WITHDRAWN

WITHDRAWN

The Chimpanzees of Gombe / Patterns of Behavior

The Chimpanzees of Gombe

Patterns of Behavior   *Jane Goodall*

The Belknap Press of Harvard University Press   Cambridge, Massachusetts, and London, England   1986

For my mother, Vanne,
for the chimpanzees of Gombe themselves,
and in memory of Louis Leakey

Contents

Abbreviations of Chimpanzee Names Used in Tables and Figures

Males

AL	Atlas
AO	Apollo
BE	Beethoven
CH	Charlie
DE	Dé
DP	Dapples
DV	David Greybeard
EV	Evered
FB	Faben
FD	Freud
FG	Figan
FR	Frodo
FT	Flint
GB	Goblin
GI	Godi
GL	Gimble
GOL	Goliath
HG	Hugo
HH	Hugh
HM	Humphrey
HX	Huxley
JG	Jageli
JI	Jimi
JJ	Jomeo
LK	Leakey
MK	Mike
MM	Michaelmas
MU	Mustard
PF	Prof
PN	Pan
PP	Pepe
PX	Pax
RX	Rix
SD	Spindle
SH	Sherry
ST	Satan
TB	Tubi
TI	Tapit
WL	Wilkie
WW	Willy Wally
WZ	Mr. Worzle

Females

AP	Aphro
AT	Athena
BB	Barbet
BM	Bumble
CA	Caramel
CD	Candy
DM	Dominie
DO	Dove
FF	Fifi
FLO	Flo
FN	Fanni
GG	Gigi
GK	Gilka
GM	Gremlin
HP	Hepziba
HR	Harmony
JO	Joanne
KD	Kidevu
KR	Kristal
LB	Little Bee
LO	Lolita
MF	Miff
ML	Melissa
MN	Mandy
MO	Mo
MZ	Moeza
NP	Nope
NV	Nova
OL	Olly
P	Pooch
PI	Patti
PL	Pallas
PM	Pom
PS	Passion
S	Sophie
SA	Sandi
SE	Sesame
SP	Sprout
SS	Skosha
SW	Sparrow
SY	Spray
VL	Villa
WK	Winkle
WN	Wunda

The Chimpanzees of Gombe / Patterns of Behavior

Introduction

All morning I had searched the Kasakela Valley for chimpanzees. At midday there had been a tropical deluge; now, an hour later, the rain was coming down in a steady, gray drizzle. Perhaps because I was wet and cold I did not see the chimpanzee until I was within 10 meters of him. Nor did he see me. He was hunched in the undergrowth at the left side of the trail with his back to me. I crouched down to watch. The chimpanzee, probably as wet and cold as I was, remained motionless. After a few moments I heard a rustle to my right and a soft call. I turned my head slowly but saw nothing in the thick bushes. When I looked back, the black shape that had been in front of me had vanished. Then came a sound from above. A large adult male stared down at me, lips tensed; with a rapid, jerky movement he shook a branch. I knew he was threatening me and quickly looked away, for prolonged staring would make him more agitated. At once I made out another chimpanzee ahead: two eyes staring in my direction and a large black hand gripping a branch. Another soft call, this time from behind. Above me the male uttered the eerie alarm call of the chimpanzee—a long, wailing *wraaa*. He shook the vegetation more vigorously and I was showered with raindrops and falling twigs and leaves. The call was taken up by other chimpanzees. All around me branches were swayed and shaken, and there was a sound of thudding feet and crashing vegetation. My instincts urged me to get up and leave; my scientific interest, my pride, and an intuitive feeling that the whole intimidating performance was merely bluff, kept me where I was. To prove myself utterly harmless I feigned disinterest and pretended to chew up leaves and stems.

Suddenly the end of a branch hit my head as the large male above me became more excited. Another chimpanzee charged through the undergrowth, straight toward me. At the very last minute he veered away and ran off at a tangent into the forest. A short while later all the black shapes had vanished. The forest was quiet again save for the pattering of rain on leaves—and the thudding of my heart.

(16 March 1961)

This photograph of William is the first closeup of a wild chimpanzee asleep in a day nest.

With David Greybeard in the early days. (H. van Lawick)

The incident just described was a major breakthrough in my study of chimpanzee behavior. I had been at Gombe (a game reserve then) for eight months. Although I had filled many notebooks with observations of chimpanzee behavior, almost all my information had been gleaned at distances of 100 meters or more. With the aid of binoculars I had learned a good deal about behaviors such as feeding and nesting, about group size and locomotor patterns, but my understanding of chimpanzee social behavior was meager. When I had tried to approach the chimpanzees more closely, they had almost always run away in fear. Gradually, however, they had become accustomed to the white-skinned upright ape who had suddenly invaded their terrain, who wore the same drab-colored clothes every day, and who never harmed them. And ultimately they dared to threaten me. That was the start of a nerve-racking few months peppered with intimidation displays, but soon the chimpanzees' aggression, like their fear, was assuaged. Some eighteen months after I began my work at Gombe, I was able to approach to within 50 meters of many of the chimpanzees without disrupting their activities. This was to lead to an amazing tolerance of me, and of other humans. Today we can move to within touching distance of a chimpanzee; he may not so much as glance up. Even now this strikes me as one of the most incredible aspects of my twenty-five years of studying chimpanzee behavior at Gombe.

It was Louis Leakey, paleontologist-cum-anthropologist, who made it possible for me to fulfill my childhood dream of studying animals in Africa. Working as a waitress during the summer months, I had saved enough money for my round-trip fare to Kenya. "If you are interested in animals you should meet Leakey," a friend told me. I did—and after I had worked for him for a year, he suggested that I set off for Kigoma in Tanzania and study the chimpanzees who inhabited the forested shores of Lake Tanganyika. But first he had to find the money

Louis Leakey.

for such an expedition. As I had no academic qualifications at the time, this was not an easy task. Moreover, there were many who felt that it was unthinkable for a young girl to set off into the African bush, unarmed, to study potentially dangerous apes. Fortunately Leakey paid little attention to his critics and eventually found seed money for me. And so it was that in July 1960 (accompanied by my mother and an African cook) I set foot, for the first time, on the sandy beach of Gombe on Lake Tanganyika.

Louis Leakey was interested in the behavior of all animals, including the human animal. He was especially curious about chimpanzees, because they are our closest relatives living today. They show many biochemical similarities to humans: the number and form of the chromosomes, the blood proteins and immune responses, the structure of the DNA. The anatomy of the chimpanzee brain is more like that of the human brain than that of any other living creature, and much of chimpanzee social behavior shows uncanny similarities to our own. Leakey reasoned that knowledge of the way of life of chimpanzees in their natural habitat might provide clues to the behavior of early man. If, as the evolutionary argument goes, man and chimpanzee once diverged from common stock, then behavior patterns existing in modern humans and modern chimpanzees were probably present also in that common ancestor—and, therefore, in Stone Age man as well.

Most modern theories concerning the life-style of our early ancestors are based on reasoning of this sort, and most refer to the behavior of the Gombe chimpanzees. Leakey's vision has been more than justified. And we can take the argument a step further. Wild chimpanzees live in an environment that is far more like that of early man than is our twentieth-century milieu. A better understanding of the adaptations of chimpanzee behavior to that environment can provide insights into the shaping of our own behavior during evolution. Aggression, for instance, seems maladaptive in present-day society. Yet if we view our aggressive tendencies in the context in which they evolved, as adaptations to cope with the hostilities of the physical environment and the competition from other humans, to ensure survival of the individual and his family or clan, we can better comprehend why we are a potentially aggressive species.

When Louis Leakey sent me off, all those years ago, he anticipated that the study might continue over a ten-year span. He knew even then that only longitudinal research would provide answers to some of the most fascinating questions. Once again he has been vindicated, except that it has been necessary to carry on even longer. For creatures who may live as long as fifty years, twenty-five years of study is still only a beginning when it comes to compiling life histories, searching for differences in early experience that might account for quirks in adult behavior, or working out kin relationships among the various members of a social group. Had my colleagues and I stopped after a mere ten years, we should have had a very different picture of the Gombe chimpanzees than we do today. We would have observed many similarities in their behavior and ours, but we would have been left with the impression that chimpanzees were far more peaceable than humans. Because we were able to continue beyond the first decade, we could document the division of a social group and observe

An adolescent female feeding.

A juvenile male brachiating.
(H. van Lawick)

the violent aggression that broke out between the newly separated factions. We discovered that in certain circumstances the chimpanzees may kill and even cannibalize individuals of their own kind. On the other side of the coin, we have learned of the extraordinarily enduring, affectionate bonds among family members, which often persist throughout life, and we have witnessed the extent to which close kin will help and support one another. We have watched adult males cooperate in a hunt, in the patrolling of their territorial boundaries, and in the protection of their females and young. Above all, we have become ever more aware of the advanced cognitive abilities of the chimpanzees that have led to sophisticated social interactions, cultural traditions, and pronounced individual variation.

I do not personally find it necessary to *justify* the study of chimpanzees by suggesting that the results will help us in our long search to comprehend human behavior—although I happen to believe that this is true. One of the most important characteristics of the human species has been our desire to acquire knowledge: through the ages this goal has led to remarkable advances in the natural sciences and in technology. Man's inherent curiosity and insatiable love of adventure led to Christopher Columbus' discovery of America; in our generation it has landed people on the moon. We do not need to justify, in terms of relevance to ourselves, the study of a creature who is surely, next to *Homo sapiens*, the most fascinating and complex in the world today.

1. The Darkness Lifts

(From Buffon, 1766)

Since therefore no one, without extraordinary delight and wonder, can look upon the living genus of Simiae, so utterly ridiculous and curious, those just mentioned, which are most like men, should be studied with open minds by experts in natural history. Wherefore one comes to wonder how it has happened that man, eager to learn, has left them hitherto in darkness, nor has wished with so little reasonableness to know the Troglodytes, who are nearest to himself . . . What, I ask, may be more conveniently an object of delight, even to Monarchs themselves, than to contemplate the animals of their home close at hand, whom we can never admire enough? Would it not be easy for the King, to whose will certainly a whole race bends, to get possession of them? It would lead not a little to Philosophy, if one were to spend a day with any of them, exploring how far human wit exceeds theirs, what distance lies between Brutish and rational discrimination; so that, moreover, light would arise for students of natural science from a complete description of them. This concerns me, for I am doubtful by what characteristic marks the Troglodytes are distinguished from Man, according to the principles of natural history; so near are some among the genera of Men and Apes as to structure of body: face, ears, mouth, teeth, hands, breasts; food, imitation, gestures, especially in those species which walk erect and are properly called *Anthropomorpha*, so that marks sufficing for the genera are found with greatest difficulty. (Hoppius, 1789, pp. 75–76)

Pan—spirit of nature, rural god of Greek mythology; so the chimpanzee genus was named. And the species and races were labeled in kind—*troglodytes*, dwellers in dark caves; *satyrus*, woodland deity, with the hooves and tail of a horse in Greek mythology, those of a goat in Roman folklore; *verus*, perhaps meaning that the existence of the creature was, after all, the very truth. These names are a legacy from the past, from the long-ago days when the anthropoid apes were known only from fragmentary tales brought back to the civilized world by sailors and other travelers who, in strange foreign lands, had heard the tales and superstitions of the natives. Aristotle mentioned "apes" more than three centuries before the birth of Christ. But it would be many more centuries before chimpanzees* were described and classified; in the interim many were the myths and legends surrounding the strange half-man, half-beast "pygmie" of the equatorial forest.

Arrival in Europe

(From Tulp, 1641)

(From Tyson, 1699)

The first chimpanzee known to have reached European shores came from Angola and was presented to the Prince of Orange in 1640. One Nicolaas Tulp, a Dutch physician and anatomist, examined and described this *Satyrus indicus*. Commenting on the false descriptions of the ancients who had credited the apes with possessing horns and hooves, he wrote, "The coming of this Indian Satyr, perchance becoming famous, may disperse their dense fog" (Tulp, 1641; cited in Yerkes and Yerkes, 1929, p. 12).

But only small patches of the fog lifted as isolated specimens of chimpanzees arrived in different European countries. Some fifty years after the publication of Tulp's description, a chimpanzee was dissected and described by a London physician, Edward Tyson. He found many physical similarities between the so-called pygmie and man, but concluded that this "Creature so very remarkable . . . [was] . . . a sort of *Animal* so much resembling *Man*, that both the Ancients and the Moderns have reputed it to be a *Puny Race* of Mankind" (Tyson, 1699; cited in Yerkes and Yerkes, 1929, pp. 14–15).

Over the next two centuries individual chimpanzees were acquired from time to time by various zoological gardens in Europe. They provided the public with much entertainment, being "very pretty Company at the Tea-Table," as one writer remarked (Boreman, 1739; cited in Morris and Morris, 1966, p. 134). Frequently succumbing to respiratory diseases, they seldom lived long. The illness of one young chimpanzee so distressed some of his devoted friends that they moved him from the zoo to a luxurious suite in one of the best hotels, then gathered around his bed as he breathed his last. "He was conscious to the end, and knew his friends, but although he had no parting words to say he expressed his feelings by looks and pressure of the hand" (cited in Morris and Morris, 1966, p. 101).

The publication of Darwin's theory of evolution in 1860 sparked new interest in animal behavior; at the same time it led to a certain amount of deliberate anthropomorphizing. If man indeed had descended from the animals, then it was desirable that his relatives be endowed with an abundance of pleasant, intelligent, and noble traits. In the 1890s R. L. Garner set off into the West African jungle where, safely barricaded into a stout cage, he was able to watch (or so we are led to believe) the behavior of chimpanzees. He also studied them in captivity and made wildly exaggerated claims about their intelligence. In 1896, encouraged by accounts of this sort, a Frenchman named Victor Meunier proposed in all seriousness an elaborate scheme for the domestication of monkeys and apes. The apes, trained to perform

* The name *chimpanzee* was first used in the *London Magazine* (September 1738, p. 465): "A most surprising creature is brought over . . . that was taken in a wood in Guinea. She is the female of the creature which the Angolans call chimpanzee, or the mockman."

a variety of menial tasks, would take the place of human servants. Chimpanzees, Meunier reasoned, would do better than mere machines, for they would be, at one and the same time, machine and mechanic. The Frenchman felt that chimpanzees would excel at the task of serving at table and that they could be taught to announce, in some way, "Food is ready." They would be trained at a special school, which would also teach them to be gardeners, nursemaids, valets, chambermaids, cobblers, sailors, construction workers, painters, guards, and so on (Morris and Morris, 1966).*

Dawn of Understanding

In 1912 the Prussian Academy of Sciences established an anthropoid station at Tenerife in the Canary Islands. It was there that the psychologist-philosopher Wolfgang Köhler carried out his classic studies with a group of chimpanzees. (His work and that of others will be described more fully in Chapter 2.) Köhler was interested in the extent to which the intellectual capacity of man and the anthropoid apes differed. He posed a variety of problems for his subjects—such as bananas that were out of reach beyond the bars or suspended from the roof—that often required the use and construction of tools for their solution. His research was carried out with the utmost scientific integrity, and with a tremendous amount of understanding of (and affection for) his chimpanzees. His 1925 book, *The Mentality of Apes*, shows that his insight into the world of the chimpanzees extended far beyond their ability to solve his problems; his descriptions of chimpanzee behavior remain among the most careful, perceptive, and important in the literature.

* In fact, there was a fairly recent account of three chimpanzees working for a year as part of the assembly line at a factory that produced foam rubber furniture. I suspect this was a publicity gimmick rather than an attempt at bending union rules to employ cheap (or chimp) labor.

Wolfgang Köhler in about 1950. (Courtesy of E. B. Newman and the Department of Psychology, Harvard University)

A group of chimpanzees working together: Konsul, Grande (*above*), Sultan, and Chica. (From Köhler, 1921)

Ioni at work in Nadie Kohts's laboratory in Moscow. (From Kohts, 1923)

Robert M. Yerkes with Panzee and Chim. (Courtesy of the Yerkes Primate Research Center, Emory University)

At the same time as the founding of the Tenerife station, a psychologist in Moscow, Nadie Kohts, acquired a male chimpanzee a year and a half old. She worked with Ioni (or Joni, as we would write) for two and a half years, testing above all his visual perception. She later repeated the tests with her own son during the first four years of *his* life.

Also in 1925, psychobiologist Robert Yerkes purchased two young chimpanzees from a sailor in Boston. They became the nucleus of the chimpanzee colony at the Yale Laboratories of Primate Biology. Subsequently Yerkes and his chimpanzees moved to Orange Park, Florida, and there established the Yerkes Primate Laboratory (which later was relocated yet again and is now the Yerkes Regional Primate Research Center in Atlanta, Georgia). Yerkes and his colleagues made gigantic strides in the understanding of chimpanzee behavior. In 1931, a young chimpanzee from the Yerkes Laboratory was raised by psychologist Winthrop Kellog and his wife in their home. For nine months Gua lived with them and their son (who was the same age as Gua); both ape and child were tested daily.

These were fertile years that provided a great deal of new and exciting information about the behavior of chimpanzees in the laboratory. But Yerkes was not satisfied: he wanted to know something of their behavior in their natural habitat. And so, in 1930, Henry Nissen from the Yerkes Laboratory set off for a two-and-a-half-month field study in French Guinea, the first venture of its kind since Garner's extraordinary exploit forty years earlier. Unfortunately, the gathering momentum of this search for understanding ended with World War II.

These early investigators delved extraordinarily deeply into chimpanzee intellectual processes—more so, I believe, than many people realize today. Köhler, after his five years of study, concluded that "chimpanzees manifest intelligent behaviour of the general kind familiar in human beings . . . a type of behaviour which counts as specifically human" (1925, p. 226). In 1943 Yerkes, summarizing the existing state of knowledge regarding chimpanzee intelligence, went even further: "Results of experimental inquiry justify, I believe, the working hypothesis that processes other than those of reinforcement and inhibition function in chimpanzee learning . . . I suspect that they presently will be identified as antecedents of human symbolic processes. Thus we leave this subject at a most exciting state of development, when discoveries of moment seem imminent" (p. 189).

From the 1960s to the Present

In actuality, more than fifteen years were to elapse before research into chimpanzee behavior again accelerated. When it did, there was a positive explosion of interest. Investigators tramped into the African forests, took chimpanzees into their homes, worked with them in trailers and enclosures, and reintroduced captive chimpanzees to the jungle. Chimpanzee subjects not only worked with traditional laboratory apparatus, they operated computers, learned languages, painted pictures, performed on stage, donated their kidneys to human patients, and orbited into space.

The proliferation of interest in behavior can be divided into four main categories: field studies, language-acquisition projects, psychological testing in traditional laboratories, and more naturalistic observations of social behavior in field enclosures.

Field Studies

At about the time that I was setting off for Gombe in 1960, under the auspices of the farsighted Louis Leakey, Adriaan Kortlandt of Holland was setting up a short field study in the eastern Congo (Zaire), and Junichero Itani from Japan was surveying the rugged country to the south of Kigoma, in Tanzania (where, after a number of brief studies, long-term research in the Mahale Mountains was eventually started and continues today under the leadership of Toshisada Nishida). Soon after this first wave Vernon and Frances Reynolds arrived for a nine-month study in Uganda's Budongo Forest, where Yukimaru Sugiyama and Akira Suzuki would subsequently work. The year 1968 saw Kortlandt and his colleagues start work in Guinea: at one of their two sites there have been a number of follow-up studies by Sugiyama and his colleagues. More recently William McGrew, Caroline Tutin, and Pamela Baldwin conducted a three-year study in the Senegal. In 1976 Michael Ghiglieri began a series of field observations on chimpanzees of Uganda's Kibale Forest. His work ended in 1981, but two years later Isabiriye Basuta commenced another study in the same area. Christophe and Hedwige Boesch have begun a long-term study in the Tai Forest of the Ivory Coast, and Tutin and Michel Fernandez are committed to studying the chimpanzees of Lope National Park, Gabon. All of these field studies are summarized in Table 1.1.

Language Acquisition

In 1821 Thomas Traill dissected and briefly described the anatomy of a chimpanzee. After examining the organs of respiration, the tongue,

Table 1.1 Major field studies of the chimpanzee throughout its range.

Country	Area	Date	Duration of study	Researchers
Guinea	Neribili	1930	2½ months	H. Nissen
	Kanka-Sili	1968–1969	Several months	A. Kortlandt, H. Albrecht, J. Koman
	Bossou	1968–1969	Several months	A. Kortlandt, H. Albrecht, J. Koman
Eastern Zaire	Beni	1960–1961	Several short field trips	A. Kortlandt
Tanzania	Gombe National Park	1960 on	25 years (continuous)	J. Goodall; many members of Gombe Stream Research Center
	Mahale Mountains	1966 on	19 years (almost continuous)	J. Itani, T. Nishida; many Japanese scientists and students
Uganda	Budongo Forest	1962	9 months	V. Reynolds, F. Reynolds
		1966	9 months	Y. Sugiyama, A. Suzuki
	Kibale Forest	1976–1981	3 field trips, 488 hours' observation	M. Ghiglieri
		1983 on	2 years	I. Basuta
Senegal	Niokolo-Koba National Park	1976–1979	3 years	W. McGrew, C. Tutin, P. Baldwin
Ivory Coast	Tai Forest	1979 on	6 years (not continuous)	C. Boesch, H. Boesch
Gabon	Lope National Park	1983 on	2 years	C. Tutin, M. Fernandez

Roger Fouts with Washoe (*on toy bike*) and Loulis. (Courtesy of R. Fouts)

and the larynx, he wrote that there did not appear to be "any reason why the [chimpanzee] should not speak." Traill went on to suggest that speech requires the power of abstraction, "which does not seem granted to the brute creation" (Traill, 1821; cited in Yerkes and Yerkes, 1929, p. 301).

To find out whether a chimpanzee *could* learn to speak, given the right environmental setting, psychologist Keith Hayes and his wife, Cathy, in 1947 adopted a month-old infant chimpanzee, Viki, and brought her up as their "daughter." Despite the fact that she lived with them and was treated as a child (until her untimely death at age six), and despite intensive language training, she learned only four words—which she "breathed" rather than spoke: *papa, mama, cup,* and *up.* In terms of language acquisition, the experiment was a failure. It now seems that there are, in fact, differences in the vocal tracts of chimpanzees and humans (in the structure of the supralaryngeal-pharyngeal area) that make it impossible for chimpanzees to produce the human vowels *a, i,* and *u,* and that they lack a certain mobility of the tongue, which humans use to change the shape of their vocal tract (Lieberman, Crelin, and Klatt, 1972; cited in Fouts and Budd, 1979).

Twenty-two years before Viki was adopted, Yerkes had suggested, "Perhaps they [chimpanzees] can be taught to use their fingers, somewhat as does the deaf and dumb person, and thus helped to acquire a simple, nonvocal, 'sign language'" (Yerkes, 1925; cited in Yerkes and Yerkes, 1929, p. 309). Not until forty years later was this approach pioneered by Allen and Beatrice Gardner, a psychologist-ethologist, husband-wife team. In 1966 the Gardners acquired Washoe, a wild-born female chimpanzee, about a year old. Except that she did not share their house with them (she lived in a trailer in their back yard), Washoe was treated like a human child. The Gardners rightly felt that an enriched environment and plenty of affection were essential for the success of their project, since it is in a social context that a human child develops language. Washoe was taught ASL (at one time known as Ameslan), the sign language of the deaf, and her vocabulary, which grew rapidly, was tested regularly, with scrupulous scientific integrity. When Washoe left the Gardners toward the end of 1970 and went to work with Roger Fouts at the Institute for Primate Studies in Oklahoma, her expressive vocabulary comprised over 130 signs.

The project made history. It is impossible to overemphasize the value of the contribution the Gardners have made to our understanding of the chimpanzee mind. The success of Project Washoe caused widespread interest, and other young chimpanzees were taught ASL, in both home and laboratory settings. The Gardners themselves launched a second project, in which four infant chimpanzees were raised in the laboratory with semideaf, fluent signers. This program has provided other insights of major significance regarding perceptual and cognitive development of young chimpanzees.

Fouts is currently working at Central Washington University, Ellensburg, Washington, with Washoe and the three survivors of the Gardners' second project. Washoe has an adopted son, Loulis, who is learning ASL without human tutelage.

One other individual should be mentioned in connection with ASL: Lucy, foster "daughter" of psychoanalyst Maurice Temerlin and his

Lucy with her human parents. (J. Carter; courtesy of J. and M. Temerlin)

wife, Jane. Lucy lived with the Temerlins for ten years and began lessons in ASL from Roger Fouts in 1972 when she was about four years old. (Fouts at that time drove around the Oklahoma countryside giving private instruction to a variety of home-raised chimpanzees; see Fouts, 1973.)

Meanwhile, investigators in more conventional laboratory settings were also pursuing imaginative research. Stimulated by the Gardners' work, David and Ann Premack, working initially with a wild-born female chimpanzee, Sarah, were the first to design an *artificial* language. This comprised pieces of plastic of various colors, sizes, shapes, and textures; the backs were metal, so that the pieces could be arranged on a magnetized board. Each piece represented a word. The trainer used these plastic symbols to ask Sarah a question; she in turn selected the correct pieces and placed them in the proper order to reply.

Soon after this project had begun, Duane Rumbaugh and his associates at the Yerkes Regional Primate Research Center devised a sophisticated computer-controlled training situation for testing the language capacities of a two-and-a-half-year-old female, Lana (Rumbaugh, Gill, and von Glasersfeld, 1973). Lana learned to use the keys of a console, each of which (there were initially twenty-five) was marked with a lexigram in Yerkish, as the language is called. All the messages Lana typed were recorded by the computer. To a certain extent, Lana could in this way control her world. Provided she made her requests in the correct word order (she spontaneously learned to monitor her sentences on a display and erase those with errors), the "machine" delivered drinks, pieces of banana, music, movies, and so on. However, it could not comply when in the middle of the night she forlornly typed, "Please machine tickle Lana period." During the daytime her trainer sat at a second console and was available to supply intervals of much-needed social contact during their computerized conversations (Gill, 1977). Other chimpanzees have since learned Yerkish, and the method is being used with some success to communicate with autistic and mentally ill humans (Parkel, White, and Warner, 1977).

Lana punches a Yerkish symbol on her console. (F. Kiernan; courtesy of D. Rumbaugh and the Yerkes Primate Research Center)

Psychological Testing

Two other major breakthroughs emanated from Yerkes. Richard Davenport and Charles Rogers proved in 1970 that chimpanzees were capable of cross-modal transfer, and the same year Gordon Gallup showed that chimpanzees *could*, whereas monkeys *could not*, recognize themselves in mirrors. (Washoe, as we shall see in the next chapter, had already identified her mirror image.)

Jürgen Döhl and Bernard Rensch have done a great deal with a gifted female chimpanzee, Julia, at the Zoological Institute in Münster. Working with extremely complex mazes and locked boxes, Julia showed that she could plan ahead in order to obtain a goal that was out of her sight.

David Premack and his colleagues, in an extraordinarily imaginative series of experiments, showed Sarah a series of videotaped scenes in order to find out whether she would understand the goals of the human

Emil Menzel with Bill, one of the "Menzel group," in 1968. (Courtesy of E. Menzel)

actor she had watched. Using Sarah and other subjects, Premack's team has investigated other aspects of the chimpanzee mind as well.

Psychologist Emil Menzel began to work, in the sixties, with a group of young wild-born chimpanzees in a large field enclosure. Many of his perceptive experiments addressed questions on the ways in which chimpanzees can communicate with one another about objects and events in their environment when these are displaced in space and in time. Menzel has made major contributions to our understanding of cognitive mapping in chimpanzees.

Captive Colonies

Of the various chimpanzee colonies in captivity the most important by far, in terms of published research, is the colony set up by Jan and Anton van Hooff at the Burgers' Zoo in Arnhem (van Hooff, 1973). Frans de Waal's recent book (1982) provides illuminating insights into the upper limits of social intelligence. The imaginative efforts of Leonid Firsov of the Pavlov Institute of Physiology led to the establishment of a chimpanzee colony that in the summer months roams a small, uninhabited island in the Pskov region of northwestern Russia. A team of scientists camps on a nearby island and visits the chimpanzees daily. Unfortunately, very few of their fascinating observations have filtered out to the West.

Other colonies established by laboratories or zoos are beginning to yield important information. The major value of such observations is that the chimpanzees can be kept under much more constant surveillance than is possible in the wild, yet have much greater freedom than in the traditional laboratory.* Our understanding of social intelligence will grow significantly as a result of continued work of this sort.

---

* For this reason I helped to establish the unique primate facility at Stanford University, where Menzel's original group and another, founded by Laurence Pinneo, were installed. Collaboration over the years between this research center and our facility at Gombe would have been of tremendous significance. Unfortunately, Stanford could not afford to maintain its chimpanzees.

And so we arrive at the present. Obviously, there has been other chimpanzee research; we have looked at only the important milestones along the road to understanding. Our debt to those who have worked long and hard before us is incalculable, for it is the accumulated results of their labors that have provided some of our most meaningful insights. Not many would be able, like Köhler, to start at the beginning, "without theoretical foundations, and in unknown territory" (1925, p. 226).

There is still much to learn. As additional aspects of chimpanzee behavior are revealed, each sets new ideas in motion and acts as a starting point for further research. It is as though we are all contributing pieces to a giant jigsaw puzzle. Slowly the picture becomes clearer, although many gaps remain. The work on captive chimpanzees contributes, spectacularly, to a better understanding of the chimpanzee mind; the field research helps to explain the adaptive value of behavior and provides insights into the complexities of social life. Because the field data concerning intelligent behavior are so often anecdotal, I shall in the next chapter attempt to synthesize, from work in the laboratory, an overall description of the complex chimpanzee mind. This can then serve as a backdrop against which to evaluate examples of intelligent behavior observed in the Gombe chimpanzees.

2. The Mind of the Chimpanzee

The Gardners with Washoe at age five. (Courtesy of R. A. and B. T. Gardner)

November 1983 The five chimpanzees housed in the psychology department of Central Washington University were going about their customary business. They may have sensed that something slightly unusual was afoot because their living quarters had been cleaned with extra thoroughness. The door of the lab opened and two visitors quietly walked in. Four of the chimpanzees stopped in midactivity and sat motionless, staring at the newcomers. The fifth and youngest looked for a moment and then began to display, banging about in the cage.

The visitors were Allen and Beatrice Gardner. They had fostered the chimpanzees who sat in stunned silence, but had not met the fifth, young Loulis. Washoe, after two or three minutes, broke the spell. Signing Trixie Gardner's name sign, she added *Come hug*. Dar, after reaching out to grab Loulis in an apparent attempt to calm him, signed Allen Gardner's name sign, as did Tatu.

Dar and Moja had only left the Gardners' fairly recently (four and two and one-half years before, respectively). It was not surprising that they should have remembered Allen's name sign. But Washoe had not seen Trixie for eleven years. (Roger and Debbi Fouts, personal communication)

Two normal humans looking at an apple tree will perceive it in approximately the same way. The trunk will look brown, the leaves green, and the apples red and rounded. If they are ripe, they will be smelled; if there is a breeze, there will be the sound of leaves rustling. The closest apples will appear larger than those on the far side of the tree, but this illusion will automatically be corrected for and the apples will be thought of as being more or less the same size.

These similarities in the perceptions of the observers are dependent upon an inherent internal programming, which sorts and arranges the complex array of elements contained in the visual, olfactory, and auditory stimuli that impinge on the sense organs in such a situation. The same observers subsequently shown a photograph of the tree will recognize it. Through an innate mechanism for abstraction and generalization they will be able to classify the actual apple tree, or a photograph of it, as a *tree*—and within the tree category they can select *apple tree*.

However, there will also be differences—perhaps very pronounced differences—in the conscious selection and interpretation of these perceptions by our two subjects. In other words, differences in the way they think about the apple tree. These conceptual differences will be greatest when the two individuals are from different cultures (particularly if one has never seen an apple before). If the two are from the same culture, conceptual differences will be largely determined by social background, occupation, and interest. One, if he is hungry, may see it as a source of food. Another, especially a small boy, may think of it as a structure for climbing. A perceptive observer may pick out all sorts of strangely shaped boughs, color combinations of fruit and leaf, patterns of branches against the sky, and so on, which may escape another observer's attention. Our small boy may notice a long-fallen bough below the tree—good for knocking down apples! The tree will be perceived differently by the artist, the forester, the market gardener, the botanist, and the person who wishes to build a house where the tree stands.

If the words *apple tree* are spoken, they serve as an abstract symbol and stimulate some kind of mental image for our two subjects. If asked about this image, they will have no difficulty describing a tree with apples, trees in general, apples in general, and so on. They will be able to imagine the taste of an apple and the feel of its skin. But the "mental experiences" of our two subjects, if unconstrained by questioning, are likely to differ even more dramatically than the way in which they perceived the tree while standing beside it. The spoken word, or any abstract symbol denoting *apple tree*, may elicit for different people very different mental images. For one it may simply be a mind picture of a tree with apples; for another, a particular apple tree where he once met his girlfriend in the moonlight; for another, an apple orchard of childhood; for another, a complex series of associations leading from Eve, the serpent, and the Garden of Eden to an atomic holocaust and the end of the world.

To what extent does the perceptual and intellectual world of the chimpanzee overlap ours? How far along the road can the chimpanzee follow us from the sight of an apple tree to the Garden of Eden? Only by working with immense patience, understanding, insight, and skill with individual chimpanzees in captivity is it possible to investigate questions that relate to the upper limits of chimpanzee intellectual ability. Let us take a look at some of the relevant research findings.

The Perceptual World

Insofar as reliable information is available, it seems that the sensory system of the chimpanzee is very similar to our own. Visual acuity, sensitivity to light, and ability to differentiate most wavelengths of the spectrum are comparable, although it seems that humans are more sensitive to the yellow-red end of the spectrum. Chimpanzees can detect differences in size and discriminate between shapes in much the same way we can; and their perception of movement is similar. They have no trouble in localizing sounds and can distinguish variations in tempo. Their sense of hearing is comparable to that of humans, although they are more sensitive to high-frequency sounds (Prestrude, 1970).

There has been no systematic investigation of chimpanzee olfaction. The chimpanzee almost certainly uses his sense of smell more than the typical civilized person, although probably no more than a person who lives close to nature—for whom the ability to detect the scent of a predator could mean the difference between life and death.

Kellog and Kellog (1933) persuaded their own son and their chimpanzee, Gua, to rank four tastes in order of preference. The human child's order was sweet, sour, salt, bitter. Gua preferred sour over sweet, but agreed on the other two. An informal observation shows that chimpanzees can spontaneously develop a taste for alcohol. The Temerlins' Lucy, when first taken to their orchard, picked apples from the branches. During subsequent visits, however, she began eating completely rotten fruits from the ground. After such sessions "she seemed happy, often laughing . . . she was getting a 'high' on the natural fermentation as the apples rotted" (Temerlin, 1975, p. 49).

Chimpanzees sometimes manipulate their own injuries in a way that would be very painful for most humans; this suggests that they may be less sensitive to pain than we are. (There is, however, a considerable variation in attitudes toward pain in different human cultures.) Chimpanzee sensitivity to temperature is probably similar to ours: they usually avoid hot sun and lie about languidly when temperatures are very high (see for instance Köhler, 1925; Fouts, 1974); when it is cold, they shiver and look miserable.

If we accept that the chimpanzee has a sensory *capacity* similar to our own, does he perceive an apple in the same way we do? Does he perceive similar patterns of roundness, straight or twisted shapes of branches, and so on? Kohts (1935), after testing her young ape, Ioni, with more than five hundred discrimination tests designed to examine the nature of his perceptions, concluded that his visual world was very like that of the human child (see also Yerkes and Petrunkevitch, 1925). An adolescent female chimpanzee was able to discriminate numerical middleness. She scored 75-percent accuracy even with as many as seventeen objects, and it seems possible that primitive counting was involved (Rohles and Devine, 1966, 1967).

The chimpanzee world is unequivocally dominated by the relations of objects in space; spatial position has clear priority over other cues such as color, shape, and size. Yerkes (1943) noted this in the following experiment. A chimpanzee was seated in the middle of a room, in each corner of which was a box of similar size and shape but of different color. The subject watched as his breakfast was placed in one of the boxes. A screen was positioned between the chimpanzee and the boxes, and the experimenter changed the location of the box containing the food. The chimpanzee consistently went to the *place* where the food had been, quite ignoring the color cues (which of course he was quite capable of perceiving). Upon discovering the box empty, the subject first searched diligently in and around it, then became angry, and eventually flung himself to the ground and cried out. The experiment was repeated with boxes that were not only of different color, but of different size and shape. The result was the same. Eventually the chimpanzee was able to learn to attend to cues other than

Figure 2.1 Use of a pole as a "ladder" to reach a suspended lure. (Adapted from Köhler, 1925)

position—but only when the delay between the setting and the execution of the problem was less than forty seconds. As a result of this predominantly spatial outlook on life, chimpanzees, as we shall see, may do better than humans in tests of place memory.

In the ASL training program photographs are routinely used for increasing and testing the sign vocabulary of the chimpanzees. If an object had been shot against a brilliantly colored background, the *color* rather than the *object* was often named (Gardner and Gardner, 1983). The Gardners also found that their chimpanzees responded far more readily to photographs of the stark trunks and branches of trees in winter, or to "Christmas trees" shot against the sky, than to pictures of trees bedecked in luxuriant summer foliage. (I wonder if an experienced chimpanzee would be able to identify an apple tree, as an *apple* tree, in winter. Many humans could not.)

Köhler has given us additional insights into the way the visual world of the chimpanzee differs from ours. Thus tables flush against the wall were not perceived as separate, movable objects, but were repeatedly bypassed by chimpanzees hunting for boxes to pull under suspended bananas. Sometimes, to solve a problem, a chimpanzee had to make a stick by tearing a piece of wood from a box. If the box was made in such a way that cracks could be detected between the boards, the chimpanzees were far more likely to see the boards as potential tools. Köhler (1925, p. 99) concluded that "visual wholes are probably more easily analyzed by the adult human than by the chimpanzee."

A problem that required chimpanzees to empty heavy rocks out of a box, so that they could move it beneath a suspended lure, proved to be very difficult—presumably for the same reason. Many of the chimpanzees never solved it, but expended great energy in trying to pull box plus stones. Even when the most gifted, Sultan, finally hit upon the solution, he removed only a few rocks—just enough to enable him, with much effort, to pull the rest.

Viki, left to herself, tended to sort things according to material property rather than function—chopsticks with pencils rather than cutlery, metal buttons with hardware rather than with bone buttons. But she could readily pick the right object for the right task, so she *was* able to perceive it in relation to function when necessary (Hayes and Hayes, 1951).

Further insights come from some of the superficially "stupid" behavior observed by Köhler. One individual, for example, searching for an appropriate tool, several times passed a small tree with a number of branches suitable for her purpose. She ignored them and tried instead to wrench a firmly attached bolt from the door. Visually, however, the bolt (which was black) stood out in contrast to the door far more than did the branches of the tree. (She did break off a branch eventually.) Köhler (1925, p. 97) goes on to describe how "optical impressions sometimes gain victory over practical considerations." When Rana, one of his least gifted animals, was trying to reach a suspended lure, her favorite method was to place a pole upright, climb rapidly, and seize the fruit even as she and the stick fell (Figure 2.1). Finding no suitable stick and several times setting a hopelessly tiny one beneath the objective, as though to climb, she suddenly picked up a second small stick and, holding the two end to end, made movements as if to climb her *optically* joined pole!

A length of rope, coiled four times around a beam—carefully, so that the coils were quite distinct and not touching each other—apparently seemed as hopeless a problem to the chimpanzees as a tangle of string does to us. Or, comments Köhler wryly (1925, p. 103), "even folding chairs to the author."

For chimpanzees, as for humans, learning and experience play a major role in the way they see their world. The home-raised chimpanzee will learn to see tables for what they are and will have no problem in moving one across the room. The shape and nature of a coil of rope will, with use, be seen more realistically. (Even folding chairs eventually become clear if you use them often enough!) A stick, initially seen as a long, straight object, can be classified as *hard* after being handled. And when it has been used to reach for food or to hit a companion, it can be classified as an *implement*.

The Processes of Learning

Learning has been defined in many ways, differing, as Hinde (1966) points out, according to the interests of the various authors. A number of ethological definitions have emphasized both that learning is adaptive and that it results in an observed change of behavior. Thorpe (1956, p. 55) defines it as "that process which manifests itself by adaptive change in individual behavior as the result of experience." Although the adaptive advantages of learning have been of crucial significance in the evolution of behavior, Stenhouse (1973) points out that the term *adaptive* should be excluded from the definition, or else it becomes logically impossible for a *maladaptive* behavior to be learned—which, particularly in our own species, is unfortunately not true. Second, certain experiences "may produce changes in internal organization . . . which are not immediately revealed by changes in behaviour" (Hinde, 1966, p. 41). These changes may then become incorporated in a *central memory store* (Stenhouse, 1973). Whether or not an experience has been "learned" will subsequently be revealed by the animal's ability, or failure, to retrieve the appropriate response from its "memory."

While biologists unanimously agree that learning is not a unitary process, controversy surrounds the actual mechanisms involved. This is not the place to enter into a discussion of the major issues inherent in learning theory: it is a fast-moving, controversial field with which I do not pretend to be fully familiar. On the other hand, a survey of the chimpanzee mind must give some consideration to learning processes, especially as more recent findings suggest that, in the higher primates, "cognition somehow emerges from and supplants the more basic, primitive associative processes of learning" (Rumbaugh and Pate, 1984, p. 586). Most of the categories used in the highly simplified account that follows are those proposed by Thorpe in 1956 and still widely quoted.

The Physical and Social Environment

There is, of course, a great deal of behavior that a chimpanzee does not have to learn (breathing and sleeping, for instance), or that appears gradually during maturation (such as many of the calls, postures, and gestures of the communication system). An extremely important aspect of early learning is the *modification* of inherent tendencies through interactions with the physical and social environment. Even though a chimpanzee raised in social isolation will spontaneously show many of the species-specific communication patterns, he must learn, through experience in the social field, the correct sequencing and contexts in which they should be performed (Menzel, 1964).

At each stage of development the youngster is ready, both physically and psychologically, to learn new behaviors, and because his phylogenetic heritage predisposes him to behave in certain ways (such as climbing, manipulating objects, being social), he will learn some things (such as skilled acrobatics in the treetops, using objects as tools, manipulating his social companions) more easily than others (such as overly complex, unnatural tests set for him by humans).

Chimpanzees acquire most of their knowledge through a complex interaction of *perceptual, trial and error,* and *observational* learning of various kinds. The extent to which an animal learns as a result of simply looking at, hearing, and smelling things in the world about him has not yet been determined, but there can be little doubt that the sorting and storing of perceptual experiences plays an important part in the gradual growth of his familiarity with the physical and social environment. A chimpanzee infant may acquire a good deal of knowledge vicariously during the months when he is firmly attached to his mother's breast, sensing her fear, excitement, or pleasure. He cannot afford to pay attention to *all* the stimuli that bombard him continually from all directions. Little by little, certain objects, sounds, or scents will come to have significance for him, whereas others will impinge on the senses but not be consciously registered. Initially he may be equally frightened by any number of events: a technician entering his cage in a white coat, an adult male chimpanzee banging on the iron wall of the next room, the scream of his mother when she is frightened, a hand appearing with a hypodermic syringe. Some of these events he learns to ignore; he *habituates* to them. Others come to have very definite implications. The sound of a key turning in the lock at a certain time heralds the arrival of breakfast and triggers much excitement; he has become *conditioned** to the sound. The sight of the syringe, if it has been used on him even once, will probably produce a fear response. But if, subsequently, it is brought in daily for several days without being used, his fear response will eventually be extinguished.

Gradually an infant acquires a mental map (McReynolds, 1962) of his surroundings. This is much easier in the confines of a laboratory, or even of a large outdoor enclosure (Menzel, 1974; Menzel, Premack, and Woodruff, 1978) than in the natural habitat where the area involved is much greater.

Some behavior is contagious. When a chimpanzee contentedly grunts as he begins to feed, or startles and rushes off with bristling hair, a companion is likely to join in the activity. This is known among ethologists as *social facilitation.* If one chimpanzee observes another

---

* Two types of conditioning are recognized: *classical* (or Pavlovian), as described, and *operant* (when chance movements are repeatedly rewarded). A chicken, for example, will eventually learn to walk in a circle if rewarded every time it takes a step to the right.

hurrying toward a tree laden with food, or rushing from an indoor to an outdoor enclosure where food has been placed, he is likely to do the same because his attention will have been directed to the food by his companion's behavior. This is known as *local enhancement*. Social facilitation will teach the chimpanzee youngster a good deal about the contexts appropriate to certain calls, what dangers to avoid, and so on; local enhancement will improve his knowledge of resources in his immediate environment.

Both positive and negative reinforcement—reward and punishment—play a crucial role in the learning process. During play and exploration the child learns more and more about the physical properties of his environment. He climbs a tree; a dead branch breaks, and he falls to the ground. Once the relationship between the quality of the branch and his tumble is appreciated, he will avoid dead branches; but it may take a number of falls. By contrast, he may learn in one trial that a fruit is delicious. During interactions with others, chimpanzee or human, the youngster learns through positive reinforcement which behaviors are acceptable (for instance, soliciting and being groomed in response; approaching with an invitation to play and being tickled) and which are unacceptable and may result in punishment (for example, biting the caretaker).

We have traced the gradual accumulation of a chimpanzee's knowledge as he matures and interacts with his environment—the molding of his acquired tendencies. In the natural environment this will produce adult behavior characteristic of his particular social group. In captivity it may produce a variety of stereotypic and bizarre behaviors (if his natural inclinations have been frustrated) or an unusually alert and well-coordinated creature (if his natural tendencies have been gently coaxed toward improved performance). Now we need to consider ways in which accumulated knowledge about situations is translated into performance.

Goal-Directed Behavior

In the laboratory an animal is often faced with a problem the solution of which, in terms of real life, is meaningless. When a chimpanzee experienced in discrimination tests is confronted with a new task (selecting the darker of two green shapes, for example), he knows from previous experience that a reward will appear if he selects *one* of the two panels. He will try either; there is no way he could make a reasoned choice. Once he learns (by trial and error) the correct response, he will be able to generalize and choose the darker of two blue—or red—objects. The ease with which he generalizes will depend on the extent of his previous experience as well as factors such as motivation, attention, and intelligence.

When a chimpanzee is attempting to solve a less contrived problem (which may range from crossing from one tree to the next, through manipulating a companion, to a Köhler-type situation in the laboratory), he is able to attempt solutions based on previous experience. Before he hits on the correct solution, he may make a number of unsuccessful moves. These were categorized by Köhler as "good" or "bad" errors. Good errors show appreciation of the essential facts of the problem—as when one of his chimpanzees, trying to reach a

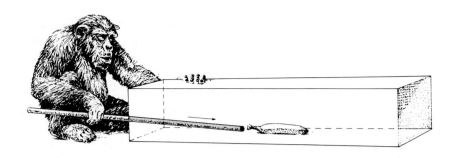

Figure 2.2 Solving the box-and-pole problem.

banana overhead, lifted a box and held it against the wall. If only it had stuck there by itself, he could have reached the lure! In the same way, Rana's construction of an optically joined stick demonstrated that she realized a longer stick was required. Bad errors, by contrast, show a complete lack of comprehension of the conditions—as when a box that was in the right place, but not high enough, was taken away and put somewhere else. Köhler also described "crude stupidities" on the part of individuals who had already demonstrated their capability for solving the problem. One of the females, for example, dragged stones about her cage in response to fruit that was outside the bars (an aftereffect of other experiments in the same place, in which stones had served as footstools).

During attempts of this sort bad as well as good errors will help a chimpanzee to become more familiar with the conditions of the problem. He may come upon the solution by chance; that is, he may achieve the goal via an accidental performance without perceiving its relevance to the problem. A young chimpanzee trying to cross from one tree to another may, having tried various methods unsuccessfully, finally move out along a branch until his weight causes it to bend down and thus provide him with a bridge. The relation of his weight (or at least his performance) to the attainment of his goal may not be immediately apparent to him. But the chances are that when the same thing happens again, he will make the necessary mental connection. The *solution to the problem* will then have been learned.

Yerkes (1943, p. 135) described the behavior of a six-year-old female, Mamo, confronted with the "box-and-pole" test. A banana was placed out of her reach in the center of a long, narrow, open-ended box, as illustrated in Figure 2.2. To obtain the reward she had to use a pole, left lying nearby, to push the banana *away* from herself until she could reach it from the far end of the box. For twelve days Mamo showed no signs of solving the problem. During her one-hour sessions she repeatedly peered at the banana, reached vainly toward it, then played around—sometimes actually *with* the pole. On the thirteenth day, as she somersaulted near one end of the box, she peered in and then suddenly went "directly to the pole and grasped it with every evidence of definiteness of purpose . . . At that very instant, and before she had approached the open end of the box with the pole, it was clear to the observer that the problem had been solved."

The spontaneous solving of such a problem is based upon the chimpanzee's perception of the relationships between things, or *insight* (Köhler, 1925). Sometimes this process may also involve a represen-

Eight-year-old Bandit, one of the Menzel chimpanzees, used to initiate escapes from his enclosure; then, when the other chimpanzees ran off to the woods, Bandit would often imitate a particular human activity he had been observing from behind the fence. (J. Mosley)

tation in the chimpanzee's mind of something not present—as when Köhler's Sultan, unable to reach a suspended treat, suddenly paused, remained motionless, then galloped off to fetch a box (which he had passed earlier in the day in another room), returned with it, and used it to reach the food reward.

The solving of new problems by observing and imitating the behavior of others, unlike social facilitation and local enhancement, also involves insight. *Imitation*, as defined by Köhler (1925, p. 190), is the ability to "understand and intelligently grasp what the action of the other means." Thorpe (1956, p. 135) proposed a very narrow definition, insisting that in true visual imitation "the pattern imitated [must] be a motor pattern that is not in the repertoire of the species."*

Some captive chimpanzees have proved their ability to imitate even *within* Thorpe's narrow definition. Viki, in the classic example, pursed her lips (a pattern not part of the innate repertoire) for the application of lipstick (Hayes and Hayes, 1951). And in the wider sense of Köhler's definition, home-raised chimpanzees have imitated actions as complex as sewing, opening bottles and pouring drinks, digging with spades—the list includes almost all of the normal actions of a person around the house.

A favorite example again involves Viki. One day she watched intently as her "mother" was laying photographs between the pages of a telephone directory to flatten them. For the next few days every available piece of paper—unopened letters, receipts, even the potholder—were all zealously pressed by Viki (Hayes, 1951). In another incident, Maurice Temerlin was ill. After having been violently sick he staggered to his bed, followed by a concerned Lucy. A few moments later she went back to the bathroom, stood upright, leaned over the toilet (as he had done), opened her mouth wide, and made gagging sounds as though trying to imitate his vomiting (Temerlin, 1975). Another interesting example of imitation was observed at the Johannesburg Municipal Zoo. Two male chimpanzees lived close together but were visually separated. One had been taught by a keeper to smoke: this, of course, delighted the spectators, who supplied him with cigarettes. The other male, who had never been taught to smoke, then began to do so; the public encouraged him to try, and he learned by watching them (Brink, 1957). Bandit, the most gifted male of Menzel's group, spent long hours watching human activities outside his enclosure. It was often he who initiated "escapes"; then, while the rest of the group bounded off to the woods, Bandit would make a beeline for abandoned tools—such as a shovel or a water hose—and use his brief time of freedom to imitate the behavior he had watched (P. Midgett, personal communication).

Washoe's adopted son, Loulis, was acquired by Roger Fouts at the age of ten months, and he and Washoe were introduced to three of the young chimpanzees from the Gardners' second project: Dar, Moja,

* Thorpe believed that imitation "seems to involve self-consciousness and the realization of another individual as resembling oneself" (1956, p. 464). Thus it was necessary to define the process in such a way as to exclude animals that were unable to form self-concepts. Observational learning is extraordinarily important in our own species—as it is in chimpanzees—and it seems best to regard Thorpe's "true" imitation as being simply the sophisticated end point of a continuum.

and Tatu. Fouts and his associates deliberately refrained from making the ASL signs in the infant's presence (with the exception of seven question signs, such as *which, who, what*). Loulis had not been taught any signs by any human at any point in his life; yet at four and a half years old, through observation and imitation, he had acquired thirty-nine signs (Fouts, Fouts, and Schoenfeld, 1984).

Tutelage

"One of the chief reasons for our use of chimpanzees as experimental subjects is their relatively great teachableness." So wrote Yerkes (1943, p. 133). This trait has been exploited to the full in the training of young chimpanzees for space travel. Morris and Morris (1966) describe in detail the conditioning procedure to which these youngsters are subjected. They must perform prolonged and elaborate operations at instrument panels, where the mental strain put upon them is calculated to be similar to that put upon a human pilot in a space capsule orbiting the earth.

The Morrises describe the fourteen-day endurance test to which a five-year-old male was subjected at the Aeromedical Research Laboratory at Holloman Air Force Base, New Mexico. Prior to the test this chimpanzee had had 1,093 hours of special training. During the test he was in a tiny cubicle, restrained by straps sewn into his nylon-mesh suit. On his feet were shoes lined with metal plates for delivering mild electric shocks if he failed to perform correctly. In front of him was the test panel at which he had to work, for reward (food and water) and to avoid punishment (electric shocks). During each twenty-four-hour period there were eight working sessions lasting sixty-six minutes each. In between, the chimpanzee was able to rest and sleep. The sessions were divided into four phases separated by intervals of one or two minutes. The first phase was heralded by a buzzer. A red light came on in one of three panel windows in front of the chimpanzee, and as long as it was there he had to press one of three levers at least once every twenty seconds to avoid a shock. When a blue light came on, he had quickly to press a different lever; if he failed he received a shock. He had to press one lever about forty-five times and the other seven times during the first fifteen minutes. During the next fifteen minutes, heralded by a green light, he had to work for water. When he pressed one of the levers at twenty-second intervals, he got a small drink from a tube above his head; if he pressed at shorter intervals (even nineteen seconds), he lost his reward. In the next fifteen minutes he worked in a similar way for food. For the final fifteen minutes he had to discriminate among three shapes, choosing the odd one each time. After that he had a well-earned rest until the whole thing began all over again. During the fourteen days the chimpanzee went through about 120 hours of these sessions. Incredibly, his performance actually improved, and by the end he was working better at his shape-discrimination test and the test for drinking and showed no deterioration on the other types of work. When he was finally released, he showed no signs of ill effect other than being a little weak on his limbs for twenty-four hours. He had not even lost weight during the whole grueling performance.

It is because a chimpanzee can be taught, and can learn through imitation because he has some understanding of the salient features of the behavior demonstrated, that he is so popular in show business. The performance of one Peter, who appeared in 1909 at the New York Theater, was described by Morris and Morris (1966). On stage he enjoyed an elaborate meal, smoked a cigar, cleaned his teeth, brushed his hair, powdered his face, tipped his keeper. He undressed, lit a candle, got into bed, and snuffed out the light. Next he got up, dressed, and—on roller skates—chased a young woman. Then came a fifteen-minute performance of trick cycling; as a finale, he drank from a tankard as he pedaled around. He then dismounted, clapped, and made his exit. The performance was purported to include fifty-six separate acts. The most renowned chimpanzee performer was of course J. Fred Muggs, who at his peak of fame had a weekly salary of £357 sterling (then about $1,000)—and that in the early 1950s!

In the natural habitat, teaching as a deliberate method employed by chimpanzee adults to modify the behavior of their youngsters probably happens only rarely—although, as we shall see, many behaviors of the mother *function* to teach her child a great deal. Yerkes (1943), however, describes how the mothers of his colony often deliberately encouraged activities such as climbing or walking.

Of particular interest in this respect is the behavior of Washoe. The teaching of ASL signs to the chimpanzees is accomplished by a mixture of molding—placing the subject's hands in the correct position—and demonstration (Gardner and Gardner, 1969). Washoe, on three separate occasions, was observed as she tried to teach her adopted son new signs. Once, upon seeing a human approach with a candy bar, Washoe began to swagger about bipedally, hair bristling, signing *food* in great excitement. Loulis, only eighteen months old, watched passively. Suddenly Washoe went to him, took his hand, and molded the sign for *food* (fingers pointing toward mouth). Another time, in a similar context, she made the sign for *gum*, but with *her* hand on *his* body. On the third occasion Washoe, apropos of nothing, picked up a small chair, took it over to Loulis, set it down in front of him, and very distinctly made the *chair* sign three times, watching him closely throughout. The two *food* signs became incorporated into Loulis' vocabulary; *chair* did not (Fouts, Hirsch, and Fouts, 1982). It is as though because Washoe herself was taught, so she is able to teach.

Factors That Influence Learning Ability

Some chimpanzees learn much more easily than others. This can be explained in part by individual variation in the genetic structuring of the central nervous system, in part by differential life experiences. We must take into account the temperament of the individual; if excitable or nervous, he is more easily distracted than if calm and confident. And of major importance is his level of motivation.

When learning a new task a chimpanzee is more likely to succeed if he is highly motivated; he will then pay attention and be less easily distracted by irrelevant stimuli. His degree of motivation is in turn affected by a variety of factors, such as the nature of the reward or punishment (if there is one), the nature of the task as it relates to his

Julia demonstrates the chimpanzee's ability to manipulate small implements as she unscrews, then removes, a screw in order to obtain food from inside the box. She shows the close attention required for problem solving. (From Rensch and Döhl, 1967)

ability to succeed, and his emotional state at the time. Köhler (1925) describes how Tschego, his oldest female, suddenly solved a problem as a result of *increased* motivation. It was her very first test: outside her cage was a pile of fruit, quite within reach if only she moved a box out of the way. For two hours she reached out helplessly, over and over again, toward the food. Suddenly some of the smaller apes began to approach. As they got closer, the danger "inspired" Tschego. She gripped the box, moved it aside, and seized the fruit before the youngsters had a chance to take it.

A second example from Köhler (1925) illustrates how *lack* of motivation can hinder the performance of a task already learned. The gifted Sultan was easily taught to gather discarded fruit peels and drop them into a bucket, thereby helping his keepers. He performed excellently for the first two days; "on the third day he had to be told every moment to go on; on the fourth he had to be *ordered* from one skin to the next; and on the fifth . . . his limbs had to be moved for every movement, seizing, picking up, walking, holding the skins over the basket, letting them drop and so on because they stopped dead at whatever place they had come to or to which they had been led" (p. 252).

The nature of a reward can also affect performance. If a food reward is too small, a satiated chimpanzee may not bother to work for it; if it is too large, he may be so eager to get it that he skimps on his task. If he is too hungry, he may be utterly transfixed by the sight of a food reward and be quite unable to put his mind to any task at all (Köhler, 1925). Pryor (1984) discusses the merits of a variable reinforcement schedule—sometimes a reward, sometimes no reward—and just occasionally, for reasons *not* connected with a better-than-usual performance (although the performance must be adequate), an extra-large, *surprise* reward. She found this "jackpot effect" extremely useful in animal training.

Chimpanzees (and other mammals) will work for rewards other than food—such as the opportunity to groom or be groomed (Falk, 1958), or to play (Yerkes and Petrunkevitch, 1925). Moreover, if a chimpanzee enjoys or is fascinated by a given task, he may repeat it over and over again just for the satisfaction he derives from the performance. Butler (1965) describes this as self-rewarding behavior. A satiated chimpanzee may continue to work at a task of this kind for even a tiny food reward (or none at all) much longer than a hungry chimpanzee is willing to work at a task he dislikes (Sultan and the banana skins) or is unable to perform. Many chimpanzees appear to enjoy drawing and painting, as discussed fully in Morris (1962). For no material reward an individual may work, for minutes on end, with intense concentration. Washoe and Moja began to draw and paint as infants (Gardner and Gardner, 1978) and their current performances are being systematically investigated (Beach, Fouts, and Fouts, 1984).

The kind of experience a chimpanzee has had, prior to being asked to perform a given task, is often crucial in determining his success or failure. Thus if a chimpanzee is to solve a problem such as acquiring a suspended fruit by hitting it down with a stick, it is of utmost importance that he has had previous opportunity to handle and play

Sultan joining two sticks to make a longer tool. (From Köhler, 1925)

with sticks. Schiller (1952) found that chimpanzees who had had *no* experience with sticks were unable to use them in test situations. After they had been allowed access to sticks in free play, however, they could transfer the knowledge thus gained to the problem situation. In the same way, Sultan was only able to fit two sticks together, thereby making a tool long enough to reach food outside his cage, *after* he had accidentally pushed one into the other during play. Before this the solution did not present itself, even when Köhler demonstrated the possibilities of the hollow stick by pushing his own finger into one end. The importance of previous experience was convincingly demonstrated in an experiment that tested the imitative abilities of Viki, of preschool human children, and of a laboratory-raised chimpanzee—all of similar age. Viki had shared many of the experiences of a typical human child; her results were much more like those of the human children than were those of the laboratory youngster—who, in fact, utterly failed most of the tests (Hayes, 1951; Hayes and Hayes, 1952).

The effect of early experience on subsequent adult learning skills is an area that, quite rightly, is receiving a good deal of attention today. There is evidence that even differences in the visual experiences of young infants can affect their ability later in life to learn tasks involving visual discrimination (Hinde, 1966). And Rumbaugh (1974) showed that social and environmental deprivation during the first two years of life had impaired at least one aspect of the higher cognitive learning processes in fourteen-year-old chimpanzees. In two-choice discrimination tests, after *reversal* of previously learned correct-incorrect responses, the performance of wild-born individuals of the same age was "profoundly superior." The two groups had been maintained in similar laboratory conditions after the first two years.

Higher Cognitive Abilities

The results of investigations into the higher cognitive abilities of creatures other than ourselves have frequently challenged the uniqueness of the human animal and thus provoked storms of protest. Fortunately, in the true spirit of science many investigators ignored their critics and continued their work.

Cross-Modal Transfer

The ability to recognize by touch what we see with our eyes, and vice versa, was until quite recently one of the attributes supposed to be unique in human beings. However, we now have clear-cut evidence that chimpanzees are capable of cross-modal transfer of this sort (Davenport and Rogers, 1970). Chimpanzees can even identify from photographs not only objects with which they are visually familiar, but those which previously they have known by touch alone (Davenport and Rogers, 1971; Davenport, Rogers, and Russell, 1975). More recently it has been shown that this ability to transfer information from one sensory channel to another is not peculiar to man and his closest relatives; it has been demonstrated also in pigtail macaques (Gunderson, 1982). I suspect that this will lead to widespread attempts to uncover the same ability in a variety of species, and that many such efforts will prove successful.

Abstraction and Generalization

Another ability, which until recently was thought to set mankind above "the beasts," is *concept formation*. In order to form a concept of something, such as an apple, it is necessary to abstract certain of its properties (such as shape, color, smell) and to make a generalization—*this* (an apple) is the same as *that* (another apple), but different from *this* (a peach). One step more, and the concept of fruit is formed—apples *and* peaches, as distinct from tennis balls. Such an exercise requires also the ability to *classify* objects according to certain key properties.

If the Gardners had not initiated the ape language-acquisition era, we might know very little about the chimpanzee's abilities in this sphere. As it is, we know a good deal, and we are learning more all the time. As the Gardners point out, Washoe's errors were in many ways even more revealing than her correct responses. For example, when asked to name a comb, she might sign *brush* but was unlikely to sign *plate*. *Plate*, however, might well be an incorrect response to a bowl—or even a cup. In other words, she was clearly classifying objects into categories. The four young chimpanzees who took part in the Gardners' second project were able to classify reliably a wide variety of different kinds of dogs as *dog*, a variety of flower species as *flower*, many kinds of insects as *bug*, and so on.* Interestingly, cars (photographs or toys) tended to be classified with animate rather than inanimate objects (Gardner and Gardner, 1983).

Washoe and other chimpanzees with language training all spontaneously generalized the use of signs, transferring them from the context in which they had been learned to new and appropriate ones. Washoe, who learned the sign *open* in connection with doors, began to use it when she wanted various containers opened, or the refrigerator, or even the water faucet (Gardner and Gardner, 1969). Lana, working with the Yerkish language, readily used stock sentences for purposes different from those she had originally been taught, in ways completely suitable to the situation at hand (Rumbaugh and Gill, 1977).

The language-trained chimpanzees have also demonstrated their comprehension of category relationships among objects. Sarah reliably distinguished between objects that were *same* or *different* and understood spatial relationships such as *on-under* and *in-out* (Premack and Premack, 1972; Premack, 1976).

When Sarah was sixteen years old, she was tested for her ability to solve analogy problems. In one series of thirty-six tests (each one unique) the problems were of the following type: the stimulus A was a saw-toothed blue shape marked with a dot; A′ was the same, but unmarked. Thus the A to A′ relation that Sarah had to grasp was removal of the dot. The B stimulus was an orange, marked crescent: the alternatives from which Sarah had to choose were a large, orange, unmarked crescent and a large, blue, marked crescent. On problems of this sort 72 percent of Sarah's choices were correct. Since each was unique, she could not have learned the solutions by trial and error; moreover, her choices for each of the first four problems were correct. These results suggest that she possessed analogical reasoning processes prior to the start of the experiment (Gillan, 1982).

Sarah was also tested on conceptual analogy problems. For example, if given that paper (A) is to scissors (A′) *same-as* apple (B) is to *either* knife *or* plate, Sarah chose knife. She chose correctly in 85 percent of such trials. Since this could have resulted from association (of apples with knives more than with plates), she was tested on problems such as lid (A) is to jar (A′) *same-as* apple (B) is to *either* knife *or* plate. Here she correctly chose plate. Gillan (1982) concluded that the chimpanzee is able to apprehend both perceptual and conceptual relations and to use these relations in *same-different* judgments.

Displacement in Time and Space

To what extent does the world of the chimpanzee extend beyond the present in time and space? In other words, how much can a chimpanzee recall of past events; how much can he anticipate—even plan for—the future?

The capability of storing information in the central nervous system, and of retrieving it as required, is of course obligatory if an animal is

* When Washoe was first introduced to live chimpanzees and asked *What those*? she signed back emphatically, *Black bugs*. Whether this tells us more about her concept of insects or of chimpanzees is not clear!

to profit from his experiences. At a purely descriptive level, it is the interaction between a given situation and the central memory store that enables an individual to arrive at a correct solution, whether his problem is the identification of a companion after an absence or a complex test situation in the laboratory.

That chimpanzees have good memories for individuals, both human and animal, who have been meaningful in their lives is well known and is convincingly illustrated in the anecdote at the start of this chapter. The literature contains many reports of chimpanzees who recognized caretakers after a year or so, and one female after four years remembered the person who had nursed her through a bad illness (Yerkes and Yerkes, 1929). Köhler (1925) remarks that one of his chimpanzees was clearly remembered by others of the group, and greeted as a friend, after an absence of four months; the oldest male of the Menzel group, Rock, was recognized upon his return after eighteen months (W. C. McGrew, personal communication). Chimpanzees also have good recall of individuals who have mistreated them (Yerkes, 1943).

The key role of location in the perceptual world of the chimpanzee has already been emphasized, and some impressive demonstrations of chimpanzee locational memory were recorded years ago by Tinkle-paugh (1932). The subjects, a young male and a young female, were led (separately) through six rooms and watched as in each room a piece of food was hidden under one of two identical containers. They were then taken back to the starting point, permitted to move freely from room to room, and allowed one choice from each pair of containers. Their scores, respectively, were 92 percent and 88 percent correct. The same individuals were tested in an even more difficult setup. As many as sixteen pairs of identical containers were arranged in a circle in a large room. The chimpanzee sat on a stool in the middle of the circle while the experimenter placed food under one container of each pair. The subject was then led to the first pair, where he (or she) made a choice. If successful, he was returned to the stool where he ate the reward before making a second choice, and so on around the circle. Both chimpanzees scored between 78 and 89 percent in all tests—better than human children.

This spatial memory enables a chimpanzee to form an excellent mental map of the important landmarks of his environment. Menzel (1974) carried out a series of trials with young chimpanzees in a large field enclosure. Each individual was carried around and allowed to watch as an experimenter hid up to eighteen pieces of food. He or she was then returned to an inside room from which the enclosure was not visible. After delays of at least two minutes individuals showed excellent recall, usually finding *all* the hidden rewards and often reaching them via shortcuts rather than following the route taken by the experimenter. Menzel (1978) subsequently tested different subjects in a different enclosure; after watching *on closed-circuit television* as the food was hidden, they performed equally successfully.

Chimpanzees not only remember where things are in their environment, they retain such knowledge over long periods. One of Köhler's chimpanzees was allowed to watch food being buried in sand, which was then smoothed over. When given the opportunity—forty-eight hours later—he ran straight to the location of the food without having seen the hiding place in the interim. More recently, Lana showed that chimpanzees can perform feats of memory of a more "academic" variety. She was readily able to store and retrieve from memory her Yerkish symbols. Furthermore, she showed a tendency to organize her recall in much the same way as humans (Buchanan, Gill, and Braggio, 1981).

Tinklepaugh (1932) carried out experiments that demonstrated anticipation. A chimpanzee who had watched a favorite food being placed under a container showed obvious signs of surprise, disappointment, and sometimes resentment if prior to his being permitted to open the container, and without his knowledge, a less-favored food had been substituted. This he expressed in bodily attitudes, facial expressions, and vocalizations. Similar responses were obtained if a large piece of food was secretly exchanged for a small piece. The Temerlins' Lucy frequently accompanied the family to their ranch for an outing. On the way they had to cross a number of bridges, which Lucy hated because they frightened her. "She sees them coming, sometimes as much as two or three hundred yards in the distance, and starts to show signs of anxiety," wrote Temerlin. One very rickety bridge scared her more than the others; a good half-mile before they reached it, she began to rock and whimper, sometimes grabbing the hand of whoever was driving (Temerlin, 1975, p. 73).

Impressive evidence of the ability to plan ahead was provided by Döhl (1968) and Rensch and Döhl (1968). A female chimpanzee, Julia, who had already demonstrated considerable aptitude in working complex mazes, was presented with two series of five locked transparent boxes each of which opened with a differently shaped key (Figure 2.3). One series of five led to a box containing a banana. The other culminated in an empty container. From two unlocked initial-choice boxes Julia was able to select the key that opened the first of the series of boxes that led to the banana; only by working backward from this goal could Julia's initial choice be made on a reasoned basis.

Julia's future goal was there in front of her. One of the females of the Arnhem colony, on the other hand, showed clear evidence of planning for a future that existed only in her mind. The incident took place soon after the weather had begun to turn cold, heralding the approach of winter. Franje, prior to leaving her indoor sleeping cage in the morning, carefully gathered an armful of straw, which she carried outside and used to construct a warm nest. She gathered the straw *before* she had felt the cold air outside, so her action apparently was based on memory of feeling cold the day before and anticipation of similar discomfort in the immediate future (de Waal, 1982).

During the long battle to toilet train Viki she was rewarded for successful performance with a candy. The late Leonard Carmichael told me of an incident that took place when he was visiting the Hayeses. Viki, on her way to the potty, paused to jump up onto the sideboard, where she took her reward from a jar. Perhaps as a result of this delay, things did not go as she had (presumably) anticipated. To her visitor's amazement a subdued and damp-bottomed Viki climbed up and returned the candy to its jar. He was the only person in the room with her.

Figure 2.3 Julia must choose the correct one of the two keys that she sees through the plastic lids of the two unlocked initial-choice boxes. This key will open the first of the locked Plexiglas boxes in the series culminating in the goal box—which contains a banana. If she makes the wrong initial choice, her key will open the first box in the second series—which leads to an empty container. (Adapted from Döhl, 1968)

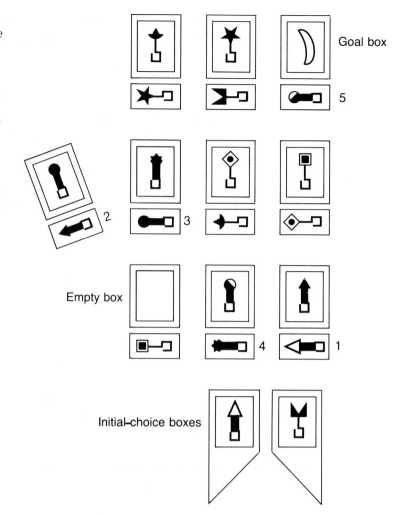

Goal box

Empty box

Initial-choice boxes

Julia at work on one of her mazes.
(From Rensch and Döhl, 1968)

There is considerable evidence relating to the ability of the chimpanzee to anticipate the social consequences of his own behavior and that of a companion, and to plan his actions accordingly. I shall return to this subject later on.

Symbolic Representation
In a review of symbolic processes in the chimpanzee, Meddin (1979, p. 99, referring to Mead, 1934; and Langer, 1957) defines symbols as "representative forms of thought: that is, symbols 'stand for' or represent something else. However, unlike concrete forms of representative thought, the referents for symbols are abstractions (synthetically constructed cognitive categories) as well as direct stimuli."

The ability to interpret a photographed image is dependent on a nonverbal type of symbolic comprehension (Davenport, Rogers, and Russell, 1975; see also Langer, cited in Meddin, 1979). As we have seen, chimpanzees have a highly developed ability to interpret photographs, first demonstrated by Köhler (1925). Viki Hayes (like other home-raised chimpanzees) used to put her ear to photographs of watches; she also gave food-barks while tapping a photograph of a chocolate bar (Hayes and Hayes, 1953). The Premacks' Sarah, as we shall see, had an extremely sophisticated understanding of scenes on video.

An extraordinary response to photographs was that of the Temerlins' Lucy. She was given a copy of the magazine *Playgirl* when she was at the height of estrus. "As she came to each picture of a nude male, her excitement visibly increased. She stared at the penis and made sounds similar to those she utters when looking at some delicious morsel, a low guttural 'uh, uh, uh, uh.' She stroked the penis with her forefinger, cautiously at first and then more rapidly. On some pictures she would first stroke the penis with her forefinger, get very excited, and then mutilate it by scratching it with her fingernail. When she finished with one picture she would turn the page and start on another . . . She did not caress or scratch any other part of the photograph." When she came to the centerfold, she spread it on the floor, positioned herself over it, and rubbed her vulva back and forth on the penis for about twenty seconds. Then she bounced up and down over the penis. Finally, having moved away, she returned and with great care directed a small trickle of urine directly onto the penis (Temerlin, 1975, p. 138).

In language-acquisition projects chimpanzees have been taught the meaning of sign, lexigram, and word symbols. These they use or respond to reliably and with high frequency. The blue plastic triangle that was Sarah's symbol for an apple apparently provided her with an appropriate mental image: she was able to ascribe properties of "redness" and "roundness" to the piece of plastic in the absence of a real apple. In other contexts Sarah was able to identify correctly properties of "blueness" and "triangularity" (Premack and Premack, 1972; Gillan, 1982).

A series of experiments was carried out with an ASL-trained youngster, Ally, to determine the relationship between his understanding of spoken words and ASL signs. Both word and sign clearly provided him with a mental picture of the object they stood for. Thus when his understanding of the spoken word *spoon* had been established (he responded correctly to spoken commands such as "Fetch a spoon!"), he was taught the ASL sign, with the spoken word as the only referent (that is, no spoon was physically present). When Ally was given the same commands as before, this time in ASL, he was able to comply successfully (Fouts, 1973).

Spontaneous combination of acquired symbols occurred in all the ASL subjects, and their first combinations (*Susan brush, There drink, Up go*, and so on) appeared at six to ten months, an age comparable to that for signing human infants, earlier than for speaking infants (Gardner and Gardner, 1980).*

Language-trained chimpanzees sometimes combine signs or lexigrams creatively in order to describe objects for which they have no symbol. Thus Washoe, when asked *What that?* as she gazed for the first time at a swan, made two signs, *water* and *bird*. It has been suggested that she was merely signing *That is water, that is bird* in succession. But this argument cannot be applied to her insistent demands for a *rock berry*, which turned out to be a Brazil nut. The Temerlins' Lucy, who also used ASL, was presented with a great array of foods—vegetables, fruits, and so on. For most of them she only had "category" signs such as *vegetable, food, drink*. Asked *What that?* she produced for some foods combination names that reflect interestingly her outlook on life. Celery was *pipe food*; watermelon, either *candy drink* or *drink fruit*; and radish, because the first she tasted was old and hot, *hurt-cry food* (Temerlin, 1975; Fouts and Budd, 1979). Lana, too, showed creative abilities, labeling a cucumber *banana which-is green* and an orange *apple which-is orange* (Gillan, 1982). Moja referred to an Alka-Seltzer as a *listen drink* and a cigarette lighter as a *metal hot* (Fouts, 1975; Gardner and Gardner, 1980).

Of particular interest is the fact that on two occasions chimpanzees invented signs. Lucy, as she got older, had to be put on a leash for her outings. One day, eager to set off but having no sign for *leash*, she signaled her wishes by holding a crooked index finger to the ring on the collar she always wore. This sign became part of her vocabulary (Temerlin, 1975). In another incident (Gardner and Gardner, 1969) a hungry Washoe forgot the sign the Gardners had taught her for *bib*. With the index fingers of both hands she traced the outline of a bib on her chest. (The Gardners corrected her, but found out later that Washoe's was, in fact, the correct ASL, which they had not known; theirs was an invented sign!)

Concept of Self

Can the chimpanzee abstract a concept of self? A self it can then look at, as an object? The subject is fully discussed by Meddin (1979). According to Mead, 1934 (cited by Meddin, p. 104), "self-consciousness arises through social interaction, in an animal which is capable of

---

* The Gardners also found that the referents for the first fifty signs that appeared in the ASL vocabularies of four home-raised chimpanzee infants and for the first fifty spoken words for eight human infants showed extensive overlap. In fact, the fifty-sign vocabulary of any one of the chimpanzees could be substituted for the fifty-word vocabulary of any one of the children without changing the degree of overlap (Gardner and Gardner, 1983).

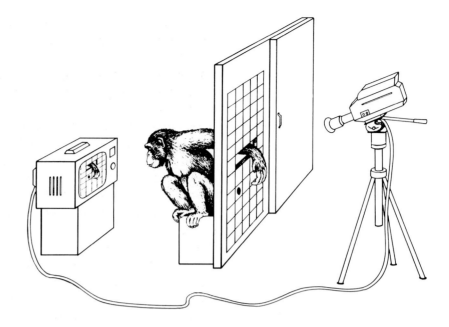

symbolic thought, and is part of an organized social group." The concept of self is a "cognitive structure, contingent on the ability of the animal to assume the perspective of others towards itself"; this, then, enables it to see itself as an object.

Washoe was the first chimpanzee to announce that she recognized herself. When asked, as she stared at her mirror image, *Who that?* she responded without hesitation, *Me Washoe* (Gardner and Gardner, 1969). That Washoe was not unique in this ability was shown experimentally by Gallup (1970, 1977). Five chimpanzees were anesthetized, then painted with patches of red dye in places that would be visible to them only when they looked in a mirror. On waking, they stared at their images and repeatedly touched the strange marks on their bodies. These chimpanzees had all previously seen their reflections in a mirror; three isolation-raised chimpanzees who had not had such experience did not react in this way to patches of paint and showed no evidence of self-recognition.

Menzel, Savage-Rumbaugh, and Lawson (1985) designed a series of ingenious tests in which two language-trained chimpanzees, Austin and Sherman, used mirrors and closed-circuit television pictures to guide their hands to objects not directly visible. Both, prior to testing, had shown unequivocal signs of self-recognition when looking at themselves in a mirror or on a television screen. For example, each shone a flashlight down his own throat while staring at his image on a monitor. Each chimpanzee, during the first series of tests, was at once able to reach around a barrier and locate a piece of food which, like his hand, he could see only via a mirror. In the next series of experiments (Figure 2.4) each was required to locate a piece of food (or, subsequently, a spot of odorless ink) placed on the outside of a door. The chimpanzee could see this object by looking into the television monitor inside his room; he could reach it by pushing his arm through a hole in the door and, keeping his eyes on the screen, orienting his hand toward the goal. This task was easily mastered even when the image was inverted, reversed, or reversed *and* inverted. In fact, when

Austin first saw this inverted image, he spontaneously turned his rump to the screen and with his head inverted looked at it through his legs. A year after these tests had been completed, with no intervening practice, the chimpanzees were required to perform the same task. Their performance was even better than before, each animal contacting the target within fifteen seconds.

The fact that Washoe, herself bathed and oiled daily, repeatedly subjected her doll to exactly the same treatment, suggests that she had a sophisticated understanding of herself as an "object." Viki, however, should have the last word on self-concept. One of her tasks was to sort a pile of mixed-up photographs depicting a variety of people and animals. She only made one "mistake." Viki Hayes was *people*. For Viki, of course, this was not a mistake at all (Hayes and Hayes, 1951).

Social Problem Solving

So far only a few investigations in this significant area have been pursued under the controlled conditions of the laboratory, and they have seldom involved more than two chimpanzees. Years ago Crawford (1937) tested two juveniles for their ability to cooperate in securing food rewards. More recently the two young language-trained chimpanzees Austin and Sherman revealed a capacity for symbolic communication. First they were taught the correct Yerkish for eleven different types of food and drink. When both were proficient, one watched as a container was filled with one of the foods. He was taken back to the room shared by the two youngsters and encouraged to ask the experimenter for the hidden food item. His companion, who had no knowledge of the contents of the container prior to "reading" the console message, was then given access to the keyboard and, in his turn, could ask for the food. Only if both chimpanzees performed correctly were they rewarded and the food shared. It turned out that from the very first they made almost no mistakes. Subsequently, when through a window between their adjoining rooms one observed the other feeding, he approached his keyboard and spontaneously requested the food he saw. His companion was encouraged to give him some (Savage-Rumbaugh, Rumbaugh, and Boysen, 1978).

Inferring Purpose in Others
During one of his experiments Köhler (1925, p. 145) recorded the following incident: "I was endeavouring to teach Chica the use of the double stick. I stood outside the bars, Sultan squatted at my side and gazed seriously . . . As Chica absolutely failed to realize what was required, I finally gave the two sticks to Sultan, in the hope that he would make things clear. He took the sticks, fitted one into the other, and did not himself appropriate the fruit, but pushed it, in a leisurely manner, towards Chica at the bars." This response, says Köhler, "clearly proved [that Sultan] really sees the task to be carried out, *from the standpoint of the other animal*" (italics added).

Finally, some fifty years later, Premack hit upon a very elegant and imaginative way of investigating the extent to which a chimpanzee can, in fact, impute mental states to others (Premack and Woodruff,

Grande climbing for a suspended banana, watched by Sultan. Note Sultan's sympathetic left hand. (From Köhler, 1925)

1978). The chimpanzee Sarah was shown a series of brief videotaped scenes of a human actor trying to cope with a variety of problems, such as attempting to escape from a locked room or shivering when an electric heater was unplugged. After watching a film, Sarah was given a pair of photographs, one of which represented the "solution" (a photograph of a key, or of a plugged-in heater). Her consistent choice of the correct photograph suggests that she "recognized the videotape as representing a problem, and understood the actor's purpose" (p. 515). In subsequent tests Sarah was presented with the same series of films, but a set of refined alternative solutions. Thus she had to choose among photos of intact, bent, or broken keys; electric cord attached, unattached, or attached but cut. Again she made consistently correct choices.

Other tests were carried out to assess whether Sarah's choice of solution would vary depending on her relationship with the particular actor. Two sets of tapes were made. In the first, her favorite keeper was faced with a problem—for example, trying to obtain bananas when his reach was impeded by a box filled with cement blocks; in the second, a disliked person, "Bill," was trying to do the same thing. As before, Sarah had to choose between a "good" and a "bad" alternative (for the above problem, the actor unloading blocks or else lying on the floor with the blocks strewn over him). There were eight tests: Sarah chose only good solutions for her favorite keeper, but six bad ones for Bill. On subsequent tests it transpired that her favorite bad photograph was the prone actor covered with blocks! The authors conclude that the chimpanzee will not fail tests that "require him to impute *wants*, *purposes*, or *affective attitudes* to another individual" (Premack and Woodruff, 1978, p. 526).

Two young chimpanzees at the Pavlov Institute of Physiology demonstrated an understanding of each other's needs. They were taught (separately) the values of three tokens of different shapes and colors; if handed to the trainer, one of the tokens produced food; one, drink; and one, toys. When both chimpanzees had reached the required level of performance, they were placed in adjoining cages. One of them, A, was hungry; he had a toy in his cage. The other, B, had eaten well and was provided with a pile of bananas. Each had a full set of tokens. Soon A extended his food token to B through the bars dividing them. B took it—and selected a banana from his pile, which he gave to his companion. Soon after this B gave A his toy token, and A handed over his toy (Schastny, reported by Popovkin, 1981). These two chimpanzees had *never* been required by their trainer to hand over food or toys, and their giving during test situations was completely unprompted.

Intentional Communication

Another series of experiments was concerned with the ability of the chimpanzee to deceive (Woodruff and Premack, 1979). To what extent could information be withheld, or false information given, when this would benefit the individual? The subject was confined to his cage and allowed to watch as food was hidden under a container. Each of four young chimpanzees learned over many trials to point out the location of the hidden food to a cooperative human, who then shared the food with his "informer." The chimpanzees also learned, after many more trials, to *withhold* information regarding the location of food from a competitive human, one who always ate the food himself. The eldest of the chimpanzees even learned to give *false* information to the competitive human.

These experiments demonstrate a capacity for intentional communication in the chimpanzee, a capacity that depends in part on the ability of the performer to understand the motives of the individual with whom he is communicating.

Intelligence and Rationality

In a recent article Mason (1982) lists some of the major characteristics of intelligent behavior per se: the ability to respond differentially to a large variety of objects and events; the tendency to modify existing knowledge in light of changing circumstances; a diversity of motives and goals; the use of variable and often indirect means to relate what is known to what is wanted. Mason concludes his list with the comment, "Evidence of planning, of foresight, of the ability to establish and work towards subgoals, to single out the essential features that define a problem, are the epitome of intelligence" (p. 134). This list might have been drawn up to characterize chimpanzee behavior.

Stenhouse (1973, p. 31) earlier defined intelligent behavior as "that which is adaptively variable within the lifetime of the individual," and intelligence as the "capacity for intelligent behavior." By contrast with instinctive behavior, which is fixed and rigid, intelligent behavior is flexible and adaptable. It is possible to learn a nonadaptive behavior but, by definition, an intelligent act cannot be nonadaptive—which is not to say that an intelligent being cannot perform unintelligent actions. Stenhouse's four-factor theory of the evolution of intelligence postulates that the oldest factor, phylogenetically, is that responsible for *sensory-motor control*, or efficiency. Next, the *central memory store* and a factor for *abstracting and generalizing information* may have evolved more or less simultaneously. Last, and most recently developed, a *withholding factor* allows an animal to refrain from responding "instinctively" to a given stimulus; an alternative response can then be performed, which may be more adaptive.

This theory enables us to trace the gradual evolution of intelligent behavior from invertebrates to humans. Thus the cockroach, able to vary its escape reaction adaptively in proportion to the time it has had to explore the area (Barnett, 1958), shows an element of intelligent behavior. Its total capacity for intelligent behavior may not rate high (by human standards), but this should not prevent us from recognizing the component it has. Showing a more advanced level of abstraction and generalization, the pigeon can be trained to peck at representations of people in photographs (Herrnstein and Loveland, 1964). And the chimpanzee, as we have seen, can choose a photograph representing the solution to someone else's problem, in accordance with his like or dislike of the person represented.

Stenhouse argues that it is the appearance and development of the fourth element, the withholding factor, that has played such a vital role in the evolution of humanlike intelligent behavior. It is a factor that may have had its origin in (and be homologous with) Pavlov's (1955) "internal inhibition." Thorpe (1956) points out that in complex tests it is often necessary for an animal to refrain from responding, and that such restraint may occur even when the action has previously been reinforced. In summarizing the performances of the young chimpanzees who learned to suppress information about the location of food, Woodruff and Premack (1979, p. 357) comment that this was "made possible, at least in part, by the subject's ability to inhibit behavioral predispositions."

Intelligence probably evolved, at least in part, in order to cope with the social pressures and problems of group living (Jolly, 1966; Kummer, 1971)—the need for an individual to adjust his own behavior and desires in accordance with the behavior and desires of others. This often leads to conflicts when, as Midgley (1978) points out, one desire must be suppressed to make way for another. It is, Stenhouse suggests, the withholding factor that enables an animal to pause when confronted by a problem, or a conflict of desires, and *choose* a solution. And it is this process of choosing that "we rightly call reasoning" (Midgley, 1978, p. 258). From a purely subjective point of view, it is when an animal pauses, as if to think, that we get the impression of intelligence at work. Köhler (1925, p. 166) noted this while he was using Sultan to demonstrate to a critical colleague the problem-solving abilities of the chimpanzee: "nothing made so great an impression on the visitor as the pause . . . during which Sultan slowly scratched his head and moved nothing but his eyes and his head gently, while he most carefully eyed the whole situation."

Humans, of course, are not merely "clever": we are a rational species. Rationality combines *cleverness*—"calculating power, the sort of thing that can be measured by intelligence tests"—and *integration*—having "a firm and effective priority system . . . based on feeling" (Midgley, 1978, p. 256). Fletcher (1966, p. 321) makes the same point when he argues that the level of intelligence shown by an individual should not only be taken to mean "the relatively unalterable capacity of the specifically cognitive, conscious, thought-processes" but also the "state of organization of his personality as a whole." Often, of course, as both Midgley and Fletcher point out, the two go hand in hand. But "intellectuals" sometimes behave in a very irrational way, whereas others, quite unable to solve complex cognitive problems, may be endowed with "sound common sense" (a judgment usually passed only when the priority system of the individual concerned happens to agree with one's own).

The structure of preferences in which our rationality is embedded is not peculiar to humans but is found also in the higher animals (Lorenz, 1963; Midgley, 1978). A chimpanzee who can, without difficulty, solve a complex series of discrimination problems, but who cannot respond appropriately to the challenge of a bigger, stronger male, would not do very well in the natural habitat. Köhler's least-gifted individual, Rana, showed many "stupidities" in her attempts to solve problems. But far more detrimental was her apparent inability to form good relationships with the other individuals of her group, since "on account of her stupidity and her dependent, dull behavior, she was for the most part *de trop*" (Köhler, 1925, p. 69).

Midgley (1978, p. 282) concludes her discussion on animal and human rationality with these words, "What is *special about people* is their power of understanding what is going on and using that understanding to regulate it" (italics added). But is this so special to humans? The Premacks' young chimpanzees were able to distinguish between the behavior of generous and selfish humans and adjust their course of action accordingly. Sarah certainly comprehended the problems faced by human actors, and she was able to regulate, in theory, the outcome of those problems. Indeed, we might almost say that she indulged in wishful thinking in her repeated choice of the picture illustrating the disliked Bill sprawled beneath cement blocks.

It has become increasingly obvious over the past decade that man does not stand in isolated splendor, separated from the beasts by an unbridgeable chasm. One by one the attributes that supposedly placed him in this exalted position have been shown to exist in "lowlier" forms of life. The slow march of evolution has progressed with measured strides from cockroach to monkey, monkey to ape, ape to man. Nonetheless, even if we differ from chimpanzees not in kind, but only in degree, it is still an overwhelmingly large degree. This we should not forget. As Yerkes (1943, p. 185) wrote, "The ape is at the beginning of a road on which man has advanced far."

Which brings us back where we started. How far can the chimpanzee travel with us on the road from the apple tree to Eden?

At the beginning of this chapter we looked at ways in which a human could perceive, use, or think about an apple tree. Let us now ask just how similar or different a chimpanzee's perceptions would be and where we have climbed beyond him to higher levels of feeling, understanding, or imagination. Of course, it will depend on which chimpanzees we take with us to look at the tree. Sultan's perceptions will differ from Lucy's; Viki will see things very differently than an individual taken from the solitary confinement of a laboratory prison; Washoe and Sarah can certainly *tell* us more about their thoughts. Sarah will inform us, if we give her bits of colorful plastic, that the apples are round and red—or green. Washoe will sign *Me Washoe food hurry hurry*. Viki will give food-grunts and leap up for an apple; Lucy will give food-grunts too, but her mind will be on fermenting windfalls and the delights of intoxication; Sultan, poor gymnast that Köhler dubbed him, will spy that stick and knock down an apple in a flash—or he may try to drag us beneath the lowest branch for use as a ladder; the laboratory chimpanzee may grunt rather hopelessly and do little else. If Washoe sees a bull chasing her adopted son across the field, she just *may* sign *Loulis hurry climb*. The laboratory chimpanzee probably will not even perceive the tree as an escape route for himself, let alone for a companion; if he has been socially deprived in infancy, he certainly will not. Sarah will not only know exactly what the fleeing individual should do, but if he happens to be an enemy of hers, she will probably try to direct him to a branch she knows will break and deposit him in the path of the enraged bull. Washoe and Moja, when they return home, may firmly mark a canvas and, when asked, label their art *apples*. This difference in perception and understanding from one individual to the next is a measure of the extent to

Chimpanzee representations of apples: *above*, drawn by Moja, who completed her drawing; *below*, drawn by Washoe, who signed *apple* when she had finished. The drawings are extraordinarily similar. Despite that similarity, the chimpanzees who drew them, when systematically tested, were found to have no regular schema for apples (or any fruit), although they did have one for other items, such as "brush." If this really is representational art, we must ask just what aspect of an apple the chimpanzee is trying to portray. (R. Fouts)

which a species can modify its inherited behavioral tendencies during the lifetime of an individual. It is not only that each individual differs; it is the truly remarkable scope revealed by the nature of the differences that is impressive.

It should by now be quite clear that the chimpanzee, given the right background and at least normal intelligence, can keep step with us much farther along the road than would have been thought possible ten years ago. I was talking not long ago to a friend about the subject of this chapter, about chimpanzees and apples. By a strange coincidence she, as a little girl, had firmly believed in a fairy who lived in the oldest apple tree in her parents' orchard. Is this a real difference between man and chimpanzee, the *imagination* that creates fairies and Eve in the Garden of Eden? Yerkes and Yerkes (1929, p. 577) wrote that there was evidence of creative imagination in the playful activities of apes who "tend to invent ways of amusing themselves and to develop frequently . . . performances which are complex and compel the attention of human observers."

Once again, Viki Hayes must have the last word. One day in the bathroom Cathy Hayes suddenly noticed Viki behaving in a most extraordinary fashion.

Very slowly and deliberately she was marching around the toilet, trailing the fingertips of one hand on the floor. Now and then she paused, glanced back at the hand and resumed her progress . . . Viki was at the pull-toy stage when a child is forever trailing some toy on a string . . . Dragging wagons, shoes, dolls or purses, her body assumed just this angle . . . She interrupted the sport one day to turn and make a series of tugging motions. That is, they would have been called tugging had there been a rope to tug, which of course there was not. She moved her hands over and around the plumbing knob in a very mysterious fashion; then placing both her fists one above the other in line with the knob, she strained backward as in a tug of war. Eventually there was a little jerk and off she went again, trailing what to my mind could only be an imaginary pull toy . . .

Viki dearly loves to "fish." Standing on the furniture, she pulls up from the floor any plaything with a string tied to it. Now from the potty she began to raise the "pull toy" hand over hand by its invisible rope. Then she lowered it gently and "fished" it up again.

Viki played the game every day, "but it never happened except around the toilet."

One afternoon, about two weeks after Viki first began to play in this way, Hayes continues:

[Viki] stopped once more at the knob and struggled with the invisible tangled rope. But this time she gave up after exerting very little effort. She sat down abruptly with her hands extended as if holding a taut cord. She looked up at my face in the mirror and then she called loudly, "Mama! Mama!" . . .

Acting out an elaborate pantomime I took the rope from her hands, and with much pulling and manipulation, untangled it from the plumbing and held out to her the rope which neither of us could see (I think). "Here you are, little one." . . . Her funny little face crinkled into a grin and she tore off around the toilet faster than ever before, dragging her imaginary toy behind her.

Out of curiosity, Cathy Hayes decided to "invent" a pull toy herself. The first day, as she walked around pretending, Viki watched in amazement, then ran to stare at the floor *where the toy would have been*. The following day this imaginary performance seemed to terrify Viki who stared, whimpered, and rocked, and finally leaped into her "mother's" arms in a very distressed state. Neither of them ever played with the imaginary toy again (Hayes, 1951, pp. 80–85). Clearly, this is an avenue of research for the future—an exploration of the uncharted realm of chimpanzee imagination.

From Laboratory to Forest

Henry Nissen, after his pioneering two and a half months observing chimpanzees in the West Africa forest, wrote that in his opinion "the ultimate capacities of these apes for complex behaviour will be found and measured in our laboratory experimental situation" (Nissen, 1931, p. 103). And it has indeed been in the controlled laboratory environment, through carefully designed experiments with captive chimpanzees, that scientists have been able to investigate the upper levels of chimpanzee cognitive ability. In such a setting, by use of judicious rewards and praise, the subject can be encouraged to exert himself over and above the effort normally required in day-to-day living. He can be provided with new experiences which, in a sense, develop his mind; he can even be given new tools, such as symbolic language, with which to express himself. And, as new cognitive complexities are revealed, more sophisticated tests can be designed to probe even deeper into the chimpanzee's mental ability. Over the sixty years since Wolfgang Köhler began his classic studies, chimpanzees seem to have been performing at ever-higher levels of sophistication.

There have been many observations of intelligent behavior among the Gombe chimpanzees. But so often they are anecdotal. And while I firmly believe that anecdotes, judiciously used, provide one of the most important keys to understanding the complexity of chimpanzee behavior, it is nevertheless comforting when solid "proof" of the cognitive ability in question is provided in the laboratory. Thus it is with profound gratitude to the patient scientists whose work I have reviewed in this chapter that I turn now to the complexities of the natural habitat, to an arena where almost all behaviors are confounded by countless variables; where years of observing, recording, and analyzing take the place of contrived testing; where sample size can often be counted on the fingers of one hand; where the only experiments are nature's own, and only time—eventually—may replicate them. Back to the forests of Africa, where it all began.

3. Research at Gombe

David Greybeard helps himself to a banana. (H. van Lawick)

1964 An extract from my diary: "Had a great idea. Perhaps we can get funds for a student—two would be better—working for their PhD's or something. They could take over when I have to go back to Cambridge. What a relief—to lose six months of Flint's life would be ghastly!"

The idea came when I was back at Gombe after a few terms at Cambridge, the second time I'd had to leave since I began the research. The first time I had packed up the whole camp, tents and all, and taken everything to Kigoma. But I had left it there, stored in the district commissioner's house, because I had known I would return: I had to. During my second absence a mycologist had looked after the camp for me, doing his own study, but keeping an eye on things along with my Tanzanian cook and helper. But neither was trained in observation methods, and Flo had given birth during those months—so much valuable information gone forever. But it would not happen again. My plan worked magically, with permission from the Tanzanian government and funding for the students from the National Geographic Society.

So it was that the Gombe Stream Research Center was conceived. I had no idea at the time that it was to become one of the best-known field stations in the world.

There have been many changes at Gombe since I first arrived there in 1960. For one thing, the Gombe Stream Reserve, as it was initially designated, in 1968 became the Gombe National Park. In the early years I observed and collected data on my own; by 1975 there was an interdisciplinary, international team of researchers working on chimpanzee and baboon behavior. The two tents that sheltered my expedition in the first years gave place to the aluminum huts and three cement-block houses of the Gombe Stream Research Center. And today the center is staffed with trained Tanzanian field assistants. Methods of data collection have ranged from recording observations in notebooks to filling in complex check sheets. But the underlying philosophy of noninterference and the building of trust between observer and subject has stayed the same.

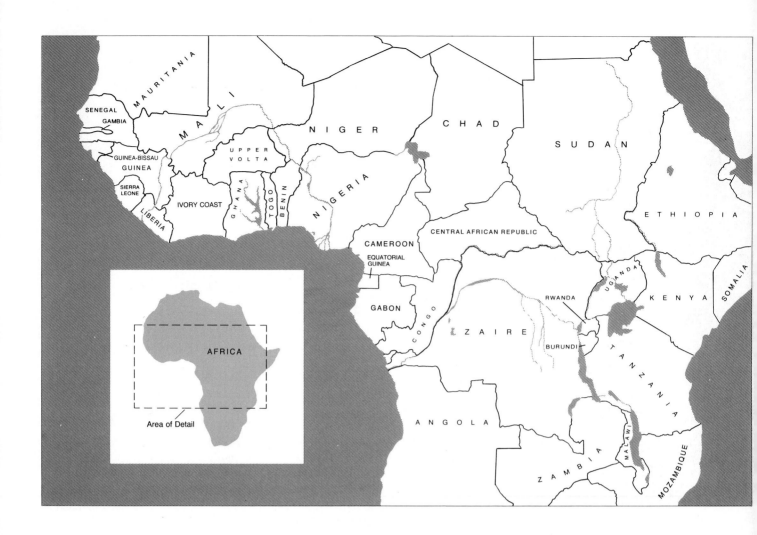

The Habitat

Chimpanzees are found in a wide belt that extends across equatorial Africa from the west coast to within a few hundred kilometers of the east coast (Figure 3.1). They are the most adaptable of the great apes, living in habitats that range from rain and montane forests to dry woodlands, and sometimes even savanna with widely scattered trees. Where chimpanzees live in low-altitude tropical rain forests, they experience little change in temperature from season to season; the humidity is always high and there may be only a few dry days during the year, although some months will be very much wetter than others. Where they live in arid areas in the north (such as the Senegal) and in the extreme southeastern limits of their range in Tanzania, there may be seven dry months in the year and considerable variation in temperature and humidity. In the Senegal midday temperatures can be as high as 42° C and drop to as low as 17° C (Baldwin, McGrew, and Tutin, 1982). In areas such as the Ruwenzoris, the fabled Ugandan "Mountains of the Moon," chimpanzees have been reported from as high as 800 meters, where temperatures, particularly at night, will be much lower (Schaller, 1963).

Gombe National Park (Figure 3.2), in the Kigoma region of Tanzania, is close to the eastern limit of the chimpanzee's range. It com-

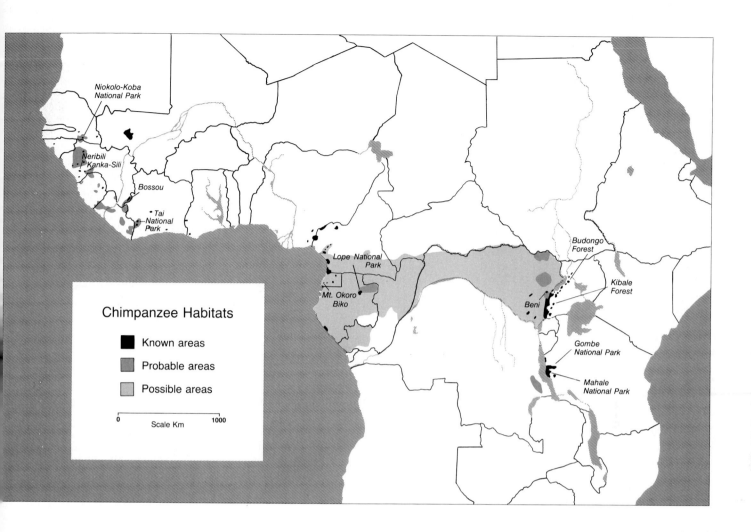

Figure 3.1 Distribution (known, probable, and possible) of *Pan troglodytes* in Africa. (Data courtesy of Geza Teleki, George Washington University, Washington, D.C., and the Survival Service Commission of the International Union for the Conservation of Nature and Natural Resources)

prises a narrow strip of rugged country on the eastern shore of Lake Tanganyika, which is a vast body of water roughly 675 kilometers long, 70 kilometers wide, and 775 meters above sea level. The hills rise steeply from the lakeshore to the crest of the rift escarpment, reaching heights of over 1,500 meters above sea level. Steep-sided valleys run from the escarpment down into the lake and these, in turn, are intersected by deep ravines draining from the ridges between the valleys. If the terrain were ironed out, the official 32 square kilometers (or 20 square miles) would probably be more than doubled.

Figures 3.3 and 3.4 show the mean rainfall per month and the mean number of rainy days per month during four selected years at Gombe. The year can be divided into the wet season and the dry season. From mid-October to mid-May there is a good deal of rain, particularly during the "long rains" between December and March. The rest of the year is dry. Because our data are organized for the most part into monthly blocks, it would be extremely time-consuming to analyze from half-month to half-month. October is therefore included in the dry season and May, with its high humidity and tall wet grass, in the wet season.

Temperatures are recorded daily inside the research building in the Kakombe Valley. Average maximum temperatures range in the wet

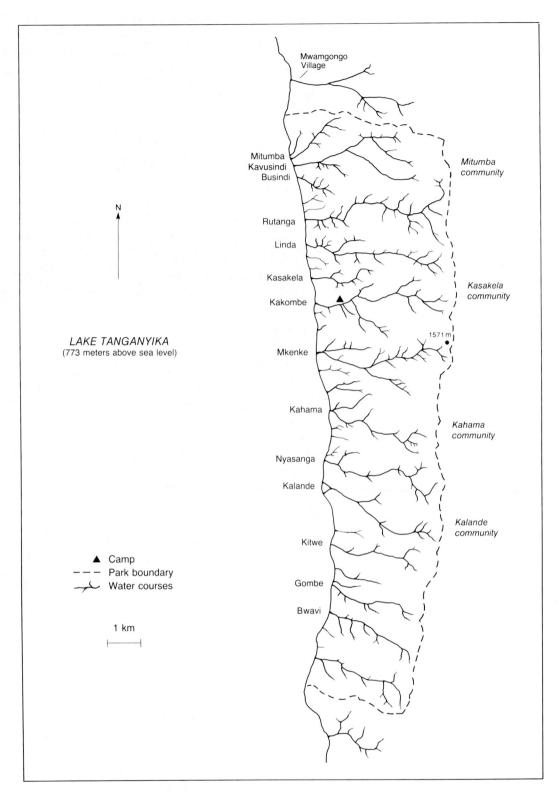

N

LAKE TANGANYIKA
(773 meters above sea level)

Mwamgongo
Village

Mitumba
Kavusindi
Busindi

Rutanga

Linda

Kasakela

Kakombe

Mkenke

Kahama

Nyasanga

Kalande

Kitwe

Gombe

Bwavi

*Mitumba
community*

*Kasakela
community*

1571 m

*Kahama
community*

*Kalande
community*

▲ Camp
- - - Park boundary
⌐⌐⌐ Water courses

1 km

Figure 3.2   Gombe National Park. The rift escarpment forms the eastern boundary.

Looking out from a high ridge over one of the steep-sided valleys. (H. van Lawick)

Each valley supports a small, clear, fast-flowing stream, fed from a watershed high up in the rift escarpment. David, carrying food, leaps across.

season from about 25° C to 26.5° C, and in the dry season from about 27° C to 30° C (with some days reaching 32° C). Mean minimum temperatures vary little, staying pretty much between 18.5° C and 21° C. August and September are the hottest months, with daily maxima regularly 30° C. Temperatures drop at the end of the rainy season, during May, June, and July.

Humidity is high, roughly 60 to 100 percent in the wet season and 30 to 70 percent in the dry season. These figures were obtained during January and August 1970 by Timothy Clutton-Brock (1972). Humidity tends to drop as temperatures rise during the day.

The temperatures recorded at the research station provide only an estimate of temperature variation in the park as a whole. The research building, with thatch on its walls and roof, is often considerably cooler than the outside slopes; at other times it is warmer, being sheltered inside from the wind. The building is situated on a fairly open slope,

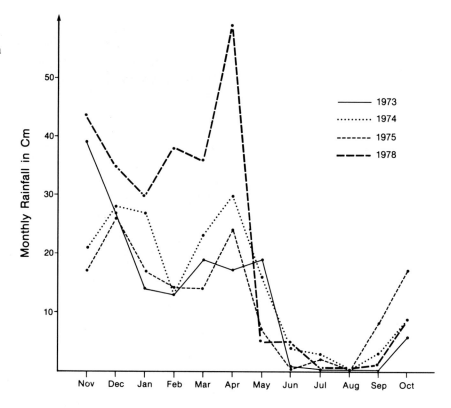

Figure 3.3 Total mean rainfall per month in three consecutive years, 1973–1975, and in 1978, a year when very high rainfall was recorded.

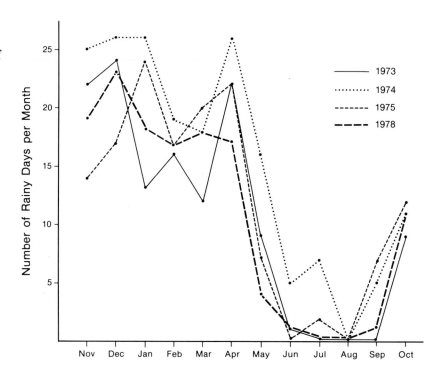

Figure 3.4 Mean number of rainy days per month in 1973–1975 and in 1978. A day was scored as "rainy" if a measurable amount of liquid was collected in the rain gauge.

so that the humidity is often higher in the densely forested valleys, but lower on the grass-covered ridges high in the hills.

In April-May and August-September winds typically are strong, starting early in the morning and dropping by midday. During the wet season they may be very blustery indeed—especially during the frequent thunderstorms. In a particularly violent storm in 1980, at least ten oil-nut palms and twenty-five large forest trees of various kinds were blown over in the lower part of one valley alone. Where three or four fell in the same area, small clearings have now appeared in the forest.

The vegetation at Gombe is of five major types (Clutton-Brock, 1972): *subalpine moorland*, along the crest of the rift escarpment; *open brachystegia* (or *miombo*) *woodland* covering the higher slopes (with no shrub layer); *semideciduous forest*, with a moderate to extremely dense shrub layer covering slopes of ridges that separate valleys, drier areas of the valley floors, and some parts of the lowermost slopes by the lakeshore; *evergreen forest*, with a moderate to dense shrub layer, mostly in valley bottoms and along streambeds; and *grassland with scattered trees*, covering many of the lowermost slopes as well as many of the upper peaks and crests or ridges between valleys. There is also the beach, composed of sand and pebbles or rocks, stretching along the very edge of the lake and supporting little in the way of vegetation, particularly during the dry season.

Trees and plants that provide food for the chimpanzees are found principally in the semideciduous and evergreen forests, but grow also in the open woodland. The grassland provides little in the way of food, and the highest peaks and ridges along the rift escarpment are rarely utilized (Wrangham, 1975).

The park today supports a population of approximately 160 eastern or long-haired chimpanzees (*Pan troglodytes schweinfurthii*). In 1960 rather large tracts of undisturbed forest stretched to the east of the rift escarpment, forming part of the extensive miombo woodlands of the Kigoma region, and chimpanzees were reported to the east and south of the park. To the north, pockets and strips of forest almost certainly linked the Gombe chimpanzees with the chimpanzee population of Burundi. Today, however, the picture is different. The land to the east of the rift "scarp" has been quite extensively cultivated, and the same is true north and south of the park boundaries. While chimpanzees still exist in patches of forest outside the park, the number and size of these refuges are gradually diminishing as human cultivation spreads. It seems likely that in the not-too-distant future the Gombe chimpanzees will be virtually isolated. A matter of very real concern is whether the population is large enough to sustain itself if this should happen.

There are seven other species of primate in the park: the olive baboon (*Papio anubis)*; the red colobus monkey (*Colobus badius)*; three species of *Cercopithecus*—the blue monkey (*C. mitis*), the redtail (*C. ascanius*), and the vervet (*C. aethiops*); and the needle-clawed bushbaby (*Euoticus elegantulus*). There are, in addition, a number of blue-redtail crosses (Clutton-Brock, 1972).

Other mammals commonly found in the park include the bushbuck (*Tragelaphus scriptus*), the bushpig (*Potamochoerus porcus*), and smaller mammals such as the civet (*Civettictis civetta*), the genet (*Genetta genetta*), the elephant shrew, and a variety of squirrels, mongooses, and rodents. The leopard (*Panthera pardus*) is there too, but is shy and seldom seen. Occasionally a serval cat (*Felis serval*) is sighted, and there have been observations of hyenas (*Crocuta crocuta*) and, once, a pair of Cape hunting dogs (*Lycaon pictus lupinuo*). In the early sixties there were at least two small herds of buffalo (*Syncerus caffer*), about sixteen animals in each; these have slowly fallen victim to poachers from the surrounding villages, and as I write only three doomed individuals remain. In the early days hippopotamuses (*Hippopotamus amphibius*) sometimes emerged from the lake at night to browse on the shore; occasionally I passed them early in the morning. But they have vanished, like the lions (*Panthera leo*), which before my time reputedly wandered through the park during the calving season of the buffaloes.

At one time crocodiles lived along the shore of the lake. I often saw two of them when I first arrived at Gombe. Today the largest reptiles found are pythons (*Python sebae*) and Nile monitor lizards (*Varanus niloticus*), which may reach 1.5 meters in length. There are many snakes, including the poisonous puff adder, night adder, bush viper, spitting cobra, Storm's water cobra, black mamba, and vine snake. Skinks, gekkos, toads, and frogs are common.

There are surprisingly few harmful or unpleasant insects and other invertebrates at Gombe. Probably the worst is a giant centipede, said to have a bite as deadly as that of a snake. Of course, there are scorpions and, near the water, tsetse flies; in the wet season there are some mosquitoes, midges, and the like.

In comparison with most other sites where forest creatures are studied, Gombe is a paradise. The almost total absence of large, potentially dangerous animals makes it unusually safe for humans to

Gombe National Park extends for some 16 kilometers (or 10 miles) along the eastern shore of Lake Tanganyika and inland only to the top of the rift escarpment. A narrow strip of country, it is intersected by numerous valleys and ravines. This is Kakombe Valley. Our camp is situated nearly a kilometer (about half a mile) inland from the lakeshore.

travel about on foot, unarmed. There is not the necessity, distracting to those engaged in field research, of keeping constant watch for danger. The crystal-clear water of the lake plunges steeply from the sandy beach to a depth of 10 meters or more before one can take ten steps; it is unsuitable as a habitat for the host snail of *schistosomiasis* (bilharzia), so one can swim in the lake with little risk. Sleeping sickness was last a problem in the area about fifty years ago and has not recurred since. Only malaria is a constant threat to the health of the humans at Gombe, and that only in the wet season.

The climate, ideal for field research during the dry season, becomes more difficult in the wet season, when books and equipment become mildewed, fungus attacks camera lenses, clothing and bedding are constantly damp, and health and morale suffer. The undergrowth becomes much thicker and all manner of thorns and brambles sprout, conspiring to hinder the zealous field-worker. But compared with, for example, the conditions endured by those studying the mountain gorillas in the Virungo volcanoes, even this is (as a visitor from that site told me a while ago) "a holiday."

Field Methods

During my early months at Gombe, as the chimpanzees gradually became accustomed to my presence, I was able to piece together, bit by bit, the overall pattern of chimpanzee life, watching through binoculars and writing everything in notebooks. Then in 1962 a mature male chimpanzee, David Greybeard, visited my base camp to feed on the fruit of an oil-nut palm that grew there. He returned each day, occasionally accompanied by another male, Goliath, until the fruit was exhausted. In the course of one such visit he took a bunch of bananas from my tent. After this I asked my African cook to leave bananas lying out; if David came, he always took them. Three months later, when the fruits of another palm in the camp ripened, David returned, sometimes accompanied not only by Goliath but by William as well. When this second palm had finished fruiting, the three males still came at times for the bananas we continued to leave out for them. Eventually other chimpanzees followed, including an adult female, Flo, and her family. When in July 1963 Flo became sexually receptive, eight mature males followed her into camp.

That was when I decided to set up an artificial feeding area. This was primarily in order to facilitate the filming at close quarters of chimpanzee behavior, for by then I had been joined by Hugo van Lawick, working for the National Geographic Society. But I was also able for the first time to make fairly regular observations on the various individuals. The number of chimpanzees visiting the feeding area, or *camp*, gradually increased. In 1964 the first infants were born to our regular camp visitors, and it became possible to study details of infant development on an almost-daily basis.

It was in 1964 that the first research assistant joined me at Gombe: the beginning of what was to become the Gombe Stream Research Center. By the end of that year a second assistant had arrived, and slowly the number grew. In 1967 the first independent project was

David Greybeard takes a banana left for him on my table in 1962.

initiated by a doctoral candidate. By 1975 there were twenty-two American, European, and Tanzanian students (undergraduate and graduate) and one postdoctoral researcher, working at Gombe—some on the behavior of baboons.

In 1968 the first Tanzanian field assistant was employed, and in the years that followed he was joined by others. Initially the responsibility of these individuals was simply to accompany students, unused to the African bush. Quite soon it became apparent that the field assistants could provide our research with an extremely important component: a core of individuals, with long-term commitment to the work (the students seldom stayed longer than eighteen months), who were totally familiar with the chimpanzees, the terrain, and the food plants. By 1970 the field assistants were receiving basic training in simple data collection (details are given in Appendix A).

The advantages and drawbacks of the banana feeding have been discussed in detail elsewhere (Goodall, 1971; Wrangham, 1974). At the beginning, a chimpanzee was in principle given bananas each time he or she visited camp. By 1967 fifty-eight chimpanzees of both sexes and all age groups were regular visitors, and some came almost daily. It was during the four-year period from 1964 to 1967 that the disadvantages of the feeding system became more and more apparent. Ranging and grouping patterns, feeding, and aggression were increasingly influenced. Moreover, because camp was at the periphery of the ranges of some individuals, chimpanzees who met frequently in order to feed on bananas would otherwise probably have associated only occasionally. The quality of the relationships of these and other chimpanzees was undoubtedly affected by this more-frequent association.

In 1968 the feeding system was extensively revised. Each chimpanzee was fed only a small number of bananas (five or six) every seven to ten days, and only when alone or when in a small group of compatible individuals. This resulted almost at once in a marked drop in

William, Goliath, and David Grey-beard share a cardboard box in my first camp.

Freud and Fifi wait for bananas, which are placed in the boxes shown here from inside the bunker. Usually the outside flap is left closed unless the chimpanzees are being fed.

The chimpanzees are weighed regularly. Bananas are placed in a tin attached to a rope that hangs from a spring balance. Here Evered climbs to take the bait. (C. Tutin)

the frequency of aggression in camp (Wrangham, 1974) and in a return to grouping and ranging patterns similar to those I had observed prior to the establishment of the feeding regime. Individuals began to visit camp much less often; some stopped coming altogether. Today chimpanzees seldom make long treks from distant parts of their range specifically in search of bananas, which means they may stay away for weeks on end. But they do visit when they are nearby, to check on the banana situation or, often, in search of other individuals. Some chimpanzees visit camp only three or four times in a year despite the fact that they almost always receive bananas when they do come.

From 1964 on, data were recorded on the interactions of the various chimpanzees who visited camp. The reproductive state of females and the health of all individuals who visited were carefully noted and, starting in 1970, they were weighed regularly.

As the chimpanzees became increasingly tolerant of their human observers, it was possible to follow a selected individual from the time he (or she) left his nest in the morning until the time he retired for the night. Over the years there have been a number of consecutive-day "long follows." The first of these was in 1968, when the pregnant Flo was followed for sixteen days in the hope of observing parturition (which, in fact, took place at night). In 1974 the top-ranking male Figan was followed for fifty consecutive days (Riss and Busse, 1977). In 1976 another chimpanzee, Fifi, was followed daily for forty-five days. In 1977 there were two marathons: sixty-nine days of watching the female Passion (to monitor her interactions with new mothers, in view of her known cannibalistic tendencies) and fifty-five days of following Melissa after she gave birth to twins. There have been a number of other follows of two to three weeks (two of them follow-ups on Figan). Most follows, however, last only one or two days, and many are less than a day.

Throughout twenty-five years of research I have paid special attention to the building up of long-term records. Individuals collecting data for dissertations or special projects were required to leave copies on file at Gombe. Students using check sheets were asked to write qualitative summaries emphasizing major points of interest. Files were maintained on various aspects of behavior such as predation, tool use, interactions with baboons, and so on. From all the available information a "character file" was compiled for each chimpanzee, with extensive cross-referencing. These records were made available to all students who worked at Gombe, in order that they might benefit from the longitudinal nature of the research.

It was in the early 1970s that the value of collaborative, cooperative field research became obvious. The pooling of data collected by a number of observers threw new light on puzzling or seldom-seen aspects of chimpanzee behavior and added significantly to our understanding of social structure. Access to pooled current data and also to back records was especially valuable for students working on ranging and feeding behavior (McGrew, 1974; Wrangham, 1975), on adolescence (Pusey, 1977), and on weaning (Clark, 1977).

In May 1975 a group of rebels from eastern Zaire crossed Lake Tanganyika in the middle of the night, captured four members of the research team, took them back to Zaire, and held them for ransom.

These young people eventually were returned safely to their parents. Nevertheless, the incident marked the end of an era at Gombe, for it was no longer considered safe for non-Tanzanians to spend extended periods of time in the park. Fortunately, it has been possible to continue the long-term research, thanks to the Tanzanian field assistants. In the first months after the students left, these men were somewhat bewildered: they did not realize how much they themselves knew, or how much they were capable of contributing to the research. It was not overnight that they assumed responsibility and became the enthusiastic and conscientious research team they constitute today: the quality as well as the quantity of data collected dropped sharply after the students left. However, by the end of 1976 the Tanzanians were beginning to collect systematically good, reliable field data (see Appendix A). Initially, I myself was only permitted to go to the center for one week at a time; I tried to get there twice in three months, but subsequently, as security has relaxed, I have been able to stay for three weeks at a stretch, every two months or so. My major emphasis at the beginning was on building up the skills of the field assistants by means of seminars, reliability checks, and lengthy discussions with each individual about his work. More recently, I have been able to spend most of my time, when in the field, collecting data myself as well as continuing to supervise the men. Two of the field assistants, Hilali Matama and Yusufu Mvruganyi, have been delegated the responsibility of day-to-day organization of the fieldwork.

These Tanzanians have had no advanced academic training, yet they undoubtedly know more about following chimpanzees through the forest, and perhaps understand more about their behavior, than most university-trained students. Throughout this book, in order to present an up-to-date picture, I have drawn extensively on data collected since 1975 (after the students left)—material that illustrates the high quality of the contribution being made by the twelve field assistants.

Relations between Chimpanzees and Observers

Reactions of the Subjects

To what extent does the presence of humans, and all that it entails, affect the behavior of our chimpanzee subjects? This is not an easy question to answer, for if we removed human observers we should not be in a position to record any changes in chimpanzee behavior that might result. However, a few comments are possible.

We have seen that the providing of bananas on a daily basis caused some changes in chimpanzee behavior. To what extent the long-term residue of those changes is still operating today is difficult to assess. Camp still represents an unnatural food source, with its supply available in one place, month after month, year after year. It is certain that if camp were not there, the chimpanzees would not visit that particular clearing with anything like the regularity with which they now pass through. Camp provides, in addition to bananas, a meeting place—an area where the probability is high that a chimpanzee will encounter his companions. A juvenile who has lost his mother, for instance, is likely to visit camp when searching for her and may wait

Following two orphans, Skosha and
Kristal, along one of the main trails.
(H. Kummer) Observing Melissa
and her family. (H. van Lawick)
The infant Flint, curious about his
human observer. (H. van Lawick)

for quite considerable periods, looking anxiously in all directions. Or he may leave, only to return after a short while several times in succession. Without doubt some individuals meet one another much more frequently than they would if camp did not exist. On the other hand, the five or six bananas to which an individual is "entitled" once every week or so have very little significance in the overall diet. (In the days of bountiful feeding I watched an adult male consume between fifty and sixty of the fruits at one sitting, after he had raided the store.) And I have already pointed out that some individuals visit camp very rarely, despite the fact that they are rewarded with bananas when they do. Some chimpanzees, even after becoming accustomed to regular bananas, have moved away and never returned, as we shall see in subsequent chapters.

We should also ask how disturbing it is for a chimpanzee to be followed through the forest by one or two humans, sometimes for days on end. Some chimpanzees show what appears to be a total lack of concern, of interest even, in the close proximity of one or more humans. Others are far more anxious. In part it depends on the behavior of the human observers: those who are insensitive to the behavior of the animal they are following, who move noisily when he is trying to listen, who make sudden movements when he is resting, who approach too closely when he is traveling, and so on, are likely to affect the behavior of their subject. The surprising degree to which almost all the chimpanzees tolerate the observers is something that developed slowly; through the years I have emphasized the importance of preserving and encouraging this remarkable chimpanzee-human relationship. If a particular subject appears nervous or irritable during a follow, the observers are instructed to fall back and watch from farther away, even if this means losing their target animal. And if a chimpanzee really wants to escape, he can do so—all too easily. Given the rugged nature of the terrain at Gombe, it is virtually impossible for a mere human to keep up with a chimpanzee who is determined to get away. Indeed, even when the target is *amenable* to being followed, observations frequently end when the subject is lost (for example, he slips easily through a patch of dense, thorny undergrowth while the humans force their way painfully after him, often on their bellies, or search for an easier route).

Individual chimpanzees differ considerably in their attitude toward humans. There is one female, Nope, who first visited camp in 1965 and has been a regular visitor ever since. Yet even today, unless she is with a number of other chimpanzees, she is too nervous to be followed. Except by me. I find her as calm and tolerant of my presence as any other habituated female. One adult male, Humphrey, often expressed dissatisfaction with his observers by throwing large rocks at them. It seemed that Humphrey was much more tolerant of male observers, and most of his threats were directed toward women. In general, chimpanzees are more likely to take liberties with humans of the female sex. Some adolescent males, for instance, clearly attempt to dominate female researchers at the time when they are struggling to dominate females of their own kind.

In the early years of the research I actively encouraged social contact—play or grooming—with six different chimpanzees (Goodall,

1971). For me personally, those contacts were a major breakthrough: they meant that I had won the trust of creatures who initially had fled when they saw me in the distance. However, once it became evident that the research would continue into the future, it was necessary to discourage contacts of this sort. Not only could such interaction distort the behavior of the chimpanzees, but it could be dangerous for the humans, for chimpanzees are much stronger than we are. Accordingly, researchers were asked not to approach their subjects closer than about 5 meters and to try to ignore, or move away from, any friendly—or unfriendly—advances that might be made. Despite our efforts to maintain a respectable distance between observer and observed, many young chimpanzees make things very difficult with their determined attempts to touch, poke, slap, lick, threaten, or try to play with the persons making observations. Usually such attentions peter out during adolescence, apart from the occasional attempts to impress observers of the female sex.

The above behaviors, extremely attention provoking when they occur, are the exceptions rather than the rule. For the most part the chimpanzees show very little interest in humans or their behavior. We have become as much a normal part of their lives as the baboons and other animals with whom they share the forest. Those of us working day after day with the chimpanzees feel, intuitively, that by and large our presence has surprisingly little effect on their behavior.

Attitudes of the Observers

Because chimpanzees exhibit behavior so remarkably similar to some of our own, many of us who have worked at Gombe over the years have developed a degree of empathy with the individuals studied. In itself, this is not a bad thing. Subtle communication cues denoting slight changes in "mood" or attitude toward other chimpanzees are more readily detected once empathy of this kind has been established, and this can aid us in the understanding of complex social processes. It is necessary, however, that we be constantly aware of the danger of anthropomorphic bias in the interpretation of behavior. Observations must be as objective as it is possible to make them. Intuitive interpretations, which may be based on an understanding stemming directly from empathy with the subject, can be tested afterward against the facts set out in the data.

It has been my practice to administer medical or other assistance to the chimpanzees in the rare cases when this was practicable. At one time some of the chimpanzees became paralyzed with a disease that was almost certainly poliomyelitis (Goodall, 1968b); oral vaccine was administered (in bananas) to as many individuals of the study group as possible. Two of the victims were shot for humane reasons. One female, Gilka, was anesthetized and a biopsy performed, in order to determine the nature of a grotesque swelling of her nose and brow; medication was subsequently administered (also in bananas), somewhat irregularly, over the next nine years (Goodall, 1983). Four times, doses of antibiotic were administered to chimpanzees with serious lacerations. An old female, Madam Bee, dying from numerous wounds, was offered bananas, eggs, and water to ease her last days. Another old female, Flo, was given eggs during the last two months of her life, and her juvenile son, who became depressed and later sick after his mother's death, was also offered various kinds of foods.

There are some scientists who frown upon such practices, believing that nature should run its course. It seems to me, however, that humans have already interfered to such a major extent, usually in a very *negative* way (habitat destruction, introduction of diseases, and so on), with so many animals in so many places that a certain amount of *positive* interference is desirable. I readily admit to a high level of emotional involvement with individual chimpanzees—without which, I suspect, the research would have come to an end many years ago.

Certainly the presence of humans, and the ongoing research, and the way the research is conducted mean that the Gombe chimpanzees are not living in an undisturbed environment. Nevertheless, each of the individuals being studied is subjected to more or less the same degree of disturbance. From early on my own special research interest has been in individual differences. Why is male A so much more aggressive than male B? Why is mother C less tolerant of her infant than mother D? To what extent are differences between adults attributable to differences in their upbringing, family life, and childhood experience?

Answers to such questions are, in a sense, equally meaningful no matter what the social and physical environment of the group under study. The important thing is to take such factors into account when making generalizations that apply to the species as a whole—especially as they relate to the ultimate causation of behaviors. This book is in essence a description of one community of chimpanzees; but in order to keep the broader picture in mind I have, whenever possible, compared the behavior of the Gombe chimpanzees with behavior reported from other study sites. It is my hope that the information will be sufficiently comprehensive to allow the reader to judge in what ways (if any) Gombe chimpanzees show behavior that is atypical of the species as a whole.

4. Who's Who

CHARLIE (B. Gray)

HUGH (G. Teleki)

One of the major benefits of this long-term study has been the accumulation of a series of life histories. Each chimpanzee has a unique combination of inherited and acquired characteristics that sets him or her apart from any other chimpanzee.* And so often, knowledge of past experiences in an individual's life helps one to understand behavior that otherwise would be puzzling.

Table 4.1 lists all the chimpanzees who have been identified and named in the course of our study, except for four elderly individuals who died before 1963: William, Horace, Lord Dracula, and Wilhelmina.

Here I would like to summarize the life histories and personalities of just a *few* of those individuals who have contributed to our understanding of chimpanzee behavior. Although these character sketches are necessarily brief, they will give perspective to the many readers who have not had the opportunity to work with chimpanzees.

CHARLIE was one of the first chimpanzees to visit my camp as an adolescent in 1963. From the beginning he was unusually fearless. HUGH, his presumed elder brother, appeared in camp the following year—a prime male with the same fearlessness and a superb physique. From the first these two males had a close, supportive relationship. Once, when Goliath (then alpha male) attacked Charlie, Hugh rushed up and displayed vigorously until Goliath stopped. By 1968, when Charlie was an adult, he and Hugh were already known for their impressive parallel charging displays; the following year the two of them together were able to intimidate all the other males.

Charlie and Hugh were the nucleus of the "southern subgroup," which ultimately split off to become the Kahama community. During

* When I began observing chimpanzees in 1960, the concept of individuality in non-human animals was unpopular in scientific circles. In fact, the first technical paper I submitted for publication was returned by a major periodical with the suggestion that a few alterations be made: where I had written "he," "she," or "who," these had been crossed out and "it" or "which" substituted. I am glad to say that the final version conferred on the chimpanzees the dignity of their separate sexes.

the process of division, from 1970 on, the Kahama individuals spent more and more time in the south. Occasionally, however, they returned to the northern part of the range, often in a compact group. When they encountered the northern males, both sides charged vigorously, drummed on trees, and called loudly. Charlie and Hugh, with their striking cooperative displays, were easily able to dominate all the northern males including—indeed, especially—the alpha male, Humphrey.

Soon after the final division of the community, Hugh disappeared. He was not old at the time and it is possible that he fell victim to intercommunity aggression, as did Charlie four years later. During those four years Charlie was alpha male of the Kahama community. The attack presumed to have led to his death was not seen; his badly wounded body was found in May 1977, lying in Kahama Stream.

DAVID GREYBEARD (H. van Lawick)

DAVID GREYBEARD does not figure prominently in this book, but he played a crucial role in the early research and cannot be omitted here. David was the first chimpanzee I saw eating meat, the first to demonstrate the use of tools, and the first to permit my close approach in the forest. It was his calm acceptance of my presence that did much to speed up habituation in the early months.

It was David Greybeard who first visited my camp and who subsequently led many others to the new source of food (bananas). He had a particularly calm and gentle disposition. When a subordinate approached him and showed submissive or nervous behavior, he almost always responded with a reassuring gesture, laying his hand on the other's body or head. He was very generous and tolerant during banana feeding, allowing others (including females and youngsters) to share his fruits. David had a close, friendly relationship with Goliath (alpha male in those early days), which continued until David's death. If Goliath, who was much more excitable, showed signs of nervousness (if, for example, I approached too closely), David would often reach out and touch his companion's groin or briefly groom him; this almost always calmed Goliath.

During periods of intense social excitement (quite common between 1963 and 1969, when there were sometimes as many as fourteen adult males in camp at the same time), David usually tried to keep out of the way. Frequently he ran to shelter beside or behind Goliath. Yet when aroused he was one of the most fearless of opponents. In fact, it was David who in 1964 led the first determined joint challenges against the new alpha male, Mike.

My own relationship with David was unique—and will never be repeated. He allowed me to groom him, and on one never-to-be-forgotten occasion, gave me a gesture of reassurance when I held out my hand to him, offering a palm nut. When David disappeared during an epidemic of pneumonia in 1968, I mourned for him as I have for no other chimpanzee before or since.

EVERED is the eldest known offspring of the timid, highly excitable Olly. He was a very early visitor to my camp, arriving with David Greybeard and Goliath in 1962. The following year he led his mother and young sister, Gilka, to the new food source.

Table 4.1 Chimpanzees who have been or still are part of the main study community at Gombe.[a]

A. Reproductive profile of the adult females

Bessie (1926?–1965)
  Bumble (1954?–*1968*)
  Beattle (1960?–1967)

Marina (1926?–1965)
  Pepe (1951?–1968)
  Miff (1956?–)
    Moeza (1969–)
    Michaelmas (1973–)
    Mo (1978–1985)
    Mel (1984–)
  Merlin (1961?–1966)

Flo (1929?–1972)
  Faben (1947?–1975)
  Figan (1953?–1982)
  Fifi (1958?–)
    Freud (1971–)
    Frodo (1976–)
    Fanni (1981–)
    Flossi (1985–)
  Flint (1964–1972)
  Flame (1968–1969)

Olly (1939?–1969)
  Evered (1952?–)
  Gilka (1960?–1979)
    m/sb (1973)
    Gandalf (1974)
    Otta (1975)
    Orion (1976)
  Grosvenor (1966)
  m/sb (1967)

Jessica (1940?–*1967*)[P]
  MacDee (1953?–1966)
  Lita (1958?–*1967*)[P]
  Jay (1965–*1967*)[P]

Sophie (1940?–1968)
  Sally (1955?–*1967*)
  SNIFF (1960?–1977)
  baby (1965)
  Sorema (1966–1968)

Sprout (1942?–)[P]
  Satan (1955?–)
  Sesame (1962?–*1973*)[P]
  Spray (1969–)[P]
  Spindle (1976–)[P]
  baby (1984)

Vodka (1943?–*1968*)[P]
  Jomeo (1956?–)
  Sherry (1961?–1979)
  Quantro (1966–*1968*)[P]

MADAM BEE (1947?–1975)
  LITTLE BEE (1960?–)[I]
    m/sb (1976)
    Tubi (1977–)
    Darbie (1984–)
  HONEY BEE (1965–)
  Bee Hinde (1971)
  WOOD BEE (1973)

Melissa (1950?–)
  m/sb (1963)
  Goblin (1964–)
  m/sb (1969)
  Gremlin (1970–)
    Getty (1982–)
  baby (1976)
  Genie (1976)
  Gyre (1977–1978)
  Gimble (1977–)
  m/sb (1984)
  Groucho (1985–)

Passion (1951?–1982)
  baby (1962?–1963)
  Pom (1965–*1983*)
    Pan (1978–1981)
  m/sb (1970)
  Prof (1971–)
  Pax (1977–)

MANDY (1951?–*1975*)
  Jane (1964–1965)
  MIDGE (1966–)[b]
  MANTIS (1972–*1975*)

Notes

Names of males are shown in green.
Small capitals designate individuals who split off to form the Kahama community or were born into that community.
Italics indicate uncertain relationship, sex not ascertained, or year when individual was last seen (may have died or emigrated).
Bracketed individuals are twins.
I = immigrant; P = peripheral.
m/sb = miscarriage, stillbirth, or vanished during first few days of life.
? = date estimated or not authenticated.

a. The age of individuals born before the study started, or during the very early years, has been estimated as follows:
  (1) By comparing photos, film, and descriptive notes on individuals age fifteen or younger when they were first seen, with chimpanzees whose age is known;
  (2) By estimating (for mothers) probable age at first pregnancy, intervals between births, and known or estimated age of eldest offspring;
  (3) By comparing photos and film of males who were older than fifteen years when first seen, with individuals who have aged during the study (the least accurate method).
Evidence in support of certain mother-offspring relationships (for offspring who were adolescent when first recognized in the study) is presented in Goodall, 1968b, p. 222. Probable sibling relationships are assumed on the basis of physical similarities, supportive relationships, and the like.

b. Kidevu is almost certainly Mandy's daughter Midge.

Table 4.1 (*continued*)

Circe (1952?–1968)
  Cindy (1965–1968)

Nope (1952?–)
  Mustard (1965–)
  Lolita (1973–1981)
  m/sb (1979)
  Hepziba (1980–1981)
  Noota (1982–)

Athena (1953?–)
  m/sb (1966)
  Atlas (1967–)
  m/sb (1972)
  Aphro (1973–)
  Apollo (1979–)
  Ariadne (1985–)

Gigi (1954?–)

Nova (1954?–1975)
  m/sb (1967)
  Skosha (1970–)

Pooch (1955?–1968)

Pallas (1955?–1982)
  m/sb (1968)
  Plato (1970–1973)
  Villa (1974–1975)
  Banda (1976)
  Kristal (1977–1983)

Caramel (1956?–)[I,P]
  Candy (1969?–)[I,P]
    California (1983–)[P]
  ⎧Castor (1976–)[P]
  ⎩Shuga (1976–)[P]

Joanne (1958?–)[I,P]
  Jageli (1971–)[I]
  Jimi (1981–)[P]

WANDA (1958?–*1975*)[I,P]
  ROMANY (1971–*1975*)[P]

Dove (1959?–)[I,P]
  Dominie (1972–)[P]
  m/sb (1977)
  Dapples (1978–1981)[P]
  Dharsi (1985–)[P]

Winkle (1959?–)[I]
  Wilkie (1972–)
  Wunda (1978–)
  Wolfi (1984–)

Sparrow (1960?–)[I,P]
  Sandi (1973–)[P]
  Barbet (1978–1982)
  Sheldon (1983–)[P]

Patti (1960?–)[I]
  baby (1978)
  Tapit (1979–1983)
  Tita (1984–)

Harmony (1960?–*1981*)[I,P]
  m/sb (1977)
  m/sb (1978)

Kidevu (1966?–)[I,b]
  Konrad (1982–)

Jenny (1966?–*1980*)[I,P]

B. Males born before the study began and/or whose mothers are not known

Name	Probable dates	Probable relationship
Mr. McGregor	1925?–1966	
Huxley	1926?–1967	
J.B.	1933?–1966	*Elder sibling of Mike*
Leakey	1935?–1970	*Elder sibling of Mr. Worzle*
Hugo	1936?–1975	
David Greybeard	1936?–1968	
GOLIATH	1937?–1975	
Mike	1938?–1975	*Younger sibling of J.B.*
Rix	1941?–1968	
HUGH	1944?–*1973*	*Elder sibling of CHARLIE*
Mr. Worzle	1944?–1969	*Younger sibling of Leakey*
Humphrey	1946?–1981	*Sibling of Melissa*
DÉ	1948?–1974	
WILLY WALLY	1949?–*1976*	*Sibling of Gigi*
CHARLIE	1951?–1977	*Younger sibling of HUGH*
GODI	1953?–1974	
Hornby	1957?–1966	
Beethoven	1969?–	Almost-certain sibling of Harmony

EVERED (H. Bauer)

Evered was obviously motivated to attain high social rank, but was at a disadvantage in that he had no close ally. His long struggle for dominance with the slightly younger Figan began when both were late adolescents. Initially Evered won these dominance conflicts. Eventually, however, Figan formed a close alliance with his elder brother and the cooperative team was able to defeat Evered conclusively. Indeed, for a while Evered became something of an outcast, spending weeks at a time wandering in the peripheral area of the community range in the north. Often, when he returned, Figan and his brother would renew their challenge, so that Evered would leave again, almost immediately. In 1975, however, Figan's brother disappeared and Evered was able to move back and enter into the social life of the community once more. During the last few years of Figan's life Evered was his most frequent companion and ally.

Between 1970 and 1983 Evered is thought to have sired more infants than any other Gombe male. During one of his long periods away in the north he was observed keeping company with an unhabituated female, presumed to be a member of the Mitumba community to the north. Even in temporary exile it seemed that he was able to take advantage of the situation!

After Olly died in 1969, Evered and Gilka were often together. Evered, if present, supported his young sister during conflicts with other community females. He shared food with her during meat eating. And, unlike other elder brothers, he was never seen to force Gilka to copulate with him. On three occasions after he had returned from one of his long absences, he spent the greater part of a day traveling peacefully with Gilka. They groomed each other, fed in the same trees, and twice nested close together. The family contact apparently was reassuring to them both.

FABEN (P. McGinnis)

FABEN, presumed eldest son of Flo, was crippled during the 1966 polio epidemic and lost the use of one arm. His young brother, Figan, took advantage of the situation, repeatedly challenged him, and became dominant. For the next couple of years the brothers associated very little, but from 1970 on they became increasingly close and supportive. Despite his paralyzed right arm, Faben learned to perform spectacular bipedal charging displays; when Figan displayed at rival males, Faben almost always joined in to help his brother. Indeed, it was Faben who helped Figan to attain alpha position.

Faben was very active in intercommunity conflicts and took part in many of the attacks on Kahama individuals. He disappeared in 1975: healthy when last seen, he may have fallen victim himself to intercommunity violence.

FIGAN was first recognized in 1961 when he was a juvenile keeping company with his old mother, Flo, and his younger sister, Fifi. All three first came to camp in 1962. The following year Faben, Figan's presumed elder brother, also began visiting camp with the rest of the family.

Figan was to become alpha male in his early twenties and even as a youngster showed signs of the qualities that would take him to the top of the male hierarchy. He was always quick to take advantage of

FIGAN (H. van Lawick)

temporary ill health in one of the older males. He even seized the opportunity to challenge Faben when his elder brother was stricken by polio in 1966. He successfully intimidated him, and remained the higher-ranking brother from then on. As he began to challenge the senior males in earnest from 1970 on, he showed increasing skill in the timing and placing of his displays.

Figan had the advantage, too, of close supportive relationships with members of his family. Flo was high ranking during his early adolescence and helped him win victories over other young males. And, having dominated his elder brother, he gradually built a close alliance with him that would be crucial to his acquisition of alpha status. Without Faben's active support Figan probably could not have seized the top position from the much heavier and more aggressive Humphrey in 1972. Figan had to overcome the slightly older Evered also, which he did with Faben's unswerving help. As described in Evered's biography, the two brothers repeatedly displayed at Evered, sometimes attacking him severely, until their rival was to some extent driven from the center of the community range.

Figan had a very excitable temperament and during tense social situations sometimes became so worked up that he would start to scream and would rush to mount or embrace a nearby companion for reassurance. Sometimes he clutched his own genitals in moments of stress. These behaviors seemed to indicate a lack of self-confidence, and many of the observers at Gombe felt that Figan would never make it to the top. Clearly, however, his high motivation and undoubted intelligence were sufficient to overcome this apparent failing.

After Faben's disappearance and presumed death in 1975, Figan temporarily lost overall control, but by 1977 was unquestionably alpha once more. In 1979 Figan was challenged by young Goblin. Up until that time Figan had supported Goblin against all the other males; when Goblin turned on his benefactor, Figan became increasingly tense and anxious. Once again he fell, temporarily, from the top position. The following year, however, four of the other senior males joined Figan in an attack on Goblin which, once again, put Figan back on top.

Figan seemed to recognize the importance of a close ally in matters pertaining to social dominance—a lesson presumably learned during his early formative years, when he had the support of his assertive mother (and probably of his elder brother too). After Faben's death Figan became friendly with Humphrey, the male from whom he had wrested the alpha position. And when Humphrey died in 1981, Figan cultivated close relations with Jomeo and his old rival, Evered.

During the last year of his life Figan was no longer the alpha male, although the senior males still showed him much respect. He was healthy when last seen, and the reason for his disappearance and presumed death in 1982 is not known.

FLO and her daughter FIFI have made extensive contributions to our understanding of maternal behavior and family relationships. Flo was another very early visitor to my camp, first appearing in 1962 along with Fifi, then an infant, and juvenile Figan. The following year her presumed eldest son, Faben, also began to accompany her to camp.

FLO (H. van Lawick)

FIFI (K. Love)

GIGI (L. Goldman)

When Flo became sexually attractive in 1963, after some five years of maternal preoccupation with Fifi, many of the community males who had not previously visited camp came with Flo—and discovered bananas. Thus Flo, like David Greybeard, played a crucial role in the early years of Gombe research.

Flo was a high-ranking, aggressive female in the early '60s, and there can be no doubt that her status and personality were strong factors in the rise to power of her son Figan, and in Fifi's present high rank. Probably because of her extreme age, Flo failed in her maternal care of her last two offspring, Flint and infant Flame. Flame disappeared when Flo was very ill, unable even to climb a tree for the night. She recovered, but seemed to have insufficient strength to enforce the independence of juvenile Flint. She continued to give in to his demands to sleep with her at night and to ride her back like an infant, until he was eight and a half years old. When she died in 1972, Flint was unable to cope and in a state of depression fell sick and died himself.

Flo holds the unique distinction of meriting an obituary in Britain's *Sunday Times*. A portion of it reads:

> Flo has contributed much to science. She and her large family have provided a wealth of information about chimpanzee behaviour—infant development, family relationships, aggression, dominance, sex—about 40,000 hours of observation if records on the different family members are pooled . . . But this should not be the final word. It is true that her life was worthwhile because it enriched human understanding. But even if no one had studied the chimpanzees at Gombe, Flo's life, rich and full of vigour and love, would still have had a meaning and a significance in the pattern of things.

Flo's surviving daughter, Fifi, has two sons, Freud and Frodo, and two daughters, Fanni and the infant Flossi.

GIGI was first recognized as a juvenile in 1963. At that time she was almost always with Willy Wally, an adolescent male some five or six years her senior. Gigi was a self-confident, assertive youngster and often provoked chimpanzees older than herself. When this got her into trouble, Willy Wally usually came to her rescue despite his timid disposition and low rank. He was frequently chased or attacked on her behalf. We suspect that the two were siblings whose mother had died.

Gigi is sterile and has come into estrus month after month ever since menarche in 1965. From the first she was sexually very popular with the adult males, despite her habit of pulling away from her partner in the middle of the sexual act. She has played an active role in several aspects of community life. She has been on many consortships with many different males, wandering with them in peripheral areas of the community range during her periods of estrus, and has formed the nucleus of many parties of adult males. Because males are more likely to patrol peripheral areas of their range when three or more are together, she has been the unwitting cause of many aggressive encounters with chimpanzees of neighboring communities.

Gigi's behavior is very like that of a male. She is large and strong for a female, and often aggressive. Her display rate is high, and she sometimes performs waterfall and streambed displays, behavior very rarely seen in other females. She is assertive in her interactions with other community members, even on occasion standing up to attacks by adult males, and since 1967 has unequivocally been the top-ranking female. She has been seen to hunt and capture prey more than any other female and is quite fearless in her determination to stand up to the defensive attacks of adult monkeys.

In 1976 she passed a blob of red, jellylike substance during menstruation, later identified as a uterine cast. Since that time she has shown irregularities in her estrous cycles and has been a less popular sexual partner for the adult males. From 1975 on Gigi has shown an interest in infants between one and a half and three years of age, and has become "auntie" to a succession of them.

GILKA

OLLY (H. van Lawick)

GILKA was an infant when I first knew her in the early sixties. She was very attractive; her face, heart-shaped with a little white beard, was elflike. Her mother, OLLY, was a timid female who usually avoided large groups. Probably this is why Gilka, when she *was* part of such a gathering, tended to be highly active. Particularly if mature males were around, Gilka would pirouette, turn somersaults, and bounce up to one after another, trying to initiate play. Olly became extremely agitated at such times, pant-grunting nervously as she made appeasing gestures toward males who seemed—to us—quite calm. Once, after retrieving her exuberant infant four times in succession, she sat on the outskirts of the group holding Gilka firmly by one wrist.

Gilka's most frequent companions during those early years, other than her elder brother, Evered, were Fifi and Figan. Olly was friendly with old Flo, and the two mothers spent a good deal of time together. But as Figan and Faben matured and began to challenge the older females, including Olly, she started to avoid even this family. Gilka was forced to spend long periods in the company of her elderly mother only. Sometimes Evered joined them, but this happened increasingly less often as he matured. At the same time, Gilka was being weaned, a depressing time for even the most confident youngster.

Next came the 1966 polio epidemic. Not only was Olly's next infant, the month-old Grosvenor, one of the first victims but Gilka herself was affected, losing partial use of one wrist and complete use of her thumb. Two years later we noticed the first slight swelling of Gilka's nose—the beginning of a fungus infection that eventually led to grotesque distortion of her face.

The following year, in 1969, Olly vanished. After this Gilka spent much time with Evered; she was also alone a good deal. During her late adolescence Gilka was not sexually popular with the Kasakela males, although she evinced the usual adolescent desire for sexual contact when she was in estrus. By this time the community had divided. Perhaps because the Kahama males showed more sexual interest in her, Gilka began spending more and more time in the south. In 1973 she was there for six consecutive months and we thought she had transferred permanently. Perhaps she would have done so except for the fact that her fungus disease, controlled before she left by regular doses of medication, escalated until by the time she returned her nose, her brow ridge, even her eyes were hugely swollen. It seemed impossible that she could see at all. Her return may have been prompted also by the beginning of the intercommunity violence.

Gilka may have been pregnant when she returned from visiting the Kahama community in 1973; but if she had a baby, we never saw it. The following year (when a course of medication had once more reduced the swellings on her face) Gilka gave birth. She was an excellent mother, but the baby vanished when he was barely one month old. The next year she gave birth again; this baby was seized, killed, and eaten by the cannibal Passion. We suspected then that her previous infant might have met the same fate. In 1976, a year after we had observed that first bizarre attack by Passion, Gilka's third infant met the same gruesome end. Gilka fought fiercely, doing her best to protect her baby, but she was smaller and weaker than Passion (who was, moreover, helped by late-adolescent Pom). In addition to losing her infant, Gilka was badly wounded. After this third assault Gilka never conceived again. Her physical condition gradually deteriorated. She developed chronic diarrhea and sores on her fingers that broke out again, worse than ever, every time we thought they had healed. Coupled with an arthritic condition in her hands, these ailments made it hard for her to move about quickly. She became increasingly emaciated and spent more and more time alone. Her most frequent companions during the last two years of her life were two other childless females, Gigi and Patti.

The bond between Gilka and her brother remained strong. True, they did not spend much time together; but when he was with Gilka, Evered was always supportive of her. In his presence Gilka was even seen to threaten Passion, whom she usually avoided fearfully. During these final years Evered provided the kind of reassuring companionship that Gilka had offered him in the days when he was undergoing the hostile attacks of Figan and Faben.

GOBLIN

GOBLIN, born to Melissa in 1964, became top male on a one-to-one basis when only sixteen years old. He achieved this status partly through his long and unusual relationship with the alpha male Figan (who supported Goblin in status conflicts with the other males), and partly through sheer determination and persistence. In 1979 Goblin finally turned on Figan and, in a series of challenges, succeeded in intimidating his onetime ally. Although Figan subsequently regained his top status (temporarily), he remained tense in Goblin's presence. Even before Figan's death, Goblin had once more become top male and by 1985 was undisputed alpha.

Until 1983 Goblin had a close supportive relationship with his mother. But when she became sexually attractive in 1983, Goblin violated the incest "taboo" that usually serves to prevent physically mature sons from attempting to copulate with their mothers. Not only did Goblin court Melissa vigorously and persistently, but actually succeeded in forcing his attentions upon her on a number of occasions. Melissa's response was initially aggressive: she threatened and even hit her son. After he had attacked her, she became fearful. Then, when her periods of estrus were over and she became pregnant, she simply ignored Goblin. After the birth of the infant, however, friendly relations were resumed. Goblin in fact has other sexual peculiarities, which will be discussed further on in this book.

GOLIATH (H. van Lawick)

HUGO (H. van Lawick)

GOLIATH was one of the first chimpanzees to allow me to approach closely in the forest, as he sat grooming with David Greybeard. By 1962 I had become familiar with the various individuals of the community and it was obvious that Goliath was alpha male. He was aggressive and had a very fast, spectacular charging display, coupled with an unusually bold disposition. If the adult males were confronted by some strange object (such as a dead python), Goliath was generally the first or only male to approach closely. When I initially offered him a banana from my hand, his response was typical: he bristled, stared, then charged past, seizing and dragging a table from the tent, and as he grabbed the banana, pushed me so hard that I all but fell.

Goliath did not relinquish his alpha position without a struggle. Even after Mike had attained top rank, Goliath still sometimes challenged him, but always lost. Goliath remained high ranking for another four years, then fell sick, lost a good deal of weight, and became very low ranking indeed. When the Kahama community split off, Goliath went with the others to the south. This was surprising, for he had never appeared to have any close links with the "southerners." He was the last of the Kahama males to visit camp, doing so in 1975, just a few months before he was brutally attacked and left to die by the Kasakela males. By that time he was very old indeed, shrunken, balding, and relatively solitary.

HUGO* was the fourth adult male to visit my camp. He appeared one day in the wake of David Greybeard and Goliath, apparently not knowing exactly where they were headed. His arrival was characteristic: he emerged from the undergrowth surrounding the clearing, took one startled look at me, and instead of fleeing (as most chimpanzees would have done) uttered a threatening call, seized a stone, and hurled it at me! Hugo was, in fact, one of Gombe's few habitual stone throwers.

In those early days Hugo was high ranking and had pronounced qualities of leadership. If it was David Greybeard who led new individuals to camp, it was frequently Hugo who led parties away. He was a very large chimpanzee, but not an overly aggressive one. Quick to threaten others, particularly in the feeding context, he was equally quick to reach out and reassure his victim afterward—sometimes almost before he had finished his gesture of aggression. Hugo formed a most unusual relationship with old Flo during her five-week estrous period in 1963 and followed her closely wherever she went. She often turned to him for comfort when she was hurt or frightened and he almost always reached out to touch or embrace her. He remained her frequent companion for two weeks after her period of estrus was over, long after the other males had lost interest. We wondered whether they might have been siblings.

Hugo retained his leadership qualities until the end. For a while he had a close relationship with the aggressive Humphrey, and he was the chosen companion of a number of young adolescent males during their early journeyings away from their mothers. Toward the end of his very long life his teeth were worn and his physique shrunken. He

* Called Rodolf in my earlier book, *In the Shadow of Man*.

and Mike, frequent grooming partners in their old age, died in the same month. Hugo's body was found during the "pneumonia" epidemic of 1975.

HUMPHREY (H. van Lawick)

HUMPHREY was a young adult male when I first knew him in 1963. He was almost always in the company of an old male, Mr. McGregor. Their facial characteristics were very similar, as were their pant-hoot calls, and they had a close and friendly relationship. McGregor would almost always hasten to help Humphrey when the younger male was threatened or attacked. It is possible that they were uncle and nephew. And, as I shall describe later, Humphrey tried to help when old Gregor was stricken with polio in 1966.

By 1969 Humphrey was very large and very aggressive. The females typically greeted him with submissive gestures before they greeted the aging alpha, Mike. The following year Humphrey attacked Mike and for a short while became alpha male himself. He held the position for only twenty months, losing it in the face of the joint challenge by Figan and his brother. Almost certainly Humphrey would not have reached the top at all if the community had not divided at that time, for he was extremely wary of the other brothers, Charlie and Hugh. When they made one of their periodic visits to the north in 1971, Humphrey always tried to avoid them.

In 1964 Humphrey developed what appeared to be an abscess inside one ear. It must have been very painful, for he would sit for minutes at a time with a finger or thumb pushed deep into the ear. Either this became a habit or the injury continued to hurt him, or perhaps the scar tissue caused "noises" in his head; whatever the reason, he often sat with one thumb in his ear right up to the end of his life. If, in fact, he did suffer from recurrent earache, it might help explain why he was unusually aggressive, attacking others—mainly females—at a high rate, and often brutally. (It is extraordinary how often, if his victim was in estrus, he managed to rip open her swelling during his attacks; it almost seems that he must have done it intentionally.) Humphrey was also aggressive to human females, frequently hurling quite large rocks at us, with uncomfortably good aim. Interestingly, he was also a very playful adult. When he approached females or adolescent males for play, however, many times they were too nervous to respond and their refusal was likely to lead to an attack.

Humphrey remained high ranking for four years after losing alpha status. During this time he became very friendly with his old rival Figan, a relationship that lasted until his death in 1981. We do not know how Humphrey died. His skull was found near the eastern boundary of the community range, and he could have been another victim of intercommunity conflict.

JOMEO and SHERRY were brothers. Jomeo first came to camp in 1964, but it was not until three years later, at the height of the banana feeding, that the rest of his family followed. When Jomeo was about nine years old, he began to challenge the community females in typical adolescent fashion. The following year, however, Jomeo twice appeared with severe wounds. It is unfortunate that we shall never know what happened, because the incidents may have had a profound

effect on his subsequent career. His adolescent swaggering at females came to a virtual halt, and he has shown a most unusual lack of motivation to better his position in the male dominance hierarchy.

Sherry began to travel extensively with his elder brother, Jomeo, even before the disappearance (and probable death) of their mother in 1968. By the middle of the following year the two brothers were almost inseparable; and when in 1971 Sherry began to challenge the females, Jomeo often ran to help if his brother got into difficulty. Perhaps as a result of this involvement, Jomeo's self-confidence increased. At any rate, he began to challenge the young male closest to him in age, Satan; Sherry often displayed up to help.

In 1977 Sherry was seen to attack Jomeo twice in one day, after which Jomeo tended to behave submissively to his young brother. The two continued to spend a good deal of time together, but less than before. Sherry's disappearance in 1979 remains a mystery. Even before Sherry vanished, Jomeo had begun to associate with the alpha male, Figan, and during the last two years of Figan's life was one of his two closest companions. He is very tolerant of young males, and the orphan Beethoven has formed a close attachment to him. I followed Jomeo one day as he wandered from one food patch to another; no less than five adolescent males trailed peacefully in his wake.

In 1978 Jomeo sustained a severe injury to his left eye. For two weeks the lid remained tightly closed and quantities of fluid dripped out. Although we thought he would lose the eye, the wound healed, leaving him partially blind and with an eye half white (from scar tissue). It gives him a sinister appearance.

A final word on Jomeo. He has always been one of the heaviest of the Gombe males, so his apparent lack of motivation to better his social rank in no way results from physical inferiority. Nor does it result from an inherent lack of aggressiveness: during the intercommunity violence Jomeo, when present, was always in the forefront of the attacks made by the Kasakela males on individuals of the smaller Kahama community. With his superior weight he undoubtedly inflicted considerable damage on the victims. Perhaps if we knew more about

LEAKEY (H. van Lawick)

MR. WORZLE (H. van Lawick)

his early history, we would better understand his interesting personality.

LEAKEY and MR. WORZLE are two males who had a very close and supportive relationship. Mr. Worzle had the distinction of having white, rather than brown, sclerotics, so that his eyes seemed very human. Leakey, several years older, also had white patches on his sclerotics. This fact, together with Leakey's friendly and protective attitude toward Worzle, suggests that they were probably brothers.

Leakey was a large, middle-aged, middle-ranked male when first I came to know him in 1963. Worzle, by contrast, was exceptionally small and extremely low ranking. He showed a number of unusual behaviors: he was one of the few males ever seen to masturbate (never to ejaculation), he sometimes threw tantrums worthy of an infant (rare in adult males), and he often whimpered when begging for food. These behaviors, plus his small size, suggest that he might have been an orphan.

Leakey's most memorable characteristic was his abnormal consorting behavior. He is the only male to date who has tried to take away two females at the same time. Interestingly, although Leakey and Worzle were never seen to go off together with a female, they sometimes took the same one (Olly) alternately—Leakey one month, Worzle the next.

Worzle died first, in 1969. He had been sick for some time with a wasting disease and had badly crippled hands (perhaps from arthritis). Leakey died the following year, probably from so-called pneumonia.

MADAM BEE and her daughter, LITTLE BEE, were regular camp visitors in 1963. A second daughter, Honey Bee, was born the year before Madam Bee lost the use of one arm during the 1966 polio epidemic. Honey Bee survived to adulthood, but Madam Bee's last two babies, Bee Hinde and Wood Bee, both died before they were a year old. Because of her ineffectual arm, it was often difficult for Madam Bee to provide adequate support for her infants, a factor that may have contributed to their deaths.

Madam Bee had always ranged to the south of camp, and it was not surprising that she joined the "southern males" when they split off to form the Kahama community. She continued to make occasional visits to camp, but they became increasingly rare. Little Bee, however, moved north on many occasions during periods of estrus; sometimes she was actually forced to leave her family by parties of Kasakela males who displayed at her if she tried to head back south. By 1975 it seemed that she had settled into the Kasakela community, although she may have occasionally returned south to visit her mother and sister.

This was the year that Madam Bee was brutally attacked by Kasakela males and left to die. Little Bee was with the party of aggressors, but in the confusion of the assault she was unobserved, and after the incident she followed the Kasakela males back to the north. Honey Bee remained with her mortally wounded mother, grooming her and keeping off the flies, until her death five days later. At irregular intervals over the next five years, Honey Bee appeared (in estrus) in

the Kasakela community range; but she has not been seen since 1980.

Little Bee almost certainly lost one infant (and possibly two) to the two cannibalistic females, Passion and her daughter, Pom: they made a determined effort in 1977 to capture newborn Tubi ("To be, or not to be . . ."). Little Bee, however, escaped and managed to hide with her infant, despite the fact that the killers searched the area for nearly an hour.

Little Bee was born with a deformed right foot, probably a club foot. She is lame and sometimes has difficulty in keeping up with a fast-moving party of adults. She also has a supernumerary nipple in the center of her chest. She has become one of the most preferred companions of the Kasakela community females, probably because she is low ranking and poses no threat. For about a year she was a favorite companion of the alpha male Figan, who may have sired Tubi during a consortship. In 1984 she gave birth to a second infant, Darbie.

MELISSA (H. van Lawick)

MELISSA, as a late adolescent, associated frequently with Mr. Mc-Gregor and Humphrey. Gregor supported both Melissa and Humphrey, and we suspect that all three were related. After the old male's death in 1966 Melissa and Humphrey seldom associated. Humphrey attacked Melissa about as often as he attacked any other female. In 1964 he pounded on her, dragged her down a slope, and slammed her into a tree; almost certainly her nose was broken as a result. But this does not necessarily refute the suggestion that they were siblings, because the brother-sister bond is seldom strong.

As a young female, Melissa was very social. She invariably joined in when other chimpanzees gave pant-hoots during general excitement, often doing a little display at the same time, stamping on the ground with one or both feet (a characteristic she retains to this day). She has always begged unusually persistently—for meat, for bananas, and, on occasion, for social grooming. If her efforts are unrewarded, she is likely to whimper or even scream.

Melissa was one of the polio victims. Afflicted in the neck and shoulders, for a while she was unable to use her arms for walking and had to travel bipedally. Eventually she regained the use of her arms, but she still cannot raise her chin properly or turn her head. All the same, she is a very high-ranking female. In part, her status is an outgrowth of the supportive relationships she developed, first with her son Goblin, now top-ranking male, subsequently with her daughter Gremlin. Melissa is a very attentive grandmother and has spent relatively more time playing with little Getty than she did with any of her own offspring.

One of Melissa's principal claims to fame was giving birth to twin sons, Gyre and Gimble. Unfortunately Gyre, always the weaker of the two, died when he was ten months old. Gimble, then also very tiny, has since picked up, though I suspect he will become an unusually small male. Melissa gave birth to two other infants, both in 1976. One of these, Genie, was killed by Passion and Pom. The other (unnamed) probably met the same fate.

MIFF (L. Goldman)

MIFF was a juvenile when first I knew her in the early 1960s. After the death of her mother, old Marina, Miff cared for her infant brother, Merlin. Although she allowed him to share her nest at night and almost never moved off without waiting for him, it seemed that she was unable to compensate for the loss of his mother. Gradually Merlin became more lethargic, played less, and developed a variety of abnormal social responses. One and a half years after the death of Marina, Merlin was skeletal in appearance. Last seen dragging a paralyzed leg, he died during the 1966 polio epidemic. Miff's elder brother, Pepe, was also affected by the disease: he was paralyzed in one arm. Like Faben, he adapted well and learned to walk considerable distances in an upright position.

Perhaps because of her early experience with Merlin, Miff has been an extremely competent mother. Her first infant, daughter Moeza, was born in 1969. Three others followed at intervals of roughly five years: Michaelmas, Mo, and, in 1984, little Mel.

Miff has often been the victim of seemingly unprovoked male aggression, mainly during periods when she was cycling. One particularly brutal assault is described in Chapter 12. It was almost certainly during an attack of this sort, when Miff was in estrus in 1977, that Michaelmas received a severe injury to his left leg. He could not use the limb at all for a number of months and we feared that he would not survive. However, by the time Mo was born the following year, he had more or less recovered, although he was still walking with a pronounced limp.

It was at this time that Passion, the cannibal, made a determined attempt to seize the newborn Mo, but was fought off by Miff, as will be described in Chapter 12. It is ironic that Mo, after surviving that early and savage attempt on her life, died when she was six years old, soon after the birth of her young sibling, Mel. Mo had become sick in 1983; she grew increasingly apathetic and emaciated, and eventually vanished in August 1985. It was during this same month that Yahaya Alamasi, working to habituate the chimpanzees of the Mitumba community in the north, "found" Moeza. She had last been seen in June 1983, badly wounded, so we thought she might have died. Instead she had become a well-integrated member of her new social group. Miff, as a late adolescent, often wandered to the northern parts of her home range during periods of estrus and subsequently has tended to spend a good many months of each year in those areas. She may even have visited, or been consorted by, Mitumba males during periods of estrus in 1977 and again in 1983. Thus it is not surprising that Moeza, as a late adolescent, also chose to mingle with Mitumba males during periods of sexual activity. Her visits gradually became longer and longer until she finally made her commitment to the north.

It seems, therefore, that Miff's grandchildren will grow up in a different community; if the northern research goes well, we shall be able to follow their progress. Meanwhile, Miff has been deprived of the support of an elder daughter and remains in her natal community with two young sons.

MIKE (H. van Lawick)

MIKE in 1963 was a low-ranking male. By judicious use of empty kerosene cans, which he hit ahead of him while charging toward his superiors, he bluffed his way to the alpha position in 1964. For some time after this Mike seemed unsure of himself and seized every opportunity to display vigorously, as though to impress whichever males were nearby; often, too, he attacked females for no obvious reason when he was tense in the presence of other adult males. When Mike took over as alpha (from Goliath), there were fourteen adult males in the community; moreover, unusually large groups of chimpanzees gathered almost daily in camp to wait for bananas. Levels of aggression were high and the situation was a demanding one for an alpha male. Without his undoubted intelligence (after all, every male had access to the empty cans; only Mike systematically used them to further his ends), his high level of motivation, and his determination to improve his rank, Mike would probably have failed to reach the exalted top position. And without another quality—for want of a better word I call it "guts"—he would not have maintained the position long, for he was subjected to a number of aggressive challenges from groups of up to five of the other adult males.

Eventually Mike calmed down and became a benign alpha. He was exceptionally generous in sharing meat, particularly with his suspected sibling, J.B. He was a zealous groomer, devoting more time than is usual in a high-ranking male to the grooming of subordinates. After J.B.'s death in the 1966 polio epidemic, Mike never acquired another ally. One of his closest friendships, in fact, was with his defeated rival, Goliath—at least until 1970, when Goliath gradually shifted his range to the south and allied himself with the Kahama males.

By this time Mike was looking old, with worn teeth and hair that had become sparse and brown. Small wonder that when a female arrived, she was likely to rush up to greet the aggressive Humphrey and ignore the aging alpha. Yet Humphrey, at that time, was still very submissive to Mike. Gradually, however, he gained confidence, and early in 1971 he attacked Mike during an arrival display. It was a decisive fight, for it conferred alpha status on Humphrey. Mike, perhaps too old to bother, made no attempt to maintain high rank and was soon showing submissive behavior to even the lowest-ranking

NOPE (L. Goldman)

adult males. He became even more of a loner: his most frequent companion at that time was old Hugo. They died within a month of each other during the pneumonia epidemic of early 1975.

NOPE, a large, well-built female, first appeared (attracted by our bananas) in 1965. She gave birth later that year to a male infant, Mustard. In the late sixties Nope spent much time in association with another powerful female, Circe (who ultimately took over the alpha female position from Flo). Nope and Circe groomed together more than is usual for females; sometimes they played, and typically they supported each other in social conflicts. We suspect, in fact, that the two may have been sisters.

Circe died of unknown causes in 1968. We thought that her infant, Cindy, would travel about with Nope and Mustard, but this did not happen. The three-year-old child, with no adult to care for her, survived for only a few weeks. After Circe's death Nope, who never formed another close relationship (except with her offspring), became increasingly solitary. Mustard, deprived of his most frequent playmate, Cindy, was forced (like Gilka) to spend a good deal of time with only his mother for company. As a juvenile and early adolescent Mustard was unusually dependent on Nope—in part, assuredly, because she did not wean him until he was seven years old. Only when he was thirteen did Mustard begin to travel regularly with the adult males.

Nope nursed her second infant, Lolita, for seven years also. Her third infant we never saw (it was either stillborn or died very early), and the next, Hepziba, vanished when she was only a year old. Nope herself appeared with fresh wounds at the time this happened, and it is possible that Hepziba was a victim of intercommunity conflict; we shall never know. Shortly afterward Lolita became very sick and she died after a few months. Thus at the beginning of 1982 Mustard was the sole survivor among Nope's offspring.

Nope has seldom been observed when sexually receptive; at such times she has simply vanished. After Hepziba's death, however, Nope remained in the Kasakela core area during her next three periods of estrus and became, each time, the nucleus of large, excited sexual groups. There was a good deal of fierce competition among the adult males for the privilege of copulating with her. The result: her last infant, daughter Noota, born in 1982.

It will be extremely interesting to follow the development of this infant. Already, at three years of age, she appears to have been affected by the asocial disposition of her elderly mother. While it is true that many females become increasingly solitary as they get older, an infant born at this period of his mother's life typically has the companionship of one, often two, elder siblings. Noota's only sibling is Mustard—and he is an adult who spends very little time with his family and who, moreover, has outgrown his adolescent playfulness.

One final point. Sometime during 1966, just after the birth of Mustard, something must have occurred to destroy, perhaps forever, the trust in humans that Nope had acquired during the previous months. Even today, almost twenty years later, it is seldom possible for observers to follow Nope unless she is with other chimpanzees. What-

PALLAS (L. Goldman)

NOVA, with SKOSHA (L. Goldman)

PASSION

ever happened, however, did not destroy the relationship that I personally had built with Nope. She is still quite relaxed and calm when I am with her—at least she is after the first few minutes, during which she apparently realizes who it is that is trailing through the forest behind her.

PALLAS and NOVA were first recognized as late adolescents in 1965. They may have transferred from a neighboring community, or they may have been attracted from some peripheral area of the community range by the then abundant supply of bananas at camp. Both females became pregnant the following year, but their babies either were stillborn or died during their first few days: we never saw them. In 1970, however, the two were more successful. Nova's Skosha arrived first; Pallas' Plato followed nine months later. The young mothers spent a great deal of time together as, indeed, they had from the first. Now that they were lactating and no longer coming regularly into estrus, they associated less frequently with other adults.

Plato, a precocious and lively youngster, died when he was three years old from some kind of gastric enteritis. And in 1975 Pallas' next infant, Villa, aged ten months, disappeared during an epidemic of what was probably pneumonia.

Nova by this time was sick and had become very thin. Perhaps as a result, she and her daughter had become increasingly peripheral and were seldom encountered in the central part of the Kasakela range. In 1975 Nova must have died, because Skosha reappeared without her. By the end of that year Skosha had been adopted by the temporarily childless Pallas. A few months later Pallas gave birth again. This baby, who disappeared during her first month of life, was almost certainly one of Passion's victims. A year later Kristal was born. Skosha, by then an integral part of Pallas' family, served as an elder sibling to the new infant. She was a much-needed playmate, because Pallas had become one of the most asocial of the Kasakela central females.

When Kristal was five years old, Pallas became sick and died. Now it was Skosha's turn to foster an orphan; she and Kristal were seldom seen apart for the next few weeks. But Kristal, depressed and lethargic after the death of her mother, vanished by the end of the year.

Pallas, as an adult, held a very low rank in the female hierarchy, but had a relaxed and friendly relationship with the adult males. Perhaps because of this she was an extremely popular sexual partner. During 1973 and again in 1976 Pallas was the nucleus of a number of large and excited sexual groups when all the community males actively competed for mating rights. (One of these occasions is described in Chapter 16.) In addition, Pallas was led away on many consortships.

Pallas was one of the most affectionate and attentive of mothers and it is tragic that, of the five infants she bore, not one has survived.

PASSION became Gombe's most notorious character. It was she, along with her late-adolescent daughter, POM, who in 1975 began to kill and eat the newborn babies of the Kasakela community. I had known Passion from the very earliest days of the study. Indeed, she figures, along with David Greybeard, in one of the very first photos ever taken

POM

of termite fishing in 1961. She did not visit camp until 1964, but during the fall of that year she became a well-known individual, for she was regularly coming into estrus and often the nucleus of large sexual groups. In 1965 she gave birth to a daughter, Pom—and thereby gave us the opportunity to observe some extraordinarily inefficient and indifferent maternal behavior.

Pom survived her somewhat rough upbringing, and as she got older a close, cooperative bond formed between mother and daughter. It was this, above all, that enabled the pair to practice their infant-killing behavior. When Passion tackled an adult female *without* Pom's help, she failed in her attempt to seize the baby. Working as a team, however, they were seen to kill and eat three tiny infants and may have taken as many as seven others over a four-year period, 1974 to 1977. Why Passion began to kill and eat babies at all is still a mystery.

From 1970 on, Passion became increasingly asocial, spending most of her time with members of her immediate family, but by 1979 this meant that she was in almost constant association with four other chimpanzees; she had two more offspring after Pom (both sons), and Pom herself gave birth in 1978. When Pom's infant died before he was three, the bond between mother and daughter became, if anything, even stronger. They were almost inseparable at the time of Passion's death in 1982 of an unknown wasting disease. Pom and her adolescent brother, Prof, between them cared for Passion's lastborn, little Pax, who has managed to survive even though he was only four years old when he lost his mother. Pom herself has not been seen since the end of 1983.

SATAN

SATAN acquired his name for the "devilish" reason that during his very first visit to camp with his family in 1964, he stole part of the manuscript of a book my mother was writing. He was a juvenile at the time, still dependent on his mother, Sprout. Satan developed into a young male with a splendid physique and today is one of Gombe's heavyweights. Although he has consistently made attempts to better his position in the dominance hierarchy, he has not shown the aggressive persistence of those who made it to the top, nor has he ever formed a lasting coalition with another male. He did manage, after a struggle of several years, to dominate the older Humphrey and to get the better of the Jomeo-Sherry alliance, but he was very submissive to Evered for many years, even though he weighed some 48 kilograms compared to Evered's 39. And today, unless he is with one of the two other senior males (Evered and Jomeo), he is completely intimidated by the 37-kilogram Goblin, and typically flees, screaming, if the young male displays at him. Nevertheless, he has played a prominent role in intercommunity conflict and has led many patrols to peripheral areas of the community range.

Satan has often taken females on consortships, most of which were unsuccessful in that they came to an end (often because the females escaped) before the most fertile period of estrus, so that Satan lost many chances of fathering infants.

5. Demographic Changes

Flo just before her death in 1972.
(H. van Lawick)

September 1972 Extract from Flo's obituary in the *Sunday Times*: "Even now a month later it is hard to believe that Flo is dead. For more than ten years this old chimpanzee has been an integral part of life at the Gombe Stream, with her torn ears and bulbous nose, her occasional spells of wild sexual activity, her dauntless, forceful personality . . . She was lying at the edge of the tiny Kakombe stream. When I turned her over, her face was peaceful and relaxed without sign of fear or pain. Her eyes were still bright and her body supple."

Flo, during her life, was a survivor. She was unaffected by the horrifying polio epidemic of 1966, although her eldest son, Faben, was crippled. She lived through a terrible attack of what was probably pneumonia, and as she lay on the ground, too weak to climb and battered by torrential rain, her lastborn infant disappeared. Flo finally died during an unusually lean dry season; by then she was old, but it will take many more years of study before we know how long it is possible for a chimpanzee at Gombe to live.

The structure of a chimpanzee community does not remain stable year after year. At Gombe, in addition to the inevitable births and deaths, some chimpanzees have immigrated into the community and others have simply vanished, leaving us uncertain whether they died or emigrated. Altogether, between 1965 and 1983, 59 infants have been born; 13 individuals have immigrated from neighboring communities; 80 chimpanzees have died, emigrated, or disappeared. Figure 5.1 shows that these increases and decreases in numbers have not been evenly distributed over the nineteen-year period.*

The various stages of the life cycle—infancy, childhood, early and late adolescence, maturity, and old age—are defined in Table 5.1 and briefly described. Table 5.2, documenting changes that have taken place from year to year in the structure of the community, shows how many chimpanzees there were at the end of each year in the different age-sex classes.

* Data from the first four and a half years of my study are less complete than those from 1965 on and have not been included in this chapter. It is virtually certain that by 1965 all members of the community, except perhaps for a few peripheral females, had been individually identified and named.

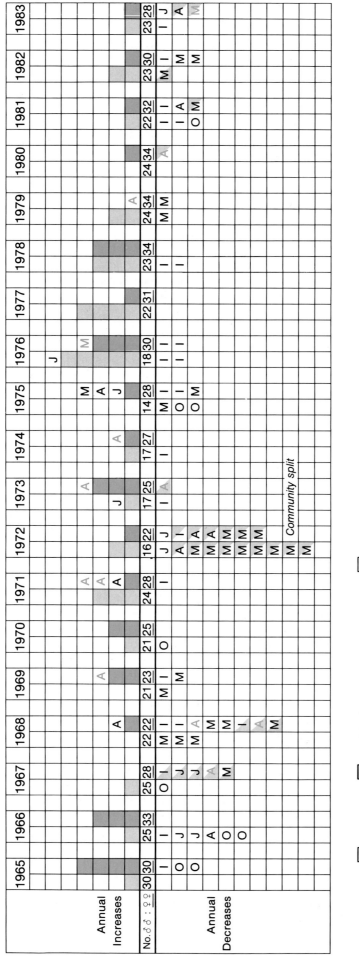

Figure 5.1 The population of the Kasakela community: annual increases caused by births and immigrations, and annual decreases resulting from deaths, emigrations, and disappearances, 1965 to 1983.

Table 5.1 Stages of the chimpanzee life cycle.

Infancy (0–5 years)
> Birth to the time when the youngster depends almost entirely on solid foods (and usually is weaned), and when riding on the mother is no longer the usual method of travel. Usually terminated by the birth of a sibling.

Childhood (5–7 years)
> Continuing close association of juvenile with the mother, but independence of her for transport and milk; youngster makes own night nest.

Early Adolescence
> (a) Males (8–12 years)
>> Gradually increasing independence; more time spent with adult males, but increasing wariness of them. More and more aggression (as testosterone levels in the blood rise) and start of struggle to dominate the females of the community.
> (b) Females (8–10 years)
>> Continuing closeness to the mother; adolescent swellings of the sex skin (irregular and unattractive to mature males); mated by infant, juvenile, and early-adolescent males.

Late Adolescence
> (a) Males (13–15 years)
>> Majority of time spent with adult males and females in estrus, but rather peripherally. By end of period usually able to dominate all adult females. (For some, *social* maturity does not occur until 16 or even 17 years of age.)
> (b) Females (11–13 or 14 years)
>> Adolescent sterility, starting with menarche and ending when the female is capable of carrying a fetus to term. Regular sexual swellings, during which she travels with adult males and is mated by them. Toward end of period some consortships formed with individual males; occasional transfers (usually temporary) to a neighboring community and mating with adult males there. Close ties with the mother retained (at Gombe, but apparently not at Mahale).

Maturity[a]
> (a) Males
>> Young Adult (16–20 years)
>>> Increasing integration into society of older males; eagerness to patrol peripheral areas of home range; attempts to raise social status.
>> Prime (21–26 years)
>>> Continued vigorous attempts to raise social status.
>> Middle Age (27–about 33 years)
>>> Glossy black of coat begins to go brown on back and legs; individuals still vigorous.
> (b) Females (14 or 15–about 33 years)
>> Raising of the family. (If reproductive cycles proceed smoothly, female may be described as *young adult* when raising first child, *prime* when raising second, and *middle-aged* after third.)

Old Age (about 33 years to death)[b]
> Gradual slowing of activity; increasing tendency to withdraw from intensive social interaction. Teeth worn or broken; thinning of hair, particularly of head and lumbar region; coat becomes browner (back of one old male liberally scattered with gray hairs).

a. The male, unlike the female, becomes *reproductively* mature (capable of fertilizing a female) before he attains *social* maturity. There is no evidence to date that males are infertile in their old age. Thus late-adolescent and old males are included in the category *reproductive maturity*.
b. Very few individuals have lived long enough during the study period to be placed in this category. The two oldest individuals at the Yerkes Center, both females, were 52.3 years and 54.8 years at the end of 1984 (R. Nadler, records of the Yerkes Regional Primate Research Center, and D. Rumbaugh, personal communication). Further years of study are necessary to determine the longevity potential of wild chimpanzees.

Table 5.2 Numbers of males and females in the six major age classes of the Kasakela community (end-of-year tallies) from 1965 to 1983.

| | Old | | Mature | | Adolescents | | | | Juveniles | | Infants | |
| | | | | | Late | | Early | | | | | |
Year	Male	Female	Male	Female	Male	Female	Male	Female	Male	Female	Male	Female
1965	2	1	13	8(3)	3	5	5	2(1)	2(1)	3(3)	4	3(1)
1966	1	1	12	9(3)	5	6(1)	2	1(1)	2	2(2)	3	5(2)
1967	0	1	14	9(2)	3	6(1)	3	0(1)	1	2	4	5(1)
1968	1	1	11	9(1)	3	3	3	3(1)	2	0	2	4
1969	2	1	11	10(1)	2	2(1)	2	2(1)	3	0	1	4(1)
1970	2	1	10	10(1)	2	2(2)	2	2	3	2	2	4(1)
1971	3	1	10	11(2)	1	3(3)	2	0(1)	3	2	5	4(1)
1972	2	0	6	12(1)	0	1(1)	3	1	1	1	4	4(1)
1973	2	0(1)	6	12(1)	0	1(1)	3	1	1	0	4(1)	7(1)
1974	2	0(1)	6	11(2)	1	2(1)	2	1	1(1)	1(1)	4	6(1)
1975	0	0(1)	5	12(1)	1	3	3	1(1)	1	1(2)	4	6
1976	0	0(1)	5	13(2)	1	3	3	1(1)	3(1)	0(2)	3(2)	6(1)
1977	0	0(1)	6	13(2)	1	3(1)	3	3(1)	3(1)	1(1)	6(2)	4(1)
1978	0	0(1)	6	16(2)	2	0(1)	2	3(1)	4(1)	4(1)	6(2)	4(1)
1979	0	0(1)	5	13(4)	2	0(2)	4(1)	3(2)	2	2(2)	7(3)	3(2)
1980	1	0(1)	6	12(5)	1	2(1)	5	1(2)	1	2(2)	7(3)	4(2)
1981	0	0(1)	6	12(4)	1	2(3)	6	1(1)	1(2)	1(2)	5(1)	4(1)
1982	0	1(1)	5	10(5)	1	1(2)	6	1(1)	4(2)	1(2)	3(2)	5
1983	0	1(1)	6	10(5)	3	1(2)	3(2)	1(2)	4	2	3(2)	2(1)

Note
The Mature column includes young adult, prime, and middle-aged. Numbers in parentheses indicate peripheral members of the community and are in addition to the other (central) community members shown. The break between 1971 and 1972 represents the community split. During 1976 two of the infants born were probably twins (one male, one female) and during 1977 two male infants were twins.

Figure 5.2 illustrates some of the striking differences that have been documented in the structure of the chimpanzee community at three different points in time—1968, 1978, and 1983. One significant change is the ratio of adult males to adult females. In 1968 this was approximately 1:1; there were 12 males and 10 females. Ten years later the picture was very different: there were 3 adult females for every male. By 1983 the males had regained a little ground and the male-female ratio was 1:2.8. A second change concerns the proportion of the community as a whole that comprises immatures—infants, juveniles, and early adolescents. In 1968 they accounted for 34 percent of the community, in 1978 they were 51 percent. Then, in 1983, 43 percent of the community was immature. This second change is, of course, the direct result of the relative increase in the number of females capable of bearing infants and raising families.

Before we consider some of the factors that affect the size and structure of any social group of animals (age of fertility, birth and death rates, causes of death), I should mention three rare events that had profound and far-reaching effects on the Gombe chimpanzee community.

(a) In 1966 there was a severe outbreak of a paralytic disease, probably poliomyelitis. Six individuals died or disappeared during the epidemic, and six others were afflicted and survived as cripples.

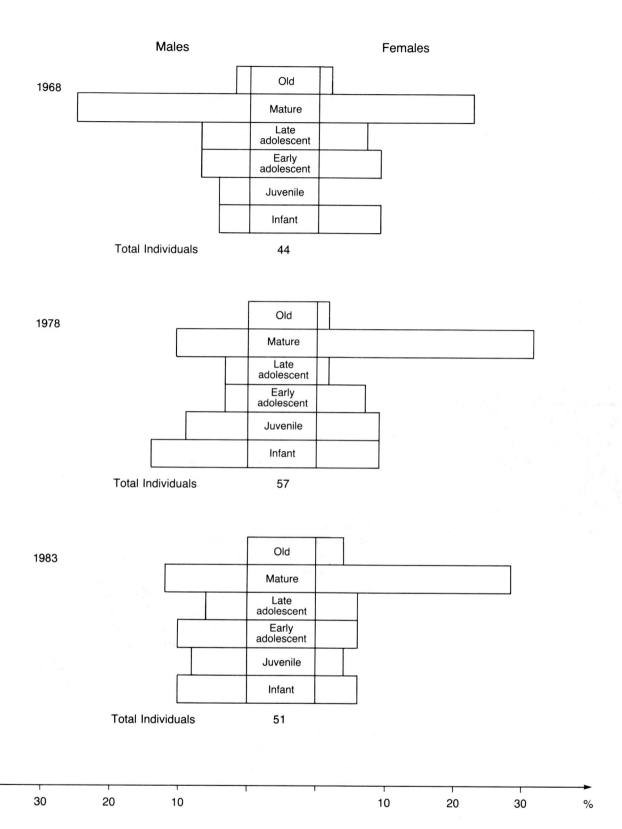

Males   Females

1968

Old

Mature

Late adolescent

Early adolescent

Juvenile

Infant

Total Individuals   44

1978

Old

Mature

Late adolescent

Early adolescent

Juvenile

Infant

Total Individuals   57

1983

Old

Mature

Late adolescent

Early adolescent

Juvenile

Infant

Total Individuals   51

%   30   20   10   10   20   30   %

Figure 5.2   Changes in population structure over time: percentage of the total number of individuals in the different age-sex classes during 1968, 1978, and 1983.

(b) In 1970 the community that I had been observing since 1960, the Kasakela-Kahama (KK) community, began to divide. By the end of 1972 two separate social groups were recognized, the Kasakela community in the north and the Kahama community in the south. This division led to fierce conflict. The Kasakela males systematically hunted down, attacked, and wounded individuals of the smaller Kahama community until all had been killed or had disappeared.

(c) In 1974 a Kasakela female, Passion, and her adult daughter, Pom, were observed to attack a community female, seize, kill, and subsequently eat her tiny infant. During a four-year period Passion and Pom were seen to kill two other infants in this way and may have been responsible for the disappearance and death of seven others. Only one infant survived during that four-year period; when the killings stopped, therefore, all the community females were either just coming into estrus after losing their babies or were pregnant anyway. This resulted in a "baby boom" in 1977–1978; the clustering of births can be seen in Figure 5.1. A second, less obvious clustering of births has occurred more recently, in 1984–1985. This, too, can be attributed to Passion's early depredations.

Age at Reproductive Maturity

Despite the long-term nature of the research, only three males and two females born during the study period have reached reproductive maturity. However, age estimates for individuals who were adolescent or younger in the early years of the study are likely to be reasonably accurate. Both males and females at Gombe mature later than chimpanzees in captivity, apparently as a result of differences in body weight (McGinnis, 1973; Pusey, 1977).

There is an acceleration in male weight gain between 9 and 12 years of age (Pusey, 1977). In captivity this acceleration correlates with a sharp rise in plasma testosterones (McCormack, 1971). Weight then increases slowly for another two years or more, until adult weight is reached at about age 16 or 17. Males at Gombe fall into two groups, heavy and light, with a mean of about 46 kilograms (100 pounds) for the heavy group and 37 kilograms (82 pounds) for the light (Pusey, 1977).

The first noticeable increase in scrotum size is apparent during the ninth year, or early in the tenth, and there is a sudden spurt in scrotum growth during a 4- to 6-month period about a year later. Ejaculation was first observed in 3 males (Wilkie, Freud, Prof) at 8.9, 9.2, and 9.4 years respectively, and it seems unlikely that it occurs much before this. In the human male, mature sperm do not appear for a few years after the first ejaculation (Katchadourian, 1976); this may be true for the male chimpanzee also. If so, males are probably not capable of fertilizing females until they are about 12 or 13, when the scrotum has reached approximately adult size. In fact, no males at Gombe are thought to have sired infants prior to their fourteenth year.

In females the first tiny swellings of the sex skin, initially involving only the labia, appear between 7.5 and 8.5 years. These gradually get larger until, when the female is between 10 and 11, she shows her first full-sized swelling and is mated by adult males. This is probably her first estrus. Menarche occurs between one and six months later.

Early-adolescent Wilkie.

The female does not show the sudden growth spurt characteristic of adolescence in the male. She reaches adult weight (32 to 37 kilograms, or 71 to 82 pounds) when she is about 19.

There is a period of adolescent sterility in the female (Hartman, 1931) between the onset of menarche and the time of the first conception leading to a live birth. At Gombe this was 1.1 to 3.0 years for 5 females (Miff, Fifi, Winkle, Gilka, Pom), but only 5 months for a sixth, Gremlin. For 5 of these 6 females, the first suspected pregnancy terminated with the birth of a live infant. The other, Gilka, may have lost an infant between 1.2 and 1.8 years after her first observed menstruation—she was not cycling when she returned from a six-month visit to the Kahama community. However, her irregular cycling may have been the result of illness (she was suffering from a fungus disease). Her first *observed* infant was conceived 2.4 years after menarche.

Of a total of 14 young females (counting the 6 mentioned above) 6 including Gilka are known or thought to have had at least one pregnancy before producing their first known offspring. However, it was not possible to tell whether these pregnancies terminated with live babies who died during the first few days of life, or whether they ended in spontaneous abortions or stillborn infants. (Abortions in young females in captivity are quite common; Graham, 1970.)

Gestation and Birth

The sexual cycle of the female chimpanzee, characterized by menstrual bleeding and periodic swelling, and deturgescence of the anogenital region, will be described in full in Chapter 16.

The gestation period for the chimpanzee averages about 8 lunar months. In the field it is not possible to calculate the length of gestation precisely, but mean gestation length for a group of 9 captive females who gave birth after single matings and normal pregnancies was 229.4 days, with a range of 203 to 244 days (Martin, Graham, and Gould, 1978).

Chimpanzee pregnancies usually lead to single births, although the twinning rate may be greater than in man (Keeling and Roberts, 1972). At the Yerkes Regional Primate Research Center there were 12 multiple births (11 sets of twins and 1 set of triplets) out of a total of 300—that is, 4 percent. Four males and 7 females contributed to these multiple births: the most prolific female, Flora, produced 2 sets of twins by 2 different males and triplets by one of them. The most prolific male, Hal, produced twins with 5 different females (one of them Flora) and was the father of her triplets (Martin, 1981). At the Washington Park Zoo in Portland, Oregon, 19 successful pregnancies produced 3 sets of twins, all born to the same female and sired (as were all her babies) by the same male (M. Yeutter, personal communication).

These multiple pregnancies commonly resulted in miscarriage, stillbirth, or survival of only one infant. Of the 12 multiple births at Yerkes, only 3 sets of twins survived, as pairs, beyond the first week (Martin, 1981). In 2 of these sets all 4 infants survived to independence; in the third, one baby died at 6 months and the other at 15 months. In the remaining 9 multiple births, one set of twins was

Melissa gazes at day-old Goblin, her first surviving infant.
(H. van Lawick)

stillborn, 2 births were premature, and 2 were infants who survived less than 3 days. In 3 instances one twin died in less than 4 days while the second lived longer—but only one of them survived beyond childhood. In the case of the triplets, one was premature, the other two lived for only a few hours. Thus of the total number of 25 babies in the 12 multiple births, only 20 percent (5) survived to adolescence.

At Gombe, 59 pregnancies resulting in live births have been recorded: this includes one pair of male fraternal twins, Gyre and Gimble. However, mothers are often not seen for 1 to 2 weeks after giving birth. There is a possibility that other twins were born but that one or both died during the early postnatal period. In 1976 an unhabituated female, Caramel, was observed at Gombe carrying two apparently same-aged infants, a male and a female. These may well have been twins, although we cannot rule out the possibility that one was adopted. One of Melissa's twins, Gyre, always the weaker, died at 10 months; the other, in 1985, is 8 years old.

The mean interval between successive births, when the previous infant was still alive, for 13 females (21 birth intervals) was 66 months (range 48 to 78). If a baby dies, the mother may give birth again in about a year; the median, for 7 females (13 intervals) was 12 months (range 10 to 16). In one of the study groups at Mahale (the M group) the duration of lactational anestrus for 4 mothers ranged from 30 to 51 months (median 45.5 months). Birth intervals for 7 females there ranged from 5.2 to 7.1 years (median 6.0) (Hiraiwa-Hasegawa, Hasegawa, and Nishida, 1983).

In captivity some chimpanzee females over 40 years have shown sporadic amenorrhea accompanied by a lack of primary follicles in the ovary (Flint, 1976). Others maintained a high frequency of menstrual cycles up to 48 years, although they conceived less often than when they were younger (Graham, 1979, 1981). One female at Gombe, Flo, showed her last sexual swelling two and a half years before her death when she was estimated to be over 40. Old age is more stressful for a chimpanzee in the wild because of parasite load, difficulty in climbing for food, exposure to heavy rain and wind, and so on. Two old females of the M group at Mahale, however, estimated as being over 40, both continued to cycle during a three-year study period, but they neither gave birth nor became pregnant (Hiraiwa-Hasegawa, Hasegawa, and Nishida, 1983).

Movement of Individuals between Communities

Even today we seldom know the precise number of adolescent and adult females in our community in any given year. Some vanish for months at a time, only to reappear later. Unhabituated females may be encountered with members of the Kasakela community; usually they are very fearful of humans. Only when the chimpanzees of the Kalande and Mitumba communities, to the south and north, have become habituated and their members are recognized, shall we be in a position to come to grips with the problem.

Immigrants and Visitors
At Gombe there are permanent transfers, or *immigrants* (females who have left their natal communities to join neighboring ones), and

temporary transfers, or *visitors* (females who visit neighboring communities for relatively short periods, usually during consecutive periods of estrus, then return to their original social group). In addition, certain *peripheral* females may continue to move back and forth between communities.

Table 5.3 gives details on twelve immigrant females.* Five of these were late adolescents; they appeared first when they were in estrus, traveling with and being mated by Kasakela males, and subsequently gave birth within the community. A sixth late adolescent was Little Bee, whose history is related below. Two of the immigrants, Caramel and Joanne, were primiparas, each accompanied by a child (they may, in fact, have been peripheral members of the community already). Joanne was in estrus and being mated by the males. Three immigrants were early-adolescent females, all of whom showed their first adult swellings and were mated by Kasakela males soon afterward. The remaining female immigrant was the juvenile daughter of Caramel.

Two of the females, Joanne and Harmony, were accompanied by dependent males: Jageli (Joanne's son) and Beethoven (presumed sibling of Harmony). Jenny, one of the early-adolescent immigrants, was caring for an infant of about eighteen months (presumed to be an orphaned sibling) when she was first seen. Two days later she was observed carrying the dead body of the infant.

The early history of one immigrant female, Little Bee, is known. She grew up in the KK community before the split. With her mother, Madam Bee, and her early-adolescent sister, she moved to the south and became part of the new Kahama community. Between 1972 and 1974 she made repeated visits to the Kasakela community during periods of estrus, returning south in between. During 1972 and 1973 she was classified as a visitor to the Kasakela group. In 1974 the Kasakela males began a series of brutal attacks on the Kahama community members. Little Bee, in estrus, was traveling with Dé, the second of the victims; after the attack the Kasakela males forcibly led her back to the north. On three other occasions there is evidence that Little Bee, when in estrus, was recruited by parties of invading Kasakela males who led her back to their range. Eventually, in 1974, she remained in the north even when anestrous and was thereafter classified as an immigrant (Pusey, 1979). The daughter of Kahama female Mandy, assuming that she and Kidevu are one and the same, followed a similar pattern.

All the late-adolescent females who eventually became immigrants first appeared as visitors. Two years after Little Bee became a permanent member of the Kasakela community, her younger sister,

* After the initiation of the banana-feeding system in 1963, "new" females appeared, one by one, as they discovered this food source. However, the fact that they were new to me does not necessarily mean that they were new to the community. Four of the females who appeared in camp during 1965 were late adolescents; they may have been transfers (Pusey, 1979), but because there is no way of determining whether this is so I have not listed them in Table 5.3 as immigrants. Three mothers with dependent young who appeared at camp during 1964 and 1965, at the height of the banana feeding, were presumed peripheral females whose core areas were some distance from camp, so that it had taken them longer to find our bananas. Two of them had (presumed) adolescent sons, who were already camp visitors before their mothers appeared. All three females left when the feeding system was changed and fewer bananas were handed out. One of them, Sprout, returned to her preferred area north of camp, where she has been encountered off and on ever since. The other two, Jessica and Vodka, may also have returned to preferred areas, but they have not been seen again.

Table 5.3 The history of twelve females who immigrated into the Kasakela community.

Age class and name	Age estimate (years)	First seen	History to first live offspring
Early adolescents			
Winkle	8/9	August 1968	When first seen (near camp), already well integrated into community. First adult swelling: July/August 1969. First infant: October 1972.
Harmony	9/10	December 1973	First adult swelling: February 1974. Last seen: 1981.
Jenny	9	June 1975	First seen with small infant, a suspected orphaned sibling (who died). First adult swelling: September 1977. Last seen: 1980.
Late adolescents			
Wanda	10/11	September 1969	Infant born: 1971. Moved south to become part of Kahama community.
Dove	11/12	February 1971	Infant born: 1972. Peripheral except when cycling.
Patti	10/11	March 1971	Seen very rarely during 1971 and 1972. Central female from November 1973 on. Lost infant 1978 (almost certainly had lost or aborted at least one before). Successful birth: 1979.
Sparrow	10/11	September 1971	Infant born: 1973. Peripheral except when cycling.
Little Bee (from Kahama)	13/14	See text	Transferred permanently to become central female after many visits back and forth after community division. Usually visited during periods of estrus. Suspected birth (lost infant): 1976. Successful birth: 1977.
Kidevu (perhaps from Kahama)	10	September 1979	Almost certainly Mandy's daughter Midge, of Kahama community (facial structures extremely similar). Infant born: 1982. Probably will become central female.
Primiparas			
Caramel	19/20	August 1975	First seen in estrus with juvenile female (presumed) offspring, Candy. Possibly already a peripheral Kasakela female. Gave birth to (presumed) twins in 1976. Seen very rarely and not included as Kasakela female until 1980.
Joanne	16/17	February 1979	First seen in estrus with late infant male (presumed) offspring, Jageli. Possibly already a peripheral Kasakela female. Infant born: 1981.
Dependent			
Candy	6	August 1975	Seen in association with (presumed) mother, Caramel. Became well integrated when cycling during 1983. Infant born: 1983.

Honey Bee, also appeared as a visitor. She was seen, very irregularly, always in estrus, during 1976, 1978, and 1979—and once, almost certainly, in 1982. Other visiting females have not been identified. During a given year there may be ten or more sightings of young females in estrus but because they are shy, it is usually not possible to get close to them. There is no way of knowing whether the sightings have been of one female, ten females, or some intermediate number.

Figure 5.3 provides additional information on the integration patterns of some of the immigrant females. Patti was an infrequently seen visitor for three years, after which she became well integrated and was soon considered to be a central female; Kidevu is following a similar pattern. (Winkle, not shown in the figure, was a central female from the time she was first seen near the camp area.) In sharp contrast, Sparrow and Dove, although they frequently associated with Kasakela individuals when cycling, did so much less often when they were anestrous and lactating. The number of days they were encountered per year correlates significantly with their reproductive condition. Harmony became well integrated for a five-year period, then disappeared and may have died. Jenny also disappeared after five years. The two primiparas, Caramel and Joanne, have remained peripheral.

During the nineteen-year period five late-adolescent females and one older, cycling, but childless female have vanished from our community. One of these, Moeza, is known to have transferred to the Mitumba community. She was encountered there two years after she was last seen (with her mother) in the Kasakela community range. She had spent long periods of time in the north, particularly when in estrus, during the year prior to her disappearance from her natal community. Sally, Bumble, Sesame, and Pom were last seen in good health and may also have emigrated. Pooch, however, disappeared during a pneumonia epidemic and was also suffering from an unhealed injury. It seems quite likely that in her case disappearance was due to death.*

All Kasakela adolescent females are thought to have made a number of visits to neighboring communities during periods of estrus; this will be further discussed in Chapter 16.

Peripheral Females
Each year, from 1965 until the present, between three and six late-adolescent and/or mature females and their dependents have been classified as peripheral. A peripheral female is one whose preferred core area is situated toward the outer limits of the community range, who associates with other community members when they are nearby,

* During the nineteen-year period reviewed it seems that whereas nine young females have transferred into the main study community, only five are thought to have transferred out permanently. Does the lure of bananas prevent young females from leaving their natal group? Probably not; it seems that these fruits, while highly prized, may quite easily be "given up." Two of the suspected emigrants, Sally and Bumble, left when banana feeding was at its height in 1967–1968 (which is also when Pooch disappeared). Moeza and Pom had visited camp and eaten bananas regularly since their infancy. (The same was true of Gilka, who abandoned camp for six months to visit the community to the south.) Moreover, some of the female visitors from neighboring social groups who have appeared in camp and discovered bananas, such as Honey Bee and Jenny, have subsequently left again.

Figure 5.3 Patterns of integration into the Kasakela community of eight immigrant females: number of days per year that these females were encountered (during follows or in camp), and number of those days that they were in estrus.

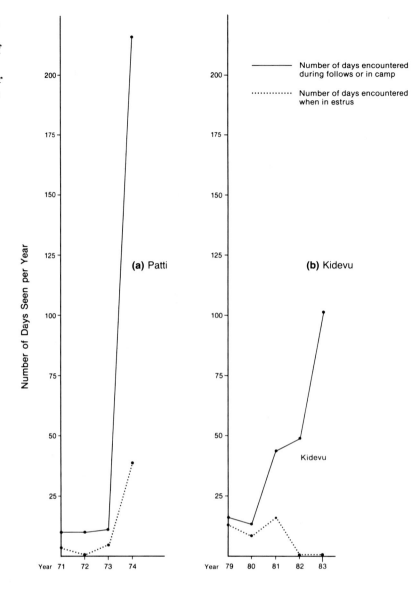

but seldom (unless cycling) travels with them when they leave her area. She is only rarely encountered during follows—perhaps two or three times in a given month, and often not at all unless the adult males are attracted to her area by some seasonal food availability.

Figure 5.3 shows the frequency with which some peripheral females have been observed between 1971 and 1983. In addition, Nova, classified as central from 1965 (when she was first identified in the study), became increasingly peripheral from 1973, perhaps as a result of ill health. She died in 1975, after which her juvenile daughter, Skosha, once more became a central member of the community.

Late-adolescent females may, as we have seen, transfer back and forth between neighboring communities and associate and mate with males of both. Even after she had been classified as a central member of the Kasakela community for two years, Harmony was encountered in estrus, apparently in consort with a neighboring male. After giving birth, the female typically makes a commitment to one of the two

Figure 5.3   (continued)

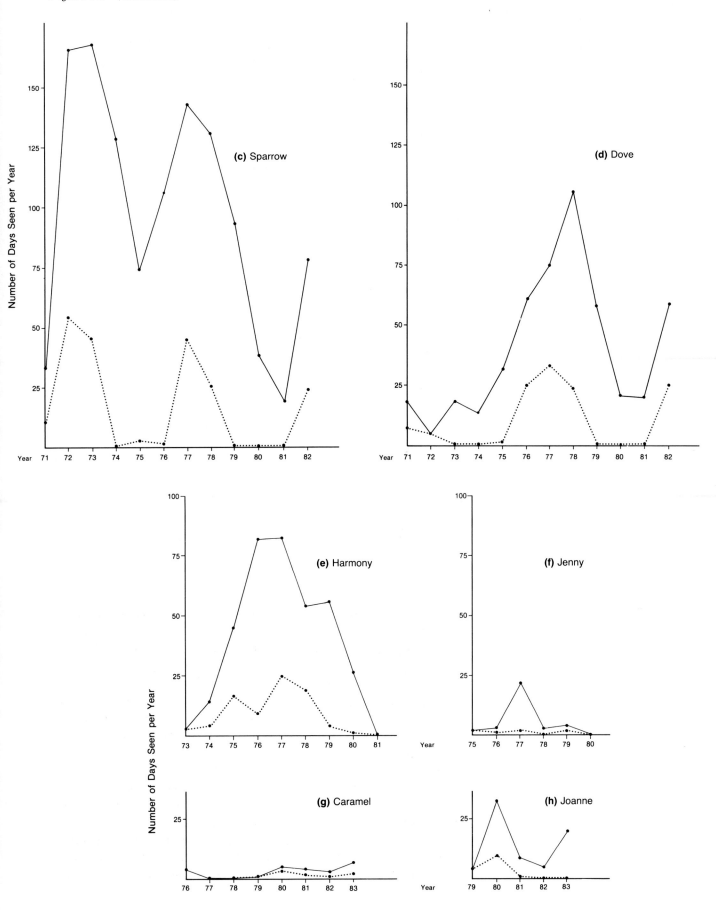

communities. If her preferred area is very close to the boundary, however, there is the possibility that, at least for a while, she may maintain ties with members of both groups. Twice young unhabituated mothers with tiny babies were encountered by Kasakela females within the Kasakela community range; the chimpanzees ignored each other. Another stranger female with a newborn met Kasakela males; there was a good deal of tension and the males surrounded her, but she was not attacked and managed to run off. This female had been a visitor the previous year (identifiable by means of a semiparalyzed leg); she has not been seen again.

When adult males encounter older neighboring females (those with at least one offspring of more than two years) there is usually severe aggression. Occasionally, however, Kasakela males and females associate peacefully with such stranger mothers—who may be peripheral members of the Kasakela community unrecognized by us, or females whose preferred areas are in the overlap zone between neighboring communities and who are peripheral members of both.

Health and Morbidity

Chimpanzees in the wild suffer from a variety of ailments. It is known that with very few exceptions they can contract the same contagious diseases as humans, but precise diagnosis is seldom possible in the field. Only two bodies, those of Flint and Plato, were sent for autopsy; and one biopsy was performed on tissue taken from an anesthetized individual, Gilka. Medication has been administered to sick individuals on a few occasions, but for the most part the various diseases and injuries sustained by the chimpanzees run their natural course. It is not easy to "interfere" without destroying the chimpanzee-human relationship.

Diseases

Respiratory. Chimpanzees quite often suffer from colds and coughs, particularly during the rainy season. Some individuals develop additional symptoms: pronounced lethargy; loss of appetite; sweating; thick, yellowy, mucous discharge from the nose; raspy breathing; deep, husky coughing, sometimes followed by gagging and swallowing. I refer to this here as pneumonia: the symptoms are similar to those described for this disease in captive chimpanzees (M. J. Schmidt, personal communication). The body of Gombe infant Plato, who had shown some of these symptoms, was sent for postmortem examination; pneumonia (along with catarrhalic enteritis) was diagnosed (D. Koehler, personal communication). In 1968 there was a bad outbreak, during which five individuals (David, Sophie, Circe, Pooch, and Flame) disappeared, and in a 1975 outbreak a ten-month infant, Villa, disappeared. In 1978 a number of chimpanzees caught colds during the dry season; the weaker of the two twins, Gyre, also aged ten months, died after having developed the husky cough and raspy breathing of pneumonia.

Individuals suffering from the symptoms described above often lay in day nests for several hours at a time, rose late in the morning, traveled only short distances, and retired early at night. Flo on two separate occasions was too ill to climb trees and lay on the ground,

Faben learned to walk long distances upright when one arm became paralyzed during the 1966 polio epidemic. (B. Gray)

Passion, four days prior to her death in 1982. She had become progressively thinner over a three-month period and was clearly suffering a good deal of abdominal pain the last few weeks of her life.

scarcely moving, for several days and nights in succession, twice in heavy rain. Her six-month-old infant, Flame, disappeared during the second of these illnesses (before we found Flo).

Poliomyelitis. In 1966 a paralytic disease broke out in the chimpanzee community; the first victims were seen in August and the last in December. It is known that the great apes are susceptible to the human poliomyelitis virus; there was a spontaneous outbreak at the Yerkes Primate Center in Florida in 1964 (H. Koprowski, personal communication). The paralytic disease at Gombe started about a month after an outbreak of polio among the human population in the Kigoma district. Two young men were afflicted in a village just beyond the southern boundary of the park. It seems almost certain, therefore, that the chimpanzees also suffered from polio. Presumably the disease spread through the park from the south; in subsequent years three partially paralyzed chimpanzees (one male and two females) were encountered in the unhabituated Kalande community to the south of the study communities.

During the five-month period in 1966, ten individuals were known to be stricken; two others disappeared and were eventually presumed dead. It was decided early in 1967 to administer oral polio vaccine to the chimpanzees (in bananas). As many as possible were given a complete course. Table 5.4 summarizes the data on the crippled victims. Those who survived adapted well to their various disabilities.

Gastrointestinal disorders. A form of gastric enteritis was diagnosed when the body of the early adolescent Flint was sent for postmortem examination. During the three weeks after his mother's death Flint, ill with chronic diarrhea, became emaciated and lethargic. Autopsy showed that his duodenum was badly inflamed and that he also had chronic peritonitis (D. Koehler, personal communication). We have seen that the infant Plato was found to have catarrhalic enteritis in addition to pneumonia.

Three other individuals (Pallas, Passion, and Lolita) also contracted some form of intestinal illness and apparently suffered considerable pain during the final three or four weeks of their lives. From time to time, as though in the grip of sudden abdominal spasms, they crouched over and remained motionless for up to two minutes. All three became increasingly slow-moving and emaciated; during the last day of Passion's life she was so weak that the effort of moving made her tremble.

Pallas had chronic diarrhea as well during the last three months of her life, a condition from which she had suffered, on and off, for ten years. All individuals have periodic bouts of diarrhea, often coinciding with a change in diet—when, for instance, a fruit crop ripens and is eaten in large quantities. Usually the condition clears up after a day or two, but persistent diarrhea lasting weeks or months sometimes occurs, particularly in old individuals.

Coprophagy, so common in captive chimpanzees, is occasionally seen in diarrheic individuals at Gombe. Pallas practiced coprophagy during bouts of diarrhea at least from 1972 until her death in 1982. Flo began to search for undigested food remains in feces (her own and that of others) when, weakened by age, she was unable to forage efficiently. In the context of intestinal illness it may be significant that both Plato and Flint, after watching their mothers eating their feces, copied the behavior. This undoubtedly contributed to their intestinal infections.

Table 5.4 Case histories of the chimpanzee polio victims.

Individual	Age class	Nature of paralysis	Outcome
Grosvenor	Infant (3 weeks)	Lost use of all four limbs during second or third day of illness.	Died same day.
McGregor	Old	Lost use of both legs; became incontinent. Moved by inching backward with hands in sitting position or pulling himself head over heels in a kind of somersault.	Shot for humane reasons after dislocation of one arm.
MacDee	Late adolescent	Lost use of both arms and hands. Moved by shuffling forward in squatting position.	Shot for humane reasons.
Merlin	Infant (4–5 years)	Lost use of one leg.	Not seen again.
Faben	Young adult	Lost use of one arm and hand. Initially moved tripedally, trailing arm on ground.	Learned to walk long distances in upright position.
Pepe	Young adult	Lost use of one arm. Initially moved in squatting position, trailing arm.	After 3 weeks started to walk in bipedal position.
Madam Bee	Middle-aged	Lost use of one arm, but retained more use of shoulder than Faben and Pepe, and could move fingers. Moved tripedally and could keep hand off ground.	Difficulty in supporting infant born 5 years after illness (during travel had no free hand). Baby died at 3 months. With next baby often walked bipedally and so could support infant.
Willy Wally	Young adult	Lost partial use of lower leg. Initially moved tripedally, with thigh flexed to abdomen.	One year later seen to use leg for three successive steps; gradually used it more often. By 1969 had progressed to walking with pronounced limp.
Melissa	Prime	Affected in shoulders and neck. Initially moved bipedally.	Badly handicapped for 2 weeks; walked normally after several months. Never regained full use of neck muscles; cannot turn head to look behind.
Gilka	Juvenile (7 years)	Partially paralyzed in right forearm and hand. Initially walked tripedally. Postmortem examination of upper limb bones showed those on left notably longer and larger in most dimensions, indicating differential usage of arms. Left metacarpals, particularly of thumb, also larger (A. Zihlman, personal communication).	Began to use bad arm for walking after 3 months. Never regained strength in wrist; never able to use thumb. Forearm in later years showed some wasting.

Note: Names of males are shown in green.

Vomiting is only rarely seen. Usually the vomit is swallowed again after it has been chewed briefly. Sometimes the chimpanzee catches the vomit in his hand before reingesting it. One individual was seen to vomit every few days for a two-month period, but subsequently recovered.

Ulcers, sores, and wasting. In 1964 Mr. Worzle, a male in his prime, got an ulcer on the dorsal surface of a broken index finger (which had set but could not be flexed). The sore never healed, other fingers became similarly affected, and by 1966 sores had broken out on his face (especially lips and ears), feet, and rump. His skin became dry and knobby in appearance. In 1967 he had difficulty walking and eventually, as all his knuckles were raw, he was forced to hobble along with the backs of his wrists on the ground. During 1968 he became very lethargic and, unable to keep up with the others, he spent much time on his own. Early in 1969 he caught a bad cold; this cleared up, but he had repeated bouts of diarrhea, his sores were often raw and bleeding, and he became increasingly emaciated. He weighed only 20.7 kilograms at his death in May 1969, age about twenty-five years.

A seventeen-year-old female, Gilka, developed chronic diarrhea early in 1977 after having successively lost to Passion three, possibly four, newborn babies. The following year Gilka developed bad sores or ulcers on the backs of her fingers and for weeks at a time had great difficulty walking. The lesions partially healed several times, then broke out again. Postmortem examination showed that several proximal and distal phalanges of her right hand were distorted by osteoarthritis (A. Zihlman, personal communication). Her weight dropped from 31.5 kilograms in 1974 to about 22.5 kilograms in 1977. The diarrhea persisted and she became increasingly emaciated, weighing only 19.5 kilograms when she died in 1979.

One other female, Nova, became excessively emaciated just prior to her death in 1975, at about twenty-one years. She had become a peripheral female during the previous two years, so no details are available.

Rhinophycomycosis entomophthorae. Gilka, whose sores are described above, had developed a small swelling on her nose in 1968, when she was about eight years old. This gradually became larger until it spread to the tissues around her eyes and brow ridge. In 1970 a biopsy was performed. Tissue culture showed that she was afflicted with a fungus disease that is quite common among people in West Africa but has not been recorded in Tanzania (Roy and Cameron, 1972). The condition was treated with doses of fanasil and potassium iodide, administered in bananas. The medication reduced the swelling, but when Gilka temporarily transferred to a neighboring community (late 1972–early 1973) and was no longer treated, her condition deteriorated. She returned with her nose, brow ridge, and tissues around her eyes hideously swollen. Both eyes were closed; she seemed to move about by sensing light and dark, and often bumped into branches. Again the swelling was reduced by medication. Pregnancy appeared to have a beneficial effect; after the birth of her infant in 1974 only her nose remained swollen, despite the fact that she was no longer receiving medication. (The sores on her fingers, her diarrhea, and general wasting were almost certainly not connected with the fungus disease.)

Gilka contracted a fungus disease (*Rhinophycomycosis entomophthorae*), which caused a grotesque swelling of the tissues around her nose and eyes. At the worst period of the disease her eyes appeared to be completely closed and she frequently bumped into objects as she moved about. (H. van Lawick)

The swelling in Olly's neck was almost certainly a goiter.
(H. van Lawick)

Humphrey began to push his thumb into his right ear in 1964, when he had an abscess just inside the external opening of the meatus. He continued to do this, on and off, for the rest of his life; this photo was taken in 1978.

Jomeo's left eye is damaged from an old injury, and his face is badly swollen as the result of an abscess under one of his upper left molars. It burst the day this picture was taken, but the swelling only gradually subsided over the next three weeks.

Goiter. One female, Olly, had a small swelling on the ventral surface of her neck when she was first identified in 1961 (about twenty-two years old). This gradually enlarged and three years later the whole ventral surface of her neck was swollen and somewhat pendulous. Goiter is a common condition in the Kigoma region, because of lack of iodine, and it seems probable that Olly's swelling was a goiter. In 1969 some further enlargement was noticed, but it is unlikely that this contributed to her disappearance and presumed death later that year.

Abscesses. In 1964 the young adult male Humphrey developed an abscess just inside one ear. It caused him much distress; he continually pushed his thumb into his ear and held it there for minutes at a time. After the abscess burst, he continued to push his thumb into his ear; four years later he began pushing a thumb into the other ear also. He continued to do this from time to time until his death in 1981.

The adult male Jomeo developed what must have been an abscess of an upper molar. Two days after he was first seen with his face frightfully swollen on one side, the abscess burst and fourteen days later the swelling had subsided.

Rashes. Numerous small pimples, starting around the nose and mouth and sometimes evident on other parts of the body (abdomen, inner surfaces of thighs and arms), typically appear at the start of the rains. These rashes are particularly common in infants up to three years of age. Some individuals are covered with a mass of spots for over three weeks.

Hemorrhoids. Flo developed an almost-certain case, which persisted for several weeks. During this time she frequently fingered the swellings and was observed to eat her feces.

Parasites. A preliminary examination of chimpanzee feces for endoparasites was undertaken in 1966 (Goodall, 1968b), followed by a more systematic investigation in 1973, when fecal specimens from thirty-two chimpanzees (from all age sex classes) were examined (File, McGrew, and Tutin, 1976). Each individual was parasitized by one or more of six helminths: *Probstmayria gombensis, Strongyloides fuelleborne, Necator* sp., *Oesophagostomum* sp., *Abbreviata caucasiac,* and *Trichuris* sp. Two ciliate species were found *(Trogoloytella abras-*

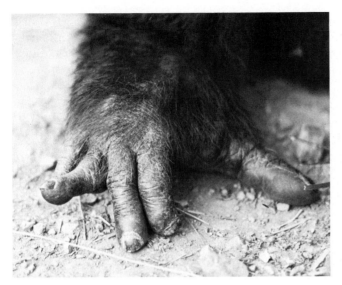

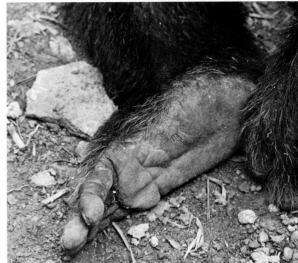

Goliath's toe was already damaged
when I first knew him in 1961.
(B. Gray)

There was no obvious scar on Leakey's foot; he probably lost the great
toe in his childhood.
(H. van Lawick)

sarti and one unidentified). None of the samples indicated heavy parasitic infection, a finding in agreement with the 1966 report. In the
earlier analysis ten samples were tested for schistosome eggs; all were
negative.

Two postmortem examinations revealed a few malaria parasites in
the blood, in the first stage of their developmental cycle.

As far as ectoparasites are concerned, the chimpanzee louse (*Pedicularis schaefi*) is probably the most common. Ticks are rapidly removed during grooming activity. Sometimes the whitish egg sacs of
the sand flea or chigger (*Tuga penetrans*) are visible on the underside
of a chimpanzee's foot; the sacs are bitten, poked, and sucked out.

Injuries

Fights between members of the same community are, for the most
part, not very severe. They usually last less than a minute and seldom
result in serious injury.

Table 5.5 lists wounds and other injuries received over seven years
by individuals of the main study community (excluding Kahama indi-

Table 5.5 Wounds and injuries received by individuals of different ages
and both sexes in the Kasakela community over seven years, 1971 and
1973–1978. Small cuts, scratches, and the like are not included.

Year	Severe				Moderate			
	♂	♀	I	Total	♂	♀	I	Total
1971	2	2	0	4	10	8	1	19
1973	1	4	1	6	5	8	7	20
1974	0	1	0	1	4	7	3	14
1975	1	0	0	1	1	8	2	11
1976	1	2	0	3	8	8	3	19
1977	0	1	2	3	1	3	0	4
1978	2	1	0	3	10	4	1	15
Total	7	11	3	21	39	46	17	102

Note: ♂ = adult male; ♀ = adult female; I = immature.

viduals after the split, and excluding cannibalized infants). The majority of these injuries were undoubtedly received during intra-community fights. The number of injuries judged serious was quite low overall (17.1 percent) and the percentages were similar for the three groups: for adult males, 15.2; for adult females, 19.3; and for dependents, 15.0.

Hands and feet were most vulnerable, for both males and females. One male sustained a broken toe during a fight; it eventually healed in a crooked position. Another, Goliath, apparently dislocated one toe, which remained out of joint for the rest of his life and often seemed to trouble him. Another male lost his great toe, and two others lost parts of their toes. Females often received deep gashes on their bottoms when they were attacked by males during periods of sexual swelling. Some of the most severe wounds and injuries recorded, from which the chimpanzees recovered, are described in Table 5.6.

Chimpanzees often lick their wounds if these are on hands or feet or other easily accessible parts. Or they may repeatedly touch the lesion with their fingers, which they then lick. Often they dab the wound with a handful of leaves, which are then usually sniffed and dropped, but may be licked and reused. Infants and juveniles have been seen to lick their mother's wounds, but an adult usually grooms carefully around the wound of a companion—though he may stare at it intently.

Table 5.7 lists the number of occasions during a two-year period (1978–1979) when chimpanzees were observed to fall. I have included only falls when the chimpanzees actually hit the ground; often they managed to save themselves by grabbing branches. Of the fifty-one falls, 31.4 percent occurred when chimpanzees trod or jumped on dead branches. Adult males fell most often during aggressive incidents, often when two were fighting each other. Twice males fell when hunting monkeys. Female aggression accounted for three of the female falls. Infants and juveniles fell most frequently during play.

The approximate distance of the falls was noted thirty-six times. Adult and late-adolescent males were the most likely to fall long distances (noteworthy in that there were relatively few males, compared to the number of females and youngsters at that time). When small infants of under two years fell, the distance was usually less than 1 meter. Older infants fell farther as they practiced locomotor skills with increasing confidence. One eighteen-month-old male fell about 10 meters and apparently winded himself.

Other falls not listed caused serious injury. Once Figan, as a young adult male, fell during a fight with two others and injured his hand or wrist, both of which swelled. He did not use the hand for the next few days and was lame for two weeks. And two males, an adult and an infant, actually fell to their deaths (see below).

One last incident occurred when the adult male Goblin swung down from a tree on a steep slope. To break his momentum, he seized hold of a small sapling, but it came out of the dry stony ground in his hand. He fell backward down the slope, grabbing at tufts of grass and other vegetation, none of which held. Finally, after tumbling some 12 meters, he managed to seize a root and brought himself to a stop—just 2 meters from the edge of a steep 15-meter drop to the shore of the lake.

Table 5.6 Selected injuries sustained by Kasakela individuals between 1964 and 1981.

Year	Individual	Age class	Nature of injury	Healing time	Context in which sustained (if known)
1964	Hugh	Prime	Bite on third toe; after 1 month end joint fell off.	1 month	Gang attack by 5 males and 1 female of own community.
1971	Passion	Prime	One eye closed or partially closed for 2 weeks. Travel very slow; eyes usually shaded with one hand. Often prone for 1 or 2 hours with eyes closed, touching eye from time to time. Runny eyes and nose. Opaque whitish patch on iris.	2 weeks, but continuously runny nose even after 10 years	Thorn penetrating eye?
1973	Fifi	Young adult (14 years)	Very deep puncture wound on head led to infection; maggots visible. Canine tooth of attacker possibly lodged in skull. Single dose tetracycline administered.	20 days	Attacked by adult male of own community.
1976	Sherry	Late adolescent (15 years)	Very deep gash on thigh. Smell of putrefaction for 1 week.	Very bad for 20 days; all right after 1 month	Attacked by older male.
1977	Pom	Late adolescent (12 years)	Multiple wounds on face, back, arms, legs, hands, and feet. Very slow movement for 2 days.	2 weeks	Unknown (reappeared in this state after absence of about a week).
1977	Michaelmas	Infant (4 years)	Possible fracture of femur or pelvis. Unable to use left leg for 10 months; thigh kept flexed against abdomen. Lower leg gradually used during climbing until use of whole limb regained.	1½ years	Almost certainly during attack on his mother (cycling at the time and subjected to frequent severe attacks).
1978	Jomeo	Prime (23 years)	Wound in left eyeball; eye tightly closed and oozing quantities of fluid. Remained closed for 10 days, emitting unpleasant odor. Pupil probably undamaged but peripheral vision apparently lost. Single dose tetracycline administered.	Some improvement after 10 days; eye open and healed after 4 weeks	Unknown.
1979	Goblin	Late adolescent (15 years)	Very deep gash 7–9 cm long in groin. Profuse bleeding for an hour. Very lame; lagged behind group next day. After 1 week much pus, which continued 2 more weeks. Became solitary much of the time.	Very bad 3 weeks, then gradually healed; all right after 5 weeks	Attacked by 5 adult males during meat eating.

Table 5.6 (*continued*)

Year	Individual	Age class	Nature of injury	Healing time	Context in which sustained (if known)
1980	Freud	Early adolescent (9 years)	Bone in foot broken or cracked, or ligament torn. Travel for 10 days with foot off ground; very lame for 3 more weeks.	Began to put weight on foot after 1 month; limped for another 1½ months	Interference when adult male copulated with his mother led to attack by male.
1980	Passion	Middle-aged	Deep horizontal gash on genital area between anus and labia, which hung down. Lame; much time spent reclining and dabbing with leaves, for 1 week. Frenzied dabbing during and after defecation and urination.	Very bad for 3 weeks; all right after 6 weeks	Suspected attack by members of unhabituated Kalande community.
1980	Pax	Infant (3 years)	Wound across groin and along penis. Almost without use of both legs first day; clung to mother's back with hands. Only a few steps per day during first week. On day 10 penis hugely swollen, much difficulty passing urine. On day 14 wound opened and began to drain.	Very bad for 2 weeks; healed after 5 weeks; scar tissue prevents normal erection of penis	Son of Passion. Wounded at same time.
1981	Freud	Early adolescent (10 years)	Deep wounds in lower back.	Difficulty in walking for 5 days; healed after 2 weeks	Attacked by adult bushpig after his capture of piglet during hunt.

Note: Names of males are shown in green.

Table 5.7 Chimpanzee falls observed during a two-year period, 1978–1979. Twenty-three falls were observed in 1978 during approximately 2,990 hours in the field, and twenty-eight the following year during about 1,975 hours.

Context	Mature		Juvenile		Infant		
	Male	Female	Male	Female	Male	Female	Total
Aggression (fights, chases)	13	3	4	0	1	0	21
Play	0	0	0	2	10	3	15
Locomotion during feeding	4	1	3	0	6	1	15
Total	17	4	7	2	17	4	51
Branch breaks	8	2	2	0	3	1	16
Distances when noted							
Over 5 meters	11	1	3	1	7	0	23
Over 10 meters	8	1	1	1	2	0	13

Disorders Following Death of the Mother

There is abundant evidence that disruption of the mother-infant bond in nonhuman primates can lead to behavioral disturbances, some of which may persist for many months (see for example Kaufman and Rosenblum, 1969; Hinde and Spencer-Booth, 1971). Rhesus (*Macaca mulatta*) and pigtail (*M. nemestrina*) infants separated from their mothers under laboratory conditions initially show increased activity, often of a searching nature, and subsequently a type of behavioral depression. The symptoms include a huddled posture, a typical "sad" facial expression, listlessness, withdrawal from social activities (particularly play), and somewhat impaired coordination. Usually the infants recover over a period of weeks, but some die with no apparent cause found on autopsy. Infant pigtail monkeys implanted with biotelemetry systems have been monitored after separation from their mothers. These studies have shown marked physiological alterations in body temperature and in heart rate and rhythm—even when, in one case, the infant was immediately adopted by a childless female who provided adequate maternal care (Reite, 1979).

Table 5.8 lists the eleven female chimpanzees who have died at Gombe over a nineteen-year period and left one or more dependent young; information is provided on the thirteen orphans. The three youngest, aged three years or less, were almost completely dependent on milk and unable, therefore, to survive without their mother despite the fact that two were adopted by an elder sibling. Six others, not yet fully weaned, were also adopted—five by an elder sibling (in Kristal's case, a "foster" sister) and the sixth, Skosha, by an adult female who may have been her biological aunt. All initially showed signs of clinical depression; they became listless, and frequency of play dropped. Kristal, just five years old when orphaned, disappeared nine months after losing her mother. Merlin lived for eighteen months. He developed a number of abnormal behaviors, such as rocking and plucking out his hair, and his tool-using skills and social responses deteriorated. Although polio was the actual cause of his death, it seems unlikely that he would have survived the approaching rains. Beattle's depression gradually decreased and a year after her mother's death her behavior seemed normal. We suspect she died a year later (because her elder sister appeared without her); if so, this was from causes almost certainly not connected with clinical depression.

Skosha, Beethoven, and Pax also got over the loss of their mothers—their behavior, that is, gradually became more normal. Pax, just four years of age when orphaned, showed few symptoms of depression. His mother had always been somewhat rejecting and lacking in affection. Orphaned infants of such mothers perhaps have a better chance of survival than those with more affectionate and tolerant mothers. Both Skosha and Beethoven, on the other hand, have shown marked retardation in their physical development. Skosha had only the tiniest and most irregular sexual swellings until she was thirteen years old, when for the first time she attracted the interest of the adolescent males. Six months later she developed her first true estrus—about three years later than is usual. Beethoven showed no scrotum growth at all until he was estimated to be thirteen or fourteen years old, and he was about fifteen when his scrotum showed the

The infant Merlin, a year after losing his mother, in the huddled position that was so typical of him at the time. The naked appearance of his lower leg resulted from hair pulling, one of several behavioral abnormalities that developed. Merlin died eighteen months after his mother, during the polio epidemic. (C. Coleman)

Table 5.8 Data on Gombe orphans between 1965 and 1983.

Mother	Offspring	Age of offspring at mother's death	Survival span[a]	Observed disorders		Comments
				Physiological	Behavioral	
Sophie	Sorema	14 months	2 weeks	Lethargic.	No further playing.	Adopted by male sib (approximately 8 years old).
Unhabituated	?Jenny's sib	?1½ years	Unknown	Unknown.	Unknown.	Adopted by presumed female sib first seen in Kasakela community carrying infant, then seen 2 days later with dead infant.
Circe	Cindy	3 years	7 weeks	Lethargic, pot-belly.	No further playing.	No sib. Traveled with various individuals of both sexes. Spent much time alone.
Passion	Pax	4 years	●	None.	Very few; whimpering during travel when left behind.	Elder sister, Pom, tried to carry and share nest, as did brother Prof. Offers refused unless frightened. Often preferred to travel with Prof.
Bessie	Beattle	4–5 years	●	Potbelly.	Decreased play; much whimpering when left behind by sib.	Adopted by female sib, Bumble, 10 years old. Often carried dorsally. In good health 1.6 years after mother's death.
Unhabituated	Beethoven	4–5 years	●	First signs of scrotum development only at 13–14 years.	Unusually violent interference when sib was mated; no sexual interest in females until age 6 or 7 years.	Adopted by presumed female sib (9 years) first seen in Kasakela community carrying infant.
Marina	Merlin	4–5 years	18 months	Lethargic, pot-belly, sunken eyes, gradual emaciation. Often blue with cold in rain. Probably died of polio.	Very little play; deterioration of social responses; rocking, hair pulling, sometimes hung upside down for several minutes. Poor coordination in tool using.	Adopted by female sib (approximately 9 years). Allowed to sleep with her, but usually rejected in attempts to ride dorsally.
Nova	Skosha	5 years	●	Potbelly first year. Almost no sign of adolescent sex swelling until 13.3 years. First adult swelling at 13.7 years.	Decreased play; unusually fearful; never seen to carry an infant.	No sibs. Traveled with various individuals, often males. Eventually adopted by adult female (close associate of Nova).

Table 5.8 *(continued)*

Mother	Offspring	Age of offspring at mother's death	Survival span[a]	Observed disorders		Comments
				Physiological	Behavioral	
Pallas	Kristal	5 years	9 months	Very lethargic.	Almost no further playing.	Traveled with foster sister, Skosha, for first months, then with unrelated adult female Miff.
Sophie	Sniff	?7 years	●	None.	None.	Adopted infant sib, Sorema.
Marina	Miff	?8 years	●	None.	None.	Adopted infant sib, Merlin.
Flo	Flint	8½ years	3 weeks	Lethargic, loss of appetite, sunken eyes. Eventual gastric trouble.	Very little play; very nervous with some big males.	Mother very old, unable to cope. Unusual dependence on her after death of infant sib at 6 months (Flint age 5 years).
Olly	Gilka	9 years	●	None.	None.	Increased frequency of association with older male sib, Evered.

Note: Names of males are shown in green.
a. ● = recovered after loss of mother.

period of rapid development that usually occurs when the young male is between nine and ten. (Another youngster, Michaelmas, who broke his leg when he was four years old, has shown similar stunting and retardation in physical development. At age eleven his scrotum is about one-half adult size.) Growth retardation (or psychosocial dwarfism) is a well-known phenomenon in human children, usually corresponding to maternal deprivation or other psychological disturbances (see for example Blodgett, 1963; Patton and Gardner, 1963; Reinhart and Drash, 1969).

Four youngsters were between seven and nine years old when their mothers died. Three of them showed few if any behavior changes and no signs of ill health (two of them cared for younger siblings). The fourth, Flint, was an exceptional case (Goodall, 1979). He was unusually dependent upon his very old mother, Flo, and after her death he spent many hours close to her body. For the next six days he showed gradually increasing signs of lethargy, although he interacted socially with a number of other individuals. His human observers lost sight of him for the following four days; when he was next seen, his physical condition had deteriorated markedly and continued to do so throughout the remaining two weeks of his life. As mentioned, postmortem examination of Flint's body showed that he had gastroenteritis and peritonitis. It seems likely that the psychological and physiological disturbances associated with loss made him more vulnerable to disease.

Studies of free-ranging Japanese macaques *(Macaca fuscata)* have shown that orphaned primiparous mothers typically mishandle their own first infants, who are likely to die (Hasegawa and Hiraiwa, 1980).

Flo in 1968, four years before her death. She was certainly over forty years old at the time. Note her very worn teeth. (B. Gray)

At Gombe only Patti has shown totally inadequate mothering, which led to the death of her week-old infant. Unfortunately her early history and family background are not known, for she transferred into the Kasakela community when she was between ten and eleven years old.

Old Age

Only eleven individuals, six males and five females, appeared old during the 1965–1983 period. Two other males and one other female classified as old disappeared prior to that time.

Three males (Hugo, Mike, and Goliath) and one female (Flo) seemed to us to be older than the other elderly individuals at the time of their death. All four became increasingly frail and emaciated during the last two years of life; the bones of the pelvis and scapular, in particular, protruded. Mike's weight dropped from 38.5 kilograms two years before death to 32.5 kilograms just before he disappeared; Hugo dropped from 38 to 28 kilograms over the same period. Flo's weight showed little change during her last two years of life, fluctuating between 22.5 and 27 kilograms. The facial hair of the males whitened, and balding occurred on the shoulders and head of all individuals. Their teeth were well worn—indeed, Flo's were almost down to the gums in 1964, eight years before her death.

All the old individuals suffered from periodic bouts of diarrhea, which persisted for a week or more at a time. The four eldest moved increasingly slowly and had difficulty climbing large trees. Whether or not the lives of these individuals were complicated by disease during their last months, there can be little doubt that old age itself was primarily responsible for their deaths.

Reproductive Disorders

One female, Gigi, reached menarche in 1965. For the next twelve years she cycled and menstruated regularly but apparently never conceived and certainly never carried a fetus to term. In May 1978 Gigi bled copiously during a period of menstruation and deposited a hemorrhagic mass. This was collected and subsequently identified (C. E. Graham, personal communication) as a uterine cast, often associated in human women with membranous dysmenorrhea (Sollereld and van Zwieten, 1978). From 1976 on, Gigi, previously very attractive sexually to the adult males, became increasingly less so; since 1977 her cycles have become irregular. Gigi has always shown reluctance to be mated and often pulls away from the male before he achieves ejaculation.

No other obvious reproductive disorders have been recorded, although injury or illness may temporarily disrupt the sexual cycles of a female. If, for example, a female's anogenital region is wounded during a period of swelling, there may be a few days of deturgescence and then a resumption of swelling. The adolescent Pooch, who disappeared six months after receiving a severe wound in her groin, showed no signs of sexual swelling during that period. The adult Melissa did not resume swelling for six and one-half months after contracting polio. After the loss of her two-year-old infant, Pom became extremely emaciated. Slowly she recovered, but did not resume cycling for eleven months after his death. Gilka's gradual decline has been described; during the last two and one-half years of her life she showed no signs of renewed sexual swelling.

Birth Abnormalities

A number of birth abnormalities have been recorded over the years, only one of which has posed problems for the chimpanzee concerned. As we saw in Chapter 4, Little Bee has a deformity of one foot—a condition similar to *Talipes equino verus*, or clubfoot, and she sometimes has difficulty keeping up with fast-moving groups of chimpanzees. One male, Worzle, had completely white (rather than the usual brown) sclerotics, and two other individuals had patches of white on the sclerotics. (One of them, Leakey, was suspected of being Worzle's sibling.) Little Bee has a supernumerary nipple in addition to her clubfoot, and her son was born with patches of dark skin on his face, still visible at six years of age.

Seasonal Influences

Effects on Health

Between 1968 and 1979 there were five severe outbreaks or epidemics of so-called pneumonia. Two of these (1970 and 1977) occurred at the beginning of the rainy season, in October (which I have been including as part of the dry season), and two (1971 and 1975) were in March. The worst, when five individuals are thought to have died of the disease, was in January 1968. Altogether there was one case per 0.9 month during the wet season, and one per 2.7 dry months. If we exclude the cases during October and consider only the really dry months, then there was one case per 15.5 months (two sickly infants).

Coughs and colds were slightly more common in the wet season (one per 0.37 wet month and one per 0.21 dry month), but dry-season colds were usually very mild compared to most in the wet season. Seasonal distribution of pneumonia and colds is shown in Figure 5.4.

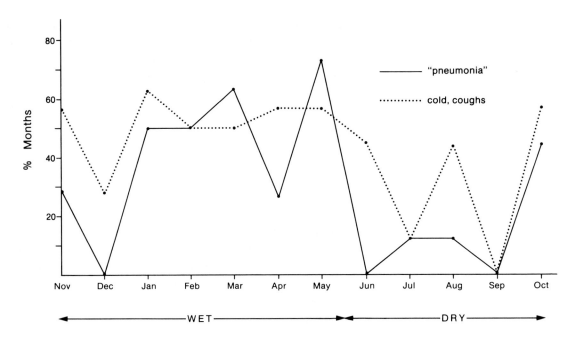

Figure 5.4 Seasonal distribution of "pneumonia" and of colds and coughs, as percentage of months over a nine-year period when there were two or more individuals showing the relevant symptoms.

Deaths resulting from illness and/or old age (and this includes victims of the polio outbreak and orphans) occurred more than twice as often during the wet season (one death every 5.8 months) as during the dry (one every 13.6 months), as is evident from the following tabulation:

	Deaths 1965 through 1983	
	Dry season (95 months)	Wet season (133 months)
Old age	1	7
Reproductive maturity	1	7
Adolescence/juvenile period	1	3
Infancy	4	6
Total	7	23

Chimpanzees shiver violently during or after heavy rain, particularly after a cold, wet night and when a strong wind is blowing. The orphan Merlin (who plucked out hair on his abdomen, arms, and legs) frequently became blue with cold during the two months of rain preceding his death. Wet, cold weather probably decreases the chance of survival of chimpanzees who are badly injured, as it is difficult for their wounds to dry and heal.

The wet season is, in general, more lush than the dry, although in most years the chimpanzees find adequate food throughout the year. Chimpanzee weights tend to be higher during the wet season. Wrangham's study (1979) spanned two dry seasons, 1972 and 1973, with an intervening wet season. He found the 1972 dry season to be a particularly difficult one with respect to food availability: chimpanzee weights were generally lower, competition for certain foods was greater, males were less likely to give food calls, they traveled shorter distances per day, and party size was smaller. Subjectively, the chimpanzees often seemed hungry.* It was during this period that the old female Flo died. (It is probably significant that the weaker of Melissa's twins died during the dry season of 1978 when, as in 1972, food seemed scarce.)

For the baboons at Gombe, who have a far more restricted range than do the chimpanzees, the effect of the season is more noticeable: often a troop ranges *farther* in the dry season and adds a new area to its range (apparently in search of supplemental food), and more of the very old members of the troop die (Gombe Stream Research Center records).

Effects on Reproduction

Mating takes place throughout the year, and there is no evidence of a birth season as such (Goodall, 1968b; Tutin, 1975). Figure 5.5b shows that the number of births during the wet and dry seasons from 1965 to 1983 was similar (one birth per 4.8 wet-season months and one per 3.3 dry-season months). The sample size, however, is still very small. Inasmuch as time of conception is estimated from time of birth, Figure 5.5a is dependent on Figure 5.5b. An assessment of the data up to 1980 (Goodall, 1983) suggested a clumping of conceptions toward the end of the wet season, but the more recent data indicate that this is probably not the case.

There is a nonrandom distribution of female estrous swellings throughout the year. Figure 5.6 shows annual distribution of swellings during four three-year periods. The first, 1972–1974, is from Tutin (1975). While the mean number of swellings per month is considerably higher during the 1975–1977 period (reflecting the fact that there were more females cycling), the annual distribution, peaking in September, is strikingly similar in all four periods. It was this pattern that led me originally to postulate a "mating season" at the end of the dry season (Goodall, 1965).

It is at the end of the dry season, too, that females most often resume full swellings after a period of lactational anestrus (excluding cases when cycling resumes after the death of an infant). This is seen in the following tabulation:

	Number of females who resumed estrous swellings	
	Mature	Late adolescent
November–January	4	4
February–April	2	2
May–July	4	2
August–October	16	2

The first full-sized estrous swellings of late-adolescent females appear to be distributed randomly throughout the year.

* Wrangham attributed this in part to the fact that whereas *Parinari curatelifola* fruits formed a major component of the diet in 1973, these trees produced few fruits in 1972. However, when Riss and Busse (1977) followed the alpha male Figan during another non-*Parinari* dry season, 1974, they found that he had no difficulty locating enough to eat. It does seem likely, though, that the lack of this succulent fruit will always have some effect on old or sick individuals, especially as it produces many windfalls, which can readily be gathered and eaten on the ground.

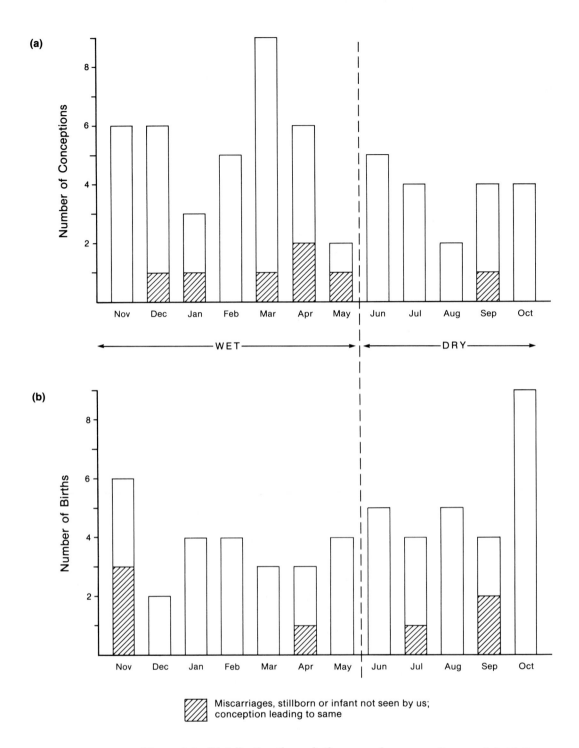

Figure 5.5 Distribution through the year of *a*, conceptions, and *b*, births, from 1965 to 1983.

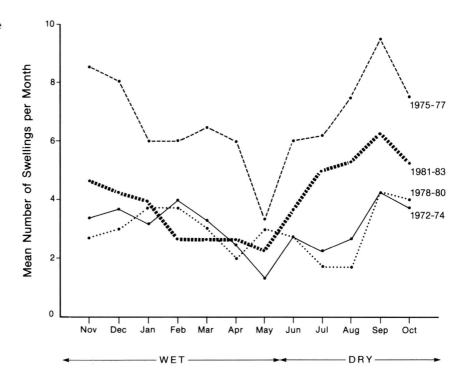

Figure 5.6 Mean number of female estrous swellings per month during four three-year periods, 1972 to 1983.

Mortality Rates

Of the sixty-six chimpanzees who have died (or disappeared and are thought to have died) in both study communities since 1963, the bodies of only twenty-four have actually been seen. Usually a chimpanzee simply disappears. However, if this disappearance coincides with an epidemic of some disease or if the individual was last seen in poor physical condition, death can be presumed. Even when no obvious signs of ill health precede the disappearance, it still is sometimes possible to be reasonably sure that death has occurred. If, for instance, a mother is seen without her small infant, we can safely conclude that the baby has died, inasmuch as it cannot survive without her milk and care. If an infant is seen traveling around without its mother and she does not reappear, it is equally plausible that the mother has died. If a fit adult male suddenly vanishes, he has probably died, whereas if a healthy young female disappears, she is apt to have transferred to a neighboring community.

Causes of Death

Our records at Gombe provide the following kinds of information with regard to causes of death: *certain* death from known, suspected, or unknown cause; and *inferred* death from suspected or unknown cause. This information is summarized in Table 5.9, which lists the individuals in each age-sex class who died or are thought to have died. Individuals from the Kahama community are included in order to give as complete a picture as possible.

The table shows that the causes of death were known or presumed for fifty-one of the sixty-six individuals who died or disappeared and were presumed dead. (Ten other individuals—five adolescent females, one young adult female, and two middle-aged females with their two

Table 5.9 Causes of death, known or presumed, for individuals of the Kasakela and Kahama communities, 1963–1983.

Cause of death	Stage of life cycle						
	Baby (under 1 year)	Infant	Juvenile	Adolescent	Young adult, prime	Middle-aged	Old
Respiratory disease	Gyre (D) Flame (W) Villa (W) Bee Hinde (D)	Plato (W)		Pooch (W)[1]	Circe (W)	Sophie (W) David (W) Leakey (W) William (W)	
Polio	Grosvenor (D)		Hornby (W)	MacDee (W) (shot)		J.B. (W)	McGregor (W) (shot)
Other illness		Tapit (W)	Lolita (W)		Gilka (W) Nova (W) Worzle (W)	Pallas (D) Passion (W)	
Orphaned		Sorema (W) Cindy (W) ?Jenny's sib (D) Kristal (D)	Merlin (W)[2] Flint (D)[3]				
Injured	Jane (W)	Pan (D)			Sniff (W) Godi (W) Dé (W) Charlie (W)	Madam Bee (D) Rix (W)	Goliath (W) Huxley (W)[4]
Killed by Passion and/or Pom	Otta (D) Orion (D) Genie (W)						
Suspected killing as above	Gandalf (D) Melissa's (W) Banda (W)						
Old age							Mike (W)[5] Hugo (W)[5] Flo (D) Bessie (W) Marina (W)
Lack of care	Patti's (W)						
Vanished; suspected death, cause unknown	Wood Bee (D) ?Sophie's (D) Hepziba (D)	Dapples (D) Barbet (D)	Beattle (D)		Pepe (D) Sherry (W) Harmony (U)	Willy Wally (U) Hugh (U) Faben (D) Olly (W) Figan (W)	Humphrey (W)*
Vanished; suspected emigration		Jay (W) Quantro (D)		Lita (W) Sally (W) Bumble (D) Jenny (D) Sesame (U)	Pom (W)	Jessica (W) Vodka (D)	

Notes: italics = body seen; ? = sex not known; D = dry-season death or disappearance; W = wet-season death or disappearance; U = season of death unknown; * = skull found, identified by teeth. Names of males are shown in green.
Contributory causes of death: 1 = unhealed injury; 2 = polio; 3 = gastroenteritis; 4 = old age; 5 = pneumonia.

Table 5.10 Injuries that led to (or contributed to) death, 1965–1981. The three babies killed and eaten by Passion and Pom are not included, nor are the Kahama victims of intercommunity violence.

Year	Name	Age	Nature of injury	Cause
1965	Jane	3 months	Fractured humerus, bone and tendons exposed; seen dead third day after appearance with injury. Carried by mother.	Unknown.
1967	Huxley	Old	Deep gash from just below eye down to lip; broke open repeatedly during 2-month period. Disappeared.	Almost certainly during fight with community male.
1968	Pooch	Adolescent	Deep wound in groin, did not use leg for walking for 6 months, sexual cycles stopped. Disappeared during epidemic of pneumonia.	Unknown.
1968	Rix	Middle-aged	Broken neck; died instantaneously.	Fell from tree during excitement of reunion between two parties.
1976	Melissa's baby	1 week	Neck seemed broken, forehead opened and bleeding, some skin torn from upper back. Carried by mother.	Possible victim of infant killers Passion and Pom.
1981	Pan	2 years, 10 months	Presumed internal injuries. Died on fourth day.	Blown from 14-meter palm in strong wind.

Note: Names of males are shown in green.

dependents—disappeared and probably emigrated or returned to peripheral core areas.) Diseases probably accounted for or contributed toward twenty-eight of the fifty-one certain deaths (55 percent). Injuries received during fighting and falls accounted for or contributed toward ten deaths (19.6 percent). The remaining causes of death were cannibalism (three certain, three inferred), loss of mother (six, although gastric enteritis and polio were contributory causes in two cases). Flo is assumed to have died of problems connected with old age, which certainly also contributed to the deaths of Hugo, Mike, Huxley, Bessie, and Marina. One infant, Patti's, apparently normal at birth, died from inadequate maternal care.

Table 5.10 lists the individuals of the Kasakela community who have died or who are thought to have died as a result of injuries. Many severe injuries were also received by Kahama individuals during attacks by Kasakela males after the community division in 1971–72. Two additional adults, Huxley and Pooch, are thought to have died after receiving bad wounds, but it seems likely that the wounds were only contributory causes of death (Huxley was an old male, Pooch disappeared during a pneumonia epidemic). There was no clue to how the 1965 infant Jane received her fatal injury. The 1976 infant may have fallen victim to Passion and Pom. Two individuals, an adult male, Rix, and a two-year-old infant, Pan, died as a result of falling from trees. Rix fell during social excitement when two parties met; he landed on rocks and broke his neck. A large, dead branch, which seemed to have recently snapped, was lying close to his body, and it was assumed that this had caused his fall (Teleki, 1973a). Pan was swept from a 14-meter palm during very strong, gusty winds. He landed flat on his

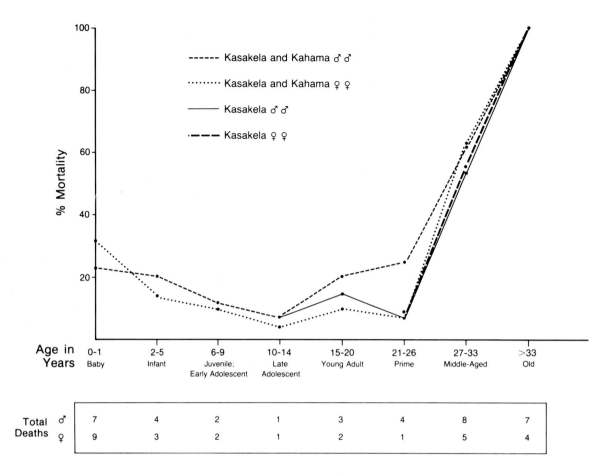

Figure 5.7 Mortality rates in the different stages of the life cycle.

back on hard, dry-season ground. Although he was able to cling to his mother when she climbed down to retrieve him, and even, after a few hours, to walk short distances, he was dead four days later. Presumably he sustained internal injuries or a fractured skull, or both.

Three babies (one to four weeks old) were seen being killed and eaten by the adult female Passion and her late-adolescent daughter, Pom. A fourth infant was found, already dead in his mother's arms, after observers had been attracted to the location by loud screaming. The nature of the baby's injuries suggested that he might have been the victim of a similar attack; possibly adult males had heard the mother's screams and arrived too late to prevent the killing but in time to prevent Passion and Pom (encountered nearby an hour later) from making off with the body (Goodall, 1977). Two other babies who disappeared at three weeks of age, and five infants who were carried to term but never seen by us, may have been additional victims of the killer females.

Adult Kasakela males killed and partially ate three infants (two and one-half to three years old) of "stranger" unhabituated mothers; it is possible that three Kasakela infants (Hepziba, Barbet, and Dapples) may have suffered a similar fate between 1980 and 1982, when Kalande males were encroaching on Kasakela territory. Cannibalism has been observed in Uganda and at Mahale, where infants were killed not only by males of different unit-groups, but also by males of the same unit-group. This will be discussed in Chapter 11.

Differential Mortality

Figure 5.7 shows mortality rates in the different stages of the life cycle based on the estimates of individual age given in Table 4.1. The years of maturity are divided into four periods: *young adult, prime, middle-aged,* and *old*. Because it was possible to follow the life histories of many Kahama individuals after the community division, they are included in the figure; but data are also presented for the Kasakela chimpanzees alone.

The infant mortality rate is initially high for both sexes (23.3 percent and 33.3 percent for male and female infants respectively during the first year of life). The death rate for the next four years of infancy is slightly lower (21.7 percent and 13.6 percent) and falls further until the 10- to 14-year age period. Of the ten males and fourteen females born during the first half of the study (1964 through 1973) 62.5 percent have survived to independence—seven of the males and eight of the females. (This does not include two female infants whose mothers vanished and whose fate is unknown.)

Once a female has reached independence, she has a good chance of surviving until the 27- to 33-year period (middle age). Young adult and prime males, however, are at slightly greater risk than females—considerably greater during the 21- to 26-year prime period if the Kahama individuals are included. This, of course, reflects the deaths due to intercommunity conflict. It seems that there may be a tendency for males to live longer than females; during the entire study period (1960 to the present) I have seen only three females (Flo, Sprout, and Wilhelmina) who looked really old, compared to six ancient-seeming males (McGregor, Hugo, Mike, Goliath, Dracula, and Hubert).

6. Communication

Passion and Pom.

November 1980   Nope is feeding, her infant sleeping in her lap. She barely glances up as Pom arrives from the east. The younger female pauses, gives two soft grunts, then begins to feed nearby also. Thirty-five minutes later we hear the pant-hoots of several males farther down the valley; I recognize Goblin's distinctive voice and Satan is there too. Pom and Nope glance toward the sounds, then continue feeding. The males call out several times over the next fifteen minutes. Pom consistently looks up but Nope pays no attention. Suddenly a female screams. Pom stares intently, turns, and with a grin and tiny squeak holds out her hand to Nope, who briefly touches her fingers. A moment later there is a renewed outburst of calling; at least two individuals are screaming urgently, one a child, and there are waa-barks of threat. Instantly Pom leaps to her feet and races toward the sound of conflict. She keeps mostly to a rough trail, so I do not get left too far behind. We run for some 500 meters. As I burst from a tangle of vines, I see that Pom has found her mother; they are grooming intently and infant Pax is close beside them. Pax and Passion evidently have been the victims of an attack, for they have tiny bleeding wounds, he on one ear, she on a toe. Prof, perhaps also involved, is in a tree above and seems subdued. Suddenly Goblin hurtles from the undergrowth, charging straight for Passion, Pom, and Pax; screaming, they leap into Prof's tree. Goblin tries to grab Pom's leg, but misses. He displays on out of sight. A moment later Satan appears and follows him. Passion and her family are left on their own.

Chimpanzees have a rich repertoire of sounds, postures, and facial expressions which function as signals* during interactions between individuals and which, in conjunction with a variety of autonomic responses and other signs indicating physiological state, emotional mood, social rank, and so on, facilitate the exchange of information

* MacKay (1972, p. 6) defines signaling as "the activity of transmitting information regardless of whether or not the activity is goal directed, [or] what impact, if any, it has on a recipient." This is similar to the definition of the sociobiologist: a signal is "any behavior that conveys information from one individual to another, regardless of whether it serves other functions as well" (Wilson, 1975, p. 595). Signals that have been modified during evolution specifically to convey information, in ethology are labeled *displays*. In this chapter I discuss the ways in which chimpanzees acquire information about one another, and about the environment from one another, as well as more direct communicative interactions between two or more individuals. I have limited the use of the

among community members. A signal, or more often a complex of signals, may be deliberately directed at another individual with the goal of changing his behavior. Thus our "Hey you, come here!" is aimed at a specific person with the expectation that he will respond in a predictable way. The same command can be conveyed without words by a beckoning movement of the arm. This may be preceded, in order to attract attention, by a whistle. An infant chimpanzee during weaning, desirous of suckling but fearful of rejection, may approach his mother until his pouted lips are close to the nipple, or he may stop a few feet away and whimper; both signals convey the same information.

If the message implicit in the signal sequence is perceived, the recipient often gives signals of his own that convey his response. The summoned human may approach obediently, or he may call out, "Wait a minute, I'm busy." The chimpanzee mother may reach to embrace her infant, allowing him to nurse, or she may discourage his closer approach by pushing him away. Sometimes the passing to and fro of signaled messages between two chimpanzees results in a prolonged dialogue during which the behavior of one or both participants may change as they decide to comply with or resist the wishes of the other. And sometimes bystanders interject signaled comments of their own. The rejected infant, after a second unsuccessful plea for his mother's breast, may throw a tantrum, provoking an irritated arm-raise threat from a nearby adult male. At this point the mother may relent and comply with her child's wishes. These sequences can be compared with the commands or pleadings, threats or obeisances, invitations or refusals, arguments or idle comments, that in human interactions are made explicit by the use of words or of nonverbal signals within the established communication repertoire of a given culture.

Some signals are not directed at a specific individual. The emergency "Mayday! Mayday! Mayday!" of an aircraft in distress is an active general-broadcast signal, an urgent appeal for *anyone* to help or take action of some sort. The loud and frenzied screaming of a chimpanzee who has been attacked may be a similar general appeal for support—although, as we shall see, it may be clearly directed at a specific ally, even if that individual is not close at hand.

Other communicative signals are designed to broadcast information in a more passive way. The physical appearance of a handsome young man, well-proportioned and athletic, advertises his vigorous masculinity to all who see him. The flamboyant sexual swelling of the female chimpanzee similarly advertises her desirable condition to any male

term *signal* to elements of behavior that appear to be *directed* acts of communication and to communicative displays such as the female's sexual swelling. Cullen (1972, p. 103) writes that a behavior must be "demonstrated to evoke a change of behavior in another individual" before it can be clearly described as a *communicative signal*.. Although most of the signals described here have, on at least some occasions, been observed to affect the behavior of others, they are so often embedded in a complex sequence of behavior that it would not be possible, without experimental manipulation, to determine the effect on the perceiver of the different components. While this could conceivably be done (at least with some signals), it becomes less important when the animals we study show such pronounced inventiveness in communicative acts. The actual signals emitted by a male during a courtship sequence show much variation, both by the same male on different occasions and by different males; the female almost certainly responds to the collective message of the various signals rather than to discrete elements within the display.

who may be interested. And just as the young man may, with clean-shaven face and wearing his best suit, make a deliberate attempt to improve his chances of acceptance for some specific job, so may the female chimpanzee. She does so by means of *directed* signals, approaching and soliciting a chosen male.

The sexual swelling, of itself, is an involuntary development, dependent upon the hormonal state of the female. She can increase or decrease its effective value as a signal by soliciting males on the one hand or shunning them on the other; but there is nothing she can do about the cyclic appearance and disappearance of the swelling that indicates, as it does, her degree of sexual receptivity. There are other actions over which the individual has little or no control and which nonetheless have a communicative function. The blushing of a human or the erection of body hair in the chimpanzee, for example, are autonomic, involuntary behaviors. They reflect internal emotional states and convey information to others regarding the *mood* of the individual concerned.

Before any meaningful interaction can take place between two chimpanzees, each must note and categorize a wealth of pertinent information regarding the appearance of the other. Thus small body size, pale skin, and white tail tuft indicate infancy; large size, big canine teeth, and characteristic genitals proclaim maleness; sex skin and breasts denote femaleness. A confident gait and manner suggest high rank; shrinking, hesitant movements reveal uncertainty or timidity. And there are all the idiosyncrasies of facial configuration, posture, gait, behavior, and perhaps odor that inform the perceiver, "This is friend, not foe; this is X, not Y." Some of these signs, such as the white tail tuft, have probably been selected during evolution for their communicative value. Others, such as male scrotum size, are by-products of other developments and function secondarily to convey information to conspecifics. Together they represent the infrastructure upon which, in a society whose members know one another as individuals, all communication is built.

A chimpanzee, like a human, can acquire a great deal of information about a given situation simply by watching the behavior of his companions and paying attention to any "messages" transmitted, deliberately or otherwise. He has a good understanding of the complex web of interindividual relationships within the community and, as we shall see in Chapter 19, is able to predict ways in which his companions are likely to react to his behavior or that of others, and to modify his own actions accordingly. This sophisticated social knowledge is the matrix in which communication is embedded.

There is another way in which an individual can acquire quite specific information about certain environmental and social situations: by paying attention to an assortment of nonintentional, nondirected behaviors of his companions that serve as incidental cues. When a man hammers a nail into the wall, he is not (usually) intending to communicate anything. His wife, nevertheless, will be informed that he is (at *last*) hanging up that picture. As MacKay (1972) points out, a person with measles is not deliberately communicating information about his health; still, his spotty face serves as a signal proclaiming his condition and enabling his fellows to avoid contamination.

When a chimpanzee bangs a hard-shelled strychnos fruit against a rock, his goal is to crack it open and eat the contents. His action, however, conveys specific information to any individual who happens to hear the banging: *a chimpanzee is over there*; and *strychnos fruits are ready for eating over there* (or at least one is). A series of soft bangs provides the additional information that *the banger is an infant* (not strong enough to make louder sounds). There is, of course, no clue to the *identity* of the unseen individual, although if, an hour before, our listening chimpanzee saw a particular mother with her infant moving in the general direction of the sounds he now hears, he may make a guess about the identity of the banger—we do not know.

In our own everyday life cues of this sort can communicate a whole wealth of information: the smell of new bread means that Aunt Cecilia is baking a day earlier than usual, which means she probably *did* get that hairdressing appointment tomorrow, which means she *won't* be able to take Johnny to school . . . and so on. The extent to which chimpanzees can make a series of complicated deductions of this sort is not yet known. Nevertheless, as the cognitive abilities of a species become more complex in the course of evolution, there is a tendency for individuals to attend and respond to increasingly complex combinations of signals, and to take into account the varying environmental situations in which they are emitted (Andrew, 1972). Certainly chimpanzees are capable of combining a number of unrelated pieces of information, and this facility gives them a more acute awareness of what is going on in their world.

Effect of Mood

A social interaction between two humans will be affected by their relationship and by the mood or emotional state of each of them. If the boss is clearly in a bad temper, his secretary is unlikely to try any frivolous chitchat; she is more likely to keep out of the way until he has calmed down. When a roommate is deeply depressed over an unhappy love affair, a sensitive girlfriend will not talk about her own flourishing romance. In the same way the response of a chimpanzee receiver to a given signal will depend, at least in part, on his relationship with the signaler and their respective moods at the time. A mild threat from a calmly feeding adult male to an adolescent may provoke a fearful response. A male of only slightly lower rank may ignore the gesture unless the signaler is in a highly aroused state, in which case he may get out of the way—unless he too is aroused, when he may retaliate in kind.

We must also realize that a given signal, while it may be clearly directed toward a specific individual with the goal of altering his behavior, may also function to alter the behavior of other individuals to whom it was *not* directed. Thus when A threatens B (with an *arm-raise* gesture), he usually attains his goal (preventing B from approaching closer). At the same time the threat may cause bystander C to move away. In other words, C learns about A's mood as a result of the communicative sequence between A and B.

There has been little systematic investigation into the emotions of the chimpanzee—not surprising in view of the difficulties inherent in such research. Some forty years ago Hebb (1945, p. 32), discussing these problems, stated, "The conclusion that emotions cannot be reliably identified is erroneous, being based on unsatisfactory scientific methods; . . . human and animal emotions can be identified in the same way." He proposed that emotions be primarily identified through their relation to temporal sequences of behavior rather than through facial expressions or any momentary state of the organism. I am not aware of any major attempts to follow through on Hebb's suggestions. Nevertheless, all those who have worked long and closely with chimpanzees have no hesitation in asserting that chimpanzees have emotions similar to those which in ourselves we label pleasure, joy, sorrow, boredom, and so on (see for example Köhler, 1925; Kohts, 1935; Yerkes, 1943).

Some of the emotional states of the chimpanzee are so obviously similar to ours that even an inexperienced observer can interpret the behavior. A youngster who hurls himself screaming to the ground, face contorted, hitting out with his arms at any nearby object, banging his head, rolling over and over, is clearly in a rage. This performance is known as the temper tantrum. Another youngster, who gambols around his mother, turning somersaults, pirouetting, and every so often rushing up to her and tumbling into her lap, patting her, or pulling her hand toward him in a request for tickling, is obviously filled with *joie de vivre*. There are few observers who would not unhesitatingly ascribe his behavior to a happy, carefree state of emotional and physical well-being. In these extremes of rage and happiness chimpanzees, like human children, tend to express their emotions with their whole bodies. Less intense feelings are revealed by more subtle changes of facial expression or body posture.

Chimpanzee adults, as well as youngsters, like small human children, tend to "act the way they feel," with little or no masking of underlying emotional state. This means that an understanding of emotion is crucial to a complete understanding of behavioral interactions in a given situation. We have a long way to go before this desirable state of affairs is reached. The emotions to which I refer in this chapter and elsewhere are broad categories based largely on intuition, and thus should not be regarded as more than tentative.

It is easy to see that chimpanzees feel *apprehension* in certain situations, reflecting their ability to anticipate an unpleasant event based on recollection of past experience. They show *fear*—fear of the strange and unknown, and social fear during intense aggressive interactions. They show *distress*, varying from that of the lost or deprived infant to the more violent emotion of a child during the peak of weaning or an adult who has been severely attacked and perhaps hurt. *Annoyance* is provoked in a higher-ranking individual by undesirable behavior in a subordinate and expressed by abrupt but not violent gestures and sounds that signal "Oh, shut up!" or "Shove off!" *Anger* is a more aggressive state, which may result in vigorous threat or attack. Combined with the thwarting of desire (or frustration)—as when a child desires to suckle and is rejected or when an adult begs persistently and is denied—anger becomes *rage*, expressed in its extreme form by the familiar temper tantrum. Chimpanzees certainly experience moods of quiet *enjoyment*, or contentment, during periods of relaxed feeding, grooming, or resting with full bellies. Far less easy to categorize are the various emotional states involved in reassurance sequences and the sharing of food, such as meat (Chapter 13). Sexual intercourse and behaviors surrounding it may be calm and peaceful, but often stimulate *sexual excitement*. The emotions underlying various levels of *social excitement* are highly complex, probably corresponding to differing levels of arousal. Expressed by means of vigorous charging displays, extravagant posturing and gesturing, and often accompanied by noisy calling, social excitement is stimulated by participating in or watching activities such as hunting, arriving at a bountiful food source, reunion between parties, rain displays, stranger contact, and so on. Probably we should add vigorous social play to the list. Depending on the particular context in which it is exhibited, social excitement is undoubtedly compounded of conflicting emotions such as pleasure and apprehension, anger and fear. Much work must still be done before the components can be isolated and the picture clarified.

Communication and Information Exchange

There are four main pathways for the transmission and reception of information: visual, tactile, auditory, and olfactory. Although a message may be conveyed through a single channel only—as when a chimpanzee receives information from sounds emitted out of his sight—most messages are transmitted through two or more of these channels. For ease of description I shall discuss each channel separately.

Visual Communication

The chimpanzee (in common with other higher monkeys, apes, and man) has elaborate facial musculature and a hairless face, presumably specialized for the production of a variety of facial expressions (Andrew, 1963). Many facial expressions, as Marler and Tenaza (1976) point out, are linked with particular vocalizations, which chimpanzees make by altering the size and shape of the mouth aperture and resonating cavities. The receivers probably learn more from the calls themselves than from these "vocalization-bound" expressions. Other expressions, illustrated in Figure 6.1, while they may also be accompanied by sounds, are sometimes given silently and more clearly function as signals.

Facial expressions play a key role in close-up communication between chimpanzees (Kohts, 1935; Yerkes, 1943; van Hooff, 1967; Goodall, 1968b; Marler, 1976), as they do in our own species (Argyle, 1972). As Altmann (1967) has pointed out, facing and looking at the addressee is probably the most common means of directing social messages. An individual can also provide information to those watching him by his visual, postural, or locomotor orientation. Subordinates, by scanning the faces of superiors, can quickly detect slight changes in expression and adjust their own behavior accordingly. I watched one infant male, Goblin, as he reached very, very cautiously toward a banana peel that lay beside a feeding adult who had dropped it; throughout the whole successful maneuver he scarcely took his eyes off the other's face.

Figure 6.1 Facial expressions of chimpanzees. (After D. Bygott)

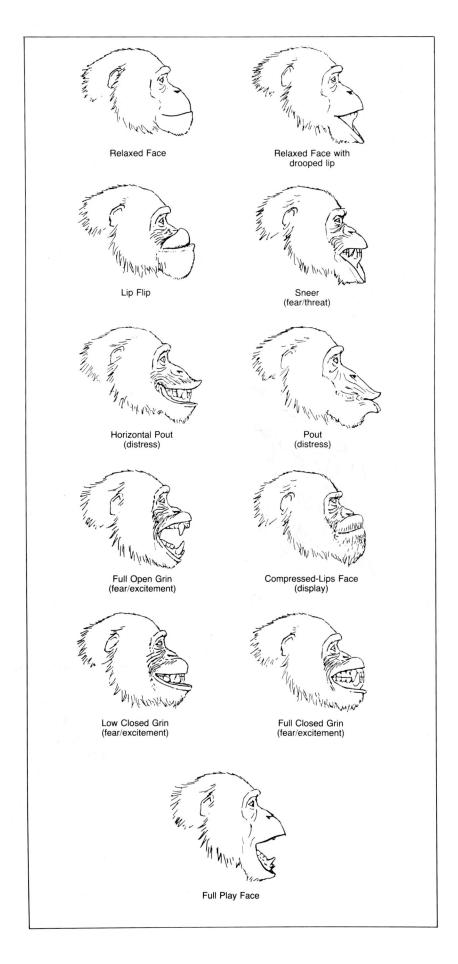

Relaxed Face

Relaxed Face with drooped lip

Lip Flip

Sneer
(fear/threat)

Horizontal Pout
(distress)

Pout
(distress)

Full Open Grin
(fear/excitement)

Compressed-Lips Face
(display)

Low Closed Grin
(fear/excitement)

Full Closed Grin
(fear/excitement)

Full Play Face

Above: Mike, squeaking after having been attacked, shows full closed grin.
Below: Leakey grins and screams during an aggressive incident.
(H. van Lawick)

In most nonhuman primates, including gorillas, prolonged eye contact between adults functions as a threat. However, in chimpanzees, as in man, relatively long bouts of eye contact may also accompany friendly interactions. In the captive colony at Arnhem the adult males, after a conflict, would sit, apparently trying to catch each other's eye, for up to fifteen minutes at a time. Once they had made and held steady eye contact, reconciliation was not far away (de Waal, 1982). When I was introduced to the home-raised chimpanzee Lucy, the first thing she did was to come and sit very close to me, and for approximately thirty seconds she and I gazed intently into each other's eyes. (This is not the place to recount what went on in my mind, nor to speculate on what was going on in hers!)

There is one facial expression which, more than any other, has dramatic signal value—the *full closed grin* (see Figure 6.1). When this expression suddenly appears, it is as though the whole face has been split by a gash of white teeth set in bright pink gums. It is often given silently, in response to an unexpected and frightening stimulus. When an individual turns to his companions with his face transfigured by this horrifying grin, it usually evokes an instant fear response in the beholders. This facial expression is ideally suited to conveying information about "danger" in situations where silence is vital for safety (as when members of a neighboring community are spotted or heard during a patrol).

Certain Gombe individuals show expressions that are never seen in others. Some animals typically allow their lower lip to droop when they are relaxed; others never do so. Some turn their upper lip up over their nostrils in the *lip flip* and make rubbing movements, whereas others do not. One mature male, David Greybeard, progressively pushed his lower lip out farther and farther as, determined to get more bananas, he inspected box after box. Hugo frequently pressed his lips together, drew back the corners of his mouth, and *mock smiled.* In no way related to the *play face* that has been referred to in the literature as *smiling* (Yerkes and Yerkes, 1929), the mock smile occurred in a variety of contexts, none of which was associated with play or good temper. Humphrey and Gigi *sneer* when suddenly alarmed.

Mr. Worzle had white sclerotics around the iris of his eyes, as do humans. It always seemed that he was unusually vigilant, for his gaze darted back and forth from side to side. In fact, such scanning is quite normal (unless the individual is very relaxed or concentrating on some task), but the white sclerotics drew attention to the movement. (Note the potential signaling value of the whites of the eyes in our own species, particularly in a person in a dark place or a dark-skinned individual.)

Body posture may also be revealing. Any change from a relaxed posture may act as an incidental signal, transmitting information to others nearby. Changes range from merely shifting position (even this can cause a young adolescent to jump nervously) through getting up and moving away, to a clear-cut directed signal such as an arm threat.

A change in the appearance or movements of an individual due to injury or ill health may affect the way he is treated by his fellows. Gross change in appearance or gait, as shown by polio victims, or

complete cessation of movement, as when an individual dies, is likely to provoke fear or aggression or both. The possession of a desired object may subtly alter the behavior of an individual and he may, accordingly, be responded to differently.

Of special importance (for human and chimpanzee observers alike) with regard to monitoring changes in the level of arousal (Mason, 1965) is erection of the body hair (piloerection, or in Gombe terminology, *bristling*). This is an autonomic behavior: a chimpanzee bristles when he is highly aggressive, or socially excited, or when he sees or hears something strange and frightening. By contrast, a chimpanzee nervous or fearful of his superiors usually has very sleek hair. Sometimes—as when a male is listening to distant calls or watching an agonistic interaction—his hair rises and falls alternately, as, presumably, he reacts emotionally to what he hears or sees. Subordinates often become nervous and move away when nearby superiors suddenly show hair erection; a bristling male is more likely to display or attack than one who is sleek.

The various postures and gestures of the nonvocal communication system are described in Chapters 12 and 13. Submissive patterns include presenting, extending the hand, crouching, and bobbing. Aggressive patterns comprise various arm-waving movements, bipedal swaggering, hunching of the shoulders, and all the components of the impressive male charging display. In general, patterns that make an individual appear larger than he is occur in situations where the actor is aggressively motivated. Thus aggressive acts may be accompanied by bipedal postures, leaping up and down and waving the arms, and by hair erection. In the charging display huge branches may be dragged or flailed, great rocks hurled or rolled, and vegetation swayed. All these acts increase the overall impression of size and dangerousness. By contrast, the submissive or frightened individual

As alpha male Mike moves past him, Figan shows submissive behavior, bobbing and uttering pant-barks. (H. van Lawick)

The compressed-lips face that typically accompanies aggression. Faben is performing a charging display. (H. van Lawick)

Alpha male Mike prepares to perform a charging display. This is a quadrupedal hunch. (H. van Lawick)

(if he is not fleeing) crouches to the ground with sleeked hair, or moves in a cautious and unobtrusive manner. He may, if he is able, remove himself altogether, or he may hide behind a tangle of vegetation.

Most communication between individuals that relies on visual cues takes place at fairly close quarters. The sexual swelling of a female in estrus, however, can function as a highly effective distance signal, alerting males who may be almost a kilometer away on the far side of a wide valley. Long-distance visibility is probably of major significance to the young immigrant or visitor female, in that she will be immediately perceived as a desirable resource before her status as

Social grooming provides relaxed, friendly physical contact. (Gombe Stream Research Center)

"stranger" can lead to violent attack. Sometimes human observers are able to recognize the characteristic form or gait of a certain individual across a valley. That the chimpanzees can also make such distinctions is suggested by observations such as the following: a juvenile who had temporarily lost his mother stopped crying when he saw her across the valley and hastened to join her.

Tactile Communication

Physical contact with others is a crucial component of messages intended to reassure or appease distressed or tense individuals. It occurs in many different social contexts, particularly greeting after a separation, reconciliation after an aggressive incident, reassurance in times of excitement, or fear. Touching, patting, embracing, and kissing are common elements in such situations. A lower-ranking individual during a reunion may not only reach out his hand toward his superior, as in supplication, but may actually make contact. The dominant, in response to submissive gestures, will in many instances respond by touching, kissing, embracing, or mounting. The lower-ranking individual may actually beg for a reassuring touch. Once the desired contact has been made, the subordinate visibly relaxes.

The friendly physical contact provided by social grooming is, as we shall see in Chapter 14, one of the most significant aspects of chimpanzee social life. Individuals with close friendly relations, such as a mother and her grown offspring or pairs of adult males, typically settle down for a long grooming session when they meet after a separation. Grooming helps to ease the tension between two adult males whose relationship is strained. And a highly aroused individual who has performed a series of charging displays, or who has been fighting, may perceptibly relax if he is groomed by others. Brief token

TOUCHING /SIM TO HUMAN

GROOMING HELPS TO EASE THE TENSION BETW. 2 MALES

grooming movements are components of many greeting, submissive, or reassurance interactions, and rapid, often ineffectual grooming may be performed by a chimpanzee in a frustrating situation (while waiting for a share of food, for example).

Tactile stimulation may also serve a calming or reassuring function when it is self-directed. Anxious or frustrated individuals scratch and groom themselves, and one male, Figan, typically held his own scrotum in response to sudden fear. In the days when the chimpanzees spent long hours waiting in camp for bananas, males quite often toyed with their erect penes; this was presumably a response to their frustration.

Physical contact is not always pleasant. The slapping, kicking, pounding, dragging, and biting that are components of aggression (and which occur too in rough play) elicit fear, screaming, and avoidance—or furious retaliation. Aggressive physical contact is sometimes the result of a disregarded signal or a wrong social response, or simply because the victim was in the wrong place at the wrong time. Scott (1958) has commented that a painful stimulus is quickly learned and hard to extinguish, and the key role played by aggressive physical punishment in the ordering of chimpanzee society must not be underestimated.

Auditory Communication

In nature, chimpanzees may be silent for hours on end, even when several are together. They may also be extremely vocal, and when two or three parties are within earshot of each other there may be frequent exchanges of noisy vocalizations. Most of their calls serve a distinct communicative function in that they may alter, often in a highly predictable way, the behavior of individuals who hear them. This is not to say that the calls are necessarily made with the intention of influencing the action of others, although they may be.

Chimpanzee vocalizations are closely bound to emotion. The production of a sound in the *absence* of the appropriate emotional state seems to be an almost impossible task for a chimpanzee. The Hayeses found that prior to intensive (and basically unproductive) training Viki was "completely unable to make any sound at all on purpose" (Hayes, 1951, p. 66). Eventually she learned to make a strangled *ahhhh* to signal her desire for a variety of objects. While her failure to produce more than four (indistinct) renderings of words was probably caused, at least in part, by differences in the ape-human vocal apparatus, her inability to vocalize at will was almost certainly an even greater impediment.

Chimpanzees can learn to *suppress* calls in situations when the production of sounds might, by drawing attention to the signaler, place him in an unpleasant or dangerous position, but even this is not easy. On one occasion when Figan was an adolescent, he waited in camp until the senior males had left and we were able to give him some bananas (he had had none before). His excited food calls quickly brought the big males racing back and Figan lost his fruit. A few days later he waited behind again, and once more received his bananas. He made no loud sounds, but the calls could be heard deep in his throat, almost causing him to gag.

It seems appropriate to relate chimpanzee calls, insofar as this is presently possible, to the underlying emotional states that appear to elicit them. Figure 6.2 is a preliminary attempt to do this. Most of the emotions listed have already been mentioned. One group of calls, however, is associated with emotions that are difficult to describe and have been tentatively labeled "sociability feelings." What emotion should we ascribe to the man who, returning after his day's work, calls out "Hello! Anybody around?" Or the apple picker who calls casually from his tree to the next, "How're you doing?" These, I believe, are comparable to the emotions that in chimpanzees generate inquiring and spontaneous pant-hoots and soft and extended grunts.

THE CALLS Chimpanzee calls are graded: acoustically most of them probably form part of a single graded system (Marler, 1969). Sound spectrograph analysis (based on data collected by Marler during three months at Gombe in 1967) led to the definition and description of fifteen distinctive calls (Marler, 1976). Since this pioneering study the Gombe chimpanzees have been followed and observed for many thousands of hours (about 58,500). It has become increasingly apparent that their vocal communication system is far more complex than had been supposed, so that despite the fact that there has been no additional sound spectrograph analysis, it seems important to extend the original list of calls.

The task of sorting by ear the various calls in the vocabulary is not an easy one: chimpanzee sound production is organized differently from our own and, for the human, it is difficult to distinguish and classify discrete calls along a graded continuum. The task is made still harder by the extreme individual variation in the production of certain calls, which in some cases seems more pronounced than differences between two types of the same main call. Nevertheless, by taking into account (a) the behavioral context in which a given call is uttered; (b) the probable underlying emotion of the caller; (c) the fact that a call, if it affects the behavior of those to whom it is directed or those hearing it, does so in a predictable direction; and (d) that it can be distinguished from other calls by experienced human observers,* it has been possible to list an additional seventeen calls (Figure 6.2). Some of these are vocalizations that were already recognized by field workers at the time Marler collected his data (for instance, Goodall, 1968b) but showed no acoustical characteristics to distinguish them from similar calls, for example, copulation pants or the huu of puzzlement. Others are divisions of major types of call, particularly the pant-hoot and the scream.

In our attempt to unravel the intricacies of chimpanzee vocal communication, we must also take into account the added complexity of combinations of calls. These, provided the chimpanzees are able to interpret them at least as well as experienced human observers (and they undoubtedly do so a great deal better), will convey additional messages.

* Four senior field assistants (H. Matama, E. Mpongo, Y. Alamasi, and Y. Mvruganyi) were interviewed with open-ended questions; their replies were in agreement, except that in some cases they favored an even greater number of divisions than the thirty-two noted in Figure 6.2.

Figure 6.2 Chimpanzee calls and the emotions or feelings with which they are most closely associated.

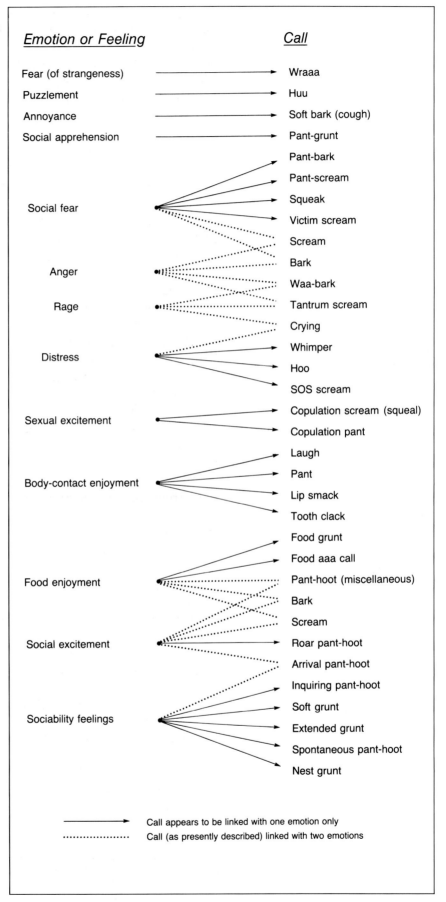

Chimpanzee calls are elicited by and/or directed toward (a) other individuals in the same party as the caller; (b) community members in different parties; (c) individuals of a neighboring community; (d) nonanimate environmental stimuli, such as a food source; (e) animals of other species. The louder vocalizations in categories (a) and (c) also transmit information to scattered members of the community. Moreover, these same calls, together with those falling in category (b), provide information to individuals of neighboring communities who chance to be within earshot. Category (c), (d), and (e) calls, in the same way, provide information to any chimpanzees who happen to hear them. To simplify matters, the vocal repertoire is described in the adjacent listing under two major headings: *intraparty calls*, those primarily concerned with communication among individuals of the same party, and *distance calls*, those that seem to be aimed primarily at transmitting information from one party (or one community) to another.

Intraparty calls. Vocalizations made by a subordinate to his superior grade from the soft pant-grunt of greeting or respect, through pant-barks, to the screams of varying pitch, intensity, and duration that are given when he is threatened or attacked. A distressed individual whimpers; if his distress is compounded by fear, and if this increases, his whimpers may grade to squeaks, then screams. If his distress is overlaid by frustration, he may end up giving the very loud and frenzied tantrum screams.

A confident aggressor is seldom noisy—apart from an occasional soft bark, he is usually quite silent during both threat and attack. When an element of anxiety or fear creeps in, however, he may utter loud waa-barks; and if the aggression becomes serious, he, along with his victim, may scream loudly.

Vocalized communication among friendly individuals comprises, for the most part, a variety of soft, gruntlike sounds. Only during social or sexual excitement are much louder sounds uttered—such as the copulation scream or squeal and loud, hysterical-sounding laughter.

Intraparty calls can serve to attract the attention of an individual with whom the caller wishes to interact, and to emphasize certain communicative gestures. They are particularly useful when chimpanzees are in thick forest where, in addition, the calls help to maintain cohesion. A party member who is temporarily out of sight of the others, but within earshot, will of course be able to learn at least something about the behavior of his unseen fellows when he hears their vocalizations. For one thing he will almost certainly be able to identify the calling individuals. An interesting confirmation of such ability occurred when a few chimpanzee calls recorded at Gombe (by Peter Marler) were played back to chimpanzees proficient in ASL. Flo's screaming, followed by Flo's pant-hoots, elicited the signed response *different-same.* A scream by Flo and a scream by Mike produced the unexpected response *same-different;* it was assumed that this meant *same kind of sound, different voice.* Other pairs of vocalizations were labeled in similar ways. The experiment was not designed to show individual differences, nor did the researcher know the identity of the callers when he played the tape (Fouts, 1974). Chimpanzee mothers recognize the screams of their own offspring

Part 1, Intraparty Calls

The *pant-grunt* is always directed by a subordinate to an individual of higher social rank. Functioning as a token of respect, it is of major importance in maintaining friendly relations among community members. The tenser the subordinate in relation to the other, the louder his pant-grunt. During a greeting between friendly individuals it is a soft, barely voiced sound; if the subordinate is apprehensive, his pant-grunts grade to *pant-barks*, a much louder call. A highly fearful individual (particularly during social excitement) may utter a series of increasingly frenzied pant-barks that, as they grade into screams, can be labeled *pant-screams*.

Whimpering is a series of softer, low-pitched sounds that may become progressively louder and higher in pitch. A chimpanzee whimpers in a variety of behavioral contexts when the constant underlying emotion is distress or need. Most commonly heard in infants, especially during weaning, the whimper occurs in older individuals when for example they are refused food by superiors, or after they have been attacked. Whimpers may grade to screams.

Fifi whimpering (the horizontal-pout face) (H. van Lawick)

The *hoo* is an isolated but distinctive part of the whimpering sequence, but has no separate identifying acoustical characteristic (Marler, 1976). The single hoo may be uttered several times in succession, but each vocalization is made separately; as a hoo sequence starts to rise and fall in pitch and volume, and when each sound is produced in temporally rapid succession, it grades into the whimper. The hoo is uttered by both an infant and (much less often) his mother when they need to reestablish physical contact—when, for example, the infant wants to ride on his mother's back during travel or when she reaches to retrieve him from a situation she perceives to be dangerous. An adult occasionally gives a hoo when begging for food or grooming.

Squeaks are short, shrill calls, usually given in a series of two to five per second (Marler, 1976). They occur in response to threats from a more dominant individual. If the subordinate becomes increasingly fearful, squeaks grade into screams, just as screams grade into squeaks as the caller gradually becomes more relaxed.

Screams are high-pitched and loud, almost always given in a series. Heard in contexts of aggression and general social excitement, they are uttered by highly stressed, fearful, frustrated, or excited individuals. Based on our interpretation of the underlying emotional state of the screaming individual, the context in which he screams, and differences in the pitch, timbre, duration, timing, and intensity of the call (as distinguished by the human ear), four kinds of scream have been tentatively identified: the victim scream, the tantrum scream, the SOS scream, and the copulation scream (or squeal). Because screams are far-carrying, they will be heard by chimpanzees scattered over quite a large area and may serve to bring an ally to the aid of the caller. The SOS scream appears to be directed toward individuals *not* in the immediate party of the caller and will be described as one of the distance calls.

(a) The *victim scream* occurs when a chimpanzee is being attacked, and can be distinguished from the many variations of screaming uttered in response to aggression because the quality of the sound (harsh and prolonged) is affected by the pounding, dragging, slamming, and so on, to which the victim is subjected. The hitting and stamping sounds made by the aggressor can also be heard.

(b) The *tantrum scream* is a very loud, harsh call, sometimes resulting in

Figan, screaming loudly, crouches in a temper tantrum. (H. van Lawick)

glottal cramps and even gagging. It occurs most commonly as part of the temper tantrum of an infant who has been rejected during weaning. An adult, after being attacked, may be torn by conflicting desires—an aggressive desire to retaliate on the one hand and fear of the aggressor on the other. He may give vent to his frustration with high-intensity tantrum screaming (he may even throw a tantrum).

(c) The *copulation scream* is described as a "clear high-pitched sound of variable length, accompanying grin" by Tutin and McGrew (1973, p. 243), who called it a *squeal*. Experienced human observers seldom have difficulty in identifying this call. For example, the field assistants sometimes write, when describing a copulation, that a female did *not* give the "sex call" but screamed as in fear.

Barks are loud, sharp sounds, usually given in long sequences with much variation in pitch. Females bark more than males. During any kind of social excitement there is likely to be a medley of different vocalizations and these will include many barklike sounds—calls that are extraordinarily confusing to human ears. Only careful analysis of a very large sample can isolate any discrete calls that may, to the chimpanzees themselves, be distinguishable from the jumble of sound.

The *waa-bark*, a loud, sharp sound, is given in a variety of agonistic contexts and is typically accompanied by an arm-raise threat or more vigorous gestures such as arm waving, running upright, and so on. The waa-bark is the comment so often interjected by bystanders during a conflict between others, and usually indicates sympathy for the victim. During a male attack on a female, other females present frequently provide a running commentary of waa-barks until the incident has ended. Often, when the victim of aggression becomes suddenly bolder after the event—as when he receives the support of an ally or as the aggressor displays away—his screams change to "defiant" waa-barks.

The *cough-threat* (or *soft bark*), a gruntlike sound uttered through slightly open mouth, is only directed down the hierarchy, by higher-ranked to lower-ranked individuals. A call that indicates slight annoyance, it functions as a mild warning to prevent a subordinate from moving closer or from doing something of which the caller clearly disapproves (such as reaching for a piece of his food).

Laughing heard during play sessions somewhat resembles human laughter. Sound spectrograph analysis shows a change from steady exhaled sound, to chucklelike pulsed exhaled sound, to "wheezing" laughter. These changes correlate with the increasing vigor of the play activity (Marler, 1976). Most laughing results from physical contact in the form of tickling with the fingers or play biting, but it sometimes occurs during chasing play. Laughing is far more common in infants and juveniles (since they play much more than adults), but when adults do play, they laugh also.

Panting, usually without voice, may occur during a grooming session or by one or both partners during a greeting (in this situation the mouth is often pressed to the body or face of the recipient). The photo at left shows a mutual *open-mouth kiss* that persisted for over thirty seconds, throughout which both male and female panted.

Copulation pants were included by Marler (1976) with other panting sounds, on the basis of sound spectrograph analysis, but the sound is so readily identifiable in the field that I have labeled it separately. In its loudest form it can be confused with laughing—and more recent sound spectrograph analysis has confirmed that the two are almost indistinguishable. However, copulation pants are generally more rapid than the panting of laughter, and often the sound lacks the voiced inhalation phase typical of laughter (P. Marler and S. Runfeldt, personal communication).

Mutual open-mouth kiss during a reunion between Humphrey and adult female Athena. She has her hand on his head.

Lip smacking and *tooth clacking* frequently accompany social grooming. These sounds, made by tongue, teeth, and mouth, are not voiced. Sudden and intense bursts of tooth clacking attract the attention of the individual being groomed and occasionally of others nearby; this sometimes functions to create new enthusiasm for a session showing signs of flagging. There is evidence that it is sometimes used with this goal in mind (Chapter 14).

Soft grunts may be exchanged when two or more familiar chimpanzees, especially family members, are foraging or traveling together. Typically one individual grunts when he pauses during travel, or when he gets up to move on, or when he hears a companion in the undergrowth nearby. One or more of the others may respond in kind. Thus these grunts function to regulate movement and cohesion among friendly individuals.

The *extended grunt*, a long, drawn-out sound that may be composed of two syllables—the *double grunt*, eh-mmmmm—is heard during rest sessions. Its significance is not clear; it may serve to proclaim identity within a party, or it may simply be an expression of contentment. The *nest grunt*, another distinctive double grunt, is given when an individual begins to look around for a suitable nest site, during the making of the nest, or as he settles down for the night.

Soft *food grunts* uttered by a relatively calm individual typically accompany the first few minutes of feeding on a much-liked food, and they may recur at various intervals during a long feeding bout. They undoubtedly function to attract the attention of party members to the resource. Both soft food grunts and the loud *food aaa call* (which functions in interparty communication) may be uttered *before* the individual arrives at a food source—even when the source is still out of sight.

One grunt, which may prove to be acoustically distinct, is given when a chimpanzee moves toward honey or driver ants (Wrangham, 1975; E. Mpongo, personal communication). These are two much-liked foods that inflict discomfort in the form of the painful bites or stings of the insects.

The *huu* of puzzlement, surprise, or slight anxiety is directed toward such things as small snakes, unknown creature rustlings, dead animals, and the like. This sound is made even if a chimpanzee is alone. The huu was included by Marler in the whimpering category on the basis of sound spectrograph analysis. However, I have never confused the huu with the hoo in the field. In response to the huu, other chimpanzees nearby almost always approach the caller and peer at the object that elicited the sound; I have never observed an approach response of this sort to the hoo of an infant (except by his mother or sibling). A chimpanzee giving the huu does not show the pouted lips that are characteristic of the hoo of distress.

Yawning is sometimes accompanied by quite a loud, breathy exhalation; yawns may be voiced as they sometimes are by humans. Tense individuals may yawn very frequently; the audible component will attract attention and perhaps provide information about the mood of the yawner.

Coughs, sneezes, hiccoughs, and loud smacking of lips during feeding all sound much the same as in humans. They may function as incidental signals, sometimes helpful in locating party members.

---

and reply even if the child is out of sight. (The same is true for other mammals—it has been demonstrated convincingly in vervet monkeys; see Cheney and Seyfarth, 1980.)

Our chimpanzee listener will also learn much from some *sequences* of vocalizations, particularly voiced comments passing back and forth between different individuals. Laughing followed by infant whimpers

Chimpanzees pant-hoot in response to distant calls. *Left to right*: Flo, Fifi, Hugo, McGregor, Melissa, and David Greybeard (partially turned). (H. van Lawick)

thus indicates a play session in which one of the playmates has become too rough. The victim's mother, feeding high in a tree, recognizes the calls of her own child and is alerted. If his whimpers change to screams, she may rush to his defense; if he utters waa-barks, she can relax, for this often indicates that the youngster has already found an ally. Similarly, the screaming of an adult, followed by squeaks and finally pant-grunts, will convey to the listener, quite as accurately as if he were observing the incident, that an aggressive interaction has taken place and that the victim has relaxed and approached the aggressor. A muffled pant-grunt will indicate a kiss.

Distance calls. If the main goal and/or function of a given call is to convey a specific message to an individual or individuals who are *not* part of the caller's immediate group, then it seems important that the information transmitted be unambiguous; the distant receiver will not only be unable to use contextual information to clarify the meaning, but he will not hear any of the quieter calls that might help him to interpret louder ones.

In animal societies whose members are organized in stable or semi-stable social groups, distance calls serve mainly to regulate intergroup spacing (the *howl* of the howler monkey, for example, or the *long-call* of the gibbon), and as a means by which a temporarily lost group member may make contact with his companions (the *waa-hoo* call of the olive baboon). In fusion-fission societies,* however, such as those of the hyena, lion, and chimpanzee, distance calls must serve a wider range of functions: they should (a) draw attention to environmental features such as the location of rich food sources or danger; (b) announce the precise location of specific individuals scattered through the home range, so that at least vocal contact can be maintained; and

---

* These are social groups whose members may separate into smaller units for feeding or other purposes, then join up again (Kummer, 1968).

(c) indicate when particular group members are in need of help from particular others.

The distance calls in the vocal repertoire list largely fulfill these functions. Distress calls—SOS and other screams—bring help from distant allies. Exchange of pant-hoots between parties provides community members with a good idea of the whereabouts of friend and foe alike; they can plan their direction of travel accordingly. When food or danger is located, the information is shared among community members through food calls and the wraaa of alarm. To what extent such vocalizations make distinctions between different kinds of food and danger, we do not know. Certainly sounds of a successful hunt are easily interpreted. But it seems likely that if there *are* other distinctions, they will relate to the *quantity* and *quality* of the food, rather than the type. After all, food desirable enough to elicit calls of delight from one individual will be relished by all. There is no need for more specific announcement. The wraaa call is heard relatively infrequently, and we do not yet know whether there are consistent variations indicating different kinds of danger. Further analysis may well reveal that there are; after all, vervet monkeys respond to a number of different alarm calls indicating, for example, *leopard*, *snake*, and *eagle* (Cheney and Seyfarth, 1980).

NONVOCAL SOUND SIGNALS Some other sound signals should be mentioned. The most important is the *drumming display*, when the chimpanzee leaps up and pounds with hands and feet against the buttresses of a large tree. This produces a sound that can carry over long distances (across a valley, for example). Drumming, like the charging display, is primarily a male activity. It is typically accompanied by pant-hoots and is frequent when the chimpanzees are traveling in large mixed parties. There are certain favorite drumming trees along chimpanzee trails, the sight of which usually triggers drumming displays from many members of a traveling party, including some females and youngsters. The drumming display is occasionally performed without accompanying calls, mostly during tense travel in the "danger zone" where the community range overlaps that of its neighbors. Many of the components of the charging display—slapping and stamping, swaying of vegetation, rock throwing—produce characteristic sounds, which serve to enhance the intimidating effect of the charging display.

Another highly distinctive sound is the swish when a small branch is shaken during male courtship displays. This is only audible over a short distance, but serves to alert a female who may not have noticed other silent courtship signals.

Loud scratching, usually up the arm or up the side of the body, is a signal that often precedes a bout of social grooming. An individual may begin scratching some distance away, and his intended partner's attention will be captured by the relatively loud, highly distinctive sound. A mother, before climbing from a tree, often pauses at a low fork and scratches, looking up at her infant. This serves as a signal; the child usually hurries to the mother and climbs aboard ready for descent. Frustrated or uneasy chimpanzees also scratch in this noisy and vigorous manner; the sound informs others about their mood.

Part 2, Distance Calls

Pant-hoots, voiced on both exhalation and inhalation, are the most commonly heard call of adult individuals and, more than any other (at least to human ears), serve to identify the caller. Field observers can reliably distinguish the pant-hoots of many individuals, and chimpanzees tested in the laboratory were able to make similar identifications (Bauer and Philip, 1983). Acoustical analysis confirmed that there are indeed consistent differences in the structure of the pant-hoots of different individuals, as well as between those of males and females (Marler and Hobbet, 1975; Marler, 1976).

Pant-hoots are given in many different situations: upon arrival at a new food source; when two parties meet; during travel, particularly from the ridges between valleys; after return from a long, tense patrol; in response to hearing the calls of another individual or party; during social excitement; and, apparently spontaneously, during feeding and, especially in the evening, during nesting. Pant-hoot choruses may break out during the night, particularly when two or more parties are sleeping within earshot, in which case the calls pass back and forth, or when a large number of chimpanzees are nesting together. It seems certain that analysis (currently under way) will reveal consistent differences in the pant-hoots given by the same individual in a variety of specific contexts. In the same way that different kinds of screams were identified, four distinctive pant-hoots are listed below.

(a) The *inquiring pant-hoot*, often accompanied by tree drumming, is given when an individual (usually a male) arrives on a high ridge, or at intervals during travel. It tends to rise in pitch toward the end of the series and is almost always followed by a pause during which the caller listens intently and (if at a lookout position) scans the surrounding countryside. A chimpanzee who hears another pant-hoot often responds by calling (usually with pant-hoots, sometimes also with waa-barks and even screams); thus the individual who initiates this question-and-answer exchange will learn both the identity and the whereabouts of those who reply. If he is on a high ridge, he may be able to monitor at least two valleys. Based on information received, he will be in a position to choose his subsequent travel route in order to join—or avoid—particular other individuals. A male who has temporarily lost contact with one or more of his companions may utter pant-hoots every ten minutes or so, then sit, listening and scanning, until he either receives a reply or, after about an hour, gives up and remains silent.

(b) The *arrival pant-hoot*, ending (probably according to individual idiosyncrasy) with rather deep roarlike sounds or higher screamlike calls, is typically given on arrival at a good food source and upon joining another party. It frequently accompanies the arrival of adult males in camp. It functions to proclaim the identity of the caller, and its inferable message "Here comes so-and-so" may be especially useful to other chimpanzees feeding in the trees, who cannot immediately see which chimpanzees are arriving below them. When pant-hoots of this nature are followed by a chorus of food calls and other sounds of social excitement, chimpanzees located elsewhere will have a very good idea of what is going on and who is around.

(c) *Roar pant-hoots* are continuous low-pitched calls given only by highly aroused individuals, rarely by females. Always accompanied by charging displays, they are most often heard during or after stranger contact, but sometimes during intense social excitement within the community.

(d) *Spontaneous pant-hoots*, uttered by peacefully feeding or even (less

Figan uttering pant-hoots.
(H. van Lawick)

often) resting individuals, are typically rather high pitched, particularly the final notes. Often, when a number of individuals call in unison, there is an almost "singing" quality. Although this call, like the inquiring pant-hoot, is likely to elicit a response (usually in kind) from those who hear it, I have given it a category of its own, not only because it sounds different (to my ears) but also because the callers do not appear to be motivated by the same need for information. After delivering their melodious chorus, the chimpanzees calmly continue feeding, with no pause for eager awaiting of response. Adult or late-adolescent males usually initiate spontaneous pant-hoots, but females and youngsters join in with gusto. Nearby females will usually respond in kind, although they often remain silent when they hear other kinds of pant-hoots.

Screams have already been described. All serve to convey information over distance, but two, the SOS scream and crying (tantrum scream interspersed with whimpering), are most obviously intended to send urgent appeals to particular individuals who are not in the immediate vicinity.

The *SOS scream* is a clear, high-pitched call, relatively long drawn out, and repeated over and over. It is made by a chimpanzee who has been attacked or severely threatened, and is an obvious appeal for help to an absent ally. When Charlie, as an adolescent, was attacked by the entire Flo family, he sought refuge up a tree and gave loud SOS screams as he scanned the opposite slopes of the valley. After some ten minutes his presumed elder brother, Hugh, charged to the rescue, scattering the members of the Flo family who had been preventing their victim's escape.

Crying, typically uttered by an infant or juvenile in distress, is a combination of loud whimpering and tantrum screaming. When a child has become accidentally separated from his mother and his anxiety increases, the screaming, interspersed with repeated bouts of whimpering, becomes louder and more frenzied. Added information is provided by the fact that the calls move, quite rapidly, from one location to another as the child searches for his mother. When she hears her offspring crying, the mother almost always hastens toward him—unless he is quite obviously moving toward her, in which case she may wait for him to join her.

The long-drawn-out *wraaa* is given when a chimpanzee comes upon a leopard or buffalo (either of which could be dangerous) or another unfamiliar (and thus potentially dangerous) creature such as a python (seldom seen at Gombe) or, in the early days of habituation, myself. Abnormal or bizarre behavior in a community member (such as the sudden cessation of movement in the male Rix, who fell from a tree and broke his neck, and in old

McGregor, who lost the use of both legs) elicited wraaa calls from highly aroused companions.

The wraaa has two functions: it serves as a distance alarm call, alerting other community members to the presence and location of danger or potential danger in the home range; and it intimidates and/or repels dangerous or potentially dangerous creatures. I speak from personal experience when I say that the combination of intimidation displays (swaggering and branch shaking) and the spine-chilling wraaa is extremely effective!

Certain calls, taken together, may be loosely referred to as *loud food calls*. When chimpanzees approach or start to feed at a desirable food source, they may, particularly when in a large party, utter a medley of pant-hoots, barks, grunts, and a loud, high-pitched vocalization, the *loud aaa*. This last is certainly specific to the feeding context; there may perhaps be distinctive *food pant-hoots* also (Plooij in Wrangham, 1975).

It is male chimpanzees who call most often and most loudly in this context (Wrangham, 1975, 1977). Females tend to utter only soft food grunts, unless they are traveling in large mixed parties, when they too may join the exuberant chorus. An individual who hears such calls often moves toward the feeding group. One of the effects of food calls, therefore, is that a rich temporary food source will be shared by community members. This is not to suggest that food calling is an altruistic behavior; no individuals have been seen to be short of food as a result of calling others to share the feast. For the most part there is enough to go around, at least for those who are the loudest callers, the adult males (Wrangham, 1979; Ghiglieri, 1984). Wrangham also suggests that, by calling, a male attracts individuals with whom he can then interact socially in ways beneficial to himself—such as females in estrus, male grooming, hunting, and patrolling companions, and so on.

When a kill has been made, an extremely noisy outbreak of calling is customary. Vocalizations, in addition to those mentioned above, include screaming and waa-barks. When other chimpanzees hear these sounds, they usually run very fast to take part in the meat-eating session that will follow. On these occasions the sounds made by the prey species probably play an important part in identifying the situation to listening chimpanzees.

Intercommunity calls. During intercommunity encounters between rival parties of males, individuals of both sides typically utter loud pant-hoots, particularly roar pant-hoots, waa-barks, barks, and screams. We cannot yet pinpoint any one call specific to this context, but there is a special intensity and "fierceness" about the cacophony of calls emitted that is unmistakable and creates in the human observer a sense of mounting excitement and apprehension. The calls, typically accompanied by displays, are given first by one community, then the other (sometimes both together), and may be bandied back and forth for an hour or more while members of one or both sides slowly retreat.

Finally there are a great many incidental signals that may be emitted by chimpanzees going about their business. They make rustling noises as they move over dry leaves or grass, and crashing sounds in the branches when feeding or playing in trees. They often eat with loud smacking of the lips. They frequently break wind. Rinds, seeds, and discarded branches make sounds when they drop to the ground. Only the cracking open of strychnos will unequivocally indicate a chimpanzee (for no other creatures at Gombe make this distinctive noise). The other sounds merely indicate the presence of approximately chimpanzee-sized creatures. When patrolling individuals hear rustling in the undergrowth, they may startle, show hair erection, grin, and embrace—only to relax if the unseen creature turns out to be a baboon or bushpig.

TOTAL SUPPRESSION OF SOUND SIGNALS Chimpanzees typically maintain silence during patrols and during consortships. Sounds are also suppressed during some hunts, although at other times a hunt is preceded by squeaks and even pant-hoots. If a chimpanzee encounters some new or alarming object in the environment, he may flee silently. When the chimpanzees suddenly came across me before they were habituated to my presence, they sometimes paused to stare before running out of sight; at other times they rushed off for a few meters and then stopped to peer at me before vanishing silently into the forest. When I surprised them in a tree, they invariably climbed rapidly and silently to the ground and then ran off. (It was *after* this stage of silent flight, when I was not quite so novel and alarming, that they sometimes approached and threatened me with their fierce wraaa calls.)

Olfactory Communication

The extent to which chimpanzees can interpret olfactory cues is not clear. Like all creatures, chimpanzees have their own distinctive species odor. Mature males, particularly during social excitement, sometimes emit a strong smell similar to very pungent human sweat, but chimpanzees sweating in the hot sun do not have this odor. Schaller (1963) detected a similar extra-strong smell, which he thought emanated from silverback male gorillas and which he also concluded was not accounted for by normal sweating.

A captive gorilla, when presented with wads of paper that had been chewed by two people with whom she was familiar, had no hesitation in correctly identifying which wad was whose (Patterson, 1979), but similar experiments have not yet been carried out with chimpanzees. Thus, while it seems likely that chimpanzees can identify an odor as being chimpanzee or nonchimpanzee, we do not know if they can proceed beyond this point and identify "friend or foe," "X or Y," from olfactory clues. Whether they can or not, they certainly appear to make considerable use of their sense of smell and pay close attention to olfactory signs in a variety of contexts such as during patrols or when searching for companions. I followed one young adult female, Pom, who had lost her mother. For three quarters of an hour she tracked her, bending to sniff the ground and vegetation every ten minutes or so, moving slowly, listening intently, and occasionally whimpering. Eventually her efforts were rewarded and she rejoined her mother—but it was not possible to estimate the extent to which her success resulted from correct interpretation of olfactory cues. Other individuals of both sexes sometimes sniff the ground, vegetation, or discarded food bits when they are searching for specific individuals.

The extent to which olfaction is important in regulating sexual response has not yet been determined, although assuredly it plays a

During a reunion Hugo inspects Passion's genital area. She embraces him and open-mouth kisses his rump. (H. van Lawick)

role. During a reunion with a female, a male often inspects her genital area. He may bend close and smell her bottom directly, or poke his finger into the vulva and then sniff the end of it. He may do this two or three times. One male, Hugo, who had formed a close possessive relationship with the sexually popular Flo, suddenly pushed her to her feet in the middle of a grooming session, two weeks after detumescence, inspected her frenziedly, then resumed grooming. Another male inspected the same female—also some days after detumescence—fourteen times in five minutes.

While inspections of this sort presumably provide a male with some information regarding the female's reproductive condition, it seems that this may not be very precise. On a number of occasions males exerted considerable energy in establishing consortships with females who were already pregnant. Moreover, although males seem able to detect the periovulatory period with some precision (this is the time when high-ranking males show possessive behavior toward females), the role played by olfaction is far from clear. Precise olfactory cues marking this period were not detected in the volatile fatty acids (copulins) from vaginal lavages of captive females (Fox, 1982). In a detailed study of captive chacma baboons, the visual channel was found to be much more important for arousing males sexually than the olfactory route (Bielert and Walt, 1982).

During reunions females quite often cursorily inspect each other's genital area; individuals of either sex sometimes touch the penis or scrotum of a male and then sniff their fingers. (In fact, when examining small dead mammals—a rat and a bat—two young females repeatedly touched the genital area, then sniffed their fingers.) When Flint was not allowed to touch his newborn sibling, he poked her with a stick and smelled the end of that. Sick individuals are sometimes sniffed repeatedly at many parts of their body. Dead chimpanzees have also been smelled carefully, as were places where a dead body had lain. Drops of blood spattered onto the vegetation during a fight may be intently smelled (and licked). Leaves a chimpanzee has used to wipe dirt or blood from himself may be picked up and sniffed by other individuals.

Signal Combinations and Contexts

It is the different ways in which signals are combined that, along with the behavioral context in which they are perceived, enable receivers to interpret correctly the meaning of a given communication sequence. For example, when a male looks toward and shakes a branch at another individual who is more than 3 meters distant, the signal indicates a desire for the other to approach. If the chimpanzees are resting, and especially if the other is a male, the signal usually implies "Come and groom me." The same signal directed toward a female during initiation of a consortship means "Follow me!" If, however, the bristling and branch shaking are accompanied by splayed thighs and a penile erection, the meaning is again different: "Come and copulate." Finally, a chimpanzee with bristling hair may shake a branch at a creature such as a snake or monitor lizard. Here the signal has still another meaning: "Go away!"

Sometimes it is *only* the context that can provide the receiver with the information necessary to interpret a given signal sequence correctly. Thus in human communication, "Go jump in the lake" means one thing when addressed by a mother to her muddy child and something quite different when it concludes a friendly argument between adults. When an adult male chimpanzee, in response to calls from across the valley, suddenly shows full hair erection, a nearby female may hurry over to him and the two may hold hands and embrace as both stare toward the sounds. If, however, the male bristles in response to the approach of a rival male, the same female is likely to rush to the safety of a tree, anticipating that there may be trouble between the two.

Chimpanzees, like other higher primates, typically pay close attention to the behavior of their companions. Much information can thus be passed on through social facilitation, so that communication signals as such often are not necessary. Let us look at a few examples. (a) As two youngsters walk along, the first one steps over a trail of ants, apparently without noticing them. His infant brother stops and begins to fish for the ants with a stick. After a few moments the elder pauses, looks back, watches the ant fishing intently, then returns and joins in. (b) An adolescent joins a party of resting adults. After a while he approaches one of them who is chewing a food wadge, and sniffs his mouth and hand intently. He then moves off, locates a patch of the same kind of food, and starts to feed. (c) An infant moves over to some bushes and stares fixedly at a spot on the ground. Twenty-five seconds later his juvenile sister looks toward him, approaches, stares at the same place, then makes a slight threat gesture; a small snake glides swiftly away.

Sometimes, of course, other members of a party will be alerted by clear-cut signals. When an individual discovers a dangerous or fearful object, for instance, he (or she), as we have seen, may announce this by a variety of vocal signals, by showing hair erection, and/or directing threat gestures toward the object. I have already described the signal function of the silent closed grin, so often given in a frightening situation. The precise location of the object will be revealed by the orientation and direction of the gaze. A sound signal may alert a companion to the availability of food. For example, two individuals, A and B, are traveling together. A climbs into a tree, B sits below and looks up. Only when he hears soft food grunts from A does B climb to join him.

Most nonhuman primates travel in relatively stable groups, so that when one individual discovers a new food source or a new potential danger, it usually becomes common knowledge rather quickly. This is not true in chimpanzee society, where an individual is often alone and may during such a time discover a new food source, a leopard den, a freshly hacked trail, and so on. Other individuals may later make the discovery just as A did; but we are interested in the extent to which A can (or will) communicate (intentionally or otherwise) his discovery to another, B. And, to the extent that he does, how does he do so? (Of course, A can be several individuals who have made a discovery together, and B in the same way can be more than one.)

First, auditory signals made by A may be detected by B, either as B passes by, or from quite far off. These may be accidental signals, such as breaking of branches or banging of strychnos, made by A during feeding. Or they may be distinctive vocalizations.

Second, B may follow A to a food source that B discovered previously; or if A's special knowledge concerns a danger, B may avoid it by following A *away*. When A, traveling now with a party of others, sets off in a certain direction with the goal of feeding on figs discovered the previous day, the others may or may not accompany him. If A is an adult male with leadership qualities or a sexually popular female in estrus, the rest of the party is likely to follow, whatever A's destination. But even if A is a low-ranking or young individual, at least some of the others may go with him if he sets off in a purposeful manner.

The coordination of movement among the individuals traveling in a small party at Gombe is a subject that needs detailed research. The chimpanzees who follow often monitor the movements of a leader and, when he (or she) starts to move on, may quickly stop feeding or grooming and hurry after. The leader usually glances toward the others as he leaves and sometimes waits if they do not immediately follow, occasionally giving a soft or extended grunt. Scratching can be a clear-cut signal in this context, indicating that he is about to go. A mother, having moved to a low branch prior to descending a tree, stops, glances at her infant, and gives slow but vigorous scratches down her side. An obedient infant responds rapidly, hurrying to cling to her for the descent.

The branch-shaking "Follow me!" signal, used by a male at the start of a consortship, occasionally occurs in other travel contexts. Sometimes the higher ranking of a pair of males, impatient to move on, shakes branches when a companion does not immediately follow. Wrangham observed this six times in eighteen months. Once Figan shook branches at Jomeo, apparently signaling that he should follow him on a hunt. And Humphrey, as described, used this signal when trying to persuade Mr. McGregor, crippled by polio, to follow him.

Compared with most other nonhuman primates, a chimpanzee has unique freedom to come and go as he wishes. At a given moment two individuals in a party, A and B, may decide to set off independently to two separate food sources. C may follow A or B—or he may remain by himself. It is sometimes difficult to pinpoint the precise reasons for C's choice. Often it will reflect C's own knowledge of the environment: he may know that there is a good food source in the direction in which A is heading and have no prior knowledge of B's source. Yet he may follow B because of B's leadership qualities or because of a special relationship with B. Or he may know that, even if he went with A to A's source, A would aggressively prevent him from feeding. If he goes with neither, he may himself know a rich source in another direction. Or he may have fed at his friend B's source that morning (and doesn't want to travel with A).

Only if one knows the relationship of A, B, and C; the nature and amount of food available at the two sources for which A and B are heading; and whether or not C knows about one, both, or neither of these sources, can one ask questions about the extent to which C can make use of information provided by A or B.

Menzel (1971, 1973, 1974) addressed this question and others connected with communication about objects, in his work with a group of young chimpanzees in a large field enclosure. He ran a series of tests, all of which were initiated in a similar way. First the group was shut in an indoor room from which they could not see the enclosure. One individual (a different one each time) was then removed and shown the location of a hidden object—which was sometimes food, sometimes a frightening thing such as a snake. The informed subject was returned to his or her companions. A few moments later the whole group was released. Menzel found that (a) the orientation and direction of travel of the informed individual, A, served as cues to his or her companions, and after a few trials the others were able to predict the direction in which the object was concealed (sometimes they even found it before A got there); (b) the *nature* of the object (food or danger) was known to the others even before the group was let out. If A had been shown a snake, for example, all emerged with signs of apprehension such as hair erection and a tendency to keep close together, move cautiously, and refrain from manual exploration of the hiding place (they threw rocks or displayed instead); (c) if *two* individuals were given different information—if A was shown a large pile of hidden food and B a small pile—the group when it emerged tended to follow A. This, Menzel concluded, was probably because A showed stronger signs of anticipatory behavior such as faster running, more vigorous signaling, or louder whimpering if no one followed. It should be pointed out that, in Menzel's setup and also at Gombe, if A makes a *deliberate* attempt to persuade B to follow him to a food source, altruistic motivation is unlikely; the persuasion most probably reflects A's desire for companionship, for whatever reason.

It is quite clear that the members of a primate group are easily able to detect the purposeful movements of a goal-oriented individual. Kummer (1968, 1971) vividly describes the complex decision making that goes on each morning before a troop of hamadryas baboons leaves its sleeping cliff for the day's march. This is another species with a fusion-fission society. During the day the troop will split into bands, the bands into clans, and the clans often into component single-male units. And so, when the troop reassembles each night, its members bring separate pieces of knowledge about food supply, predators, and so on. This perhaps is why in the morning the adult males, followed by their females, begin to move tentatively in different directions. Like the pseudopods of an amoeba, bulges appear around the periphery of the troop, some persisting and growing a little, others being withdrawn as the males who initiated them return toward the center. Finally, one of the senior males appears to make a decision and sets off. His purposeful walk is very different from the tentative movements made by the peripheral males, and soon the whole troop is on the move. At this stage it seems that all know where they are headed, for different adult males replace one another at the head of the troop.

On one occasion the departure from the cliff was unusual in that a single adult male suddenly set off with a unique swinging gait, ignoring the pseudopod, which had begun to move in the direction most usually taken by the troop. Almost at once the entire troop followed him, although he moved in a rarely taken direction. It so happened that on this particular morning the normal route was blocked by a flooded river (Kummer, 1968). Whether there were signals other than that purposeful and unique swinging gait we do not know; probably that, in itself, was sufficient. On three other occasions when males set off with this same gait, they too galvanized the troop into following them; but the reasons for their behavior could not be determined.

To what extent can a chimpanzee in nature convey information to his fellows about specific events that he has taken part in or observed? Can he convey more than "Something very bad (or good) happened to me"? If a male chimpanzee traveling near the periphery of his range encounters a party of his neighbors and is attacked and wounded, how much of this information can he share with other members of his community? As a result of communicative signals emitted by the male himself, probably very little. Nevertheless, those individuals he encounters upon his return may learn a great deal about the nature of his ordeal from some of the following cues:

Sounds of the conflict—the screaming of a known community
    member along with the aggressive calls of strangers
Direction from which the sounds came, and approximate location
Direction from which the victim arrives
Fearful, apprehensive behavior of the victim (wide grin, screaming,
    embracing upon reunion)
Nervous glances from time to time toward the scene of the
    encounter
The nature of his wounds and where on the body they occur
Pant-hoots or drumming of strangers continuing, on and off, in the
    direction from which the victim came
Possibly the odor of strangers still clinging to him

Because chimpanzees are capable of combining pieces of information
separated in space and time in an orderly and meaningful way, it
seems likely that community members can to some extent find out
about events of this sort that occur within earshot. On the other hand,
if after a few days our victim meets an individual who at the time of
the attack was far away and out of earshot, it seems most unlikely
that there would be any exchange of information concerning the incident—except (from the healing wounds) that there had been a fight
of some sort.

One female, Passion, and her daughter, Pom, began to kill and eat
the babies of other community females, as will be described in detail
in Chapter 12. They were only seen to do this when the mothers of
the victims were with dependent young only. There is no firm evidence
that any individuals other than the mothers who were attacked ever
knew about the bizarre behavior. On the other hand, the mothers
were, it seems, able to communicate at least their fear and/or dislike
of Passion to some other community members. When Gilka (who had
already lost one infant to the killer) screamed as Passion approached
to stare at her newborn, two adult males, one after the other, rushed
over and attacked Passion. In fact, Passion's behavior led to one
extremely interesting communication sequence. This involved Miff, a
female who had saved her baby from Passion but who remained extremely fearful of and hostile toward Passion for many months after
the attack. When the infant was still very tiny, Miff encountered
Passion and at once fled, screaming, until she met two adult males,
some hundred meters away. Miff then turned back toward Passion,
her screams became aggressive waa-barks, and glancing over her
shoulder at the males, she started back the way she had come. The
males, responding to her solicitation for help, followed. When they
reached Passion, the males displayed—and Passion fled. Miff quite
obviously had already gained some information regarding Passion's
abnormal behavior, probably as the result of a previous, unobserved
encounter.

Nonverbal "Dialects"

Chimpanzees raised in social isolation during their first two years of
life show a number of characteristic communication signals (postures,
gestures, and calls of the sort described), but these appeared in

Goblin and Melissa, socially grooming, each hold an overhead branch with one hand.

Two adult chimpanzees at Mahale demonstrate the hand clasp while grooming each other. This pattern has never been observed at Gombe. (C. Tutin and W. C. McGrew)

incomplete sequences and inappropriate contexts when the isolates were introduced to normal chimpanzee youngsters (Menzel, 1964). This means that certain behavioral patterns are inherited, and if the youngster grows up in a normal social environment, he will learn, as a result of trial and error, reward and punishment, how and when to use them. In other words, as a result of his experiences with others, his repertoire of species-specific gestures, postures, and calls will gradually be patterned and organized into functional sequences, which will appear in contexts traditional for his group.

Because of the plasticity of behavior shown by the apes, which is associated with the complexity of the brain and the significant role played by experience during the long period of maturation, individual chimpanzees tend to develop a variety of idiosyncratic behaviors. Thus we would expect that chimpanzees in different geographic localities might show variation at the group level in communication patterns—the equivalent of local dialects. That this is so is gradually emerging from comparison of data from different field and captive studies.

Three examples will illustrate. At Gombe, when two chimpanzees are socially grooming, each often holds an overhead branch with one hand while grooming his or her companion with the other. At Mahale, 160 kilometers to the south, and in the Kibale forest of Uganda, a pair of grooming chimpanzees often sit, each clasping the hand of the other at a point above their heads (McGrew and Tutin, 1978; Ghiglieri, 1984). At Gombe, a mother reaches back with a characteristic *climb-aboard* gesture, which signals her infant to get onto her back. In a film of chimpanzees in West Africa made by Adriaan Kortlandt, I saw an adult male make precisely the same gesture to another adult male in a situation of extreme social excitement (the group was faced with a stuffed leopard); in response the second male mounted the first. The end result was a typical dorsoventral embrace—common at Gombe in similar contexts—but the West African male used the climb-aboard signal in a context that has never been seen to elicit the gesture at Gombe. Chimpanzees at Mahale and at Bossou, Guinea, have incorporated the distinctive *leaf clipping* into their courtship displays, in which a male picks a large green leaf and rapidly bites off small pieces with exaggerated jaw movements, until only the midrib is left (Nishida, 1980). This element has never been seen at Gombe.

If we include differences observed in captive groups, our list of culturally acquired dialects is much lengthened. For example, the bipedal swagger, seen almost entirely in aggressive contexts at Gombe, was frequently used as an invitation to play among adolescent and adult chimpanzees of two captive groups, at the Stanford Outdoor Primate Facility (now defunct) and at Lion Country Safaris in Florida. In almost all chimpanzee groups for which data are available, submissive presenting is one of the most common signals in females and youngsters; in the Florida group it was seen only three times during an intensive two-month study (Gale and Cool, 1971).

As more information regarding social behavior becomes available from different field and captive studies, if filmed sequences are available for comparative purposes, and if investigators collaborate and visit one another's study groups, many differences of this sort undoubtedly will be unveiled.

The dialects presumably arise when idiosyncratic behavior developed by one individual is imitated by others. At Gombe a juvenile, Fifi, suddenly showed *wrist-shaking*, a behavior that had appeared spontaneously in a captive individual (Gardner and Gardner, 1969). Fifi used it when threatening an older female. A younger individual, Gilka, was with Fifi at the time. The following week not only was Fifi seen to wrist-shake again (in a similar context), but Gilka too used the gesture. Subsequently Gilka wrist-shook frequently, in a variety of contexts. Fifi also continued to use the new pattern, but infrequently and usually only as a threat. During the following year the gesture was used by both youngsters less and less often and finally vanished from their repertoire. Another time an infant began to inspect female bottoms by poking them with a little twig, then sniffing the end—instead of using his finger in the usual way. Within two weeks another infant was doing the same thing. This pattern, like wrist-shaking, was gradually extinguished. But the two examples show how a new communicative signal, or an existing signal used in a new way, can spread from an "inventor" to his or her companions.

If a new communication signal is to be incorporated into the repertoire of the group as a whole, more is required than that other individuals imitate the pattern—if the signal is to convey a message, then that message must be *understood* by both sender and recipient. When the new pattern bears close resemblance to other signals used in the same context, this will not be difficult. Fifi's wrist-shaking, being similar to the flapping and other kinds of arm-waving that serve as threats, was probably correctly interpreted by the recipients. The climb-aboard signal used by the male in West Africa is almost identical to the command given by the mother and learned during infancy. The leaf clipping is an *addition* to the typical courtship pattern; if a female, under experimental conditions, responded to that element alone, it would undoubtedly be through long association of leaf with penis. Of particular interest was the novel courtship display of an adolescent captive male chimpanzee of the Menzel group (Tutin and McGrew, 1973). Typically, courtship comprises many elements of aggressive behavior, and when young Shadow began to seek sexual intercourse with the older females using these aggressive poses, they turned and attacked him. So Shadow invented an extraordinary idiosyncratic display in the sexual context: he stood upright, manually everted his lower lip, hooked it over his chin in the *lip flip*,* and gazed toward the female of his choice. The females quickly learned to respond appropriately to this strange performance. How? An erect penis, of itself, is not sufficient indication of a male's sexual interest, for it occurs frequently in contexts other than sexual arousal. Presumably it was the combination of erect penis and direction of gaze. Whatever it was, it worked; and it serves as a perfect example of the inventiveness, plasticity of behavior, social awareness, and ability to communicate that makes the chimpanzee the most fascinating of all the primates to study—except, of course, *Homo sapiens* himself.

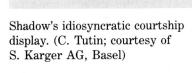

Shadow's idiosyncratic courtship display. (C. Tutin; courtesy of S. Karger AG, Basel)

* This lip flip is different from the Gombe lip flip illustrated in Figure 6.1, in which the upper lip is turned inside out over the nose. At Gombe this often occurs in the context of self-grooming and almost certainly communicates nothing. Thus Shadow "invented" the facial expression, as well as the courtship display itself.

7. A Unique Society

Fifi and her family.

July 1977   Five days from the social calendar of adult female Fifi

Day 1   Early in the morning Fifi and her two offspring, Freud and Frodo, are still lying quietly in their nests. The three chimpanzees who have slept close by, Satan, Sparrow, and four-year-old Sandi, rise and leave. For the entire day Fifi and her family move about on their own. Fifi grooms and plays with her sons, and Freud interacts frequently with his infant brother—carrying, playing with, and grooming him.

Day 2   During the first part of the morning the Fifis are alone. At 1100 hours they arrive in camp, where they meet four adult females and their families. Freud is preoccupied with the sexual swelling of late-adolescent Pom. The Fifis, who have received no bananas, travel and feed for just over an hour with two of these families, the Winkles and the Melissas. From 1400 on, Fifi and her sons are alone. Altogether they are with others for only three of the twelve daylight hours.

Day 3   The Fifis spend most of the morning alone, apart from brief encounters with two different families and the young male Goblin. In the afternoon, however, Fifi joins a large party. A colobus monkey is hunted and caught, and Fifi begs successfully from Evered, sharing her spoils with Freud. After she has finished, Fifi waits patiently while Freud plays wildly with other infants. The family then spend a peaceful afternoon on their own. But that evening they nest among the other members of the same big gathering. They have spent nine and a half hours by themselves.

Day 4   Once again Fifi and her family remain in their nests while most of the other chimpanzees set off in the morning. After that, apart from feeding on galls with Sparrow and Sandi for two and a quarter hours, the Fifis are alone all day.

Day 5   Another day spent almost entirely on their own. There is one brief meeting (about thirty-five minutes) with Passion and her family, when Freud is again fascinated by Pom's sexual swelling. The Fifis spend yet another night on their own, Frodo sharing his mother's nest, Freud in his own small nest close by.

The chimpanzee community is a *fusion-fission* society, the term first applied by Kummer (1968) to hamadryas baboons. It is an unusual type of organization. Most populations of nonhuman primates are made up of a number of relatively closed social groups, which may be one-male groups, multimale groups, or occasionally monogamous pairs. The fusion-fission arrangement, in which portions of the parent group may on a *regular* basis separate from and then rejoin the rest, is highly formalized in the hamadryas baboon and the gelada (Crook, 1966; Kummer, 1968; Dunbar and Dunbar, 1975). These species are organized into multilevel societies: the hamadryas troop, for example, which travels as a whole for part of the time, in fact comprises several rigidly closed one-male units that separate during daily foraging, particularly when food is scarce, but recombine later in the day. A number of troops usually congregate at night, in the sleeping cliffs, to form huge bands. In spider monkeys (Klein and Klein, 1973) and long-tailed macaques *Macaca fascicularis* (Mackinnon and Mackinnon, 1978) the organization seems to be somewhat less rigid, in that groups of various sizes may congregate for feeding. The baboons at Gombe sometimes show considerable fragmentation during daily foraging, and again the distribution of individuals among subgroups is quite flexible.

Fusion and fission in chimpanzee society are carried to the limits of flexibility; individuals of either sex have almost complete freedom to come and go as they wish. The membership of temporary parties is constantly changing. Adults and adolescents can and do forage, travel, and sleep quite on their own, sometimes for days at a time. This unique organization means, for one thing, that the day-to-day social experiences of a chimpanzee are far more variable than those of almost any other primate. For example, although there may be fragmentation of a baboon troop at Gombe, each member of that troop is virtually certain to encounter the same individuals day after day, and to visit similar areas of the home range on a daily basis, with gradual shifts as seasonal food supplies change. Even the place and time of encounters with neighboring troops are relatively predictable. Members of kinship groups remain together even if the troop fragments, so that help and support are at hand during most social conflicts.

The situation is completely different for a chimpanzee, who rarely sees *all* the members of his community on the same day and probably *never* sees them two days in succession. He may travel one day in a large, noisy, excitable gathering; the next day, completely by himself. He may feed peacefully with a small, compatible party in the morning, then join fifteen other chimpanzees, after a successful hunt, in the late afternoon. He may be one of six males competing for the same female one week, and associate with one female, far from any other males, the next. He may spend one day at the very center of his core area and move out to a far-flung boundary on a patrol the next.

In chimpanzee society some individuals meet on a fairly regular basis, others but seldom. There is always uncertainty: it is never possible for a chimpanzee, upon wakening in the morning, to be quite sure whom he will encounter during the day. Each male is likely to encounter all the other adult males of his community several times a week (unless one of them is away on a consortship) and, in addition, many of the females. A female, on the other hand, although she may be very social and encounter many individuals during a week when she is in estrus, becomes more solitary when she is no longer cycling. Some anestrous females, particularly those whose core areas are in the central part of the community range, encounter many individuals (even if briefly) on a regular basis; more peripheral females encounter far fewer individuals per week, with some females seeing one another only a few times during an entire year. These major differences in frequency of association obviously affect the relationships among community members; at the same time, as we shall see, the nature of the relationships among the different individuals can affect the patterns of association. The society is extraordinarily complex: without a high level of intelligence, chimpanzees would not be able to cope with the uncertainties and tensions so often engendered by the constantly changing social environment.

In order to present a realistic picture of day-to-day variation in social experience, I include in Appendix B exact translations from the Kiswahili* of two full-day follows on the same individual, the adult male Jomeo. The accompanying travel and association charts make clear the major differences between the two days. On 9 May 1981 Jomeo encountered one adult female and a juvenile: he spent all but forty-five minutes of the day, and the nights of 8 and 9 May, alone. By contrast, on 14 October 1982, Jomeo at some time was within 100 meters of all the Kasakela independent males and all but four of the adult females; he was not alone at all. On 9 May one social interaction was recorded for Jomeo; on 14 October he was involved in twelve agonistic interactions and seven grooming sessions with a total of eight individuals. On 9 May only two chimpanzee vocalizations (by the lone female he met) was heard during the entire day; on 14 October Jomeo himself vocalized at least twenty-two times. The total distance covered on 9 May was about 1.4 kilometers; on 14 October the males traveled some 7.2 kilometers, moving out to the northern periphery of their range. I myself followed three different mother-infant pairs in the course of 14 October. Word Picture 7.1 summarizes these observations.

The Nature of a Society

The form of an animal society has evolved over time as a result of the interplay between environmental pressures and the adaptive strategies developed by members of the species to cope with those pressures. Each individual coming into the world with its own inherited genetic package will, through its contact with the environment and its behavioral interactions with conspecifics, contribute to the special structure of its society. At the same time, its own inborn tendencies will be molded and modified by the behavior and relationships of other group members—by the social structure in which it lives.

We have, then, a double challenge: we need to understand the behavioral interactions and relationships of the members of an animal society if we are to comprehend the social structure of that species;

* Kiswahili is the correct name for the language common to the East African countries, usually referred to as "Swahili."

and we need to understand the social structure to appreciate the constraints it imposes on the expression of individual behavior. Only when a given species has been studied in the natural habitat over a period of years is it possible to come to grips with so complex an issue.

Hinde (1976, 1979) has provided a conceptual framework within which information regarding social structure in a variety of primate species can be analyzed, evaluated, and compared. He suggests that we describe the observed *surface structure* of a given society in terms of the kind, quality, and patterning of the *relationships* among the various members. These relationships in turn are described by reference to the kind, quality, and patterning of the behavioral *interactions* of the individuals concerned. At each of these levels our task is to search for fundamental principles—rules that underlie the interactions and relationships upon which the structure is based.

The surface structure is likely to be stable over time, maintaining its organization despite the deaths of constituent members. Youngsters growing up in the social environment will develop patterns of interaction and relationships similar to those of their elders. Thus the relationships of individuals within the society are, as Hinde (1976, p. 8) puts it, influenced by membership in the group, which "has properties that are more than the sum of the constituent relationships." Moreover, although environmental constraints or demographic changes may lead to temporary changes in the patterning of relationships, a society may be able, through social tradition, to revert to previous patterns when conditions permit (see Kummer, 1971). At the same time, when we are dealing with an animal capable of performing innovative acts and able to acquire behavior through observing and imitating his companions, we may expect to find variation from one group to another in the social traditions themselves. Even so, we must try to make generalizations that *will* apply equally to other groups of the same species—to find patterns that are independent of aberrant or idiosyncratic behavior and that will ultimately enable us to come to grips with the *structure* of that society.

Patterns of Association
The most deep-seated principles underlying chimpanzee community structure are those concerned with sex differences in sociability and in the choice of companions. Males are more gregarious than females and prefer each other's company, except when females are in estrus. Females are less sociable and spend most of their time with their own offspring—except when cycling, at which time they become very sociable. This basic pattern is molded into the characteristic ever-changing fusion-fission society by (a) environmental and demographic factors, such as the changing nature of the food supply or the number of males in the community at a given time; (b) monthly and yearly variations in the number of cycling females; and (c) social factors, such as dominance reversals or intrusion of neighbors into the home territory. Each of these factors will affect at any given time the size and composition of parties, the frequency and duration of association of the various members of the community, and the probability of who chooses to associate with whom.

A Chimpanzee Gathering, 14 October 1982

It was a clear, sunny day with a light breeze. This was the end of the dry season and the vegetation (except in the valleys) was fairly open, allowing for good observation. I began my follow of Melissa and her adult daughter, Gremlin, with their two infants, Gimble and Getty, in camp. At 0905 Melissa led her family (Gimble walking, Getty riding) toward the loud calls that for some time had been heard to the east, farther up Kakombe Valley. At 0930 the Melissa family joined two other females, Fifi and Little Bee, and their families. Gremlin displayed, running along the ground, stamping her feet, hair bristling, then pant-grunted submissively as she approached and briefly groomed Fifi. Gimble also greeted Fifi, then began a boisterous play session with the youngsters Frodo and Tubi.

At 0957 Melissa and her party joined forces with the males and the other females. There was an eruption of pant-hoots, barks, and pant-grunts before things finally calmed down.

At 1021 the aggressive Satan charged past Melissa (who rushed out of the way, screaming) in a magnificent display, one of a series that had been directed toward Melissa's mature son, Goblin. Followed by Jomeo, the two field assistants, and (presently) Goblin, Satan vanished down the slope through the tall grass.

Melissa stayed with Gremlin and adult female Miff, who was in estrus (and who became my second target). Athena, also partially swollen, was nearby. Seven adolescent males stayed back with these females—with almost continuous erections. From time to time one of the young males copulated with one of the females; each time one or more of the infants interfered, running up and touching or pushing at the male concerned. Levels of arousal were high and during the wild play of the youngsters many squabbles broke out. Athena's juvenile daughter, leaving a play session when it became rough, made a small nest on the ground, in which she rolled about playfully. She began to tickle herself in the neck with her fingers, laughing loudly.

Infant Mo, age three and a half, was being weaned. Each time she tried to suckle, Miff rejected her (at least initially). Mo whimpered and then screamed loudly and insistently for minutes on end, adding to the tension in the group.

There were occasional outbursts of calling from Jomeo's direction; each time, the females and youngsters gazed intently toward the big males, often responding with pant-hoots or waa-barks, then moving on toward the rest of the gathering, feeding on the way. Pant-hoots behind us indicated that another group had also temporarily separated itself from the males.

At 1230 Jomeo moved slightly apart from the others and fed quietly; soon he rejoined the gathering as all moved on, calling and displaying. During a rest period Miff lay stretched out on a rock while other adults groomed. It was hot and the wind had dropped. Nearby some of the youngsters played with juvenile and infant baboons.

Soon the males moved again, following Satan. Miff and the females followed along in the rear. We all climbed high up toward the rift escarpment. The terrain became more open, rough, and rocky; the grass was shorter, sparser, and finally nonexistent. Ahead of us Jomeo and the males climbed into tall trees. Some of the females joined them; others, including Miff, moved higher still and began feeding on delicious *muhandehande* fruits (*Uapaca nitida*). After a while the males (and the field assistants) joined us

Adult male Jomeo, target of the follow of 14 October 1982, is in the foreground grooming Dominie, adolescent daughter of Dove, who sits to the right. In the background are two of the Tanzanian field assistants, E. Mpongo and G. Paulo. (H. Kummer)

Part of the gathering of 14 October 1982, feeding on young leaf buds near the rift escarpment. (H. Kummer)

and soon the whole gathering, widely spread out, was feeding on these fruits, moving leisurely from tree to tree.

At 1318 I changed targets again and began following the two orphans, Skosha and Kristal. Pant-grunting as Satan climbed into their tree, they moved to greet him. Up on the skyline, 50 meters above us, I saw Jomeo and the young male Atlas heading north, followed by the field assistants. Satan climbed down and wandered after them, and one by one the others followed. An adolescent male, Beethoven, out of sight of his superiors, performed a spectacular, hair-bristling, charging display (directed at no one, as though "practicing"), then moved on after the others.

By 1415 the only chimpanzees visible, other than my targets, were Pom and her young siblings (who had recently lost *their* mother, too). Calls from the large gathering grew fainter and fainter as the members pushed northward; there was silence as they moved down into Linda Valley. Pax, age four and a half, tried to get Kristal to play, but she refused. Gradually the two families drifted apart. I followed Skosha and Kristal back into Kakombe Valley, leaving Pom and her family the only chimpanzees on the upper Kasakela rift.

Later in the day (we learn from the reports of the field assistants) other females dropped out as the main party headed farther and farther north. Only two females, both of whom were in estrus and both of whom had core areas to the north of the community range, stayed with the males after 1615.

One of the things that was so striking that day (as on other similar days) was the sheer exuberance of the chimpanzees. If they had been humans, we would not hesitate to say that they were having tremendous fun. There was a good deal of noise and much play, mating, and grooming. There were wild displays with a lot of screaming, but no serious fighting. The females, who tended to stay back feeding when the males moved on, were clearly attracted by pant-hoots and other calls. Typically they called too, stopped feeding, and moved on to rejoin the action. It was a carnival atmosphere.

The sociograms that constitute Figure 7.1 show, for each member of the community, the frequency with which he or she was observed in association with each other individual during one year, 1981. It is a somewhat rough method of looking at association, because the target individuals (on whose associations the measures are based) were followed for unequal lengths of time, but it does permit inclusion of those chimpanzees who were rarely or never followed. Moreover, for all its shortcomings, the information presented in the sociograms compares favorably with data for the same year derived from more precise target information on the same individuals, which we shall consider later. The raw data from which these sociograms have been compiled are given in Appendix D.*

The scores have been divided into six different orders of association: first-order associations link those individuals seen together most frequently, and sixth-order those seen together very rarely.

First-order associations, shown in Figure 7.1a, are the closest of all—those between mothers and their dependent young. The association links between youngsters up to seven or eight years of age and other community members are virtually identical to those of their mothers and are not treated separately.

Second-order associations are shown in the same sociogram. Three link mothers and their adult or adolescent daughters, and one links an adult female, Pallas, with her adopted daughter, Skosha.

Third-order associations are also shown in Figure 7.1a. Two link mothers with their adolescent sons, three are between pairs of adult males, one is between a pair of adult females (Patti and Gigi). The other four link two of the adolescent males, Jageli and Beethoven, to the adult males. It can be seen that Beethoven, who is an orphan, associated very closely with three different adult males during the year.

Fourth-order associations, shown in Figure 7.1b, link other males (including adolescents), adolescent Jageli to his mother, and many of the males to the sterile, cycling female Gigi. (Gigi has no offspring and has always associated closely with the adult males.) The other cycling females were either unhabituated and seldom seen (Sprout and Kidevu), or adolescent and not highly popular with the males. The adult female Athena is linked to her son Atlas and to the adolescent Beethoven.

Fifth-order associations (Figure 7.1c) link many more community members. Here we see additional links between adult and adolescent males, as well as the first links between adult males and nonrelated adult females, and between nonrelated females. Less sociable females (still only linked to their immediate families) were Passion, Nope, Pallas, Winkle, and Joanne (a rarely seen peripheral female). Miff and Athena were more sociable; in Miff's case, partly because her daughter Moeza was in estrus during the year.

* In 1973–1975 a start was made at Stanford University on computerizing much of the Gombe longitudinal data, particularly the association patterns of target individuals. Unfortunately this program came to an end when the Gombe-Stanford affiliation was broken off after the kidnapping of Stanford students. Since then I have had neither the resources nor the ability to reinstitute such an effort. All data presented here, therefore, have been laboriously hand analyzed.

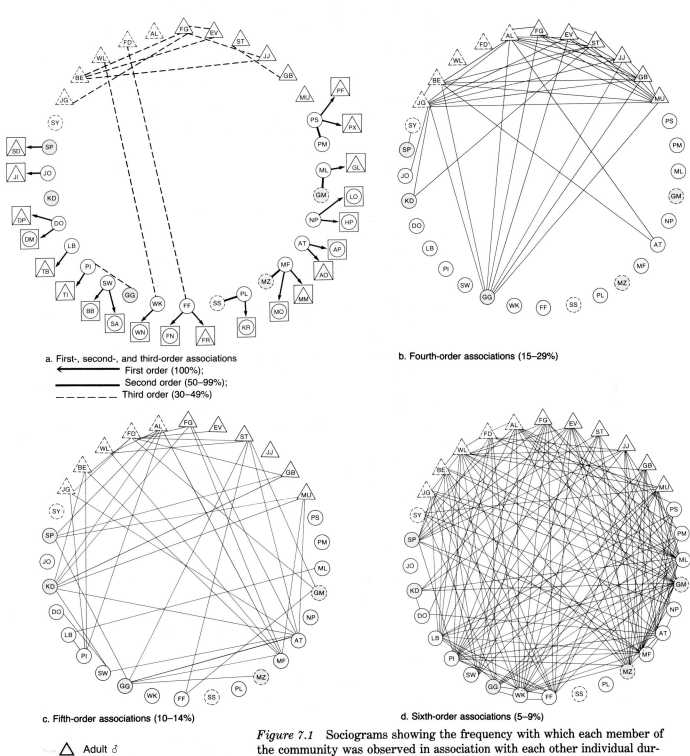

a. First-, second-, and third-order associations

First order (100%);

Second order (50–99%);

––––––– Third order (30–49%)

b. Fourth-order associations (15–29%)

c. Fifth-order associations (10–14%)

d. Sixth-order associations (5–9%)

△ Adult ♂

△ Adolescent ♂

◯ Adult ♀

◯ Adolescent ♀

◯ Adult ♀ in estrus

◯ Adolescent ♀ in estrus

▣ ♂ offspring

◎ ♀ offspring

Figure 7.1 Sociograms showing the frequency with which each member of the community was observed in association with each other individual during 1981. Data are from all follows on all individuals (*out-of-camp data only*). A matrix (Appendix D) was compiled by totaling for each individual the number of quarter-hours (or portions thereof) that he or she was seen with each other community member. From this matrix a crude measure of association was calculated for each dyad (A and B) by the following formula:

$$\frac{A + B \text{ together}}{(A \text{ without } B) + (B \text{ without } A) + (A \text{ and } B \text{ together})} \times 100.$$

The resulting measures of association have been divided into six different orders of association ranging from 5 to 100 percent. Dyads seen together less than 5 percent of the time are not included.

Sixth-order associations (Figure 7.1*d*) bring in almost all remaining individuals. Only two adult females were outside the association network at the 5-percent level, Pallas and the peripheral Joanne. Joanne may well have associated more often with the others than is indicated in this sociogram; she is very shy and usually leaves a group when human observers appear. Pallas, however, is a central female, a regular camp visitor; the diagram gives a true picture of her extremely asocial disposition during that period of her life.

These sociograms, for all their crudity, illustrate overall patterns—the frequent association between mothers and elder offspring, especially daughters, and between males; the higher levels of association of males with females who are cycling than with those who are anestrous; and the remarkably low frequency of association with other community members shown by some anestrous females. All these patterns will be detailed more precisely in subsequent sections.*

Composition and Size of Parties

Chimpanzees typically travel, feed, and sleep in parties of five or less (excluding infants and juveniles). The mean party size for adult males was between four and five both before and after the community split and before and after the days of intensive banana feeding (Goodall, 1968b; Bygott, 1974; Wrangham, 1975; Bauer, 1976). For five anestrous females, median party size was 1.6 (excluding dependent offspring) during 1971–72 (Bauer, 1976). Females in estrus move in parties ranging in size from two during a consortship to very large gatherings.

There are a number of different types of party. Halperin (1979) has discussed party types under different headings, but those that follow seem to me to be the most logical categories:

All-male party—two or more adult and/or adolescent males

Family unit—a mother and her dependents, with or without older offspring

Nursery unit—two or more family units, sometimes accompanied by nonrelated childless females

Mixed party—one or more adult or adolescent males with one or more adult or adolescent females, with or without dependents

Sexual party—mixed party in which one or more of the females is in estrus

Consortship—exclusive relationship of one adult male with one adult female (estrous or anestrous) when they travel on their own, with or without her dependents or older offspring

Gathering—group comprising at least half of the members of the community, including at least half of the mature males (actually, of course, nothing more than an extra-large, mixed, usually sexual, party)

Lone individual—completely alone.†

* Much of the artwork in this volume has been designed to illustrate differences in behavior: between males and females; between wet season and dry season; between one year and another. In these figures I have tried to provide a visual impression of the *patterns* of these differences rather than emphasize their detail.

† A good deal of the published work on chimpanzees is concerned with ecological topics, such as feeding and ranging behavior, and party size has traditionally been limited to independent adults. However, Wrangham and Smuts (1980) scored as "alone" mothers who were associating with their *adult* daughters and vice versa. In this volume, the word *alone* means alone.

A small mixed party responds with pant-hoots to similar calls from across the valley. *Left to right*: Miff, Mo, Humphrey, Michaelmas (*in background*), Jomeo, and Patti with her infant.

With this overall pattern as background, we can turn now to a consideration of the factors that influence the size and composition of parties from day to day and month to month. It must be remembered that this exercise will result in a measure of oversimplification, because in practice environmental and social factors are interwoven with the unique personalities of the chimpanzees themselves. Four key variables are the following:

Nature of food supply. The nature and availability of seasonal foods exert a continuing and highly significant influence on association patterns. When food is scarce and found only in patches, the chimpanzees are apt to move about in small parties or alone; when there is much food, whether spread out or concentrated (in space and/or time), many individuals tend to feed together (Wrangham, 1977). At the height of the banana feeding, party size increased dramatically, but dropped back to previous levels when the system was changed (Wrangham, 1974). Seasonal variation in food supply has a marked influence on the fragmentation of hamadryas baboons, gelada, spider monkeys, and baboons at Gombe (Crook, 1966; Eisenburg and Kuehn, 1966; Kummer, 1968; Klein and Klein, 1973; A. Sindimwo, field data).

Demographic. When there are many infants of similar age, their mothers often move about together and nursery parties tend to be larger than at other times. The greater the number of adult males in the community, the larger the parties that move out to patrol the boundaries of the home range.

Danger. As yet there is no clear-cut evidence that predator pressure leads to an increase in party size in chimpanzees, but it seems likely. Kummer et al. (1981) found that hamadryas baboons at the Saudi Arabia study site moved in smaller aggregations than did those in Ethiopia; they suggest that this was the result of much lower predator pressure in Saudi (almost none in some places). The frequent splitting of baboon troops at Gombe into subgroups (sometimes of only three or four individuals) may also be a response to the lack of danger from predators.

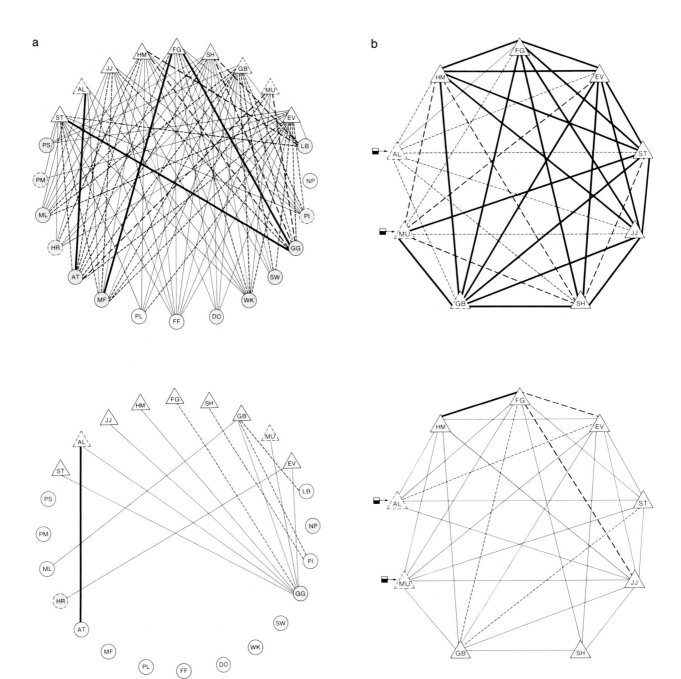

c

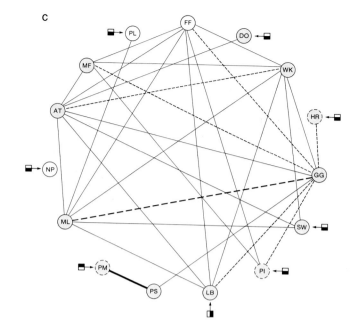

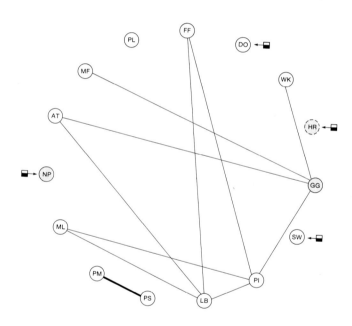

Figure 7.2 Sociograms showing levels of association between pairs of *a*, males and females, *b*, males, and *c*, females, in two years, 1977 (*above*) and 1979 (*below*). These data have been extracted from the target information on all individuals (over thirteen years of age) who were followed for seventy hours or more during each of the years. The indexes indicate the proportion of time that A spent with B during all follows on both. Individuals for whom it was not possible to compile an association index (because they were not followed for enough hours) appear on these sociograms if they scored higher than 5 percent in the index of a more frequently followed chimpanzee. Association scores below 5 percent are not included.

———————	50–99% association
- - - - - - -	40–49% association
— — — — —	30–39% association
———————	5–29% association

△	Adult ♂
△ (dashed)	Adolescent ♂
○	Adult ♀
○	Adult ♀ in estrus
○ (dashed)	Adolescent ♀ in estrus
▢	Less than 70 target hours during the year

Availability of females in estrus. The presence or absence of cycling females is without doubt the single most significant factor in the overall patterning of the chimpanzee community from year to year. A sexually popular female in estrus, unless she has been taken on a consortship, will be surrounded by most or all of the community males and many of the females (some as a result of the sexual motivation of their juvenile or early-adolescent sons). The alpha male Figan was followed for fifty consecutive days in 1974; his party size was significantly greater when a female in estrus was present (Riss and Busse, 1977). When several females are in estrus simultaneously, party size is often even larger. An example is a gathering in late 1976 that persisted for eight days. Excluding infants and juveniles, there was an average of 11.3 chimpanzees per day (range 9 to 14). Five swollen females were part of the gathering for varying lengths of time; anestrous females joined and then wandered off again, some of them several times. The nucleus was the sexually popular Pallas, at the height of estrus. When she began to detumesce, the seven males, who had remained with her continuously, gradually dispersed.

Year-to-Year Changes

Figure 7.2 illustrates association patterns among the Kasakela chimpanzees during two different years, 1977 and 1979. The two years were selected because they represent extremes for one highly important variable, the number of cycling females. In 1977 fourteen females were cycling during all or part of the year, compared with six in 1979. Some of these females are not shown in the figure; two in 1977 and three in 1979 were peripheral or visiting and seldom seen, so that their association scores with target individuals were very low. Nevertheless, they had some influence on the association patterns of the community. If all these females are taken into account, the mean number of females in estrus per month was 6.3 in 1977 and only 1.9 in 1979.

As the sociograms show, the presence of many sexually attractive females led to dramatically increased levels of association among males and females. Moreover, males also spent more time with each other, and among females there were higher levels of association too. It was the frequency of gatherings throughout the year that resulted in the overall increase of sociability. Anestrous females, as we have seen, are often attracted to the "excitement" of a gathering; even if they are not, their youngsters may be, and mothers are sometimes influenced by the desires of their offspring. All but the least sociable members of the community were involved in the riotous sexual jamborees of 1977.

The difference in overall levels of association among community members in the two selected years is dramatic, but we should not forget that the underlying principles of social organization remain unchanged. Males may be more or less gregarious at different times, but they are always more gregarious than females. And while females may associate with other females more in some years than in others, they always do so far, far less than males associate with males. There may be more gatherings in some years than others (depending on the availability of sexual partners and the nature of the food supply), but there will always be some (because chimpanzees are inherently attracted by social excitement) and they will always be attended more often and for longer by males than by females. This difference in sociability between the sexes has far-reaching effects on many aspects of chimpanzee behavior and we should examine it more closely.

Sociability

Males versus Females

Adult males are more sociable than adult anestrous females in all the following measures: the frequency with which, during a given period of time, an individual meets with nonfamily members; the number of such others with whom he or she associates; and the duration of the associations formed (Bauer, 1976; Halperin, 1979; Nishida, 1979; Wrangham and Smuts, 1980). Wrangham and Smuts analyzed Gombe data from 1972 to mid-1975 and showed that males spent more *party days* (whole days in association with others) than did anestrous females, who in turn spent more days *with family only* than males did *alone*. On *mixed days* (when part of the day was spent with others) males spent less time alone and joined parties for longer than anestrous females.

Figure 7.3 shows the "sociability webs" of (*a*) four adult females and (*b*) four adult males. All individuals over eight years old are included in these webs, infants and juveniles only when their association index with the target was greater than 50 percent.

Three major differences in the association patterns of adult males and adult females are immediately apparent: (1) the center rings for males are empty, whereas for some females such as Passion they may be occupied by as many as four other individuals; (2) the male webs include many more individuals in the 10–39 percent rings than those of the females; and (3) the female webs show a much greater year-to-year variation than those of the males. The pattern of social life for the two sexes is obviously very different. When a male starts to leave his mother at the age of about eight years, he will for the rest of his life spend many hours by himself. This is evident from the "alone" scores in the outer sections of the male webs. A female, when first she starts to travel independently of her mother, will also spend time by herself. But from the time when, two or three years later, she gives birth to her first infant, she is unlikely (unless that infant dies) to spend *any* time completely alone for the rest of her life. At Gombe the bonding between mothers and their adolescent and adult daughters is very strong indeed, as can be seen from a glance at Passion's web. The close-knit family circle at the heart of a female's web of relationships provides her, as she gets older, with companionship, grooming, play, and—most important—support during agonistic interactions with other community members.

The major change in the association patterns of a female during years when she is cycling is also revealed by these webs. This is particularly striking for the more asocial females such as Passion and Pallas. Interestingly, a mother may become more sociable not only when *she* is cycling, but also during her daughter's cycles, as we see in Miff's web for 1981.

Figure 7.3 (overleaf)
Sociability webs of *a*, four adult females, and *b*, four adult males, compiled from the association index of each individual, expressed as a percentage of the total time he or she was followed as a target that year. Each segment of a web represents one year, and the number of hours the individual was followed as target is indicated. Each symbol stands for one chimpanzee of a particular age-sex class (the symbol changes as that individual progresses to the next age category); close relatives are identified. The concentric rings represent different degrees of association, those near the center being the higher. Association scores below 5 percent have been omitted.

For the females, the shaded segments represent years during which the target female was cycling; the number of her estrous swellings during each of those years is given. The outer ring shows the percentage of target hours that she was with members of her immediate family only. The innermost ring (100-percent association) is enlarged below the web for clarity.

In the case of the males, the outer ring shows the percentage of target hours per year that the male was alone.

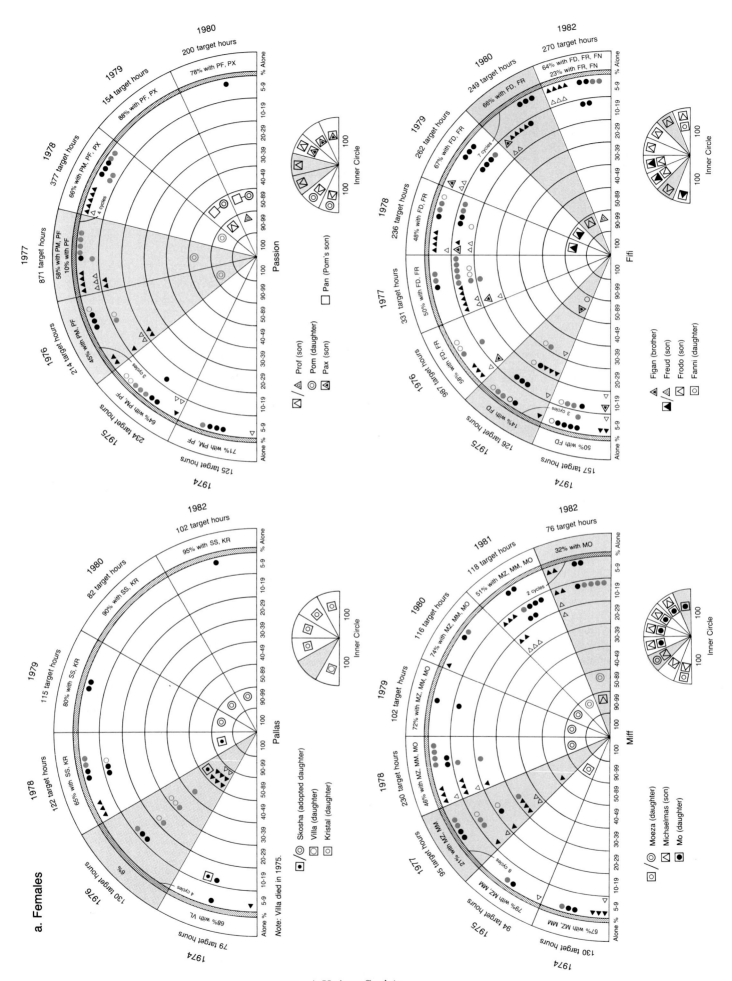

a. Females

Note: Villa died in 1975.

Pallas

Skosha (adopted daughter) ◉
Villa (daughter) ◻
Kristal (daughter) ◻

Inner Circle

Passion

Prof (son) ◁ / ◁
Pom (daughter) ◉
Pax (son) ◁

Pan (Pom's son) ◻

Inner Circle

Miff

Moeza (daughter) ◉ / ◉
Michaelmas (son) ◁
Mo (daughter) ●

Inner Circle

Fifi

Figan (brother) ◁ / ◁
Freud (son) ◀ / ◁
Frodo (son) ◁
Fanni (daughter) ◉

Inner Circle

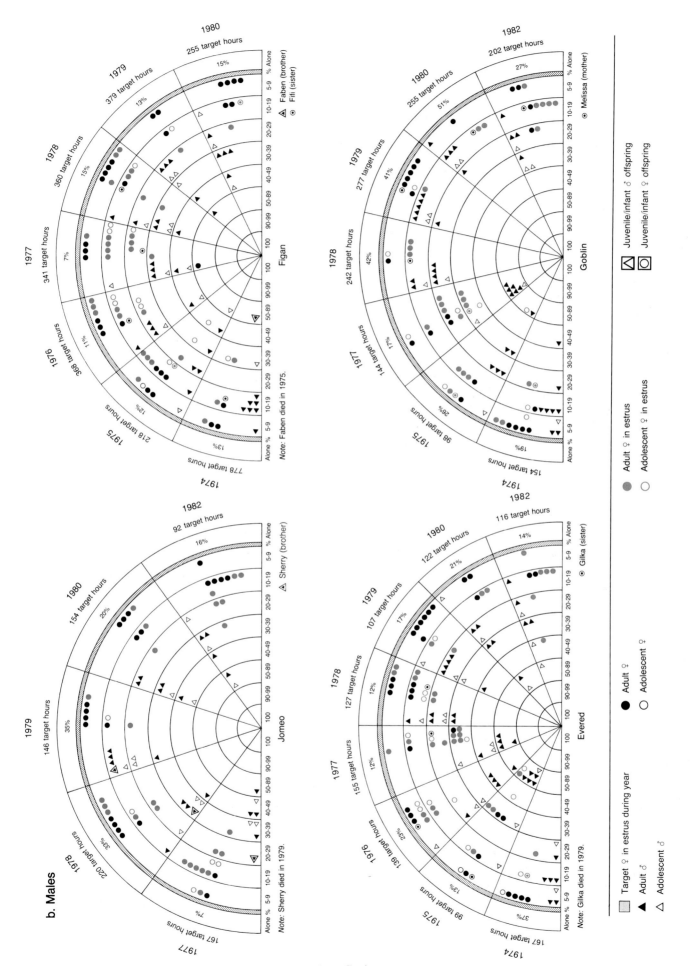

b. Males

Note: Faben died in 1975.

Note: Sherry died in 1979.

Note: Gilka died in 1979.

The adult males of a community, although they compete vigorously for social dominance, are for the most part relaxed in one another's company and spend a good deal of time grooming, feeding, and traveling together. In this all-male party Hugh is groomed by his presumed sibling Charlie. The older Hugo is in the background. (B. Gray)

The webs of the adult males show a much more uniform pattern from year to year, the major changes being the extent to which they associate with females. The increase in male-male association scores during 1977 (the "sexy" year) is evident in the webs of the four males, and association levels remained high during 1978 when there were still a good many cycling females in the community. (By contrast, Fifi, the only female in this sample who was anestrous during 1977, shows only a slight increase in association with males, at the 10–19 percent level.) Some of the fluctuations in association patterns for the males reflect changes in dominance status (Evered, for instance, in 1974 and Goblin in 1978 and 1979; these will be discussed in Chapter 15).

While the webs illustrate the overall association patterns of the adult males and the *central* females of the community, they do not provide information about the peripheral and/or shy females (or other individuals on whom there have not been enough follows). Another way of measuring sociability is to monitor the frequency with which different individuals take part in gatherings, some of which last a week or more, with chimpanzees arriving and leaving at different times. Community members have the opportunity at such gatherings to meet and interact with many other individuals—to play, groom, display, make noise. In a sense, gatherings are the hub of chimpanzee social life.

Figure 7.4 shows the extent to which adult males, adult females, and youngsters participated in observed gatherings over a ten-year period. Only gatherings that were attended by at least half of the community, including at least half of the adult males, and that persisted over at least two days, have been included. This analysis again

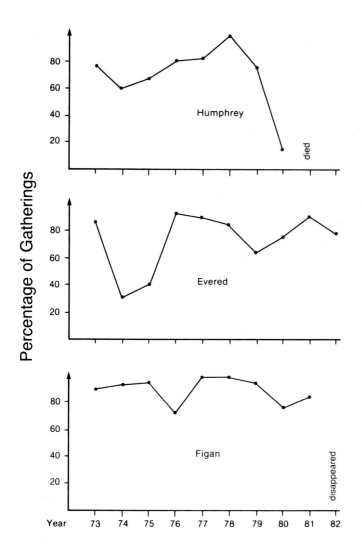

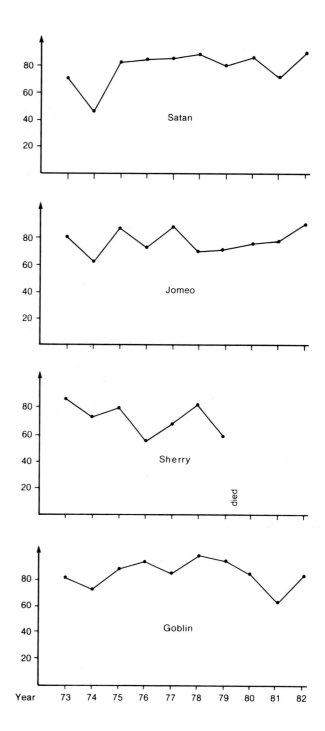

Figure 7.4 The frequency with which the adult males (*above*) and the adult females (*next pages*) of the Kasakela community participated in gatherings over a ten-year period. For each of the years reviewed the number of days that each individual attended gatherings was counted, and his or her overall participation was expressed as a percentage of the total number of gathering days for that year. The graphs of offspring are plotted with those of their mothers.

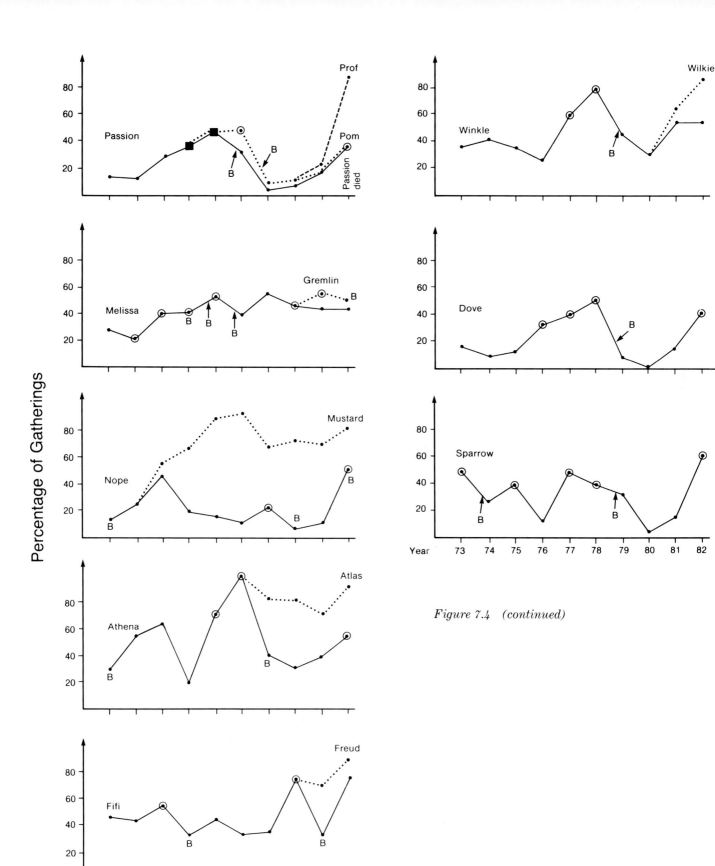

Figure 7.4 (continued)

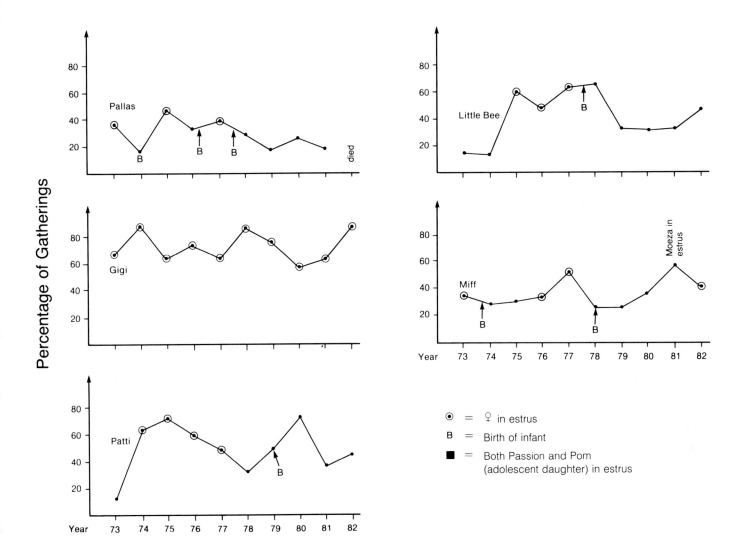

Percentage of Gatherings

⊙ = ♀ in estrus

B = Birth of infant

■ = Both Passion and Pom
(adolescent daughter) in estrus

enables us to compare all members of the community; even shy peripheral females are much less fearful of human observers when they join large gatherings.

As expected, males scored higher than did anestrous females; they joined more such gatherings on more different days. (They also stayed for longer periods of time after joining than did females, but this is not shown in the figure.) Fluctuations in male attendance reflect, for the most part, extended absences from the community range during consortships. This accounts for the lower attendance of Satan in 1974, Evered in 1979, and, at least to some extent, Sherry in 1976. Males may also become slightly less social after loss of dominance status; this accounts for the drop in attendance seen for Evered in 1974–75, for Figan in 1980, and for Goblin in 1981. Sherry lost a major dominance conflict in 1976 and, again, this may have been partly responsible for his decreased participation in gatherings that year.

The females show greater individual variation and considerably more fluctuation than do the males. They attend many gatherings during years when they are cycling unless, like Nope in 1979, they are led away on lengthy consortships. The three least sociable of the central females (Passion, Pallas, and Nope) and the two most peripheral of the habituated females (Dove and Sparrow) seldom attended gatherings except during cycling years. The ever-cycling, sterile Gigi not unexpectedly showed a level of attendance similar to that of the adult males.

Figure 7.4 also gives information on differences in sociability of males and females as they go through adolescence. Young males begin to attend many gatherings without their mothers from age ten (Wilkie and Freud) to twelve (Atlas). The two young females (Gremlin and Pom) did not show this sudden surge of independence.

Changes over the Life Cycle
A little more detail on this subject will serve to emphasize the difference between the sexes.

The male. The extent to which a male of eight years old or less associates with individuals outside his immediate family circle depends largely on the personality and sociability of his mother. True, the child can in a limited way extend periods of association with other families (by refusing to follow his mother when she leaves), but it is the frequency with which she *joins* others that is the significant factor. The five- or six-year-old male is usually very eager to join any party where there are (a) other youngsters to play with, (b) females in estrus, or (c) any kind of social excitement involving the adult males (such as loud calling, displaying, and drumming). Until he is eight or nine years old, however, he is unlikely to leave his mother to indulge these desires; the frequency with which he actually does join such parties will depend almost entirely on whether his mother shares his inclinations, or whether he can "persuade" her to follow him. For example, in response to the excited calling of a party he may start toward it, then pause and look back at his mother. If she does not follow, he often begins to whimper. He may go farther and stay away for ten minutes or so—but if she still does not come, he usually returns (often still crying). Some mothers comply more readily than others with their son's desires (see also Pusey, 1983). A glance at the sociability webs of asocial Passion and social Fifi, and at the frequency with which these two females joined gatherings (Figure 7.4) gives some idea of the difference in the early social experiences of their two sons, Prof and Freud.

As adolescence progresses, the male spends less and less time with his mother and more and more time with nonfamily members, particularly adult males and females in estrus. During late adolescence the male tends to become somewhat peripheral, spending a good deal of time by himself (Pusey, 1977; Halperin, 1979). Figure 7.5 shows time spent alone (during follows), plotted by age, for seven Kasakela males. We see that Goblin's solitary hours rose during late adolescence, leveled off, then gradually dropped as he became integrated into adult male society. Once maturity is reached, there is little variation in the "alone" scores for the males, and none of them subsequently rose higher than 45 percent, except Sherry prior to his death.

Figan's graph shows that he spent very little time alone at any age. When he was a young adult, in 1972 and 1973, he scored lowest of all

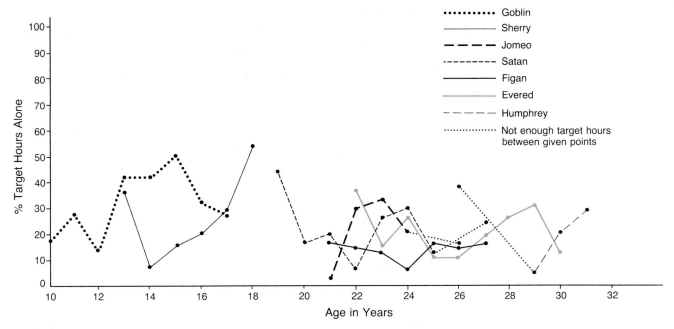

Figure 7.5 Percentage of target hours spent alone by seven adult males at different ages.

the males in Halperin's (1979) *lone-male* category—a mere 2 percent. (This was calculated by sampling target-male party size every hour, for follows that were not less than seven hours.) In part Figan's high sociability was a result of his frequent association with his elder brother, Faben, but even with this taken into account, Figan's "alone" score was still lower than that of other males. During the fifty-day follow in 1974, Figan's total "alone" score was 16 percent (Riss and Busse, 1977). His brother was still alive at that time but, as Figure 7.5 shows, Faben's death in 1975 did not lead to an increase in Figan's time alone.

Some males may become more solitary during old age. In 1972–73 the two oldest males in Halperin's sample, Mike and Hugo (estimated to be thirty-four and thirty-six respectively), scored rather differently. Mike had 54 percent of his points in the lone-male category, while Hugo was more sociable and had only 29 percent there. Data from an unpublished manuscript (Polis, 1975) show that this difference in sociability between the two persisted until the last three months of their lives, at which point Hugo became much more solitary, with scores similar to Mike's. During the final year of life the frequency with which each attended gatherings dropped sharply: from 79 percent to 42 percent for Mike, and from 74 percent to 44 percent for Hugo. Neither of them attended any of the thirteen gatherings that were observed during the last three months of their lives.

The female. The early social experiences of the infant and juvenile female also depend on the sociability and personality of the mother. Like her male counterpart, the female child can, to some extent, prolong contact with other family units. However, as a juvenile and early adolescent she does not exhibit the eagerness to participate in gatherings that we have observed in the young male—not until she enters late adolescence at about ten years of age. This period, characterized by recurrent sexual cycles and receptivity to the males, is

Adolescent females often leave their mothers during periods of estrus to travel with the mature males. This is Fifi, age about eleven years.

likely to be intensely social, and she often leaves her mother during estrus to travel with the males. Her association patterns become dramatically different from those of the late-adolescent male. He tends to become peripheral, she plunges into the social life of the community. Moreover, it is at this time that she begins to visit neighboring communities and, whether she transfers permanently or returns to her natal social group between periods of estrus, she enlarges her circle of acquaintances in a way not possible for the male at any stage of his life. If a female does transfer, late adolescence will be even more social; probably to avoid hostility from resident females, she spends much time, even when anestrous, in association with adult males (Pusey, 1979). If she remains in her natal community, however, and if her mother is still alive, she spends most anestrous periods with her family.

After giving birth to her first infant, the female typically becomes less gregarious. She continues to associate frequently with her mother (if the latter is still living), but spends a good deal of time with her infant only. This stage of comparative peace once more erupts into bursts of sociability with the resumption of estrous cycles. Figure 7.6 shows the "with family only" scores for Kasakela central females and clearly demonstrates the effects of estrus; during years when the various females cycled, they spent less time "with family only." The mother's sociability can also, as we have seen, be affected by her daughter's estrous condition. During one year in which Gremlin was cycling, the percentage of occasions when Melissa was encountered (during follows of other individuals) in a party with others was greater when her daughter was in estrus.

	Melissa in party	
	Gremlin estrous	Gremlin anestrous
January to June	86%	47%
July to December	87%	50%

These data point to increased levels of sociability during Gremlin's periods of estrus. Similar data were obtained during the long 1977 follow of Passion. During one two-week block, when Pom was anestrous, the family spent 83 percent of the time on its own. When Pom was in estrus, during a second two-week block, this figure dropped to 45 percent. During the first block, Passion encountered other individuals or parties on ten occasions. Over the two weeks the nonfamily member with whom she associated most was an adult male; she was with him for 12 percent of the 168 hours of observation. During the second block, Passion encountered other individuals or parties on nineteen occasions: she spent 33 percent of the 168 hours with a different adult male.

Increased contacts of this sort may be caused by (a) the daughter's reluctance to leave her mother during estrus (which means that males are attracted to the family) or (b) the mother's reluctance to allow her daughter to leave her (which means she will follow as the young female seeks out the males). Probably both factors are operative, to a degree determined by the type of relationship between mother and daughter.

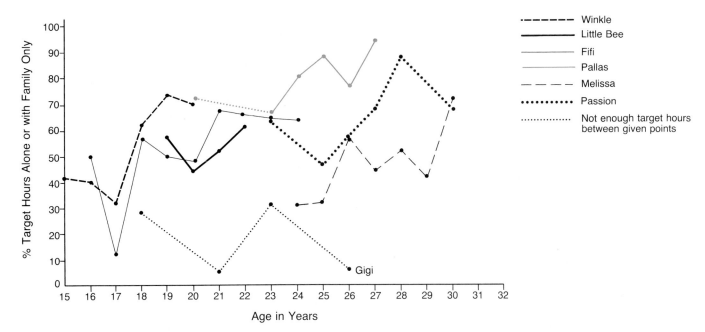

Figure 7.6 Percentage of target hours spent alone, or with family only, by seven central females at different ages.

From left to right: Fanni grooms her brother, Frodo, who is groomed by his mother, Fifi, who in turn is groomed by her adolescent son, Freud.

As a female gets older, she demonstrates a gradually increasing tendency to spend time with family members only. This is shown quite clearly in Figure 7.6. This decrease in gregariousness probably results from the fact that her growing family provides her with social benefits that she does not, therefore, need to seek elsewhere.

One of the most striking aspects of chimpanzee society, in comparison with other nonhuman primate societies, is the relative autonomy enjoyed by every individual after the age of about nine years. Once the young male breaks the psychological bonds of dependency that for so many years have attached him to his mother, he is free at almost all times to join—or to avoid—any other individual. While this is true to some extent for the female also, it seems that her psychological dependence on her mother is stronger; moreover, it is sometimes difficult, or impossible, for her to leave if she is taken by a male on a consortship, because he will actively try to prevent her departure. But this is the exception rather than the rule. A chimpanzee can almost always withdraw from an undesirable social situation and either remain alone for a while or seek a more congenial companion. In startling contrast to the vast majority of nonhuman primates, chimpanzees of both sexes can and do *regularly* go about their affairs with privacy. For the male, this means complete privacy; for the female, it is privacy shared with her immediate family.

This freedom in choice of companions, travel route, activity, and so on, plays an important role in reducing stress, particularly for males. At the same time the constant separations and reunions, at varying time intervals, impose unusual demands on social intelligence and awareness and exert strong pressures for flexibility of behavior. For an adult male, the presence of an ally may be of crucial importance to the outcome of a dominance conflict, but at any given time a much-needed partner may be not just at the far side of the troop (as is true, for example, in baboon society), but in a different valley, quite out of earshot. And he may be away for days or even weeks. His companion must search for alternative ways to cope with his rival. A female faces similar problems as her offspring grow increasingly independent and leave her alone more and more often, just as they become old enough to be effective as allies.

The nature of the society into which he is born inevitably has a marked influence on the development of an infant. The age, rank, and personality of the mother is crucial to certain aspects of infant socialization in all primate groups. For the chimpanzee youngster, other highly significant factors are the extent to which his mother is *social* or *asocial* and, far more than for infants of other species, his *position in the family*. A firstborn with an asocial mother may spend hour upon hour with only his parent for company. Thus her personality (affectionate or brusque, restrictive or permissive, playful or serious) plays a uniquely important role in his development. Subsequent infants have the advantage of built-in playmates; an infant male has a role model in an elder brother.

In most primate societies the infant has the benefit of a relatively stable social group. A young baboon, for example, is constantly surrounded by other infants and juveniles and by various relatives who, in addition to the mother, are likely to take its part in agonistic encounters. It is not difficult for the youngster to learn the probable outcome of his actions with respect to his companions. The chimpanzee mother, however, associates from day to day with different community members in different combinations. For her infant the task of learning the dynamics of the social scene is highly complex. He must find out, through experience, that if he hurts a playmate not only will the outcome be affected by the rank of his own mother relative to that of his playmate's mother, but also by the presence or absence of elder siblings and other relatives of either or both families. We must remember that a central female sees other community members regularly, a peripheral one less often. Entry into the society of adult males will be somewhat different for the sons of two such females: one will have his mother readily available in times of stress, the other often will not. This reality affects the kinds of relationships the young males build with the other males.

We have now reached the point where we can turn to a more detailed examination of the relationships that exist among the different members of the community, relationships upon which the structure of the community ultimately depends.

8. *Relationships*

Gilka grooms her brother, Evered.
(E. Tsolo)

June 1975　We suddenly notice Evered sitting in the bushes to the north of the camp clearing; he has not been seen for many weeks. On previous occasions he has been fiercely attacked on his return from an absence by brothers Figan and Faben. No wonder he looks ill at ease and glances nervously around, startling at every rustle. After twenty minutes he stands, hair on end, staring to the east. Then he relaxes; it is his young sister, Gilka. He moves forward and Gilka hurries up, giving soft pant-grunts. They embrace, then settle to groom. Presently Gilka leaves him and approaches one of the feeding boxes. She has not had bananas for twelve days, so we put some out for her. Evered, hair bristling, rushes up; it is weeks since *he* had a banana. Any other female would have fled, screaming—and probably would have been attacked anyway. But Gilka calmly takes her share while Evered watches; then she moves aside and he takes his. Evered eats faster, and she has several left when his are gone. He sits close, watching each mouthful, then reaches out and is allowed to take her peels. They leave camp. Gilka travels slowly, for she is very pregnant. Evered waits patiently whenever she pauses to rest. They are on their own for the remainder of the day, grooming, feeding side by side in palms, and nesting in the same tree at night. Perhaps it is the relaxation and comfort he finds in her familiar, nonthreatening presence that will give him courage next day to face his bitter rival Figan.

The chimpanzees of Gombe have been studied for a quarter of a century. We have watched youngsters mature and mature individuals grow old. We have noted the pronounced individual variation in behavior that gives each chimpanzee his or her unique personality. And we have observed and recorded literally thousands of interactions between hundreds of dyads. We can turn now to this wealth of data and search for rules underlying the development of different kinds of relationships between different individuals. What, for example, are the characteristics of relationships between adult males and infants? How will such relationships change as the youngsters mature? What is the effect of specific social and environmental factors (such as sexual competition or a lean dry season)? Or, to turn the question around, to what extent does the nature of a given relationship (friendly or hostile) influence the way that the chimpanzees involved behave in particular contexts (during a reunion, for example)?

A relationship can best be described in terms of the interactions between the individuals concerned. Hinde (1976) provides a framework, suggesting that we note the *type*, the *quality*, and the *frequency*, both absolute and relative, of such interactions. The *number of different types* of interactions is meaningful (within the mother-infant relationship, for example, are included suckling, transport, grooming, play, protection, and many others, whereas two adult females may be involved in only an occasional reunion). The degree to which interactions are *reciprocal* (A grooms B, B grooms A) or *complementary* (A solicits, B grooms) must also be taken into consideration. Another significant factor is the extent to which the behavior of two individuals *meshes*—that is, the extent to which their goals are similar or, if different, how readily one defers to or competes with the wishes of the other.

Chimpanzees (like other group-living mammals) tend to behave in varied and predictable ways to members of the several age-sex classes. These tendencies are different for the two sexes and change with age. For example, a juvenile male plays with other youngsters; he watches and imitates adult males. When he becomes adolescent, he may still play with his peers, although the quality of his play is likely to be much rougher, but he also becomes involved in aggressive, competitive interactions with them; still fascinated by adult males, he nonetheless directs submissive and fearful behavior toward them. As an adult, he associates and interacts frequently with other adult males: he grooms with all, cooperates with some, competes with others for status and access to females. He is (usually) tolerant and friendly toward the infant and juvenile males who watch and imitate him. And so on.

These tendencies, in part inherited, are to some degree molded by the social context. As we have seen, the frequency with which the infant plays with peers depends on the kind of mother he has; it also depends on demographic factors—the number of other infants available as playmates. The quality of his play may depend on the rank of his mother, for sons of high-ranking mothers are likely to be more assertive and rough than those of more subordinate females.

From early infancy a male typically shows sexual interest in a female who is in estrus, but the exact form of that interest varies. As an infant, he will sniff and touch any attractively swollen bottom. When he is older, he will try to achieve intromission and make thrusting movements. In maturity, the nature of his sexual response will be determined by the female's age and popularity as a sexual partner, his rank in the male hierarchy, the other individuals who are present, and the like. The frequency with which a male *interacts* with a given female depends, at least in part, on the extent to which he *associates* with her; this in turn depends on their genetic relatedness, her sociability and age, the position of her core area, and sometimes the age and sex of her offspring.

A female infant shows little interest in sexual swellings and throughout her childhood is much less fascinated with adult males than is her male counterpart. Young males and young females alike are attracted to small infants, but the female is likely to be less rough in her play than the young male. The adolescent female may become suddenly more aggressive toward other females, but she does not systematically challenge them as does the young male, she performs far fewer charging displays toward them, and she seldom threatens those who rank higher than her mother. She is less likely to form close and enduring friendly relationships with nonrelated adults than is the male. And when she does, her less social nature, leading to lower levels of association, makes such bonds less obvious. A number of friendly relationships between adult females are disrupted when the son of one reaches adolescence and begins to challenge his mother's friend. If the latter threatens or attacks him as a result, his mother will typically support him—and the lower-ranking of the females may then start to avoid her former companion.

At the deepest level, then, the patterning of relationships is based on the age and sex of the individuals concerned, their degree of familiarity, and their relative social status. Variations in the social context in which interactions occur can profoundly influence their nature. During times of great tension normally friendly and highly tolerant family members may even attack one another. Moreover, the outcome of an interaction between two individuals (and sometimes the course of their relationship) may be significantly altered by the participation or intervention of a third. In chimpanzee society triadic interactions (as well as those involving greater numbers of individuals) play a highly significant role, and there are many well-established and durable alliances. Often two allies will unite with the common goal of intimidating or defeating a third.

All of these social variables must be taken into account in any attempt to untangle the dynamics of social relationships within the chimpanzee community. Moreover, it is important to reiterate that every chimpanzee is a unique individual with a special combination of inherited and acquired characteristics, a fact that will inevitably affect the nature of *every* relationship to a greater or lesser extent.

Major Categories

There are three types of relationships—friendly, sexual, and unfriendly. In addition, there are some that are difficult to categorize, and as a temporary measure (until more sophisticated data collection and analysis permit better understanding), we describe these as "neutral." Let us look briefly at each category.

(a) *Friendly Relationships*. Affiliative behaviors heavily outweigh aggressive ones, both in quantity and in quality. Three subdivisions deserve special comment:

Caretaker—typically the relationship of mother to child. It is characterized by high levels of association, protection, assistance, and support, as well as many affiliative behaviors.

Friend—the strongest and most enduring friendly relationship, characterized by *two-way* affiliative, supportive interactions. The caretaker relationship of mother and child usually leads to friendship between mother and older offspring, particularly (at Gombe) between mother and adult daughter—the strongest of all bonds among adults.

The adult brothers Faben and Figan enjoyed a close supportive relationship from 1968 until Faben's disappearance and presumed death in 1975. (L. Goldman)

Friendships also exist between adult brothers and some pairs of non-related males, usually of very different social rank.

Follower—relationship between a youngster (usually an adolescent male) and a particular adult male. The young male watches, imitates, occasionally grooms his superior, and is, by and large, well tolerated by him. The older male sometimes supports the youngster during interactions with others.

(b) *Sexual Relationships*. These may be confined almost entirely to the sexual act with attendant courtship behaviors and, often, some aggression. Or they may include high levels of relaxed association, as during consortships (which may last for over a month and may occur at times when the female is anestrous). The sexual relationship is basically a complementary one, and the extent to which the sexual desire of the male meshes with the receptivity of the female (toward males in general or one male in particular) will determine whether it can *also* be included in the category of friendly relationships. Some sexual relationships are definitely unfriendly.

(c) *Unfriendly Relationships*. Aggressive or avoidance behaviors outweigh friendly, affiliative ones. There are two basic types of unfriendly relationships:

Competitive—between individuals of any age and either sex, during periods when they are competing for status. This sort of relationship is most obvious between pairs of adult males, one of whom is striving to better and the other to maintain his social rank; and between an adolescent male and adult females, during his struggle to dominate them. Sometimes a similar relationship develops between an adolescent female and some of the older females. Competitive relationships are also evident among infants or juveniles of similar age, particularly males; they are expressed in the frequency with which play bouts lead to aggressive incidents. Other individuals are often drawn into competitive interactions as coalition partners of one or both of the contestants.

There often comes a point when a relationship between males becomes ambiguous and tense, usually just after a reversal in dominance has taken place. At this stage the individuals concerned may show high levels of both aggressive and friendly behavior. Once the two have adjusted to their new relative positions, the relationship, whether it is basically friendly or unfriendly, stabilizes.

Hostile—characterized by very high levels of aggression, usually in one direction, and by fear and avoidance on the part of the subordinate. Resident females, for instance, are hostile toward new immigrant females. In its extreme form hostility can lead to death, as when adult males attack individuals of a neighboring community.

(d) *Neutral Relationships.* There are some relationships, particularly between adult females, that cannot be labeled either friendly or unfriendly. The individuals concerned are very rarely seen to interact. Sometimes this is because they are seldom observed together, but there are instances of females who seem to ignore each other completely even when they spend hours together. Occasionally the nature of their relationship can be determined by pooling records of their interactions over several years. But this kind of determination is not always possible either, so for present purposes I have listed nebulous relationships of this sort as "neutral."

One final point. It has become increasingly clear that while a *change* in the frequency of interactions between two individuals often indicates a change in their relationship, frequency per se is not a useful measure. When two individuals interact very little, it can mean either that their relationship is hostile and one avoids the other, or that they are very relaxed and tolerant of each other and have no need to interact frequently. Menzel (1975) describes pairs of chimpanzees who virtually never groomed or played and were not often seen in physical contact; yet they were seldom more than 9 meters apart and defended each other in fights. Better ways must be found of quantifying the more subtle aspects of chimpanzee communication if we are to understand fully the complex relations among chimpanzees, particularly female chimpanzees, in the wild.

Change over Time

Because relationships among the members of a social group are determined in part by underlying principles of age and social rank, all must change over time as individuals mature. Some of these changes are gradual, such as a mother's increasing rejection of a child during weaning, with a decisive push toward independence at the birth of the next infant. Others are more sudden, as when a young female comes into estrus for the first time. Relationships may be altered also by unpredictable events, such as the crippling of some individuals at Gombe during the polio epidemic or the bizarre attacks of infant killers Passion and Pom.

Notwithstanding this pattern of continual change, the fundamental nature of some associations remains relatively constant. Among family members, for example, relations remain basically friendly throughout life, despite inevitable ups and downs. But the relationship of two male siblings is likely to run a rather different course from that of two nonrelated males. Let us first consider the brothers—one, say, five years older than the other. When the baby is born, the elder child is initially allowed to touch him. Then he is permitted to groom, play with, and finally carry the infant. He is usually very protective of his small sibling. Their play, gentle at first, becomes rougher as the younger brother grows up. Interactions become less frequent when the elder of the two begins to spend more time away from his family. The relationship, however, remains friendly and supportive, and they often feed, patrol, or nest together. It seems that all younger brothers may, at some point, become the higher ranked: during this period of dominance reversal there will be tension and some aggression between the two, but friendly relations are soon resumed. A friendship of this kind can have far-reaching consequences for a male eager to rise in the dominance hierarchy, because brothers typically support each other during social conflicts and almost never take sides against their sibling.

The relationship of the two nonrelated males (one, again, five years older than the other) will be different in many ways. Some of the elements that characterize their early interactions may well be similar: the elder may groom, play with, even carry and protect the younger. But he will do these things far less often, because he will not have the same opportunity. He will spend far less time with the younger male and the degree of familiarity will be much less. When the younger of the two begins to leave his mother, he is unlikely to follow the unrelated adolescent; they will not do things together or support each other. And when, as is likely, the younger male challenges the older, their relationship may become very unfriendly, with much aggression and probably some serious fighting. They may eventually become more friendly again, but this can take years rather than months. Nor will either ever be able to *count on* the active support of the other during social conflict with a third individual.

Commentaries 8.1, 8.2, and 8.3 describe, for two well-studied individuals (Figan, alpha male for ten years, and his sister Fifi) the relationships that appeared to be most meaningful to each of them over a twenty-year period. These accounts give some sense of the complexity and diversity of chimpanzee relationships in the natural habitat.

Figan's Relationships
Commentary 8.1 describes Figan's relationships with three other males:

(a) Faben, his brother—a stable and enduring friendship, disrupted by a brief unfriendly period when the younger Figan challenged Faben and became dominant over him.

(b) Humphrey, nonrelated—initially unfriendly, subsequently hostile as Figan wrested alpha position from the older male, finally friendly.

(c) Evered, nonrelated—initially ambiguous, then intensely hostile for several years, and finally (toward the end of Figan's life) friendly.

Figan's relationship with one other male, Goblin, is briefly mentioned, because it illustrates a different pattern. Initially friendly, that relationship became tense and then very hostile until Figan's disappearance and presumed death.*

* Much of the material concerning Figan's associations with these males has already been published (Goodall, 1968b, 1971; Riss and Goodall, 1977; Riss and Busse, 1977); I have supplemented these sources with unpublished analyses by David Riss, as well as the recent data.

Some of Figan's Meaningful Relationships with Other Males

Figan and Faben   When Figan was a juvenile, still dependent on his mother, Flo, he spent a good deal of time with Faben. In those days Faben was obviously the more dominant of the brothers, but very few aggressive incidents were observed; the two enjoyed a relaxed relationship and often played. Then, in 1966, Faben was stricken with polio and lost the use of one arm. Figan, taking instant advantage of the situation, began to challenge his brother. Once, as Faben groomed Flo in a tree, Figan swayed the branches so violently that Faben, not yet adjusted to his disability, was shaken, screaming, to the ground. It was after this that Faben was first observed being submissive to his young brother, greeting him with pant-grunts. After this Figan's aggression toward Faben stopped; but so, for a while, did their close relationship. They were only occasionally seen together, usually when each had sought the company of their mother in the family unit.

During 1969 things changed: the brothers began to travel and groom together often, and sometimes displayed in unison at other males. Eventually, in 1973, Faben gave his allegiance wholeheartedly to Figan and they became close allies. It was thanks to Faben's support that Figan was eventually able to become alpha male of the Kasakela community—a position he almost lost when his brother vanished (and presumably died) in 1975.

Figan and Humphrey   Humphrey had a very aggressive nature and Figan, as a youngster, usually avoided him. On the few occasions in the mid-sixties when Figan joined Humphrey and Faben in a play session, Figan was clearly nervous. By 1970 Faben as well as Figan had become extremely submissive to Humphrey; Figan still kept away from the high-ranking and aggressive male whenever possible, but even so Humphrey was seen to attack him twice that year (and Faben four times).

Early in 1971 Humphrey became alpha male. Figan still seemed to be very submissive to him, but Humphrey must have sensed the potential threat of the younger male; he very often displayed in his presence (Bygott, 1974). At that time Faben had not yet allied himself with his brother—indeed, he quite often traveled with Humphrey and once actually supported him against Figan (R. Wrangham, personal communication). During that year Figan gradually became more confident. In October 1972 Figan and Humphrey almost certainly had a serious fight. Observers, hearing screaming and waa-barks, rushed to the scene and found both males with fresh wounds; they were in a state of intense social excitement and continued to scream and display for some time (Bauer, 1976).

By April 1973 Figan and Faben had become close allies; Faben almost always joined his brother's displays. This was the key factor in Humphrey's downfall. Figan seldom challenged the older, heavier male unless Faben was nearby. The repeated displays of this powerful alliance had a marked effect on Humphrey, who became increasingly tense and fearful. The final showdown came early in May. One evening Figan began to display wildly in the treetops where a large group of chimpanzees, including Humphrey, was settling for the night. Suddenly Figan hurtled down to attack Humphrey in his nest. Both males fell to the ground and Humphrey ran off screaming. For good measure Figan attacked a second time, just as Humphrey had settled in a second nest. The outcome was the same. Faben was present; he did not actively help Figan but provided implicit support. (Observer: D. Riss.)

Figan wrist-bends and leans away as high-ranked Humphrey passes him.

Faben seeks reassurance from his younger brother, Figan, during social excitement. (H. van Lawick)

After that evening Figan was dominant over Humphrey for the rest of his life. Gradually, during the next two years, a close friendship developed between them. It was to Humphrey that Figan turned for reassurance when, after Faben's death, the other males sometimes joined forces against him.

Figan and Evered The victory over Humphrey did not confer alpha status on Figan. He still had to dominate the other contender, Evered—and in this endeavor Faben's support again was crucial. Figan and Evered had been childhood playmates. When, as sometimes happened, their play became too rough, it was Figan, the younger, who ran off screaming. Flo, higher ranking than Evered's timid mother, Olly, initially hurried to support her son, but by 1966 Flo was wary of Evered, now a late adolescent, and seldom intervened on Figan's behalf.

The first serious fight observed between Figan and Evered was in 1970. It happened when Evered had just reappeared after a long consortship. Figan and Faben raced toward him, chased him into a tall tree, and attacked him. Flo, Fifi, and Flint charged about below, uttering fierce waa-barks and roar pant-hoots. Finally Evered fell some 11 meters, then fled, screaming; blood poured from a split cheek. (Observer: D. Bygott.) After that Evered, for the first time, became submissive to Figan; nonetheless, Figan was still uneasy in Evered's presence. For the next two years their relationship was tense. Then, early in 1972, Evered with the support of another male attacked and bested Figan. Following this incident there was even more tension when they met, and they displayed vigorously whenever they were in a party together. It seemed, though, that they both tried to avoid a major confrontation; neither charged directly *at* the other, merely in the general vicinity of his rival. Once they displayed back and forth, first one and then the other, for nearly an hour before they stopped, obviously exhausted.

In December of that year we thought Evered might regain the upper hand; he joined with Humphrey and the two attacked and wounded Figan.

Faben, though present, did not help his brother. During the next four months, however, the Figan-Faben alliance was sealed and this was Evered's undoing.

In May 1973, four days before Figan's decisive win over Humphrey, the brothers began to persecute Evered. The first observed incident started with an unprovoked attack by Figan, after which the brothers displayed repeatedly below Evered's tree. Their victim became increasingly nervous, whimpering and even grooming an infant as he sought reassurance; he was unable to escape until the brothers moved off. His final defeat came three weeks later. It began when Figan and Faben heard Evered's pant-hoots. They raced toward the sound and, finding Evered in a tree, began to display below. Figan then climbed into an adjoining tree and sat, while Faben moved up the steep slope until he was level with Evered. Now close to their victim, both suddenly sprang at him. All three fell to the ground and Evered, screaming, fled with the brothers in pursuit. For the next hour the two charged back and forth below Evered as he cowered in another tree. (Observer: D. Riss.)

From then on Evered showed extreme submission to Figan, sometimes going out of his way to do so. Rising early one morning, for instance, he sat for twenty-five minutes below Figan, who was still in his nest; when Figan finally looked down, Evered pant-grunted and bobbed frantically. Figan continued to display at high frequency whenever Evered was nearby, as though to emphasize his status. And when Faben went off on a thirteen-day consortship, Figan occasionally climbed high in a tree and stared around, giving loud SOS screams; he was obviously calling to his brother.

When Faben returned, and despite Evered's submissive behavior, the brothers resumed their persecution of him; they seemed to be trying to drive him from the community. Another attack was seen in 1974. This followed a forty-minute period during which Figan and Faben sat calmly below Evered (who had again fled up a tree) and groomed each other. When they finally rushed up after him, the assault was fierce and again Evered fled, screaming. Once more he vanished for an extended period to the north. In August that year another attack on Evered was seen. The brothers were joined by Humphrey and Gigi, and Evered sustained some minor wounds.

In June 1975 Faben disappeared for good. With his ally gone, Figan lost much of his confidence, while Evered became less fearful and no longer spent weeks at a time in the north. Sometimes he held his ground when Figan displayed toward him: then the two males would embrace, kiss, and pat each other vigorously on the mouth and scrotum, both screaming loudly. The other males now began to take advantage of Faben's death; they sometimes joined forces against Figan, who usually ran to Humphrey for comfort. For a nine-month period Figan could not be described as alpha. Gradually, however, he regained the top position, though never again was he in the position of supreme power that he had enjoyed while Faben was alive.

After Humphrey's death in 1980 Figan turned for companionship and support to his longtime enemy, Evered, and also to Jomeo. The three became very friendly and often supported one another against the challenges of young Goblin. Goblin had been Figan's follower as an adolescent, and for years Figan had supported him in his conflicts against others. Then, when Goblin had risen almost to the top of the male hierarchy (as a result of Figan's help), he turned on Figan and challenged his alpha status. This will be described in detail in Chapter 15.

Relations between the Sexes

Figan, Fifi, and Flo Both Figan and Fifi continued to spend much time with Flo even after reaching adolescence, which means that they saw a good deal of each other as well. Flo and Fifi were almost always very relaxed in Figan's company, even after he was full grown. Only twice was he ever seen to attack Flo; both episodes involved mild poundings and took place at the height of the banana feeding. Flo seemed not at all fearful of her son, but rather enraged; she screamed in fury, then gave loud waa-barks as he displayed away. Once, when Figan was about twenty years old, Flo was sitting directly in the path of one of his charging displays. Other females, screaming, raced to escape, but old Flo continued to sit calmly. As he approached, she ducked her head and he leaped right over her and went on his way.

As a youngster Figan had been supportive of Fifi. In adulthood, however, he seldom helped his sister, rarely shared meat with her, and did not often groom with her. (In these ways, he was very unlike his contemporary, Evered, who had a much warmer relationship with his sister, Gilka.) Fifi was seldom afraid of Figan; like Flo, she remained quite calm during many of his displays, although she sometimes climbed a tree to get out of his way. She was less submissive to Figan than to any other adult male; indeed, when Figan was followed for fifty days in 1974 she did not pant-grunt to him once. At that time she associated with him more than did any other anestrous female (Riss and Busse, 1977). In 1976, when Fifi was followed for forty-five days, Figan was her most frequent male companion. The actual amounts of time involved, however, were not large—about 73 hours out of a total of 563 daylight hours in the 1974 long follow of Figan, and 62 of 540 daylight hours in 1976. Until his death Figan continued to be one of the three males with whom Fifi associated most.

Figan never made any attempt to copulate with his mother, but he did several times force his sexual attentions on Fifi, as described in Chapter 16.

Figan and Two Nonrelated Adult Females There were two females, in addition to his mother and sister, with whom Figan spent a good deal of time from 1974 until his death: Gigi and Melissa. Figan associated with the ever-cycling Gigi almost every time she was in estrus from 1965 on unless (as was often the case) she was taken away by another suitor. The longest consortship of Figan's life was probably when he and Gigi were gone for thirty-nine days in 1972. His close association with Melissa dates back to the late sixties, when he several times led her off on consortships, accompanied by infant Goblin. During years when she was anestrous, Melissa was often seen with Figan; they had a calm and relaxed relationship. The two did not interact a great deal, but this is typical of a male's relationship with anestrous females; Figan was involved in more grooming sessions with Melissa than with his sister.

Fifi and Three Nonrelated Males One male who had a major impact on Fifi's life was Leakey. In 1967 and 1968, when she was a late adolescent, he took her off on seven consortships for a total of sixty-seven days—and tried unsuccessfully to lead her away on a good many other occasions too. Although Fifi was sometimes reluctant to follow Leakey away from her familiar range, she showed no such unwillingness to respond to sexual advances of a more immediate kind. Like many other adolescent females, she would hasten to a male at the least sign of sexual arousal. One day, as Leakey groomed with a male companion, Fifi, after reclining and gazing intently at his limp penis, reached out and began to tweak it with her fingers. She presently achieved the desired result and instantly turned around and solicited copulation. She was, however, totally ignored.

As an adult, Fifi has spent a good deal of time with two childhood playmates, Satan and Evered. She has groomed with these two males more than with any others. As far back as 1968 Fifi's character file described Satan as her "special" male companion. Her relationship with him has always been more colorful than that with Evered, characterized by higher frequencies of interaction in many contexts: pant-grunts and other submissive behaviors on her part, a certain amount of aggression (including attacks) on his, and quite frequent play sessions initiated by Satan. Often, too, Fifi is allowed to share Satan's meat.

Commentary 8.3

Some of Fifi's Meaningful Relationships with Other Females

Fifi and Flo   Fifi's only really close bond with another female has been with her mother. When Fifi was an infant, Flo was an affectionate, tolerant, and playful mother. She was protective and always supported Fifi in social conflicts. During Fifi's late adolescence she was almost always with Flo. Only during her periods of sexual swelling did Fifi leave her mother to travel with the adult males. After giving birth, Fifi continued to spend a great deal of time with Flo. They still supported each other and were frequent grooming partners. During the last few months of Flo's life, however, Fifi was with her less often. By this time Flo was old and frail and confined her travel to a tiny area in Kasakela Valley. It was the lean dry season of 1972 and Fifi, with an infant to care for, presumably felt the need to move farther afield to find enough to eat.

Fifi and Winkle   Fifi's relationship with Winkle was initially hostile. Winkle appeared in the Kasakela community as an early adolescent and was often involved in aggressive incidents with resident females (Pusey, 1977). Flo was among Winkle's most zealous persecutors, and the ten-year-old Fifi eagerly supported her mother. Gradually, as Winkle was accepted by the Kasakela females, she became more aggressive. If Fifi threatened her, Winkle retaliated and Flo, as she became older, supported Fifi less and less often. After Flo's death in 1972 Winkle and Fifi often traveled together. Over the next few years a strong play bond developed between their infants, Freud and Wilkie. Neither youngster had an elder sibling to play with, so when they were alone with their mothers they continually pestered them for attention. When Fifi and Winkle met, however, Freud and Wilkie rushed to play. The two mothers were then able to feed or rest in peace; it is undoubtedly no coincidence that they spent much time together.

By 1979, however, Freud at age eight was becoming rougher; his play sessions with Wilkie, a year younger, more and more often ended in fighting. Winkle always hastened to her son's support, and aggressive incidents between the two mothers resulted. At the same time, Fifi had resumed cycling and spent much time with adult males. As a result she and Winkle associated far less.

Three years later, however, they once more became frequent companions. Winkle's daughter, four-year-old Wunda, was fascinated by Fifi's year-old infant, Fanni. Even if Winkle wanted to leave Fifi and travel in another direction, she often gave in when her daughter, wanting to stay with Fanni, refused to follow. In 1983 Winkle was Fifi's second most frequent companion, and Wunda and Fanni were very close playmates.

Fifi and Patti Patti too was an immigrant. She appeared in 1973 and Fifi, about fifteen years old, was one of the three females most aggressive to her. They spent little time together until 1975 when both were cycling and often traveled in the same sexual parties. The following year, when Fifi was first pregnant and then nursing a small infant, she and Patti continued to associate a good deal. This was almost entirely owing to Patti's persistence. Again and again Fifi, when she saw Patti approach, displayed toward her vigorously, hair bristling, stamping, and swaying vegetation; three times this behavior ended with an attack. Patti ran screaming from two of these encounters, but once she turned and fought back and it was Fifi—heavily pregnant—who came off the worse. The reasons behind Patti's determination to associate with Fifi are not clear, although one factor may have been the strong sexual attraction between herself and five-year-old Freud. And Freud's obvious desire to stay with Patti may have influenced his mother in her growing tolerance of the younger female. Toward the end of that year, after Fifi had given birth, she and Patti cooperated once to drive off an adolescent male. And they were even observed to embark on quite a vigorous play session, initiated by Patti (who had tried to play with Fifi before, but failed).

In 1977, when Freud was six, he began (somewhat precociously) to challenge low-ranking females, including Patti. If Patti retaliated, and usually she did, Fifi hurried aggressively to her son's defense. Patti then began to associate less frequently with Fifi and her family. But the following year, when Patti was pregnant, she became fascinated with two-year-old Frodo. Freud continued to threaten Patti, but now if she retaliated, Fifi, preoccupied with her infant, was less likely to intervene.

Fifi and Patti continued to associate at relatively high levels. The quality of their relationship, however, was ambiguous, as illustrated by the following incident. One day in 1978, Freud, hair bristling, swaggered toward adult female Sparrow, flailing a large branch. Sparrow, squeaking, ran to Patti and Little Bee, who sat nearby. The three females gave loud waa-barks. Patti displayed vigorously at Freud, who ran to Fifi. Patti continued her display until Fifi, with Frodo clinging to her belly, squeaked and moved hastily out of the way, then sat whimpering, Freud beside her. A few minutes later two more adult females joined the party. Freud again swaggered toward Sparrow. Instantly Patti, who had moved almost out of sight, charged back in a bristling display directed at Freud. He rushed to his mother again and Fifi once more retreated, whimpering. Then, after just two and a half minutes, Patti wandered up to Fifi and began to tickle her bottom. For almost a minute Fifi ignored her, but suddenly a grunt of laughter broke from her and she began to tickle Patti's neck. They played, rolling over and laughing, once chasing around a bush, for nine minutes.

Fifi and Little Bee These two females grew up in the same community but had little to do with each other because their mothers seldom associated. Not until 1978, when Fifi had two offspring and Little Bee one, did the two families begin to spend time together. Frodo was two years old then, just a year older than Tubi; they were highly compatible playmates. Freud, too, often joined their play and although he was rough with the older youngsters like Wilkie, he was gentle with the younger ones. Little Bee, when moving about with Fifi, was temporarily freed from Tubi's insistent demands for her attention.

By the end of 1979 Freud had intensified his displays against community females. Little Bee, low ranking and crippled by her clubfoot, became increasingly intimidated. At the same time Frodo began to play roughly with the younger infants. When Tubi was subjected to such play assaults, he retaliated vigorously, rushing at and trying to bite or hit Frodo, interspersing his screams with waa-barks. Frodo usually avoided these furious charges, but there were times when Tubi hurt him; then he too began to scream. As Little Bee hurried to rescue her son, she sometimes encountered Fifi bent on avenging Frodo; this often resulted in aggressive incidents between the mothers. The two families gradually spent less time together. And their association further decreased when Fifi resumed estrous cycles in 1980.

The following year, when Fifi gave birth again, she and Little Bee resumed their interrupted association until, in 1982, it was Little Bee's turn to resume cycling. She traveled with the males in sexual parties that Fifi, with her small infant, tended to avoid. Once again the two spent less time together. But in 1983 Little Bee again was Fifi's most frequent companion.

The paucity of interactions between adult females even when they are friendly is well illustrated by Fifi's relationship with Little Bee. In 1978, for example, when Fifi associated very frequently with Little Bee, a search through *all* the records, including the detailed mother-infant reports, revealed a total of only ten interactions between them, six of which were mildly aggressive. From 1976 to 1982, despite the many hours they spent together in camp, to all intents and purposes they did not groom each other at all; Little Bee groomed Fifi once, for five minutes.

Fifi and Gigi These two females grew up together. As juveniles and early adolescents, they played often; as adults, they have spent a good deal of time together. Their relationship has always been fundamentally friendly, and Gigi was one of Fifi's preferred grooming partners. From 1975 through 1977 Gigi was Fifi's most frequent companion (sharing the position with Patti in 1975). Several factors were probably responsible: (a) both were cycling and often traveled in the same sexual parties; (b) Fifi was still spending a good deal of time with Figan, and Gigi was Figan's preferred female companion; (c) Freud was highly attracted sexually to Gigi; and (d) in 1977 Gigi's fascination with small infants made Frodo very appealing to her.

After this period Fifi and Gigi spent less time together. Gigi continued to travel in sexual parties while Fifi spent more time with her family; and Frodo lost the dependent, tractable qualities that Gigi finds so irresistible. Not until 1982, when Fifi's next infant was a year and a half old, did Gigi once again begin to spend time with Fifi's family.

A characteristic shared by most male chimpanzees is the preoccupation, from adolescence on, with maintaining and bettering their social rank, and many of their interactions are devoted to this end. Change in the relative dominance rank affects not only the type of relationship between two males, but also the extent to which they associate. Figure 8.1a depicts variations in the amount of time that Figan spent with other adult males (during Figan follows) over an eight-year period. Males who were with him for less than 10 percent of the total number of target hours per year have not been included. The figure shows how Figan's associations altered in relation to the events described in the Commentary. The death of his brother Faben led to a change in Figan's social status, which resulted after a few months in an increase of association with Humphrey and, subsequently, Jomeo. The marked decrease in his association with Goblin from 1979 on mirrors the intense hostility that developed between them, just as the gradual increase in time spent with Evered from 1976 on reflects the ending of their period of enmity. The amount of time that Figan and Satan spent together fluctuated considerably. From 1976 until Goblin's challenge in 1979, Satan was the most serious contender for the alpha position. The relationship between Figan and Satan was tense, and this probably explains the yearly fluctuations in their association.

Commentary 8.2 includes descriptions of Figan's relations with four females: Flo, his mother, and Fifi, his sister, with both of whom Figan had a close and friendly bond; Gigi, his most frequent sexual partner; and Melissa, whom he took on several consortships when he was young.

For the most part, relationships between adult males and nonrelated adult females are patterned by the rhythms of the female estrous cycle. Even those females with whom the males interact very little during anestrous periods become drawn into social life when they are swollen. Figure 8.2a shows how the amount of time that Figan was seen to spend with his preferred females varied from year to year in accordance with this principle. For the most part he was patient and tolerant in his relations with the females of his community, and his sexual interactions with them were almost always friendly.

Fifi's Relationships

Commentary 8.3 gives an account of Fifi's relations with five other females: Flo, her mother; Winkle and Patti, two females slightly younger than herself who immigrated into the Kasakela community during their adolescence; and Little Bee and Gigi, who grew up with Fifi in the Kasakela community. Little Bee left, temporarily, to join her mother and sister in the newly formed Kahama community, but returned a few years later.

Only with Flo did Fifi maintain a relationship that was basically friendly and supportive at all times, characterized by high levels of association and grooming. Fifi's interactions with Winkle and Patti when they first appeared in the community were hostile. Patti persevered in her attempts to establish a friendly relationship with Fifi and, to some extent, was successful; but Fifi and Winkle, while no longer hostile, are not particularly friendly either. Fifi's relations with Gigi and with Little Bee have always been quite friendly.

Females are preoccupied, from about age thirteen on, with raising their families, a task that is interspersed, every three to five years, with periods of sexual activity. Relationships with other females are colored by the stage of their estrous cycles and the age and sex of their offspring, factors that affect (among other things) their level of sociability. The frequency with which two females associate has a meaningful influence on their relationship; equally, the nature of their relationship itself is likely to affect the amount of time they spend together.

Figure 8.1b shows the adult females with whom Fifi has associated most frequently from 1974 (two years after her mother's death) to 1982. If we compare it to Figure 8.1a, which provides comparable

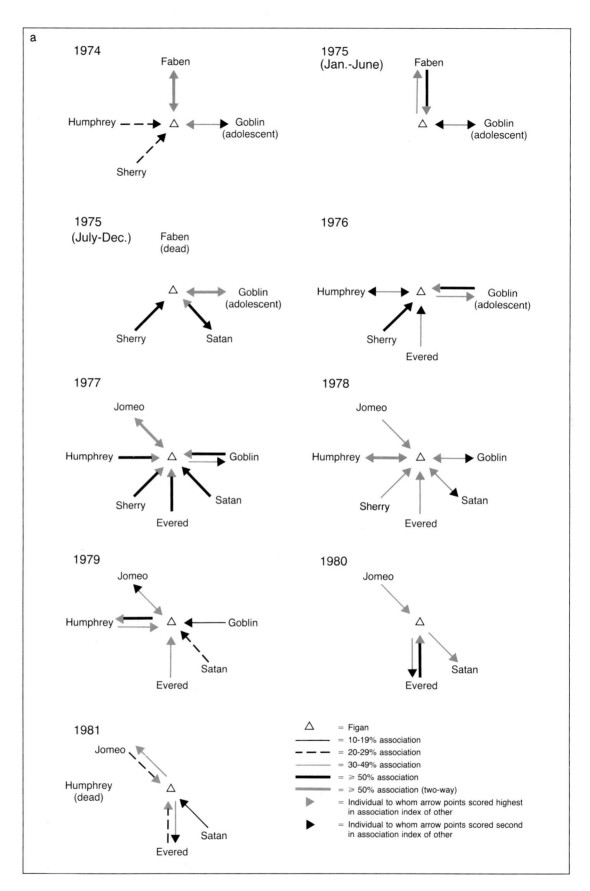

Figure 8.1 Association patterns of *a*, Figan with adult males (including Goblin as an adolescent), 1974 to 1981, and *b*, Fifi with adult females, 1974 to 1982.

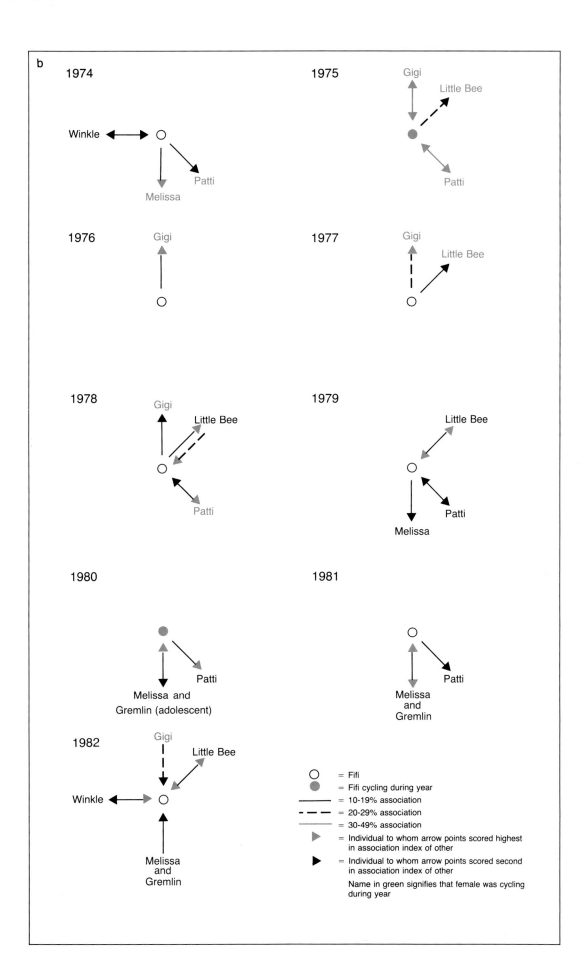

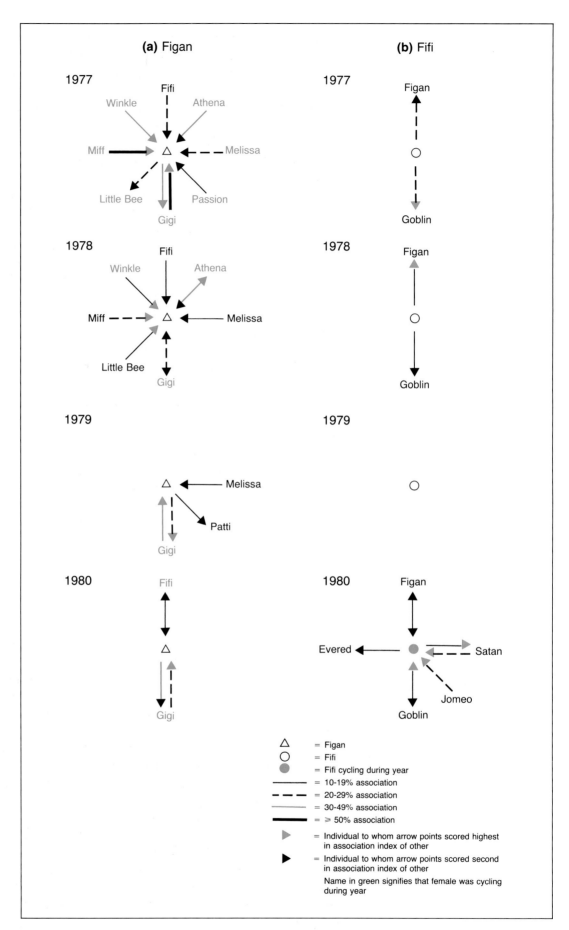

Figure 8.2 Association patterns from 1977 to 1980 of *a*, Figan with adult and late-adolescent females, and *b*, Fifi with adult males.

information for Figan, two striking differences are apparent. First, Fifi spent far less time with even her closest female companions than did Figan with other males; second, there is much less consistency, from year to year, in Fifi's choice of associates—a reflection of the continually changing nature of the reproductive cycles and family structures of Fifi and her companions.

Many of the adult Kasakela females do not even appear in Figure 8.1*b*; they never spent enough time with Fifi, at least during observation hours, to qualify. Even though it is often not possible to define relationships characterized by extremely low levels of association, in some instances the infrequent association is probably a result of deliberate avoidance. This was certainly true for Fifi and Passion. Fifi almost never chose to travel with Passion and her family; when they were together, it was usually because they met in camp or were part of the same large gathering. Any interactions were usually unfriendly. As we shall see in the next section, after the death of Passion, who was higher ranking than Fifi, the relationship between Fifi and Passion's daughter became actively hostile.

In Commentary 8.2 are described Fifi's relationships with four males: Figan, her eldest brother, with whom she was always very relaxed; Leakey, her first consort partner when she was a late adolescent; and Evered and Satan, her most frequent grooming partners, both of whom have taken her on consortships. With a fifth male, Humphrey, Fifi had an unfriendly relationship.

The relatively small amount of time that Fifi spent with adult males during her anestrous years is illustrated vividly in Figure 8.2*b*. During one year, 1979, there was not one adult male with whom Fifi was seen for as much as 10 percent of her total (262) target hours. Once again a comparison with Figure 8.2*a*, which shows Figan's association with adult females over the same time period, illustrates the sex difference in sociability.

The Social Networks of Two Individuals

Now that we have looked, in a general way, at some of the more significant relationships in the lives of two chimpanzees at Gombe, we should try to examine, in a little more detail, the way that a chimpanzee relates to *all* the other members of the community. There are three sources of Gombe data from which it is possible to glean information: observations recorded during follows when he (or she) was the target, information recorded when other individuals were the targets, and information recorded daily in camp. In order to gain as complete a picture as possible of the relationships of each chimpanzee with every other, we need to combine these data.

I have attempted to do this for two individuals, Goblin and Fifi. As shown in Table 8.1, both were quite extensively observed during the year 1983. I should like to have compiled information on Figan, but he had vanished by 1983, a year selected for present purposes because the data were already organized in a way that made it relatively simple to extract the desired information. Let me use Goblin to explain the procedure:

Table 8.1 Number of hours in 1983 that Goblin and Fifi were observed.

	Hours observed	
Source of data	Goblin	Fifi
Target of follows	314.00	179.00
Mother-infant study	0	31.00
Observation in camp	159.50	165.25
Follows of other individuals	613.00	476.50
Total	1,086.50	851.75

(a) From all follows when Goblin was the target, the number of times he interacted with each other adult or late adolescent was calculated. A behavioral sequence (A approaches and pant-grunts; B bristles; A squeaks and extends hand toward B; B swaggers and attacks; A screams and flees, then approaches for reassurance) was scored as one interaction. A grooming session was scored as one interaction, whether it lasted for one minute or forty.

(b) The total number of hours that Goblin was observed in association with each other individual during Goblin follows was calculated.

(c) For each Goblin-other dyad the number of interactions per Goblin observation hour was calculated.

(d) Details of all Goblin's interactions with others were carefully extracted from camp reports and from follows of other individuals.

(e) All grooming involving the target chimpanzee was recorded during follows. The time spent by Goblin in grooming and in being groomed by each other individual was calculated as a percentage of the time he spent with that individual.

The precise data from Goblin follows are given in Table 8.2: the number of hours that he was seen in association with every other individual, and the number of interactions recorded between each Goblin-other dyad.

Figure 8.3a depicts Goblin's social network during 1983. It was compiled from the data shown in Table 8.2 plus information extracted from the other sources indicated above. Goblin appears at the center of the network. Other individuals (over twelve years of age) are represented by circles of differing sizes and at various distances from Goblin. The circles closest to him represent individuals with whom he associated most frequently: the largest circles are those of the chimpanzees with whom he was observed to interact most often during follows. (The size of each circle depends upon the number of interactions between the target and the individual represented by the circle, per target hour.) The position of each circle indicates the sex of the individual represented and the type of relationship Goblin had with him or her. Males appear in the top half of the figure: those with whom Goblin was friendly are clustered to the left, those with whom he was unfriendly to the right, and those near the twelve o'clock mark are the individuals with whom Goblin had a tense, ambivalent relationship. In the lower half of the web, where Goblin's relationships with the females are plotted, those with whom he had friendly and unfriendly relationships are positioned to the left and right, as for the

Table 8.2 Number of hours during follows of Goblin (314 target hours) and Fifi (210 target hours) that they associated with each other individual of the community, and number of interactions recorded between each dyad in 1983. Except for Fifi's three offspring, immature individuals under twelve years of age are not included. Each category (males, females, and offspring) is listed in decreasing order of age.

Community members	Goblin		Fifi	
	Hours together	Number of interactions	Hours together	Number of interactions
Evered	103.0	81	12.0	6
Satan	119.5	74	13.0	6
Jomeo	73.0	60	10.5	2
Goblin			15.5	7
Mustard	72.0	21	5.0	2
Atlas	122.0	15	9.5	2
Beethoven	134.0	11	12.5	1
Jageli	141.5	23	12.0	1
Melissa	68.0	5	13.0	5
Nope	9.5	1	1.5	1
Athena	30.5	3	0	0
Gigi	98.5	20	21.0	1
Miff	35.0	10	1.5	2
Fifi	73.0	12		
Dove	29.0	4	3.5	1
Winkle	87.5	24	23.0	4
Sparrow	30.0	6	2.0	1
Little Bee	47.5	22	30.0	5
Patti	19.0	20	12.5	3
Sprout	27.5	5	0	0
Pom	34.0	9	3.5	1
Kidevu	60.0	9	0	0
Gremlin	69.0	7	14.5	1
Candy	41.5	1	0	0
Skosha	53.5	3	4.5	1
Fifi's offspring				
Freud			111.5	38
Frodo			170.0	45
Fanni			210.0	140

males. Females who cycled during the year, with whom Goblin had sexual relationships, are indicated. Females with whom he interacted very little and with whom his interactions were neither particularly friendly nor unfriendly are clustered around the six o'clock mark and labeled "neutral." Thus the size of a circle and its distance from the center is dependent on precise information from target follows, whereas the position of a circle—to the right or left—takes into account all data collected during the year.

Finally, Figures 8.4a and 8.5a, drawn from precise Goblin follow information, show relative scores: the former gives the number of interactions per hour based on the number of hours that each Goblin-other dyad was observed, and the latter gives the percentage of time spent grooming relative to the time that each pair was seen together.

Figures 8.3b, 8.4b, and 8.5b present the same information for Fifi, again based on Table 8.2. In her case two individuals under age

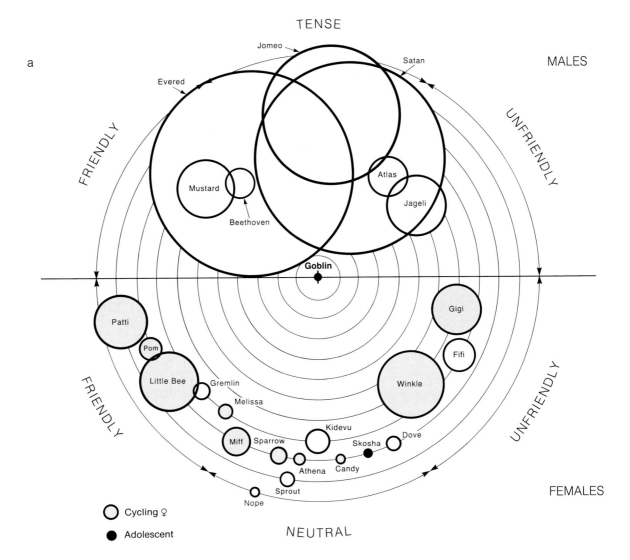

Figure 8.3 The 1983 social networks of targets *a*, Goblin, and *b*, Fifi, who appear at the center of their respective networks. Each circle represents a different individual of the Kasakela community. Males are shown in the upper half, females below; the type of relationship (friendly, unfriendly, tense, or neutral) is indicated around the edge. The distance of the circles from the center indicates the percentage of total hours that the target was observed with each other individual, those nearest to the center being the most frequent companions. The size of each circle depends upon the number of interactions between the target and the individual represented by the circle, per target hour.

thirteen have been included—her sons, Freud and Frodo. Before we compare the two social networks, I should emphasize one thing: the number of interactions that actually took place during the follows inevitably exceeded those recorded. The analysis does not include the many subtle glances, soft grunts, inconspicuous avoidances, and so on that characterize chimpanzee communication. These are quite often described in the reports, but they are not recorded reliably—and even if attempts were made to do so, many would be missed. Moreover, while periods marked "bad observation" have been subtracted from the total number of hours, some of even the more conspicuous interactions will undoubtedly have been missed. Nevertheless, I believe that by taking into account all the data recorded during the year, the picture presented here is reasonably true to life. Certainly the differences between the two networks reflect a real variation in the ways that males and females relate to other individuals in their society.

Male and Female Compared

Table 8.2, which gives Goblin's and Fifi's association scores with the other individuals of the community, highlights the pronounced sex difference in sociability. A comparison of the two social networks

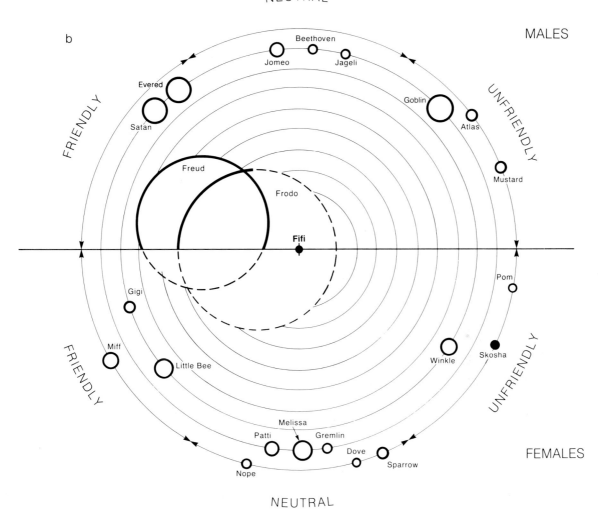

presented in Figure 8.3 shows this more vividly: the circles on Fifi's network (excluding those representing her two sons) for the most part are smaller and farther from the center than the circles on Goblin's network. The two circles representing Freud and Frodo, however, are not only large, but are closer to the center than any on Goblin's network. If we were to include a circle representing infant Fanni, twenty-one months old at the start of the year, it would be three times larger than Frodo's. And even more than is the case with other dyads, many interactions between a mother and her infant go unrecorded during follows (except in the special mother-infant study, data from which will be presented elsewhere). Indeed, we might say that in some ways a mother is interacting with her small infant all the time. Thus if we compare for Goblin and Fifi the *total* rate of interacting with others per target hour, Fifi's score would be much higher than Goblin's. Based on the numbers in Table 8.2, their scores are about equal; 1.5 interactions per hour for Goblin, 1.4 for Fifi. But if we compare rate of interaction with *nonrelated* individuals of age thirteen or more, then Goblin's score is much higher; Fifi's is only 0.3 interaction per target hour. Moreover, Figure 8.4 shows that Goblin's rate of interaction with most other individuals, based on the number

a. Goblin

b. Fifi

Interactions per Hour of Association

Adult ♂
Adolescent ♂
Close kin ♂
Cycling ♀
Close kin (Melissa)
Adult ♀
Association 4 hours or less
Peripheral ♀

Evered, Satan, Jomeo, Goblin, Mustard, Atlas, Jageli, Beethoven, Melissa, Athena, Gigi, Miff, Winkle, Little Bee, Patti, Pom, Sprout, Sparrow, Candy, Fifi, Nope, Dove, Kidevu, Gremlin, Skosha, Freud, Frodo

Adult ♂♂ — Adolescent ♂♂ — Cycling ♀♀ — Other ♀♀ — Offspring of hours each dyad was observed together, exceeds Fifi's in almost all instances (not including the five chimpanzees with whom Fifi was observed for less than four hours during the year).

Figure 8.5 shows that during follows, with the exception of Fifi and her offspring, neither Goblin nor Fifi groomed or received grooming very frequently. Individuals vary greatly in the amount of time spent in social grooming, and these two chimpanzees are not industrious groomers. During his follows Goblin was only involved in grooming for 2.5 percent of the total time; Fifi, if we include sessions with her offspring, for 5.3 percent of the time. Goblin only groomed three individuals for more than 3 percent of the time that he was with them—the three senior males; Fifi groomed Satan and Freud for over 5 percent of the time they were observed together. Goblin, at some time during the year, was observed to groom with all but six of the adults of the community; Fifi was never seen to groom with the three youngest males, nor with seven of the females who are represented in her social network.

Same-Sex Relationships

GOBLIN AND THE MALES   During 1983 Goblin was working to attain complete dominance over the senior males, Evered, Satan, and Jomeo. His overall relationship with them was tense. In Figure 8.3a, however, Evered's circle appears on the more friendly side. This reflects the fact that Goblin was not observed to attack Evered, whereas he was seen to attack both Satan and Jomeo—three and four times respectively. There were times too when Goblin and Evered seemed very relaxed when they were together. Only if interactions are recorded in much finer detail will it be possible to quantify impressions of this sort. The two younger adult males, Mustard and Atlas, as well as the two late adolescents, Jageli and Beethoven, associated with Goblin quite frequently. Mustard, although definitely subordinate to Goblin, was less frenzied in his submissive acts than Atlas or Jageli; he groomed with Goblin quite often, and twice cooperated with him in hunting. Atlas, Jageli, and Beethoven were mostly seen with Goblin in sexual parties: it was females in estrus who drew them together, and Goblin was intolerant of his subordinates at such times, often threatening them if they approached too closely. Beethoven usually kept well out of Goblin's way and so avoided provoking him; Atlas and Jageli were less cautious and were threatened more often.

FIFI AND THE FEMALES   Fifi spent more time during the year with Gigi, Little Bee, and Winkle than with other females. Her relations with Gigi and Little Bee, and also with Miff, were relatively friendly, but with Winkle aggressive incidents often flared up. And Fifi was definitely hostile in her dealings with Pom and Skosha. Never friendly toward Pom, Fifi was free to express her dislike after the death of Pom's mother, who ranked higher than Fifi. Similarly, after Skosha's "foster mother," Pallas, died, Fifi (along with Melissa and her daughter Gremlin) repeatedly threatened, chased, and sometimes attacked Skosha; it seemed that they were trying to drive the adolescent right out of the community.

Figure 8.5 Percentage of time during which *a*, Goblin, and *b*, Fifi, groomed other individuals relative to the number of hours each dyad was observed together during follows in 1983. Also shown is the percentage of time during which Goblin and Fifi were groomed. Although Goblin and Fifi were not seen to groom during the 314 hours that Goblin was followed, Fifi was observed to groom Goblin during the 210 hours that she was the target.

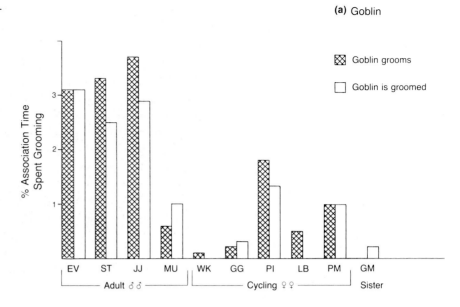

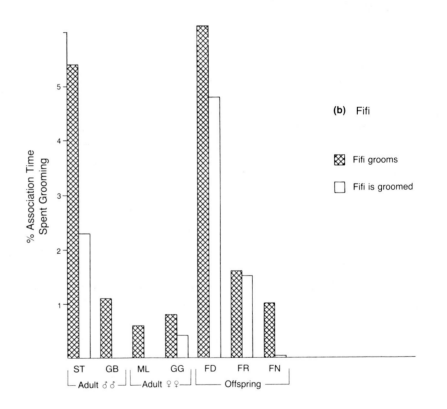

The other females are clustered in the "neutral" zone of Fifi's network. This is because of the difficulty, already discussed, of defining the nature of relationships between individuals who actually interact but rarely. Such interactions as were recorded between Fifi and these females were often aggressive in nature: they threatened one another during feeding, or when their infants squabbled. But against this recorded list of threats, displays, and even the occasional attack, one must set the long hours when they associated peacefully but were not seen to interact in any very obvious ways.

Goblin sits with his sister Gremlin, who cradles her newborn infant, Getty.

Bonds between mothers and their offspring may last throughout life. Here Melissa is seen with her twins and her grown son, Goblin (age thirteen years). As they hear sudden calls up the valley, they embrace each other.

Male-Female Relationships

GOBLIN AND THE FEMALES   As Figure 8.3a shows, many females came into estrus during 1983. Goblin was impatient in the sexual context, quick to show aggression when a female did not immediately obey his summons. In particular, he had many altercations with Gigi, resulting from her frequent attempts to avoid his sexual advances (as she often avoids other adult males). With Little Bee, Pom, and Patti, Goblin's relations were much more harmonious. These females usually responded quite rapidly to his sexual overtures, and levels of grooming between them were relatively high (Figure 8.5a). He was most relaxed and friendly with Patti, whose 1979 infant, Tapit, he is thought to have sired.

Of special interest is Goblin's relationship with his mother, Melissa, who came into estrus in 1983 for the first time in six years. During both her periods of sexual swelling, her son was seen to try to copulate with her. Melissa was initially aggressive in her response, but after Goblin actually attacked her when she avoided his courtship, she became fearful. This sequence of events had a profound effect on their relationship, which until that time had been friendly and supportive.

Goblin seldom interacted with anestrous females, either during follows or at other times. His relationship with Fifi was basically unfriendly: he sometimes threatened her, she showed submissive or avoidance patterns, he was never seen to reassure her, and grooming was rarely observed. Goblin's relationship with his sister Gremlin was fairly relaxed and friendly. His levels of association and interaction with other noncycling females were so low that it was not possible to define their nature with any precision, so I have categorized them (in the same way as many of the Fifi-female relationships) as "neutral."

FIFI AND THE MALES   A glance at Figure 8.3b shows that Fifi interacted with Satan and Evered, her two long-term favorites, more than with the other adult males. Satan was her most frequent adult nonrelated grooming partner (Figure 8.5b), and she was twice seen to play with him during the year. Although she did not groom with Evered during follows, she did at other times (during follows of other individuals, and in camp). Her relationship with Goblin was, as we have seen, basically unfriendly. Fifi was more relaxed with the senior males (including Goblin) than with the younger Mustard and Atlas. She had been able to intimidate them up until 1982 and now that she could no longer do so, she was quick to respond submissively to any sign of aggression. In this way, and by actually avoiding them, she was able to escape retribution.

Adaptive Value

An understanding of the adaptive value of relationships will help to explain why animals live in groups, rather than pursuing solitary lives. In the course of evolution natural selection has ruthlessly eliminated behaviors that have led to reduced levels of survival and reproductive success. (This does not mean that every action performed by an individual must necessarily contribute to his or her genetic survival, for behavior that is neutral will be tolerated.) However, behavior as deep-rooted and prevalent as the formation of close friendly relationships among group members must have decisive evolutionary advantages for the individuals concerned.

Kummer (1979), in a valuable conceptual approach, suggests that we look at the relationships in a social group from the point of view of a given individual, A. Any group member, B, with whom A interacts, is viewed as a potential resource for A. Interactions are seen as mechanisms for building or changing particular aspects of the relationship (such as strengthening a friendship, or decreasing hostility). In order to maximize B's potential value and minimize his potential harm, A must monitor B's behavior. He must try to assess his qualities (aggression, timidity), his tendencies (to attack or protect A, to avoid or follow A), and his availability (spatial proximity, extent to which other individuals may interfere with B's behavior). A can try to alter B's tendencies—for instance, by grooming or showing other affiliative behaviors to strengthen B's friendly responses; or by showing submissive behaviors to placate B if he threatens A; or by attacking and defeating him so that he is no longer dangerous to A. If B is stronger than A and overtly hostile, A must avoid B, at least for the time being, though he may attempt to improve B's tendencies toward him at a later date should this seem profitable. From this perspective interactions with B can be viewed as investments in a social bond that either now or in the future will be beneficial to A.

Seyfarth, Cheney, and Hinde (1978) stress the value of looking for the long-term benefits of interactions in relationships. Thus a female juvenile baboon grooms with high-ranking, nonrelated females, and interacts with their infants, more often than with those of lower rank. This tendency can be interpreted as a step toward establishing friendly relations with those individuals who will potentially be the most useful to her in the future—because a female may derive beneficial consequences from a good relationship with a high-ranking female in terms of access to scarce resources (Weisbard and Goy, 1976). A male juvenile, by contrast, mostly grooms low-ranking females. Because he will eventually dominate *all* females, there would be no obvious benefit in cultivating good relations with those of high rank; furthermore, he is likely to transfer out of his natal troop during adolescence (Packer, 1979). The low-ranking females he grooms are those he is most likely to mate (unchallenged by high-ranking males); thus his behavior, unlike that of the female, is directed at *immediate* benefits. When a young female chimpanzee first transfers into a new community, she may make repeated efforts to establish a good relationship with one or more young, potentially high-ranking females. Cases in point are Winkle's and Patti's persistent association with Fifi, despite the repeated aggression this provoked. These determined efforts to reduce levels of unfamiliarity were eventually successful. Quite clearly it is important that the immigrant be accepted by the stronger, higher-ranking females, in that the likelihood of continuing powerful alliances against her will thereby be diminished.

Behavior that at the time seems irrelevant can often be understood when viewed as part of a long-term investment. A young hamadryas male who had recently acquired two young consorts led them back and forth across the sleeping cliff in the evening, repeatedly looking over his shoulder. He was presumably testing their following responses, or building up their following tendencies, or both (Kummer, 1968, 1979). A male chimpanzee may attack (often severely) a cycling, nonswollen female for no apparent reason. This may be an attempt to shape her behavior, for it seems that she is then more likely to follow him on a consortship when she comes into estrus.

Kummer (1979) points out that the potential benefits of a given relationship may be ecological and/or social. To this I would add that there may also be emotional or psychological benefits. The advantages gained by close bonding among family members encompass all three categories.

Ecological Benefits

Kummer stresses the importance of determining the ecological benefits that one individual, A, can derive from his or her relationship with another, B. In the list that follows, relating specifically to chimpanzees, I have *excluded* benefits derived by an infant from its mother.

(a) B can make loud food calls, which will attract A to a food source, perhaps one previously unknown to A. This benefit can be important, particularly in times of food scarcity. Or the calls may simply motivate A to feed at a source already known to him by directing his attention to it at that particular time. On one occasion the adult male Satan, seemingly in response to loud food calls from Goblin, joined him in a tree where there were a few ripe fruits. Shortly after the two males left, a baboon troop stripped the tree of its remaining food. Satan might have missed that particular meal if Goblin's obvious delight had not attracted his attention.

(b) B can catch mammalian food and share it with A. Sometimes B may share plant foods with A, but this is rare except in mother-infant relationships and is unlikely to be a significant benefit, at least at Gombe.

David Greybeard, screaming, is threatened by an adult male baboon during banana feeding in 1962. When the baboon lunges toward him, he runs to seek reassurance contact from his close friend, Goliath. Eventually Goliath joins the fray in support of David.

(c) B can improve A's food-getting skills through attracting A's attention to a certain aspect of the environment (say, a tree with colobus monkeys), through social facilitation (B starts to hunt, A follows suit), or through observational learning on the part of A, who may watch B and subsequently utilize similar techniques. Other food-getting methods are mostly learned during infancy from the mother or an elder sibling.

(d) B can drive competitors from A's food or help A to do so. David Greybeard, when repeatedly threatened and hit by an aggressive male baboon who was trying to seize his bananas, kept running to solicit help from his friend Goliath. At first Goliath ignored David, but eventually he responded. Sometimes a lower-ranking male will aggressively threaten and even attack chimpanzees who approach to beg for a share from a higher-ranking companion. Of course he himself begs for a share, but many of the others are driven off and the possessor of the meat is able to eat more, and eat in more peace.

(e) B can warn A about danger (by calls or by startle reactions and avoidance), or protect A, or physically remove A from danger. Chimpanzees often learn that a snake is present from the response of an individual who has spied it. Adolescents and juveniles sometimes gather up infants who are too close to aggressive baboons or other danger.

(f) Much information about travel routes is, of course, learned from the mother, but additional topographical knowledge is acquired when an adolescent male A, or an immigrant or cycling female A, follows an adult male B. When traveling with adult males, A will also learn more about the community boundaries.

(g) B can clean A's skin and remove parasites.

(h) B can care for A when A is sick, keeping flies from wounds and grooming, thereby removing fly eggs or maggots from festering sores.

Social Benefits

The extent to which social benefits may lead to increased individual or inclusive fitness is often difficult to assess. Among the most obvious benefits are the following:

(a) B can be A's mate, and bear and raise A's child, or B can impregnate A, depending on their respective sex.

(b) B can protect A from threat or attack by a conspecific, as when Figan displayed toward older males who turned on Goblin when he challenged them.

(c) B can cooperate with A to intimidate or attack an opponent, as Faben repeatedly joined Figan in his persecution of Evered.

(d) A can acquire added competence in social interactions by watching B and doing as he does, as Goblin imitated Figan's early-morning displays.

(e) A can learn appropriate social behavior as a result of being disciplined by B. Adolescent Goblin, for example, screamed during a patrol and was threatened and hit by Humphrey; the adult males were maintaining vocal silence in a potentially dangerous situation.

(f) If A and B are infants or juveniles, A during play sessions with B will learn some of the social manipulations so important during triadic interactions in later life. Evered, for instance, learned that Flo might run to Figan's aid if their play got too rough. A also gains practice in fighting skills during play sessions of this sort.

In some relationships the social benefits conferred by one member of a dyad on the other are clear-cut: if, for example, A is an infant and B is his mother, or if A is a male and B a higher-ranking male who consistently supports his lower-ranked friend. Sometimes, however, the benefit for one member of a dyad is not immediately obvious. If we turn the last example around, what does the higher-ranked male gain socially from helping his lower-ranked friend? Particularly in cases when, in doing so, he risks being hurt himself? If he is helping a close relative, as when one sibling helps another, then the supportive act may increase his *inclusive fitness* through the mechanism of *kin selection* (Haldane, 1955; Hamilton, 1964). This evolutionary theory is based on the fact that natural selection is concerned, above all, with the survival of genes. Thus, even if a male gorilla is killed defending his group (Fossey, 1983), provided that his action enables close relatives to live, many of his genes, present in them, will survive. In this way self-sacrificial behavior can sometimes be selected for.

If the higher-ranking male is not a relative, his behavior is usually explained in terms of reciprocal altruism (Trivers, 1971): you help your friend today in the hope that he will help you tomorrow. This will be discussed in Chapter 13. There is an additional advantage that one individual can gain from supporting another, whether that other is kin or nonkin, higher or lower ranked: between them they may decrease the reproductive chances of a rival (Hinde, 1984).

What of the follower relationship between an adult and an adolescent male? The advantages to the adolescent are obvious; fascination for the older male assists the younger in loosening childhood dependency on his mother, and his exposure to male behaviors (such as hunting and patrolling) provides opportunity for observing and learning these patterns. Moreover, he becomes familiar with areas of the home range that he might never visit with his mother. The gradually increasing amount of aggression that the older male directs toward him as he matures will teach him about male temperament, and he will become more and more adept at finding ways of avoiding trouble and raising his status in the adult male dominance hierarchy. How does the older member of the partnership benefit? Probably very little. But then, for the most part the relationship imposes no burden upon him, either: it is almost entirely initiated and maintained by the follower. Furthermore, by facilitating the tendencies of the youngster to associate and form closer bonds with the males of his community and to acquire knowledge of territorial behavior, the older male helps to mold an individual who will be better fitted in the future for defending the community range and its resources against encroachment by neighbors. This development may be extremely important to the descendants of the "teacher," since, as we shall see in Chapter 17, a powerful and cohesive group of adult males is sometimes necessary for the survival of the community.

Emotional or Psychological Benefits

Emotional distress can precipitate and exacerbate a variety of illnesses in humans and animals, including a rather wide range of medical problems. In our own species sudden danger, whether actual or anticipated, typically leads to changes in endocrine and autonomic processes and hence to a wide variety of visceral disorders. These may trigger readily detectable emotional distress. Significant elevations of adrenocortical hormones in such situations are quite persistent when the anxiety continues unabated; they drop substantially when the distress is alleviated (Handler, 1970).

For humans the presence of a companion in a time of stress can be extremely beneficial. Embracing, stroking, and other forms of physical contact often help to calm us when we are very upset, just as the infant is quieted when his mother picks him up and comforts him. The presence of supportive companions can crucially affect human efforts in coping with a variety of stressful events—such as disruption of social ties (Hamburg, Elliott, and Parron, 1982)—and can protect against clinical depression (Brown, Bhrolchain, and Harris, 1975). Single people who go to high-risk locations in the course of their work (such as areas of terrorist activity) tend to find it more difficult to cope with fear than do those who are married or who have close friends, and they often leave before the end of their contracts (M. Gould, personal communication). In fact, very close relationships are sometimes *formed* between people who cope, together, with a dangerous or distressing experience.

Emotional bonding or attachment begins, in an evolutionary as well as an ontogenetic sense, with the crucial mother-infant bond. There is abundant evidence in the literature on both nonhuman primates (Kaufman and Rosenblum, 1969; Hinde and Spencer-Booth, 1971) and humans (Spitz, 1946; Bowlby, 1973) that disruption of this relationship, even when the infant is physically well cared for in the absence of the mother or other primary caretaker, results in serious emotional distress that is likely to leave lasting scars. The infant will suffer, at the same time, from underlying physiological disturbances (Reite, 1979). Two of the six young chimpanzees at Gombe who lost their mothers when they were between four and six years of age (that is, they were *nutritionally* independent) died within eighteen months.

In the chimpanzee, as in man, the period of emotional dependency on the mother has been extended considerably beyond childhood. Not only the infant, but also the juvenile and the young adolescent, will show distressed searching behavior, accompanied by whimpering and sometimes crying, if he or she becomes separated from the mother. Even an older offspring, who often *deliberately* leaves his mother, may become distressed if he loses her *accidentally*. Fully mature females, with infants of their own, sometimes search for their own mothers for hours at a stretch, whimpering from time to time. The weaning period followed by the birth of a new infant can cause quite serious emotional disturbances in youngsters (between four and five years of age)—decrease in frequency of play, and marked increase in whimpering and contact with the mother. These symptoms probably parallel feelings of rejection in human youngsters during similar childhood crises.

The abnormal relationship between Flo and her son Flint is relevant here. As we have seen, Flint was still dependent on his mother, riding on her back and sharing her nest at night, when he was eight years old. And by this time Flo herself was to some extent dependent on her son. If they came to a fork in the trail and each took a different direction, Flo was as likely to whimper, turn, and follow Flint, as he was to give in to her. The final anomaly was Flint's extreme depression after Flo's death, followed by his illness and death three and a half weeks later.

Throughout childhood the chimpanzee youngster typically seeks comfort from his mother in times of stress. As he spends more time farther away from her, he seeks reassurance from other individuals, as I shall discuss in Chapter 13. But he is likely to run first to his mother if she is there. Figan, even as a late adolescent, often ran to touch or briefly groom Flo when he was upset; usually she reached out to reassure him. The adolescent male who has begun to travel with the mature males often leaves a group after a period of intense social excitement (particularly if he himself was involved in any aggressive episodes) and rejoins his mother and siblings. Apparently he derives comfort from the stable family environment, in part because of the very high level of familiarity and probably also a feeling of security derived from the days of infancy. And if a family member cannot physically side with him in time of trouble, at least he or she will almost never join others against him.

The close and supportive relationship between adult brothers, such as that between Figan and Faben, is clearly of great social benefit to both. For Figan, it resulted in the attainment of alpha status; for Faben, the protection of a high-ranking ally. When Faben died, it seemed that Figan, after his long reliance on Flo, overlapping and followed by his friendly relationship with Faben, found it psychologically necessary to establish a close bond with another individual. He selected his deposed rival, Humphrey—presumably because by then Humphrey was the least threatening of the adult males. The difference in Figan's relationship with Faben and Humphrey is of interest: whereas Faben *actively* supported Figan during social conflicts, Humphrey, even after their friendship was assured, almost never did. But he did not join the others against Figan, and so he represented a measure of security. Figan's emotional dependence on Humphrey was evident when on three separate occasions he whimpered and screamed, on and off, for periods of up to forty-five minutes as he searched for the older male after they had become separated during foraging. And when Humphrey died, Figan forged the final friendship in the series with his onetime bitter rival, Evered.

Family Benefits

In chimpanzee society the closest bonds, as we have seen, are those between family members—between a mother and her offspring, and between siblings. Both members of such a pair benefit (Trivers, 1972), but in different ways, at different times, and in different contexts. Let us examine the advantages first of all from the point of view of the offspring. For him the long period of maternal care brings with it

Flo looks at her first grandson, Freud, cradled in Fifi's arms. Flint grooms his sister. (H. van Lawick)

Flo embraces her adult son Faben and tries to shelter behind him as adult male J.B. charges past. Flo and Faben are both screaming. (H. van Lawick)

obvious benefits in terms of protection, warmth, food, and opportunity to learn the various skills necessary for adult life.

If the mother is high ranking, this is beneficial for both her sons and her daughters, for they then have a good start in their long struggle to attain high rank. The juvenile or adolescent female gains valuable experience when allowed to play with, groom, carry, and protect her infant siblings—experience that enables her to care more competently for her own first child. Bonds that develop between the siblings themselves during these years are likely to endure, particularly those between brothers; this may well be crucial in determining social rank in later life. And, for both males and females, the presence of a sibling may be helpful and comforting in a variety of situations. For a secondborn infant and any born thereafter, the presence of older siblings adds to the protective and supportive atmosphere in which the child grows up. Often they help to care for the infant, even to the extent of adopting it in the event of the mother's death. For the infant male, an elder brother acts as a built-in role model from whom he learns a great deal about male behaviors.

From the point of view of the mother, the presence of older offspring provides her with companionship, grooming partners, and—often—support. They relieve her to a degree of her otherwise constant preoccupation with a small infant, by playing with, grooming, and otherwise diverting him. As they grow older, her sons and particularly her daughters become ever more useful as coalition partners with whose help she may rise to a higher position in the female hierarchy. Most important, by successfully rearing healthy offspring who are themselves able to reproduce, the mother increases her own overall inclusive fitness.

Finally, if we consider the network of relationships within the family as a whole, close kin, by helping one another, may increase the chances for high reproductive success, at the same time increasing the inclusive fitness of family members to the detriment of nonrelatives. In terms of the ultimate benefits of evolutionary biology as well as the immediate benefits of day-to-day life, a female's investment in her offspring can yield high dividends for herself and for her close kin.

9. Ranging Patterns

Adult males on a patrol sit close. (C. Busse)

September 1979 All morning we have been climbing steadily higher. Now we are not far below the rift escarpment. The chimpanzees ahead are close together and silent, their hair bristling slightly as they gaze into a narrow ravine. They are nervous, for males of the powerful Kalande community have been seen nearby. Satan moves a few steps forward, hair now fully erect. Suddenly he rushes back to Jomeo and Goblin and all three embrace with wide grins of fear. They are quite silent. From the bushes ahead a stick snaps; the three are off, noiselessly racing back to the north. I try to follow, but soon lose them. Nor can I find any sign of chimpanzees in the ravine. So I climb up until I can see all around. Below, stretching to the invisible coastline of Zaire in the west, is the blue water of Lake Tanganyika, to the north the territory of the Kasakela community. I gaze across Mkenke Valley to the well-known valleys beyond: Kakombe, heart of the KK community before the division; Kasakela; Linda; Rutanga; and finally Mitumba-Kavusindi Valley, the northern boundary of the park. Down near the Mitumba beach is the little area where Evered so often takes his consort females. I look southward; Kahama and Nyasanga, the two valleys the Kahama males claimed for themselves, scene of the bitter hostility between the separated parts of the original community, taken over now by the Kalande chimpanzees from the next valley to the south. Beyond Kalande, the hazy ridges of Kitwe, Gombe, and Bwavi, where it is unlikely that any of our known chimpanzees have ever roamed. I sit for an hour, but there is no sight or sound of a chimpanzee. I start the downward climb.

Group-living animals share an area of land within which they forage, sleep, raise their young, and go about their other daily activities. The size of this home range depends on many factors. Among the more important are the size and food requirements of the animal, the number of individuals in the group, the density of the surrounding population, and the type of habitat. Environmental factors, which may themselves cause differences in group size and structure, can underlie variations in the size of the range of different species, and of groups of the *same* species in *different* areas. As we shall see, chimpanzees living in the relatively lush environment of Gombe have smaller home ranges than do those inhabiting harsher and more arid parts of Africa.

At Gombe chimpanzees and baboons share the same environment; the chimpanzee community, with its fusion-fission organization, ranges over a larger area than does the more cohesive baboon troop, which has about the same number of (slightly smaller) individuals but with a somewhat more catholic diet. Animals do not use all parts of their range equally all the time, particularly when the area covered is large. One portion, the core area (Kaufmann, 1962), is likely to be utilized extensively; other locations may be visited only occasionally—when, for example, certain foods are seasonally available there.

Influence of the Individual

Unlike most primate species, chimpanzees follow no regular route in their daily search for food. Nor do they return to well-used sleeping sites each night; they construct their nests close to where they have had their last meal of the day.* Within the home range of the community, each adult chimpanzee is free to choose where he will go and the routes he will follow to get there. To some extent, however, his choice will be influenced by the movements of the other community members.

Some chimpanzees, more than others, have the ability to affect the travel patterns of their companions. Because of the fluid nature of chimpanzee society, there is no single overall leader. Almost any adult, or even adolescent, may at one time or another lead a small party and determine the direction of travel. In a sexual party it is often the female in estrus who controls the pace of travel, even if she plays no part in deciding the direction. It is common to see several or all of the adult males of the community waiting patiently under her tree, gazing up from time to time as she continues to feed or to groom or nurse her infant. Once, after patiently waiting for twenty minutes, Satan walked off for 15 meters. He paused, looked back, sat for a while self-grooming, and finally returned to the other males. Twice more he set off in this fashion, only to retrace his steps. Eventually he settled into a grooming session; but the instant the female started to descend he moved off again, leading the party in his chosen direction. In the consortship situation too, once the male has led the female to his chosen location, it is typically she who determines the pace of travel during the days that follow.†

When a family unit travels on its own, the mother normally is undisputed leader—unless she is with a late adolescent or adult son, in which case she often follows him. There may be disputes, as when an infant or juvenile male desires, against the wishes of his mother, to join nearby chimpanzees; some mothers give in much more readily than others. As the child gets older, disputes about who should go where and when often end with mother and offspring simply going in different directions.

Some chimpanzees have qualities that set them apart as leaders. This was first apparent in David Greybeard. Between 1961 (when

* Sometimes nest sites are used for several days in succession, by the same or different individuals, particularly when the location is close to a rich crop of food; the sites may be used again the following year when the same food becomes available.

† Even the male hamadryas baboon, who usually herds his female aggressively if she strays from him, meekly *follows* her when she is in estrus (Kummer, 1968).

Goliath, then William, followed David to camp) and 1965 (when the last peripheral individuals appeared) about 75 percent of the newcomers arrived for the first time with David. He possessed all the characteristics that seem to be the most significant for leadership. He was calm, tolerant, and remarkably nonaggressive, yet at the same time he was self-confident and, when thwarted, determined to get his own way. He was quick to reach out and reassure nervous or fearful subordinates, and his quiet approach and grooming often served to relax socially aroused high-ranking males also.

Hugo was another leader. More excitable than David, he was nevertheless quick to deal out reassurance on occasions when he did become aggressive. When Winkle and Patti transferred into the Kasakela community as adolescents, they often traveled in Hugo's company. As he got older, young males first beginning to leave their mothers often moved about with him.

Bygott (1974), defining a leader as the individual walking first in the line of travel, found that in 1971 there were no individuals who invariably led, but that those who did so most frequently were three middle-ranked males—Evered, Faben, and Hugo (quite old by then)—and the young, high-ranking Figan. Humphrey, alpha male at that time, seldom led his companions, and Mike, the top-ranked male before Humphrey, was something of a loner.

Figan, even as an adolescent, showed an unusual power of leadership (as I shall discuss in Chapter 19) that characterized him throughout his reign as alpha. From 1979 on Figan became increasingly reluctant to travel to peripheral areas in the south. Four times during such excursions he was seen to stop and then head back north. On each occasion the others also turned back and followed him. In marked contrast, when Humphrey, who had always been reluctant to travel in the south, left parties that were headed in that direction, the others almost always went on without him. Many younger males, however, are eager to make contact with chimpanzees of neighboring communities. Individuals such as Satan and Sherry typically pushed forward during excursions, boldly leading the way toward possible excitement, and there were many occasions when the rest of their party followed.

To a considerable extent, therefore, an individual chimpanzee can influence the ranging patterns of the community as a whole.

Daily Travel Patterns

A number of factors influence the distance traveled in a day: (a) seasonal distribution of foods, distance between major crops being particularly important; (b) health—sick individuals do not travel far; (c) weather—travel is curtailed when it pours rain; (d) consortships—a male-female pair travel very short distances once they have established their consort range; (e) the activity of the previous day—after a long excursion there is a tendency to travel a much shorter distance the following day. During Wrangham's (1975) two-year study the longest distance covered in a day was 10.7 kilometers by an adult male, Hugh, who traveled only 2.4 kilometers the following day.

Males and females utilize the home range of the community in different ways (Wrangham, 1975, 1979; Wrangham and Smuts, 1980).

Adult male Hugh, walking.
(B. Gray)

Table 9.1 Kilometers per day traveled during long follows. A dash indicates that the follow did not last all day; the chimpanzee was lost to the observer's sight for some time.

Subject and dates of follow	Day 1	Day 2	Day 3	Day 4	Day 5	Day 6	Day 7	Day 8	Mean
Figan									
7.16.74 to 7.23.74	6.2	4.1	3.6	3.0	4.2	3.3	4.0	4.8	4.2
7.24.74 to 7.31.74	2.5	2.5	2.4	4.4	2.5	3.3	—	3.5	3.0
8.1.74 to 8.8.74	3.0	3.4	4.1	3.3	—	5.0	3.0	3.8	3.7
8.9.74 to 8.16.74	2.9	3.6	2.9	5.0	3.5	2.7	2.9	2.6	3.3
7.19.77 to 7.26.77	3.8	4.8	4.6	4.0	5.7	7.5	—	4.8	5.0
Goblin									
8.19.83 to 8.26.83	7.0	2.5	4.8	—	4.6	6.0	6.2	4.0	5.0
Passion (pregnant)									
8.4.77 to 8.11.77	3.7	2.9	3.3	2.5	1.3	2.7	2.9	3.5	2.9
8.12.77 to 8.21.77	1.9	3.7	2.8	1.9	2.4	2.0	2.3	2.5	2.4
8.22.77[a] to 8.29.77	2.9	—	1.5	2.0	2.6	3.5	1.3	3.1	2.4
9.11.77[b] to 9.18.77	3.1	3.5	4.5	2.1	2.0	4.1	4.3	5.2	3.6
Pom (anestrous)									
8.2.83 to 8.9.83	1.4	3.4	1.8	1.3	3.4	2.5	1.4	2.9	2.3
Fifi									
6.3.76[c] to 6.11.76	1.4	2.3	1.9	1.5	1.5	1.6	1.4	2.5	1.8
7.5.77[d] to 7.10.77	—	1.4	1.6	2.5	—	1.5	1.6	—	1.7
3.22.81[e] to 3.29.81	0.7	0.7	1.3	1.0	0.6	0.9	1.8	0.3	0.91
Winkle									
1.21.79 to 1.28.79	3.2	1.0	1.4	0.6	1.3	0.9	1.3	1.0	1.3
1.1.79[f] to 1.8.79	0.8	0.6	0.8	0.8	0.6	1.2	1.0	2.4	1.0
Gremlin (estrous)									
8.21.81 to 8.30.81	1.9	2.1	2.6	3.8	4.1	3.7	—	3.2	3.1
Evered/Winkle (consortship)									
5.4.78 to 5.12.78	1.3	2.2	1.0	0.4	0.8	1.2	1.1	1.2	1.0

a. Pom anestrous.
b. Pom estrous.
c. Infant birth 6.1.76.

d. Infant age one year.
e. Infant age about one week.
f. Infant birth 12.29.78.

Males typically travel farther in a day than do females (a mean distance of 4.9 kilometers versus a mean of about 3.0 kilometers). Males also range more widely than females, tending to visit each of the boundary areas of the home range once every four days or so, whereas females, at least when anestrous, spend a great deal of time in their core areas. The distance traveled by a female will be affected by her reproductive state: she ranges farther than usual when in estrus and associating with the adult males, less during late pregnancy and for the first few weeks after giving birth. Her movements may be influenced by the age and sex of her offspring, as we shall see.

Over the years at Gombe there have been, as already described, a number of long follows when target individuals were observed on consecutive days for periods ranging from one to nine weeks. These follows (carried out by dint of considerable physical exertion and dedication on the part of the observers) have resulted in unique data on the ranging patterns of the Gombe chimpanzees. Figures 9.1 through 9.4 are maps on which are traced the daily travel routes of six adults and a consort pair over eight-day periods (and one of six days) selected from these follows. For Passion and Winkle, two sets of data are shown; for Figan, three sets. The maps enable us to see how different seasons or different circumstances affect individual travel patterns. Details of mean daily distances covered during these follows, as well as during an additional long follow on the young female Gremlin, are shown in Table 9.1.

Figure 9.1 illustrates the travel routes of two adult males, Figan and Goblin, during dry-season months. Figan, alpha male at the time, was followed for fifty days in July–August 1974 (Riss and Busse, 1977). From this marathon, two blocks of eight days have been extracted (Figure 9.1a and b). They represent two extremes: the eight-day period when Figan traveled most widely and the period during which he traveled least. In 1977 Figan was followed for fourteen days: the eight days selected for Figure 9.1c include his longest journey (on the sixth day) and his shortest. Goblin was followed for eight days in 1983 (Figure 9.1d).

The travel patterns of Figan in 1977 and of Goblin in 1983 were similar. The mean distance covered per day was the same for both, 5.0 kilometers. In 1974 Figan traveled less widely, and his mean daily distance for the entire period was 4.4 kilometers. Wrangham (1979) also noted differences in the mean daily distances traveled during successive dry seasons: in 1972 the mean for eight males was 3.9 kilometers; the following year, for six males, it was 6.2 kilometers. Wrangham suggests the shorter daily mean in 1972 may have resulted from scarcity of food; the following year was more bountiful. In 1974 food was not scarce (Riss and Busse, 1977), yet Figan's mean daily distance was similar to that of the males in 1972.

During the 1974 follow Figan made three excursions to the northern boundary, two to the east, and one to the south—an average of one excursion per eight days. Only two of these visits to peripheral areas resulted in patrolling behavior (see Chapter 17). During the fourteen days he was followed in 1977 Figan made six excursions, two to the north, one to the east, and three to the south. On five of these occasions he and his party, upon reaching the farthest point of their journey, settled down to feed; on the sixth, the party patrolled in the

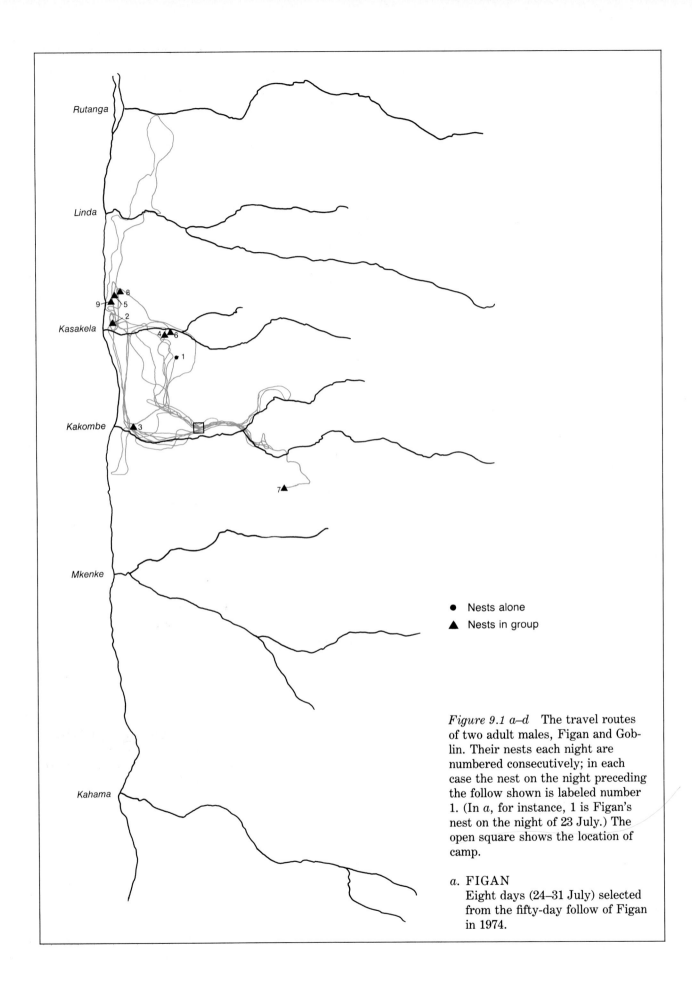

Figure 9.1 a–d The travel routes of two adult males, Figan and Goblin. Their nests each night are numbered consecutively; in each case the nest on the night preceding the follow shown is labeled number 1. (In *a*, for instance, 1 is Figan's nest on the night of 23 July.) The open square shows the location of camp.

a. FIGAN
   Eight days (24–31 July) selected from the fifty-day follow of Figan in 1974.

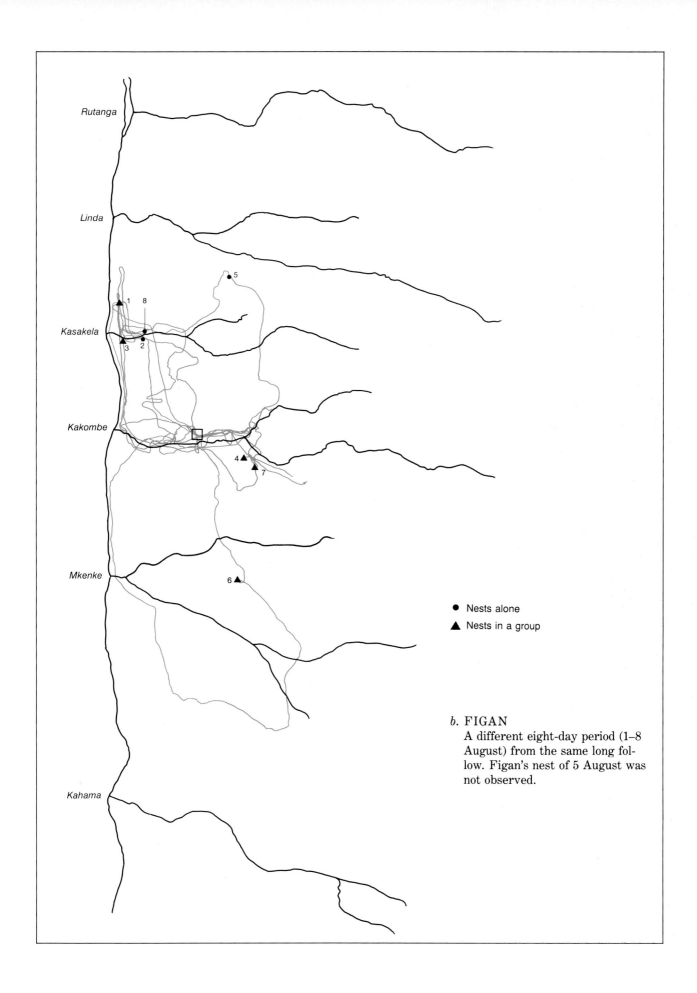

Rutanga

Linda

5

1 8

Kasakela

3 2

Kakombe

4

7

Mkenke

6

● Nests alone

▲ Nests in a group

b. FIGAN
A different eight-day period (1–8
August) from the same long follow. Figan's nest of 5 August was
not observed.

Kahama

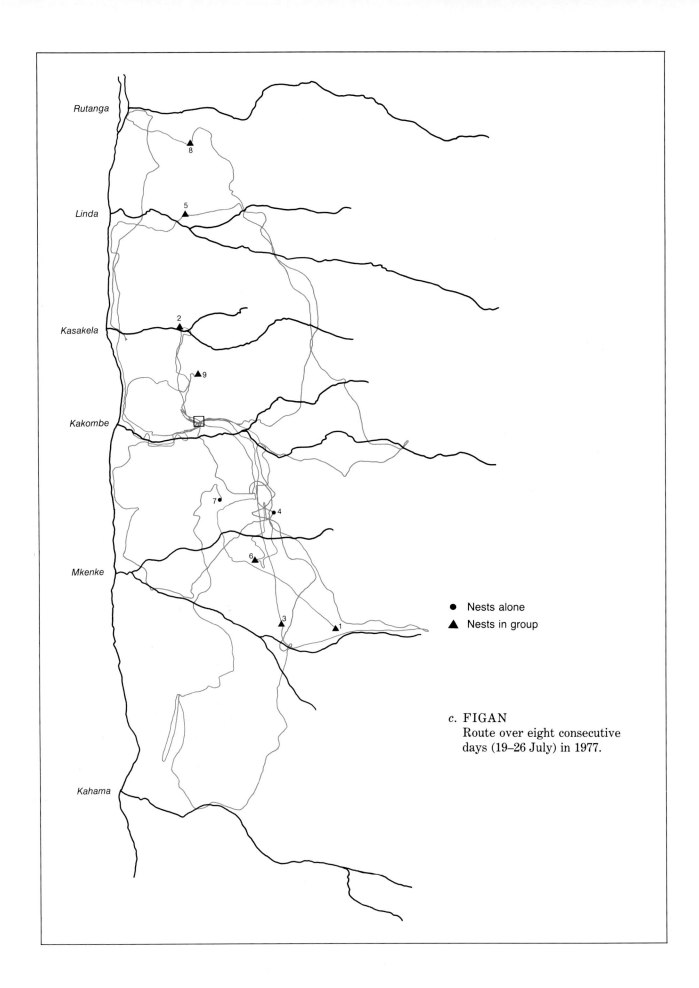

Rutanga

Linda

Kasakela

Kakombe

Mkenke

Kahama

● Nests alone
▲ Nests in group

c. FIGAN
Route over eight consecutive
days (19–26 July) in 1977.

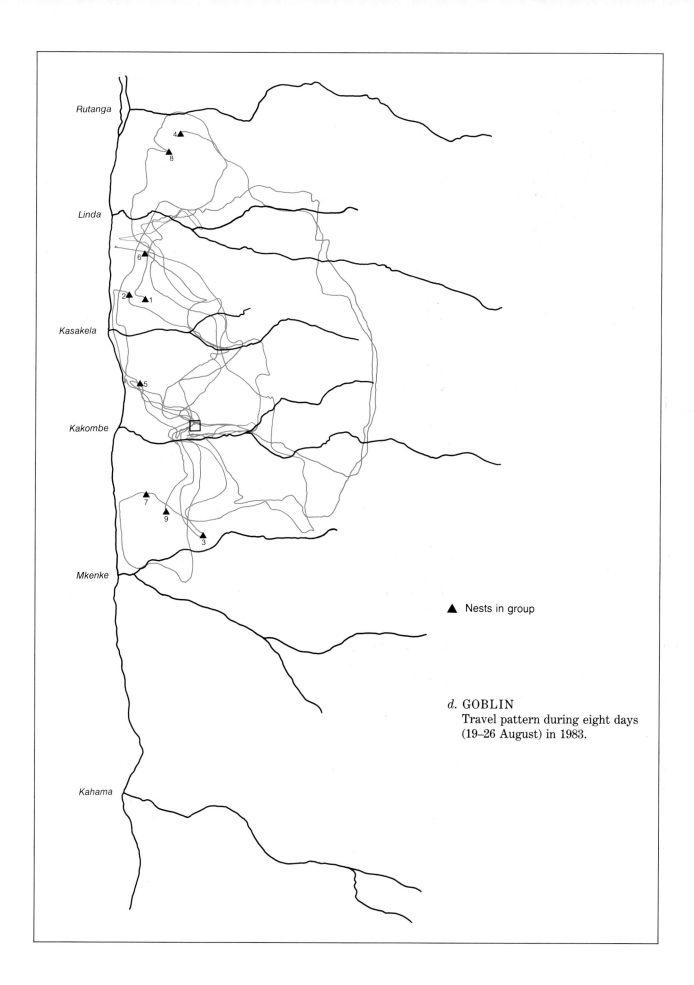

d. GOBLIN
Travel pattern during eight days
(19–26 August) in 1983.

A young female in estrus follows her male during a consortship.

southern overlap zone, moving cautiously and gazing toward the neighboring community from tall trees. They displayed vigorously upon return to their own core area after some four hours of silence. Goblin in 1983 made excursions three times to the northern overlap zone, once to the northeast, and once to the southern periphery, where he and his party patrolled.

Figure 9.2 shows the seven-day and eight-day ranges of two anestrous females, Fifi and Winkle. Both were accompanied by juvenile sons: in addition, Fifi was carrying a year-old infant and Winkle a three-week-old baby. The two females journeyed less extensively than either Figan or Goblin, remaining in relatively small areas; the mean distances traveled per day were low—1.7 kilometers for Fifi and 1.3 for Winkle.

Figure 9.3 shows two eight-day ranges for the anestrous female Passion. When her late-adolescent daughter, Pom, was in estrus (Figure 9.3*b*), Passion traveled more widely (mean distance per day 3.6 kilometers) than when Pom was anestrous (mean 2.4 kilometers). Figure 9.3*c* shows the eight-day range of Pom in 1983, after her mother's death. The area over which she ranged had shifted to the north because of intercommunity conflict, but the distances traveled were similar to 1977, when she was anestrous and with her mother (Figure 9.3*a*).

The final map in this series (Figure 9.4) shows a typical consort range, the one established by Evered and Winkle for ten days in May 1978. The mean distance traveled per day, once the range had been established, was 1.3 kilometers. The farthest they moved in a day was 2.2 kilometers; once it was only 0.4 kilometer.

Individual Patterns over Time

Year Range

The year range of an individual is the sum of all areas visited during a given year. Each chimpanzee is, of course, observed for relatively few days per year. Thus the recorded year range is bound to be smaller than the actual year range; in other words, there will be places beyond the plotted boundaries that a chimpanzee visits when he is not being followed. Wrangham (1975), using a grid of 100 meters × 100 meters superimposed on one of the standard maps used to plot chimpanzee travel in the field, found that the number of squares in which an individual was seen gradually increased relative to the number of hours he was observed during a year. After about 250 hours of observation there was a leveling off, when it could be assumed that the recorded range was an approximation of the real range.

The year range of a healthy adult male is larger than that of a healthy anestrous female. In 1972–73, the areas utilized by the adult males (all of whom were followed for 250 hours or more), ranged from 9 to 12 square kilometers, with a median of 10.3. For three anestrous mothers, also followed for the requisite number of hours, the median year range was 6.8 square kilometers, ranging from 5.8 to 7.0. For cycling females, who travel with the males during periods of estrus, the year range is of course larger: for three such females the median year range was 9.8 square kilometers, ranging from 8.3 to 11.0 square kilometers (Wrangham, 1975, 1979).

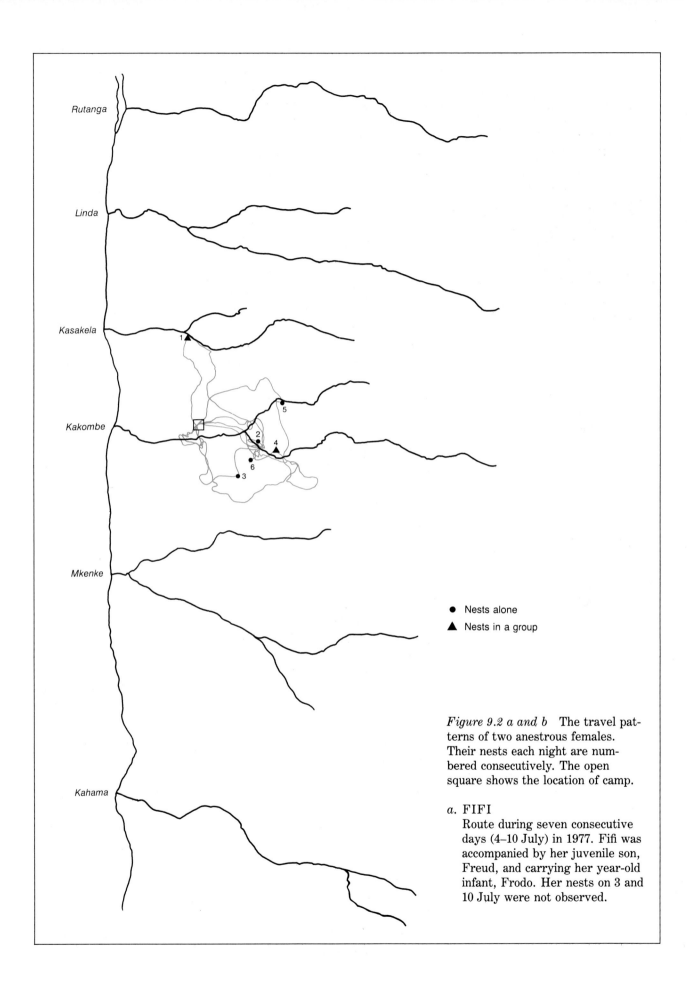

Rutanga

Linda

Kasakela

Kakombe

Mkenke

Kahama

● Nests alone
▲ Nests in a group

Figure 9.2 a and b The travel patterns of two anestrous females. Their nests each night are numbered consecutively. The open square shows the location of camp.

a. FIFI
Route during seven consecutive days (4–10 July) in 1977. Fifi was accompanied by her juvenile son, Freud, and carrying her year-old infant, Frodo. Her nests on 3 and 10 July were not observed.

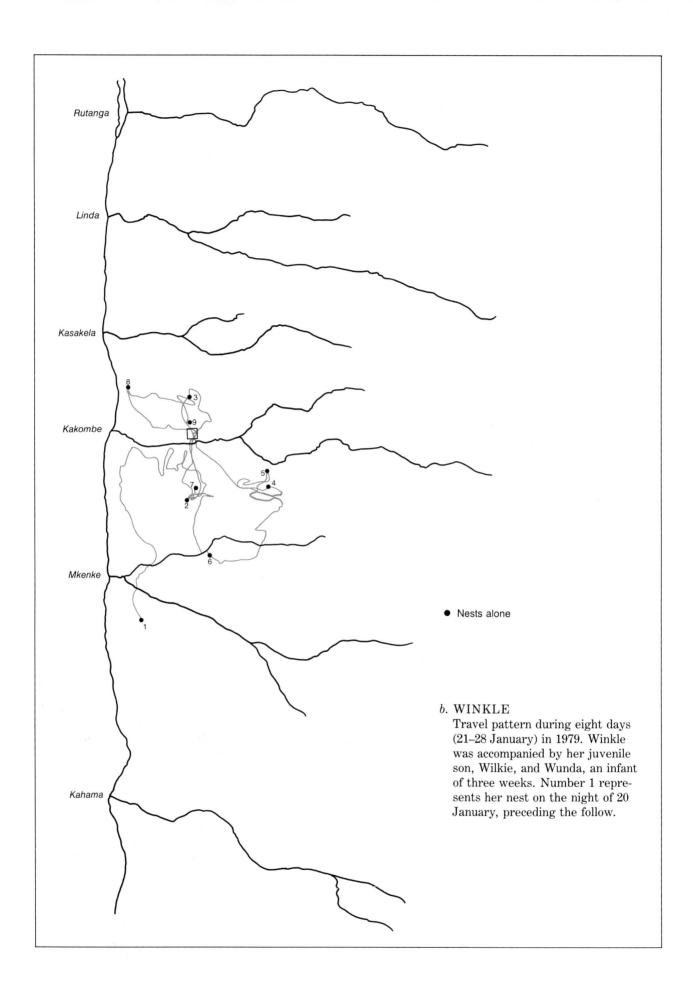

Rutanga

Linda

Kasakela

8
3
9

Kakombe

5
4
7
2

6

Mkenke

1

● Nests alone

b. WINKLE
Travel pattern during eight days
(21–28 January) in 1979. Winkle
was accompanied by her juvenile
son, Wilkie, and Wunda, an infant
of three weeks. Number 1 represents her nest on the night of 20
January, preceding the follow.

Kahama

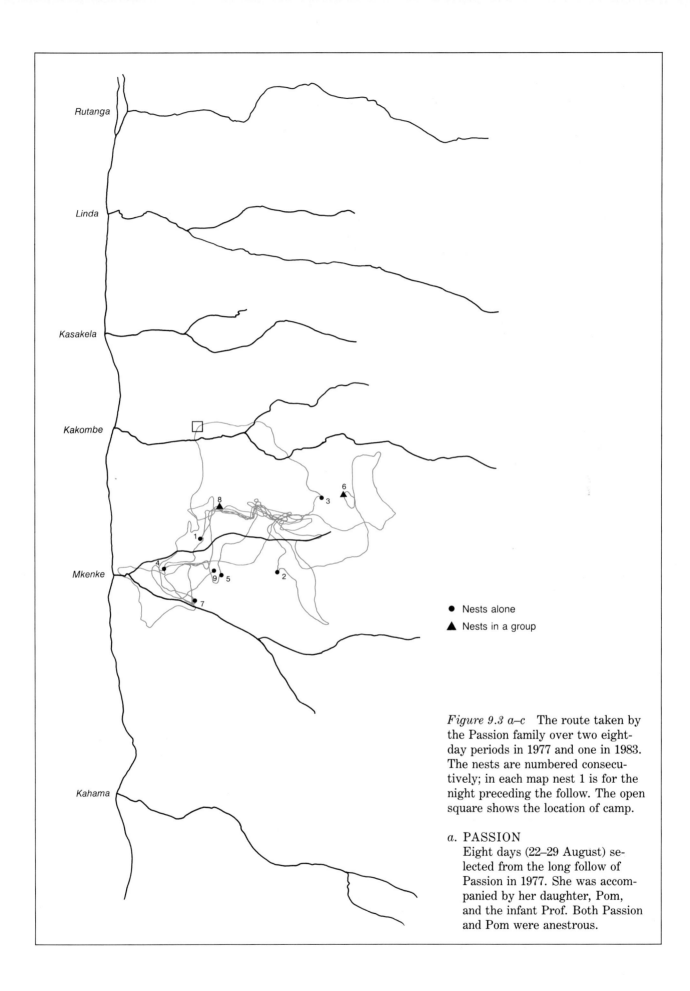

Figure 9.3 a–c The route taken by the Passion family over two eight-day periods in 1977 and one in 1983. The nests are numbered consecutively; in each map nest 1 is for the night preceding the follow. The open square shows the location of camp.

a. PASSION

Eight days (22–29 August) selected from the long follow of Passion in 1977. She was accompanied by her daughter, Pom, and the infant Prof. Both Passion and Pom were anestrous.

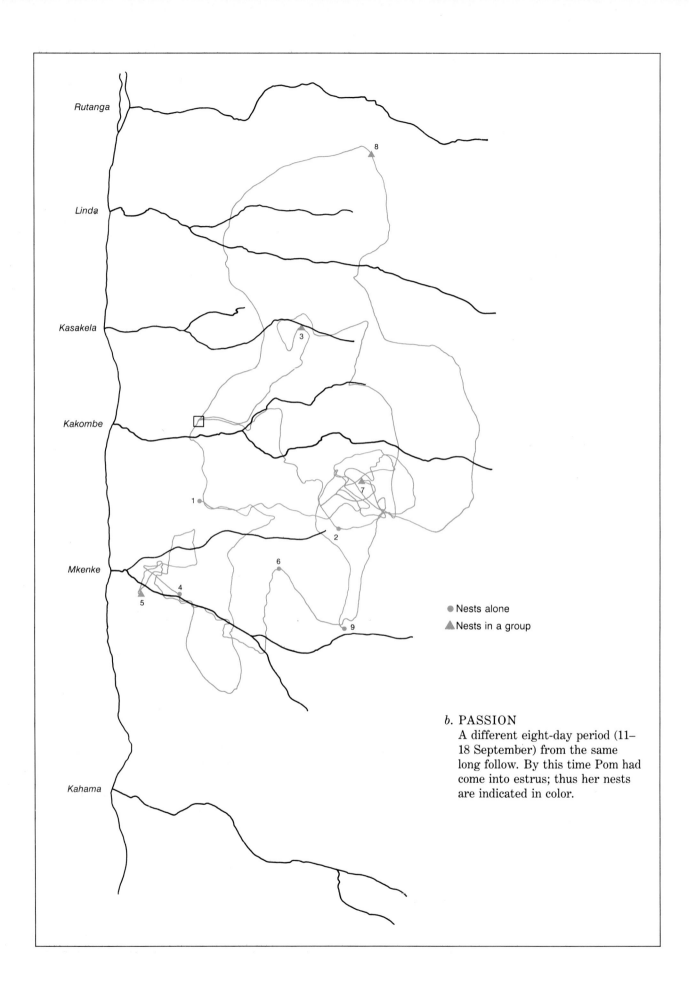

Rutanga

Linda

Kasakela

Kakombe

Mkenke

Kahama

● Nests alone
▲ Nests in a group

b. PASSION
A different eight-day period (11–18 September) from the same long follow. By this time Pom had come into estrus; thus her nests are indicated in color.

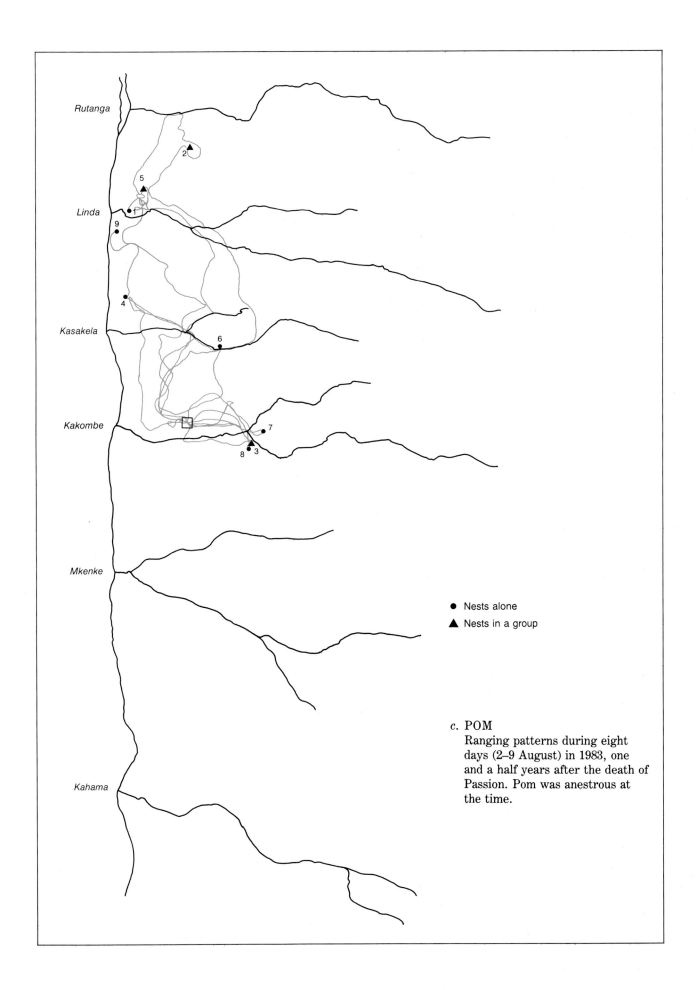

Rutanga

2 ▲

5 ▲

Linda
1 ●

9
●

4
●

Kasakela

6 ●

Kakombe
□
7 ●
8 ▲
● 3

Mkenke

● Nests alone

▲ Nests in a group

c. POM
Ranging patterns during eight days (2–9 August) in 1983, one and a half years after the death of Passion. Pom was anestrous at the time.

Kahama

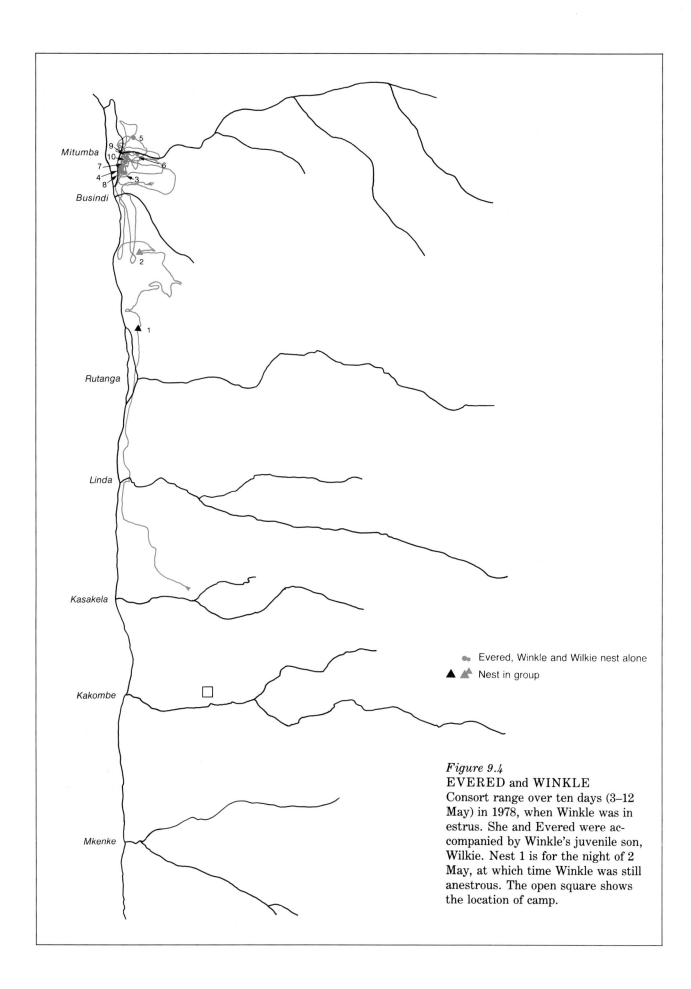

Figure 9.4
EVERED and WINKLE
Consort range over ten days (3–12 May) in 1978, when Winkle was in estrus. She and Evered were accompanied by Winkle's juvenile son, Wilkie. Nest 1 is for the night of 2 May, at which time Winkle was still anestrous. The open square shows the location of camp.

Mitumba

Busindi

Rutanga

Linda

Kasakela

Kakombe

Mkenke

Evered, Winkle and Wilkie nest alone

Nest in group

An adult female in her nest, formed by bending the branches over onto the base platform. (H. van Lawick)

Figure 9.5 shows the 1978 year ranges for the alpha male, Figan; for a cycling female, Athena; and for two anestrous females, Miff and Passion. The year ranges for all three classes (adult male, cycling female, and anestrous female) were larger than they had been during 1972–73. Figan's range was 14.0 square kilometers; Athena's was 13.6 square kilometers, and Miff's and Passion's were 9.2 and 8.2 square kilometers respectively. The increase in size was because the Kasakela community had extended its range to the south after annexing the territory of the Kahama community.

The preferred areas of individual females are spaced out through the community range, although there is extensive overlap: no part of her range is exclusive to any female, and some share virtually the same core area. Figures 9.5c and 9.5d show that Miff spent much time in the northern part of the community range, whereas Passion preferred the south. This pattern was already established during Wrangham's study in 1972–73. Places where mothers sleep, particularly when they are with their offspring only, unaccompanied by nonrelated adults, are good markers of preferred areas. Of the fifty-two nest sites plotted for Miff, forty-seven (90.4 percent) were north of Kasakela Stream, and 93.8 percent of the sixteen family-only nest sites were in the north. Passion's preferred area was very different. Of the sixty-one nest sites plotted for her, forty-five (73.8 percent) were south of Kasakela Stream, and 84 percent of the thirty-nine family-only nest sites were in the south. In fact, Passion traveled more extensively than usual in the north that year while accompanying her daughter, Pom, who was cycling and moving about with the adult males.

Life Range

Obviously the ranging patterns of an individual change in relation to his or her age. A male juvenile still traveling with his (anestrous) mother will have a smaller year range than when he is older and spending most of his time with adult males. A young female's range will increase in the same way when, during periods of estrus, she travels more widely with the mature males of her community. Sooner or later she will accompany one of these males on a consortship to a peripheral area of the home range, and thereby will probably increase her range still more. If she then transfers, temporarily or permanently, to a neighboring community, additional acreage will be added to her range. After she has given birth and made a commitment to one of the two communities, her range will become smaller. Little Bee had a range of about 15 square kilometers in 1974–75 when she was moving back and forth between the Kasakela and Kahama communities (Pierce, 1978). In 1978, when she was nursing an infant, her observed range was 8 square kilometers. The year range of an old individual is likely to decrease still further: Flo's was estimated at only about 3 square kilometers during the year prior to her death (based on about 150 hours of observation; Wrangham, 1975).

For a male chimpanzee, the total area traveled during his lifetime is likely to correspond to the range of his community, changing over time only as the community range changes. A female, if she transfers even temporarily from one community to another, will move over a larger area during her lifetime than a male.

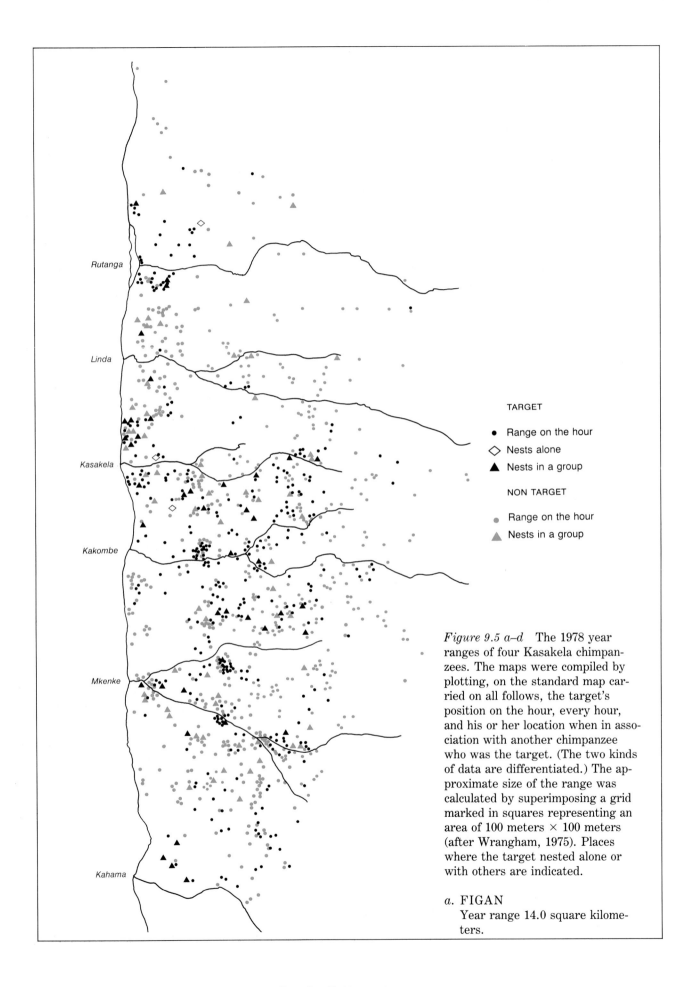

Rutanga

Linda

Kasakela

Kakombe

Mkenke

Kahama

TARGET

• Range on the hour

◇ Nests alone

▲ Nests in a group

NON TARGET

• Range on the hour

▲ Nests in a group

Figure 9.5 a–d The 1978 year ranges of four Kasakela chimpanzees. The maps were compiled by plotting, on the standard map carried on all follows, the target's position on the hour, every hour, and his or her location when in association with another chimpanzee who was the target. (The two kinds of data are differentiated.) The approximate size of the range was calculated by superimposing a grid marked in squares representing an area of 100 meters × 100 meters (after Wrangham, 1975). Places where the target nested alone or with others are indicated.

a. FIGAN
    Year range 14.0 square kilometers.

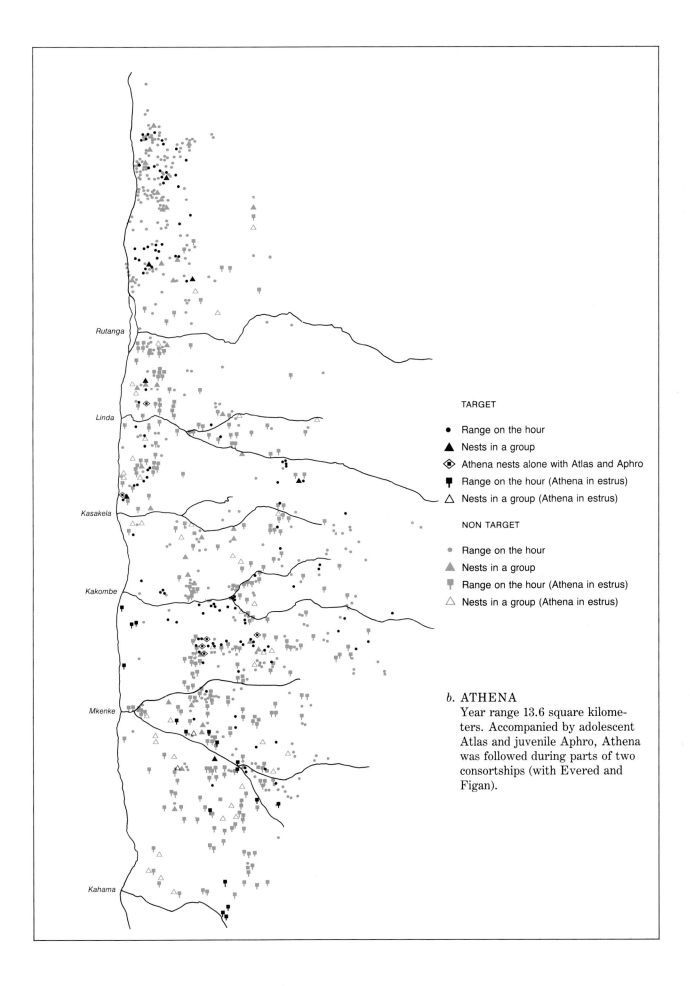

Rutanga

Linda

Kasakela

Kakombe

Mkenke

Kahama

TARGET

● Range on the hour
▲ Nests in a group
◈ Athena nests alone with Atlas and Aphro
▉ Range on the hour (Athena in estrus)
△ Nests in a group (Athena in estrus)

NON TARGET

● Range on the hour
▲ Nests in a group
▉ Range on the hour (Athena in estrus)
△ Nests in a group (Athena in estrus)

b. ATHENA
Year range 13.6 square kilometers. Accompanied by adolescent
Atlas and juvenile Aphro, Athena
was followed during parts of two
consortships (with Evered and
Figan).

225   *Ranging Patterns*

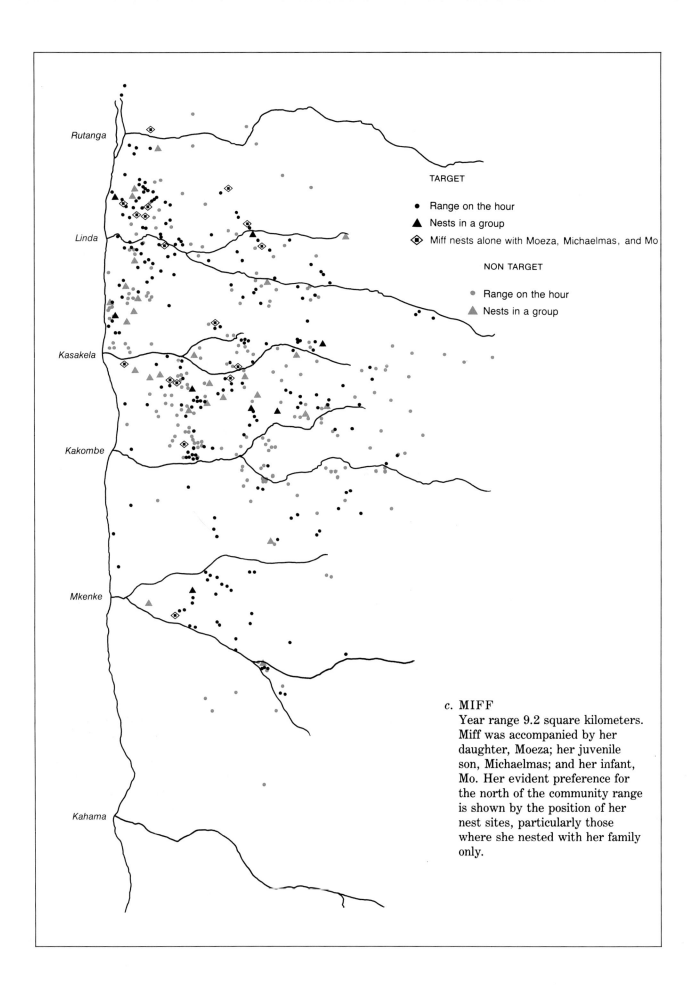

TARGET

● Range on the hour

▲ Nests in a group

◈ Miff nests alone with Moeza, Michaelmas, and Mo

NON TARGET

● Range on the hour

▲ Nests in a group

Rutanga

Linda

Kasakela

Kakombe

Mkenke

Kahama

c. MIFF

Year range 9.2 square kilometers. Miff was accompanied by her daughter, Moeza; her juvenile son, Michaelmas; and her infant, Mo. Her evident preference for the north of the community range is shown by the position of her nest sites, particularly those where she nested with her family only.

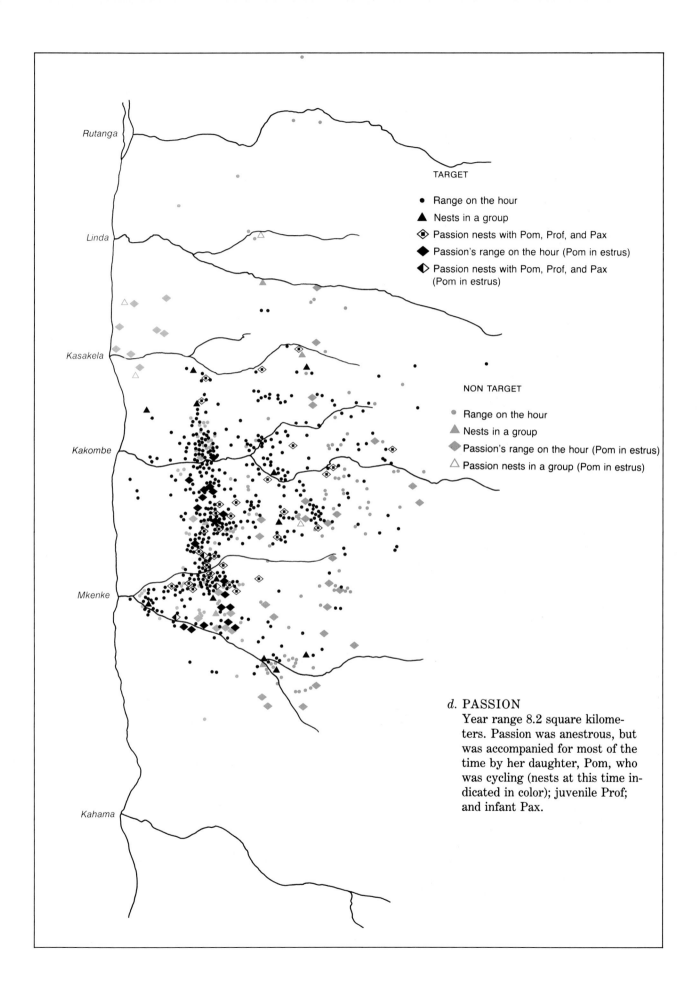

Rutanga

Linda

Kasakela

Kakombe

Mkenke

Kahama

TARGET

● Range on the hour

▲ Nests in a group

◈ Passion nests with Pom, Prof, and Pax

◆ Passion's range on the hour (Pom in estrus)

◗ Passion nests with Pom, Prof, and Pax (Pom in estrus)

NON TARGET

● Range on the hour

▲ Nests in a group

◆ Passion's range on the hour (Pom in estrus)

△ Passion nests in a group (Pom in estrus)

d. PASSION
Year range 8.2 square kilometers. Passion was anestrous, but was accompanied for most of the time by her daughter, Pom, who was cycling (nests at this time indicated in color); juvenile Prof; and infant Pax.

Community Range

The sum of all areas visited by each adult male of the community during a given year is referred to as the community range (Wrangham, 1979). This will be approximately the same size as the ranges of its adult male members, but considerably larger than those of some of the anestrous mothers. Just as the actual year range of an individual is likely to be larger than the observed year range, so the community range in a given year is apt to be larger than the area in which the chimpanzees were actually seen.

The community range must be large enough to support the males and their females and young. It is not sharply delineated from that of neighboring communities, although to the north and south, streams often serve as rough boundary lines. The chimpanzees sometimes cross these boundaries into the range of an adjacent community, but usually show signs of apprehension when they do so. Thus an area the equivalent of at least one valley can be referred to as an overlap zone, utilized at different times by members of neighboring communities. In general, as we shall see in Chapter 17, this arrangement is not a friendly one.

The size of the home range at Gombe changes over time primarily as a result of changes in the number of the community's adult males from year to year. When there are many males, they may be able to enlarge their range at the expense of weaker neighbors; when there are few, they may lose ground and their range may shrink. In the early sixties, when there were as many as fourteen adult males in the KK community, their range was estimated as not less than 24 square kilometers; in 1981, when there were only six adult males (compared with at least ten in each of the unhabituated communities to the north and south), the Kasakela range was only 9.6 kilometers.

Figure 9.6 provides an overview of changes that have been mapped in the community range from 1971 (when, for the first time, chimpanzees were regularly followed for long distances) through 1982. The fission of the KK community in 1972 resulted in an unequal division of the home range; the Kasekela community in the north was the larger of the two and occupied the greater area. The figure shows the increase in the Kasakela core area after the attack on the last of the Kahama community chimpanzees in the south, in 1977. It shows, too, the subsequent shrinkage starting in 1979 caused by the invasion of the strong Kalande community from the south, as well as the apparent reluctance of the Kasakela males to utilize the zone of overlap with the Mitumba community in the north. Finally, we see the more recent expansion of the community range, to both south and north, in 1982; this growth may have been due to the presence on many Kasakela patrols of up to five late-adolescent males. More detail will be forthcoming in Chapter 17.

Other Chimpanzee Populations

Until chimpanzees have been identified individually and followed as they travel about their community range, their patterns of movement and the overall size of the area utilized by a particular social group can only be estimated. Nevertheless, it is clear that in some areas chimpanzees travel much more widely than they do at Gombe.

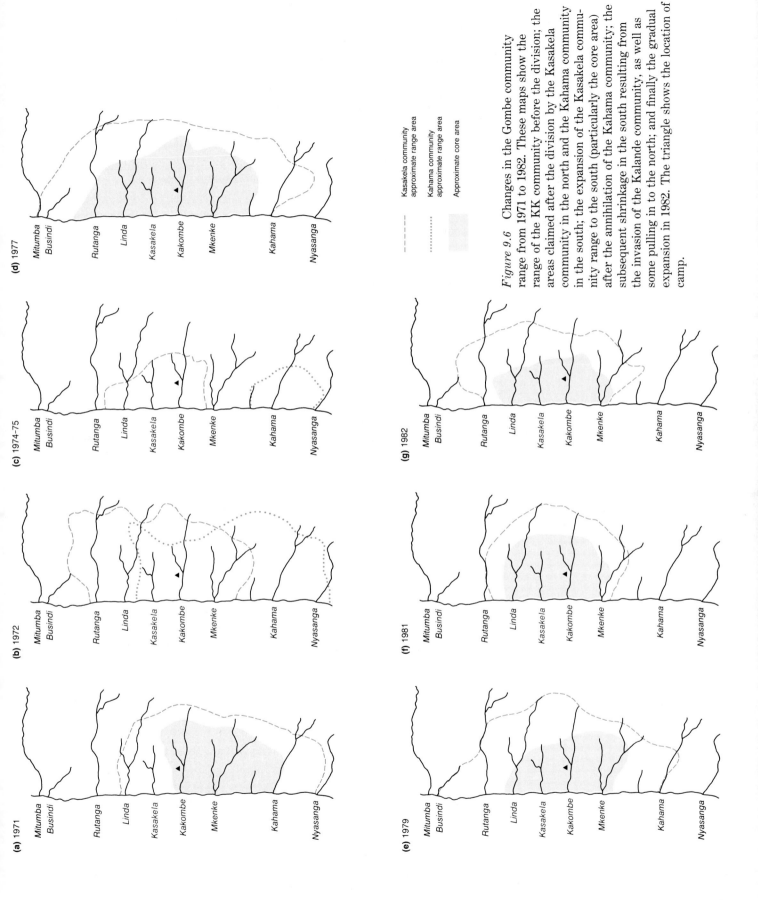

(a) 1971

(b) 1972

(c) 1974–75

(d) 1977

(e) 1979

(f) 1981

(g) 1982

Mitumba
Busindi
Rutanga
Linda
Kasakela
Kakombe
Mkenke
Kahama
Nyasanga

- – – – – Kasakela community
approximate range area

· · · · · · Kahama community
approximate range area

Approximate core area

Figure 9.6 Changes in the Gombe community range from 1971 to 1982. These maps show the range of the KK community before the division; the areas claimed after the division by the Kasakela community in the north and the Kahama community in the south; the expansion of the Kasakela community range to the south (particularly the core area) after the annihilation of the Kahama community; the subsequent shrinkage in the south resulting from the invasion of the Kalande community, as well as some pulling in to the north; and finally the gradual expansion in 1982. The triangle shows the location of camp.

Two families travel along the lake-shore, a part of the range seldom utilized. Youngsters learn traditional family travel routes during their long years of dependency.

The Mount Asserik chimpanzees in the Senegal occupy the hottest, driest, and most open habitat yet known for chimpanzees (McGrew, Baldwin, and Tutin, 1981). In this region there are no barriers to limit the range of the chimpanzees (except a river, which can be crossed at the end of the dry season), nor do there appear to be other communities of chimpanzees in the vicinity. During the wet season these chimpanzees range widely, exploiting outlying patches of food: the total area utilized has been estimated at between 278 and 333 square kilometers. Only at the end of the dry season is their wandering limited by a shortage of running water and perhaps by the extreme heat of the open woodland country (Baldwin, McGrew, and Tutin, 1982). Large home ranges have also been estimated for chimpanzee communities in relatively arid areas in western Tanzania (Suzuki, 1969; Izawa, 1970; Kano, 1971a, 1972).

At Mahale, where the conditions are more similar to those at Gombe, the community range of the K group, when there were five or six mature males, was about 10.4 square kilometers (Nishida, 1979). The range of the large M group (eleven mature males in 1982; Hiraiwa-Hasegawa, Hasegawa, and Nishida, 1983) is at least 33 square kilometers (Kawanaka, 1982b). Once a year, toward the end of the dry season, this group migrates to the northern part of its home range, the overlap zone between the two communities. Often on the very day when M-group individuals arrived, the K group migrated to the northern part of *its* home range. After four or five months M-group chimpanzees moved back to the south, vacating the overlap zone, which was again utilized by K group. (This pattern was eventually influenced by the feeding station set up in the overlap zone.) The migratory route from the southern border of M-group range to the overlap zone in the north is about 5 kilometers (Kawanaka, 1982b). This seasonal movement is very similar to that shown by the Kalande community at Gombe, which moves northward during the wet season (during the peak of the *Landolphia* fruiting season) and retreats southward during April or May.

The movements of chimpanzees within their community range are dictated by a variety of environmental and social factors that may differ from one locality to another; the distribution of food and water; the availability of females in estrus; the size and movement of neighboring communities; the presence of predators or other dangers. Traditional travel routes are passed on by mothers to their offspring, by adult to adolescent males, and by residents to immigrants. Community members may wander over a huge range, as in the Senegal, curtailed only by shortage of water in the driest months, or they may be restricted to a tiny area by the encroachment of human cultivation, as is the case with a small remnant population north of Gombe (A. Seki, personal communication). The long-term studies at Gombe and Mahale have provided unique information regarding changes in ranging patterns over time.

10. Feeding

An infant watches, fascinated, as his mother eats unripe strychnos fruit.

September 1980 As Passion and her family approach the strychnos tree, Pom, who is leading, turns, pant-grunts softly, and touches her mother's brow. She runs on ahead; Passion, with a small sound of threat, rushes after. They reach the tree together. Passion climbs at once and three-year-old Pax follows, but Pom, after gazing at the sparse crop, moves on. Passion feeds for ten minutes, banging the hard fruits against the trunk to crack them open, then picking out the flesh with her lips. The fruits are unripe and bitter and as she feeds, she salivates copiously. Pax picks a few but is too young to be able to open them. Presently Passion climbs down, carrying four unopened strychnos, and sits to eat them in comfort. Pax begs and eventually gets a tiny piece. As he chews, a dribble of saliva trickles down his chin; instantly he picks a blade of grass and dabs at the sticky juice. As he continues to feed, he repeatedly wipes his mouth with tiny pieces of grass. He uses nine of these napkins in the five minutes it takes him to consume the morsel. Passion does not clean herself at all and at the end of her meal has a mess of saliva and fruit juice all over her chin, chest, and hands.

The Gombe habitat, with its permanent water supply in every valley and its mosaic pattern of forest, woodland, and grassland, supports a wide variety of plant and animal species. These in turn provide an adequate—usually plentiful—food supply for the chimpanzees. Food tends to be more abundant during the lush wet season, but healthy chimpanzees have no difficulty finding enough to eat during most dry seasons. Occasionally, however, crops are poor at this time and chimpanzees, particularly the old and the sick, may suffer as a result.

On most days chimpanzees spend close to half their waking hours (an average of 47 percent) feeding, and much of the rest of their time (13 percent) moving from one food source to the next (Wrangham and Smuts, 1980). There are two major feeding peaks, one in the morning between 0700 and 0900 hours, and the other between 1530 and 1930. Wrangham (1977) recorded a third peak, around 1300 hours. In fact, food can be eaten at any time—even (rarely) during moonlit nights.

It has been possible to identify 184 items of vegetable food eaten by chimpanzees from 141 species of trees and plants (Wrangham, 1975). Of these, 26 were seen to be eaten on only one occasion. Of the remaining 158 items 48 percent are fruits, 25 percent are leaves and leaf buds, and 27 percent comprise a miscellany of seeds, blossoms, stems, pith, bark, and resin. Chimpanzees supplement their vegetable diet with a variety of insects, birds' eggs, birds, and small and medium-sized mammals. (Hunting will be discussed separately in the next chapter.) Chimpanzees also consume various substances such as soil, perhaps for their mineral content.

Figure 10.1, compiled from full-day follows during 1978 and 1979, shows the proportion of time spent by target individuals in feeding on different food types. There was little variation in the percentages. Fruits were the largest single component of the diet in most of the twenty-four months; leaves were eaten very frequently in all months; seeds scored high in three months (May to July 1978); and insects were consumed in large amounts during November of both years.

Their feeding behavior suggests that chimpanzees crave variety in their diet. After feeding on one type of food for a while, they often move on to feed elsewhere on a different kind of food—even when food is still available at the first source. Wrangham (1977), in his detailed study of feeding behavior in adult males, found that the variety selected was similar from month to month. He estimated that in a typical month the males fed on forty to sixty different food items, with the number of different foods per day throughout the year averaging thirteen.

The foods eaten by target chimpanzees are recorded routinely during every follow (Appendix A). These data, however, cannot be compared with Wrangham's; his lists of food plants were far more comprehensive. He was concerned with feeding per se and noted items such as leaves, seeds, and other parts of small herbs that tend to be overlooked by other observers. He was also more aware of changes in food choice when his targets were feeding high up in tangled vegetation. Although it is not possible to make a detailed comparison of male and female diets, it seems that females may eat a wider variety of leaves per month and eat them slightly more often than males. This difference is now being investigated. Also, as we shall see, there are major sex differences in feeding on animal proteins.

Seasonality

Some foods can be regarded as staples in that they contribute to the overall diet throughout the year (though more in some months than in others). Most major foods, however, are seasonal and are eaten in large quantities only during certain months. Some crops can be harvested, quite regularly, every year at about the same time; other species produce bumper harvests every other year, and only insignificant amounts in between. Most erratic are "plagues" of caterpillars, which when they do occur are eaten in very large amounts.

These patterns are illustrated in Figure 10.2, which shows the percentage of total feeding time per month that target chimpanzees

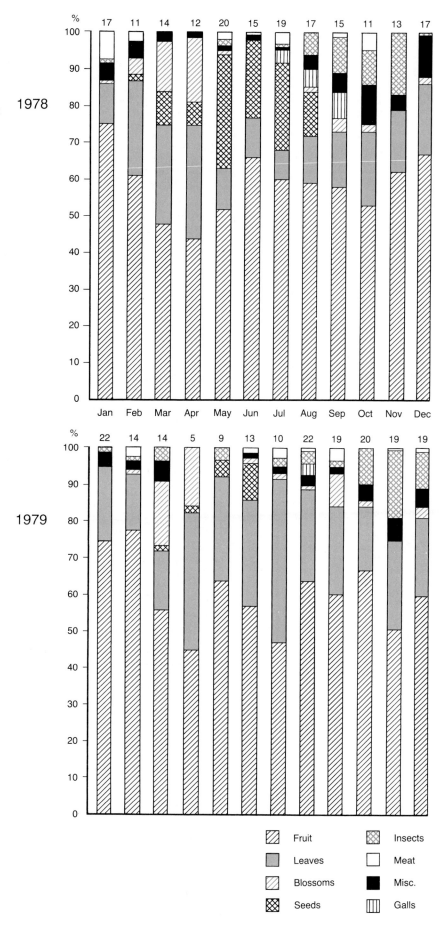

Figure 10.1 Percentage of the overall chimpanzee diet constituted by various types of foods in 1978 and 1979. The data are for males and females, from full-day follows only. The number of follows in the month is indicated at the top of each column.

1978

1979

Fruit
Leaves
Blossoms
Seeds
Insects
Meat
Misc.
Galls

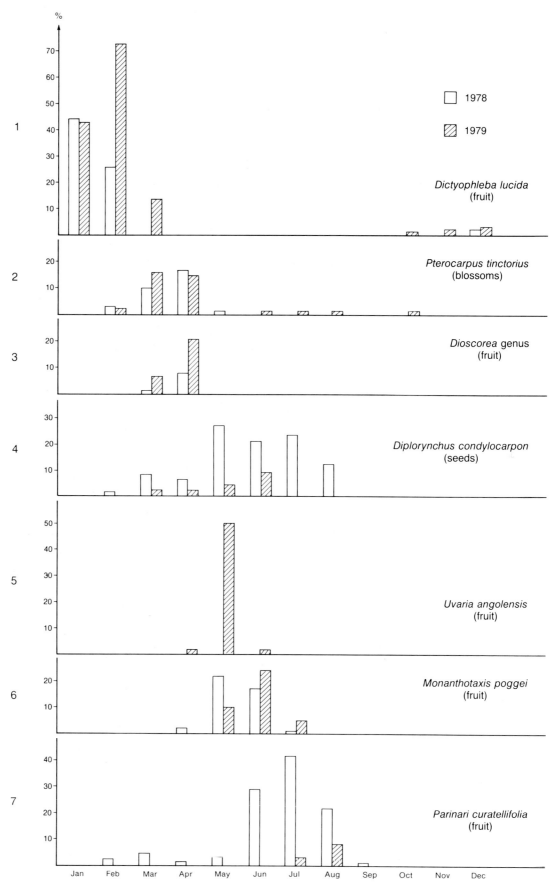

Figure 10.2 Percentage of total feeding time per month over the two-year period 1978 and 1979 that target chimpanzees spent in eating seventeen major foods.

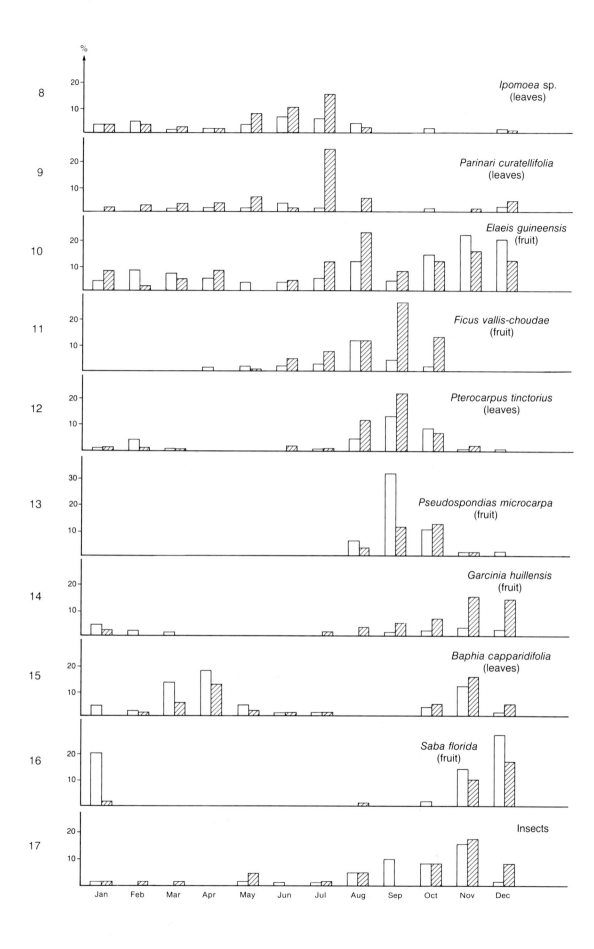

spent in eating seventeen major foods over a two-year period (1978 and 1979).* My arbitrary criterion for including a food is that it was eaten for at least 15 percent of the total feeding time during at least one of the twenty-four months. Of the sixteen plant foods, ten were clearly seasonal. Four (nos. 8, 9, 10, 15) contributed to the diet throughout the year, but were eaten more extensively in some months than others. Of the six foods (nos. 4, 5, 7, 9, 11, 13) that contributed more substantially to the diet during one of the two years, one (no. 5) was not available at all in 1978 and one (no. 7) was available only in very small amounts in 1979. While insects of one sort or another were eaten during all but six of the twenty-four months, the different types were markedly seasonal and only termites made significant contributions to the diet in terms of feeding time.

Finding Food

During the fifty-day follow of the adult male Figan he never appeared to search for food, but typically returned to foods eaten on the same or a previous day. Sometimes as he traveled he came upon "new" foods—that is, those which had become ripe or ready to eat since he was there before; these were then incorporated into his diet (Riss and Busse, 1977). Sometimes the attention of a chimpanzee is directed to a new source when he sees fallen fruits on the ground. He may eat one or two of these, gaze into the foliage above, and if he sees more fruits, climb and start to feed. At other times his attention may be attracted to a new source when he sees or hears chimpanzees feeding there. Or he may observe the feeding behavior of other species such as baboons. Once, for example, a female chimpanzee heard a sound overhead, looked up, saw a juvenile baboon feeding (on palm nuts), and at once rushed up the tree and displaced him. Probably monkeys and noisy birds such as hornbills and touracos also reveal the whereabouts of ripe fruit. Three times a food supply was apparently discovered during boundary patrols near the periphery of the home range. The following day the chimpanzees returned and fed there again.

An individual may be led to a food source that is new to him when he travels with a companion who has previously discovered it. Or one who is perhaps more observant than he. For example, a young male walked right past a trail of driver ants, *Dorylus (Anomma) nigricans*, without appearing to notice them. When his companion, who was following, stopped and began to eat them, he returned and watched, then ate some himself.

Upon climbing into a well-laden food tree, a party of chimpanzees is likely to utter loud food calls (combinations of pant-hoots, food aaa calls, barks, and even screams). A single male may also call in this way. Other chimpanzees, hearing such calls, may then be attracted to the food. After a successful hunt a party of chimpanzees usually calls even more loudly. Screams, pant-hoots, and barks uttered in this context appear to convey quite specific information to nearby chimpanzees, who are likely to show signs of excitement such as hair erection or grinning and embracing, and then run toward the site of

* Figure 10.2 gives the Latin names of the various foods; in the follows and elsewhere they are referred to by their Kiswahili and Kiha names.

the kill. Calls made by baboons after *they* have captured a prey convey similar information to chimpanzees, who usually race to seize the meat.

Chimpanzees have excellent mental maps of their home range and know the location of many of the food resources in it. Wrangham (1975, 1977) has presented examples of the ease with which individuals can return unhesitatingly to certain foods. During the termite season some females visit the same termite mounds daily, often approaching them in a different order and from a different direction. They may select a grass stem for subsequent use as a tool, even when the heap is at a distance of 100 meters or more and out of sight. There are times when a party of chimpanzees utters anticipatory soft food grunts or food aaa calls for up to three minutes before arriving at a known food source, such as a fruiting tree. When two females are traveling together, the subordinate may suddenly pant-grunt, or squeak and grin, and touch or embrace the other before hastening toward the desired food. It seems likely that chimpanzees in the wild are able to remember the whereabouts of food sources over considerable periods of time, probably from one year to the next.

Methods and Time Spent

Eating the juicy *Saba florida* fruit.

Upon arrival at a food tree, chimpanzees may simply climb up and start to feed. At other times they will carefully inspect the available food, sniffing or tasting fallen fruits, then gazing up before finally climbing. Once they start to feed, they may show a good deal of selectivity. Fruits may be chosen on the basis of color, degree of softness, or odor. Thus fruits that have ripened to become red, purple, or yellow are chosen unhesitatingly rather than green ones. Some fruits remain green when ripe, or are eaten when only partially ripe and still green in color: the chimpanzees often squeeze each one before selecting or rejecting it.* Or they may test a fruit by gently pressing it with their teeth, then sniffing it; perhaps the bruising of the skin intensifies the smell. Relative size is also likely to play a role; a small-sized fruit among a bunch of larger fruits is often unripe and is frequently ignored.

Some foods, such as blossoms and leaves, are typically stripped directly off their twigs (usually by mouth, sometimes by hand), chewed, and swallowed. However, leaves of one of the fig species, *Ficus urceolaris*, are usually picked individually by hand, piled together, and then folded over before being chewed, as described by Wrangham (1977). Young leaves of *Aspilia* sp. are tested, then one by one broken off with the mouth. They are subsequently pressed and rubbed against the palate without being chewed (Wrangham and Nishida, 1983).

Some fruits, after being picked, are squeezed or rubbed in the mouth in the same way. Seeds, which may be bitter or even poisonous, can thus be swallowed or discarded without being bitten into; the muscular lips and heavily ridged palate of the chimpanzee are probably

* One female arrived in a fig tree late in the evening. It was too dark to start feeding, but she moved rapidly from branch to branch for one and a half minutes, feeling two or three fruits per branch. After nesting nearby, the following morning she rose early and immediately began feeding at the last of the sites she had tested the previous evening.

Melissa peers over her nose at a wadge of seeds and skin from which the juice is being expressed. After sucking a while longer, she will discard the wadge. (H. van Lawick)

Gremlin tears a section of pith from the frond of an oil-nut palm. Her infant, Getty, chews the discarded fragments.

a specialization for the crushing of fruit in this manner (Wrangham, 1977). The process results in the formation of a wadge of skin, seeds, or fibers, which the chimpanzee may squeeze and suck for ten minutes or more until the last juices have been extracted. While squeezing such a wadge between lips and incisors, the chimpanzee from time to time pushes it out on his protruded lower lip and peers at it over his nose. Sometimes he holds a wadge in one hand while continuing to feed until a second wadge has been formed; then he relaxes, sucking at each in turn. When he moves off after a bout of feeding, he often holds a wadge in his mouth for a while and continues sucking as he travels.

When feeding on very soft foods such as pulpy or overripe fruit (also eggs and meat), a chimpanzee sometimes picks leaves and adds them to his mouthful, forming a wadge that is squeezed and sucked as described. Although these leaves are sometimes swallowed, they are customarily discarded after the juices have been expressed.

Many kinds of seeds, fruits, and pith require skillful manipulation and good oral dexterity before the food can be extracted and eaten. Small seedpods may be put directly into the mouth and opened with lips and teeth, the seeds extracted, and the pods ejected. Some seeds are eaten from pods that are less than a centimeter long. When extracting seeds from the tough fibrous seedpods of *Diplorynchus condylocarpon*, the chimpanzee cracks each pod with his molars, then places it upside down between his incisors. Finally he pulls it apart using his thumb and forefinger. The seeds are picked out with the lips, chewed, and swallowed. The tough outer rind of some large fruits is pierced with the canines before being torn open with hands and teeth. The hard shells of the tennis-ball-sized *Strychnos* sp. are smashed against the trunk or branch of a tree, or taken down to the ground and hit against a rock, after which the shell is pulled apart with hands and teeth. The fruits of the oil-nut palm are often difficult to acquire, in that each one must be carefully poked out from between the spiny tips of the florescence.

Pith from the frond of the oil-nut palm is eaten quite extensively during the dry season, and each feeding bout involves a considerable expenditure of energy. First the chimpanzee must bend down a frond, then peel off the very hard, spine-ridged outer covering. This process

Prof with a piece of palm-frond pith that he has prepared for eating.

Prof feeding on *Diplorynchus condylocarpon* seeds: cracking the pod, using his incisors and canines (as a rule, the Gombe chimpanzees use their molars); tearing open the cracked pod; and picking seeds from the sticky pod.

Melissa uses her teeth to scrape resin from a treetrunk.

Melissa stuffs her mouth with *Harungana* berries.

is initiated with the canines and finished with the hands—or even hands and *feet*—and teeth. Eventually about half a meter of pith, looking rather like peeled sugarcane, is broken off. The end is crushed with the molars, then a bite-sized piece is torn off, chewed, and sucked. The fibers are eventually discarded. Chimpanzees also chew and suck on the fibers of the dried male flower cluster of the palm, after it has fallen to the ground. When the chimpanzee tears the stem apart, a cloud of blackish dust is created, and the fibers themselves are somewhat black in appearance. They too are discarded after being chewed and sucked. Chunks of wood fibers from dead palms may be torn off and sucked; sometimes these are quite dry, at other times they are damp and smell slightly of rotten wood and fungus. Cardboard or cloth, sometimes stolen from humans, is torn, sucked, and wadged in a similar way. Bark is pulled off with teeth and/or hands, then chewed. Sometimes rather large sheets are taken from the trunk, and the inside is scraped with the incisors. Resin is licked from treetrunk or branch as it oozes out, or if it has hardened with exposure to the air, it is scraped off with the incisors and/or canines, or picked off with the fingers.

For the most part chimpanzees eat food where they find it. Sometimes, however, usually toward the end of a long feeding bout, an individual will take a food-laden branch or a collection of large fruits to a comfortable and/or shady spot in a tree or on the ground, and there feed at leisure. Hands and mouth may be stuffed with fallen fruits, after which the chimpanzee sits down to eat in comfort. If a party moves on before one of its members has finished feeding, he may tear off a food-laden branch and carry it with him to eat during the next pause.

The actual time spent feeding per day is to a large extent related to the types of food eaten, because some foods require more processing time than others. This was demonstrated by Riss and Busse (1977) during their fifty-day Figan follow. During the first twenty days Figan spent much time feeding on two types of food requiring long processing, *Diplorynchus condylocarpon* seeds and palm nuts, and on the berries of *Harungana madagascariensis*, which are very small so that large numbers must be plucked (by mouth) from countless sprays before an appreciable amount can be ingested. These three foods contributed to the high proportion of time that Figan spent in feeding (67.5 percent). During the next thirty days he fed extensively on foods having short processing times (fruits of *Ficus* sp., *Pseudospondias microcarpa*, and leaves of *Parinari curatellifolia*) and spent only 49.8 percent of his time feeding.

Figure 10.3 shows the percentage of time spent in feeding, per month, by adult males and females who were targets of full-day follows during 1978: the graphs for the two sexes are similar. Reference to Figure 10.2 shows the foods that were most frequently eaten each month. The longer time spent feeding in May, June, and July was obviously the result of the seasonal availability of *Diplorynchus condylocarpon* seeds and *Monanthotaxis poggei* fruits: the former must be extracted from hard pods; the latter are spread out over large areas and gathered during lengthy periods of foraging as the chimpanzees wander from bush to bush plucking the small, bright yellow fruits.

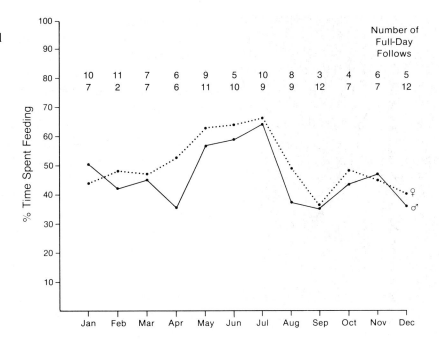

Figure 10.3 Percentage of time spent in feeding by adult males and females during the year 1978. The data are from full-day follows only.

Time spent feeding per day may also be affected by a variety of social factors. Bygott (1974) and Wrangham (1975) found that adult males tended to spend less time feeding when they were in large parties than when they were on their own or with one or two compatible individuals. Wrangham and Smuts (1980) analyzed the Gombe data further and found that females were less affected by the presence of others than were males, in that their time spent feeding in and out of parties was similar. Other factors that tend to depress time spent feeding are tensions occurring on the first day of a consortship, intense competition among males for a popular female in estrus, and patrol activity.

Ill health can affect the time spent feeding, for very sick individuals eat little or nothing; the same is true of new mothers.

Size of Party

For the most part chimpanzees forage in small parties. Riss and Busse (1977) found that the mean size of Figan's party (excluding youngsters under eight years of age), when he fed on nine of the eleven foods he ate most commonly, was between 2.2 and 4.4. In 1977 he was followed for fourteen days, also in the dry season; his party size then was between 2.0 and 4.4, when he fed on ten of the eleven most frequently eaten foods (four of which were common to both years).

The effect of different types of foods on the size of party is difficult to assess because of other variables, notably the presence or absence of females in estrus. Nevertheless, there are certain characteristics of foods that do tend to increase or decrease the probability of a large feeding party—for instance, whether the food is distributed widely or is concentrated in a small area.

Feeding in an oil-nut palm. (H. van Lawick)

Five major types of food source may be distinguished at Gombe:

(1) *Widely distributed, few feeding sites; food supply frequently available.* The fruit of the oil-nut palm can be found all year round, since each tree has its own individual cycle. Palm trees are widely distributed throughout the lower altitudes of the park. The fruits ripen in compact bunches close to the trunk, just above the lower fronds. Usually only one bunch per tree is ready for eating at any given time, so that there is only room for two adult chimpanzees to feed comfortably, one sitting at each side of the bunch on the base of the fronds. Sometimes two bunches will ripen together; even so, only four adults can feed at once. Quite often two or three trees in a cluster of palms bear ripe fruit simultaneously, in which case more chimpanzees can feed at the same time. Termite mounds are somewhat similar, because there are seldom more than three or four productive tunnels in one mound and sometimes only one.

(2) *Many feeding sites in isolated trees; food supply available over several weeks.* Some very large trees may become laden with their crop and many chimpanzees can feed together, even though no similar food occurs within 500 meters or more. As the crop gradually decreases, fewer and fewer individuals can obtain a good meal at the same time. Examples of this type of food are figs of *Ficus kitablu* and galls on both leaf buds and mature leaves of *Chlorophora excelsa*.

(3) *Widely distributed crop, many feeding sites; food supply available for month or more.* Some foods are found in abundance on stands of trees, shrubs, or vines that have a wide distribution in the community range. Foods in this category can be harvested over periods of up to four months as the fruits, seeds, or blossoms gradually become ready for eating. There are many of these major seasonal crops and they become, one after the other, important food items during succeeding months. Examples are *Landolphia* fruits, *Parinari* fruits and leaves (eaten at different times), *Diplorynchus* seeds, and blossoms and leaf buds of *Pterocarpus tinctorius* (also eaten at different times). Some (like *Monanthotaxis* and *Harungana* fruits) are found on low shrubs; the chimpanzees wander among them, feeding as they go and occasionally pausing when they come to a rich patch.

(4) *Highly preferred; very localized in space and time.* *Dalbergia* buds, for instance, are harvested from a very large tree that offers many feeding sites, but the buds are only available for a three- or four-day period. Another food source, even more localized in time, is meat. This offers few feeding sites if the captor keeps all or most of the prey, but more if the carcass is widely divided (see next chapter).

(5) *Dispersed, in patches of suitable habitat such as streambanks, quite widely through the range.* These foods often grow amid other species and usually do not constitute a high percentage of the chimpanzee diet in any one month. Many of these foods are the leaves of small vines or herbs, such as *Asystasia gangetica* or *Ipomoea* sp.

A sixth source is our *banana-feeding area*, where this highly prized food is offered at regular intervals throughout the year. This source, of course, is different from any of the natural food resources of the Gombe chimpanzees. In other areas wild chimpanzees raid human plantations for, say, papaya (Kortlandt, 1962) or grapefruit (Albrecht and Dunnett, 1971), but those fruits are available as they ripen and

Mother and infant (Athena and Apollo) feeding on *Pterocarpus* buds. Note how the long chimpanzee arms are used to reach food at the end of a slender branch.

access is not artificially controlled. Initially the Gombe banana-feeding area offered multiple sites, as a number of boxes were filled and many chimpanzees fed at the same time. Since 1969 it has been the practice to feed the chimpanzees when they arrive alone or in small compatible groups, and thus the area now offers few feeding sites.

There are, of course, food resources with other characteristics, but most of those which form a significant part of the chimpanzee's diet can be fitted into one of the above categories. In general, party size is smaller when chimpanzees feed on category 1 foods than when they feed on foods of category 2 or 3. Thus Wrangham (1975, 1977) found that party size was smaller when chimpanzees fed on palm nuts (category 1) than when they fed on *Harungana* berries (category 2).

Wrangham also noted that chimpanzees who had fed together tended then to travel together, at least for a while. When rich crops of category 3 foods are available, therefore, chimpanzees often remain in quite large parties, even when they are feeding on other food types in the vicinity. More analysis needs to be done on the size of foraging parties as it relates to *major* food resources during the various seasons of the year.

One way of examining the effect of different food types on party size is to compare data for the same individual, feeding on the same foods, in different years. Table 10.1 shows that for the five foods that Figan ate in fairly large amounts during the 1974 and 1977 follows, his party size was very similar. Of interest is the sudden availability, in both 1974 and 1977, of *Dalbergia* buds. In both years Figan fed in these trees for about two hours a day for three consecutive days; both times his mean party size was higher than for any other type of food eaten during the follows.

Table 10.1 Mean size of parties in which target chimpanzees fed on different foods. Data are from long follows only (fourteen consecutive days or more). The number of individuals feeding was scored every half hour during all feeding bouts whenever the food was eaten during the course of the follow. The total number of half hours of feeding on each food is given in parentheses. Bracketed figures signify that there were fewer than five feeding bouts on that particular food.

| Food | Category | Figan party size[a] | | Fifi party size[b] | | Gremlin party size |
		1974 (50 days)[c]	1977 (14 days)	1976 (28 days)	1977 (14 days)	1981 (15 days)
Palm nuts	1	2.2 (108)	2.5 (31)	1.4 (28)	1.2 (8)	2.7 (30)
Galls on *Chlorophora*	2	—	5.1 (36)	[5.0 (4)]	1.7 (78)	2.5 (12)
Diplorynchus seeds	3	3.5 (59)	3.2 (12)	1.5 (8)	1.5 (12)	—
Harungana berries	3	4.3 (45)	3.0 (15)	[4.0 (3)]	—	—
Figs (*Ficus vallis-choudae*)	3	4.3 (11)	4.8 (6)	2.9 (17)	—	4.0 (12)
Dalbergia buds	4	7.4 (37)	8.8 (16)	—	—	—
Ipomoea leaves	5	4.4 (33)	3.4 (17)	3.8 (15)	1.0 (14)	3.1 (13)

a. During a seven-day follow in 1976 Figan's mean party size when feeding on *Diplorynchus* seeds was 3.7 and when feeding on *Harungana* berries, 4.0.
b. During a six-day follow in 1978 Fifi's mean party size when feeding on palm nuts was 1.1.
c. Data are from Riss and Busse, 1977.

Table 10.2 Mean size of parties in which target males and females fed on two types of food when estrous females were present and when they were absent. The data are taken from Gombe travel and association charts.

| Condition | Feeding on *Uvaria* | | Feeding on *Diplorynchus* | |
	Party size	No. parties	Party size	No. parties
Target males				
Estrous female present	8.8	24	9.0	7
No estrous female present	3.6	42	1.1	7
Target females				
Estrous female present	17.0	1	8.5	11
No estrous female present	2.1	33	3.2	18

The table also gives information taken from the long follows on Fifi (anestrous) during the same time of year (dry season) in 1976 and 1977. Mostly Fifi fed with only one or two other individuals (excluding her juvenile son) whatever the category of food. Bauer (1976), analyzing our 1972 association charts (Appendix D), found that the median party size (excluding youngsters under eight years of age) in which chimpanzees fed on a given food type tended to be larger for males than for anestrous females.

Table 10.1 includes information on the size of the party in which a young female in estrus, Gremlin, fed on some of these foods. Her party size tended to be larger than that of Fifi, who was anestrous when these data were collected. In fact, the presence of a female in estrus usually overrides the factor of food type in determining party size. I analyzed data over a two-month period when fruits of *Uvaria angolensis* were being eaten extensively in 1977, and another two-month period when *Diplorynchus* seeds were in season. The results, given in Table 10.2, show that for both foods the mean party size for target males was considerably higher when sexually attractive females formed part of the group. This factor also affected the size of parties in which target anestrous females fed (on *Diplorynchus* seeds): the mean was considerably higher when an estrous female was present.

The introduction of large-scale banana feeding at Gombe had profound effects on many aspects of chimpanzee behavior, including party size in the feeding area (Wrangham, 1974). To summarize, the number of chimpanzees in the camp area gradually increased, from 1963 on, as more and more chimpanzees discovered the new food source and tended to congregate there for longer and longer periods, in hopes of another handout. It was not unusual, at the height of the banana feeding, for parties of over twenty individuals to be waiting around in camp. By 1970 the size of parties in camp had decreased. Although, as Wrangham and Smuts (1980) have shown, the party size of target males and females tends to increase during the time they visit camp, this does not appear to have any significant effect on party size outside the camp area.

Competition for Plant Foods

When two or more chimpanzees are feeding together there may be competition, either for actual food objects or for feeding sites. Wrangham (1975) found that *overt* competition (one individual displacing another or showing threat or attack behavior) was not common. He observed approximately one such incident every twenty hours. However, he inferred competition from instances like the following: (a) individual B sat, not feeding, until more dominant chimpanzee A left his feeding site, at which point B took over; (b) A sat in a site surrounded by food in abundance, which he could obtain simply by reaching out; B, on the other hand, had to move all over the tree, picking the few available fruits from various branches; or (c) B, in a large party, did not feed at all, and it appeared that all the feeding sites were occupied.

In a rough analysis of the overt competition for plant foods (excluding bananas) that was recorded during ten months (March to December) in 1979, I included all incidents, whether or not they involved the target individual. The data cannot, therefore, be treated statistically and are intended merely to illustrate the kinds of competitive behaviors most often seen, the contexts in which they occur, and the age-sex classes between which they are likely to arise.

Of the total 104 competitive incidents, by far the most (50 percent) were between mothers and their five- to eight-year-old offspring; 18.3 percent were aggressive behaviors directed by mature females at juvenile offspring of other females, and 9.6 percent were episodes between siblings. Of the remainder, nine incidents were directed by adult males at adolescents or juveniles and seven involved aggressive interactions between adult females. (Competition over plant food is, in fact, one of the two most common causes of aggression in adult females; see Chapter 12.) Only twice during the time period was an adult male seen to threaten (mildly) a female, and only once another adult male. Fifteen straightforward displacements were described in which one individual approached another, who promptly left the feeding place. Probably this is the most common overt competitive behavior, but it is less likely to be recorded than the more obvious aggressive patterns. By far the most frequently observed of these was the arm-raise threat accompanied by the soft bark (48 percent of the total number of incidents). Other patterns included the seizing of a branch containing food that was just in front of another member of the feeding party, sometimes just as the latter reached out for it (recorded nine times); racing ahead to get to a desirable food site after noticing another individual headed in that direction (recorded seven times); and aggressive contacts such as pushing another away from food, hitting, and actual fighting such as stamping, sparring, or grappling. Fighting was only recorded eight times, four times between adult females, twice between adult and adolescent females, once between an adolescent male and an adult female, and once between siblings. On one occasion an adult male, Humphrey, threw a rock at a juvenile when she headed toward a fallen fruit about 4 meters away from him.

When bananas were provided daily at camp in large amounts, there was a dramatic rise in the frequency and the severity of food competition as well as in the numbers of chimpanzees visiting the camp at any one time. When we changed the feeding system, aggression and party size in camp both dropped. During the ten-month period in 1979 there were only three fights in camp over bananas—twice between two females, and once between a mother and her juvenile son. On five occasions adult males charged after adult females and took their bananas; the females dropped them before physical contact had been made, and the incidents did not lead to fighting.

Water and Minerals

Fast-running, clear water flows down each Gombe valley from the watershed on the rift escarpment. Chimpanzees often pause to drink when they cross a stream, particularly in the dry season when they may drink two or three times in a day. In the wet season they sometimes do not drink all day. Chimpanzees crouch down and suck up stream water with their lips, usually for only a few seconds at a time. Water may also be drunk from hollows in treetrunks; if it cannot be reached with the lips, a leaf sponge may be used (Chapter 18). After a dry-season rain shower chimpanzees, particularly youngsters, often pick up dead leaves and suck off the raindrops; they may also suck or lick raindrops from their own hair.

At one site in the park, outside the range of the habituated community, chimpanzees have been seen to eat soil from a cliff face. A number of hollows have been scraped in this cliff, and baboons (as well as bushbuck and a variety of birds) have also been observed feeding there. I watched the behavior five times in 1961: the chimpanzees remained at the site for about thirty minutes per session, alternately scraping at the soil and licking it from the palms of their hands. Analysis of a sample showed that a small amount of halite, or sodium chloride, was present.

Fifi drinks from a stream, as four-year-old Freud plays with the rocks. Youngsters sometimes spend minutes at a time dabbling in the water.

Frodo tears off a mouthful of fibers from a dried male flower cluster of the oil-nut palm.

Analysis of a dried oil-nut palm flower cluster showed the fibers to be unusually rich in potassium—4.2 percent, which is much the same as beet molasses. Other minerals present in small amounts were magnesium—0.83 percent, calcium—0.51 percent, phosphorus—0.18 percent, and the barest trace of sodium—0.013 percent.

All individuals of the study community eagerly lick a block of mineral salts that is occasionally offered in camp in lieu of bananas. Moreover, the chimpanzees today continue to lick at a cement ledge where in 1967 a python skin was salted for preservation. In an effort to discourage the behavior, the ledge has been liberally soaked in kerosene once and in used motor oil twice—to no avail!

Quite frequently chimpanzees enter the empty huts of fishermen and spend up to ten minutes feeding on ash from the cooking grates. Human urine is sometimes licked, a behavior seen also in the mountain gorillas (Schaller, 1963). Once a pan of photographic chemicals was inadvertently tipped onto the ground, and during the next few days a number of chimpanzees spent up to thirty minutes nibbling at the contaminated soil.

Coprophagy

Coprophagy occurs only rarely at Gombe. One adult female, Pallas, was observed to eat her feces on many occasions (from 1971 until her death in 1982), usually when she had diarrhea. For the most part she merely picked out and ate undigested food material such as small whole fruits or seeds. Her offspring, after watching intently, sometimes ate a little themselves. A juvenile male, Frodo, was seen to eat his feces (twice the whole stool) five times during one month; then he discontinued the habit. During one dry season (1981) there was a sudden increase in observed coprophagy when six individuals showed the behavior (four of them several times). In each case they fed on undigested food remains. When the very old female Flo was no longer able to climb thick-trunked trees or travel far, she sometimes picked food remains from her own feces and those of others. Her juvenile son imitated this behavior.

Insects and Insect Products

Insects are probably included in the diet of chimpanzees throughout their range.* There are reports of insect eating from all areas where chimpanzees have been systematically studied, even as long ago as the pioneering study of Nissen (1931). Moreover, captive chimpanzees catch and eat insects with relish (Köhler, 1925), even home-raised individuals (Kohts, 1935). Table 10.3 lists those species which at Gombe are eaten regularly (six), occasionally (at least five), or incidentally (probably accidentally along with plant food). The table also lists the products of insect industry, such as honey and galls, which are regularly consumed. Many of the same species, or closely allied ones, are eaten by the chimpanzees of Mahale, although cultural differences exist with respect to selection and methods of feeding.

Figure 10.4 shows the percentage of total feeding time per month that target chimpanzees fed on insects of any kind over a two-year period. To get some idea of the actual frequency with which insects are eaten, I made a rough analysis of the data from *all* follows (not only full-day follows) and included insect-eating bouts seen in individuals other than the target. Of the 289 days that chimpanzees were followed for six hours or more in 1979, insects were seen to be eaten on 44.6 percent. In 1980, chimpanzees ate insects on 46.4 percent of 278 days. Figure 10.4 indicates the months in which insect eating was observed most frequently.

Figure 10.5 breaks down the data on insect eating, during full-day follows only, to show the seasonal distribution of feeding on different species of insects. Termites (*Macrotermes*) obviously contributed significantly to the diet during October and November and were also eaten quite extensively during March and August of 1979 and September of 1978. The peak in May, in both years, was when the winged form of *Pseudacanthotermes* was eaten. Weaver ants were eaten between August and October in both years. There was less seasonality in feeding on driver ants. In some years caterpillars are eaten extensively during January or June (two different species), but in 1978 and 1979 they were consumed in modest quantities.

Termites

Macrotermes bellicosus is the most important insect in the Gombe chimpanzee diet, in terms of the proportion of total feeding time (87.4 percent of the total time spent feeding on insects in 1978 and 86.3 percent in 1979) and the overall frequency with which it is eaten (58.6 percent of days when insects were observed to be eaten in 1979 and 63.6 percent in 1980). The main termite "fishing" season, as indicated

---

* Insects, while they contribute animal protein to the diet, may also be eaten for their essential amino acids. Termites may be selected for their lipid content; the unsaturated fatty acids in animal lipids are perhaps as necessary to the chimpanzee as amino acids (Hladik, 1977).

Table 10.3 Species of insects eaten at Gombe and Mahale.

Order	Family (common name) and genus	Species eaten at— Gombe	Species eaten at— Mahale
Isoptera	Termites		
	Macrotermes[a]	*M. bellicosus*	*M.? herus*, B group
	Pseudacanthotermes	*P. militaris*	*P. militaris*
			P. spiniger
Hymenoptera	Ants		
	Oecophylla (weavers)	*O. longinoda*	*O. longinoda*
	Dorylus (Anomma) (drivers)	*D. nigricans*	—
	Crematogaster	*C.* sp.	*C. clariventris*, K group
	Camponotus (carpenters)	?	Five species
	Monomorium	?	*M. afrum*
	Wasps		
	Blastophaga	Various, in figs	Various, in figs
	Polistes (paper wasps)	Unidentified species (once)	—
	Others?	Larvae possibly eaten from dead wood	—
Lepidoptera	Caterpillars		
	Unidentified	From leaves of *Annona senegalensis*	Also caterpillars of two unidentified moths
	Unidentified	From leaves of *Baphia capparidifolia*	
Coleoptera	Beetle grubs		
	Longicorna?	Unidentified species	—
	Cerambycidae?	—	Unidentified species
Orthoptera	Crickets		
	Acridoidae	—	Unidentified species

Insect Products

Order	Family (common name) and genus	Species eaten at— Gombe	Species eaten at— Mahale
From Hymenoptera	Honey		
	Apis (honey bees)	*A. mellifera*	Unidentified species
	Trigona (stingless bees)	—	Two species
	Anthophoridae (mining bees)	—	Unidentified species
From Hemiptera	Galls		
	Phytolyma	*P. lata* on leaves of *Chlorophora excelsa*	*P. lata* on same leaves
	Paracopium	*P. glabricorne* on flower buds of *Clerodendrum schweinfurthii*	
From Isoptera	Termite clay		
	Pseudacanthotermes	Possibly *P. spiniger*	*P. spiniger*

a. *Macrotermes* are also eaten in the Mount Asserik region of the Senegal (McGrew, Tutin, and Baldwin, 1979) and at three sites in Mbini (Río Muni), West Africa (Jones and Sabater Pi, 1969). There is no evidence that these termites are eaten in the Budongo Forest of Uganda, although they almost certainly occur there (McGrew, Tutin, and Baldwin, 1979). A translocated group of chimpanzees in Gabon typically ignore *Macrotermes* mounds, even though insects form a high percentage of the diet (Hladik, 1977). An unidentified species of termite is also eaten in Guinea (Sugiyama and Koman, 1979) and West Cameroons (Struhsaker and Hunkeler, 1971).

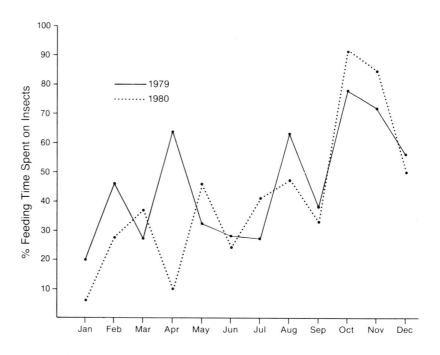

Figure 10.4 Percentage of total feeding time per month that target chimpanzees fed on insects (of any kind) over the two-year period 1979 and 1980. Data are from follows at least a half-day long.

Flo feeding on termites (*Macrotermes bellicosus*). (H. van Lawick)

in Figure 10.5, is November. This is the time when the reproductives leave their nests to form new colonies. When chimpanzees see these winged "princes" and "princesses" swarming, they often run to the nest and catch them as they emerge. Baboons, monkeys, many small mammals, birds (and humans) also feed on the reproductives, which are large (about 2.5 centimeters long) and nutritious. The chimpanzees also feed on the soldiers and (much less often) the workers, using simple tools to extract them from their underground tunnels. This termite-fishing behavior has been described in some detail (for instance, Goodall, 1970) and will be further discussed in Chapter 18. Briefly, the chimpanzee, with index finger, second finger, or thumb, scrapes away the plug constructed by worker termites to seal a passage entrance. Next, a grass stem or other suitable instrument is pushed down the passage. After a pause it is withdrawn, carefully, so as not to dislodge insects that have gripped the tool with their mandibles. Any soldiers are picked off with lips and teeth and chewed; workers are usually wiped off as the tool is pulled through the fingers and cleaned, ready for the next insertion. Soldiers that fall to the surface of the heap are picked up with the lips, between thumb and forefingers, or with a characteristic mopping movement of the back of the hand and wrist. A passage yielding only a few termites may, nevertheless, be worked for minutes on end, particularly by females during the dry season. On such occasions worker termites are almost always eaten as well as soldiers.

Pseudacanthotermes militaris are eaten much less frequently. The only records are from the end of the rainy season (May and June), when the reproductives leave the nests and form huge swarms that, for a couple of hours, eddy around the tops of tall trees. Adult chimpanzees sometimes capture the insects as they leave their underground nests. Three times chimpanzees (adolescents twice and a juvenile once) climbed up and reached from high branches to grab handfuls of the insects as they flew around the treetops.

Ants

Two species of ants are eaten quite frequently at Gombe. Weaver ants construct arboreal nests by drawing leaves together and binding them with silk produced by their larvae. The chimpanzee selects a nest, rushes from the site (to avoid the aggressive response of uncaptured insects, which inflict painful bites), and holding the nest tightly with one hand, crushes it with vigorous downward squeezing movements of the free hand or foot. This action brushes off the ants swarming on the surface and prevents their crawling up the chimpanzee's arm. The nest is then gradually pulled apart and the insects (many of which have been partially crushed) are picked off with the lips as they are revealed. Often, after the juices of the ants (which taste hot and acid) have been extracted, the remains are discarded.

The manner in which chimpanzees collect driver ants has been described in detail by McGrew (1974). First the underground nest is opened with rapid digging movements of one or both hands. Then a fairly long stick or *wand*, broken from a nearby sapling or shrub, is inserted into the nest. After a few moments, when the ants have swarmed about halfway up the tool, it is withdrawn and swept through

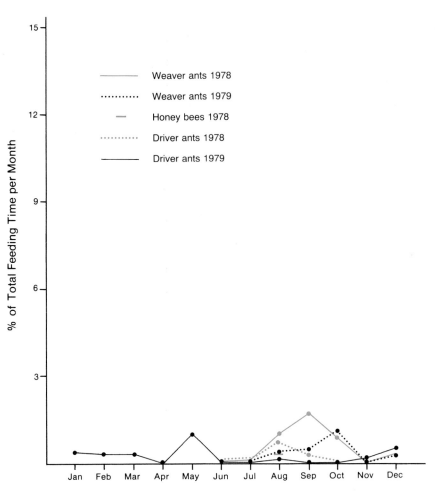

Figure 10.5 Seasonal distribution of time spent in feeding on various insects over the two-year period 1978 and 1979. Ants and honey bees appear above, termites on next page. The data are from full-day follows only.

the free hand; the mass of insects is then rapidly transferred to the mouth. Two chimpanzees, McGregor and Pom, varied the technique by sweeping the tool through their mouths so that the mass was picked off with the lips. The insects are eaten with rapid and frantic chewing movements, presumably in order to crush the ants as quickly as possible because their bites too are very painful. Twice we have observed females, after opening nests, reach in and extract large handfuls of eggs and grubs before commencing to dip in the usual way.

When a nest is disturbed, the ants swarm rapidly to the surface and fan out defensively, climbing up anything they encounter. When chimpanzees feed on the ground, they stand as far away from the nest as possible. Sometimes they insert their stick, retreat, and when it is covered with swarming ants, hurry in to seize it before stepping back to feed. Often, however, a chimpanzee climbs to a strategic position above the nest and dips from comparative safety. In this case the wand must be transferred to a foot after withdrawal, because one hand is occupied with holding onto the tree. I watched one female, perched on a sapling, become increasingly frenzied as more and more ants swarmed up the trunk. She swept them down, first with one

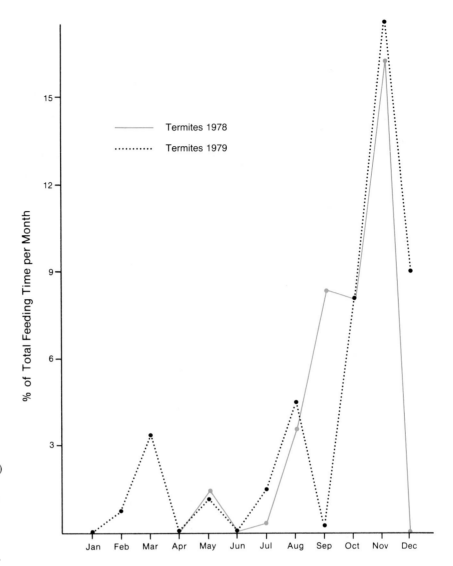

Termites 1978

Termites 1979

% of Total Feeding Time per Month

Jan Feb Mar Apr May Jun Jul Aug Sep Oct Nov Dec

(a) Miff pauses as she feeds on weaver ants (*Oecophylla longinoda*) to pick off those that are crawling up her wrist. In her right hand she holds the remains of the ants' crushed leaf nest. (b) Miff's four-year-old daughter, Mo, reaches for some of the leaves of the ants' nest. (c) Mo picks ants off the leaves she took from Miff and eats them. Michaelmas, Miff's juvenile son, has also taken some of his mother's ant-covered leaves.

Feeding on driver ants (*Dorylus nigricans*). Miff keeps as far as possible from the swarming, vicious ants while inserting her slender tool into their nest.

foot, then the other, stamping on the trunk, slapping at her feet and waving them frantically in the air; but she continued to dip for a total of ten minutes. Such behavior is quite typical. McGrew describes occasions when chimpanzees pushed over nearby upright saplings in order to reposition them for use as aboveground perches. He also relates an instance when a chimpanzee, suspended from an overhead branch, fell directly into an ant nest when the perch broke!

McGrew found the average duration of an ant-dipping episode to be 18.9 minutes. He estimated (by collecting a mass of ants similar in size to that consumed regularly by a chimpanzee) that the average dip weight is about 0.45 gram (the mass contained 292 ants) and that the mean intake for an adult during a dipping session is about 20 grams.

The whereabouts of driver ant nests is often known and the same nest may be visited on a number of successive days. Sometimes a chimpanzee visits a nest that has been abandoned (for driver ants, unlike termites, are quite migratory); after a few digging movements the chimpanzee gives up and moves away.

An arboreal ant, *Crematogaster* sp., was found three times, in quite large numbers, in fecal samples analyzed during the 1964–65 study period. Wrangham twice saw chimpanzees picking these ants off dead branches, and I watched a female break open dead wood and consume the eggs, larvae, and adults of the crematogaster nest that she found inside. Occasionally chimpanzees feed on other ant species. In 1978, for example, this was seen five times. Three times these ants were described as small and reddish black, and they may well have been crematogaster. Twice the ants were black with very large heads and described as big and fierce; probably these were a species of *Camponotus*, or carpenter ants.

Caterpillars

Two kinds of caterpillars, from unidentified species of moths, are eaten extensively in some years. One spins a web on the underside of *Annona* leaves; during bumper years (1964, 1981, and perhaps 1973) chimpanzees fed on them for as much as three hours a day. During two weeks in May–June 1981, target chimpanzees (eleven females and three males) fed on these caterpillars for 11 percent of the 126 hours they were followed. The second type of caterpillar lives on leaves of *Baphia*. It is eaten in January and February, but again only sporadically except during certain years. (It formed 1.5 percent of the target male diet in January 1973, according to Wrangham, 1975, and was eaten extensively in January–February 1978.) Other kinds of caterpillars are eaten occasionally: an adolescent female ate a brown stick caterpillar, and an infant squeezed out and ate the inside of a long green stick caterpillar, then played with the skin.

Other Insects

Twice chimpanzees ate the large white grubs of a species of beetle, possibly a longicorn; they were chewed with leaves. Occasionally dead branches were torn open and some kind of larvae picked out and eaten. Once a female chimpanzee seized the nest of a paper wasp (*Polistes*) and ate the whole thing. Presumably there were eggs and larvae inside.

Sometimes insects are eaten incidentally. A variety of *Blastophaga* is ingested when the chimpanzees feed on figs. Sometimes ants are found in figs, in which case they will be eaten too. An assortment of other small insects and their eggs are undoubtedly consumed along with the leaves that form such a high percentage of the chimpanzee diet.

On a number of occasions chimpanzees have been seen to eat the eggs of *Tuga penetrans* (commonly known as "chigger"), which forms egg sacs beneath the skin of the host—in chimpanzees and humans usually on the sole of the foot or between the toes. These eggs appear to be a special delicacy. Juvenile Prof watched intently as his mother, Passion, poked at a chigger on her foot. As she broke the skin and the white eggs began to ooze out, he pushed in and sucked them. Passion allowed this for a moment. When she pushed him away (to suck herself), he struggled to return and, when prevented, threw a tantrum. Another mother, Miff, after watching her adult daughter Moeza work on a chigger, suddenly grabbed at the affected foot. Moeza rushed away and was chased by her mother twice around a tree before she managed to climb up and consume her chigger eggs in peace!

Honey, Galls, and Termite Clay

The chimpanzees sometimes raid beehives, usually in the dry-season months (Figure 10.5). It is not clear whether these are all nests of the same species. Some are constructed on the underside of branches or on treetrunks, some are in hollows of trees, some are underground. However, all identifiable specimens of bees found in the 1964–65 feces analysis were those of the honey bee (*Apis mellifera*).

During the three-year period 1978 to 1980 chimpanzees were seen to raid hives on twenty-one separate occasions. Three times the raiders simply sat eating honey, surrounded by bees at which they merely slapped from time to time, albeit rather frenziedly. Two small infants, in fact, who were clinging to their mothers, whimpered and burrowed their faces into their mother's breasts. Afterward the females spent some time pulling stings from themselves; one mother was also seen to pull them from her infant during a grooming period. Nine times, after seizing one or two handfuls of honey, the chimpanzees were driven off by swarms of bees, and nine times they ran off without *any* honey. It is uncertain whether the extent to which they are prepared to put up with the pain inflicted by stings depends upon the amount of honey available or its accessibility. A total of sixteen different individuals were actively involved in the raids (six males, ten females). Four of them were observed on different occasions (a) running off without honey and (b) staying, ignoring the bees, and feeding. One female ran once, but stayed on three other occasions. Thus it appears that the decision to flee or stay cannot be explained by differential individual tolerance.

Often larvae and a few workers are eaten along with the honey when the chimpanzees seize and chew the actual honeycomb. The grubs are a delicacy for the people of many tribes, and although the chimpanzees probably seek the honey for its own sake, they undoubtedly appreciate the grubs also.

Two types of galls are eaten. One is formed by *Phytolyma lata* on the leaves of *Chlorophora excelsa*. The chimpanzees pick one or several leaves, remove the galls with their lips, then drop the leaves, neatly punctured with round holes. The other gall is caused by a lace bug (*Paracopium glabricorne*) in the flower bud of *Clerodendrum schweinfurthii*. The deformed flower head, about the size of a human fist, is picked and eaten when a chimpanzee happens to pass by during travel. When chimpanzees eat galls, they inevitably eat the insects that have formed them, but almost certainly are seeking the gall substance itself. Typically, during gall formation additional tannins are produced, but at the same time starches are turned to sugar by the actions of the insect (Frost, 1959).

Some galls are eaten by humans (for example, the "gall of sage" produced by a species of *Aulax* in the Near East), and one produced by a species of *Callirhytis* is used for animal feed in the United States (it contains 63.6 percent carbohydrate and 9.3 percent protein). It is possible that the galls on *Chlorophora* are highly nutritive, because they are an important food source, especially in September. Wrangham found them to rank as the fourth most frequently eaten food in September 1972 and the most eaten in 1973. In September 1978 galls again were the most frequently eaten food by males. The reason for the low percentage of galls in the female diet that year is not clear; in other years females spent long periods of time eating them. In 1977, during the long follow of Passion, she fed on galls for 15.8 percent of her total feeding time in August and 10 percent in September. The same galls are eaten in the Mahale Mountains during August and September (Nishida and Hiraiwa, 1982).

About once a day, as they pass termite mounds (usually of *Pseudacanthotermes militaris*), chimpanzees pause to pick off and eat small amounts—not more than would fill a walnut. Analysis of samples of termite clay collected by R. Wrangham revealed substantial quantities of potassium, magnesium, and calcium and traces of copper, manganese, zinc, and sodium. There was wide variation in the amounts of the different minerals present in samples taken from different termite mounds. These minerals are present, often in larger amounts, in some plant foods and it is not yet clear exactly why chimpanzees (and other primates) feed on termite clay. It may be to neutralize tannins and other poisons present in plant foods (Hladik, 1977). The clay is eaten by humans in rural areas in many parts of the world (M. Latham, personal communication). Sold in markets in most parts of Tanzania, it is eaten by the Waha people of the Kigoma region, particularly by pregnant women from the fourth month on. They apparently are attracted by the smell of the clay at this time and eat a few lumps after lunch and/or the evening meal (H. Matama, personal communication).

Sex Differences in Insect Eating

Accumulated evidence shows that females at Gombe are more insectivorous than males. Even in the very early days of my study it was evident that females spent more time fishing for termites than did males (Goodall, 1968b). Analysis of the association charts of July 1972 to January 1974, and reexamination of data from the 1964–65 feces analysis, showed that the Gombe females also ate more termites and weavers, and possibly more driver ants, than did males (McGrew, 1979, 1981).

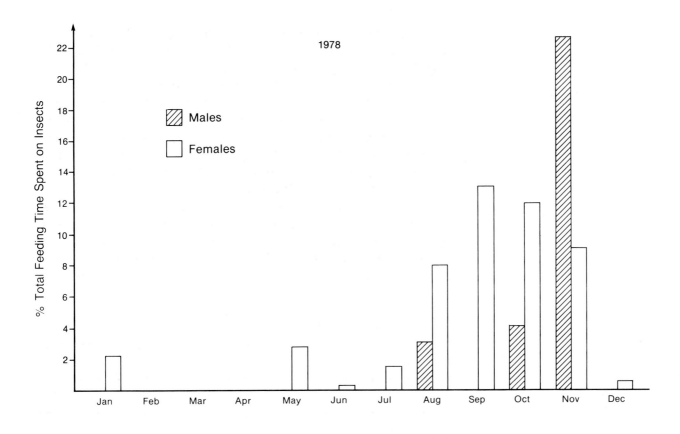

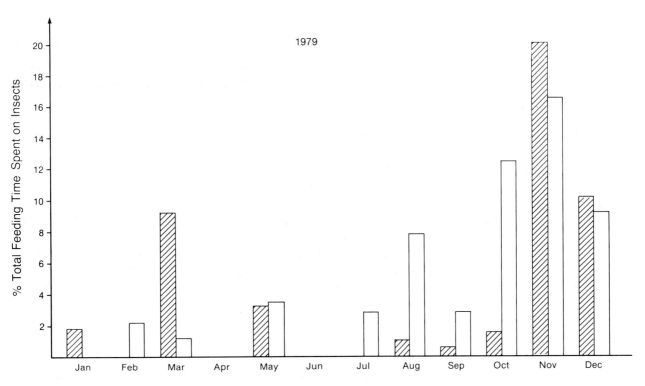

Figure 10.6 Percentage of total feeding time per month that insects (of any sort) were eaten by males and females during full-day follows in 1978 and 1979.

Figure 10.6, which analyzes insect eating in males and females, is a breakdown of the data shown in Figure 10.2 from the association charts of 1978 and 1979. In both years females spent a greater proportion of their overall feeding time in consuming insects than did males: for thirteen of the nineteen months that insects were eaten, females scored higher, and they ate insects during eight months when males were not seen to eat any. Moreover, females ate at least some insects on 42 percent of the days they were followed during the two years, whereas the corresponding figure for males was 27 percent.

One surprising fact emerges from the above data: in both years male scores were higher than female scores during November, the peak of the termite-fishing season. Initially I thought that perhaps this was an artifact of sampling procedures. However, analysis of insect eating during the same month in three additional years produced similar results, as shown in Table 10.4. Why, if females are more insectivorous than males, should they apparently eat fewer than males during the period of greatest abundance?

Table 10.5 has been compiled from all observations of termite feeding when complete sessions were observed. The information is divided into two categories: sessions at a termite mound that yielded at least some termites, and those that were absolutely unproductive. The first category is further divided into time spent searching and time spent eating. The information is not precise because although the interval is recorded from the time an individual arrives at a heap to the time he begins to feed on termites, pauses during feeding (as a new passage is sought) are noted only if they are longer than about two minutes. The number of sessions recorded for each sex per month is raw data, uncorrected for the fact that there are more females than males in the community or for the different amounts of time that the two sexes were observed. Nevertheless, the table gives a reasonably good idea of the pattern of termite fishing over a two-year period. We see that while females search for and feed on termites throughout the year, males tend to confine their termite fishing to the peak season, November and December.

In the dry season, when termites are in the lower levels of their nests, females spend long periods searching for them and, when they find a productive heap, long periods feeding on them. So far, precise data on the number of termites per tool, or the percentage of times that any termites at all are captured per probe, have not been collected. However, it is not uncommon for a female at this time of year to continue fishing for an hour or more even when she is getting only a few termites every ten minutes. The longest session recorded during the two years was in May 1980, when Passion (with Pom and their offspring) worked a mound for a total of four hours, twenty-nine minutes. Unfortunately, we do not know Passion's total intake during all that time, but it is certain that she was catching very few termites.

During October, when the first rains fall, the females become even more persistent in their fishing endeavors. As Table 10.5 shows, the percentage of occasions when a mound is productive increases during this month—and the percentage of time spent in searching drops. In fact, the females are often catching more termites during productive sessions than they were before, although again no systematic data collection or analysis has been undertaken.

Table 10.4 Percentage of full-day follows on males and females during November of five different years in which the chimpanzees were seen to eat insects.

	Males		Females	
Year	No. follows	% when insects eaten	No. follows	% when insects eaten
1976	7	28.9	20	22.5
1977	5	16.7	20	4.5[a]
1978	6	23.6	7	9.1
1979	10	19.8	10	16.8
1980	7	24.3	7	19.5
Average		22.7		14.5

Note: Male-female differences are significant by the Mann-Whitney U test (one tailed), $p < 0.05$.

a. This unusually low score was due to the large number of follows on one female who had just given birth to twins. If 1977 is omitted, the female average becomes 17.0 percent.

Table 10.5 Sex differences in termite-fishing behavior over a two-year period. The data are from all follows on adult males and females. Figures in italics are from one bout only.

	Females								Males							
	Positive bouts			Negative bouts					Positive bouts			Negative bouts				
Month	No.	Mean length (minutes)	Range (minutes)	No.	Mean length (minutes)	Range (minutes)	% bouts positive	% time spent searching	No.	Mean length (minutes)	Range (minutes)	No.	Mean length (minutes)	Range (minutes)	% bouts positive	% time spent searching
1979																
Jan.	0			0					0			0				
Feb.	1	*45*		1	*5*			*11.1*	0			0				
Mar.	3	52	19–152	0				13.9	1	*2.5*		1	*2*			
Apr.	0			0					0			0				
May	4	59	41–80	3	10	1–17	57.1	22.8	0			0				
June	1	*40*		1	*5*			*20.0*	0			0				
July	1	*79*		0				*16.9*	1	*10.0*		0				
Aug.	6	34	11–101	6	11.5	1–27	50.0	33.4	1	*15.0*		1	*1*			
Sept.	9	54.4	9–172	19	5	1–12	32.1	37.2	1	*15.0*		1	*1*			
Oct.	26	40.7	2–100	22	5	1–25	54.2	28.6	4	9.0	5–10	0				
Nov.	37	31.1	3–77	6	10	1–16	86.0	6.9	43	33.9	5–134	7	2	1–15	86.0	3.8
Dec.	20	26.0	2–50	6	6	1–7	76.9	6.9	9	37.0	7–70	4	1.5	1–8	30.8	5.0
1980																
Jan.	6	19.5	17–105	5	7	2–25	54.5	12.7	1	*60.0*		1	*1*			
Feb.	8	42	10–97	5	4	4–6	61.5	18.5	3	30	30–57	1	*6*			
Mar.	14	36	1–157	12	14	1–46	53.8	22.4	0			0				
Apr.	17	31.5	10–156	18	45	1–70	48.6	37.5	1	*20*		1	*11*			
May	2		16–269	0					0			0				
June	7	72	11–208	6	7.5	3–28	53.8	31.8	0			0				
July	8	38.5	18–230	9	6	1–36	47.1	34.1	0			0				
Aug.	7	67	34–180	6	12	5–16	53.8	26.1	0			0				
Sept.	9	60	30–106	10	9	3–32	47.4	31.6	1	*33*		1	*1*			
Oct.	26	41.4	16–192	13	14.7	1–27	65.8	28.6	3	7.7		—				
Nov.	63	21.1	6–52	17	5.7	1–13	78.8	5.8	50	28.8	4–90	16	4.8	1–3	75.8	5.1
Dec.	8	22.2	5–51	10	3.6	1–7	65.6	17.9	19	22.2	10–43	10	3.6	1–6	65.5	7.9

Table 10.6 Sex differences in frequency of visits to termite heaps and time spent working at them. The data are from forty-one full-day follows during October and November 1979.

Sex and month	No. follows	No. days heaps inspected and/or worked	No. heaps investigated	% heaps productive	Mean no. minutes per heap		Mean no. minutes per day	
					Working/feeding	Feeding	Working/feeding	Feeding
Females								
October	13	10	29	44.8	55.3	29.3	85.0	71.9
November	10	9	29	79.3	24.3	20.7	66.7	62.0
Males								
October	8	3	3	100.0	7.7	7.7	7.7	7.7
November	10	9	21	85.7	32.7	31.8	79.4	77.3

Table 10.7 Length (in minutes) of bouts of feeding on weaver and driver ants during 1980, and on driver ants (from McGrew, 1979). Also shown are bouts of feeding on caterpillars (from *Annona* leaves) during 1980.[a]

Food	No. feeding bouts		Mean length (minutes)		Range (minutes)		No. individuals contributing to bouts	
	Males	Females	Males	Females	Males	Females	Males	Females
Weaver ants, 1980	14	57	8.9	8.0	2–17	2–21	7	12
Driver ants, 1980	25	35	8.9	12.0	1–35	1–44	7	12
Driver ants (McGrew, 1979)	8	19	13.6	15.5	5–32	3–48	—	—
Caterpillars, 1980	3	25	27.7	32.2	8–45	4–115	2	6

a. The male-female difference for weaver and driver ants in 1980 is not statistically significant (χ^2 test, $df = 2$).

Table 10.6, compiled from full-day follows only, shows clearly that females investigated more termite heaps during October than did males, that they spent longer inspecting the heaps, and that when they found a productive heap they spent considerably longer feeding than did the males. Only Goblin and Mustard were observed to fish for termites during October. Both selected tools on their way to the heaps; perhaps they had encountered females feeding at these heaps on a previous occasion and knew they were productive. But they spent an average of only about seven minutes per heap. As soon as their efforts were unsuccessful, they gave up; whereas the females persisted, very patiently, for up to one hundred minutes.

In November the reproductive termites are ready to emigrate. There is great activity within the nest; the workers are busy opening, then sealing, the passages; the soldiers gather to defend the nest from intruders. Suddenly it becomes very easy to catch termites, and to catch them in abundance. It is at this time that the adult males capitalize on an abundant rich and nutritious food source. Table 10.6 shows that during both years males spent longer at each heap, and longer feeding per heap and per day, than did females. The females actually spent, on average, less time termite fishing than during October. They may, however, have been eating far more, merely consuming the insects at a much faster rate. Both males and females wasted relatively little time searching at nonproductive heaps. If a preliminary investigation suggested that the termites were not "biting," the chimpanzees simply moved on to another heap.

Data are currently being collected on length and type of tool and approximate yield of termites per session. This information will highlight individual differences in technique both within and between the sexes. Perhaps analysis of these data will help to explain the intriguing difference between male and female termite-fishing behavior in November. A possible explanation is that because males so seldom eat termites during off-season months the insects, suddenly incorporated into the diet, represent a "new" food, an exciting and tasty delicacy to be hunted for and consumed with gusto. Or perhaps (although this seems unlikely) males, having had less practice, catch fewer termites per tool than females and therefore need to work longer to achieve the same amount of nourishment.

The sex difference in eating weaver and driver ants is less clear-cut, mainly because the behavior is seen less often (Figure 10.5). Moreover, weaver ants are frequently eaten high in the trees, where bouts may be overlooked. McGrew's analysis of the 1964–65 feces data showed that a higher percentage of female samples than male samples had remains of weaver ants, but for driver ants there were not enough samples to come to any conclusion. However, more recent data from full-day follows showed that during 1978 and 1979 the percentage of days when ants of either species were eaten was almost the same for males and females. If figures for the two years are combined, females were seen to eat ants on 10.3 percent and males on 9.2 percent of the days on which they were followed.

Bout lengths for males and females feeding on weaver ants were very similar, as shown in Table 10.7. The mean length for bouts of driver ant dipping, however, was slightly longer for females than for males (12.0 and 8.9 minutes respectively). McGrew's (1979) results gave longer mean bout lengths for both females (15.5 minutes) and males (13.6 minutes). It may be significant that during 1978 and 1979 females accounted for four out of the five sessions that were over thirty minutes long.

McGrew suggested that females were more highly motivated to dip for driver ants when they had the opportunity than were males; that the proportion of individuals who actively consumed ants versus those who merely watched was significantly different. The more recent observations did not show such a clear difference: females dipped on 90.6 percent of fifty-five occasions that they were observed in the vicinity of a nest and males on 76.9 percent of thirty-nine occasions.

Looked at another way, during 78.3 percent of the forty-six occasions when the behavior was seen (in 1979 and 1980) *all* adults who were present dipped. On six occasions one or more females did not participate; on six occasions one or more males did not participate.

Even though there may be little difference in the motivation of adult individuals of the two sexes once they have arrived at nests, only females have been recorded visiting the same site more than once in a day. McGrew saw one female return to the same place four times in a single follow. He also described an incident in which a female passed a column of ants in the morning while she was traveling with a male, then returned later in the day on her own, searched the area for eight minutes, found the nest, and fed on the ants.

Years when caterpillars were eaten in large quantities were few and far between and it is not clear whether there is a sex difference in caterpillar feeding. During the May–June period of plenty in 1981 there were too few full-day follows on adult males to compare their scores with those of females. Certainly females ate caterpillars in great quantities. Indeed, one female ate them for 42.2 percent of her total feeding time during one day and 31.9 percent during another. The next time there is a caterpillar fiesta at Gombe, we shall try to get an adequate sample of follows on both sexes.

In summary then, females not only eat insects more frequently than males, but tend to spend longer periods feeding on them. The major difference seems to be that females have more persistence, or patience, than males. They are prepared to spend several hours fishing at a termite mound even when they reap little reward (in volume consumed, at any rate). And while they probably do not eat substantially more drivers, they are, again, willing to spend longer at the nests, and check them out more frequently than do the males.

Birds and Small Mammals

Hugo with a domestic hen's egg. He will crack the shell and then wadge all with leaves. (H. van Lawick)

The first report of the eating of birds comes from the London Zoo, where in 1883 an adult female, Sally, used to catch and devour a pigeon almost every night. At Gombe birds are not eaten very often, as we shall see in the next chapter. Eggs and young of the following species have been recorded as prey: at least two types of weavers, *Ploceidae* sp.; francolins, *Francolinus squamatus*; palm-nut vultures, *Gypohierax angolensis*; helmeted guinea fowl, *Numida melaegris*; and green pigeons, *Treron australis* (Goodall, 1968b; Wrangham, 1975). Only twice have chimpanzees at Gombe been observed to kill and eat adult birds, both of which were injured. One was an African pied wagtail, *Motacilla aguimp*, and the other a harrier hawk, *Polyboroides typus*.

At Mahale chimpanzees also take eggs and fledglings. The one adult bird that was captured and eaten, a white-browed coucal, *Centropus superciliosus*, was also injured (Nishida and Uehara, 1983). Both at Gombe and at Mahale the young of domestic chickens, *Gallus gallus*, have been caught and consumed (Goodall, 1968b; Nishida and Uehara, 1983).

Often a chimpanzee raids a nest after his attention has been attracted by the calls of nestlings or the movements of adult birds. Once Figan, as an adolescent, stared over the valley as a pair of palm-nut vultures landed on the nest they had just constructed. Figan set off and presently climbed into the tree, about a kilometer away, and inspected the (empty) nest. He paid little attention as the adults repeatedly dived at him. At Mahale, however, an adult male chimpanzee who attempted to take eggs or fledglings from the nest of a crowned hawk eagle (*Stephanoaetus coronatus*) gave up in the face of

The early adolescent Wilkie, feeding on a fledgling bird that he caught as it flew just above the ground.

the adults' aggression (Takahata, Hasegawa, and Nishida, 1984). Juvenile and early-adolescent chimpanzees typically investigate birds' nests when they come upon them during foraging.

Only twice at Gombe have small mammals been captured. Both times the hunters were adolescent females: Pom seized young squirrels and Gremlin caught a rat and ate it, head first. (Often rodents are disregarded; Flint, for instance, ignored a family of newborn rats, and several times adult chimpanzees have been observed to hit at a mouse or rat without attempting to kill it.)

Cultural Differences

Variation in food items selected by chimpanzees in different areas are, for the most part, related to availability. At Gombe, for instance, Kahama community chimpanzees feed on *Morus lactea* fruits, but none of these trees are found in the Kasakela community range (Wrangham, 1977). There is also evidence that food items commonly eaten by one chimpanzee community are ignored by another despite their availability—or they are eaten in a different manner. The oil-nut palm (*Elaeis guineensis*) grows at Gombe and Mahale, having been introduced from West Africa. At Gombe, the chimpanzees feed on the fruits, pith, dried flower stems, and dried or rotten wood fibers. Indeed, as we have seen, the fruits are probably their single most important food throughout the year. Mahale chimpanzees have never been seen to feed on any part of this palm (Nishida et al., 1983). In Liberia (Beatty, 1951) and at Bossou, Guinea (Sugiyama and Koman, 1979), chimpanzees crack open the hard seed of the oil-nut palm and eat the kernel (it has not been reported whether or not the flesh is also part of the diet). In the Tai National Park, Ivory Coast, the oil-nut palm is present, though not abundant; only the pith has been seen to be eaten during three years of study (C. Boesch and H. Boesch, personal communication).

Evered chewing on a piece of dead palm wood.

A recent paper (Nishida et al., 1983) documents other differences in the vegetarian feeding behavior of the Gombe and Mahale chimpanzees. Sixteen different foods from nine plant species present in both areas are eaten *regularly* at Mahale and either have not been seen to be eaten at all at Gombe, or only a few times. Six different foods from three plant species present in both areas are eaten at Gombe but have not been recorded as part of the diet of the Mahale chimpanzees. One of the Mahale foods (fruits of *Cordia millenii*) and one of the Gombe foods (fruits of the oil-nut palm) constitute very significant parts of the chimpanzee diet in the area where they are eaten.

An interesting difference is seen in wood-eating behavior. The Gombe chimpanzees, as mentioned, occasionally chew on dead-wood fibers (from the oil-nut palm, for instance). At Mahale, although wood eating is not common, the chimpanzees sometimes chew or lick the dry wood of trees such as *Ficus capensis* and *Garcinia huillensis*. They visit some trees so regularly that huge "caves" are formed. When feeding, the chimpanzees creep into these hollows so that only their rumps protrude (Nishida and Uehara, 1983). Behavior of this sort has never been observed at Gombe, even though both species of tree are present and chimpanzees feed on the fruits.

At Mahale the chimpanzees have been observed feeding on the bark (inner, outer, or both) of twenty-one different plant and tree species (Nishida, 1976): thirteen of these are present at Gombe, but the barks of only three (*Brachystegia bussei*, *Pseudospondias microcarpa*, and *Sterculia tragacantha*) are eaten at both localities. Chimpanzees at Gombe eat four other types of bark, two from trees also present at Mahale (*Ficus vallis-choudae* and the oil-nut palm) and two not listed for Mahale. The Mahale chimpanzees have occasionally been seen to bite off and consume pieces of root, but not to dig for them (Nishida and Uehara, 1983); this behavior has not been recorded at Gombe.

At Mahale the chimpanzees sometimes lick rocks lying in streambeds or on the shore of Lake Tanganyika (Nishida and Uehara, 1983). This practice has never been seen at Gombe, although it is common among the baboons there.

Another difference between the chimpanzee populations of Gombe and Mahale lies in the species of ants selected as food and the frequency with which they are eaten. Weaver and driver ants are found at both localities: weavers are eaten by the chimpanzees of Gombe and Mahale, but drivers by the Gombe chimpanzees only. Arboreal ants (mainly *Crematogaster* of several species, at least five species of *Camponotus*, and the small *Monomorium afrum*) are eaten very frequently at Mahale (Nishida and Hiraiwa, 1982) but only a few of these species are consumed at Gombe, and those only very rarely.

Even a preliminary comparison of feeding behavior in the two areas has revealed some differences in feeding methods. At both sites small, rough-surfaced leaves (such as *Ficus urceolaris*) may be piled up in the hand before chewing; at Gombe they are sometimes collected and piled up in the mouth. The hard seedpods of *Diplorynchus condylo-carpon* are usually cracked open with the molars at Gombe, with the incisors and canines at Mahale. A more thorough investigation will almost certainly reveal many other variations in the feeding techniques of these two chimpanzee populations. (Differences in the use of tools for feeding are described in Chapter 18.)

Some of the above dissimilarities in the diet of the Gombe and Mahale chimpanzees may result from intraspecific variation in plant chemistry, which can affect food selection even in a single population (R. Wrangham, personal communication). Most of the differences, however, are probably due to tradition. An infant chimpanzee, from the age of about five months, tends to watch closely when his mother is feeding; he may put his face very close to hers and sniff the food she is eating. Once he begins to eat small pieces of solid food (at about four to six months), he usually chews the food on which his mother is feeding, taking some from her hand or mouth or from nearby food clusters. He also samples foods eaten by siblings, peers, or other individuals who happen to be in the group. Infants do sometimes pick and chew plant materials that adults have not been seen to feed on, but such experiments, at least sometimes, are discouraged by their mothers or other individuals. When a year-old infant began mouthing the seed of a *Garcinia* sp. fruit (which the adults spit out or swallow without chewing), his mother flicked it from between his teeth (Wrangham, 1975). When three-year-old Gilka picked up a piece of papaya and began to lick it, her mother at once snatched it, smelled it, then threw it away. And when Flint, aged seven months, nibbled a tiny piece of a biscuit I was eating, his elder sister, Fifi, hurried up to him, sniffed the remainder, and vigorously hit it away. At Mahale, females on two occasions snatched from their infants' mouths, and discarded, leaves of plants that were not part of the normal adult diet. Interventions of this sort undoubtedly serve to maintain the traditional food preferences of a community.

Occasionally an infant, having once tasted a new food and presumably found it palatable, will try it again. In 1965 Goblin, aged nine months, repeatedly sniffed, licked, and then played with the skin of a mango that I had left lying on the ground. That was the first time I had seen a chimpanzee show any interest in a mango fruit, although mango trees (*Mangifera indica*) are quite common along the shoreline and foothills in the Kigoma region and were introduced from India about a hundred years ago. Three years later Goblin was observed to pick and eat a partially ripe mango and he consumed an unripe one in 1972 when he was eight years old (Wrangham, 1977). He has not been seen to eat these fruits again, but other individuals have. In 1979 an adult male and two females with their three offspring all fed on fallen, unripe mango fruits (the male was on his own, the females and young were together). The following year two other females spent some thirty minutes eating unripe mangos, and on a different occasion the infant of a fifth female ate several of the fruits. Thus it seems that gradually mangos are being incorporated into the diet of the Kasakela community. While we do not know whether this is in any way connected with Goblin's experiments as an infant, it does seem that youngsters, with their more flexible behavior and their predilection for exploration, are the most likely age class to introduce a new feeding tradition. Without doubt infants and juveniles were directly

Food habits of the community are learned during infancy. Frodo tastes the food in his mother's mouth. His elder brother, Freud, watches closely.

Feeding traditions are passed on through mechanisms of social facilitation and observational learning. One-and-a-half-year-old Getty observes his mother's six-year-old brother, Gimble, feed on leaves. Getty then samples the same food, while Gimble watches him.

responsible for initiating many of the new feeding behaviors that spread through a troop of Japanese macaques (see for example Kawamura, 1959).

In 1982 the first observation of mango eating was recorded at Mahale; the fruit was picked and consumed by a five-year-old. It will be interesting to see whether other Mahale chimpanzees will imitate the behavior (Takasaki, 1983). Perhaps at both study sites mangos will eventually become part of the normal diet of the adult chimpanzees.

11. Hunting

Intense excitement just after a kill. Three adult males grin and scream as they crouch over a carcass. (H. van Lawick)

November 1981 It is late afternoon when the chimpanzees—six adult females and their young—encounter bushpigs, about eight adults and some piglets foraging and grunting in the tangled undergrowth. Melissa reaches to touch Fifi and both give excitement grins. Hair bristling, all move stealthily forward. Gigi and ten-year-old Freud are the first to give chase. The pigs, startled, stampede in confusion and Gigi grabs a piglet. It squeals in fright and instantly three of the adult pigs turn and charge the hunter, uttering furious roarlike snorts. Gigi drops her prey and leaps into a tree, screaming. All is confusion: bodies crash through the undergrowth, adult pigs roar and snort and toss their manes; the chimpanzees with waa-barks and screams take refuge in the branches above. Every so often one of the hunters leaps down to try again.

Five minutes from the start of the hunt young Freud catches another piglet, but almost at once he is seized by a huge sow. Dropping his prey, he screams in terror and pain. The other chimpanzees do nothing except bark and scream and shake the vegetation in threat—until suddenly one of them charges straight at the desperate Freud. The sow turns to this new challenge, releasing Freud, who drags himself into some low branches. He is bleeding heavily. Gigi leaps from the wicked jaws only just in time. The adult pigs continue to mill around, but the hunt is over and the chimpanzees sit quietly. After another fourteen minutes the pigs leave, and only then do the chimpanzees climb silently to the ground and move in another direction. Freud, still bleeding, limps slowly after them.

The hunting, killing, and eating of medium-sized mammals is probably a characteristic behavior of chimpanzees throughout their range—though more common in some areas than others. Meat eating has been recorded at three study sites in western Tanzania, in Uganda, and in the Senegal and Ivory Coast. Two groups of chimpanzees that have been rehabilitated into the wild, in the Gambia and in Gabon, have also been seen hunting, killing, and eating prey. Table 11.1 lists the prey species recorded to date from these areas and gives full references. The observed prey range in size from mice, rats, and small birds to a more-than-half-grown bushpig estimated to weigh not less than 20 kilograms. A variety of primates, including human infants,

Table 11.1 The prey species of chimpanzees in different parts of their range.

Species	Western Tanzania		Uganda[c]	Senegal[d]	Ivory Coast[e]	Ex-captive	
	Gombe[a]	Other[b]				Gambia[f]	Gabon[g]
Primates							
Colobus							
Colobus badius	•	•			•	•	
C. polykomos			•				
Baboon							
Papio anubis	•						
Cercopithecus							
Cercopithecus mitis	•	•	•			•	
C. ascanius	•	•				•	
C. aethiops		•				•	
Prosimians							
Galago sp. (unidentified)						•	
G. senegalensis		•		•			
G. crassicaudatus		•					
G. aleni							•
Perodicticus potto						•	
Chimpanzee							
Pan schweinfurthii	•	•	•				
Human							
Homo sapiens	•						
Ungulates							
Bushbuck							
Tragelaphus scriptus	•	•					
Suni							
Nesotragus moschatus		•					
Bushpig							
Potamochoerus porcus	•	•					
Blue duiker							
Caphalophus monticola		•					

account for about half the prey species. Cannibalism has been reported at two sites in Tanzania—Gombe and Mahale—and also in Uganda.

Table 11.2 lists the various mammals and birds that have been caught and/or eaten by chimpanzees, or of which the remains have been found in their feces, during twenty-two years of research at Gombe. The first fifteen years of the study are divided into five periods (determined by the data-collection methods at the time). Much of the material pertaining to these periods has already been published. The remaining seven years are listed separately. While the table is not designed to provide information on the frequency of meat eating at Gombe, it does show that the behavior has been recorded throughout the study period and that meat is not an uncommon food. It also gives some idea of the relative frequency with which the different kinds of prey have been consumed. The data are more precise from 1970 on, when more hours were spent following target individuals.

Red colobus monkeys are by far the most frequently eaten prey: they account for 59 percent of the total number (374) of medium-sized

Table 11.1 (continued)

Species	Location						
	Western Tanzania		Uganda[c]	Senegal[d]	Ivory Coast[e]	Ex-captive	
	Gombe[a]	Other[b]				Gambia[f]	Gabon[g]
Rodents							
Mice and rats							
Unidentified species	•						•
Cricetomys eminii		•					
Squirrel							
Funsciurus sp.	•						
Protoxerus stangeri		•					
Insectivores							
Checkered elephant shrew							
Rhynchocyon cernei		•					
Carnivores							
Mongoose (white tail)							
Ichneumia albicauda		•					
Myracoidea							
Hyrax							
Heterohyrax brucei		•					

a. Goodall, 1963, 1968b; Teleki, 1973c, 1981; Wrangham, 1975, 1977; Busse, 1977, 1978; McGrew, 1983; Wrangham and Bergmann-Riss, in press.
b. Japanese scientists have carried out extensive studies in western Tanzania since 1966. Hunting behaviors listed here were first observed in the Kasakati Basin and other areas near the shores of Lake Tanganyika (Izawa and Itani, 1966; Kawabe, 1966; Kano, 1971b) and in the Mahale Mountains (Nishida, 1968; Nishida, Uehara, and Nyundo, 1979; Kawanaka, 1982a; Norikoshi, 1983; Takahata, Hasegawa, and Nishida, 1984).
c. Suzuki, 1971; Ghiglieri, 1984.
d. McGrew et al., 1978.
e. Boesch, 1978.
f. Brewer, 1978.
g. Hladik, 1973.

Table 11.2 Numbers of different prey species observed to have been caught and/or eaten by Gombe chimpanzees (or of which remains were found in their feces) during twenty-two years of observation. Numbers in parentheses represent number of predatory incidents involved when this was different from the number of prey captures—that is, in 1970–1971 six bushpig young were seen to be consumed, captured during four hunts.

Year	Number of months	Prey								
		Red colobus	Bushpig	Bushbuck	Baboon	Monkey (redtail/ blue)	Chimpanzee	Rodent	Bird	Unidentified
1960–1963	33	4	3	2	0	1	0	0	8	5
1964–1967	45	6	4	5	4	1	0	0	3	4
1968–1969	24	3	2	1	12	0	0	0	0	1
1970–1971	24	5	6 (4)	2	1	0	1	0	0	1
1972–1974	36	42	9 (5)	9	1	4	0	1	4	1
1975	12	18 (12)	5 (2)	3	0	0	2	1	0	0
1976	12	19 (13)	10 (6)	3	0	0	2	0	1	0
1977	12	37 (23)	10 (5)	6	0	0	0	0	0	0
1978	12	32 (32)	9 (7)	6	0	0	0	0	4 (3)	0
1979	12	25 (17)	2 (2)	5	3	0	1	0	2 (2)	0
1980	12	16 (13)	0	4	2	0	0	0	0	0
1981	12	14 (8)	6 (5)	3	2	1	0	0	4 (2)	1
Total, 1960–1981		221	66	49	25	7	6	2	26	13

prey from 1970 to 1975, and 66 percent from 1976 to 1981.* The two *Cercopithecus* prey species are encountered far less frequently than the red colobus at Gombe; moreover, the size of their troops is smaller and they are more agile than the colobus—which, as they crash about in the canopy overhead, are very obvious even from a distance. For these reasons *Cercopithecus* fall victim to chimpanzees less often.

Chimpanzees encounter baboons more frequently than any other primate species. The number of five-minute observation points when baboons were visible or audible within 200 meters of target male chimpanzees in the 1972–73 period was calculated as 4.3 percent, compared to 0.68 percent for red colobus and 0.36 percent for redtail (Wrangham and Bergmann-Riss, in press). Nevertheless, with the notable exception of 1968, baboons were preyed upon much less frequently than colobus and during some years were probably not caught at all. Similar numbers of young bushpigs and bushbucks were observed to be eaten per year; as we shall see, many of the fawns were taken over from baboons and not captured by the chimpanzees themselves.

It is seldom possible to get satisfactory pictures of meat-eating sessions high in the trees. It is even more difficult to get pictures of the actual hunt; in fact, insofar as the Gombe chimpanzees are concerned, none exist. From two examples of hunting behavior I have therefore provided word pictures for this chapter, taken from the reports of the field assistants who observed the incidents. I have translated them almost word for word, with a little added explanation and occasional slight condensation. I trust these descriptions will bring hunting behavior to life for those who have never seen it.

Methods

To some extent all hunting may be opportunistic and happen only after the sighting of suitable prey. Nevertheless, there is a difference between a capture that occurs when a chimpanzee simply happens upon and seizes a certain prey (such as a bushbuck he presses to the ground) and a capture that follows an extended period (up to two hours) of watching, following, and chasing intended prey. On several occasions chimpanzees saw monkeys across a valley, perhaps half a kilometer distant, and at once set off to hunt them. Sometimes, after

---

* The impact of chimpanzee predation on the red colobus monkey population at Gombe is obviously significant. T. Clutton-Brock (personal communication to C. Busse, 1977) estimated that there were between three hundred and five hundred red colobus within the range of the Kasakela and Kahama chimpanzee communities. If so, then in the 1973–74 period, *observed* chimpanzee predation resulted in colobus mortality of between 8 and 13 percent. Because these figures do not take account of *unobserved* predations, the actual number of monkeys killed was inevitably larger. Wrangham and Bergmann-Riss (in press) calculate that the fifteen males of the two study communities, between 1970 and 1975, killed an estimated 30 to 50 percent of the colobus population—if, that is, Clutton-Brock's estimated population size was correct. They suggest that such high predation rates could not be sustained by a prey species, and that perhaps the colobus density is in fact greater. In any case, these monkeys certainly suffer severe annual loss from chimpanzee predation.

a kill has been made, an individual who has been unable to acquire a share moves away from the prey and starts hunting again.

A single chimpanzee may hunt successfully on his or her own; often, however, hunting is a group activity. A lone chimpanzee who has made a kill usually keeps silent, whereas loud (and distinctive) calls typically break out at the climax of a successful group hunt. Thus the latter are far more likely to be detected (both by nearby chimpanzees and by human observers). During some hunts there is clear evidence of cooperation between two or more individuals, but the methods differ depending on the kind of prey.

Arboreal Monkeys

Red colobus monkeys are found in large multimale troops of up to about eighty individuals, with a preponderance of females (Clutton-Brock, 1972). Smaller groups and occasional lone individuals are also seen. When chimpanzees who are being followed by humans encounter colobus, the monkeys typically utter their high-pitched alarm calls and move off through the canopy. However, it is difficult to be certain that this is not, at least in part, a response to the presence of the observers (they react in the same way to the approach of humans when no chimpanzees are present). In the early days of the study I observed twenty-three encounters between chimpanzees and colobus when my presence did not affect the behavior of the monkeys because I was on the opposite side of the valley, watching through binoculars. Sixteen times chimpanzees and colobus fed within 30 meters or so of each other and the monkeys showed no concern.

When chimpanzees first see a troop of colobus nearby, they almost always show some interest, even if they only pause and gaze at the monkeys. Sometimes they grin, utter small sounds (grunts or squeaks), and reach out to touch or embrace one another—behavior that is almost always followed by hunting. On a few occasions parties of chimpanzees traveled rapidly toward a colobus troop and began to hunt after hearing the monkeys call from distances of up to 500 meters away.

Before starting to hunt in earnest, the chimpanzees often spend time on the ground gazing up into the canopy, or walking below as the monkeys leap above them. It seems that they are assessing the situation: the availablility of suitable victims, particularly mother-infant pairs and small juveniles; where they are located; and probably the arboreal pathways along which the monkeys would be able to escape. Wrangham (1975) observed that chimpanzees seldom hunted when they encountered monkeys moving through unbroken canopy, but usually did so when the canopy was broken, particularly when the monkeys were feeding in tall, emergent trees from which escape was difficult. Sometimes the chimpanzees make this preliminary assessment while they are still undetected by their prey, sitting silently and peering through the foliage at a monkey troop above. At other times they make no effort at concealment prior to hunting and, indeed, may utter loud pant-hoots before climbing to chase. The relative frequency of the stealthy versus the open approach is undoubtedly influenced by the presence of humans, who are likely to be seen as they move through the forest and thus attract the monkeys' attention to the would-be hunters. There would be little point in the chimpanzees' trying to hide once they had been detected.

Most observed hunts took place when two or more chimpanzees were together, although they also hunted, sometimes successfully, when on their own. In 1973–74, nineteen hunts by lone male chimpanzees were observed, of which six were successful: four of these captures were made by one Kahama male, Sniff, whose capture rate was unusually high (Busse, 1977). During the four-year period 1976 to 1979, only nine hunts by lone males were recorded. Twice the male concerned, Satan, was lost to view while still hunting, so the outcome was unknown. The other hunts were all unsuccessful. (Satan did catch one infant, but was forced to release it, unharmed, when adult male colobus attacked him.) Twice during these four years lone male chimpanzees remained quietly near troops of colobus, once for nearly two hours, but began to hunt immediately when they were joined by a second male.

Sometimes only one member of a party climbs and chases monkeys, but often two or more do so at the same time—after the same or different prey. Others remain below, following the hunt on the ground, or climb and watch from the branches. In mixed male-female parties, females often remain somewhat peripheral to the main activity, although this is by no means always so. After a bout of unsuccessful chasing the chimpanzees may give up and move off, but sometimes they remain, watching the monkeys for periods of up to one and one-half hours, and then hunt again.

From 1973 to 1981 the percentage of observed hunts in a given year that were successful ranged from 33.3 to 78.0 (with a median of 41.4). As indicated in Table 11.3, more than one colobus may be captured in a single hunt. Sometimes two are seized more or less simultaneously. At other times chimpanzees who are at first unsuccessful continue to hunt after one capture has been made—sometimes after trying, but failing, to acquire meat from the possessor of the first prey. Table 11.4 provides information on the frequency and success of multiple hunts. On three occasions individuals who were *already in possession of meat* nevertheless hunted again, successfully. Once, as Figan sat feeding on an infant he had snatched from its mother, he intently watched other chimpanzees who were still hunting. Suddenly he handed his entire carcass to Humphrey (who had been begging), joined the hunt, and caught a second monkey—which he then ate. Humphrey once made a kill while still holding a large piece of meat in one hand; and Sherry, holding in his mouth the remains of an infant that he had killed and partially eaten, captured an adult female (its mother) when she ran toward him (avoiding another hunter).

On 130 occasions between 1975 and 1981 the age class of colobus victims was recorded. Seventy-eight percent of the victims were infants and juveniles, and of these, 41 percent were infants snatched from their mothers. Of the twenty-three adults captured, 57 percent were mothers of these infants; six were males, three of which had been attacked as they tried to defend their females and young.

Sometimes monkeys fall during hunts. They may be knocked to the

Table 11.3 The success of male chimpanzees in observed hunts of red colobus monkeys.

Year	Observed hunts	No. of hunts when at least one prey captured	Percent success
1973	14	7	50.0
1974	14	11	78.0
1975	32	12	37.5
1976	29	12	41.4
1977	33	20	60.6
1978	30	19	63.0
1979	40	16	40.0
1980	28	11	39.0
1981	24	8	33.3

Table 11.4 Frequency of colobus monkey hunts when multiple kills were made.

No. of kills per successful hunt	Year							
	1973–74	1975	1976	1977	1978	1979	1980	1981
2	6	5	1	8	5	4	2	3
3	1	0	1	2	1	1	1	1
4	0	0	0	0	1	0	0	0
5	0	0	1	0	0	0	0	0
Percent successful hunts in which multiple kills were made	38.9	41.7	23.1	43.5	50.4	29.4	23.1	50.0

ground by chimpanzees trying to capture them (or their infants); or they may fall when a branch breaks, or when making too large a leap—particularly after a long pursuit when they are probably tired. Between 1975 and 1980, thirty monkeys fell: twenty-two of these, seventeen juveniles and five adults, were captured; four juveniles and four adults managed to escape. Another three monkeys, two adults and a juvenile, jumped to the ground when fleeing during hunts; one, an adult male, escaped. Fourteen of those killed were seized by arboreal hunters who hurled themselves down after their prey, but the others (almost half) were captured by chimpanzees already on the ground, watching the progress of the hunt from below. Individuals who remain on the ground during a hunt may well be anticipating the fall of a monkey.

Adult male colobus monkeys, on many occasions, actively defend troop members from the predatory attacks of one or more chimpanzees. Busse (1976) found that half of the successful predations recorded in the 1973–74 period occurred within seven minutes of the initial encounter and he suggests that the reason may have been that the monkeys were still widely dispersed and potential victims often far from the protection of the adult males. As a hunt progresses, adult males quickly move toward the chimpanzees and usually threaten them vigorously, shaking branches and bobbing up and down, uttering their alto threat calls. In the face of colobus male aggression, chimpanzees almost always retreat, often whimpering or screaming, and climb rapidly (sometimes precipitously) to the ground.

Once, as three adult male chimpanzees (Figan, Jomeo, and Goblin) started to hunt, three male colobus bounded down toward them, shaking branches and barking. The chimpanzees, screaming, quickly left their tree. The colobus followed them to the ground, and the entire party of chimpanzees (which also included an adolescent male and a mother with her offspring) turned and fled! When the aggressive monkeys climbed into the trees again, the chimpanzees began to move quietly along the ground, looking up at the large and very spread-out monkey troop above. Five minutes after being chased, Figan again climbed to hunt. At once a male rushed toward him, shaking branches and uttering threat calls. Figan whimpered and quickly climbed down. As the chimpanzees sat below, still looking up, three male colobus (probably the same three) leaped down to the lowest branches of the tree only a few meters above the chimpanzees' heads and threatened them. Once again, the hunters were routed, this time for good. (Observers: H. Mpongo and Y. Alamasi.)

In 1965 I watched old Mr. McGregor as he was chased along the ground for about 10 meters, screaming loudly, with a single male colobus bounding after him. Another time one adult male colobus successfully defended several juveniles who were in a tall tree, their escape routes cut off by four chimpanzees. Every time one of the hunters approached, he was threatened by this male and retreated; after one and a half hours the chimpanzees gave up and went away (Busse, 1976).

On nineteen different occasions male colobus actually attacked chimpanzees as they hunted, leaping onto them and biting at their backs or around their scrota. When this happened, the hunters almost always retreated. Sherry was considered the "bravest" and twice continued hunting in the face of vigorous defensive action. Once he turned and drove off two males who had leaped at him and were biting around his rump; only when they charged back and attacked him yet again did he give up. Another time, when he was after a mother-infant pair, a single male charged at him and jumped right onto his back. Sherry, screaming loudly, managed to shake him off and raced on after his prey. The defending male did not give up, and as Sherry seized the mother and grabbed her infant, the colobus male again leaped at him, biting his loins. Finally Sherry, screaming louder than ever, was forced to abandon the hunt and the infant was rescued. (Observers: H. Matama and E. Mpongo.)

There were six other instances when colobus males were able to rescue troop members who had been captured. Sherry had to relinquish another infant he had taken from its mother, and on different occasions he, Figan, and Gigi each had to let go of females they had grabbed prior to snatching their infants. Gigi was forced to relinquish a young female she had captured and Satan, who had cornered a young male against a wide treetrunk, turned and fled when six other males charged to the rescue of their companion. Twice colobus males

Flo begging as Mike tears a colobus monkey carcass. (B. Gray)

were aggressive to male chimpanzees who were *not* hunting; one went up to two chimpanzees on the ground and gave one of them, Figan, a shove (whereupon he was chased up a tree) and another leaped some 6 meters onto Jomeo's head as he sat, all unsuspecting, below (Wrangham, 1975).

After a kill has been made, the mothers of infant victims sometimes remain nearby while the prey is being eaten. One mother tried to pull her infant away from Evered as he fed on it, but was chased off by a female chimpanzee. All but one of the adult females killed or wounded by the chimpanzees were mothers who had had infants seized from them and had not immediately run off. One of these mothers even tried to reach her infant (who was being eaten) while she herself was being killed.

Blue monkeys and redtails, as shown in Table 11.2, have not often been recorded as prey at Gombe. Only five actual *hunts* on blue monkeys have been observed, of which two were successful; the others failed in the face of aggression from the male monkeys. Only one redtail hunt was seen when Faben (despite his paralyzed arm) chased a juvenile along the ground and seized it (Wrangham, 1975).

Bushpigs
Bushpigs are nocturnal and lie in dense thickets during the day. Very rarely, when it is gray and damp, they move about in the daytime. Observation conditions in the thick undergrowth are poor, but it was possible to make fairly accurate counts on fifty-five occasions between 1970 and 1980, and it seems that at Gombe the pigs mostly move about alone or in small groups known as sounders. Up to six piglets were seen in the same sounder, sometimes of two different age groups.

Over the entire twenty-two years sixty-six piglets were seen being consumed by the chimpanzees on forty-five different occasions—1.5

piglets per successful hunt. From 1972 to 1981, when more long follows enabled us to get a clearer picture of predatory behavior, over fifty searches or hunts were observed, of which twenty-two were successful. On ten other occasions chimpanzees were found feeding on freshly killed piglets. Most victims were small, still in their striped natal coats, but three times they were older: the largest was estimated at just over half adult size, the other two just under.

When chimpanzees see or hear pigs in the undergrowth, they almost always stop to investigate. If one or more adults run off, the chimpanzees check the undergrowth carefully; six times they found and seized piglets. Once an adult male, Mike, having disturbed a sow, searched for a few minutes but failed to find her three small piglets. (The sow returned and moved them eight hours later; Wrangham, 1975.) Another male, Satan, emerged from a thicket with two striped piglets, leaving a third. Bushpigs construct untidy nests (mounds of matted grass) where very small piglets are concealed; when chimpanzees come across these, they search them intently, pulling away handfuls of grass and sniffing all around. Three times, nests were occupied. Once a sow ran off and two piglets were caught. Another time an adult charged out, leaving a piglet that managed to escape, racing after its (presumed) mother. The third time only an adult emerged, and diligent looking revealed no piglets.

Pig hunts are almost always very difficult to observe. For one thing, they usually take place in thick undergrowth. For another, human observers are normally not very close, partly because there is a certain amount of danger from an infuriated adult bushpig, but mainly because the pigs for the most part are afraid of humans; the farther away the observers are, the less likely it is that their presence will distort the sequence of events. Occasionally a chimpanzee creeps up to a sounder and steals a piglet before the prey is aware of his presence (this probably happens far more often when noisy humans are not around). After the adult pigs have seen the chimpanzees, they either run off with the young or gather defensively around them. Twice adults have been seen pushing piglets with their snouts, trying to move them away from chimpanzees.

Once adult pigs have been alerted to danger, capture of the prey becomes much more difficult. Typically the hunters charge toward the sounder, displaying, stamping on the ground, and waving vegetation. The adult pigs retaliate, grunting fiercely and running at the predators, who hastily leap into the safety of trees overhead. There they continue to display, uttering loud waa-barks and pant-hoots, swaying the vegetation, and every so often making another charge toward the prey. During the confusion, as chimpanzees and pigs mill about, one or more piglets are often caught. Three times adult chimpanzees surrounded thickets in which pigs had taken refuge, charging in, one by one or two together, until eventually one of the hunters managed to seize the prey. Once chimpanzees encountered two adults, a three-quarters-grown youngster, and a small piglet. The pigs kept quite still while the chimpanzees surrounded them, the three large ones facing outward, flanking the piglet. The chimpanzees did not start to hunt immediately, but glanced at one another, and every so often gave food calls. After approximately two minutes one of the hunters, Mike, moved toward the pigs and threw a rock, which hit one adult on the snout. Still they did not run, and no other chimpanzee approached. Only when the prey broke formation, startled perhaps by the closer approach of the observer, did the chimpanzees move closer: one minute later adult male Hugo was seen with the dead piglet, up in a tree. (Observer: F. Plooij.)

A total of twenty-seven hunts have been observed and of these, 66.7 percent were successful. On the remaining nine occasions it was almost certainly the defensive, protective behavior of the adult pigs that saved their young. Sometimes an adult made repeated charges, grunting and roaring, shaking its mane, tossing its head, causing the hunters to rush up trees again and again. During one hunt the adults actually rescued two piglets that had been captured, as described in the anecdote at the start of this chapter. (Observers: H. Mpongo and H. Matama.) Probably rescues of this sort are much more common when humans are not there to distract and frighten the bushpigs.

Bushbuck Fawns

A total of forty-nine fawns are known to have been wholly or partially consumed by chimpanzees since 1960. Only a few, however, were seen to be actually captured by chimpanzee hunters. Many were known or suspected to have been stolen from baboon predators. Table 11.5 provides information on the capture and/or eating of bushbuck fawns.

The seven known captures occurred when chimpanzees came across fawns during travel. Bushbuck young, like those of many antelope species, rely for protection from predators on camouflage, immobility, and minimum body odor. Thus captures took place without prolonged hunts, when chimpanzees suddenly noticed the prey. Humphrey in 1979 saw a fawn when it was about 35 meters from him, raced over, seized it, and ran off. The fawn bleated a few times; the mother snorted and ran toward it, then bounded away as other chimpanzees and human observers approached. Another capture occurred when the victim was heard bleating. After a few seconds the oldest male in the chimpanzee party, Hugo, seized the fawn; other males, who were screaming and embracing in excitement, hurried up to the prey (Wrangham, 1975). A third fawn was captured in 1977 by a late-adolescent male, Goblin, who was on his own. The other four captures were made by females. One was by Gigi, who during travel suddenly stopped, stared around, moved very cautiously through the grass for nine minutes, pausing frequently to sniff the vegetation, then seized a very tiny fawn. Passion found and killed two fawns within a week (in 1977), both thought to be only two or three days old. And Fifi seized an older fawn as she traveled with a mixed party; other chimpanzees hurried over and took part in the killing of the prey.

We do not have an accurate picture of the behavior of the mother bushbuck during predatory incidents, because these antelopes are very shy and always run from human observers. However, the second time I ever observed chimpanzees eating meat, in 1961, was on the other side of a valley. I heard excited calls and watched through my binoculars as chimpanzees consumed a fawn. The mother remained below the tree for an hour, by which time it was dusk and I could no longer see. Whenever a chimpanzee (or baboon, for this prey had

Table 11.5 The capture and/or eating of bushbuck fawns, 1960–1981.

Method of capture	Known	Suspected	Total
Caught by chimpanzees	7	7	14
Stolen from baboons	14	13	27
Undetermined			8
Chimpanzee found with remains	2		
Male chimpanzee joined party carrying fresh kill	1		
Remains in chimpanzee feces	5		
Total			49

almost certainly been taken from baboons) climbed to the ground, the doe charged at it, chasing it back into the tree. On only two other occasions were mother bushbucks seen to run to the defense of their young; after charging toward the captors, they saw the human observers and ran off.

On a number of occasions when an adult bushbuck was startled by the approach of chimpanzees (and humans) and bounded off, barking, the chimpanzees clearly showed predatory interest, moving toward the place where the bushbuck had been and searching the grass. Possibly it was the presence of the mother that attracted Humphrey's attention in the 1979 kill.

Baboons

The first time that chimpanzees were observed to eat a baboon was in 1964. Before this, however, I had observed three definite and two probable baboon hunts (one in 1961, two in 1962, and two in 1964). Table 11.2 indicates that, with the remarkable exception of 1968, not more than three baboons per year have been seen to be captured, and for six years (1973–1978) *none* were seen to be captured. Morever (apart from one incident, where two male chimpanzees moved toward a young baboon, hair erect, then embraced each other), no hunting at all was seen from 1976 to 1978.

Table 11.6 Circumstances of infant and juvenile baboons during eighteen successful and thirty-two unsuccessful hunts by chimpanzees, 1968–1981.

Circumstances of prey	Infant with—				Older infant or juvenile		Success	Failure	Total
	Mother		Adult male						
	+	−	+	−	+	−			
In troop	8	7	0	2	2	7	10	16	26
In small group	1	9	1	1	2	4	4	14	18
Isolated	1	1	2	0	1	1	4	2	6
Success	10		3		5		18		} 50
Failure		17		3		12		32	

Note: + = successful hunt; − = unsuccessful hunt.

A total of twenty-five young baboons are known to have been eaten by the habituated chimpanzees (and two by unhabituated individuals in the south) since the beginning of the study. One other infant was captured and killed, but retrieved by its mother before being eaten. Of the twenty-two prey whose ages were known *precisely* (eight) or *approximately* (fourteen), 77.3 percent were black infants (those under six months old). Victims were seized sometimes from the midst of their troops, sometimes from a smaller subgroup, and sometimes when they were on their own or with just one caretaker (Table 11.6). Fifty hunts, of which eighteen were successful, were well observed. In many of these instances the attention of the chimpanzees was first aroused by various distress calls made by infant baboons. Sometimes this was during weaning incidents, sometimes when the infants were being carried by adult males, who often handled them roughly. On four separate occasions chimpanzees tried (but failed) to seize the same three-week-old infant when she screamed as an adult male took her from her neglectful mother—twice when she was two weeks old and twice the following week (Ransom, 1972).

The tactics employed by male chimpanzees when hunting baboons have been described in detail by Teleki (1973c). Sometimes two or more (occasionally just one) male chimpanzees rushed toward an intended victim and seized or tried to seize it on the run. At other times several males approached their prey slowly, and there was often a high level of coordination and cooperation among them. In 54 percent of the well-observed hunts the chimpanzees concentrated on mother-infant pairs, and 55.5 percent of the well-observed captures were infants seized from their mothers. Once an intended victim, or its mother, or another nearby individual became alert to the possibility of danger, there were loud calls of fear (from infant or mother) or threat (from adult males). These sounds often attracted other baboons, who ran to the defense of the endangered youngster. Sometimes the ensuing interspecific aggression resulted in absolute pandemonium, with chimpanzees and baboons (often including females) screaming, roaring, barking, and lunging. During skirmishes of this sort the combatants sometimes engaged in physical conflict, standing up and hitting out at one another. On six occasions male baboons leaped onto chimpanzees and appeared to be slashing with their canines at the chimpanzees' backs. An adolescent male, Mustard, was only released when an adult male, Evered, displayed toward the scene so that the baboon ran off.

After such incidents human observers always looked for signs of injury in the chimpanzees, but only once was an open wound seen (Wrangham, 1975). Baboon males have large and powerful canines; in other areas they have been known to mortally wound leopards caught attacking troop members (Goodall, 1975a), and it is puzzling that they almost never hurt chimpanzee hunters. Nevertheless, many times chimpanzees were clearly prevented from seizing their intended prey by intensive mobbing and harassment from adult male baboons. Once Figan (in 1969) seized a black infant from her mother, then raced screaming into thick vegetation with a number of male baboons after him. He reappeared without the infant, who was later seen being carried by her mother; but her head was crushed and she was dead.

Mike (in 1968) also lost an infant when a male baboon jumped onto his back. As Mike reared up and twisted to dislodge his opponent, the infant flew from his grip and was lost in the long grass—but was captured by another male chimpanzee who ran off with it (Ransom, 1972; Teleki, 1973c).

After a capture some of the baboons who have been involved typically follow the chimpanzees to the site chosen for the eating of the meat. One juvenile was not killed until thirty minutes after his capture; every time he uttered a small sound, there was an outburst of baboon vocalization from below the tree, aggressive advances from some of the males, and occasional brief fights between chimpanzees and male baboons (Ransom, 1972; Teleki, 1973c). Once a victim is dead, however, male baboons soon lose interest and leave, whereas a mother often remains longer at the site. Three times mothers tried to pull away their infants who were being eaten; they were threatened and chased off by nearby chimpanzees, but returned and stayed near the feeding hunters. One mother remained close to the site of the kill for a total of four hours: most of this time she was on her own, although an adult male stayed with her for a while.

Baboon hunts tend to be much shorter in duration than predatory attacks on colobus monkeys. When attempts to seize baboon victims fail, the hunters usually give up quickly; presumably once a troop has been alerted, it is not worthwhile for the hunters to continue. Only twice have two attempts (on different victims) been made during the same encounter. In 1968 four males attempted to seize an infant from an adult male baboon. They failed, but three minutes later the same chimpanzees rushed out of sight into the vegetation and when they emerged, one of them, Hugo, was grasping a ten-month-old infant (Teleki, 1973c). In 1981 four adult male chimpanzees ran toward a mother-infant baboon pair and one of them, Satan, seized the infant and began to eat, sharing the meat with his companions. Twenty-five minutes later a female with a black infant passed nearby. Figan got up and moved cautiously toward her. She at once began to scream and run, and an adult male baboon lunged at Figan, roaring and grunting. Figan avoided him and ran after the female for nearly 100 meters before giving up. (Observers: R. Fadhili and H. Matama.)

Usually at least two chimpanzees take part in baboon hunts, but occasionally a single male tries on his own. In 1968 Mike approached a mother-infant pair who sat with an adult male; the mother (who had been victimized before) screamed and ran off, and Mike did not persist. Figan hunted alone when he seized and killed the black infant that he lost when pursued by adult male baboons. Only twice have male chimpanzees who were entirely on their own been seen to make attempts. Evered in 1969 approached two mothers who sat, with their infants, close to an adult male. Evered grabbed at one of the infants but failed to seize it; the male, joined by another, chased after him through two trees (Teleki, 1973c). In 1981 Humphrey made a determined attempt to capture a six-month-old infant who, with his mother, was far from the rest of the troop. A few moments after the two climbed a tree above him, Humphrey moved slowly toward them and appeared to try to conceal himself under some thick undergrowth. He then watched them for a few minutes as they fed, before moving closer. Eight minutes from the time they appeared, he suddenly

Leakey and Mike with a baboon kill. The victim's mother watches. (B. Gray)

charged up the tree toward them. The mother screamed, seized her infant, and leaped away. Humphrey swung down, then climbed another tree for which the baboon was headed. She, however, jumped to the ground and raced off; he followed for 30 meters before giving up. (Observers: Y. Mvruganyi and Y. Alamasi.)

The yearly fluctuation in observed frequency of predation on baboons, with its extraordinary peak in 1968, followed by the gradual decline and apparent cessation, and then the renewed interest from 1979 on, is extremely interesting. The 1968 peak, occurring as it did during the height of the banana feeding, was clearly linked with the abnormal congregation of members of the two species in camp. Wrangham (1977) suggests that this high predation rate may have been simply a result of increased opportunity; certainly chimpanzees and baboons have spent less time at the feeding area since 1970. Nevertheless, chimpanzees have continued to visit camp frequently, and the baboons of at least one troop pass through every day. Even in 1978, eight years after the change in the feeding system, chimpanzees and baboons were together for over 123 hours in camp. Baboons are encountered during most follows, and in the same year individuals of the two species spent at least fifty-four hours within 100 meters of each other during follows out of camp (usually rather closer, as they were often feeding on the same foods). Thus lack of opportunity does not explain the fact that for so long no baboons at all were seen to be caught.

Ransom, who studied the baboons during the 1968–69 period, suggests that their increased vigilance may have played a considerable part in the reduction in numbers of hunts. He comments that during the first few occasions when he watched predatory incidents, the target mother-infant pairs and the associated adult male baboons seemed oblivious to the chimpanzees' intentions. Then, as more kills were recorded, the baboons became increasingly alert to signs of danger. In January 1969 an adult male baboon was seen for the first time to chase a mother-infant pair away from a chimpanzee who was showing predatory interest. Mothers of small infants, particularly those who had already been involved, began to react more quickly to the approaches or gazes of interested male chimpanzees (Ransom, 1972). During 1969 the ratio of successful to unsuccessful hunts dropped sharply, and it was after this that baboon hunting decreased in frequency and finally appeared to come to a halt.

Why, then, the sudden renewal of interest in 1979? Perhaps it was not until there had been a fortuitous, opportunistic kill that baboons were once more perceived by the chimpanzees as a possible source of food. It may be significant that all seven of the successful kills between 1979 and 1981 were from *unhabituated* troops (Table 11.7). Habituated baboons, during encounters with hunting chimpanzees, are not distracted by the presence of human observers and so can concentrate their efforts on repelling the predators. Unhabituated baboons are typically nervous and usually run away. Two of the three observed 1979 kills occurred as the troop, including the adult males, ran off, and the chimpanzees were able to snatch infants from their mothers.* After these relatively easy captures the chimpanzees once more began to hunt habituated baboons; during 1980 and 1981 eleven such attempts were observed. These elicited extremely aggressive and rapid defensive action from the adult male baboons, and the chimpanzee hunters were routed. During this period, however, four infants from the habituated troops disappeared and it is possible that one or more were victims of chimpanzee predatory attacks. More recently, in 1984, two infants from these troops were observed to be killed and eaten by chimpanzees (one was an orphan who had just lost her mother).

Human Infants

Prior to my arrival at Gombe there were two reports of chimpanzee attacks on human infants in the area. One of these took place outside the national park (near the Manyovu Road, to the east). An African woman was gathering firewood when a male chimpanzee suddenly appeared, leaped at her, and seized the infant from her back. The woman was injured; the infant was dead when recovered and had been partially eaten (Thomas, 1961). The second incident took place in the park (a game reserve at the time) near Nyasanga Beach. A six-year-old boy was minding his baby brother when a male chimpanzee rushed at him and seized the baby. The child gave chase, whereupon the chimpanzee dropped the baby, which survived, pulled the older child to the ground, and began biting at his face. The screaming alerted his

These injuries were sustained when the victim, age about six years, rescued his infant brother, who had been seized by a male chimpanzee, apparently for food. (H. van Lawick)

* It seems that chimpanzees are unlikely to attack an animal that remains quite calm and does not run. One predatory incident involved a blind infant baboon. The two brothers, Jomeo and Sherry, approached a mother-infant pair, who were about 15 meters from the only adult male baboon in sight. None of the baboons showed any interest in the chimpanzees. Jomeo stared at the infant, his hair erect. He moved to within 2 meters and began swaying a branch and hitting his hand forcefully on the ground. When he had been doing this for five minutes and the baboons were still ignoring him, he gave up and moved away. (Observer: C. Packer.) When Mike in the incident already described threw a rock at an adult pig which stood, defending the young, it was probably an attempt to make him run off. Spotted hyenas (*Crocutta crocutta*), when hunting large animals (such as zebras or wildebeeste), show great reluctance to attack if the prey stands its ground (Goodall and van Lawick, 1970).

Table 11.7 Observed predatory attempts on young baboons of habituated versus unhabituated baboon troops, 1968–1981.

Year	Successful		Unsuccessful	
	Habituated	Unhabituated	Habituated	Unhabituated
1968	10	0	13	0
1969	1	1	6	0
1970	1	0	2	0
1971	1	0	1	0
1972	2	0	1	0
1973	0	0	1	0
1974	0	0	3	0
1975	0	0	2	1
1976	0	0	0	1
1977	0	0	0	0
1978	0	0	0	0
1979	0	3	0	0
1980	0	2	4	2
1981	0	2	7	1

mother who, along with other women, ran at the chimpanzee, which fled. This child also survived, but with a badly mutilated face. No further predatory attacks on human infants have been reported since.

Cannibalism

Between 1971 and 1984 six infant chimpanzees were seen to be killed and/or eaten by members of the study community: the information is summarized in Table 11.8. Most of the cases have already been described in detail (Bygott, 1972; Goodall, 1977).

Three times the victims were infants of "stranger" females, estimated as being one and one-half to two and one-half years old. In cases 1 and 6, parties of Kasakela males were seen as they savagely attacked the mothers of the infants; during the fighting the infants were seized by the aggressors. In case 3 it was the sound of fierce fighting that led the observers to the scene—where they found three adult males feeding on the body of the infant, but did not see the mother. Kasakela males were seen to seize another infant during a similar attack on a stranger female. This infant was not killed and eaten, however, but carried by four different Kasakela males (including a four-year-old juvenile, Freud) and later abandoned.

The victims in cases 2, 4, and 5 were all small babies (under a month old) of Kasakela females and were seized, after fierce struggles, by the Kasakela female Passion, twice helped by her late-adolescent daughter, Pom. These two females were observed as they *tried* to capture two additional small infants, and Passion on her own made a determined attempt to seize a third. Another newborn infant may also have been killed by one or both of these females. Observers, attracted by sounds of conflict, came upon a party of Kasakela males with Melissa, who was clutching a dead newborn infant. His forehead had been bitten into (which was how Passion and Pom had killed

Table 11.8 Cannibalistic events observed at Gombe, in chronological order.

Case no.	Date	Age and sex of victim	Community	Captors	Cannibals
1	September 1971	1½–2½ years, sex unknown	Other	Humphrey	Humphrey, Mike, Jomeo
2	August 1975	3 weeks, female—Gilka's	Same	Passion	Passion, Pom, Prof, Skosha
3	October 1975	1½–2½ years, male	Other	Adult males	Jomeo, Satan, Figan
4	October 1976	3 weeks, male—Gilka's	Same	Passion, Pom	Passion, Pom, Prof, Skosha
5	November 1976	3 weeks, female—Melissa's	Same	Passion, Pom	Passion, Prof
6	March 1979	1½–2½ years, sex unknown	Other	Sherry, Evered	Sherry, Evered, Jomeo, Mustard

known victims). When the two cannibalistic females later joined the party, Melissa kept very close to one of the adult males. It seems likely that the intervention of the adult males, alerted by the mother's screams, had rescued the infant—but only after it had been killed. During the four-year period 1974 to 1977 two other infants vanished during their first month of life (Gilka's Gandalf and Pallas' Banda) and three other mothers, known or suspected to have been pregnant, may also have lost small babies. Throughout the same period only one mother, Fifi, was able to raise her infant. It is possible that Passion and Pom were responsible for some or all of these additional infant deaths.

Whereas the killing of infants by the adult males (cases 1 and 6) seemed to be a *consequence of the attacks on their mothers* (similar equally brutal attacks on stranger females were seen on many other occasions), Passion and Pom attacked the mothers of their victims *only in order to acquire the infants* as meat. Once they had possession of the babies, no further aggression was directed toward the mothers. Indeed, when Melissa approached the killers, who were then eating her infant, Passion reached out and embraced her.

Cannibalism has also been reported among chimpanzees in Uganda (Suzuki, 1971) and Mahale in Tanzania (Nishida, Uehara, and Nyundo, 1979; Kawanaka and Seifu, 1979; Kawanaka, 1981; Norikoshi, 1982; and Takahata, 1985). The Uganda victim was a newborn infant; its capture was not seen, but it was eaten by adult males. Two of the Mahale victims were newborn (one and one-half and two months) and were killed by members of the same community—in these cases, thought to be adult *males*. The other two victims were one and one-half and three years old respectively; one was killed and eaten by males of the same community, the other by males of a neighboring community. All four of the infants were males.

Table 11.8 (continued)

Chimpanzees present who did not eat[a]	Observer	Comments
2 adult males, 1 female, 1 infant	D. Bygott F. Plooij	Severe attack on mother by all males; infant killed by eating.
None	E. Tsolo H. Matama	Killed by biting into frontal bones; continuous feeding for 5 hours.
1 adult male, 1 adult female	R. Bambanganya	Figan and Jomeo found in possession of dead and bleeding infant.
2 adult females, 2 immatures	E. Mpongo G. Kipuyo E. Tsolo	Passion and Pom fight Gilka; Pom kills infant by biting forehead; continuous feeding for 5 hours.
Pom, Melissa, and Melissa's juvenile daughter	L. Lukumay R. Bambanganya	Severe fight with mother; Pom kills infant, then watches as Passion and Prof share meat; mother and juvenile daughter stay and watch.
5 adult males, 2 adult females	J. Athumani Y. Alamasi	7 males attack mother.

a. Individuals who arrived after the killing but did not stay are not included here, nor are victims' mothers unless they remained.

Cannibalism is well documented for many carnivorous species such as the lion (Schaller, 1972), the leopard (Turnbull-Kemp, 1967), and the hyena (Kruuk, 1972), but I have only been able to find a few references to cannibalism in nonhuman primates. On two occasions an adult male redtail seized, killed, and ate infants of the troop he had recently taken over (Struhsaker, 1977). An adult male chacma baboon (*Papio ursinus*) was observed to seize an infant from its mother, run off with it, and start to feed on the body (Saayman, 1971). At Gombe one female baboon ate part of her own infant, which had died of severe injuries (she nibbled at the flesh around one of the wounds and ate some of the protruding intestines), and the body of a premature infant was seized by another female and consumed. (Observer: A. Sindimwo.) Fossey (1984) found the remains of a gorilla infant in the feces of two other gorillas of the same group, a female and her son. It seems that cannibalism does not often occur in nonhuman primates: in most cases of infanticide described for langurs (Hrdy, 1977), gorillas (Fossey, 1984), and baboons (Collins, Busse, and Goodall, 1984) the victims were not eaten. However, in view of the clear-cut predatory behavior of chimpanzees, it should perhaps not be surprising that they, like humans, sometimes eat the flesh of victims of their own species.

Cooperation in the Hunt

I define cooperation as the action of two or more chimpanzees, directed toward the same goal at the same time, the sum of which makes it more probable that that goal will be attained. When two chimpanzees chase the same monkey, it is more than likely that each has as his goal the catching of the prey for himself—not some altruistic notion of helping his companion. Nevertheless, the end result is the same.

The most sophisticated examples of cooperative hunting have been observed during predatory attacks on baboons. A number have already been described (Goodall, 1968b; Teleki, 1973c). As mentioned, it was rare for chimpanzees to hunt baboons on their own: indeed, in a number of instances one chimpanzee appeared to try to alert his companions to the possibility of a capture. For example, in 1980 Goblin, after intently watching a group of infant baboons playing, moved over to a tree where Satan and Evered were grooming. He stood below, hair erect, and looked from them to the baboons. Satan and Evered climbed down at once, Goblin and Satan embraced, grinning and squeaking, and all three slowly approached the young baboons. On that occasion they were unsuccessful, as they were chased off by adult males. At other times a chimpanzee started toward his intended prey, then paused, looked back at another male, and sometimes reached out to touch him; often the two then hunted together.

Occasionally, when a potential baboon victim was partially isolated from its troop, three or more adult male chimpanzees carefully positioned themselves so as to block off escape routes while one of their number climbed toward the prey (Goodall, 1968b; Teleki, 1973c). One example, taken from relatively recent observations, illustrates the behavior. In September 1979, all but one of the Kasakela adult males were traveling in Lower Mkenke Valley when they came upon a female baboon with a tiny black infant. She was feeding in an oil-nut palm and appeared to be quite on her own. Goblin grinned, squeaked softly, and reached to touch Satan. All six males had their hair erect. When the baboon noticed the chimpanzees, she stopped feeding and gazed toward them. After approximately half a minute she began to show signs of unease; she backed away from the predators, gecking softly. Jomeo, moving very slowly, left the other males and climbed a tree close to the palm. At this point the female baboon began to scream, but she did not run off. When he had climbed to a branch level with her and about 5 meters away, Jomeo stopped. He stared at her, then began to shake a branch—possibly to try to make her run. She screamed even louder, but apparently no other baboons were within earshot. Two minutes later Figan and Sherry, also moving slowly, climbed two other trees. A male chimpanzee was now in each of the trees to which the baboon could have leaped from her palm; the other three, still looking up, waited on the ground. At this point Jomeo leaped over onto the palm. The baboon made a large jump into Figan's tree, where she was easily grabbed and her infant seized. The mother ran off 6 meters or so, where she remained, screaming, then uttering waa-hoo calls for the next fifteen minutes while the chimpanzees consumed her infant. (Observers: A. Ibrahim and S. Rukemata.)

She is not the only baboon to have been trapped in this way. One of the 1968 captures took place when Charlie climbed into a palm where an adult male baboon fed, an infant clinging to him ventrally. Charlie slowly leaned across the tree, seized the infant, and quickly climbed down to where other chimpanzees waited below (Ransom, 1972). And on another occasion, in 1969, I watched Figan follow a small male juvenile from one palm crown to an adjacent one, then back again, three times in succession. The baboon became more and more frightened and the chase faster and faster, until the quarry took a huge leap into a neighboring tree, where he was almost caught by a second male chimpanzee who was waiting on the ground. However, an adult male baboon charged to the rescue and the victim escaped.

During the intense harassment of chimpanzees by adult male baboons that typically follows a successful capture, the presence of other male chimpanzees may well be essential if the captor is to maintain possession of the prey. After Humphrey had seized an infant during one of the 1979 hunts, the male baboons were extremely fierce in their offensive tactics, but the other five chimpanzee hunters repeatedly charged them, flailing vegetation and (one of them) hurling many rocks. The male baboons were thus unable to concentrate on Humphrey. (Observers: H. Mpongo and R. Fadhili.)

Cooperative tactics have also been observed during hunts on colobus monkeys. Busse (1977) noted only two instances, during the 1973–74 period, when more than one chimpanzee chased the same monkey at the same time—and on both occasions pairs of brothers were involved. Subsequently, however, there have been many observations of cooperation between nonrelated males. Once, for example, Sherry and Goblin chased the same juvenile colobus. Twice in the course of this long hunt Goblin climbed rapidly down from the tree, raced ahead along the ground, then climbed another tree directly in the path of the fleeing monkey. The second time this happened, the monkey was caught by Sherry. (Observers: H. Mkono and J. Athumani.) Another time Figan and Evered chased the same prey. Figan raced ahead (through the branches), causing the monkey to swerve; Evered leaped at the prey but missed, and Figan then made a capture. (Observers: H. Mkono and J. Athumani.) One of the best examples of cooperative hunting was seen in 1969 when five chimpanzees surrounded one male colobus in a tall tree. As two of them ran toward it, it jumped to the next tree, but at once a chimpanzee waiting there rushed to the same branch, cutting off its escape. The monkey was forced to return to the original tree, where again its escape was blocked. This continued for approximately fifteen minutes, after which one of the hunters managed to seize the (probably exhausted) prey. (Observer: D. Bygott.) Another cooperative colobus hunt is described in Word Picture 11.1.

Cooperation has been observed also during bushpig hunts. There was one occasion when Humphrey, Evered, Figan, and Satan encountered a sow with a half-grown young. The pigs took refuge in a thick clump of tangled vegetation. The four hunters surrounded the thicket, darting in singly or in pairs while the others prevented escape. After approximately ten minutes one of the males managed to seize the prey, the other three converged, and the adult (perhaps because of the presence of the human observers) ran off. It seems unlikely that a single chimpanzee could have captured such a large prey on his own—and had he done so, he probably could not have killed the victim nor escaped the charges of the sow. Another time Figan, after gazing intently into a thicket where a sow and piglets had run, looked back at Jomeo and gave the characteristic branch shake that is normally used to summon females during consortships. Jomeo at once hurried over, both males entered the thicket, and a piglet was captured. (Observers: H. Matama and R. Bambanganya.)

Mention must be made also of the very close cooperation of the infant-killers, Passion and Pom. It is true, Passion on her own was able to seize Gilka's infant in 1975—but Gilka was small and weak,

The Hunting of Colobus Monkeys

19 July 1981. Observers: H. Mpongo and G. Kahela.

The target chimpanzee is the adult male Evered. He is traveling with the alpha male, Figan, two adolescent males, Atlas and Beethoven, two females, Gigi and Miff, and Miff's offspring. At 0744 this small party encounters a colobus monkey troop. At once Figan and Evered stop and stare up at the monkeys with hair erect. Figan seizes Evered around the waist in an embrace. Uttering soft squeaks, all the chimpanzees begin to run along the ground, watching the monkeys which, giving their alarm calls, start to leap away through the trees. At 0745 the two adult males, along with Atlas and Beethoven, climb and chase a small juvenile. After a short time they appear to give up and return to the ground. Two minutes later, however, Figan and Evered climb and chase again. As Evered rushes after a mother-infant pair, three male colobus run toward him, threatening, shaking branches, and calling. Evered gives a loud waa-bark and then screams as he is chased back to the ground. Figan continues to hunt in the trees.

At 0749 Evered, Atlas, and Beethoven climb again. One colobus, a large juvenile, is isolated in a tall msebei tree about 10 meters away from the nearest members of its troop. As the chimpanzees approach, the intended victim screams and the hunters begin to utter food aaa calls. They pause as they get close to the tree, watching the prey intently. Then Atlas and Beethoven move closer. The colobus leaps away, and Beethoven gives a loud waa-bark and follows. Figan now joins the chase.

At 0750 the colobus makes a huge jump to the next tree, landing on very slender branches. Figan cannot follow and stops. The other three hunters, however, race after, following different arboreal routes; Evered manages to get ahead of the prey, leaping into its flight path. It stops and again the hunters pause, gazing at their quarry, and giving loud food aaa calls.

At 0754 Beethoven goes up to the colobus, which is clinging to the very end of a branch. He approaches slowly, giving tiny squeaks interspersed with waa-barks. As he gets to within about a meter, he gives a very loud waa and shakes a branch; the monkey again makes a big leap, landing in a neighboring tree. There it encounters Figan, who has circled around and who now rushes after it. At 0755 Figan is about 2 meters from the prey; he makes a grab for its tail, but misses. The colobus leaps to a slender branch. Figan follows, the branch breaks, and both Figan and the prey fall about 10 meters to the ground. The monkey runs off but Figan gets up rather slowly, as though shaken by the fall. Atlas and Beethoven hurtle down to the ground and chase after the monkey, while Evered remains in the tree watching. Moments later the prey once more leaps into the branches, pursued by the two adolescent hunters, and at 0756 Atlas manages to grab it. It screams loudly and two male colobus run to the rescue—but Atlas avoids them, leaping away with the prey. Beethoven at once runs up to Atlas, grabs hold of the monkey, and bites at its face, screaming as he does so. Figan, moving very slowly, approaches uttering loud food aaa calls. He takes hold of the prey, breaks open its head, and starts to feed on the brains. Beethoven, meanwhile, chews at its nose and lips while Atlas tears flesh from its back. All three chimpanzees have been screaming; now they become quiet.

At 0803 Figan suddenly seizes the carcass from Atlas and Beethoven, hair bristling. Atlas at once begins to scream loudly, until Figan reaches out and lays one hand first on his back, then on his groin. Atlas is now quiet. After a few moments Evered approaches, closely followed by Gigi, Miff, and her offspring. They all pant-grunt loudly to Figan and begin to beg. Atlas is chewing at one leg; Beethoven whimpers as he sits close, watching.

At 0808 Pallas and her family arrive. Evered, who has had no meat, erects his hair, and even as Pallas pant-grunts submissively to him, leaps onto her and attacks, stamping on and hitting her.

At 0812 Figan disembowels the prey and lets Atlas take all the viscera. At 0815 Evered, still unsuccessful, again attacks Pallas, who escapes, rushes down to the ground, screaming, and leaves the scene of the kill altogether, followed by her offspring. At the same time the mother Athena arrives; she climbs toward Figan, pant-grunting, only to be met and vigorously attacked by Evered. She screams, gets away from Evered, and once more moves toward Figan, who has the prey.

At 0820 Figan leaps away from the individuals clustered around him and, flailing the carcass with one hand, displays along the ground, then climbs another tree. Everyone follows him and they all start to beg again. Figan tears off an arm; Miff takes it and moves on, followed by Michaelmas and Mo, who beg from her. She gives a small piece to Michaelmas and lets Mo chew at the end of the bone. Figan then allows Gigi to take a large portion—one leg with part of the rump. Immediately Evered attacks Gigi, but she hangs on grimly while screaming very loudly indeed. Figan displays in their direction and Evered stops fighting Gigi, but follows her as she moves off. He sits near her, with his hair on end, but does not attack her, nor does he try to beg from Figan.

At 0900 Jomeo, Goblin, Jageli, and Prof arrive. Jomeo and Goblin display vigorously below the meat-eating clusters. Jomeo rushes up to Gigi and attacks her, pounding on her, then seizing the meat and pulling. Gigi, however, hangs on tightly and Jomeo lets go. Meanwhile Goblin is approaching Gigi, moving slowly with hair on end. Gigi, still screaming, climbs up and away from him, but Goblin follows, still moving slowly, and begins to beg. Five minutes later he is still begging very persistently, moving around when she turns her back, holding the meat, but not attempting to take it by force; at this point she allows him to share. Athena's daughter, Aphro, who has been close to Figan since the family arrived, now gets the remains of the leg Figan has been eating. Suddenly, at 0906, Gigi relinquishes her hold on her piece of meat and Goblin moves off with it. All at once Melissa is noticed, with her son Gimble. They have climbed up to beg from Goblin, along with Prof and Jageli.

At 0916 the target chimpanzee, Evered, finally leaves the scene, having had no meat at all.

---

having sustained partial paralysis of one arm and hand during the polio epidemic, and was suffering also from a fungus infection of nose and brow. When Gilka was again attacked the following year, her fungus infection was greatly improved and she put up a much harder fight. (She also knew what to expect.) She might even have saved her infant if she had had to battle with only one adversary. But during that attack, and also the 1976 attack on Melissa, Passion and Pom worked as a superb team: Passion, heavier and stronger, grappled with the mother, while Pom pulled at the baby. When Passion attacked Miff on her own in 1978 (Pom was pregnant and did not join in), she was unable to seize the infant, despite a long and very violent struggle during which both females fell some 6 meters to the ground.

Pom (*left*), Passion, and Prof share the flesh of Gilka's newborn infant. (E. Tsolo)

In summary, then, although chimpanzees can and do hunt some prey very successfully on their own, the presence of other hunters is often beneficial and in some cases, particularly during hunts of baboons and large bushpig young, may even be essential to the success of the hunt. In some instances too, the hunters clearly are aware that they are unlikely to succeed on their own—as when Figan summoned Jomeo to join him in the pig hunt.

The Kill

Chimpanzees kill their prey by (a) biting into the head or neck, (b) flailing the body so that the head is smashed against branches, rocks, or the ground, (c) disemboweling it, or (d) simply holding it and tearing off pieces of flesh (or entire limbs) until it dies. When several chimpanzees converge upon a small animal, it may be literally torn to pieces within moments of capture. Small prey, such as infant or juvenile colobus, black infant baboons, and striped piglets are usually killed by eating. Since the brains of such victims are almost always consumed first, death is very quick. The chimpanzee, often holding the victim by its neck, bites into the head, crushing the frontal bones. Thirty-one of the thirty-four infant or juvenile colobus whose capture was well observed were killed in this way (two were flailed and one was disemboweled).

Prey such as an adult monkey or large bushpig youngster sometimes present problems. The chimpanzee is not equipped with the teeth of a carnivore and often has trouble tearing open the skin of the victim.

However, a combination of smashing against treetrunks or rocks and tearing at limbs (sometimes breaking them) usually ensures that large victims are rendered fairly immobile within five to ten minutes, although they may not actually die until later. Between 1975 and 1980, seventeen adult (or large adolescent) colobus were captured and eaten, and two more were captured and badly wounded, but not killed. In fifteen of these incidents adult male chimpanzees were involved. Six times the victims were dispatched within five minutes of capture—they were torn apart by two or more male hunters, throttled, or disemboweled. Three times adult males joined adolescents or females who had been struggling with full-sized monkey prey for ten minutes or more, and quickly killed the prey. One victim, however, was flailed, stamped on, and bitten for nearly ten minutes by Jomeo and his brother Sherry before it finally died, and another time Jomeo, although he quickly incapacitated the adult colobus he had captured, continued to eat it for forty-seven minutes until it finally died.

Three times adult males, after making a few attempts at smashing or disemboweling full-sized monkeys, left them and did not return, and once Humphrey and Figan ripped open the abdomen of a pregnant female, removed the fetus, then as they ate lost all further interest in the mother. These four monkeys were all finally killed and eaten by adolescent and female chimpanzees. Adolescent males were actually involved in eight prolonged struggles with adult monkeys, and twice gave up before the victims were dead, without eating any meat.

Two examples will illustrate these somewhat gruesome episodes. In 1977 Jomeo, after a brief struggle with an adult male colobus in the trees, pulled his victim to the ground by the tail, then displayed, dragging the monkey. The colobus kept seizing hold of vegetation, but was not strong enough to break Jomeo's grip. His captor then let go and the monkey lay, squeaking faintly, but seemingly unable to walk. Adolescent Freud approached and bit at his hand, but retreated as the monkey moved violently. Five minutes later Jomeo's brother Sherry approached, took hold of the monkey's tail, dragged him a short way, let go, then turned around and bit at his face. Seizing the tail again, he displayed vigorously, flailing the monkey three times against trees and rocks. Jomeo immediately returned and also displayed, dragging and flailing the victim. Once again the monkey was left lying on the ground, still not quite dead. Another adolescent male chimpanzee approached, stared, then bit off part of his genitals. Freud returned and bit off some more, and Jomeo completed the castration. At this point, nine minutes after his capture, the monkey was dead. (Observers: R. Fadhili and H. Matama.)

In 1980 Mustard chased and caught a female colobus with a fairly large infant. As he tried to seize the infant, he and the female fell to the ground and the infant escaped. The mother also ran off, hotly pursued by Mustard, who chased her for 20 meters or so, then yanked her tail and displayed with her. She, like the male, grabbed hold of vegetation to no avail. Mustard paused, seized her again and displayed vigorously, flailing her against the ground, then stamping on and hitting her. After this he sat down, turned the exhausted and battered female onto her back, and tried to bite into her belly. He failed, and screamed as she bit his hand. Once again he stood up and hit at her,

then leaped onto her and stamped, still screaming loudly. His screams alerted Evered and Figan, who ran up and took over. Figan ripped open her belly, then bit at her face, while Evered tore off a leg; at this point she made her last sound and died. (Observer: H. Mkono.)

Most of the baboon infants and juveniles whose death was observed died quickly. Often several adult male chimpanzees rushed up and tore the body of the victim apart. One ten-month-old infant, however, was consumed by only one adult male and was still alive and calling feebly for forty minutes after his capture. The three large bushpig young took between eleven and twenty-three minutes to die as they were slowly torn apart; the largest gave his final scream when Humphrey tore out his heart.

Piracy and Scavenging

Piracy

On twenty-seven separate occasions chimpanzees have been seen feeding on bushbuck fawns which they were known or suspected to have seized from baboons. The data can be summarized as follows:

Observation	Times seen
Baboon seen actually killing fawn; chimpanzee seen taking it	2
Chimpanzees race toward excited baboon calls, seven times mingled with bleating of dying fawn; chimpanzees seen to take prey	12
Target chimpanzee races toward baboon calls; observers arrive some minutes later and find target flailing body of fawn (disemboweled), surrounded by baboons	1
Outbursts of chimpanzee and baboon excited calls lead target chimpanzee and/or observer to chimpanzees, feeding on fawns, surrounded by baboons	11
Outburst of chimpanzee/baboon calling heard; chimpanzee rushes toward sounds; later appears with part of bushbuck kill	1

Ten of the known or suspected seizures (37 percent) were made by females; one of them is described in Word Picture 11.2. A typical example of piracy by an adult male occurred in October 1978. At 0810 hours Humphrey and Satan heard a sudden uproar of baboon calling; with hair bristling, they grinned, squeaked, and embraced, then raced toward the sounds. When the observers caught up with them five minutes later, Humphrey already had the kill, which he had taken into a tree. The mother bushbuck was still in the vicinity, charging at baboons (of Camp Troop) who were searching for fallen scraps on the ground, but she left when the humans arrived. One adult male baboon sat close to Humphrey and Satan, who were sharing the meat. Every so often he threatened them, showing his canines in a huge yawn and flashing the pale skin of his eyelids, whereupon one or both of the chimpanzees retaliated, raising their arms in threat and giving waa-barks. About an hour later Humphrey, startled by the sudden appearance of another male chimpanzee, dropped his portion (most of the rump and both hind legs). He quickly started down after it, but

a young male baboon, Hector, was quicker and raced off with it. He was hotly pursued by three chimpanzees (two adolescent males and Gigi) who had joined the group and had also been waiting for meat. Four hours later Gigi was seen again—carrying the skin of the fawn! (Observers: A. Seki and H. Mkono.)

Baboons at Gombe have been seen killing, or feeding on, bushbuck fawns on at least thirteen occasions that did *not* involve chimpanzees. They also prey on bushpig young (seen twice), fruit bats, rats, birds, lizards and gekkos, and frogs. When chimpanzees came upon a male baboon eating a guinea fowl, one of them, Mike, chased him and seized most of the prey (Morris and Goodall, 1977). In fact, on only one occasion did a chimpanzee make *no* attempt to take meat from baboons: when an adult female, Winkle, encountered three Camp Troop males, one of whom had the remains of a bushbuck kill. Possibly there was not enough meat for her to make the effort. On two occasions female chimpanzees raced toward the sounds of excited baboon calling, only to find in each case that the commotion was caused by fierce male-male rivalry over females in estrus (interesting examples, which show that chimpanzees cannot always correctly interpret baboon vocalizations).

Once baboons were present when chimpanzees were hunting piglets. As Hugo made a capture and flailed the body of his victim against the ground, an adult male baboon rushed at him and threatened. Unfortunately he belonged to an unhabituated troop and ran off when he noticed the observers; possibly the baboons were hunting at the same time. There were two other occasions when, just after chimpanzees had killed piglets, baboons arrived and hunted about for scraps. And once, very early in the study, I found David Greybeard feeding on a piglet, closely watched by an adult male baboon. There were a number of aggressive incidents between the two, including a bout of actual fighting. By contrast, baboons were only present during six out of all the occasions when chimpanzees fed on colobus monkeys—and not even on those occasions did they show any interest in the consumption, or search for scraps. Very recently, however, an adult male baboon twice tried to take colobus meat from a male chimpanzee, and spent hours searching for and eating scraps under the tree in which the chimpanzees were feeding.

Only once has a baboon been seen to succeed in snatching meat from a chimpanzee. The young male Hector (of Camp Troop) watched closely as Passion and her family fed on a bird of prey. When Prof, Passion's juvenile son, finally managed to get a share from his mother, he moved a short distance from her to eat. Hector at once followed, made a lightning grab, and rushed off with a wing that represented most of Prof's prize. Prof screamed in rage as he rushed back to Passion and made futile threats in the direction of Hector. (Observers: E. Mpongo and Y. Alamasi.) It will be surprising if other incidents of this sort (baboons "stealing" from chimpanzees) are not recorded as time goes on.

Scavenging
Only ten times in the history of the research at Gombe have we observed the chimpanzees feeding on meat that they found on the

The Seizure of a Bushbuck Prey from a Baboon

10 November 1981. Observers: H. Mpongo and G. Kahela.
At 1412 hours Melissa, with her daughter Gremlin (who is pregnant) and infant son Gimble, hear the sounds of baboons making a kill. At once they move rapidly toward the uproar. They arrive to find one adult male baboon tearing at the meat of a freshly killed fawn, while other males threaten him, slapping their hands on the ground, showing their canines and the whites of their eyelids as they yawn, uttering fierce-sounding roarlike grunts. Slowly Melissa and Gremlin move toward them. The baboon, dragging his prey, moves a short distance away, then starts to feed again.

At 1414 Melissa and Gremlin suddenly run at the baboon with loud waa-barks, waving their arms. (Gimble has climbed into the tree and is following overhead.) As the baboon threatens, showing his canines and lunging toward the chimpanzees, Melissa stops, whimpering. Abruptly she seizes a thick dead branch and, with hair erect, hurls it toward the baboon, who utters a roarlike grunt and leaps aside; the missile does not hit him. Melissa then displays, swaying the vegetation wildly, leaping up and down, gradually approaching the baboon more closely. All at once he drops the meat and lunges at her, making contact and seeming to bite her arm. Melissa, with loud waa-barks, flails her arms and hits out with her fists. The baboon turns and retrieves his kill. Melissa follows him, then stops and watches as he continues to drag the prey. (Gremlin meanwhile has retreated to a distance of 3 meters.)

At 1415 Melissa and Gremlin again move toward the baboon, waving their arms and waa-barking as he lunges toward them. He now starts frenziedly tearing at the rump of the prey. As he feeds, Melissa and Gremlin watch; Gimble climbs down and joins his mother.

At 1420 the baboon suddenly runs toward Gimble, who screams and races back up a tree. The baboon returns to his kill, tears at it, then seizes it in his mouth and moves off with it as Melissa again starts to display toward him, shaking vegetation vigorously and uttering loud waa-barks. As the baboon retreats, his prey becomes entangled in the vegetation. He pulls but cannot move it, so tears off a piece while Melissa continues to sway branches. She is not approaching him and is whimpering as she watches. Suddenly the baboon rips off a piece of flesh and runs away. Melissa at once races to the prey and seizes hold of it; but the baboon promptly returns and pulls the meat from her, again with a roarlike grunt and showing his canines. Melissa, screaming, hangs on determinedly. Gremlin again retreats. This time she climbs a tree and moves toward Melissa overhead, accompanied by Gimble. Both are giving waa-barks, and Gremlin starts to wave and shake branches just above the baboon. Melissa still has hold of the carcass, and now starts to climb with it toward Gremlin. Suddenly the baboon lets go and Melissa puts the carcass over her shoulder, holding it by a front leg. From below the baboon looks up, shows his teeth, then leaps up after the chimpanzees. At this Gremlin seizes a dead branch, breaks it off, and flails it at the baboon. He stops and sits. Gremlin throws the branch at him; it misses. Immediately the baboon lunges toward Melissa, threatening her. But now, at 1425, she ignores him and starts to feed, Gremlin and Gimble moving close and joining in.

Throughout these incidents other baboons have been milling about nearby, grunting and threatening, but they have never actually clashed with Melissa, the possessor of the fawn being left to defend the meat himself. Now, as the chimpanzees feed, the dispossessed baboon sits close by and continues to threaten. But at 1440 two other female chimpanzees arrive and cluster around Melissa, begging, and at this the baboon leaves the tree. With the rest of his troop he begins to search the ground below for fallen scraps.

At 1510 the baboons finally leave; the chimpanzees are still eating meat. At 1740 Melissa moves off with the remains of the carcass—head, neck, and chest—on her back. At 1750 when Gimble pulls at the meat, Melissa tears off a piece and leaves the rest. Gimble tries to drag it after him, but fails. Gremlin comes up at this point and, giving food grunts, drops the piece of meat she has and drags the carcass as she follows Melissa. Finally, at 1800, Gremlin detaches a piece of meat and leaves the rest of the carcass on the ground.

---

ground. Four times this was remnants of previous chimpanzee kills. Twice the finders returned to feed on pieces that they themselves had abandoned earlier the same day. Once Gilka, who had been present at a meat-eating session without herself getting a share, returned the next day and found the body of an adult female monkey who had been badly wounded during the hunt, but left alive. And another individual appropriated half a monkey, which other chimpanzees had abandoned a few hours earlier. Five other colobus remains that were scavenged may also have been left from previous chimpanzee kills; indeed, two were found quite near areas where there had been hunts the day before. While the chimpanzees at Gombe usually eat the whole carcass, there are occasions—especially when multiple kills have been made and there is plenty of meat for all, or when only one or two individuals are involved in the eating of meat—when pieces are abandoned, presumably when all are satiated. (Although sometimes chunks of meat are taken to bed for eating the next day—or perhaps for midnight feasting?)

One case of scavenging certainly did *not* involve the remains of a chimpanzee hunt. H. Matama watched what appeared to be an inter-troop conflict during which an adult male colobus seized and bit a juvenile male. The youngster fell to the ground and had bled to death within ten minutes. The body was brought back for me to look at; I was taking some photographs when suddenly a black hand appeared in the frame and the monkey was gone! Adolescent female Gremlin had approached silently and appropriated the body. It was taken from her immediately by Jomeo, who had arrived with her, and the small party (which included Gremlin's mother) spent most of the rest of the day feeding on it.

On other occasions chimpanzees ignored the remains of a bushbuck fawn that had been abandoned by baboons earlier in the day, and also the still-warm body of a guinea fowl. When we put out the body of an adult bushpig (which had died of spear wounds), the many chimpanzees who encountered it in camp showed fearful behavior, staring with erect hair and giving small huu calls. Many then sniffed the ground all around, also the trunks and even the branches of nearby trees. It seems likely they were searching for signs of a leopard, the only predator in the area that could kill an adult pig (other than a human, but it seems unlikely that chimpanzees would search for signs of humans up trees).

At Mahale there have been four observations of scavenging. Twice the carcasses were adult female blue duikers, twice adult bushbucks. One of the duikers had been dead for some time and was beginning to decompose; after being carried, then dropped, by an adolescent male chimpanzee, it was eagerly snatched up when offered to a group of chimpanzees by researchers and almost totally consumed by fourteen chimpanzees in two hours. The second duiker was found by research personnel. Its body was taken back to camp and skinned; chimpanzees consumed the viscera and other scraps that were thrown out.

Both bushbucks had almost certainly been killed by leopards. Their bodies were discovered by the chimpanzees after they had spent some time intently sniffing nearby vegetation, even (as at Gombe) climbing a tree and sniffing the trunk (Hasegawa et al., 1983).

Eating Meat

Chimpanzees tear off chunks of meat with their teeth and hands, sometimes using their feet too when strength is required for dividing the carcass. Almost always each morsel is chewed up together with a wadge of leaves, sometimes dead ones. These wadges, although they may be swallowed, are usually discarded along with any unwanted portion of the meat, such as pieces of bone or skin. Large bones are usually cracked open and the marrow extracted. Chimpanzees suck, chew, and may even swallow small bones and bone fragments, as well as pieces of skin, especially individuals who have been unable to acquire juicier portions. They also chew the leaf-meat wadges that have been discarded by their luckier companions.

Meat is eaten slowly. One old male, Hugo, fed on the body of a large infant baboon for almost nine hours. When he finally abandoned the carcass to the other chimpanzees present (who had been able to obtain only very small scraps before), the head, arms, legs, and part of the torso still remained.

When eating small prey, such as infant colobus or piglets, a chimpanzee typically begins with the head. He bites open the skull, sucks the blood, then consumes the brain. This was the method employed in thirty-two out of the thirty-four well-observed infant or small-juvenile colobus prey, most bushpiglets, and the three bushbuck fawns when only one or two individuals were present at the capture. When an older prey is involved, the viscera are typically eaten first (as they were in the remaining two colobus infants).

When Passion in 1977 captured a newborn bushbuck fawn, the stages of eating were well recorded. After biting down hard on its face, destroying both eyes and causing copious hemorrhaging, she bit off one ear, which she ate with leaves. She then broke open the skull and fed on the brains. Her late-adolescent daughter, Pom, after first drinking blood from the mouth and ears, began feeding on the hooves, as did the infant Prof. Pom then disemboweled the fawn and fed on the viscera, while Prof ate brains beside his mother. The family next began feeding on fleshy parts of the chest, shoulders, and rump. Four hours and fifty minutes later the carcass was abandoned, the legs (somewhat depleted) and backbone held together by skin.

Three adult males with a baboon kill. Hugo has most of the carcass; Leakey reaches up for a length of gut.

A few days later Passion caught another fawn. Again she first bit into its eyes, drank the blood that poured out, and then, as before, fed on the brain. On this occasion, too, Pom first fed on the hooves, then the viscera. Passion still retained the carcass when she nested thirty minutes after darkness fell, three hours and forty minutes after the capture. The next morning she did not feed from the remains but carried them until 0840, when she abandoned them. (Observers: H. Mkono and Y. Alamasi; Y. Mvruganyi and H. Mkono.)

The heads of older prey animals are probably left until the end because their skulls are harder and more difficult to open than those of small infants. Teleki (1973c, p. 144) describes how three male chimpanzees (Mike five times, Hugo and Leakey once each) opened seven such skulls: five baboons, one bushbuck, and one adolescent colobus. Six times the chimpanzees used their canines and incisors to break into the skull through the top of the cranium. They bit down "at times with a force that caused the entire body . . . to tremble." In the remaining case the chimpanzee enlarged the foramen magnum. When the hole was large enough, he inserted two or three fingers and scooped out the brain. I observed this performance on two other occasions. Once a chewed mass of leaves was used to clean out the inside of a brain case (Wrangham, 1975).

Teleki noted that the brain appeared to be a desirable item, and that although other parts of the carcass were readily shared, the brain never was. Wrangham (1975) observed no such preference and saw the brain being shared on several occasions. However, it is certainly true that the brain was eaten first in almost all cases when the thickness of the skull did not make its opening difficult, and from a preliminary survey of the data it seems that when a carcass is divided, the possessor, if he is high ranked, often retains the head end for himself. Once Figan caught a juvenile colobus a few moments before Humphrey made a second capture. With small screams of excitement Figan raced to where Humphrey sat and, clutching his own prey, seized hold of Humphrey's also. He did not try to wrest it away from the older male, but bit open the head and consumed it, together with the brain, while Humphrey disemboweled the infant and ate the viscera. After this Figan ate the brain of his own monkey, and Humphrey moved off with the headless body of his! (Observers: G. Kipuyo and R. Bambanganya.)

As mentioned, chimpanzees may suck and drink blood after biting into the head of their prey. Once Goblin, having punctured the neck of an adult colobus, drank blood (perhaps from the jugular) for approximately one minute. When disemboweling an adult colobus, Miff drank blood five times from her cupped hand. During meat eating drops of blood are systematically licked off leaves, branches, or the ground by individuals who have been unable to get portions of the kill. Spilled blood from wounded chimpanzees is sometimes consumed by their companions. Once three different individuals spent up to thirty minutes licking Melissa's blood as it dripped onto the leaves below from a wound in her genital swelling. An adolescent female, Skosha, threw a tantrum when the leaves she was licking were pulled away from her by an adult female (Melissa's daughter), and subsequently Skosha broke off a blood-spattered twig and carried it for over 100 meters, occasionally pausing to lick the leaves.

The fecal content of the large intestines is often consumed with apparent relish. Once, when a female colobus had been captured, Humphrey repeatedly poked one finger into her anus and licked off the feces. Another time, when Humphrey attempted without success to seize a baboon infant, he spent the next ten minutes carefully consuming every scrap of the trail of "fear dung" she had left as she made her escape.

During the year when chimpanzee feces were regularly examined, we could tell immediately when chimpanzees had been eating meat, as the samples were full of hair, bones, even lumps of flesh. One sample yielded a monkey finger, another an ear, and a third an incredible five inches of tail, bone and all! One morning, after Mike had been eating bushbuck meat, he picked pieces of flesh out of his own dung and ate them.

Two of the three recorded instances at Gombe when adult males killed chimpanzee infants of stranger females (Table 11.8, cases 1 and 3) resulted in some very bizarre behavior during the ensuing cannibalism (Bygott, 1972; Goodall, 1977). In case 1 only a few of the individuals present ate, or tried to eat, any of the flesh, and only Mike and Humphrey ate for more than a few minutes (Humphrey for twenty and Mike for ninety), although three others (Jomeo, Figan, and Satan) each carried the body around. Abnormal patterns included repeatedly charging with and flailing the body, poking at and examining the body, pounding on the chest or head with fists, and playing with or grooming it. (Similar strange behavior was seen in the Budongo Forest cannibalism; Suzuki, 1971.) Moreover, relatively little had been consumed of infants 1 and 3 when they were abandoned. By contrast, the four males feeding on infant 6, and Passion and Pom when feeding on infants 2, 4, and 5, showed normal meat-eating behavior, as did the male cannibals in the four Mahale Mountain cases (Nishida, Uehara, and Nyundo, 1979; Kawanaka, 1981; Norikoshi, 1982; Takahata, 1985).

We shall see that during the intercommunity attacks on the Kahama individuals some of the attack patterns were similar to those shown by chimpanzees killing or eating adult monkeys.

Distributing Meat

A chimpanzee traveling by himself or herself can, as we have seen, hunt successfully. In that case he or she quietly consumes the meat alone—unless another individual happens to pass by and notices the hunter, when he will almost certainly approach and try to obtain a share. However, when a capture is made during a group hunt, especially when this involves a number of adult males, intense excitement usually breaks out as those present converge on the successful hunter. An adolescent or low-ranking male captor, or a female, is likely to lose possession of the carcass within moments of capture. During the 1973–74 period seven colobus were observed to be taken forcibly within two minutes of capture (Busse, 1976). Thus it can never be assumed that an individual found in possession of a carcass is the real captor; he may merely be a successful thief.

During the first few minutes after a kill has been made, before the excitement has died away, those who are present (at least the males) often manage to obtain quite large portions. They rush up to the successful hunter, seize hold of the prey, and pull. Sometimes competitors tear off and move away with large portions of viscera or entire limbs; at other times they start to feed on the carcass along with the captor. During this initial division the forest is filled with screams, barks, waa-barks, and pant-hoots—sounds that alert other chimpanzees in the area, most of whom hurry to join the meat eating.

Meat is a highly coveted food and often there is intense aggressive competition around a kill. This aggression comprises (a) attacks on possessors of meat by those who have none, (b) attacks or, more usually, displays or threats by possessors toward individuals trying to share their prey, and (c) attacks or threats directed by those who have not managed to acquire portions toward lower-ranking individuals who are also trying to get some meat. Wrangham (1975) analyzed the frequency of aggression during the first half-hour of nineteen well-observed predations between 1970 and 1973. He found that of the twenty attacks made on individuals in possession of meat, the aggressor was successful only three times. From 1975 to 1977 individuals with meat were seen to be attacked on fourteen occasions. Again, only three fights resulted in success for the aggressor—and two of those were attacks by males on females. Twice the disputed meat was dropped during the conflicts and taken by "scrap-hunters" below.

Hugo crouches over his prey as
Mike displays past and Leakey
screams in excitement. (B. Gray)

A begging cluster gathered around
Mike and his colobus kill; Fifi holds
one leg and Flo is eating a small
piece she has detached from the carcass. Flint pushes in between Flo
and Mike. Faben is not begging.
(B. Gray)

As Wrangham has pointed out, an individual in possession of meat is
usually well able to defend it by crouching over it and protecting it
with arms and body (much as a mother protects her infant when under
attack). Moreover, the possessor is usually highly motivated to retain
his prize. Even a low-ranking male may hold onto his prey in the face
of intense aggression. Once, for example, alpha male Figan attempted
to wrest a colobus monkey kill from young, low-ranking Sherry; despite four violent attacks he was unsuccessful.

Begging, described in detail elsewhere (Goodall, 1968b; Teleki,
1973c), is the way most chimpanzees try to get some meat for themselves. Their success or failure will depend on a variety of factors,
such as the amount of meat involved, the amount the possessor has
already consumed, and the relative age, rank, and relationship of the
two individuals. Some males are more generous than others, some
more tolerant, some more aggressive. Those who beg may reach out

to touch the meat, often glancing at the face of the possessor as they do so, as though to gauge his mood—or they hold out a hand, palm up, toward him, or reach to his mouth for the wadge he is chewing. When a low-ranking individual begs, the possessor may quite ignore him or her, at the most turning his head away from an outstretched hand. Or he may push the hand away, turn his back, or give a mild threat. Sometimes, especially if he is surrounded by a cluster of begging chimpanzees, he seems to lose patience and suddenly, hair erect, displays wildly through or over them, often hitting at or stamping on them as he passes, flailing the kill or holding it in his mouth. Then he settles down in another place and starts to feed again. Before long, though, the cluster usually gathers around him once more. Meat is typically consumed in a tree, and Wrangham (1975) has suggested that a likely reason for this is that far fewer can cluster around a possessor on a branch than one sitting on the ground. In fact, it is in the context of frustrated attempts to acquire meat that so many aggressive interactions occur. Word Picture 11.1 describes this kind of frustration and redirected aggression.

There are, however, many occasions when begging is successful and supplicators may be allowed to take, or may actually be given, pieces of varying size. I shall discuss this phenomenon in Chapter 13.

Success in Hunting and Meat Eating

Adult Males

Wrangham (1975) observed no obvious differences in hunting ability among the adult males. Even Faben, with his paralyzed arm, caught a baboon, a juvenile redtail monkey, and a piglet; and Willy Wally, with his semiparalyzed leg, twice caught a colobus. Busse (1977) also noted little individual difference, although the oldest and the youngest males during his study period (1973–74) had lower success rates than those in their prime.

Table 11.9 is a summary of observed hunting participation and success of the Gombe adult males from 1973 to 1981. Only colobus hunts are included here, because of the described variation in the skills required to catch a monkey versus, for example, a piglet. The table shows that capture rates for Figan and Sherry were consistently high from 1973–74 on for Figan, and from 1976 on for Sherry (who prior to this, as an adolescent was not seen to catch any colobus). The other males showed more variability.

Table 11.9 does not include information on cooperative success in hunting. If two males hunted together, a success is scored for the individual who actually captured the prey; if both seized it at approximately the same time, both are scored as succeeding. Table 11.10, therefore, has been compiled to show the percentage of well-observed hunts when males chased and captured monkeys on their own and when they captured in cooperation with another male or males. I have included under the heading of cooperation (a) joint chases and captures and (b) occasions when one hunter hit a monkey to the ground and caused it to fall, enabling another chimpanzee below to seize the prey. The table shows that some males, such as Figan and Sherry, captured

Table 11.9 Capture rates of the seven Kasakela males when hunting red colobus monkeys, 1973–1981. Capture rate (one monkey per x hunts) is calculated on the basis of number of well-observed hunts at which a male was present and caught a prey.

	Number of hunts at which present						Capture rate						
Male	1976	1977	1978	1979	1980	1981	1973–74[a]	1976	1977	1978	1979	1980	1981
Humphrey	11	12	14	19	6	5	7.5	5.5	4.0	14.0	4.8	6.0	0
Evered	9	11	8	12	8	14	6.5	0	3.7	2.7	0	1.6	7.0
Figan	19	23	20	30	22	14	4.0	2.5	4.6	4.0	6.0	3.7	4.6
Satan	14	16	12	13	10	13	15.0	7.0	3.2	3.0	6.5	3.3	6.5
Jomeo	15	14	12	17	11	8	4.0	15.0	4.7	12.0	8.5	3.6	8.0
Sherry	13	11	15	17	[b]	[b]	0	2.1	3.6	3.8	4.3	[b]	[b]
Goblin	14	10	11	19	7	12	0	0	5.0	2.2	0	0	4.0

Note: The differences among males (excluding Sherry) in capture rate are significant by the Friedman two-way analysis of variance test (one tailed), $p < 0.001$.
a. These data are from Busse (1977), who does not provide information on the number of hunts attended.
b. Sherry died in 1979.

Table 11.10 Colobus monkey prey acquisition during a seven-year period (1975–1981) for seven Kasakela males (Humphrey died in 1981 and Sherry in 1979). Only well-observed hunts are included.

	Humphrey	Evered	Figan	Satan	Jomeo	Sherry	Goblin
Captures observed	10	10	35	15	11	22	16
Thefts observed[a]	1	2	5	8	0	0	0
Total acquisitions	11	12	40	23	11	22	16
Prey lost to others	1	0	0	0	1	2	5
% cooperative captures	45.5	16.5	17.5	8.5	45.5	27.5	31.25
% independent captures	45.5	66.5	70.0	56.5	54.5	72.5	68.25
% stolen from others	9.0	17.0	12.5	35.0	0	0	0

Note: The differences among males in cooperative captures are significant (χ^2 test, $df = 6$, $p < 0.001$); they are not significant for independent captures.
a. Includes only prey seized from the captor almost immediately (before he had a chance to eat more than a few mouthfuls).

most of their prey on their own, whereas others, such as Humphrey and Jomeo, were more likely to make kills when they hunted with others. In fact, each of Humphrey's cooperative captures occurred when he seized prey that had been hit to the ground or had fallen as a result of hunting by other chimpanzees; Jomeo, on the other hand, actually chased and seized prey together with other individuals (twice his brother Sherry).

It should be emphasized that the amount of meat actually eaten by a male does not necessarily depend on his success as a hunter. Adolescent and young mature males (as well as females) may sometimes catch prey but be unable to retain possession. Table 11.10 also shows the number of monkeys caught by each of the six males that they lost, immediately or soon after capture. Of the sixteen monkeys Goblin was seen to capture between 1975 and 1981 he lost five (31 percent) to older male chimpanzees within a very short time after catching them. Sherry lost two of his kills (9 percent) and Humphrey and Jomeo each lost one infant. The table does not include instances where a successful hunter lost *part* of his prey to others, or shared large amounts with them.

One way in which the individual males appear to differ is the extent to which they attempt to steal newly captured prey from other chimpanzees. As Table 11.10 shows, Satan was the greatest exponent of this type of meat acquisition—eight (35 percent) of the whole monkey carcasses that Satan was seen eating during the seven years were seized from others immediately after capture. Four times Satan took kills from females, three times from Goblin, and once from Sherry. He also made very vigorous attempts, which failed, to take other kills from Sherry (who received help from his brother Jomeo) and Gigi (who clung to her meat despite the battering she received).

Sometimes an individual who has made a successful capture runs off with the prey, undoubtedly anticipating that he may lose it if he remains. Humphrey was seen to behave thus for the first time in 1968, when he unobtrusively snatched up an infant baboon dropped in tall grass as its captor battled with an adult male baboon. Presently observers spotted Humphrey quietly eating the prey on the other side of the valley. He made off with three colobus monkeys in 1979 and another in 1980 (three times the prey had fallen to the ground during chases by other chimpanzees above him). Once he rushed off with a bushbuck fawn. The others who had been with him at that time ran in the direction he had taken, stopping from time to time to sniff the ground or to gather scraps where Humphrey had paused to disembowel his prey, but they never caught up with him.

Sometimes individuals "disappear" during a hunt. Quite possibly they have made a capture and quietly moved away, thereby ensuring undisputed possession and a peaceful meal. Jomeo vanished in this way on a number of occasions when, in the early '70s, he was seldom able to maintain possession of a kill in the presence of older males. Wrangham (1975) describes one occasion when Jomeo, after exerting considerable energy in killing a colobus, lost his prey before he had taken even one mouthful. He licked a few drops of blood from the leaves, gazed around, and vanished. Once when Jomeo and Sherry had cooperated in the capture of a baboon infant, they hurried off silently with their prey; when other chimpanzees arrived at the capture site, having been alerted by the sounds of conflict, they were unable to find either captors or kill.

Just as success in capturing prey does not necessarily mean that an individual will eat a lot of meat, so it cannot be assumed that the unsuccessful hunter will go hungry. Some individuals, as we have seen, are highly skillful at acquiring meat from others—by snatching the carcass or part of it, by sharing with the possessor, or by begging for pieces. Even an individual searching for scraps below an ongoing feast in the branches may get a sudden windfall. During a fight, for example, a large piece of meat may be dropped. Table 11.11 gives a rough idea of the success rate in meat acquisition of the seven adult males during two years. There are two measures of success—the percentage of occasions when a male attended a well-observed meat-eating session that he was seen (a) eating meat at all and (b) eating large amounts of meat (at least a whole arm or leg). The table shows quite clearly that the older males (Humphrey, Evered, and Figan)

Table 11.11 Number of well-observed meat-eating sessions during 1978 and 1980 at which different adult males were present and the percentage of those occasions when each was observed to eat (a) some meat (including large amounts) and (b) a large amount of meat. This includes prey captured by the individual scored.

Male	No. of sessions at which present	% sessions at which meat was eaten		No. of times individual made kill
		Some	Much	
Humphrey	35	90.5	62.5	2
Evered	27	79.0	76.0	6
Figan	46	85.0	78.0	13
Satan	33	77.5	42.5	7
Jomeo	28	58.0	50.5	2
Sherry[a]	16	75.0	56.0	7
Goblin	31	31.5	10.5	0

Note: The differences among males in sessions at which much meat was eaten are significant by the χ^2 test (one tailed), $df = 6$, $p < 0.001$.
a. Sherry's scores are for 1978 only.

Gigi and Miff gaze at a colobus troop in the canopy above.
(C. Busse)

scored higher on the second measure (eating a lot of meat) than did the younger ones. Humphrey's 1978 scores were higher than his later ones; he was seen eating some meat at every session (100 percent) and a lot of meat during 71 percent of them. This compared with 81 percent and 54 percent in 1980, the year before his death. Wrangham (1975) and Busse (1977) also found that older males were more successful in obtaining meat than younger ones.

Begging behavior has not yet been analyzed in detail. However, a preliminary survey of the data reveals clear-cut differences between the adult males in the extent to which they are *prepared to beg* from their companions. Thus, from 1977 to 1981 Humphrey and Jomeo begged very consistently at almost all meat-eating sessions they attended. Figan, on the other hand, rarely begged (although he snatched, or tried to snatch, kills by force on a number of occasions). This is not a characteristic of alpha males; when Mike was alpha, he showed no reluctance to beg. Goblin almost *never* begged: in 1978 he was seen to do so twice, in 1979 not at all, in 1980 twice (once from a female, Gigi). Both Figan and Goblin, when in the presence of meat-eating individuals, tended to display vigorously at the scene of activity, then to sit some distance away. Goblin, on a number of occasions, moved off 20 to 30 meters and began feeding on fruits or other vegetable foods. These individual differences in readiness to beg are of interest and need careful investigation in the future. Similar differences, as we shall see, are found in the females.

Adult Females

While it is certainly true that adult males hunt far more frequently than do females, recent observations show that females at Gombe hunt more often, and eat much more meat, than was previously supposed. Until 1976 females were only rarely the subject of full-day or consecutive-day follows: the new information on hunting has largely been acquired by rectifying that bias.

The first time females were observed to kill was in 1970, when Pallas and Athena, traveling together, each captured a young bushpig.

Table 11.12 Hunting scores of Kasakela females from 1974 to 1981.

Prey	Caught and eaten	Caught and lost	Stolen and eaten	Stolen and lost	Participated in killing
Colobus	13	9	2	1[a]	6
Bushbuck	4	0	10[b]	0	0
Bushpig	7	3[c]	0	0	0
Chimpanzee	3	0	0	0	0
Rodent	1	1[c]	0	0	0
Bird	4	1	0	0	0

a. Stolen from adolescent male chimpanzee, lost to adult male.
b. Ten prey stolen from baboons, two of which were lost to other female chimpanzees.
c. Lost to mother.

Table 11.13 Differential involvement of seven female chimpanzees in various types of predatory activity. Numbers in parentheses are prey lost after capture or seizure.

Prey species	Gigi	Fifi	Passion	Miff	Athena	Pallas	Melissa
Colobus							
Captured	8 (3)	3 (2)	0	1 (1[a])	0 (1)	0	0
Stolen from chimpanzee	0	2[b]	0	1[a]	0	0	0
Helped to kill	0	2	0	2	1	0	0
Bushbuck							
Captured	1	1	2	0	0	0	0
Stolen from chimpanzee	1	0	0	0	0	1	0
Stolen from baboon	3	0	0	4	4	0	2 (1)
Bushpig							
Captured	2 (1)	0	2	0	1	1 (1)	0
Stolen from chimpanzee	0	0	1[c]	0	0	0	0
Chimpanzee							
Captured	0	0	3	0	0	0	0
Rodent							
Stolen from chimpanzee	0	0	1[c]	0	0	0	0
Bird							
Captured	0	0	1	1	0	0	1

a. Taken by Miff from adolescent male; kill seized by adult male.
b. Taken by Fifi from juvenile son.
c. Taken by Passion from adolescent daughter.

The next observed female kill was at the end of 1973, when Gigi captured another piglet. Table 11.12 summarizes data on female prey acquisition from 1974 to 1981. During this eight-year period females were seen to capture or steal, and then eat, at least part of forty-four prey animals. In addition, they captured or seized another fifteen prey, but subsequently lost them (they escaped, were rescued by conspecifics, or seized by other chimpanzees—twice by other females). Females also participated in six inefficient killings of adult colobus monkeys.

Table 11.13 provides details of the involvement in predatory activities of individual females. All these females were not observed for similar amounts of time, so the table cannot be used to compare their performances. Nevertheless, if we consider only females who were followed *extensively* during the eight-year period (Fifi, Passion, Melissa, and Pallas), some interesting differences are apparent. Fifi, for example, captured or acquired seven colobus monkeys; the other three, none. (Passion and Pallas, in fact, were never observed even to *hunt* colobus.) Only Passion (together with Pom) was observed to show predatory interest in infant chimpanzees. Gigi, although not one of the frequently followed chimpanzees, nevertheless participated in more of the observed predatory incidents than did any other female (as shown in Table 11.13).

A female traveling in a family unit with her dependent young, or in a small female party, is more likely to be successful in obtaining and retaining prey than when she hunts in a mixed party. Thus six of the twelve colobus, five of six piglets, three of four fawns, and all of the thirteen fawns stolen from baboons were obtained in the family unit or small female party context. One of Fifi's kills was made just two weeks after she had given birth in 1976. She managed to chase a mother-infant pair in a small colobus troop for over 200 meters, supporting her own tiny infant with one hand. She caught hold of the mother and seized the baby, which she was already eating when her five-year-old son, Freud (who had been whimpering as he struggled to catch up), joined her to share the meat. Fifi and her family hunted successfully on their own on two other occasions (once it was Freud who actually seized the prey; Fifi, who was right beside him, appropriated it at once but shared the flesh with him). Another time Fifi and Freud were found eating a freshly caught infant colobus; no other chimpanzees were nearby and it was assumed that they had captured it. The Miff family also hunted colobus successfully.

Much of the observed female predatory behavior (including unsuccessful hunting) was, however, in the context of mixed-party hunts. Here we can more easily compare the involvement of the different individuals. Table 11.14 shows the number of times that each female was present when adult males were hunting colobus (during well-observed hunts) over a three-year period, and the percentage of those occasions that the females were seen to participate actively. I have included only instances when they were seen to climb trees and chase a prey, although it could be argued that they were also hunting when they ran along below. The number of times that a female was *present* merely reflects her level of sociability and cannot be regarded as any indicator of predatory interest. The percentage of occasions present that she was seen to *hunt* does indicate predatory interest; the table reveals some clear-cut differences among the females, even though during the confusion of the hunts some female participation undoubtedly was overlooked.

Gigi tops the overall scores, hunting on 67.6 percent of the occasions that she was present over the three-year period. The females at the bottom end of the scale (Pallas, Passion, Little Bee, and Winkle) were never observed to hunt at all, although their adolescent and juvenile offspring frequently did. It should be added that these four females were not observed to hunt monkeys even when no male chimpanzees were present.

Of the nineteen prey of all kinds that female chimpanzees captured in mixed parties (excluding prey that escaped), seven were almost immediately appropriated by adult males—who often then shared the meat with the female captors. When Fifi, in a large mixed party, captured a bushbuck fawn, she held it to the ground, screaming in excitement. Miff ran over and also grasped the prey, screaming, but neither female made any attempt to kill it. Within a few moments Figan and Evered raced up and quickly killed and tore apart the prey; Fifi and Miff each retained a large portion. (Observers: H. Matama and E. Mpongo.)

Ten of the twelve mixed-party prey that were *not* appropriated by adult males were captured by Gigi. Once again, her behavior is of special interest. On a number of occasions she maintained possession of her meat despite determined assaults by adult males. Once, for example, she caught a large juvenile colobus when it fell or jumped to the ground during a mixed-party hunt. Satan instantly leaped down, chased after Gigi, and attacked her vigorously as she crouched over the prey. She managed to escape and rushed up a tree with her prey. Satan followed and, displaying through the branches, once more caught and attacked her. Both fell some 10 meters to the ground. Gigi picked herself up and ran, with Satan in close pursuit. At this point she had some respite, for Satan displayed off in the direction of a second kill being eaten nearby. Eight minutes later, however, he was back (empty-handed) and once more raced after Gigi and attacked her; again both fell at least 10 meters. Even as the assault continued on the ground, Goblin charged over and attacked Satan. Gigi escaped and, still clutching her prey, climbed back into the branches. Satan went back to the second kill again, but at this point Sherry arrived, charged toward Gigi, grabbed the monkey, and hung on tightly. So did Gigi; both pulled, screaming loudly. At once Satan charged back and attacked them, stamping and hitting on their backs, but they continued to cling to the prey. Satan then concentrated his assault on Sherry and, as he lifted him bodily off the meat, Gigi was again able to climb a tree with the kill. At once Satan raced after her. As she fled, she crashed to the ground. Sherry, waiting below, grabbed the prey and this time managed to tear off a large part. Satan turned on Sherry (taking the meat from him) and Gigi, for a while at any rate, was left in peace to eat a little of her hard-won spoils.

Other females besides Gigi sometimes showed surprising tenacity when adult males attacked them for their meat. Once in 1974, when Melissa had part of a fawn that she had taken from baboons, a number of other chimpanzees, attracted by the sound of conflict, arrived on

Table 11.14 Number of times each central female attended mixed-party colobus hunts in three years, 1977 to 1979, and the percentage of these that she climbed trees and actively participated in the hunting.

Female:	Gigi	Fifi	Miff	Sparrow	Patti	Melissa	Athena	Winkle	Little Bee	Passion	Pallas
Number of times attended colobus hunt:	34	17	14	9	17	8	14	17	13	6	4
Percent involvement:	67.6	53.0	50.0	44.4	40.0	25.0	14.0	0	0	0	0

the scene. Melissa was brutally attacked first by Humphrey and then by Figan. She clung to the meat through both these onslaughts, but lost it when Humphrey attacked her for the second time. At this point Gigi somehow managed to gain possession of the carcass—and *she* maintained her hold on it during the aggressive attacks of three different males, one of whom dragged her some 15 meters down the rocky slope. (Observers: D. Anderson and C. Busse.)

When in 1975 Athena had possession of the remains of another fawn (also taken from baboons), she clung to it persistently throughout a prolonged attack by Figan, during which they both fell from a tree. The fight continued, unseen in the undergrowth; then Athena reappeared, still clutching her meat. Figan followed her slowly back up into the tree and appeared to calm her somewhat with a reassurance touch. After a few moments he made a lightning grab that Athena apparently was not expecting. As he made off with the meat, she threw a tantrum worthy of an infant. Figan managed to wrest a colobus kill from Gigi in a similar way, seizing it suddenly after a session of peaceful begging. On that occasion Gigi, who had turned her back to him, whirled around, saw the adult female Nope (who had also been begging), attacked her fiercely, and hurled her to the ground.

Twice females actually took prey from young adult males. In 1973 Gigi snatched a monkey from Jomeo, but it was almost instantly reappropriated by other adult males. And in 1977 Miff managed to take a colobus from Mustard: it was taken from her immediately by Satan (who shared with her).

On six occasions when adult colobus fell or were hit to the ground during hunts, female chimpanzees took part in the attempts to dismember or disembowel them. Once Miff tried repeatedly and unsuccessfully to bite into the stomach of an adult male; when Nope arrived she was able to disembowel the victim almost immediately. When Freud and Mustard had dragged and flailed an adult monkey for thirty-nine minutes, it was Miff who tore off the first limb and started the process that led to the death of the victim. On another occasion Sparrow entered the abdominal cavity through the anus, pulling out a length of intestines.

Three times female chimpanzees caught adult female colobus, but then seemed unsure of what to do. Patti seized a mother whose infant had been captured, bit her leg, but then let her go; the monkey moved away, seemingly unharmed. Patti grappled with another adult female and hit her to the ground. She followed her victim, but seemed afraid to touch her. They looked at each other for a moment, then Patti moved off and the monkey escaped. The strangest of these captures occurred when Harmony encountered two colobus monkeys on the ground, late in the evening. She at once ran toward them. One quickly climbed a tree, but the other seemed sick and made little attempt to escape. Harmony seized her victim, clasped her tightly, bit her—then let her go. The monkey, an adult female, began to move away; Harmony followed, caught and held her, then once more released her. Twice more this strange sequence was repeated, after which it was almost dark and Harmony left the monkey and made her nest. (Observer: E. Tsolo.)

Athena with a bushbuck fawn.
(L. Goldman)

During colobus hunts female chimpanzees, like males, were often chased by male monkeys. As Patti followed a hunt along the ground, a male colobus jumped down and threatened her. As she turned to avoid him, he leaped after her and bit her fully swollen bottom. Another time, when Gigi was hunting alone, a male bounded aggressively toward her. She stamped, waved her arms, and gave loud waa-barks, but he continued to approach. She stood firm, and as he jumped at her and bit her arm, she hit out at him. He finally ran off, and Gigi chased him for 5 meters, slapping and stamping on the branches and giving fierce waa's. This is one of the few instances when a chimpanzee was seen to put a defending male colobus to flight.

Females were often present during baboon hunts, but were never observed to participate in events leading up to a capture. (Fifi was once the victim of adult male baboon aggression during a hunt, but escaped unscathed.) However, as we have seen, females, either single-handedly or in pairs, may seize newly captured bushbuck fawns from adult male baboons. Because male baboons do not typically share meat at Gombe, these fawns did not represent a "troop possession," and thus their appropriation did not trigger the mobbing response that followed predatory attacks on baboon infants. Even so, the female chimpanzees who made these seizures faced a good deal of vigorous aggression from the males who were in possession at the time. Two such incidents have been described elsewhere (Morris and Goodall, 1977); Word Picture 11.2 relates another.

Females, like males, differ in their abilities to acquire meat at group meat-eating sessions. They beg much more frequently from males than from other females. This is mainly, of course, because males are

Table 11.15 Percentage of ninety well-observed group meat-eating sessions (including at least one adult male) attended by each female during five years, 1972–1973 and 1978–1980, and the percentage of those occasions that each female was seen to eat (a) some meat and (b) a significant amount of meat. Females are listed in order of success in acquiring large portions.

Female	% meat-eating sessions attended	% sessions at which meat was eaten	
		Some	Much
Gigi	59.5	72.5	54.0
Miff	29.5	63.0	34.0
Winkle	30.0	64.0	31.5
Melissa	20.0	52.0	31.0
Fifi	28.5	45.5	29.0
Athena	35.0	41.5	28.5
Nope	23.0	42.5	18.0
Pallas	18.0	25.5	14.0
Little Bee[a]	22.0	26.5	7.0
Patti[a]	23.5	12.5	6.0
Passion	19.0	28.0	0

Note: The differences among females in sessions at which much meat was eaten are significant (χ^2 test, one tailed, $df = 10$, $p < 0.001$).
a. Scores are for sixty sessions, 1978–1980 only.

When Mike turns his back on the begging Flo, she moves around the trunk of the tree in order to continue her solicitations. (P. McGinnis)

more often in possession of large portions of meat, but also because, if females do beg from other females, they are seldom rewarded. When Gigi captured a bushbuck fawn, Miff, Moeza, and Michaelmas, who were with her at the time, begged persistently, whimpering for minutes on end and even throwing tantrums. They were not given even the tiniest scrap and finally moved off and left Gigi with her prey.

Table 11.15 summarizes the data on female success during ninety well-observed meat-eating sessions, thirty during 1972–1973 and sixty from 1978 to 1980. Females showed some variation in the frequency with which they attended these sessions and, like the males, in their persistence and success in obtaining meat. Gigi, as expected, scored highest on all measures. This is partly because of her success in hunting and maintaining possession of her *own* prey, and partly owing to her success in begging from males. Back in the 1968–69 period Gigi, while she was not observed to eat meat particularly often, once acquired the largest portion recorded for any female (Teleki, 1973c).

By contrast, Passion attended only 19 percent of the sessions, was only seen eating meat during less than one third of these, and was never seen with large amounts. In the 1968–69 period her attendance at meat-eating sessions was similar to that of most other females, but she was not seen eating meat once, nor even begging. During the more recent years the few occasions when she did participate were when large amounts of meat were available—and she ate very little. In 1978, for example, she attended eight sessions and participated in three. Once she took a small piece of meat from the carcass, once she picked up a scrap from the ground, and once she begged and was rewarded (her *only* observed beg). During 1979 and 1980 she was only present during five sessions and was not seen to eat meat, or try to eat meat, at all. Yet this lack of involvement in group sessions certainly does not imply a lack of predatory interest: witness her consumption of fawns, piglets, and infant chimpanzees.

The other asocial female, Pallas, likewise attended very few sessions and was seldom observed eating meat. The two young immigrants, Little Bee and Patti, showed similar low scores. The older central females were present at about half of the sessions, were seen eating some meat during about half of those they attended, and obtained significant portions at about one third.

The reproductive condition of females may affect their success in obtaining meat. Teleki (1973c) found, in 1968–69, that some females attended several sessions but only participated when in estrus. He analyzed 236 instances of females begging: 132 of those involved females who were fully swollen, and they scored 69 percent success. The other 104 begs involved anestrous or partly swollen females; they scored 40 percent success. When Gigi got her one large portion (most of the viscera of a baboon infant), she was fully swollen.

While similar trends are apparent from a preliminary survey of more recent data, reproductive stage seems to make little difference for some females. Fifi showed the same high levels of success during both cycling and lactating phases. Her high-level involvement in predatory incidents may be connected with her frequent exposure to meat-eating behavior as an infant and juvenile. (Her mother, Flo, attended almost all the recorded meat-eating sessions in 1968–69 and succeeded in getting meat at most. She was experienced and persistent at begging, and was not intimidated for long by attacks from frustrated male bystanders: she ran off screaming, but was soon back again. In the two years before her death she appeared at only two meat-eating sessions. Both times she acquired large portions of meat, once pushing in with the males around the carcass.) Passion's daughter, Pom, however, has frequently hunted colobus, despite the fact that her mother was *never* seen to do so.

The factors involved in female hunting ability and female participation in different types of predatory activity are complex and cannot be correlated with any degree of assurance to age, rank, or upbringing. It is another fascinating area for further research.

12. Aggression

David Greybeard seeks reassurance from his ally, Goliath.

December 1963 Goliath arrives in camp alone, late one evening. Every so often he stands upright to stare back in the direction from which he has come. He seems nervous and startles at every sound. Six minutes later three adult males appear on one of the trails leading to camp; one is high-ranking Hugh. They pause, hair on end, then abruptly charge down toward Goliath. But he has vanished silently into the bushes on the far side of the clearing. For the next five minutes the three crash about the undergrowth, searching for the runaway, then they emerge and are given bananas. As they sit feeding, Goliath's head peeps out from behind a big treetrunk, some distance up the opposite slope. He quickly ducks back into hiding when one of the three looks up. They all sleep nearby; we hear them call a few times during the night.

Early next morning Hugh returns to camp with his two companions. A few minutes later Goliath charges down, dragging a huge branch. To our amazement he runs straight at Hugh and attacks him. The two big males fight, rolling over, grappling and hitting each other. It is not until the battle is already in progress that we realize why Goliath, so fearful the evening before, is suddenly so brave today: we hear the deep pant-hoots of David Greybeard. He appears from the undergrowth and displays in his slow, magnificent way around the combatants. He must have joined Goliath late the evening before, and even though he does not actually join in the fight, his presence provides moral support. Suddenly Goliath leaps right onto Hugh, grabbing the hair of his shoulders, pounding on his back with both feet. Hugh gives up; he manages to pull away and runs off, screaming and defeated.

Early field studies of chimpanzees (including my own) gave rise to the myth of the gentle, peace-loving ape. As more data on chimpanzee behavior have been collected over the years, at Gombe and elsewhere, this myth has gradually been dispelled. It is true that chimpanzees, particularly when traveling in small, compatible groups, may maintain peaceful relationships for hours or days. Nevertheless, they can easily be roused to sudden violence, particularly during social excitement. While most fights do not lead to wounding, some certainly do—particularly those directed at individuals of neighboring social groups.

The term *agonistic* was first used by Scott and Fredericson (1951) to cover sequences of attack, threat, defense, escape, and appeasement. To this list we should add, when considering chimpanzees, reassurance (Goodall, 1968b). Agonistic behaviors, as Brown (1975, p. 40) puts it, are "functionally related to intraspecific, competitive situations; intricately related motivationally and physiologically; and tend to occur together in space and time." In describing agonistic behavior in the chimpanzee, I group attack, threat, and defense under the heading *aggression*. *Attack* is when one individual makes physical contact with another in a behavioral context that does not appear friendly, and in response to which the recipient typically flees or ceases to perform the activity that triggered the aggression. He may also show a variety of submissive or appeasing gestures, or he may fight back. *Threat* is noncontact behavior that elicits a similar response in the threatened individual. Nagel and Kummer (1974, p. 159) include only attack—"massive physical impact"—in their definition of aggression, but I choose to include threat also, partly because threat displays often derive from attack patterns (see for example Marler and Hamilton, 1966; Eibl-Eibesfeldt, 1979), and partly because some actions classified as threats (because they do not involve physical contact) probably only miss being defined as attack because of the skill of the victim in avoiding the aggression of the actor.

Chimpanzees, like many other primates, are capable of skillful manipulation of their companions: an individual may form a coalition and, through the strength of his partner, or through their combined strengths, threaten or attack a rival (Kummer, 1971). For the most part aggressive behavior serves to *maintain* or *increase* distance between individuals (and between communities) through flight or avoidance, although in some species (such as the chimpanzee or the hamadryas baboon) it can serve also to *reduce* distance, as in aggressive herding (Kummer, 1968) when a male threatens or attacks a female who then follows him. For the most part, however, interindividual distance is decreased (and bonds maintained) as a result of the appeasing, submissive, and reassurance components of agonistic behavior, to be discussed in the next chapter.

Aggressive Patterns

Threat

Like the majority of animals, chimpanzees solve more disputes by threat than by actual fighting. Despite the fact that most fights do not appear harmful, physical attack can be dangerous for the aggressor as well as the victim and is best avoided. Gestures and postures of threat are listed below in order of increasing intensity and increasing likelihood that they will culminate in actual attack:

The *head tip*, a slight upward jerk of the chin
The *arm-raise* threat, a rapid raising of the arm; usually forearm only, elbow bent
Hitting toward the threatened individual with an overarm throwing movement
Flapping, a variant of hitting toward; rapid slapping movements in the air, in the direction of the objective, given mostly by females

Hugh threatens a female with the bipedal swagger during banana-feeding excitement. (H. van Lawick)

The *sitting hunch*, when the shoulders are raised and the arms are held out from the body, either to the side or in front

The *quadrupedal hunch*, when the back is rounded and the head pulled in between the shoulders

Swaying branches, either vigorously or with short jerking movements

Throwing rocks, branches, or other loose objects

Flailing, or hitting toward with a stick or branch

The *bipedal swagger*, when the individual sways from foot to foot, arms akimbo

Running upright toward the opponent, often waving both arms in the air

Charging fast at the other.

Two or more of these patterns may be combined, especially in the charging display, which will be described separately below. Vocalizations associated with threatening behavior range from the soft bark (which typically accompanies the first three patterns), to the bark and waa-bark (when threats are more vigorous), to screaming (in high-intensity threats when the threatening individual is both highly aroused and fearful).

The more intense threats are almost always accompanied by hair erection. This autonomic behavior, of itself, serves as a nondirected threat, alerting other chimpanzees to the aggressive mood of the individual concerned. If this is a high-ranking male, a subordinate may detour around him or leave the vicinity altogether, so that additional directed, threatening gestures may be unnecessary.

The actual form of threat used depends on a variety of factors such as the context, the level of arousal of the interactants, their relative dominance status, and the particular individuals involved.

The charging display is a highly effective form of bluff that enables a male to appear larger and more dangerous than he perhaps is. J.B. (*right*) typically swayed vegetation. Faben (*below*) almost always displayed in a bipedal posture because of his paralyzed arm. (H. van Lawick)

The Charging Display

Most often performed by males, but sometimes by females also, this is the most dramatic threat. Display elements include erection of hair; running quickly or slowly along the ground, sometimes upright; dragging or flailing branches; throwing rocks or other loose material; slapping the ground with hands and stamping with feet, or both alternately; leaping up to hit and stamp at a tree (drumming display); seizing low branches or palm fronds and swaying them vigorously from side to side; leaping from branch to branch in a tree and showing exaggerated brachiation. Different males tend to have their own characteristic displays: Hugh and Pepe typically ran upright and occasionally beat their chests; Goliath charged very fast indeed; and Bygott (1974, p. 56) describes one male, Godi, who performed "prolonged displays during which he stamped around in slow motion and threw literally dozens of sticks and stones randomly into the air."

It was Bygott who first pointed out that there are two distinct types of display, vocal and nonvocal. The vocal display is accompanied by pant-hoots and often by the noisier elements of the display pattern, such as slapping and stamping on the ground or on treetrunks. This display is seldom directed at any particular individual; if an attack does occur it is usually of the hit-in-passing variety. The nonvocal display, by contrast, is frequently directed at specific individuals. It often includes one or more of the visual elements of the display pattern, such as bipedal swaggering or waving the arms. Often the nonvocal display is performed several times in succession; each performance may be directed at the same individual and the displayer may charge past a little closer each time. This display is more likely to be followed by an attack than any other threat (Bygott, 1974). The performer seems to lose certain inhibitions. Thus a male, usually gentle and protective toward small infants, may seize, drag, or throw one during such a display; an adolescent male may charge very close to a senior male to whom he is usually deferential, and may even hit him.

Attack

Attack patterns comprise hitting, kicking, stamping on, dragging, slamming, biting, scratching, and grappling. During an attack a male may leap onto his victim's back and stamp with his feet. A smaller individual may be lifted bodily and slammed to the ground, or dragged. Female chimpanzees are more likely to grapple, rolling over and over, sometimes pulling at each other's hair and scratching. Occasionally, two or more aggressors gang up on a single victim and attack simultaneously or consecutively.

According to my definition, aggressive hits or pushes must be included as attacks. Intuitively, however, it seems that such minor aggressive acts should be considered separately. Therefore observed attacks are recorded in three categories: level 1 is made up of aggressive hits and pushes, or quick kicks, delivered in passing; level 2 includes aggressive pounding, dragging, and so on, but lasts less than half a minute and does not lead to serious injury; level 3 involves a much more violent attack that lasts over half a minute, and during which serious wounds may be inflicted. Throughout this volume, unless otherwise stated, *attack* refers to levels 2 and 3 only.

The following tabulation shows the total number of observed attacks made by adult males during four years, the percentage of these that were level 3, and the percentage of the level 3 attacks that led to serious injuries (one or more deep, bleeding gashes). There is remarkably little difference in the percentages from year to year.

	1976	1977	1978	1979
Total fights	92	127	93	124
% level 3	15.2	16.5	16.1	13.7
% level 3 that were serious	28.6	14.3	26.7	23.5

There is a special form of attack, level 4, that is typically directed at chimpanzees of neighboring communities. It is a brutal assault on a single victim by two or more (up to six observed) adults, usually males, lasting more than five minutes. Such attacks have been seen on twenty-one occasions. All resulted in extremely severe wounding. Six lasted between fifteen and twenty minutes; these victims were known (once) or presumed (five times) to have died from their injuries. Only two *intra*community attacks lasted more than five minutes; both were perpetrated by the infant-killer Passion.

Responses

Responses to acts of aggression range from precipitous flight to fairly calm avoidance, or even no response at all. Youngsters (and very occasionally adults) may throw temper tantrums. The victim may scream loudly, squeak, or whimper. During an actual attack a victim may crouch passively, screaming; struggle to escape; or turn and retaliate. After the aggression the victim may run off or direct a variety of submissive, or *appeasing*, behaviors toward the aggressor. These behaviors may, in turn, lead to gestures of *reassurance*, usually involving physical contact, from the aggressor to the victim. Sometimes the victim seeks reassurance contact of this sort from a third individual who happens to be nearby. This whole complex of appeasement and reassurance will be discussed fully in Chapter 13.

Like the act of aggression, the precise response of a given victim depends on a variety of factors: the intensity and form of the aggressive pattern, the behavioral context in which it occurs, the individuality of the interactants, their relative size and rank, and their state of arousal. Thus a mild threat delivered by one chimpanzee to another of comparable rank may elicit no response at all; the same pattern directed toward a much lower ranked individual may elicit a submissive or fearful response. A juvenile without his mother typically responds to threats with more fear than if she were present. A female who would usually flee the aggressive charge of a high-ranking male may actually leap toward and hit him if he has seized and is displaying with her infant. Many females are less nervous and submissive when they are in estrus. And there are numerous other environmental and motivational factors to take into account.

In general, aggressive behavior directed toward a subordinate serves to inhibit the behavior that elicited the threat or attack in the first place, but here again the behavioral context and motivation of the interactants must be taken into account. A female who moves too close to a male as he feeds on fruits in a tree will usually, in response to his arm threat and soft bark, squeak and move away to feed elsewhere. The same female who approaches the same male to beg for meat may, in response to a similar threat, scream and show submissive patterns; yet, after a few moments, she may continue to try to obtain a share of his prey. Indeed, she often returns to beg again even after he has attacked her quite violently and chased her out of the tree. A four-year-old infant who feeds too close to his mother will usually move away quite calmly, or at most whimper or squeak a little, in response to her soft bark or gentle push. But the same aggressive patterns elicited by his approach to suckle will often provoke a tantrum during which the child may actually hit his mother and then continue to try to reach her nipple.

Coalitions and Snowballing

Coalitions

The victim of an aggressive act may, as I have described, seek reassurance contact from a third individual. Such contact may serve to calm him (or her) and may terminate the episode. This is not always the case, however: sometimes, while the contact reduces the manifestations of fearful behavior, screams or squeaks then give way to aggressive waa-barks and, particularly if the third individual C is higher ranking than the original aggressor A, the victim B may then direct threatening behavior toward A. This sequence of events was first described as *protected threat* by Kummer (1957). B, in addition to directing threat behaviors toward A from the security of his position close to the higher-ranking C, may also solicit C's support in an act of joint aggression against A. He will look repeatedly from A to his potential ally C, while continuing to threaten A; he may maintain or renew physical contact with C, or he may move toward A, then look back to see whether he is, in fact, going to be supported. This behavior may result in the formation of a coalition when B, together with his ally C, engages in an aggressive interaction with A (who may retreat in the face of a joint challenge, even if both B and C individually rank lower than he does).

The success or failure of an individual in gaining support depends on numerous factors, the most important of which are the relations between B and C, and between C and A. As we have seen, close bonds may develop between certain dyads: successful coalitions very often comprise a pair of this sort. In particular, family members are likely to help one another.

Snowballing

There are many occasions when an aggressive interaction between two chimpanzees leads to the involvement of others:

(a) The victim of an attack flees the aggressor, then seizes an innocent bystander and attacks him or her; this is *redirected aggression* (Moynihan, 1955).

(b) After attacking one victim, the aggressor may almost immediately seize and attack a second.

(c) An individual, after watching an aggressive encounter, may charge over and join in, either attacking simultaneously with the first aggressor, or taking over when he stops.

(d) The same sort of intervention may take place on behalf of the victim (particularly when this is a relative).

(e) A third or even fourth individual may join in, and we may speak of a *gang attack*.

(f) One aggressive incident may be followed almost immediately by another, involving two quite different actors.

(g) In all of the above, still other individuals may become involved in response to active solicitation by the victim (or victims) as described in the preceding section. De Waal (1982) recounts an incident in the chimpanzee colony at Arnhem that initially involved a fight between two males, but eventually drew in fifteen bystanders.

Figure 12.1 represents schematically the various courses that may be followed by chimpanzees participating in an aggressive episode.

Causes

In order to investigate why chimpanzees are aggressive, it is necessary to delve into successively deeper levels of behavior. At one level—the immediate cause—an act of aggression is sometimes easy to understand: A tries to snatch B's fruit; B attacks A. This is a simple case of competition for food. But why didn't B attack A when A snatched B's fruit yesterday? Why didn't B attack D, who is the same sex and age as A, when D took B's fruit? Why did E, who is the same sex and age as B, only threaten A when A snatched E's fruit? It is when we ask why the same individual behaves differentially on different occasions, or why different individuals respond differentially to the same situation, that the investigation becomes more complex—and more interesting. Each chimpanzee has his or her unique set of genetic and environmentally acquired characteristics, which combine to create a calmer or more fearful—or more aggressive—personality.

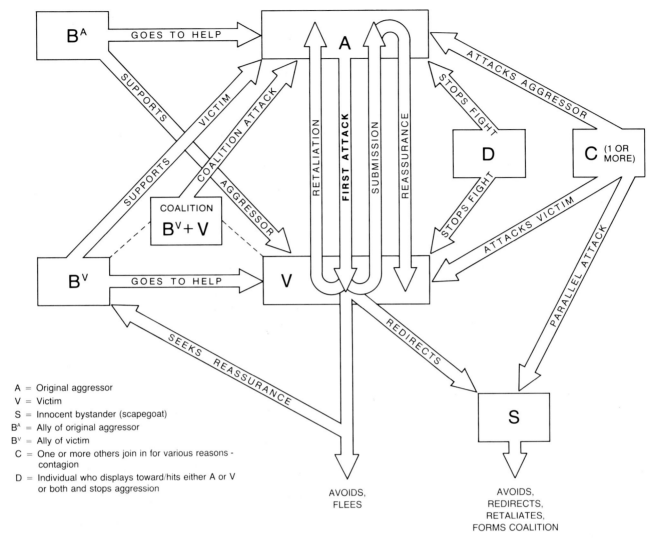

A = Original aggressor
V = Victim
S = Innocent bystander (scapegoat)
B^A = Ally of original aggressor
B^V = Ally of victim
C = One or more others join in for various reasons -
    contagion
D = Individual who displays toward/hits either A or V
    or both and stops aggression

Figure 12.1 Schematic representation of the various behaviors that may take place after an attack.

An event that occurred in the recent past, such as winning or losing a major fight, can have a temporary or permanent effect on the aggressiveness of an individual. Moreover, a number of environmental factors—ecological (the weather) and social (the number of chimpanzees present)—will affect the behavior of all individuals, making it more or less likely that they will become aggressive in a given situation. It is *these* factors that I want to discuss before moving on to examine the sex differences in various aspects of aggressive behavior and the adaptive function of different aggressive acts.

Immediate Causes

The emotions that give rise to acts of aggression appear to range from mild irritation and annoyance to extremes of rage and hostility. Fear may also serve to trigger acts of aggression (see for example Hebb, 1945). Table 12.1 lists the kinds of situation most likely to lead to aggressive responses of threat or attack among the Gombe chimpanzees.

Table 12.1 Immediate causes of aggression.

Immediate cause	Comments
b does not give appropriate response to A's request/command	A "punishes" explicit "disobedience" or coercively molds b's behavior
B tries to take object from A	A is competing for food resources (or a tool or a "toy")
B/b (a male) courts/copulates with female c (A is a male)	A/A is competing for mating rights/punishing implicit disobedience
b (a female) copulates with male c (A is a male)	A is punishing implicit disobedience/competing for mating rights
b (a male) copulates with female c (A is a female)	A is competing for sexual favors with female c
b (a female) copulates with male c (A is a female)	A is competing for sexual favors with female b
b grooms/plays with/sits by c, who is friendly with A	A is competing for social access or protecting a relationship
b "teases" A	A punishes b
b inadvertently/accidentally disturbs/aggravates/startles A	A punishes b's mistake
B threatens/attacks a, who cannot escape	a is defending himself/herself
B/B threatens/attacks A/a	A/a is retaliating
B is merely present (B's behavior is irrelevant)	A is challenging a rival or competing for dominance
B + B are merely present	A/A is challenging rivals/emphasizing his rank
B grooms/sits near C/C (who is another male of similar rank)	A is preventing formation of/breaking an alliance or possible alliance
B/b threatens/attacks A/a's close kin/subordinate friend	A/a is protecting/helping victim
B/b threatens/attacks A's ally C/c (B is subordinate to ally C/c and/or rival of A)	A/A is supporting ally/taking opportunity to jointly challenge
B aggressor or victim in ongoing fight	A is taking opportunity of "stealing a march" on rival
b₁ + b₂ quarreling/fighting near A	A is stopping aggression/emphasizing rank/restoring social harmony/protecting both combatants
b is nearby; A is frustrated in attaining goal	A is redirecting aggression onto scapegoat
b is nearby; A has been watching ongoing aggression	A has been influenced by social facilitation or contagion
X (familiar individual, rank irrelevant) shows abnormal behavior/looks strange	A is initially afraid of X; X has become aversive to A
X (rank irrelevant) is a member of a neighboring community	A expresses inherent dislike of strangers

Note: In each case A, A, a is aggressor anB, B, B, b is recipient of aggression
B is dominant to a; B is *just* above/below A in rank; b is subordinate to A
C is a third individual dominant to a; C is a third individual who ranks *just* above/below A; c is a third
individual subordinate to A.

Punishment, Retaliation, and Challenging

Aggression can be triggered during any interaction between two individuals when a *conflict of interests* is involved: in males competing for mating rights or meat; in females squabbling over food; in youngsters wanting to nurse during weaning, or competing with each other for the opportunity to play with an infant. The aggression may be labeled punitive or coercive when it is directed by a dominant to a subordinate, retaliatory or challenging when the subordinate is the aggressor. Retaliatory aggression may be directed away from the more dominant individual, who (by his actions) caused the aggressive feelings, onto a subordinate scapegoat.

PUNISHMENT AND COERCION   By means of a direct threat or attack, a dominant individual can express disapproval of a subordinate's behavior. If, for example, B, because he wants a share of A's food, continues to approach despite A's bristling hair, A may threaten or attack. Or if B does not move away from a feeding site when A approaches with the obvious intention of taking it over, A may also react aggressively. There are many times when a subordinate ignores or fails to comply with a specific request or demand by a superior. Coercive aggression may result when the superior, by means of threats or attacks, tries to enforce his wishes. Thus a male may repeatedly display around and eventually attack a female who does not respond correctly (with the crouch-present) to his courtship gestures. He may continue these tactics until she submits. When a female refuses to follow a male at the start of a consortship, the result, as we shall see, may be a particularly severe level 3 attack.

B's behavior in the above examples may be viewed as explicit "disobedience," for which he (or she) has been "punished" by A. Implicit disobedience, too, is punished. When a high-ranking male is showing possessive behavior toward a female in estrus (Chapter 16) and a subordinate male attempts a clandestine copulation, either the male or (more likely) the female may be attacked. A female who replies to calls of other males when her consort male has been trying to lead her (secretively) away may also be punished.

Hebb (1945) and de Waal and Hoekstra (1980) have described the deliberate "teasing" of superiors, which led to aggressive punishments. At the Arnhem colony juveniles often approached older individuals who were sitting quietly and threw sand or sticks at them while stamping noisily on the ground. This was one of the most common causes of aggression in the adult females there. At Gombe, youngsters between three and five years of age sometimes dangle above resting adults and kick at their heads and shoulders; often this leads to annoyed threats, particularly from some of the adult males.

There are many occasions when a subordinate inadvertently bothers or startles a superior. Köhler (1925, p. 248) was the first to describe noisy social disturbance as a cause of aggression in chimpanzees. Commenting on the behavior of his oldest animal, he wrote: "All her life, peace was Tschego's essential need. When a noisy quarrel broke out among the other animals and came near her, she always grew angry, sprang up, stamped her foot, and struck out with her arms at the disturbers of her peace." If one of the others came too close, she would seize one of his hands and give it a hard bite. At Gombe youngsters who bump into—or fall onto—individuals who are peacefully resting nearby may be punished. Noisy submissive behavior, particularly the frenzied pant-barking and bobbing of an adolescent male toward his senior, can also provoke an irritable aggressive response.

RETALIATION   There are, of course, many occasions when a subordinate becomes annoyed or enraged by the behavior of a superior. At such times he may direct threatening calls and gestures at the aggressor—but usually only when he judges the distance between them great enough for him to do so with impunity. Females and youngsters often behave thus after being attacked by adult males; they rush up trees, give "defiant" waa-barks, and flap with their hands from the safety of the branches. An adult male, after being attacked, may charge after the higher-ranked aggressor, uttering tantrum screams and slapping at the ground in obvious rage. But if his superior stops and turns toward him, he will usually stop chasing immediately.

Sometimes in response to an attack a subordinate will turn and fight back. This type of retaliation may occur when the victim is unable to escape an aggressor who is holding and biting him or dragging him along the ground. Pain and fear are likely to be components of his aggressive response, which in this case may be viewed as self-defense. Mostly, however, fighting back occurs when the victim is quite close in rank to the original aggressor, or when an ally of the victim is close by or actually helping him. When play becomes rough, particularly that between older juveniles and early adolescents, one of the partners is likely to get hurt. At this point he (or she) may run off screaming or retaliate by attacking the other, and the play session ends in fighting.

FRUSTRATED REDIRECTION   There are many occasions when a subordinate becomes quite clearly enraged by the behavior of a higher-ranking individual who prevents him from getting his own way. But unless he is very close in rank, he may not dare to express these feelings in direct retaliation. If, for example, lower-ranking A wants a prized food, such as meat, that is in higher-ranking B's possession, and he begs for a share but is repulsed by B, or if he is inhibited from even approaching B because he is afraid of B's aggressive reaction, then A will become frustrated. Frustration is a condition arising when something interferes with the attainment of a desired goal (Dollard et al., 1939). If B's rank were only *slightly* higher, A might attempt to take the disputed object by force. But in the present situation he dares not try, and this too will frustrate him. He may instead threaten or attack an innocent bystander, lower-ranking C. His aggression is caused by his inability to compete successfully with higher-ranking B for the desired object. There is then the possibility that C in turn will attack another scapegoat, D. C's aggression resulted from a frustrated desire to retaliate in kind—because he did not dare to attack his aggressor, A.

In the above example the attacks by A on C and C on D are easy to ascribe to redirected aggression. However, it is not always possible to identify this kind of attack because it can be considerably *displaced in time.* I can best illustrate this with an actual example. The target chimpanzee was the young male Goblin, who at the time had a tense relationship with Satan. During the morning there was a total of four aggressive interactions between the two, involving fighting and much displaying. Both males showed very high levels of arousal. An hour after they had separated, Goblin (on his own) encountered a female and her family; he attacked all three, one after the other. These were quite severe attacks—not the brief chase-and-thump variety that so often characterizes male-female reunions. Some forty-five minutes later Goblin, once more on his own, encountered another female; he subjected her, also, to a severe attack. It certainly seems reasonable to suppose that at least the *violence* of these four attacks was related, psychologically, to the intense rivalry between Goblin and Satan noticed that morning. On many other occasions males have been observed to attack females or adolescent males fiercely, and for no obvious reason, on arrival. Many of these may have been the result of prior "unsatisfactory" dominance conflicts or other frustrating experiences, not seen by us.

Before leaving the subject of redirected aggression we should note that a frustrated individual (usually a male), rather than rush off and attack a scapegoat, may perform a charging display, during which he will stamp on the ground or treetrunks (or human observers!), wave branches about, and perhaps throw rocks. These, clearly, are all aggressive patterns, even though directed at inanimate objects. The performance of such a display usually reduces the displayer's level of arousal (Goodall, 1968b, 1971) or, as van Hooff (1973, p. 199) puts it, "acts as an outlet for social tensions." A temper tantrum serves a similar purpose.

CHALLENGING   The aggression we have considered thus far is that caused by particular activities of other individuals that directly conflict with the needs or desires of the aggressor, or that are aroused by aggression directed toward him. There are other aggressive acts that appear to be elicited in A by the *mere presence* of another individual, B—what B happens to be doing at the time plays no part in triggering A's aggression. This pattern was commonly observed in the Arnhem colony, where a male would approach his "prospective opponent from a distance without any apparent relation to the latter's present behavior" (de Waal and Hoekstra, 1980, p. 932). The sight of B does not always elicit an aggressive response from A, but it does on many occasions and in many different contexts. Often an aggressive incident of this sort starts during a period of social excitement, particularly during reunions and arrival at food sources. At other times the aggression begins when B is sitting peacefully minding his own business. In the Arnhem colony there were occasions when an aggressor "quietly and menacingly prepared himself for the encounter by searching for a stick or a heavy stone," even when his opponent was *out of sight* (ibid.).

Because A's aggressive impulse on these occasions is not obviously correlated with changes in B's behavior or in the environment, we can assume that it is due to a change that has taken place internally, and that this change has been brought about by B's seen or known presence. To put it very simply, the sight or thought of B, at that particular moment, puts A in an "aggressive mood." One could probe deeper and search for reasons that make A feel aggressive at one time and not another; to do this, it would be necessary to monitor A's behavior for long periods in order to evaluate the various factors that might influence his mood.

Almost all of the aggression that falls into this category is initiated by an individual who occupies a position just *below* B in social rank. It is important to reemphasize that these are not just isolated incidents: A does not direct aggression toward B just once, but many times and in many contexts. It is obvious that A is actively challenging B. Aggression of this sort, which appears to be motivated on A's part by a desire to intimidate B, almost always leads in the long term (sometimes after countless such incidents) to a reversal in dominance rank between A and B. Once this reversal has taken place, the relationship tends to become more relaxed. But a young or low-ranking male, working his way up the hierarchy, will for some time be unable to dominate *pairs* (or larger numbers) of the senior males when they are together—even if he can intimidate each of them when they are on their own. Just as the mere presence of a single rival, ranking slightly above A, may trigger an aggressive conflict, so may the mere presence of two or more senior males. At Gombe two individuals in particular disrupted pairs or peaceful groups of senior males again and again with vigorous displays. These were Mike and Goblin. Mike, during his struggle for the alpha position, sometimes displayed directly toward groups of up to ten males, most of whom ranked higher than he did. Because he enhanced these displays with noisy empty kerosene cans, the others were easily intimidated. They scattered, and often after a while would gather around Mike and groom him. When Goblin, in his turn, displayed toward two or more senior males, coalitions against him were a frequent result, and often he was attacked. But eventually (as we shall see in Chapter 15), through persistent repetition of these aggressive tactics, he, like Mike, was able to assume the top-ranking position.

Even after one male has unequivocally dominated another, he may from time to time challenge him, with displays or actual attacks, for no obvious reason—especially during a reunion. Alpha males in particular often cause confusion as they charge through peacefully resting groups, sometimes hitting or mildly attacking a senior male in the process. This aggressive behavior may be interpreted as a strategy for reemphasizing and thus maintaining high rank.

The "desire" to dominate is, in fact, one of the most common causes of aggression among adult males at Gombe. Because it occurs in so many contexts, however, and because of the absence of specific eliciting causes, it is not always easy to define. While many of the conflicts involve intimidation tactics only, they can lead to quite serious fighting. Adolescent males challenge adult females in this way as they begin their long struggle to subordinate them before turning their attention to the lower ranking of the adult males. There is, too, a certain amount of competition for social rank among females at Gombe but, as will become apparent, it is very infrequent in comparison with competition between males.

Intervention in the Affairs of Others
Sometimes as a chimpanzee watches an interaction between two or more others, he hurries over and intervenes or joins in aggressively. If the context is agonistic, this action may be in response to an appeal for help by one of the interactants. Or it may be simply the nature of the interaction and the identity of the participants that prompt his behavior. Such interference may be divided into intervention in interactions that are friendly in nature (including sexual) and those that are aggressive.

TO PREVENT FRIENDLY CONTACT Nagel and Kummer (1974, p. 178) have described the development of relationships between previously unfamiliar geladas. In this situation every individual "tended to intervene whenever it saw friendly behavior between two others." The authors conclude that the final social structure that emerged over time "was a direct effect of the success with which pair relationships were aggressively suppressed by other group members."

At the Arnhem colony, the young male Luit displayed repeatedly toward the alpha male on many occasions when the latter groomed or sat by one of his female supporters. Although this often provoked coalitions against Luit, the frequency of contact between Luit's rival and the females gradually decreased, and eventually a reversal in rank of the two males took place. De Waal (1982) labels this behavior "alliance breaking."

I have already described how, at Gombe, both Mike and Goblin, as they strove to attain top rank, disrupted peaceful pairs or groups of senior males again and again.

Other aggressive interventions appear to be caused by feelings of sexual and social competitiveness which, if we were describing human behavior, we should label jealousy or envy. We have seen that an adult male who has established a possessive relationship with a female

(a) Adult male Hugo takes bananas. Melissa seeks reassurance contact while her infant, Goblin, pushes in between Hugo and the adult female Olly. (b) As Melissa reaches for a banana, Hugo attacks her mildly. Goblin, at the left, watches and Athena, gathering her infant, rushes out of the way. (c) Hugo returns to his banana box. Melissa approaches, screaming, with full open grin and extends her hand; Hugo responds by touching. Goblin also approaches. (d) Melissa moves closer, now squeaking, and Goblin, with a slightly nervous grin, turns to watch as Olly also returns, uttering pant-grunts. (B. Gray)

c

d

is likely to intervene aggressively if a subordinate male tries to copulate with her. And, occasionally, a female will threaten or attack one member of a copulating pair. Once, for example, when Passion was in estrus, she solicited a young male who ignored her. When he subsequently began to copulate with her daughter, Pom, Passion ran up and slapped him; he screamed and displayed away.

Now and then a juvenile threatens or hits a youngster who tries to initiate play or grooming with his (or her) infant sibling. Melissa sometimes threatened (and once attacked) individuals who, by starting to groom with her adult son, Goblin, thereby deflected his attention from herself. And resident females occasionally threatened immigrant females who were grooming with community males.

In the context of meat eating, an individual unable to obtain a share often threatens or attacks those who are begging from the possessor of the kill—irrespective of whether those who beg are successful. These actions should probably be classified as redirected aggression, perpetrated by an individual who dares not confront the possessor. Sometimes, though, feelings of "envy" may be involved—when the victim has been successful in his begging.

TO PROTECT   A mother invariably rushes to rescue her infant—and often her older offspring, too—when she sees that the child is being intimidated or hurt by another individual. Often this intervention is a response to a direct appeal for help. If the distress of an infant is due to real danger—as when an adult male is displaying with the youngster—the aggressive, defensive response of the mother (who will leap at and attack the male no matter what his rank) is undoubtedly compounded by feelings of fear. Probably they are feelings similar to those aroused when she herself is in danger. On other occasions, when the individual hurting or frightening her child is less intimidating to the mother (such as when an older youngster is being overly aggressive in play), her aggression may be more an expression of annoyance.

There are many examples of active intervention by adults and young alike when the victim of a fight is a close family member. Infants may hurl themselves at adult males who are attacking their mothers or siblings. And even nonrelated individuals, of any age and either sex, are likely to rush to try to protect an infant in serious trouble, directing threatening behaviors toward the cause of the distress. I shall describe some of these incidents in the next chapter.

TO SUPPORT A FRIEND OR ALLY   In many instances an individual A will hasten to the support of ally C during a conflict between C and a third party B. Here again the aggressive intervention may be prompted by an appeal for help—when, for example, A's ally C is subordinate to B. As a result of the intervention, A and C together may be able to turn the tables on B, even if B ranks above each of them on a one-to-one basis.

At other times A joins the conflict between B and C even when ally C is dominant to B and clearly winning the fight before A joins in: A's participation merely adds emphasis to C's victory. Interventions of this sort may be motivated by a desire to help ally C, or by a desire to conclusively intimidate B, when B is a rival. (And, whether ally C is winning or losing the fight, A's support will function to strengthen their coalition relationship.)

TO ATTACK A RIVAL OPPORTUNISTICALLY   Sometimes an individual seizes the opportunity provided by an ongoing conflict to rush up and hit or stamp on a rival B, at a time when B is fully occupied and unlikely, therefore, to retaliate. The rival may be the winner—as when Satan rushed up and stamped on Goblin a few times as the latter attacked an adolescent male. (Satan at the time was being subjected to repeated challenges by the younger Goblin.) Or the rival may be the loser—as when, in the Arnhem colony, a subordinate female "helped" the alpha male against her high-ranking rival (de Waal and Hoekstra, 1980). In these cases the interventions may be viewed as "sneaky" challenges.

TO STOP THE AGGRESSION Sometimes an individual who ranks higher than both participants may charge toward, threaten, or mildly hit one or both of them, thereby ending the incident. This sort of intervention, sometimes called impartial "policing" (Kurland, 1977), occurs most often at Gombe when a male displays toward two females who are fighting or squabbling, thereby ending their dispute. A higher-ranking male may break up an attack by a lower-ranking male on a female. And a mother sometimes intervenes in a noisy dispute between her offspring, with threats or slaps directed impartially at either or both. The motivation underlying these interventions is not yet clear. The arbiter may dislike the disturbance and act to stop it. Or he may act from feelings of protectiveness toward both of the combatants. Or, perhaps, he may seize the opportunity to assert his rank by impressing any individuals who happen to be watching.

Contagion

I have described a number of different motives that may prompt a chimpanzee to take part in an aggressive interaction between two others. Sometimes, however, an individual seems to join in for none of these reasons: he may suddenly charge over and take part in a fight that, as far as the observer can tell, has nothing to do with him. Or for no apparent reason he may, after watching an aggressive episode, initiate a parallel incident with an innocent bystander.

Aggression of this sort can lead to serious fighting because more than one other individual may join in. In one incident a community male, Hugh, was simultaneously attacked by five adult males and one female, received a number of wounds, and lost one toe as a result. Usually the victims of this kind of aggression are females; if they have infants, these are likely to be attacked also. An unusually violent episode began when Miff, in estrus, joined a group of four males. The alpha male, Figan, approached her, hair erect, and as she pant-grunted submissively he suddenly began to attack her. During the assault he seized her four-year-old infant, Michaelmas, from her back and hurled him away. When Figan left Miff and displayed off through the trees, Michaelmas jumped back onto his mother. At that moment the young male Goblin displayed up and also attacked Miff, throwing her to the ground and stamping on her. He too pulled the infant from his mother and threw him some 3 meters. Michaelmas landed near an adolescent male, who at once seized him by one foot and began to display with him, dragging him over the ground. As Goblin ceased his attack, the powerful and aggressive Humphrey charged up and took over, subjecting the unfortunate female to the most severe pounding of all. When he finally displayed away and Michaelmas managed to escape and return to his mother, Miff just sat, making small squeaks, her bottom bleeding from a gash inflicted by Humphrey. (July 1977; observer: H. Matama.)

A second attack of this sort began at the start of heavy rain, when the three adult males present began to display. Satan was the first to approach the victim, the adult female Pallas. He leaped into her tree and attacked her severely. After a time she was able to pull away from him and ran, screaming loudly. Satan charged off, continuing his display, and briefly attacked one of the other two females present. Meanwhile, Faben climbed Pallas' tree, chased after her, and began to pound on her. He was closely followed by his younger sibling, the alpha male Figan, who leaped down onto Pallas and the two of them fell 3 meters or so to the ground. Pallas raced off, but Figan pursued and tackled her again, pounding and stamping on her for fifteen seconds. Faben charged up and joined in. Pallas escaped once more, and fled up a nearby tree. By now Satan had reappeared, and he caught her and continued the assault as the brothers displayed away into the forest. Soon Satan followed them, leaving Pallas, squeaking softly, with blood streaming from a deep gash on her mouth. (May 1973; observer: D. Riss.)

In both of the above incidents it is possible that each of the male aggressors, given the environmental stimulations (arrival of a female in estrus, heavy rain), might have attacked independently of the others; if so, the first attacks merely served to focus attention on suitable victims. There are other occasions, however, when a second aggressor, who was sitting quite calmly before, gets up and charges over to join a nearby attack. Fortunately for the victims, such multiple-aggressor attacks are rare. During a five-year period there were only three observations of fights in which more than two aggressors converged on a single individual. In each case three males attacked a single female.

It is possible that the observation of fighting may in itself stimulate aggressive behavior in the onlooker through the mechanism of social facilitation. If this is so, a chimpanzee, his level of arousal raised, may join a fight simply for the sake of fighting—we might almost say "for the fun of it." It is known that in some situations aggression is attractive. Rats and monkeys, when they have electrodes implanted in areas of the brain that stimulate aggressive threat, will continue to stimulate themselves, and rats can learn to press levers when the reward is the opportunity to fight with other rats (Eibl-Eibesfeldt, 1979). Young adult males in a number of primate species may actually seek out aggressive encounters (at least with neighbors), as we shall see in Chapter 17. And in our own species, a study on the effect of TV violence concluded that "violence is valued, wanted, enjoyed" and that "watching dramatized violence may actually lead to subsequent aggressive behavior" (Gilula and Daniels, 1969, p. 403).

Fear of the Unfamiliar
In a few rare instances we have observed aggressive acts that are directed toward victims who, by their abnormal behavior, appear to frighten the aggressors. The most notable examples took place during the 1966 polio epidemic when, as described, three individuals became partially paralyzed and as a result developed bizarre movements. When the other chimpanzees saw these cripples for the first time, they reacted with extreme fear; as their fear decreased, their behavior became increasingly aggressive, and many of them displayed toward and even hit the victims. (For a complete description, see Goodall, 1971, chap. 17.) When adult male Rix fell from a tree and broke his neck, group members showed intense excitement and anxiety, displayed around the dead body, and threw stones at it; they also directed many aggressive acts at each other (Teleki, 1973c). Similar fearful behavior followed by outbursts of aggression occurred when male

chimpanzees first saw their reflections in a large mirror. One male repeatedly brandished a large stick, then turned to a companion with a wide grin of fear and embraced him.

At Lion Country Safaris an adult male, removed from his companions for a short while and returned in a semianesthetized condition, was set upon by the others, male and female alike. He was quite unable to defend himself, and almost certainly would have been killed but for human intervention (M. Cusano, personal communication).

There are other aggressive responses that may also stem from feelings of aversion following fear, such as those directed toward some creatures of different species (pythons, for example), which will be described in a later section. The sight of "stranger" conspecifics from neighboring communities may elicit feelings of fear, particularly when the strangers are adult males. Chimpanzees typically show fearful behavior when they travel in the unsafe peripheral zones of their community range, startling at the sound of a twig breaking or at a sudden rustle in the undergrowth. Obviously they are well aware of the danger inherent in a surprise encounter with a strong group of hostile neighboring males. When patrolling individuals return to the "safe" area of the home range after a long period of stealthy travel, they sometimes engage in a series of vigorous charging displays, hurling rocks and branches and sometimes attacking subordinate scapegoats (Goodall et al., 1979). Such performances may well serve as outlets for tensions built up during travel in the danger zone.

Encounters with Strangers
Köhler (1925, p. 246) describes the introduction of a young female chimpanzee to his Tenerife colony. For some weeks she had been housed a few meters away from the others, well within their sight. Thus she was not totally unfamiliar when she was finally placed in with the group. At first the others gazed at her in "stony silence." Then one of them uttered "the ape cry of indignant fury" (a waa-bark?), which was taken up by all the rest. "The next moment the newcomer had disappeared under a raging crowd of assailants, who dug their teeth into her skin, and who were only kept off by our most determined interference." Köhler goes on: "She was a poor, weak creature, who at no time showed the slightest wish for a fight, and there was really nothing to arouse their anger, *except that she was a stranger*" (italics added).

This aggression apparently was prompted by what appears to be an inherent dislike or "hatred" of strangers. It takes some time, usually, for chimpanzees to overcome this aversion: all the group members continued to harass the young Tenerife female for several weeks. At Gombe, as we shall see, resident females may show quite severe aggression toward young immigrants for some months after they have joined the new community.

It may be significant that the worst assaults on members of neighboring communities at Gombe were perpetrated on individuals who were not completely strange to the aggressors. The Kahama chimpanzees had associated with the Kasakela males for years before the community division and had only become estranged a few years before the attacks. A compounding factor here perhaps is that when a chimpanzee rejoins his or her companions after a long separation, levels of arousal are apt to be high. In the case of family members or other individuals with close, supportive bonds, this excitement is expressed by embracing, holding, patting, screaming, and so on. But when there is any kind of competitive dominance relationship between the returning individual and one or more of those whom he meets, it seems that the same high level of arousal is likely to be expressed in violent hostility. Certainly there was unusually severe aggression when Evered rejoined Figan and Faben during the hostile stage of their relationship. In one captive colony (at Holloman Air Force Base in New Mexico) the most severe aggression observed was when individuals were reintroduced after periods of prolonged absence (Wilson and Wilson, 1968).

Behavioral Contexts
Any discussion of the immediate cause of an aggressive act must, of course, be concerned with the behavioral context in which the aggression occurs. In fact, it is not always possible to determine the precise cause, because a multiplicity of factors may be involved—and because the incident may be the outcome of an event that took place some time earlier. Moreover, when a number of chimpanzees charge about, displaying and screaming, and when aggression snowballs, it can be extremely confusing for a human observer, especially when the undergrowth is dense or the actors are high in the trees. Often, therefore, it is more practical and more accurate to categorize aggressive acts with regard to the contexts in which they occur rather than try to assign an immediate cause for each. Sometimes one label, such as "competition for food," is descriptive of both context *and* cause, but this is not always the case.

Most male attacks occur at times of social arousal, such as during reunions, arrival at food sources, hearing distant calls, and so on, when party size is greater than four or five adults. Bygott (1974, 1979) found that of eighty-three attacks by males during one year, 39 percent occurred within five minutes of a reunion. During a five-year period (1976–1980) the percentages of observed male attacks that took place in this context were remarkably similar (26, 17, 29, 25, and 29 percent respectively). However, knowing that a male attacks in the *context* of reunion does not tell us much about the *cause* of his aggression. Appraisal of the attackers and victims of these reunion fights reveals that they occur in the following situations:

(1) The attacker, A, and his male victim, B, are known to be rivals and A ranks just below B;

(2) A ranks just *above* B and is fighting in retaliation after one or more displays directed toward him by B;

(3) A attacks a female or a much lower-ranked male when a higher-ranked rival has just arrived;

(4) A, who has been for some time in a group with a higher-ranked rival, attacks a female just after she arrives in that group;

(5) A has been for some time in seemingly peaceful association with a known rival, and when they encounter C, A's ally of long standing, A suddenly displays at and attacks B;

Charlie hits Passion during an arrival display as he charges past. (Gombe Stream Research Center)

(6) A, during a charging display that does not appear to be directed at anyone, briefly leaps onto, pounds a few times, and leaves B—who is any individual of either sex who ranks lower than himself.

In the above, (1) and (2) might be described as straight dominance conflicts, (3) and (4) are probably aggression triggered by the sight of the rival but redirected onto a scapegoat, while (5) suggests either that A was feeling aggressive toward his rival all along, but did not dare attack until his ally arrived, or that the presence of the ally precipitated a change in his perception of his rival. An attack of type (6) seems to be almost like part of the charging display; subjectively one has the impression that whether the displayer stamps on a chimpanzee or a treetrunk or swats at a human who happens to be in the way makes little difference to him. (Many of the aggressive acts performed by males during other forms of intense social excitement can be interpreted as status rivalry.)

Most of the aggression at Gombe takes place in the following behavioral contexts: reunion, social excitement, competition for food, sexual competition, protection, and "no obvious context." There is a significant sex difference in the frequency with which males and females show aggressive behavior in these major contexts, as we shall see in a subsequent section.

Interspecific Aggression
Various aggressive patterns, ranging from threat to fierce attack, may be directed toward animals of other species. Often this aggression occurs in the context of hunting. The behavior surrounding the actual killing of a prey animal may or may not include aggressive feelings. In many cases the hunter probably feels no more aggressive toward his victim than the man who catches and kills a fish for his supper. (Indeed, both chimpanzees and humans often do not actually kill a prey that cannot escape and poses no immediate threat.) But when

the prey fights back, as does an adult colobus monkey, particularly if the chimpanzee hunter is bitten, his subsequent actions may well be prompted by retaliatory aggression.

Competitive aggression over food between chimpanzees and baboons has already been described in Chapters 10 and 11. And once, when bushpigs and chimpanzees were feeding on the same fallen fruits, a young male chimpanzee hurled a rock at the pigs and they ran off.

Play between young chimpanzees and young baboons is common at Gombe. Quite often it becomes highly aggressive. Sometimes the chimpanzee has the upper hand and chases off his playmates with much stamping, slapping, and hurling of rocks and sticks; at other times it is he who is routed by the baboons. These incidents occasionally end in actual attacks, during which the opponents, screaming loudly, may hit or grapple with one another. Females of both species are then likely to intervene aggressively in defense of their young, and nearby adult males may act protectively.

A few encounters between chimpanzees and large predators (lions and leopards) have been observed at both Gombe and Mahale: the chimpanzees sometimes brandished branches, hurled rocks and sticks, and performed intimidation displays. When confronted by a stuffed leopard (in field experiments performed by Adriaan Kortlandt and his colleagues), the chimpanzees displayed aggressively and some of them actually attacked the dummy with "clubs." These acts will be described in Chapter 18. I was subjected to a good many aggressive intimidation displays before the chimpanzees became accustomed to my presence, and one adult male boldly thumped my head as I lay motionless on the ground (hoping thereby to allay his fear). Yahaya Alamasi and Sufi Matama, currently working to habituate the chimpanzees of the Mitumba community in the north, have been the objects of a number of hostile displays, but none has involved physical contact.

Small predators may actually be assaulted and incapacitated, as will be recounted in Chapter 18. A variety of aggressive behaviors, such as swaying of branches, hitting with sticks, and throwing, may be directed at creatures such as pythons and monitor lizards.

Factors Affecting Level of Arousal

A relaxed and calm chimpanzee is less likely to threaten a subordinate who approaches to feed nearby than one who is tense and socially aroused. As mentioned, hair erection is a useful indicator of the state of arousal—a chimpanzee whose hair bristles all over the body is far more likely to show intense forms of aggressive and/or fearful behavior than one with sleek hair.

There are a number of ecological and social factors which, in general terms, affect the state of arousal and thus the likelihood that a chimpanzee will behave aggressively in a given situation. An important ecological factor is the nature and abundance of the *food supply*—which to a large extent is determined by the season of the year. Probably the most significant effect of seasonal variation is on the size of the foraging party. When food is relatively scarce, particularly if the sources are widely scattered, chimpanzees usually travel in small

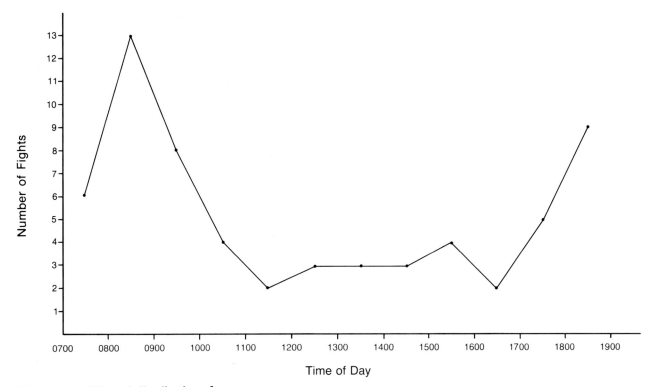

Figure 12.2 Diurnal distribution of fights during the hottest months of the year 1978, July through September.

parties of two or three compatible individuals; in this situation aggression is rare. When food is abundant in large stands of fruiting trees, many individuals forage together and aggressive incidents are relatively frequent. And when a highly prized item (such as meat) is in short supply, aggressive competition will be at its height.

The *weather* and *temperature*, again largely determined by the time of year, can also affect the expression of aggressive behavior. Very heavy rain, typical of the last three months of the wet season, tends to depress all activity; the chimpanzees sit huddled waiting for it to end. However, at the *onset* of heavy rain there may be much wild displaying among the males, during which aggressive incidents often take place. Similar displaying occurs also when a sudden very strong wind springs up.

During the hottest part of the day, between about 1200 and 1500 hours, the chimpanzees usually spend time resting; aggression, particularly the more violent kind, is less frequent. Figure 12.2 shows that frequency of attack during the hottest months of the year (July, August, and September), when midday temperatures seldom fall below 30° C, is highest in the cooler morning and evening periods.

When chimpanzees are sick, they sometimes seem to be especially irritable and often direct mild forms of threat, such as the arm raise or head tip, at individuals who approach them too closely or annoy them in other ways. Pain can elicit an immediate aggressive response (Plotnik, 1974). It tends to increase irritability: a mature male who had a broken toe repeatedly threatened noisy youngsters who played nearby.

There are two major social factors that affect the frequency and intensity of observed aggressive behavior. The first of these is *party size.* An increase in the number of associating chimpanzees correlates positively with increased rate of aggression per individual (Wrangham, 1975; Bygott, 1979). This is partly because in a large group two or more relatively incompatible individuals are apt to be present, and partly because large aggregations so often form in response to situations likely to increase the level of arousal (after a successful hunt, for instance, or because of the presence of one or more females in estrus). In situations of this sort, when the threshold for aggressive behavior is lower, aggression is more likely to be triggered by seemingly trivial causes—such as an adolescent male passing too close to a grooming senior. Moreover, there are more individuals to become involved in aggressive chains as a result of the snowballing effect.

The second social factor is the current state of the *male dominance hierarchy* (Chapter 15). During periods when the rank order of the adult males is relatively stable, there is less fighting than at times of change. During two years when the current alpha male was being challenged, there was much tension among the adult males. It often erupted into charging displays and fighting during times of social excitement—as when a large party arrived at a rich food source, when two parties met, or when a resting party got up and started to move on. The males typically redirected their aggression onto adolescent male and female scapegoats. A similar increase of aggression, and redirection against females, was noted in the top-ranked males of a baboon troop during a period of social instability after the defeat of the top alpha male (Sapolsky, 1982).

Individual Differences

In all creatures the acquisition and modification of behavior is colored by experience during the individual life cycle. As we move up the evolutionary scale, such experience plays an ever more important role.

Early Experience

In common with other group-living primates, chimpanzee infants are exposed to acts of aggression from a very early age. Sometimes this is at first hand, for even a mother carrying a tiny baby may be attacked, and although she typically crouches over it at such times, the infant may be hurt. Attacks, even when not performed on the mother, are noisy and obvious, as are charging displays; usually other individuals in the vicinity, including infants, watch such events intently—unless they are too busy rushing out of the way.

It has been shown that human children, when given the opportunity to watch a variety of aggressive behaviors, will subsequently imitate the patterns they observe (Bandura, 1970). An infant chimpanzee, too, after intently watching a charging display, may perform some of the same patterns. A three-year-old male (Atlas), for example, ran to the security of his mother as an adult male gave pant-hoots preceding a charging display, then watched as the male ran, slapping the ground with his hands, stamping with his feet, ending his display by jumping up and drumming with his hands on a treetrunk. When the adult male had moved away, the infant left his mother, ran a short distance with much stamping of his feet, then paused near the drumming tree. He gazed at it, approached, and—very cautiously and gently—hit it twice with his knuckles. A female of similar age (Fanni) also watched from her mother's arms as a male displayed. She then went to the spot and several times stamped her own feet on the ground. The slap-stamp elements of the charging display, in particular, are often incorporated into sessions of locomotor play by infants of both sexes.

The young chimpanzee gradually acquires the aggressive patterns of the adult, partly as a result of the maturation of species-specific behaviors, partly from actual experience of aggressive actions, and partly as a result of observational learning. Adult females provide important models of agonistic patterns for small infants of both sexes, particularly the mothers of the infants concerned. An elder sibling may also be important. One young male, Frodo, from the age of three years not only watched his elder brother intently, but often "helped" him when he aggressively challenged young females. Adult males become increasingly important as models for juvenile and adolescent males. From the age of about nine years, when a young male begins to leave his mother, he may select one particular adult male and associate with him very frequently (Pusey, 1977). A follower relationship of this sort between young Goblin and the alpha male Figan will be described in Chapter 15. Goblin learned a number of Figan's display and attack techniques, which he used to good advantage.

The personality of the mother (especially whether she is sociable or relatively solitary) and her dominance rank have an effect on the aggressive behavior of an infant. If she is sociable, the infant will spend more time with other chimpanzees and have more opportunities both to watch and to experience aggression. The infants of high-ranking females tend to be more assertive than those of low-ranking ones, presumably because they learn early that their mothers will not only come to their help, but are likely to protect them successfully. Older siblings will protect them in the same way. Aggression may thus be repeatedly rewarded with success, and both sons and daughters of high-ranking mothers are likely to rank high when they become adults. Of course, infants of low-ranking females are sometimes assertive too—but with less rewarding results.

Historical Factors
There are incidents that we might call "historical" in the lives of animals, as of people, which have lasting effects on subsequent behavior. As a result of our many years of observation at Gombe, we have been able to record a number of these incidents; some of them are relevant to the better understanding of individual differences in aggressiveness. Three examples might be mentioned, all of which are described more fully elsewhere in this book.

Mike, after being soundly beaten in a fight with Humphrey, not only lost his alpha status, but from that moment ranked lower than almost all of the other males. Sherry, after being badly wounded during a dominance conflict with Satan, not only became highly submissive to Satan, but from that point on seemed much less motivated to rise to a top position in the male hierarchy. Following the cannibalistic attacks of Passion on other Kasakela mothers, some of them continued to avoid the killer for the rest of her life (she died seven years after the first attack was seen). Miff showed fearful and (when she had male support) aggressive behavior toward Passion throughout this time.

The above incidents, and others like them, were fortuitous observations in that they enabled us to interpret subsequent agonistic behavior that otherwise would have been puzzling. Inevitably, many similar historical incidents go unrecorded, leaving us with puzzle after puzzle, the solutions to which lie locked in the experiences and memories of the individuals concerned.

Genetic and Physiological Factors

It has been possible through selective breeding to produce "aggressive" and "gentle" strains of dogs (Scott and Fuller, 1965). Probably some aspects of aggressiveness are hereditary also in chimpanzees, but the interaction of genetically determined and environmentally acquired behavior in this species is so complex that it is difficult to come to grips with the problem. We do know, as a result of experiments in which chimpanzees were raised for the first six months of their lives in social isolation, that many of the motor patterns associated with aggression are inherited—although when these individuals were later introduced to social groups, innate patterns appeared as incomplete sequences and in inappropriate contexts (Menzel, 1964).

A key variable in any aggressive interaction is the sex of the interactants. Many studies of free-ranging primates have provided clear evidence that males in general show higher frequencies of aggressive behavior, and tend to be more violent, than females (see for instance Kummer, 1968; Lindburg, 1971; Saayman, 1971; Poirier, 1974; Galdikas, 1979). Certainly this is true of the chimpanzees at Gombe. Bygott (1979) found that the adult males were significantly more aggressive than the females, particularly with respect to frequency of attack. Further analysis (see Table 12.2 later in this chapter) confirms that the frequency of attack by adult males is significantly higher than for females.

From an evolutionary point of view, males can afford to be more aggressive, and more violently aggressive, than females because, at least among the higher primates, they are not directly involved in caring for the young. Females, on the other hand, must for many years conduct their affairs with infants clinging to their bodies; frequent outbursts of violence would scarcely be in the best interests of these infants (and in many cases would be positively harmful), so that such behavior is likely to have been selected against.

In a number of primate species males, particularly where they are known to be the more aggressive sex, are also larger, heavier, and probably stronger than females. However, the main reason for the difference in aggressiveness between males and females lies in the underlying neurological and endocrinological mechanisms involved in the expression of the behavior. Very little work in these areas has been done with chimpanzees: the following comments refer mostly to rhesus monkeys and to rats.

It is generally accepted that the hypothalamus plays a key role in the regulation of aggressive behavior (see Hunsperger, 1956), receiving and integrating messages from other parts of the limbic system,

Searching for chimpanzees in the early years of the study. The peaks of the rift escarpment are in the background. (H. van Lawick)

Getty playing on the spring balance. (C. Boehm)

Pom as an adolescent, sitting on the banana trench.

Above: Beethoven, Prof, and Pax (foreground) in a fig tree overlooking Lake Tanganyika.
Below: Watching baboons on the lakeshore. (C. Boehm)

Goblin with a colobus monkey kill.
(C. Boehm)

Eighteen-month-old Fanni has her
eye on Fifi's partially chewed
palm nut.

Above: Gigi and Fifi pant-hoot in response to distant calls.
Year-old Frodo remains silent.
Below: Goblin grooms alpha male Figan.

Gremlin carries Getty in the
ventral position.

Jomeo.

Above: Flo plays idly with her
nine-month-old son Flint.
Below: Satan pauses in his grooming
of Humphrey as both look toward
an arriving chimpanzee.

Above: Pom fishing for termites.
Below: Females and offspring
at termite nest.

Above: Two young males, Tubi and
Jageli, wait for Gigi, in estrus.
Below: Hugo performs one of his
majestic, slow-motion charging
displays. (H. van Lawick)

Above: Getty and Gremlin.
Below: Winkle, Wolfi, and six-year-old Wunda.

Above: Darbi and Tubi.
Below: Fifi looks at her infant brother, Flint, secure in his mother's arms. Adolescent Figan grooms Flo. (H. van Lawick)

A mixed party resting and grooming. (H. van Lawick)

and sending out messages via circuits running through the midbrain, which then stimulate different portions of the spinal cord to produce patterns of attack and the emotional expression of aggression (Smythies, 1970; Flynn et al., 1970, cited in Konner, 1982, p. 190). Brain stimulation experiments in rhesus monkeys confirm that the hypothalamus plays an important role in the expression of aggression in this species (Perachio, 1978).

It has been shown in rats and mice that there are structural differences between the brains of males and females, and that these differences lie in the hypothalamus. At birth the brain is undifferentiated between the sexes, but circulating levels of the male gonadal hormone testosterone, *immediately after birth,* stimulate development of the male pattern in the hypothalamus of the male infant. If a male is castrated immediately after birth, he retains the female pattern. If a female is treated with injections of testosterone at a comparable age, she develops the male pattern.

Although brain structure differences of this sort have not to date been demonstrated in primates, it is well established that circulating levels of gonadal hormones (especially testosterone) during fetal development have important and lasting effects on some aspects of the behavior of rhesus monkeys. The gonads of the male rhesus start to produce testosterone around the forty-fifth day of fetal life; but androgens, which are found in the blood of the male infant immediately after birth, disappear after two days. That it is this prenatal exposure to testosterone which causes the typically higher levels of aggression in male infants (see for example Harlow, 1965; Goldfoot and Wallen, 1978) is suggested by a number of experiments. In one (Chamove, Harlow, and Mitchell, 1967) a group of rhesus, separated from their mothers at birth, were raised in social isolation for the first three years of their lives (thus effectively eliminating rearing variables, such as differential treatment of the sexes by mothers or peers). The isolates were then paired with randomly sexed infant monkeys. All the male isolates directed significantly more hitting at their companions than did the females (who in fact showed more nurturing behavior). In another experiment a male castrated at four months continued to show levels of aggression quite comparable to those of intact males throughout the juvenile period (Goy, 1966). And when adult female rhesus were treated with injections of testosterone from the thirtieth through the ninetieth days of their pregnancies, four androgenized females were born; these pseudohermaphrodites were tested between two and five months of age, and showed levels of aggressive behavior midway between those of normal males and normal females of the same age.

Having established that androgens play a vital role in the *initial* differentiation of male and female aggressiveness, and that this difference persists through the juvenile stage despite the absence of gonadal hormones, we should next investigate the relationship between aggressive behavior and male sex hormones in the adult. That there is a relationship has been clearly established: in a variety of studies on rhesus males those individuals with the highest levels of plasma testosterone tended also to be the most aggressive individuals (Gordon et al., 1979). However, the nature of this relationship is not clear. When testosterone levels of intact males were artificially raised, there was no corresponding increase in aggression in rhesus (two males only in a pilot study; Gordon et al., 1979) or chimpanzees (three males; Doering et al., 1980). A number of male rhesus monkeys were castrated (as adults) and then returned to the seminatural habitat of Cayo Santiago in Puerto Rico; they continued to show aggressive behavior and sometimes were able to fight and defeat larger intact males (Wilson and Vessey, 1968). Castrated males in the laboratory also continued to direct aggression at male companions (Perachio, 1978).

The relationship between androgens and dominance rank has also been investigated: plasma testosterone levels corresponded positively with rank in an *all-male* group of rhesus (Rose, Holaday, and Bernstein, 1971), but no significant correlations were found in studies of *heterosexual* groups of rhesus (Gordon, Rose, and Bernstein, 1976; Perachio, 1978) or Japanese macaques (Eaton and Resko, 1974).

Of particular interest is the finding that *social events* may have a direct effect on plasma and androgen levels. Male monkeys with access to sexually attractive females showed a marked rise over baseline levels of plasma testosterone in rhesus (Rose, Gordon, and Bernstein, 1972) and in *dominant* talapoin monkeys (Keverne, Meller, and Martinez-Arias, 1978). Male rhesus also showed a rise in plasma testosterone levels after they had decisively won fights; depressed levels were characteristic of males who had lost fights, or who had been exposed to severe stress upon introduction to established groups, when they were repeatedly harassed by several individuals (Rose et al., 1974). In the talapoins it was further shown that *subordinate* males did not show increased plasma testosterone levels when caged with sexually attractive females as long as the dominant male of the group was also present, although levels did rise when they were caged with females on their own.

Sapolsky (1983) has examined rank-related endocrine differences in olive baboons in the field. During a period when the dominance hierarchy was stable, the seven top-ranked males were less aggressive than the nine lower-ranked males, and they did not show particularly high testosterone levels. However, during a period of instability (after wounding of the alpha male) the six top-ranked males became the most aggressive and had the highest levels of testosterone. Sapolsky comments that the entire endocrine system of these males became less efficient: their cortisol stress response was impaired, and they were no longer capable of elevating testosterone levels following stress.

Thus it now appears that the level of plasma testosterone may *reflect* rather than *determine* events that are neurally mediated. Keverne, Meller, and Martinez-Arias (1978) suggest that social events may influence the hypothalamic amines, which then modify aggressive (and sexual) behavior and gonadotropin release; this sequence will result in changes in the level of plasma and androgen levels. Perachio (1978) also stressed the role played by the hypothalamus in controlling the release of testosterone and as a site of endocrine and behavioral interactions.

Table 12.2 shows that during two separate years target males attacked other individuals significantly more frequently than females did. The attack rate of different individuals varies widely:* for the males in this sample, the range was one fight per 27 hours to one per 207 hours, with a mean of one fight per 62 hours; for the females, the range was one fight per 47 hours to none in 230 hours, with a mean of one fight per 106 hours. Bygott (1974) found that the most aggressive male in 1971 (the alpha, Humphrey) had an attack rate of one fight per 9 hours; the median for the fourteen males (before the community division) was one fight per 33 hours. These higher scores were probably due to a variety of factors: the large number of males, the extremely tense situation between them at the time of the community division and the change in alpha males, and the typically larger party size. Moreover, Bygott followed for shorter time periods, often starting when the sound of calling attracted him to a large party, whereas the more recent data are taken from full-day follows.

Table 12.2 also shows the frequency with which males and females were victims of attack by others. Target males attacked three to four times more frequently than they were attacked in both years. This was true for the female targets in 1978. In 1976, however, females were attacked more frequently, probably because of instability in the male dominance hierarchy, as we shall see.

Table 12.3 shows that during 1978 a significantly greater percentage of the aggressive interactions initiated by adult males led to attack (29 percent) than was the case for females (17 percent). Fifteen (16.1 percent) of the attacks by males were severe level 3 ones; only two (7.4 percent) female-initiated fights were severe.

In addition, the table shows the major contexts in which aggression was observed during 1978. About one third of total male aggression, as well as one third of male attacks, occurred in the context of reunion. Other major contexts for male attacks were meat eating, sexual, social excitement, and "no obvious context." As already discussed, male aggression in the framework of reunion, social excitement, and "no obvious context" is often triggered by dominance rivalry. All of the twenty attacks made by adult males on other adult males in these three contexts were almost certainly dominance conflicts. The attacks perpetrated by adult males on adult females in the "no obvious context" category almost all involved cycling females who were not fully swollen. This puzzling aggression will be discussed in some detail in Chapter 16.

The picture for the adult females is very different. There were two major contexts that elicited aggressive behavior—feeding (on plant foods) and protection. The table shows that roughly half (48 percent) of female attacks occurred in these contexts and, as for the adult males, about a quarter (22 percent) for "no obvious context." Many of

---

* It would not be appropriate to compare differences in attack rate between individuals from these data. As we have seen, the likelihood that a chimpanzee will attack is highly dependent on environmental variables and it would be necessary to analyze individual aggressive behavior in relation to these variables to obtain a meaningful picture of individual differences.

Table 12.2 Frequency of attacking and being attacked during target hours (not less than 100 hours for any individual) for males and the most frequently followed females during two years, 1976 and 1978. Follows when the targets were quite alone, or with dependent offspring only, are excluded.

Sex	Year	Number of target hours	Frequency of attacking	Frequency of being attacked	Observation hours per attack by target	Observation hours per attack on target
Males	1976	829 (6 males)	16	4	51.8	207.3
Males	1978	1,570 (7 males)	23	8	68.3	196.3
Females	1976	897 (6 females)	9	7	99.7	128.1
Females	1978	1,647 (7 females)	15	5	109.8	329.4

Note: Males made attacks significantly more frequently than females.
1976: χ^2 test (one tailed), $df = 1$, $p < 0.001$.
1978: χ^2 test (one tailed), $df = 1$, $p < 0.005$.

Table 12.3 Frequency of aggression (including attacks) and attacks by males and females in various contexts during 1978.

	Overall aggression				Attacks			
	Total incidents		Percentage		Total number		Percentage	
Context	Males	Females	Males	Females	Males	Females	Males	Females
Reunion	105	8	33	5	28	1	30	4
Social excitement	36	2	11	2	8	1	9	4
Plant food	18	56	6	35	3	6	3	22
Meat	47	8	15	5	17	2	18	7
Sexual	46	3	14	2	13	0	14	0
Protection	13	56	4	35	4	7	5	26
No obvious context	45	10	14	6	16	6	17	22
Miscellaneous	9	16	3	10	4	4	4	15
Total	319	159			93	27		

Note: Significantly more male aggression resulted in attack than was the case for females: χ^2 test (one tailed), $df = 1$, $p \leq 0.05$.

these episodes were probably due to dominance rivalry, but the picture is far less clear than it was for the males.

In order to investigate these male-female differences more closely, I have broken down the information presented in Table 12.3 in two ways. Table 12.4 shows the actual number of aggressive acts (including attacks) and the number of level 2 and 3 attacks that were *directed* by, first, adult males and, second, adult females against members of the different age-sex classes. It should be borne in mind that most females spent far more time with infants, juveniles, and adolescents than did adult males. Adult males associated most often with adult and adolescent males and with females in estrus. The table shows that adult males directed 82 percent of their overall aggression and 88 percent of their *attacks* against other adults. That females were victims most often is not surprising, partly because they are a less risky proposition and partly because there are more of them; for both of these reasons females are more likely to become scapegoats in instances of redirected aggression. The amount of adult male aggression that was directed toward youngsters is insignificant and will be discussed elsewhere (from the point of view of the youngsters, for whom

Severe attack on Melissa by Evered. There was no obvious reason for this attack, which lasted for about forty seconds. Melissa was in the flat phase of her cycle. The infant Gremlin clung ventrally throughout. (B. Gray)

343 *Aggression*

Table 12.4 Distribution during one year (1978) of the total aggressive acts (including attacks), and of attacks only, directed toward the various age-sex classes by adult males and adult females.

Direction of aggression	Number of aggressive acts	Number of attacks
Directed by adult males—		
Toward adult males	109	30
Toward adult females	154	52
Toward adolescent males	22	4
Toward adolescent females	13	2
Toward juvenile males	12	3
Toward juvenile females	9	2
Total	319	93
Directed by adult females—		
Toward adult males	7	1
Toward adult females	53	18
Toward adolescent males	54	2
Toward adolescent females	33	5
Toward juvenile males	8	1
Toward juvenile females	4	0
Total	159	27

it was not so insignificant!). Females directed similar percentages of aggression toward other adult females and toward adolescent males (33 percent in each case) and 21 percent toward adolescent females. (In fact, about a quarter of their overall aggression was directed toward their own offspring of various ages.) Many (66.7 percent) of the observed female *attacks* were directed against other adult females.

Table 12.5 shows the number of aggressive acts and attacks that were *received* by adult males and adult females from members of other age-sex classes. Almost all of the aggression received by the males was, not surprisingly, from other adult males, and all but one of the attacks. About half (55 percent) of all aggression received by females was perpetrated by adult males, and over half (67.5 percent) of the actual attacks. The relatively high 15 percent of aggressive acts received by females from juvenile males was entirely due to the belligerence of one youngster, Freud.

Figures 12.3 and 12.4 present the same data separated into the various behavioral contexts in which the incidents were observed. Adult males (Figure 12.3*a* and *b*) were aggressive to other adult males most often in the contexts of reunion and other social excitement—which, as we have seen, often involved dominance conflicts. The males also attacked one another frequently in the sexual context: a total of nine females cycled during the year, and tension often ran high in large sexual parties.

Males were aggressive to females considerably more often during reunion than in other contexts, but their attacks on females were distributed fairly evenly through reunion, meat eating, sexual behavior, and "no obvious context." In almost all contexts the *proportion* of aggression directed toward females (about one half) and other adult males (about one third) was very similar—both for overall aggression and for actual attacks. In the social excitement context, however,

Table 12.5 Distribution during one year (1978) of the total aggressive acts (including attacks), and of attacks only, received by adult males and adult females from the various age-sex classes.

Direction of aggression	Number of aggressive acts	Number of attacks
Aggression toward adult males—		
By adult males	109	30
By adult females	7	1
By adolescent males	1	0
By adolescent females	0	0
By juvenile males	1	0
By juvenile females	2	0
Total	120	31
Aggression toward adult females—		
By adult males	154	52
By adult females	53	18
By adolescent males	24	5
By adolescent females	5	0
By juvenile males	42[a]	2[a]
By juvenile females	1	0
Total	279	77

a. All these aggressive acts were perpetrated by one precocious youngster, Freud.

David Greybeard attacks the adolescent female Pooch. (H. van Lawick)

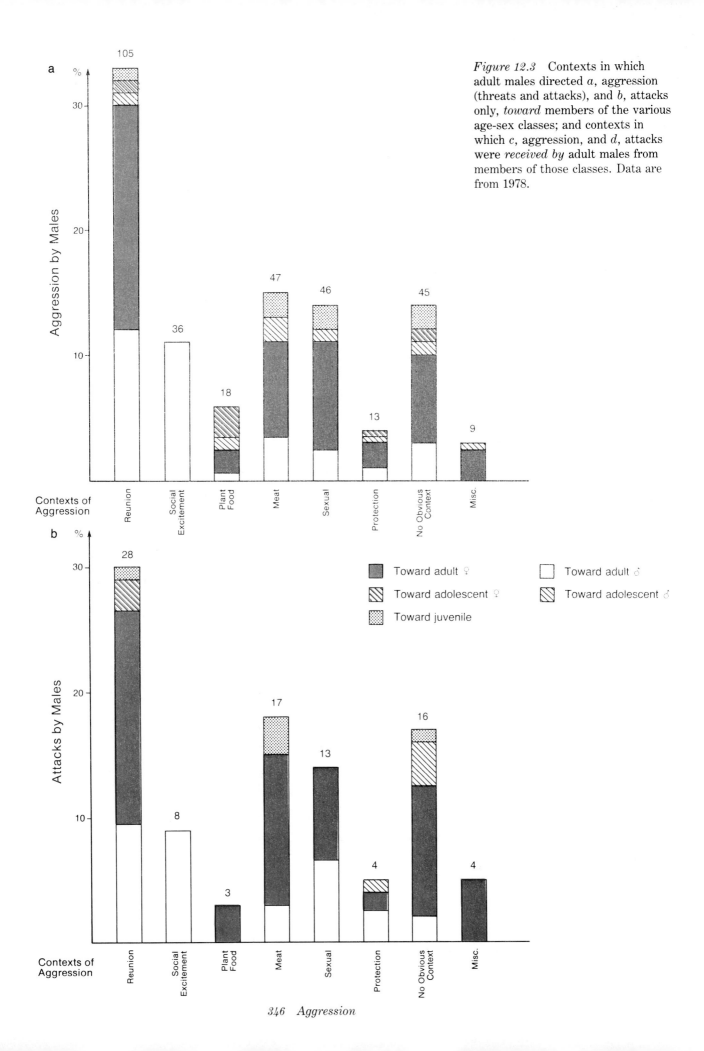

Figure 12.3 Contexts in which adult males directed *a*, aggression (threats and attacks), and *b*, attacks only, *toward* members of the various age-sex classes; and contexts in which *c*, aggression, and *d*, attacks were *received by* adult males from members of those classes. Data are from 1978.

Legend:
- Toward adult ♀
- Toward adolescent ♀
- Toward juvenile
- Toward adult ♂
- Toward adolescent ♂

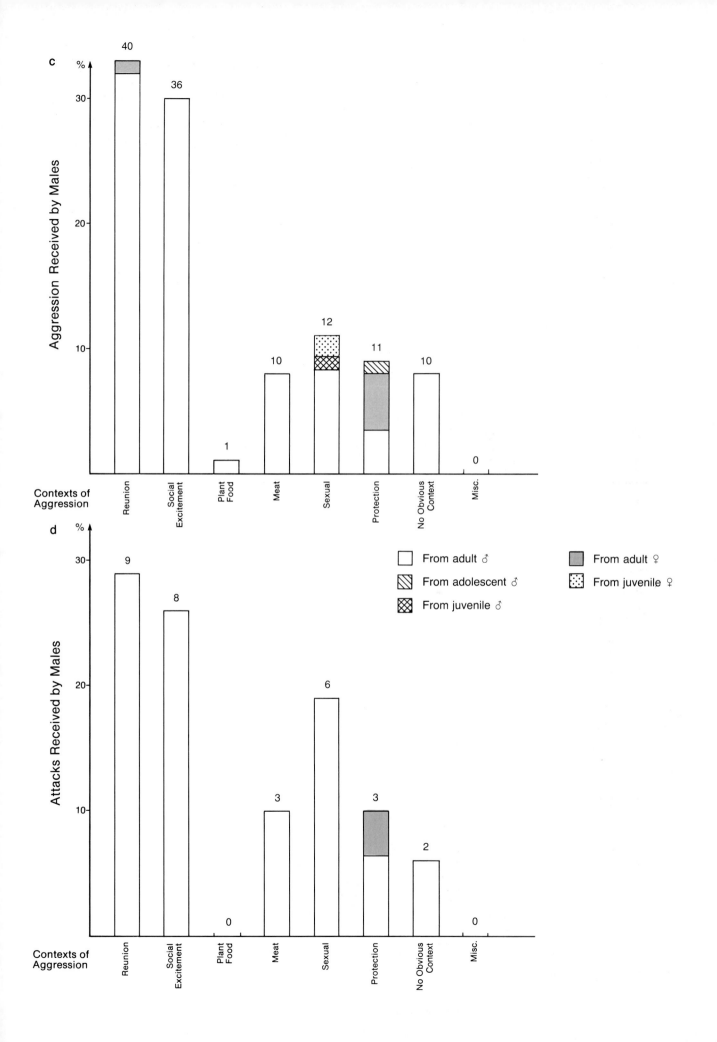

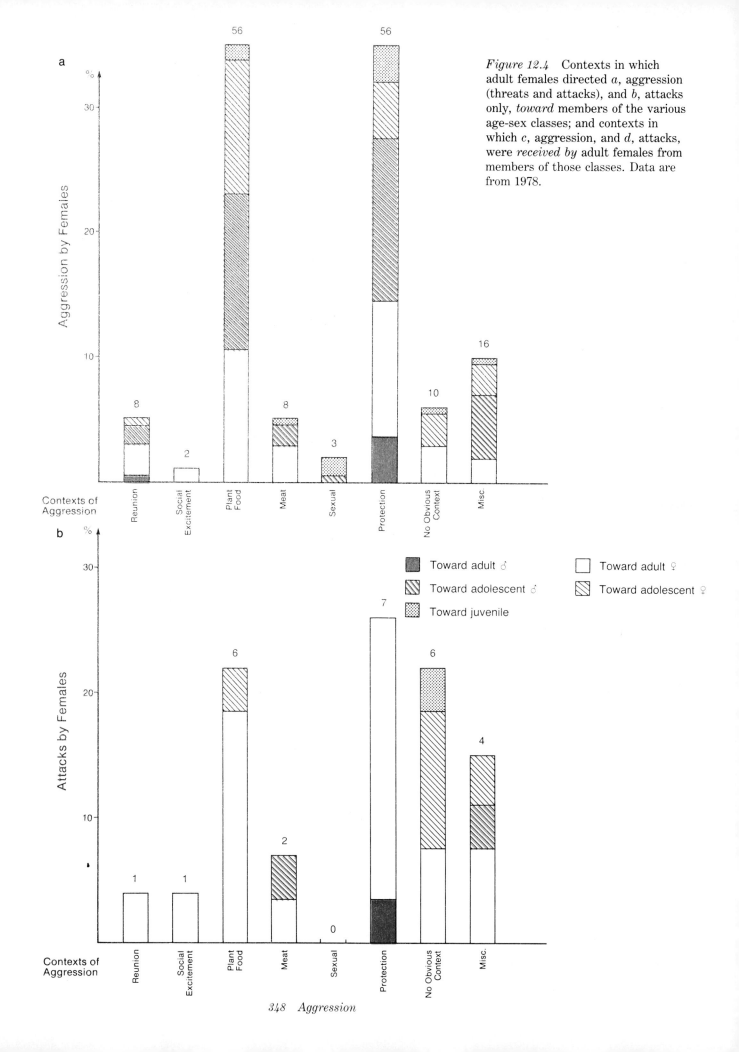

Figure 12.4 Contexts in which adult females directed *a*, aggression (threats and attacks), and *b*, attacks only, *toward* members of the various age-sex classes; and contexts in which *c*, aggression, and *d*, attacks, were *received by* adult females from members of those classes. Data are from 1978.

Legend:
- Toward adult ♂ (solid dark)
- Toward adult ♀ (white)
- Toward adolescent ♂ (hatched)
- Toward adolescent ♀ (hatched)
- Toward juvenile (dotted)

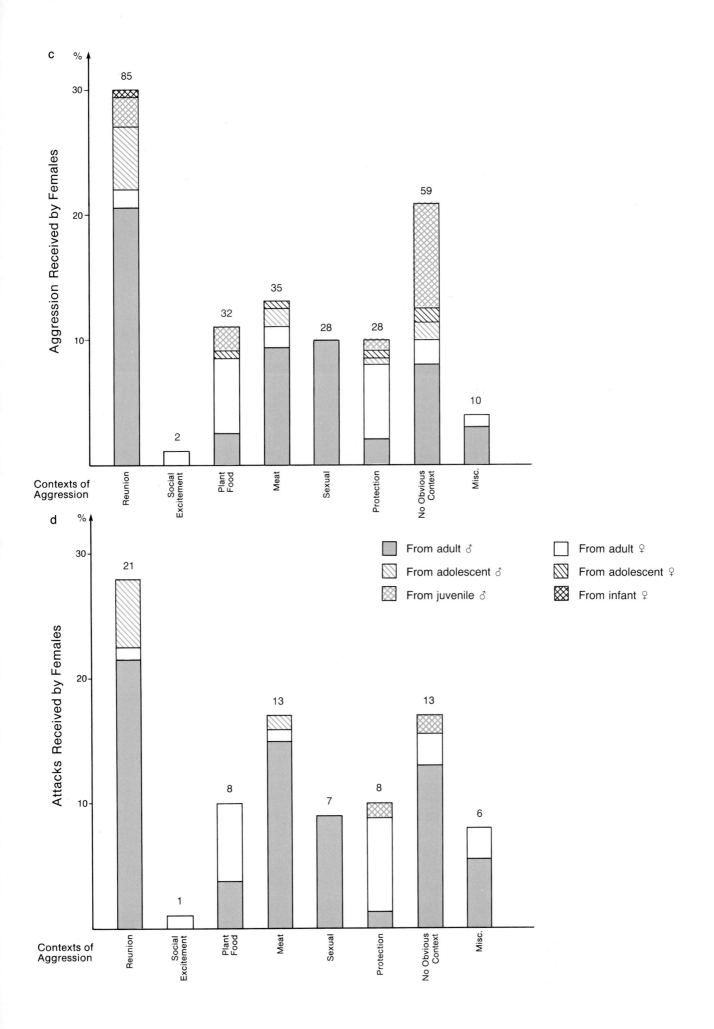

aggression was directed only against other males—almost certainly, all the incidents represented dominance conflicts.

In most contexts, obviously, adult males were victimized almost entirely by other adult males (Figure 12.3c and d). The only exceptions were in the sexual context, when youngsters interfered aggressively in copulations, and in protection, when mothers tried to rescue their offspring from male aggression. During the period reviewed this was the only context in which a female was seen actually to attack a male (once).

Thirty-two percent of all aggressive interactions between adult females (Figure 12.4a) were in the plant feeding context. Here perhaps I should give an example: Two females and their young are feeding on the tips of long vine stems. One of them starts to pull on such a stem. The tip is up near the other female, who feeds at a higher site in the tangled vegetation. As she sees the vine stem moving down past her, she stares, her hair begins to bristle, she seizes the stem and yanks it toward her, then begins to swagger bipedally. The offender squeaks, reaches her hand up, and after a moment the aggressor calms down and briefly touches the outstretched fingers. (The aggressor eats the disputed shoot.) The relatively high proportion of *overall* aggression directed against youngsters in this context is not mirrored in the distribution of attacks (Figure 12.4b), almost all of which were on nonrelated adult females.

The percentage of female-female aggression in the protection context was the same as in the plant food context (Figure 12.4a). A typical example: As two infants play, one gets hurt, screams, and looks toward his mother. She arm-raises and soft-barks at the playmate, who screams in his turn. At this, the other mother (who holds similar social rank) charges up, hair bristling, and the two females spar briefly, hitting at each other and screaming. Then they calm down and the infants continue their game.

Of the fifty-six aggressive incidents initiated by females in the plant feeding context, 31 percent occurred when mothers threatened their offspring (including adolescents) when they tried to feed too close. And all fifty-six instances of aggression in the protection context were either when a mother went to the aid of her child or when (five times) an adult sibling helped a younger one.

The contexts in which females were most often victimized by males (Figure 12.4c and d) were reunion, meat eating, sexual, and "no obvious context." All the juvenile male aggression in this last context was, as mentioned, perpetrated by Freud; transitional between childhood and early adolescence, he was precociously challenging many of the adult females.

During this particular year adult males were observed to initiate fifteen serious level 3 fights. Eight of these were directed against other males, five of whom were wounded quite badly—two during meat-eating incidents; two when the alpha male, Figan, attacked them during social excitement; and the fifth and worst, when Goblin challenged Figan and precipitated an attack against him by five adult males. The seven adult females violently attacked by males were not visibly badly wounded (although they were in other years), but they were probably severely bruised.

Cannibalistic Attack on Melissa and Her Infant, November 1976

At 1710 Melissa, with her three-week-old female infant, Genie, clinging in the ventral position, and followed by her six-year-old daughter, Gremlin, climbed to a low branch of a tree, and sat. After one minute Prof, Passion's juvenile son, appeared. He approached Melissa, peered at, and then sniffed her infant. Melissa gave loud pant-grunts and cradled Genie more closely. Prof was followed by his eleven-year-old sister, Pom. As she stood nearby, Melissa reached to touch her and then, supporting Genie with one hand, she climbed from the tree into an oil-nut palm above. As she reached the crown, disappearing temporarily from the observers' sight, Passion (who had presumably been there all the time) suddenly leaped at her. Melissa, screaming loudly, half climbed, half slid down the trunk with Passion close behind. Pom, Prof, and Gremlin also climbed rapidly to the ground.

The fight that followed lasted just under ten minutes. Passion and her daughter cooperated in the attack; as Passion held Melissa to the ground, biting at her face and hands, Pom tried to pull away the infant. Two minutes after the start of the fight, blood was pouring from Melissa's upper lip. At one point Passion snatched the baby away but Melissa seized it back, biting Passion's hands. Passion, leaping around, seized Melissa from the rear and bit deeply into her rump (the wound actually penetrated the rectum just above the anus). Melissa, ignoring this savaging, struggled with Pom. Passion then grabbed one of Melissa's hands and bit the fingers repeatedly, chewing on them. Simultaneously Pom, reaching into Melissa's lap, managed to bite the head of the baby. Melissa still held on, and Passion seemed to try to turn her over. Then, using one foot, Passion pushed at Melissa's chest while Pom pulled at her hands. Melissa, still clinging to the baby, bit Passion's foot while Pom held and bit one of her hands. During the entire fight all the participants screamed loudly. Finally Pom managed to run off with the infant. At this point Gremlin—who had been trying to help her mother throughout the fight, but had been repeatedly pushed out of the way—hurled herself at Pom, and Melissa managed to retrieve the infant; but almost at once, Pom got it back and ran off. Pom climbed a tree with the corpse (for the baby is thought to have died when Pom first bit into its forehead), and Passion and Prof followed. Melissa tried to climb also, but fell back, seemingly exhausted, as a small dead branch broke. She watched from the ground as Passion took the body and began to feed.

Three minutes later Melissa managed to climb up. Passion moved away, leaving some entrails on the branch; Melissa picked these up and sniffed them, then dropped them and moved after the killers. As she approached, she began to scream again and once more Passion avoided her. Melissa, whimpering now, followed.

Fifteen minutes after the loss of her infant, Melissa again approached Passion. The two mothers, in silence, stared at each other; then Melissa reached out and Passion touched her bleeding hand. As Passion, with Prof, continued to feed on the infant, Melissa began to dab at her wounds. Her face was badly swollen, her hands lacerated, her rump bleeding heavily. At 1830 Melissa again reached to Passion, and the two females briefly held hands. Melissa continued to watch, lying on a branch, tending her wounds. At 1842 Melissa again approached; Passion embraced her and continued feeding. She reached out her foot toward Melissa, who touched it. As darkness fell, the chimpanzees were still there, Passion and Prof feeding, the others watching. (On this specific occasion Pom never tried to eat any of the flesh, although she had on both previous occasions.)

Two days later when Melissa was next seen, she was still for the most part walking upright to keep her lacerated hands off the ground. Her other wounds were still raw and swollen. She did not fully recover for another four weeks.

There were two level 3 fights observed between adult females, one of which led to serious wounding. This was when Miff was fighting to save her infant from the predatory attack of the cannibal, Passion. During the fight, which lasted over four minutes, both females twice fell from a tall tree, almost to the ground, then climbed back up and continued the battle. Eventually Passion, hindered perhaps by her own four-month-old infant, gave up and retreated. Miff remained in the tree uttering very loud waa-barks. Three minutes later an adult male displayed up, hair erect. Passion by then was nowhere to be seen. Miff did not remain with the adult male, but traveled with her daughter and juvenile son for the rest of the day, occasionally dabbing at the wounds on her face and hands. The infant did not appear to be harmed at all. (Observer: H. Matama.)

It is in this extraordinary context that the most severe female-female fighting has been recorded. Between 1975 (when Passion was first seen to seize, kill, and eat the newborn infant of another community female, Gilka) and 1978 (when the incident just described occurred) a total of five such attacks were observed, of which three were successful. Gilka lost two infants, and Melissa one; Passion was helped in three of the attacks by her late-adolescent daughter, Pom. Five other infants disappeared during this period, and elsewhere (Goodall, 1977) I have presented evidence suggesting that they too may have been victims of Passion and Pom. The incidents are summarized in Table 12.6.

As we have seen, females typically are not so violently aggressive as males. That they can, on occasion, show extremely fierce attack behavior is illustrated by the incident described in Word Picture 12.1—the November 1976 seizure of Melissa's infant by Passion. This was the longest of her attacks and provides the best example of the cooperative behavior shown by Passion and her daughter that ensured the success of their enterprise.

Adult male chimpanzees at Gombe have also been seen to attack, kill, and partially eat infants, but in all cases so far these were infants of females from other communities. In the Mahale Mountains, adult males of M group are known (once) and suspected (three times) of killing and eating infants of their own community (Nishida, Uehara, and Nyundo, 1979; Kawanaka, 1981; Hiraiwa-Hasegawa, Hasegawa, and Nishida, 1983; Takahata, 1985).

In this chapter I have shown that males have a higher frequency of aggression than females, that a greater percentage of their aggressive interactions include physical attack, and that more of those attacks are severe. Moreover, the contexts in which aggression is most likely to occur are, for the most part, different for males and females. There is one further difference. The male, as we have seen, becomes increasingly aggressive throughout adolescence. However, once he has established himself in the male dominance hierarchy and subordinated the adult females of his community, he will not experience major changes from year to year in the frequency of his involvement in aggressive interactions. Certainly there may be more aggression when the male hierarchy is unstable: there will be more tension in male-male relationships, more social excitement, and more redirected aggression. Even so, for the average male, the contexts in which he instigates and receives aggression will not vary greatly. Every year he will spend much time traveling with other males and mixed parties of males and females. These parties will encounter others, at which times there will typically be high levels of arousal and aggression. Every year he will go on boundary patrols, encounter strangers, compete for meat. And every year there will be cycling females to compete for, to lead away aggressively on consortships. His social life will follow a similar pattern, and there will be a similar pattern of aggressive involvement. Only during certain very specific years—if, for example, he challenges the alpha male or loses a close ally of long standing—will there be a *major* difference.

For the female, however, the picture is completely different. As a late adolescent she will travel with mixed parties, she will cycle regularly, she will direct a good deal of aggression toward other females, and she will frequently be a victim not only of the adult males, but of the females also. Then, having become pregnant and given birth, she will become less social, spending increasingly more time on her own or with her mother. As a result, her level of aggressive involvement will drop dramatically. As her offspring gets older, however, the mother will increasingly become involved in the squabbles of childhood—an involvement which, as we have seen, will lead her into conflict with other mothers of the community. Then, quite suddenly, she will resume estrous swellings. Once more she will be a sought-after sexual partner, once more she will travel in big mixed parties, with the resulting increase in aggressive involvement with the adult males. This period will end with her second pregnancy, and her aggressive interactions will again, for the most part, be with other mothers. But if her firstborn is a son, she will soon be drawn into a whole new series of conflicts as he challenges the adult females of the community during his adolescence. Again and again, especially as he begins to direct his efforts to subordinating the more aggressive of these, his mother will be involved in quite serious conflicts on his behalf. Eventually this is likely to raise her own status, because her son is, in turn, likely to help her as he becomes bigger and stronger.

Functions in Chimpanzee Society

Aggression is part of the complex network of social relationships within the chimpanzee community and, along with the other patterns of agonistic and friendly behavior, it plays its role in structuring chimpanzee society. It is helpful to an understanding of the multifaceted function of aggressive behavior to first look at the role it plays in infant socialization. Much of the early behavior of an infant is shaped

Table 12.6   Observed attacks on mothers of newborn infants by Passion[a] and Pom.

Date	Description of attack
August 1975	*Mother:* Gilka *Infant:* Three-week-old female *Duration of fight:* One minute *Injuries sustained by mother:* None *Attack sequence:* Gilka sits in camp. Passion and her family arrive. Passion after a few moments suddenly charges at Gilka, who flees, screaming loudly. Passion chases for about 60 meters, catches, and attacks her. Gilka clings tightly to her infant, but when Pom runs up to join Passion, Gilka suddenly lets go. Passion puts baby into ventral position, chases Gilka away, sits down, and deliberately bites into frontal bones of baby, killing it instantly. Gilka returns cautiously, gives waa-bark, then leaves. Observers: H. Matama and E. Tsolo
October 1976	*Mother:* Gilka *Infant:* Three-week-old male *Duration of fight:* Two to three minutes *Injuries sustained by mother:* Lacerations from biting on both hands and one foot, and on brow *Attack sequence:* Pom approaches Gilka and lies near her. Five minutes later Passion approaches. Pom moves toward Passion, then both charge Gilka, who flees, screaming. Pom runs ahead and cuts her off. Passion seizes Gilka and flings her to ground. Gilka crouches over infant. Pom attacks Gilka, while Passion holds onto and chews infant and Gilka hits at and tries to bite Passion. Passion then bites Gilka's brow, which bleeds. Attack continues. Passion bites Gilka's hand hard. Aggressors turn Gilka onto her back, Pom seizes infant and runs. She kills him the same way Passion killed infant (above). Gilka races after Pom and grabs at infant; Passion follows and attacks, biting Gilka's hand and foot. Gilka, bleeding badly, continues to fight. Passion and Pom travel off fast. Gilka follows a short way, then gives up. Observers: G. Kipuyo and E. Mpongo
November 1976	*Mother:* Melissa *Infant:* Three-week-old female *Duration of fight:* Nine to ten minutes *Injuries sustained by mother:* Numerous gashes with much bleeding and subsequent swelling of face; very deep gash on rump, which penetrated rectum (feces oozed out of wound); deep lacerations of both hands. Walked bipedally the following day. Condition very poor for ten days; healed in about a month *Attack sequence:* See Word Picture 12.1 Observers: R. Bambanganya and L. Lukumay

by mild aggressive responses to inappropriate behavior—first by the mother, subsequently by other members of the community. As the child gets older, aggressive punishment becomes more severe. Attacks, especially from adults, are extremely distressing to infants and juveniles and because, as Scott (1958) points out, fear is easily learned and hard to extinguish, incidents of this sort rapidly teach youngsters which behaviors should be avoided, at least in the presence of which individuals.

A most important function of maternal aggression is to promote independence. A three-year-old infant is allowed to suckle, to ride on his mother's back, and to share her food, but during the fourth and fifth years maternal intolerance increases and the child is gradually forced to walk by himself and to feed at a greater distance from her. In some mother-child pairs weaning is accompanied by frequent aggressive incidents (Clark, 1977).

Table 12.6 (continued)

Date	Description of attack
August 1977	*Mother:* Little Bee *Infant:* One-week-old male (Tubi, survived) *Duration of fight:* One to two minutes *Injuries sustained by mother:* Little Bee next seen after one week; no obvious wounds *Attack sequence:* Pom and Prof climb up to Little Bee, who is in palm tree. Pom touches and sniffs infant. Pom repeatedly looks down at Passion, who sits below. Little Bee becomes uneasy, moves out along frond, uttering tiny squeaks. Five minutes after arriving, Passion suddenly rushes up. I yell at her but she pays no attention. She and Pom leap after Little Bee, who is screaming and fleeing through the branches. Little Bee leaps wildly into the next tree. Passion and Pom race after her, and a fight ensues out of sight high in the tree. During this Little Bee escapes. Passion and Pom search for nearly an hour before giving up and moving away. Observer: J. Goodall
May 1978	*Mother:* Miff *Infant:* Two-week-old female (Mo, survived) *Duration of fight:* Four to five minutes *Injuries sustained by mother:* Gashes on brow and upper lip *Attack sequence:* Passion, with four-month-old infant clinging ventrally, arrives as Miff sits. Prof approaches and Miff presents to him. Two minutes later Passion and Pom approach Miff who pant-grunts, presents, then moves away. Passion races after her and grabs at baby, but Miff swings around, screaming, hits toward Passion, then races on. (Pom does not help; she is pregnant, also intimidated by the field assistants, who threaten her.) Passion chases Miff up tall tree. They grapple and Miff bites Passion's hands. Both fall. Miff rushes up again, with Passion following. Again they grapple and again Passion falls. She climbs up and attacks a third time, then gives up and leaves. Observers: E. Mpongo, H. Matama, and R. Bambanganya

a. Passion never sustained any wounds, despite the fact that Gilka, Melissa, and Miff were all seen to bite her.

Aggression, as we have seen, compels attention, and youngsters watch fights and displays with interest. Perhaps to some extent this underlies the fascination shown by young males in the behavior of adult males (Goodall, 1971). When these youngsters first leave their mothers, it is the big males with whom they choose to associate, despite the inevitable risk that they will become the targets of male aggression. An early-adolescent male, on occasion, will actually leave his mother to travel with a male who has just attacked him severely.

The child assimilates a great deal about aggression (and social dominance) during play sessions. As he (or she) gets older, play is increasingly likely to become very rough and, in fact, often ends in aggression. The child thus not only learns facts about the strengths and weaknesses of his playmates and the dominance position of his own mother in relation to theirs, but he also develops skills in fighting, bluffing, and forming alliances. These skills, once they are mastered, will enable him to compete in a society in which the ability to defend his "rights" and raise his status by fighting may be important. For the young male, successful domination of a female, which is accomplished by means of displays and fighting, functions at least sometimes to ensure that she will follow him in consortships. And if she refuses, he will display at her and sometimes attack her severely until she does. Here, as when the male hamadryas herds his females with the *neck bite* (Kummer, 1968), aggression serves to reduce the distance between individuals. The pair may then remain together, with no further aggression, for periods of several weeks. Mild fights and displays directed toward young cycling females from other communities function to recruit them into the male's own social group.

The rank of a female, as we have seen, may be of crucial importance to the future status of her child, whether male or female; the offspring of high-ranking females have a good chance of becoming high ranking themselves—and it is the more aggressive females who have the higher status.

Male-male aggression, at least as it operates within the community, functions primarily to establish and maintain a dominance hierarchy. But once a reasonably stable ranking order has been worked out, the benefits are obvious—not only to the more dominant but also, as Lack (1966) has pointed out, to the subordinate. Knowing that it is useless for him to compete for resources with those who rank higher, the lower-ranking individual does not waste time and effort, or run the risk of injury, but goes off to try his luck elsewhere. Chimpanzees are highly excitable, and especially when a number are together, trivial disputes may erupt into aggressive interactions and displaying; but as long as the hierarchy is stable, serious fighting seldom occurs. The minor attacks and the wild charging displays with all their elements of aggression function to relieve social tensions and thus to minimize the physiologically undesirable components of stress.

Often the role played by aggression is not immediately obvious, simply because it operates at second hand through the dominance hierarchy it helped to create. Thus when many individuals climb into a large fruiting tree, there may be no overt aggression, yet the dominant individuals will be seen to occupy the best feeding sites, while subordinates move about picking one fruit here and another there, or simply waiting on the ground (Wrangham, 1979).

And so, gradually, the complex functions of aggression are disclosed. But they reveal only half the picture. We cannot hope to understand the role played by aggression in the ordering of chimpanzee society without consideration of the equally powerful forces of social attraction. It is the interplay between those two opposing forces, aggressive hostility and punishment on the one side and close and enduring friendly bonds on the other, which has led to the unique social organization that we label a community.

13. Friendly Behavior

(H. van Lawick)

July 1974 Observer Eslom Mpongo followed Madam Bee as she headed slowly for Kahama Stream. Her two daughters, young adult Little Bee and adolescent Honey Bee, were far ahead along the trail that led to a stand of *Saba florida* vines with their large, lemonlike fruits. Madam Bee looked old and sick. Her arm, paralyzed by polio, dragged and several half-healed wounds were visible on her back, head, and one leg. It was very hot that summer, and food was relatively scarce so that the chimpanzees sometimes had to travel considerable distances from one feeding place to the next. Again and again Madam Bee stopped to rest. When soft food calls indicated that the two young females had arrived at the food site, Madam Bee moved a little faster; but when she got there, it seemed that she was too tired or weak to climb. She looked up at her daughters, then lay on the ground and watched as they moved about, searching for ripe fruits. After about ten minutes Little Bee climbed down. She carried one of the fruits by its stem in her mouth and had a second in one hand. As she reached the ground, Madam Bee gave a few soft grunts. Little Bee approached, also grunting, and placed the fruit from her hand on the ground beside her mother. She then sat nearby and the two females ate together.

Chimpanzees, as we have seen, are capable of very violent behavior. Aggression, particularly in its more extreme form, is vivid and attention catching, and it is easy to get the impression that chimpanzees are more aggressive than they really are. In actuality, peaceful interactions are far more frequent than aggressive ones; mild threatening gestures are more common than vigorous ones; threats per se occur much more often than fights; and serious, wounding fights are very rare compared to brief, relatively mild ones. In this chapter we shall investigate the peaceful side of chimpanzee nature—the mechanisms that serve to maintain or restore social harmony and promote cohesion among community members.

It is virtually impossible to overemphasize the importance of amicable physical contact in chimpanzee relations. Stemming from long years of dependency, when the mother's embrace soothes the hurts and fears of childhood, chimpanzees have a deep-rooted need for contact with a friendly conspecific in times of physical or emotional stress. That such contact is psychologically beneficial has been demonstrated. Infant chimpanzees tested in a laboratory were considerably less distressed by mild electric shocks when held by a familiar caretaker (Mason, 1965). In the same study social grooming served to reduce levels of arousal in both the groomer and the individual groomed. Indeed, social grooming, providing as it does the opportunity for long periods of relaxed, harmonious contact among adults, is probably the single most significant element in the repertoire of friendly behaviors and will be discussed separately in the next chapter.

In Fear or Excitement

When a chimpanzee is unexpectedly frightened or startled—by an unaccustomed sound, by a fight breaking out near him, by an unusual sight—he typically seeks physical contact with a companion by touching, embracing, mounting, or kissing. When David Greybeard first saw his reflection in a mirror, he grinned widely in fear, his hair bristled, and he abruptly reached out to four-year-old Fifi, who was nearby, and drew her into a close embrace. Slowly the grin left his face and his hair sleeked.

Similar contact-seeking behavior occurs during high excitement. When a kill has just been made, the onlookers often embrace, kiss, and pat one another, screaming loudly, before rushing toward the successful hunter to try to acquire a share of the meat. When two females were unexpectedly confronted with an enormous pile of bananas, they flung their arms around each other's neck and pressed their open mouths to each other's shoulders while uttering excited

Flo (*left*) and Passion embrace during banana-feeding excitement. (H. van Lawick)

Excited anticipation as Faben (*foreground*) and Charlie wait for bananas at the trench. Both are squeaking, as is Dé (*on trench*), who also touches his penis. (B. Gray)

Leakey and Hugo embrace and kiss in excitement as Mike starts to feed on the infant baboon he has just caught. (B. Gray)

In response to calls from another group, Mandy and adolescent Jomeo pant-hoot. Their physical contact reduces the level of arousal. (H. van Lawick)

Hugh and Charlie during banana excitement. (B. Gray)

food calls before they took a single fruit. In the days of frequent banana feeding the chimpanzees sometimes became very tense and excited in anticipation of the opening of a box, and often embraced, mounted, pressed open mouths on each other, and patted one another vigorously, especially on the mouth and scrotum. Faben usually turned his rump toward a companion in such a context, then, as the other reached to touch him, made thrusting movements, bouncing his genitals up and down against his comrade's palm. His brother, Figan, sometimes held his *own* genitals in moments of sudden fear or excitement, particularly if no other chimpanzee was close by. As he grew older he self-touched less frequently, but still did so in moments of extreme tension (during a boundary patrol, for example) throughout his life. Other males occasionally touched their own penes briefly when excited.

Submission and Reassurance

Submissive behaviors comprise a variety of nonaggressive postures, gestures, and calls that are directed up the hierarchy, from a low-ranking to a higher-ranking chimpanzee. They occur not only when the subordinate has been threatened or attacked by the other, but even when he (or she) approaches or is approached by, passes or is passed by, a higher-ranking individual who has *not* shown overt signs of aggression toward the gesturer. The patterns of submission (detailed in Goodall, 1968a) include *presenting* or turning the rump toward, most commonly female to male; *crouching* and *bowing; bobbing* (facing the dominant while extending, then flexing, the arms); *kissing* (with lips pouted forward to barely touch the other or with mouth wide open and pressed against the other); *embracing; mounting* (a close embrace from the rear sometimes accompanied by thrusting pelvic movements); and a variety of hand and arm movements, such as *reaching toward* (usually with palm up), offering the back of the wrist, touching or grooming some part of the body, and *bending away* (the subordinate flexes elbow and wrist, at the same time drawing one arm close to the body and leaning slightly away from the other individual). These postures and gestures are accompanied by pant-grunts, squeaks, whimpers, or screams, depending on the level of distress of the subordinate (which will be influenced, among other things, by the level of arousal of the dominant).

Pant-grunts, bending away, and touching are common responses to the approach or passing of a relaxed high-ranking individual; if he shows signs of arousal (such as hair erection), the submissive response will probably be more intense—even when the subordinate is not, in fact, being threatened. Figan as an early adolescent invariably rushed to pant-bark and bob frenziedly whenever the high-ranking Hugh came anywhere near him. Many adolescent males since that time have shown this kind of tense, submissive behavior toward particular high-ranking males, sometimes but not always toward the current alpha. Fully mature males sometimes behave in the same way; Evered did during the hostile period of his relationship with Figan (Chapter 8).

The higher-ranking individual, although he (or she) may ignore submissive behaviors of this sort, or may even threaten or attack the subordinate, is most likely to respond by touching, patting, kissing, embracing, or grooming the other. Any of these friendly responses tend to calm and relax the subordinate and thus serve to reestablish harmonious relations between the individuals concerned.

Sometimes, as we have seen, a chimpanzee who has been threatened or attacked seeks reassurance contact not from the aggressor but from a third individual, using postures and gestures of the submissive repertoire. The solicited individual may offer reassurance in the same way as an aggressor. An adult male challenged by another male often runs, screaming, to a third and establishes contact with him. Often both will then scream, embrace, mount, or groom each other while looking toward the original aggressor. This, as described earlier, is how a victim tries to enlist the help of an ally. There are occasions, however, when it seems that the primary goal is to establish reassurance contact—as when fourteen-year-old Figan, after being attacked by a rival, went to hold hands with his mother. Or when Melissa, after a tense interaction with a high-ranking male, approached and held her hand toward old Mr. McGregor (perhaps her uncle).

Occasionally, even when a conflict appears to have been resolved as a result of reassurance contact and the subordinate has relaxed, the underlying resentment may not have been appeased. The adolescent Evered, for example, attacked Pooch during banana feeding. She screamed and presented, Evered reached to touch and briefly groom her, and the two then ate side by side. All was peaceful for a few minutes until the old male Huxley, Pooch's close friend and supporter, arrived with bristling hair. The moment she saw him, Pooch, made bold by the presence of her ally, turned and with fierce waa-barks attacked Evered.

Despite the fact that a subordinate may harbor a grudge for a while, there is no doubt that reassurance contact plays a major role in conflict resolution and, along with social grooming, enables chimpanzees to relate to one another in an extremely relaxed way most of the time. Thus Mike, soon after attacking old Flo very severely (causing her to bleed from two wounds), subsequently responded to her submissive gestures by embracing and patting her until she stopped screaming. He then groomed her for ten minutes, after which Flo was so relaxed that she actually slept with her head almost on Mike's foot.

Sometimes an individual mediates an agonistic interaction between two others. In a typical example, Frodo and the younger Tubi are wrestling. Frodo gets rough and Tubi whimpers, then screams. Enraged, he hurls himself at Frodo and bites hard. Now Frodo screams. Little Bee, with a nervous grin, reaches to calm Frodo. As she does so, she glances at Frodo's mother, Fifi, higher ranking than herself (and very likely to intervene on her son's behalf should the quarrel continue). On another occasion Little Bee may ignore Frodo and reach to touch Fifi, in a direct attempt to appease her. I described in Chapter 4 how the nervous mother Olly often directed submissive behavior toward the adult males with whom her infant was playing, even though they showed absolutely no signs of aggression.

Submissive behavior directed *up* the dominance hierarchy can be readily understood when considered in relation to the deep-seated need for reassurance contact experienced by an emotionally or physically distressed chimpanzee. When such behavior (essentially an

Melissa with her newborn infant, Goblin. After a tense encounter with a young male, she now approaches and reaches toward her probable sibling, Humphrey. Mr. McGregor, also thought to be a relative, is present, as is David Greybeard.

Alpha male Mike reassures Melissa as she approaches with submissive squeaks. (H. van Lawick)

Faben seeks reassurance from Goliath after being attacked by Mike. Leakey is grooming Goliath. (H. van Lawick)

Figan, after being attacked by Goliath, approaches him, screaming, with a submissive crouch. Goliath calms him with a reassurance gesture. Figan's screams change to whimpers and he begins to relax. (H. van Lawick)

appeal for friendly contact) is ignored by a nearby higher-ranking male who has just threatened or attacked him, an adolescent may continue to crouch, screaming or whimpering, looking toward the aggressor and sometimes holding out his hand in a begging gesture. He may even throw a tantrum.

This need for contact with the aggressor sometimes results in a conflict situation (Hinde, 1966): a subordinate who has been attacked may approach the aggressor in a series of zigzags as he alternately moves toward, then turns away from, his intensely desired, intensely feared goal. Figan once followed Goliath in this zigzag fashion for at least 50 meters after a particularly savage attack. When Goliath finally sat, still bristling, Figan stopped and turned as if to flee; then gradually, screaming loudly all the time, he backed toward the big male, whom he watched over his shoulder. When he finally arrived within arm's reach he remained motionless, crouched flat to the ground and screaming. Goliath patted him gently for over half a minute until gradually Figan quieted and relaxed. There are times when the fearful subordinate stops *beyond* arm's reach, then leans forward and extends a hand to the other—who by reaching toward the supplicant is just able to touch, pat, or hold the outstretched hand. Occasionally, of course, a subordinate is unable to overcome his fear; he may approach to within 5 meters or so, then rush away, or stop and move no closer.

Reassurance behavior may be triggered by the chimpanzee's apparent dislike of noisy submissive acts. A subordinate who seeks reassurance after aggression usually screams or squeaks loudly while reaching toward or crouching in front of a superior. Some form of friendly contact serves to soothe him. The following is a typical example of reassurance behavior that functions to restore a harmonious atmosphere. Goliath, second-ranking male at the time, sat eating a large pile of bananas. Adolescent male Pepe approached to within arm's reach, then crouched, whimpering and squeaking, while looking at the older male and three times reaching his hand toward him. After a few moments Goliath began to pat Pepe on the face and head. When

Young adult male Pepe approaches Goliath with pant-grunts and wrist bend. Later he sits close to the older male and watches him feed. (H. van Lawick)

he stopped, Pepe reached slowly for a banana. At the last moment he jerked his hand back and began to scream, although Goliath had made no threat. After thirty seconds the high-ranking male again began to pat Pepe on the head, and fifteen or twenty seconds later Pepe quieted. Goliath stopped his patting and Pepe, watching the older male intently, picked up a few fruits and moved away with them. Both sat peacefully feeding about 8 meters apart. Goliath, by means of reassurance gestures, had brought about a more peaceful atmosphere, one that he undoubtedly preferred. Thus his behavior, which functioned to calm Pepe, also served his own best interests.

While reassurance behavior may in many cases be selfishly motivated, we must not forget that the main patterns of reassurance—patting, embracing, kissing—are almost certainly derived from maternal gestures, which evolved to comfort and calm the child. In like manner, the emotions aroused in the dominant individual by the submissive actions of a subordinate—who is, after all, behaving much like a fearful child—may derive from maternal concern for offspring. The postures and gestures themselves, many of which are inborn characteristics, are channeled and reinforced during early learning experience in the social group. I shall return to this point a bit later.

Of course, there are many occasions at Gombe when the victims of aggression do not seek reassurance contact from the aggressor or from any other individual. Either they move some distance away from the aggressor and gradually calm down, or they leave the social group and go off by themselves.

Table 13.1 lists the number of aggressive incidents between adults that were observed in camp during three different periods, two months only during 1966, and all months of 1978 and 1979. Although much more aggression was observed in 1966 (as a result of heavy banana feeding, large parties in camp, and many more adult males), a similar percentage of aggressive incidents led to reassurance in all three periods.

The table also shows that when the aggressor was an adult male, both male and female subordinates were equally likely to seek reassurance from him, and that male aggressors responded positively to about half of the submissive gestures directed toward them. Aggression directed by adult females toward other adult females was less likely to lead to reassurance, and females were much less likely to seek reassurance from female aggressors than from males. When they did, it was often granted. These differences between the sexes reflect the lack of a clear-cut dominance hierarchy among the Gombe females, to be discussed in Chapter 15.

De Waal and Roosmalen (1979), working with the chimpanzees of the Arnhem colony, made a detailed study of submissive and reassurance behavior after conflict, labeling it *reconciliation*. Here, too, reassurance was sometimes sought from third parties, but opponents preferentially sought body contact with each other. The gestures used by the Arnhem chimpanzees were similar to those observed at Gombe, but it was obviously still more important in the captive situation (where the chimpanzees could not escape from one another) that quarrels be speedily settled. Tense relationships prevailed throughout the group until opponents had become at least temporarily reconciled.

Tense reunion between Passion, who grins and squeaks, and Faben. Three-year-old Pom watches intently. (P. McGinnis)

Table 13.1 Episodes in which aggression (between adult males, between males and females, and between females) was followed by reassurance. Also shown is the frequency with which reassurance was sought after an aggressive incident, and the degree of success of such appeals. Data are from three separate periods, two months in 1966 and two years, 1978 and 1979.

Type of aggression	Number of aggressive incidents	Percentage reassurance given	Percentage reassurance sought	Percentage of appeals successful
Male to male				
1966	41	31.5	24.5	60.0
1978	58	15.5	5.2	66.5
1979	104	21.0	15.5	62.5
Male to female				
1966	44	20.0	20.0	50.0
1978	105	27.5	21.0	41.0
1979	99	27.0	27.0	44.5
Female to female				
1966	19	5.5	5.5	100.0
1978	38	8.0	8.0	0
1979	33	3.0	0	0

There were even occasions when adult females assisted rival males to reestablish friendly contact, as I shall describe in Chapter 19.

The methods used for conflict resolution and control of aggression in other captive groups are impressive also. At Lion Country Safaris in Florida, for example, when the alpha was working himself up for an aggressive display, the third-ranking male regularly approached and walked backward in front of him, in an upright position, with a play face. Although the alpha sometimes ignored him, and occasionally attacked him, the strategy often led to a play session between the two males (Gale and Cool, 1971). In two other groups, originally housed in small cages (at the Washington Park and Zurich zoos), the dominant males showed what to me seemed incredible self-restraint in situations that without question would have led to fighting at Gombe. Social mechanisms of this sort for conflict resolution and inhibition of violent aggressive patterns constitute an extremely important area for future research.

Greeting

When chimpanzees meet after a period of separation, there is often some kind of interaction between them that we may define as a *greeting*. The exchange may be friendly or unfriendly; interactions in this context range from a grunt and a quick glance at a newcomer, through a delighted embrace, to a vicious attack. The type and intensity of greeting is determined by many social and environmental variables, the most important of which pertain to the kind of relationship between the individuals concerned, their moods, and to some extent the length of time for which they have been separated. Two females who seldom associate may, if they meet in a large fruiting tree, barely acknowledge each other's presence (indeed, they sometimes appear to ignore each other completely); but if one of them arrives and joins a meat-eating cluster where the other is begging, they may fight. A mother and her adult daughter who have been foraging separately for a few hours may merely look at each other and give a few grunts when they meet; but if they have been separated for a week or more, they are likely to fling their arms around each other with grunts or little screams of excitement, then settle down for a session of social grooming. A female may approach a high-ranking male with soft pant-grunts and reach to touch him if he is relaxed and peaceful; but if he has erect hair and has, for instance, recently been involved in an aggressive incident, she may rush halfway toward him with uneasy

Fifi greets Hugo with a kiss. (H. van Lawick)

Mr. Worzle shows the sitting hunch as Flo approaches with apprehensive pant-grunts. (H. van Lawick)

Hugo stamps on the old female Marina during a reunion. Marina is crouched protectively over her three-year-old infant, Merlin. As Marina sits, screaming, Hugo immediately embraces her. Merlin's foot is visible as he continues to cling to his mother.

pant-barks, then lose her nerve and race up a tree. An adult male may reach out and idly touch the genital area of an anestrous female who presents to him, inspect and groom her vigorously if she is showing some signs of genital swelling, or copulate with her if she is in estrus. The nature of a greeting may be determined by interactions that took place when the two individuals were previously together; an adolescent male who left a party after being attacked by a high-ranking individual is likely to be especially fearful of him when next they meet. And, as described in Chapter 12, many of the threats or attacks that occur during reunions are the result of redirected aggression, the recipient being merely a scapegoat.

One point should be made here. Quite often, during an excited reunion sequence (when, say, two parties of chimpanzees meet at a food source) low-ranking individuals are attacked during charging displays. These incidents, accompanied by bristling hair, stamping, hitting, and piercing screams, may be interpreted by the chimpanzee actors in a rather different way than they are by human observers. In our own society some greetings, particularly between active and healthy adolescents, can be quite boisterous, and we all know the too-hearty thump on the back. The greeting illustrated above shows adult male Hugo stamping vigorously on the frail back of old Marina, who crouches over her infant, screaming at the top of her lungs. Having hit and stamped a few times, and without breaking contact, Hugo embraces and then grooms his "victim." I am not suggesting that Marina enjoyed being attacked—merely that it may not have seemed as bad to her as it did to the human onlookers.

Greeting behavior, made up as it is of elements of submissive, aggressive, reassurance, and sexual behavior, serves to reestablish the nature of existing relationships, in that the subordinate typically indicates with clear-cut submissive signals his recognition of the higher rank of the other. Tension generated by reunion with a social competitor (unless of an extremely hostile nature) is decreased by reassurance contact, and further relaxation is made possible by the long sessions of social grooming that may follow reunion of adult males.

<table>
<tr><td>

Tolerance

</td><td>

Mothers are, for the most part, extremely patient with their infants, although some are more long-suffering than others. Thus one infant may be permitted to clamber all over his mother, grabbing onto her cheek hair or her ears, whereas another infant will be firmly prevented from taking such liberties. Sometimes an infant ignores the obvious signals indicating that his mother is ready to move on to a new location. Some mothers resign themselves to waiting patiently until the game, termite-fishing session, or whatever is finished. Others are less lenient and may seize the child and drag him away, or simply set off without him. And some mothers are far more tolerant than others during the long months of weaning.

Small infants are usually permitted to climb over adult males who are resting, to sit close beside them as they feed, and even to share their food. Often it seems that a male cannot resist reaching out to draw an infant into a close embrace, to pat him, or to initiate gentle play. At no time does he demonstrate his tolerance of infants more strikingly than in the sexual context. During adult copulations infants frequently interfere, rushing up and reaching to touch or push at the male's face, particularly his mouth, and sometimes jumping onto the female's back as they do so. An infant may be quite calm as he interferes, but if his mother is involved he often becomes agitated and may actually hit at the male during copulation. Juveniles and even young adolescents sometimes interfere when their mother is in estrus. Sometimes two, occasionally three, youngsters interfere at the same time, so that the male can barely see his sexual partner through the tangle of intervening bodies. Yet almost always he tries to ignore it all. Turning his head from one side to the other in an (often vain) attempt to avoid his tormentors, he gets on with the business in hand. The photograph on page 468 shows such a situation. (It is perhaps fortunate for the infants that the whole performance is over in a few seconds, before the male's patience can be too sorely tried.)

Occasionally a male threatens an infant who interferes, with an arm-raise gesture and soft bark. He may be slightly more aggressive toward an older child, hitting out at him during the copulation or giving him a few thumps afterward. Goblin developed what could almost be described as a phobia in this context;* his aggressive reaction serves to highlight the extraordinary degree of patience shown by the other males.

Usually adult males are highly tolerant of the noisy behavior of ultrasubmissive adolescents. Even when a youngster actually impedes

</td></tr>
</table>

* At first Goblin merely threatened or mildly hit at infants who approached when he was copulating. But by the time he was sixteen years old, his punishments were unusually severe: that year (1980) he attacked nine-year-old Freud (who interfered when Goblin copulated with his mother) so severely that he apparently broke a bone in Freud's foot or ankle. And the following year it seemed that Goblin was unable to ejaculate if an infant came anywhere near him during copulation; he left the female, chased the youngster away, then returned and began to court the female all over again.

Pom greets Evered.

the progress of a high-ranking male, moving backward in his line of travel with exaggerated bobbing and noisy pant-barks, his senior seldom reacts with more than a slight arm-raise threat. Once Hugh was actually forced to move in a zigzag fashion in order to proceed at all when Figan behaved in this way. Pom, from late adolescence on, had a very tense relationship with adult males. Again and again as she approached to greet a senior male, she crouched in front of him with frenzied pant-barks and made curious dabbing, punching gestures toward his face. Generally the male merely turned his head away.

Males are often very lenient also in the context of feeding. It is not uncommon to see an adult male share a fruiting cluster of palm nuts with one or even two other adults of either sex. Of course, to a large extent, the degree of indulgence the male shows in this situation depends on his relationship with the individual concerned, and those who approach are likely to be those with whom he is friendly. The others will seldom try to join him. The tolerance of a feeding male to the very close approach and/or persistent begging of others is, as we shall see, remarkable. In contrast, females in the feeding context are usually highly intolerant of the close approach of subordinates, including their own offspring.

Social Play

Play is the hallmark of childhood and will be described fully elsewhere. With its component of friendly physical contact, play undoubtedly is important in the development of friendships that may persist throughout life. Figan and Pooch, for example, often played together as juveniles. During their late adolescence they went off together on what was, for each of them, almost certainly the first consortship. Before Pooch disappeared in 1968, she went on five other inferred consortships; four of these were with Figan. Often the two of them associated peacefully between Pooch's periods of estrus.

A chimpanzee infant has his first experience of social play from his mother as, very gently, she tickles him with her fingers or with little nibbling, nuzzling movements of her jaws. Initially these bouts are brief, but by the time the infant is six months old and begins to respond to her with play face and laughing, the bouts become longer. Mother-offspring play is common throughout infancy, and some females frequently play with juvenile, adolescent, or even adult offspring. Flo, aged at least forty years, often joined in games with her offspring when, for example, they were chasing around a tree—even if it was simply to reach out and grab at their feet as they passed her. Flo's daughter Fifi plays with her family in a similar way. Other mothers (such as Passion and Melissa) are far less playful (Goodall, 1967).

When an infant is between three and five months of age, the mother usually allows other youngsters to approach and initiate mild play: often the first overtures of this sort are made by an elder sibling. Gradually, as the infant matures, he plays with more individuals and sessions become longer and more vigorous. By the time the child is three years old, play quite often ends in screaming and aggression as one playmate gets too rough and hurts the other. The frequency of play peaks between the ages of two and four years, then decreases during weaning (Goodall, 1968b; Clark, 1977).

When two infants meet after a separation, they typically embrace, and this often leads directly to tickling and mild wrestling. A play session between two older individuals may commence when one of them approaches with the *play walk*—back rounded, head slightly bent down and pulled back between the shoulders, and movements exaggerated. One of the two then reaches to tickle or thump the other. A child attempting to initiate play with an adult male sometimes approaches, then sits, facing away and leaning slightly forward in the *back present*. This is the signal often used to initiate interspecific play, with young baboons and also with humans in the days when four of the youngsters (Fifi, Figan, Pooch, and Gigi) occasionally played with us. The invitation is made more specific when the chimpanzee reaches back and makes scratching-tickling movements with the fingers on his shoulders. An adult male may approach a youngster with the play walk, then kick back with one foot as he moves past: sometimes this leads to a chase around a clump of vegetation. Play between adults frequently starts with *finger wrestling*, as one reaches to the hand or foot of the chosen partner and makes gentle pulling, tickling motions.

At the start of a session between two youngsters, or during a pause, one of them may pick a play twig, approach the other and, brandishing this in the air or holding it in his mouth, gambol past his playmate, who typically gives chase and makes a grab for the twig.

Play, while it gradually becomes much less frequent during adolescence, does persist into adulthood. The first time I saw adult males playing together was in 1961. David Greybeard and Goliath lay stretched out in the shade. Suddenly Goliath reached over to David's hand and began to tickle it. David responded and a bout of finger wrestling began. For the next two minutes, apart from the movement of their hands, both males lay quite still, David gazing into the bushes, Goliath looking up through the branches of the tree. Suddenly David rolled over and dug his fingers into Goliath's groin. Goliath reciprocated, tickling David between shoulder and neck; soon both were

Freud uses the back present.
(L. Goldman)

Figan, chewing a wadge, grabs at Hugo's foot as the adult male kicks back in play. Figan later chased Hugo around the tree. (H. van Lawick)

Fanni watches as Fifi and Satan play.

laughing. After three minutes of this they got up and Goliath chased David five times around the trunk of a tree. Then, as if by mutual consent, they stopped and began a grooming session that lasted for twenty-one minutes.

Play between adult males and females is almost always initiated by the male. He may use the finger-wrestling technique, or he may make tickling movements under the female's chin. Humphrey, perhaps more aggressive toward females than any other male, was also one of the most playful. Often, however, when a female responded to his overtures, he became rough; if she pulled away or squeaked in fear, he was apt to attack her. Many females, not surprisingly, were reluctant to accept his invitations to play. Flo, however, almost always reciprocated. She played quite often with other adult males too. Perhaps because of this, Fifi as a juvenile often romped with the big males; today, as an adult, she is still a playful female.

Adult females only rarely play with one another, and many have never been seen to do so. Sterile Gigi does so more often than the others, perhaps because she has no family of her own. Her most frequent playmates over the years have been Fifi and Patti. Once all three of these females played together; Freud, six years old at the time, tried to join in but they were more interested in grabbing and tickling one another's feet.

An analysis of play data during one year (1967) showed that females with offspring played more than other adults. Some of the adult males were observed playing—mostly with infants or young juveniles—quite regularly, others very rarely indeed. The more playful males went through phases when they played relatively frequently, while at other times they did so far less or not at all. Humphrey, for example, was seen playing (with individuals of all age-sex classes) on fifty-four occasions during January and early February, but for the rest of the year never more than five times in a given month and for five months not at all. Today, now that the chimpanzees no longer spend hours hanging about camp waiting for bananas, adult play is observed less often. Nevertheless, when food is plentiful and particularly when the chimpanzees come together in large gatherings, adult play is often seen. Word Picture 13.1 describes one such occasion.

Food Sharing

A chimpanzee infant younger than two and a half to three years is usually allowed to share his mother's food with little or no resistance on her part. He may take bites from items in her hand, take chewed food directly from her mouth with his lips, or take (or be given) pieces in his hand. As he gets older, his mother becomes increasingly reluctant to share with him and this often results in long bouts of whimpering, sometimes ending in a tantrum, if the infant does not get his way. Mothers of youngsters in their fourth and fifth years generally continue to share foods that are difficult to process, but refuse requests for items the child can acquire for himself (Silk, 1978). There is, however, a distinct difference in the generosity of the various mothers. Some consistently refuse to share with a three-year-old even those foods that are virtually impossible for the child to obtain otherwise, such as strychnos fruits (which infants younger than four cannot crack open), whereas other mothers share strychnos with their five- or six-year-old offspring.

At Gombe adult chimpanzees rarely share plant foods. Excluding bananas, Wrangham noted only five instances during his eighteen-month study; four times this involved strychnos fruits. McGrew (1975) analyzed the sharing of bananas: 86 percent of the observed cases involved family members, almost always mothers sharing with infants. Most of the remaining instances were adult males sharing with females (and, occasionally, their infants).

By contrast, meat, which is not only highly prized but also available in relatively small amounts, is almost always shared (unless one individual manages to escape with a whole carcass). A possessor shares meat for a variety of reasons. For one thing, as Wrangham (1975) points out, the individual may have more than he needs, particularly

Dove holds bananas away from her son, a year and a half old.

(H. van Lawick)

Play Behavior at the Height of the Landolphia *Season*

16 January 1978. Observer: J. Goodall.
Many of the Kasakela chimpanzees and I with them had spent the morning in a large gathering, moving from one patch of yellow and red fruits to the next. At midday the adults lay down to rest or settled to groom, while the youngsters played. I had completed a five-hour infant record and was sitting near Humphrey. Presently, at 1315, eleven-year-old Atlas approached and began gathering a huge armful of dead leaves. These he dumped on the ground close to Humphrey, then tried to stand on his head in the middle of the pile. Perhaps this was an invitation to play; at any rate, Humphrey grabbed at Atlas, then set off around a treetrunk, his companion in hot pursuit. Suddenly Humphrey stopped and kicked back forcefully, so that Atlas quite literally flew through the air—but rushed back for more play. From time to time the chasing stopped and the two wrestled and tickled. Both were laughing loudly; indeed, Atlas sounded quite hysterical. At 1320 Gigi arrived and sat watching (which caused a break in play as Atlas raced over, copulated with her, then ran back to Humphrey). Several times during the session not only Atlas, but also thirty-two-year-old Humphrey, turned somersaults. The game ended at 1327 when Humphrey moved off to investigate some screaming just out of sight.

I followed Humphrey and came upon Fifi and her family. She was sprawled out, in her usual indolent fashion, nursing Frodo. Nearby, Figan groomed Pallas. At 1334 Freud (seven years old) came running from one of the juvenile play sessions and back-presented to Figan, looking over his shoulder with a play face. Figan, not usually a playful individual, instantly stopped grooming and seized Freud's foot. This was the start of a twenty-three-minute play session during which Freud and Figan both turned somersaults, wrestled, tickled, and chased each other around Fifi. From time to time infant Frodo made little daring sorties to grab at a flailing arm or leg, then dashed back to reestablish contact with his mother. Fifi, from her reclined position, occasionally tickled Freud or Figan, and even nonrelated Pallas joined in once for a few moments. At 1355 Figan began cantering in a wide circle around two trees and a clump of bushes, kicking back at Freud, who was close behind. Five, six, seven times they circled, and Fifi, whenever they passed, grabbed at her brother's heels. Abruptly Figan broke off and walked away. Freud followed a short distance, then returned to Fifi and Frodo. Soon after this the whole gathering moved on.

after he has been feeding for a time, and it is not worth his while to keep moving from and/or threatening others who are eager for portions. There are occasions when the solicitations of begging chimpanzees make it all but impossible for the possessor to feed; at the very least they are a source of irritation. If he dispenses pieces of meat, the recipients usually move off, sometimes followed by others. Even if the possessor only relinquishes chewed leaf-meat wadges, social harmony will temporarily be restored and he will be able to eat a few mouthfuls in peace.

Quite often a supplicant may be allowed to share the carcass with the possessor, or to break off a portion. Leaf-meat wadges are frequently deposited in the outstretched hand of a begging individual or are transferred directly from mouth to mouth, usually after the meat

Flo begs for the wadge of meat that Mike is chewing. (B. Gray)

Flo, whimpering, begs from Mike. She is eventually successful, for he allows her to take the wadge directly from his mouth into hers. (B. Gray)

Fifi begs for a wadge from her brother, Figan.

eater has extracted all he wants but sometimes after only a few chews. Occasionally the possessor breaks off a portion of meat and places it in the outstretched hand of a suppliant. On a few occasions possessors gave meat to individuals close by who were not overtly begging, although their intense interest in the food was obvious. This happened most often at the end of long meat-eating sessions, when a male sometimes handed over the entire remains of the carcass to the individual closest to him. Frequently this was a female.

Sometimes after a lengthy and unrewarded session of begging, the suppliant starts to whimper or (especially an infant or juvenile) throws a tantrum. Once, when Goliath (in 1968) had the freshly caught body of a baboon infant, Mr. Worzle begged persistently, following the possessor from branch to branch, hand outstretched and whimpering. When Goliath pushed Worzle's hand away for the eleventh time, the lower-ranking male threw a violent tantrum: he hurled himself backward off the branch, screaming and hitting wildly at the surrounding vegetation. Goliath looked at him and then, with a great effort (using hands, teeth, and one foot), tore his prey in two and handed the entire hindquarters to Worzle.

McGrew (1975) found that some males were more likely to share bananas than others. Similar differences in generosity were obvious in meat eating, but so many variables are involved that it has not been possible to rank individual males precisely. It is clear, however, that in male-to-female sharing Evered scored high (as he did in banana sharing); Figan was particularly responsive to adult males; Humphrey was reluctant to share with males other than Figan; and so on.

Adult females, with the exception of mother-daughter pairs, rarely share with each other. When a female has possession of a carcass (or a large portion of one), other females seldom beg from her; when they do, they are hardly ever rewarded. This response is independent of the amount of meat available. When Gigi had an entire bushbuck fawn, she refused to share with the Miff family, even though they begged repeatedly and threw tantrums before giving up; and Miff herself, having captured an infant colobus, would not allow Nope even the smallest morsel. This sex difference can probably be explained by the fact that a female eats a large portion of meat relatively less often than a male does, so it is more of a treat for her. Moreover, nonrelated adult females in the wild support and help one another far less often in other ways than is the case for males. As we shall see, a female is likely also to refuse a request for grooming that comes from another female.

Food sharing is commonly reported among the members of captive groups. Köhler (1925, p. 255) describes how an individual, in response to begging, "may suddenly gather some fruit together and hold it out to the other or take the banana which he was just going to put in his mouth, break it in half and hand one piece to the other." When, "in the interests of science," young Sultan was for several consecutive days shut up without his supper, he became frenzied in his appeals to old Tschego as she sat nearby, eating her own meal. He whimpered, screamed, stretched his arms toward her, even threw bits of straw in her direction. Eventually she responded by gathering a pile of food and taking it over to him. She fed him in this manner for five succes-

Melissa takes a piece of fruit that Mike holds toward her in response to her whimpering. (H. van Lawick)

sive evenings. She then entered the flat, anestrous stage of her cycle and ignored him—until she began her next swelling, when she once more supplied him with food.

Another anecdote comes from the Arnhem colony. The adult female Oor, feeding from a leafy branch, ignored the whimpering of her begging infant. After a few moments Oor's best friend approached, picked up the branch, broke off a piece for the infant and for herself, then handed the largest part back to Oor (de Waal, 1982). In another captive group (at the Washington Park Zoo) the one adult male, Bill, shared occasionally with all the others (three mothers and their infants, all sired by him) but most often with his only son, Goliath. During feeding Bill routinely took the fruits he preferred from the central pile and moved off with them. Goliath was the only individual allowed to take from this supply. He also took food from his father's mouth and hands. Once a day the chimpanzees were given milk, poured directly into their mouths. Bill sometimes took a large mouthful, retaining it as he moved away. One day Goliath ran up to him holding out his hand. Bill picked up the infant, transferred milk directly into his mouth, then held him ventrally until he swallowed, This became customary, and Bill would repeat the procedure two or three times (from the same mouthful) if Goliath continued to beg. A few days later Goliath's mother shared milk with the infant in the same way (King, Stevens, and Mellen, 1980). Yerkes (1943) observed a similar incident when a female was offered fruit juice from a cup through the bars of her cage. Like Bill, she filled her mouth, then deliberately walked across her cage to that of her neighbor, a female with whom she was on friendly terms, and transferred the juice to her mouth. She returned for more juice and took that also to her friend. The process was repeated until the cup was empty.

In one study of food sharing it was found that chimpanzees tended to share more readily food items that were less valued (Nissen and Crawford, 1936). The authors concluded that a positive response to begging was often a means of getting rid of a so-called noxious stimulus. While this is undoubtedly the explanation for a great many instances of food sharing, it is also true that chimpanzees unquestionably understand the desires of a begging individual; the frequency of the behavior, along with the obvious volitional nature of some giving, provides a sound biological basis for the emergence of altruistic food sharing in our own species.

Helping and Altruism

Altruism, biologically defined, is behavior that benefits another individual at some cost to the altruist. The cost, at times, may be high—an altruist may even lose his life. In any discussion of altruistic behavior it is important to realize that there are two very different ways of measuring this cost. First, the *immediate* cost: a wound, for example, received when helping an ally in a fight. Second, the *ultimate* genetic cost: any decrease in the proportion of the altruist's genes (relative to the genes of competitors) in the gene pool of the future that results from his altruistic behavior. Benefits can be measured in the same two ways.

It is necessary also to consider separately altruistic acts that are directed toward close kin, who share many of the altruist's genes (the precise number being related to the degree of kinship, with full brothers and sisters sharing half, sons and daughters sharing a quarter, and so on), and those that are directed toward individuals who are not closely related, or not related at all, who share few or none of the altruist's genes.

Before considering the evolution of altruistic behavior, let me give some examples of helping and altruism in chimpanzee society. An early report is that of Savage and Wyman (1843–44, p. 386). They describe how a female chimpanzee jumped hastily down from a tree when approached by hunters, but then returned for her infant. She "took him into her arms, at which moment she was shot." At Gombe a mother will risk severe punishment by attacking an adult male who is harming her child (during a charging display, for instance). Melissa even leaped at alpha male Mike as he dragged her infant during excitement; he let go of the infant and attacked Melissa. Another time, when the high-ranking and unusually aggressive adult male Humphrey attacked the adolescent Little Bee, her mother and younger sister both hurled themselves at the aggressor—who fled! There are many occasions when an infant or juvenile tries to help his or her mother when she is being attacked. Sometimes this merely involves following at a distance and uttering loud waa-barks, but there are occasions when a child will actually hit or bite the aggressor, even when this is an adult male. Word Picture 13.2 provides a detailed example of helping behavior among the members of the Melissa family.

Adults of either sex are likely to go to the aid of their mothers if they happen to be nearby. Goblin once ran some 200 meters when he heard the loud screams of his mother, Melissa, who was being attacked by another female. When he arrived, he displayed toward and attacked his mother's aggressor. Adult males often support their younger siblings, especially brothers, during aggressive incidents:

Helping Behavior among Family Members

15 August 1983. Observer: Y. Alamasi.
This incident took place one summer when Melissa was in estrus. For the first part of the morning she had foraged peacefully with two of her offspring, five-year-old Gimble and his adult sister, Gremlin. At 1013 Satan approached the small family party. Hair bristling and penis erect, he began to court Melissa, staring up at her as she fed and shaking branches. At once she climbed down to him and, with pant-grunts, crouched for mating. As he began to copulate and she to squeal, her daughter suddenly jumped to the ground and with her year-old infant clinging to her belly leaped toward the pair and hit at Satan. The vegetation was thick, and the observer thought that Gremlin had perhaps mistaken the sexual act for one of aggression. Be that as it may, Satan at once stopped mating Melissa and turned to attack her daughter. Gremlin, screaming, crouched protectively over her infant as Satan stamped on her, pounded her with his fists, then dragged her through the undergrowth. At this point Gimble ran at Satan with waa-barks and hit him repeatedly on the head and shoulders. Satan, ignoring this puny assault, continued to attack Gremlin. Then Melissa, who had rushed up a tree, returned to the fray and, along with Gimble, hurled herself at the big male. Satan instantly turned on Melissa. Seizing hold of her by one arm and the hair on her back, he actually picked her up off the ground and threw her some 3 meters. Melissa once more seized the opportunity to escape. As she ran, screaming loudly, Satan charged after her and little Gimble followed behind, still uttering waa-barks, as did Gremlin from the safety of a tree. Melissa also took refuge in the branches. Satan did not follow but stopped below her tree, glaring up, hair still bristling. Presently he began to shake branches again, summoning her to him. For one and a half minutes Melissa refused to obey; then she climbed down and, still screaming, crouched before him. Satan carefully inspected her genital area, then wandered away. Soon afterward Melissa and Gremlin settled down to a twenty-minute grooming session as the two youngsters played nearby.

thus Faben and Jomeo frequently hurried to help Figan and Sherry, respectively. Evered was very supportive of his grown sister, Gilka, and almost always intervened on her behalf if he was in the vicinity when she got into social difficulties. Other adult males too occasionally aided their sisters.

Nissen (1931, p. 88), during his pioneering field study, described how he once sat transfixed as a very large male chimpanzee, after staring intently in his direction, suddenly charged toward him. "When about 30 feet away he stopped, picked something up, and ran back up the incline: he was carrying a young chimpanzee, perhaps three years old." Nissen surmised that the youngster, moving through tall grass, had been unaware that he was headed for potential danger. Because chimpanzees were hunted in that area, the male ran considerable risk. We do not know whether he was closely related to the infant. He may have been the biological father, but even if he was, it would not have influenced his behavior, because fathers do not recognize their own young. There are a good many examples of chimpanzees running risks for individuals *known* to have no close relationship with the altruist.

As we have seen, it is not unusual for a male to support an ally who is being threatened or attacked by a third, more dominant individual. Sometimes this is for his own opportunistic ends, but not always. Adult females occasionally help one another in the same way. Twice chimpanzees ran to the rescue of individuals who were being attacked during hunting: Evered charged to the aid of Mustard who was being held to the ground by an infuriated male baboon, and Gigi enabled Freud to escape from the clutches of a bushpig. Chimpanzees are mostly too agile and too smart to be endangered by animals of other species; if this were not the case, we should probably see more altruistic behavior in this context.

Perhaps the most dramatic examples of helping nonrelatives come from two captive chimpanzee colonies. Because they cannot swim, chimpanzees are usually afraid to enter water and it is therefore possible to confine them to islands surrounded by water-filled moats. Washoe spent some time (at Norman, Oklahoma) on an island ringed with an electric fence. One day a three-year-old female, Cindy, somehow jumped this fence. She fell into the moat, splashed wildly, and sank. As she reappeared Washoe leaped over the fence, landed on the narrow strip of ground at the water's edge, and, clinging tightly to a clump of grass, stepped into the water and managed to seize one of Cindy's arms as the infant surfaced again. Washoe was about nine years old at the time; she was not related to Cindy and had not known her for very long (R. Fouts and D. Fouts, personal communication). At Lion Country Safaris in Florida a number of rescues or attempted rescues have been observed. Two involved mothers and their own infants. One adult male rescued an infant; a second male is thought to have drowned during an unsuccessful attempt to rescue another infant. Once two adult males made a dramatic effort to pull a third who was drowning from the moat (M. Cusano, personal communication).

The emergence of helping behaviors among family members can be understood in the framework of the evolutionary theory of kin selection (Haldane, 1955; Hamilton, 1964), as briefly outlined in Chapter 8. This theory makes it possible to calculate the genetic costs and benefits that can be incurred when an altruist helps individuals who are related to him in different ways. An altruistic act that imposes a heavy cost on the actor's ultimate genetic success will, of course, be selected against and will not have a chance to become part of the repertoire of helping behaviors of the breeding population to which the individual belongs. On the other hand, however costly the immediate result (severe wounding, for instance), it will be tolerated by natural selection provided it does not impair the ultimate reproductive fitness of the altruist. An individual who sacrifices his life for others can actually incur an overall net genetic *benefit* if by so doing he saves the lives of, say, three siblings, all of whom survive and perpetuate their half-shares of his all-important genes.

I believe (and see also Boehm, 1981) that this same mechanism of kin selection can explain the emergence of altruism that is directed toward nonrelatives. Let me trace the hypothetical evolutionary pathway that may have led to the emergence of "selfless" altruism in the higher social animals.

In those species whose young go through a period of relative helplessness and dependency, selective pressures have required that biological parents (at least one, of either sex) devote a considerable amount of time and energy to child rearing. This effort can be considered an investment (Trivers, 1972). The payoff, for the parents, is the number of offspring who will be able to compete successfully with other, nonrelated individuals. In other words, it is not enough for female A to produce a larger number of offspring than female B unless her offspring, collectively, produce more youngsters than do those of female B. During the evolution of the higher primates the period of immaturity was gradually prolonged so that increasingly heavier demands were placed on parental nurturing behaviors. And because the female had to devote ever more time and energy to the raising of her young, the relationship between mother and child was strengthened and became more meaningful.

In the higher primates affiliative bonding in early life is of crucial importance for the developing youngster. The disastrous effects of severe disruption of the mother-child bond are well documented (see Chapter 5). Even when no obvious behavioral abnormalities are apparent in the adult, evidence is accumulating which suggests that certain vital functions may be seriously and permanently affected by disturbances in the caretaker-child relationship during infancy. Adolescent rhesus monkeys who had been separated from their mothers for not more than three weeks during infancy were more fearful and less adventurous than controls (Hinde and Spencer-Booth, 1971). Chimpanzees taken from their mothers during the first year of life were less able to concentrate on problem-solving tasks than were those who had had normal relationships with their mothers (Rumbaugh, 1974). In our own species a youngster who has enjoyed a close, affectionate bond with a primary caretaker in early life is likely to be more "persistent and effective in facing environmental challenges on its own" than one who had a less satisfactory bond (Sroufe, 1979, p. 837). Moreover, the deprived individual is more likely to show high levels of stress during crises in adulthood (Brodkin et al., 1984). Because the well-adjusted individual with effective coping strategies is likely to do a better job of child raising than the deprived individual, affiliative bonding mechanisms can be selectively strengthened.

I have already outlined in Chapter 8 the benefits likely to accrue to chimpanzee mothers and their offspring as a result of mutually supportive behaviors, with the resultant increase in inclusive genetic fitness for all family members. Thus even though the costs incurred through nurturing behaviors may in individual cases be high, for the most part the benefits will outweigh them. And as Brown (1975) points out, while parents often run the risk of being hurt when battling on behalf of their offspring, they are, in fact, seldom badly injured. Natural selection is quite amenable to risks and can only step in on occasions when risks are realized—as when the parent's reproductive potential, or ability to raise existing offspring, is damaged. These scattered hard-luck cases will be insignificant in the statistics of the evolutionary process because the overall net benefits are so great.

The strong selective pressures acting on familial nurturing behaviors have produced individuals with well-developed tendencies to care for, help, and protect. For maximum gain in reproductive fitness, of course, these helping behaviors should be directed exclusively toward kin, the closer the better. But among the higher social mammals, who recognize one another as individuals and much of whose behavior is regulated by learning and experience, attitudes toward kin are shaped, to a large extent, by the degree of *familiarity* of the individuals concerned, and depend on close and prolonged association. It is only logical, therefore, that helping behaviors will on occasion be extended to familiar individuals even when they are *not* very close kin. Patterns of comfort and reassurance, helping and sharing, that have emerged over thousands of years in the context of the mother-child and family relationship and that are firmly embedded in the genetic endowment, may be released not only by the distress or pleas of biological kin, but by similar appeals from unrelated but highly familiar individuals.* Midgley (1978, p. 136), discussing societies of higher animals, comments on "the power of adults to mutually treat each other as honorary parents and children," and this is a central theme in both Wickler's *Sexual Code* (1969) and Eibl-Eibesfeldt's *Love and Hate* (1971).

Because altruistic behavior toward nonkin is dependent, genetically, upon firmly established helping behaviors among close kin (Boehm, 1981),† we do not have to argue a separate evolutionary mechanism for its selection. And, in the higher social mammals, the development of close affiliative and supportive bonds with individuals outside the immediate family circle can be much expanded through tradition.

In a society of animals who recognize one another, who form lasting relationships, and who have demonstrably long memories, an individual who helps another in time of trouble can often expect payment in kind. Reciprocal altruism (Trivers, 1971), which was briefly discussed in Chapter 8, provides, as it were, a bonus to the altruist—added reward for an action that he might have performed anyway, even without the prospect of reward at a later date. An adult male olive baboon, who is unlikely to be closely related to other adult males in his troop, is most apt to solicit help from the male who also requests help from him (Packer, 1977)—in other words, his coalition partner. Chimpanzees can readily remember who helped them (as well as who sided against them) from one occasion to the next, even when the occasions are separated by long time intervals. Yerkes (1943, p. 32)

* Of interest here is Satan's response to the outstretched hand—the appeal for reassurance—of a *stranger* female: he instantly moved from her, then scrubbed the place of contact with leaves. On other occasions, too, community males ignored the submissive gestures of strangers (Chapter 14).

† Boehm (1981, p. 173) has described this evolutionary mechanism for the emergence of altruism directed toward nonkin as "parasitic selection." A single set of genes, he suggests, could permit simultaneously "an eminently selfish behavior, relatively well prepared by genes, which brings large benefits in terms of individual relative fitness, and a similar but functionally quite different and more labile behavior, which results in modest genetic losses due to altruism." The selfish, or *basic*, behavior is relatively compulsive or innate (such as a mother protecting her infant, whatever the cost); the second, *emergent* behavior, is optional (such as Gigi rescuing Freud from a pig). In those cases where altruistic behavior toward a nonrelative imposes high costs in terms of individual fitness, they can be borne, in the same way as are the occasional costly or fatal sacrifices of biological parents, by the overall net gain in the context of helping close kin.

Orphan Merlin is nervous as he approaches Charlie during a reunion. Charlie embraces the child, then Merlin touches Charlie's mouth in greeting. (N. Washington)

observed that a chimpanzee "may clearly express devotion, and its appreciation may be lasting, for memory of kindness is demonstrably long." Among the chimpanzees of the Arnhem colony failure to reciprocate in kind was sometimes punished. An adult female, who had previously helped one of the adult males in an aggressive incident with his rival, turned on him and actually hit him when he later refused to respond to her appeal to him for support (de Waal, 1982). Thus the old adage "one good turn deserves another" is, without doubt, deeply rooted in our primate heritage.

Mellen (1981), writing on the evolution of love, traces its origin through the helping behaviors we have just discussed. He suggests a complex of genes for various kinds and degrees of affection, leading to generalized tendencies to form emotional attachments with others.

Love can take many forms, and we use the word freely and loosely in our speech (we love our spouse and our children; we love God; we also "love" music and a friend's new house—and ice cream). I am concerned here only with the emotions that accompany very close bonding, such as sympathy, tenderness, joy, understanding. In Chapter 17 I suggest that chimpanzee intercommunity violence foreshadows human cruelty and warfare; here we should ask whether chimpanzees also show behaviors that might be considered precursors of some aspects of human love. I have already described aspects of affiliative bonding that unquestionably are milestones on the path: the greetings between close friends, the pleasure in relaxed physical contact, the sharing of food, and the helping and soothing of those in distress.

Köhler noted that the individuals of his group sometimes showed highly impulsive behavior that was not always directed toward concrete material advantage. As an example he cited the exuberant greetings the chimpanzees gave him when he first arrived in the morning: the kissing and embracing lasted for some time before a single mouthful of food was taken. Once, hearing plaintive calling at night, Köhler went out in torrential rain and found two chimpanzees accidentally locked out of their sleeping quarters. He forced open the door and stood aside to let them in. "Although the cold water was streaming down their shivering bodies on all sides (and although I myself was standing in the middle of the pouring torrent), before slipping into their warm, dry den they turned to me and put their arms round me, one round my body, the other round my knees, in a frenzy of joy" (Köhler, 1925, p. 250).

Köhler also described the concern shown by the young female Rana when her small companion, Konsul, became fatally ill. As he lay on the ground, Rana approached and invited him to accompany her, but he hardly moved. "She grew attentive, first lifted his head, and then, putting her arms around the little fellow, carefully lifted his weak body and seemed by her bearing and her look so deeply concerned that there could be, at that moment, no doubt whatsoever as to the state of her feelings" (1925, p. 242).

I told in Chapter 1 of the psychoanalyst Temerlin and his family, who spent ten years with the female chimpanzee Lucy. Temerlin (1975, pp. 164–165) wrote that Lucy's affection for her human family was "sufficiently intense and enduring that I would not hesitate to call it love. It is always there, side by side with a protectiveness and concern for us that is touching and tender to see." If Temerlin or his wife was ill and vomited, Lucy would run to the bathroom and stand by them, trying to comfort them by kissing them and putting her arms around them. Once when this failed to alleviate the symptoms, Lucy slammed down the lid of the toilet, hitting it while screaming loudly. If Jane Temerlin was sick in bed, Lucy would "show tender protectiveness,

Lucy looks at a scratch on the forehead of Jane Temerlin and signs, "That, that hurt." (J. Carter; courtesy of J. and M. Temerlin)

(a) Olly cradling her month-old infant, the first victim of the supposed polio epidemic, just before his death. His arms and legs are paralyzed. (b) Olly continues to carry the infant for three days after his death, but she no longer cradles him or shows any other nurturing maternal behavior. (H. van Lawick)

Orphan Pax (age five years) with his siblings, Prof and adult female Pom, one year after the death of their mother.

bring her food, share her own food, or sit on the edge of the bed trying to comfort her." And if Jane was emotionally upset, Lucy noticed it immediately and tried to console her. A captive male who attacked a human companion in rage subsequently showed signs of distress; with his fingers he tried to close the lips of the wounds (Lorenz, 1963).

At Gombe, concern for sick, injured, or distressed individuals is observed primarily among members of the same family. Mothers care tenderly for their sick infants. When Olly's month-old infant lost the use of his arms and legs because of polio, she supported him carefully and each time he screamed she cradled him at once, solicitously arranging his limbs so they would not be crushed. The moment he died (or, at least, lost consciousness and no longer cried out), she began to treat the body in an extremely careless way, flinging it onto her back, holding it by one leg, and allowing it to fall to the ground.

On six occasions an elder sibling or a presumed sibling (twice males) cared for a younger one when the mother died. These caretakers showed a great deal of concern for their charges and without exception were highly protective. The juvenile Sniff carried his fourteen-month-old sister wherever he traveled during the two weeks she survived the death of their mother. Several times when he tried to rescue the infant during the excitement generated by banana feeding, Sniff was buffeted or attacked by one of the adult males. The youngest of the orphans, Passion's four-year-old son, Pax, preferred to walk and sleep on his own, but when, as often happened, the pace of travel was too fast for him, he began to whimper. For at least the first eight weeks either his adult sister, Pom, or his ten-year-old brother, Prof, returned and tried to carry him, but their offers were persistently (and inexplicably) refused. Indeed, when Pom tried forcibly to place him on her back, he threw a tantrum and clung to the vegetation. Pom became extremely upset, holding her hand toward him and whimpering as she

Prof Rescues His Brother, Pax

21 November 1982. Observer: J. Goodall.

It was midafternoon on a cool, gray day. Goblin, top-ranked male, sat near Miff, who was in estrus. Only a few other chimpanzees were present: Miff's juvenile son, Michaelmas; her four-and-a-half-year-old daughter, Mo; eleven-year-old Prof; and his young brother, Pax. (Their mother had died the year before, when Pax was only four years old.) Pax approached Miff, stood behind her, and shook a small branch, courting her. She paid no attention. Presently he moved closer and attempted to copulate with her. Not even looking around, Miff kicked Pax, sending him tumbling back, head over heels, into the vegetation. Screaming loudly, grabbing now at the hair of his head, now at the leaves around him, hurling himself from side to side, Pax threw one of the most violent tantrums I have ever seen. Miff ignored him; Goblin, probably irritated by the commotion, began to bristle, glaring toward the screaming infant. At this point Prof approached. He stood for a moment, looked from Pax to Goblin, and seemed to weigh the situation. Then he went over to his small brother, seized him by one wrist, and firmly dragged him away through the undergrowth for about 8 meters. Only when Pax stopped screaming did Prof release his hold; then the two brothers left the other chimpanzees and went off by themselves. Almost certainly Prof, anticipating an aggressive reaction from Goblin, had saved Pax from worse punishment than rejection by an uncooperative female.

persisted in her efforts to help—to no avail. It was the same at night. He cried softly as he constructed his own small nest close to Pom's; she whimpered too, reaching out and begging him to join her, but he never did. After a year Pax became very closely bonded to Prof. Word Picture 13.3 illustrates Prof's concern for the well-being of his young brother.

In 1975 the adult female Pallas (who had lost two infants of her own) adopted five-year-old Skosha, daughter of her closest companion. (Pallas may in fact have been a biological aunt.) Pallas showed the utmost patience with Skosha, often waiting half an hour or more for her, carrying her, sharing food, and reassuring the child during her frequent outbreaks of screaming. Pallas gave birth again, but died when her child, Kristal, was five years old. Skosha, still an integral part of the family, adopted her foster sister. She was an affectionate caretaker, and it was through no lack of concern on her part that Kristal became sick and died a year later.

When nine-year-old Freud hurt his ankle (probably a broken bone), he was unable to put his foot to the ground and initially moved very slowly. His mother, Fifi, typically waited for him when he stopped to rest, but sometimes she moved off before he was ready to hobble after. Three times when this happened Freud's four-year-old brother, Frodo, hung back, looked first at his mother, than at Freud, and began to whimper. He continued to cry until Fifi stopped once more, and eventually the family moved on together. Frodo showed his concern in other ways too, sitting close to his brother, gazing at the injured ankle, and grooming him.

Skosha with her five-year-old foster sister, after the death of Kristal's mother.

Miff adopted her infant brother, Merlin, after Marina died in 1965. (E. Koning)

After being fatally attacked by the Kasakela males, Madam Bee lay, scarcely able to move, until she died five days later. During this time her adolescent daughter, Honey Bee, remained nearby, grooming her mother and keeping away the flies. And when old Mr. McGregor was crippled by polio and unable to walk, he too had a faithful attendant in the younger Humphrey (possibly his nephew). As I have described, Humphrey, in defense of his friend, not only dared to attack the higher-ranking and powerful Goliath, but stayed near Gregor for several hours each day throughout the last two weeks of the old male's life.

Care of the sick, as mentioned, is rarely observed among unrelated chimpanzees at Gombe. Gregor was shunned by those individuals whom he approached for social grooming (Goodall, forthcoming). When Fiji had a gash in her head, which became infected, those individuals to whom she presented the wound for grooming seemed fearful and moved away. In a captive chimpanzee group, however, the situation is different—perhaps because members of the group so often are raised together and are as familiar as close kin in the wild. Köhler's (1925) chimpanzees zealously squeezed pus from wounds and removed splinters, and Yerkes (1943) and Fouts (1983) have described similar behavior. Miles (1963) observed an adult male remove a speck of grit from his companion's eye in response to her whimpering and solicitation. One adolescent female of Menzel's group not only groomed her companion's teeth, but even extracted a deciduous molar that was loose (McGrew and Tutin, 1972). As Köhler (1925) noted, such manipulations are undoubtedly prompted in some measure by a fascination with the lesions themselves. The results, however, are beneficial and it is easy to see how such behavior could have led to altruistic health care in humans.

In order to feel compassion, an individual must have some understanding of the wants or needs of the sufferer. That chimpanzees have the cognitive ability to empathize, at least to some extent, has been shown by Woodruff and Premack (1979): Sarah consistently chose "good" solutions for problems faced by a friend and "bad" ones for a person she disliked. At the start of this chapter I described how Little Bee took food to her ailing mother. It was not the only time she was seen to help Madam Bee in this way. On two other occasions she climbed down to the old female with her mouth and one hand full of palm nuts and laid the fruits from her hand beside her mother. Quite clearly she had some understanding of the needs of the old female. This ability to empathize, so highly developed in our own species, prompts much altruism in humans. If we know that another, especially a close relative or friend, is suffering, then we ourselves become emotionally disturbed, sometimes to the point of anguish. Only by helping (or trying to help) can we hope to alleviate our own distress. Was Little Bee, I wonder, motivated by a similar kind of emotion? Whatever the answer, it is evident that chimpanzees have made considerable progress along the road to humanlike love and compassion.

14. Grooming

Satan relaxes as he is groomed.

July 1981   Satan is traveling alone through the forest. Suddenly he comes upon Goblin, who is resting on the ground some 30 meters ahead, peacefully chewing on a wadge of palm nuts. These two males have been, for some time, actively competing for dominance rank. As soon as he sees the younger male, Satan stops short, hair bristling. Goblin catches sight of Satan at about the same time and quickly sits up, all his hair erect. For the next forty-five seconds neither male moves. Goblin looks directly at Satan, who avoids the other male's gaze. Abruptly, hair still on end, Satan walks up to Goblin, who stays where he is, his hair bristling even more. He seems to swell up to twice his size. Satan, as he arrives within arm's reach of Goblin, suddenly turns and presents his rump for grooming. Goblin responds immediately with quick, vigorous movements. After thirty seconds Satan turns to face Goblin and the grooming becomes mutual. At first it is very tense on both sides with loud tooth clacking, but gradually the partners calm down. Twenty-one minutes later the two travel off and feed together, temporarily very relaxed in each other's company.

Social grooming permeates virtually every aspect of chimpanzee social life, serving many functions above and beyond skin care. It provides the opportunity for long bouts of relaxed, friendly physical contact— which, as we have seen, is crucial for reducing stress in a variety of situations. Grooming thus plays a vital role in regulating relationships, especially in restoring harmony in the wake of aggression. Even self-grooming (from which, presumably, social grooming stemmed; see Reynolds, 1981) often takes place during times of tension and helps to calm an anxious or frustrated individual.

Because grooming enters into so many aspects of social behavior, it is mentioned in almost every chapter of this book. The one detailed study of social grooming at Gombe deals with adult males only (Simpson, 1973). Other publications discuss grooming as it relates to agonistic relationships among adult males (Bygott, 1974, 1979), reunions (Bauer, 1976, 1979), adolescent behavior (Pusey, 1977, 1983), sexual behavior (McGinnis, 1973, 1979; Tutin, 1979), and weaning (Clark, 1977). My own early monograph (Goodall, 1968b) includes descriptions of grooming patterns and techniques.

Social grooming can be an enjoyable activity for both the groomer and the groomed. Chimpanzees have learned novel tasks when the reward was a session of being groomed—and also, a little less enthusiastically, when the reward was an opportunity to groom (Falk, 1958). Viki Hayes was never more relaxed and calm than when she was groomed by her human foster parents. Not only did she groom them, frequently and intently, but if they tried to prevent her from doing so (when, for example, she was picking at an abrasion on their skin), she became very upset, pushing at their hands and sometimes throwing a tantrum (Hayes, 1951).

Techniques

At Gombe one has only to watch a relaxed grooming session to see that the chimpanzees derive considerable pleasure from the activity. The individual being groomed may sit, even lie, in a comfortable position while a companion works over him. Sometimes the groomer lazily runs a forefinger through the other's hair, his own eyes drooping. Or he may work in a very businesslike manner, pushing and pulling the other into different positions to gain access to various parts of the anatomy, parting the hair with his fingers, removing particles of dry skin, and so on. From time to time the groomer comes upon a tick or louse.* With a quick lunge that may almost knock his client off balance, he seizes the affected part and, tooth clacking loudly, picks off the prize—which he then chews with exaggerated movements of the jaw. His companion watches the whole operation with fascination. Such moments, unpredictable and obviously stimulating, probably help to maintain the interest of the groomer during a long session; they provide, in a sense, the "jackpot effect" so effective in animal training (Pryor, 1984).

A session is defined here as any bout of grooming longer than (an arbitrary) two minutes and not interrupted by breaks of more than two minutes. Shorter bouts often represent token grooming, part of the submission-reassurance sequence already described. Some sessions involve two individuals only, others many more. The largest number of participants recorded was twelve, including two infants (who played more than they groomed). During a long session involving a number of individuals, some may stop grooming and wander away while others may approach and join in. Often the group divides into a number of smaller grooming clusters.

A session is likely to last longer if both partners groom, either reciprocally (one after the other) or mutually (both at the same time). Table 14.1 shows that sessions between adult males were, on average, the longest, followed by those between mothers and adult offspring. Grooming sessions between adult females who were not closely related usually lasted less than five minutes. The longest grooming session ever recorded involved a mother and her young adult son; it continued for two hours and forty-five minutes.

Pepe self-grooming. He uses both fingers and lips to part his hair and expose the skin. (H. van Lawick)

* The chimpanzee louse *(Pedicularis schaefi)* has the habit of freezing into immobility when exposed to light, thereby becoming extremely difficult to detect (Kuhn, 1968, cited in Ghiglieri, 1984). There is thus a need for the very close inspection of the skin that characterizes chimpanzee grooming behavior.

Table 14.1 Duration of grooming sessions between same-sex and mixed-sex dyads and between larger clusters in camp during 1978. Scores for two mothers and their adult offspring are shown separately.

Age-sex combination	Pairs of groomers		Three or more groomers	
	Mean duration (minutes)	No. of sessions	Mean duration (minutes)	No. of sessions
Adult males	25.9	24	35.7	6
Adult males and females	13.5[a]	61	29.9	47
Adult females	6.3[b]	36	22.6	7
Passion and daughter Pom	19.3	42		
Melissa and son Goblin	16.6	10		

a. Excluding mother and adult son.
b. Excluding mother and adult daughter.

Pepe is tooth clapping as he grooms a companion. Note the precision of his grip. (B. Gray)

As one would expect, grooming has its own set of signals. A chimpanzee wanting to initiate a grooming session may simply approach a chosen partner and start to groom. Or he may present his rump, back, or (occasionally) bowed head—those parts that are difficult or impossible to see during self-grooming. At the same time he may deliberately scratch the presented part. Sometimes he stops about 3 meters away, raises his arm, and, usually holding onto an overhead branch, scratches with vigorous downward movements from elbow to chest while gazing at the other.

In response to such solicitations the chosen partner can respond in one of four ways: by grooming, by countersoliciting, by ignoring the request, or by moving away. The choice made will depend on a variety of factors such as the relationship between the two, the other individuals present, and whether the individual chosen is already grooming another. If the solicitor is ignored, he may give up and move away, he may groom the other himself, or he may solicit more vigorously— by emphatic scratching, whimpering, or touching the other's arm or hand. These last two entreaties are seen most often in the context of

Fifi whimpers and touches her mother's arm as Flo grooms Flint. Then, as Fifi raises her arm, still whimpering, Flo responds to her daughter's solicitations. Flint now starts to scratch, a request for his mother to resume grooming *him*. (H. van Lawick)

family grooming. Indeed, one offspring, trying to divert his mother's attention from a sibling, may take her hand and pull it toward himself. Or he may physically place himself between the grooming pair. Another gesture commonly seen in infants and juveniles consists in holding one arm up in the air; this is sometimes accompanied by whimpering and scratching. This may be the origin of the *hand-clasp* posture so often adopted during mutual grooming by chimpanzees at Mahale (McGrew and Tutin, 1978) and in some areas of Uganda (Ghiglieri, 1984).*

After a period of solo grooming, a groomer often pauses and requests his partner to take a turn—by vigorous scratching, by deliberate repositioning and presenting of some portion of his body, or by

* Curiously, while the hand clasp occurred in 38 percent of the (nonmaternal) sessions of grooming recorded for the chimpanzees at Kanyawara, it was not observed at all at Ngogo, 10 kilometers away (Ghiglieri, 1984).

A large grooming cluster. Hugh (*foreground*) is groomed by Hugo (*right*) and alpha male Mike, while Goliath and a second male groom Mike. Faben is grooming Goliath as Sniff, an adolescent male, grooms Faben. Figan grooms the adult female Pallas (*extreme right*). (H. van Lawick)

using one of the other signals mentioned above. Hugo occasionally shook branches, hair beginning to bristle, if a lower-ranking male grooming partner ignored more typical requests. And Melissa, after six minutes of whimpering, rocking, and scratching, once actually gave old Mr. McGregor a hard shove with one foot.

There is, at both Gombe and Mahale (Goodall, 1968b; Nishida, 1980), the curious tradition of *leaf grooming*, which occurs most often in the context of social grooming. The chimpanzee picks one or more leaves and, peering at them closely, grooms them most intently, sometimes lip smacking at the same time. Leaf grooming invariably attracts much interest from those nearby who gather round to watch; the leaf groomer may then be groomed by one or more of his audience. Thus it may serve to revive the flagging interest of a companion during a session or to divert his attention from another partner. Sometimes, though, an individual who is already being groomed suddenly reaches out to seize and groom leaves: often this is a prelude to leaving the grooming session altogether. There are occasions, too, when a solitary individual leaf grooms—and (subjectively) it looks like a form of doodling. The behavior, presumably (at least in origin) a displacement activity, is not yet properly understood.

Changes with Development

Immediately after giving birth, the chimpanzee mother licks and grooms her baby (Davenport, 1979; Goodall and Athumani, 1980; Fouts, Hirsch, and Fouts, 1982). During the first few months of an infant's life, bouts of maternal grooming (at Gombe) usually last for only a few minutes, or even seconds—and often occur intermittently during self-grooming. Some mothers, however, groom their babies for longer (Nicolson, 1977) and prolonged bouts are usual in captivity (Davenport, 1979).

As infants grow older, the duration of maternal grooming sessions increases. By the time they are nine to twelve months old, however, infants are often restive when their mothers seize them in order to groom. They struggle to escape, or make vigorous attempts to change the nature of the activity from grooming to play. Sometimes they succeed, particularly when they have mothers who incline to playfulness themselves. If not, they may reach out and idly play with leaves, branches, or anything they can get hold of while submitting to the mother's wishes. After a few minutes most infants relax and become quiet under the gentle manipulation of maternal fingers. No part of the anatomy is overlooked by a businesslike mother. Flo—and Fifi after her—favored meticulous attention to the ears and genitals, persisting despite the wriggling of their infants. Ear grooming appeared to cause discomfort, whereas genital grooming often elicited laughing (and erections in male infants).

Infants start to make brief grooming movements during the second or third month, but do not become proficient in adult grooming techniques until they are about a year and a half old. The behavior appears to be innate. Viki Hayes, who had never had an opportunity to watch other chimpanzees, began to "groom" the pattern on her bedclothes when she was eight weeks old. At six months she showed appropriate

Athena grooms her two-year-old son, Apollo.

Old Mr. McGregor grooms the early adolescent Pooch.

hand and finger movements, and at twelve months she began to lip smack and tooth clack (Hayes, 1951). During early attempts to groom, a Gombe infant may appear very "serious," but after a few correct movements he often abandons the idea and ruffles the hair wildly, sometimes playfully slapping the recipient of his attentions.

As the infant matures, grooming begins to occupy a more important place in the scheme of things. When he is about five months old, he may for the first time approach his mother and solicit grooming, climbing onto her lap and stretching out, side or belly uppermost. A mother almost always responds to an invitation of this sort. As the infant's own grooming skills increase, grooming sessions with the mother get longer, and individuals outside the immediate family sometimes groom him also (Goodall, 1968b).

Prior to weaning, a mother typically grooms her infant (of either sex) more than the infant grooms her. After the birth of the next baby, sessions between the mother and her juvenile offspring become longer, and the frequency of both mutual and reciprocal grooming increases. Still, throughout childhood it is the mother who does most of the grooming. During early adolescence a sex difference appears: a daughter, but not a son, tends to groom her mother more than she herself is groomed (Clark, 1977; Pusey, 1983). The early-adolescent female grooms very little outside her close family circle, although she may sometimes groom other mothers—particularly those with small infants, which fascinate her. When she begins to cycle, she occasionally participates in grooming sessions with nonrelated males. Initially her male partners tend to be adolescents like herself, but after her first estrus (when she is about ten years old) she will sometimes be groomed by adult males (Pusey, 1977). Even so, for as long as she maintains close ties with her mother (which may be for life), her grooming partners for the most part are members of her immediate family.

The adolescent male engages in grooming activities with nonrelated individuals more often than his female counterpart. In some ways, the changes over time in his grooming behavior reflect not only the changes in his association patterns, as he spends more time away from his family, but also his changing relationships with other members of the community, particularly his mother and the older males. He has, from infancy on, groomed adult males when they were peaceful and relaxed, and now he does so more often. Indeed, he may even solicit grooming, pausing in his work to stand, rump toward his senior, glancing (hopefully) back over his shoulder. By and large, adult males ignore such requests, and after a few moments the youngster abandons his pose and continues to groom. From the age of nine or ten years the young male becomes increasingly apprehensive of the adult males; he often watches, enthralled, as they groom together, not daring to join in. At the same time, as he gains in stature, the adult females whom he has been grooming for some years (particularly when in estrus) begin to reciprocate. During late adolescence he joins the grooming sessions of the senior males more and more often. When, at thirteen to fifteen years old, he *regularly* takes part in their grooming sessions and is *regularly* groomed by one or more of them, it is an indication that he has attained social maturity.

There are certain general principles, or rules, that underlie choice of grooming partner, making it more likely that A will groom B rather than C, that C will groom A rather than B, and so on. A monkey or ape is most likely to select as a grooming partner an individual with whom he (or she) already has a close, supportive relationship, or with whom a better relationship would be advantageous. And the chosen partner will reciprocate, or not, for similar reasons (see for instance Seyfarth, 1980).

An adult male tends to groom higher-ranking or older male partners more than he is groomed by them (Simpson, 1973; Bygott, 1974, 1979). The benefits to be derived from a good relationship with a higher-ranking male are obvious: older males may prove important allies against the challenges of younger males working their way up the hierarchy. In a society in which most or all of the adult males cooperate to patrol territorial boundaries, it is clearly beneficial if *all* are on relatively good terms, and male chimpanzees do in fact spend time grooming males who are lower ranking and younger than themselves.

If the relationship between two males becomes tense, they are likely to groom each other more often. Bitter hostility between them, however, can virtually extinguish grooming. Thus at the peak of their antagonism (in 1973–74), apart from a few grooming movements made by Figan on Evered during the excitement of a hunt, no grooming at all was recorded during eighteen months of intensive study (Riss and Goodall, 1977). In 1976, after Faben had vanished and Evered had returned to the social life of the community, he was seen to groom Figan only once, for four minutes; and very little grooming was seen in 1977 and 1978, even though by then their relationship was no longer hostile. When in 1979 the senior males began to support one another against Goblin, Figan and Evered groomed often; and in 1980, when they had a close friendly relationship, levels of grooming between them shot up and remained high for the rest of Figan's life. At the same time, Figan and Goblin were never seen to groom again after

David Greybeard grooms Hugo who grooms Flo who grooms Fifi.
(H. van Lawick)

the gang attack on Goblin (led by Figan) at the end of 1979. To some extent, therefore, yearly changes in levels of grooming between pairs of males reflect changes in their relationships.

The adult female grooms most with members of her own family. Obviously she benefits from maintaining close relations with them, for it is they who will subsequently provide her with support and, particularly in the case of a daughter, companionship. She grooms very little with females who are not close kin, even those with whom she associates most frequently and with whom she has a friendly relationship. Females in captivity groom one another more. They also interact more and frequently form stable, mutually supportive bonds (Köhler, 1925; de Waal, 1982). Often, of course, they are as familiar with one another as are mothers and daughters at Gombe.

If a female can establish relaxed relations with the adult males, she will undergo less nervous tension and stress. She tends to groom more with the older males who, as we shall see (Chapter 16), are likely to be her most frequent sexual partners during the fertile phases of her reproductive cycles. The female, however, does not need to work at maintaining the sexual relationship as such: physiological mechanisms, such as her swelling and perhaps pheromones, do this for her. It is the male who must put effort into his relationship with her, because her cooperation may be critical to his overall reproductive success. In one captive group the sole adult male groomed females significantly more when they were a quarter swollen than at other phases of the estrous cycle. This suggests a mechanism for influencing the female's preference in his favor prior to the more competitive situation likely to arise as she reaches her maximum swelling (Rapaport, Yeutter-Curington, and Thomas, 1984).

Because grooming sessions last longer when both partners groom, the reluctance of certain age-sex classes to reciprocate helps to explain differences in grooming distribution among the various members of the community. Thus while females respond readily to requests for grooming by offspring of all ages, they groom longer when the child is old enough to reciprocate. The low level of grooming among adult females at Gombe (other than those who are very closely related) can be explained in part by the reluctance of most females to reciprocate when groomed by another female. After a few minutes the groomer either gives up or herself solicits grooming from her partner, probably without success.

Six Profiles

Figures 14.1 and 14.2 are grooming profiles of three adult males and three adult females during one year, 1978. These profiles show, for each chimpanzee, the time in camp spent grooming and being groomed by other individuals (a) relative to the time that each pair spent together and (b) as a percentage of the total time spent grooming and being groomed during that year. These histograms, while they represent the behavior of only six individuals during only one year, nevertheless illustrate many of the underlying principles that dictate partner choice.

The three males in Figure 14.1 were selected because they show three different patterns. Jomeo very clearly groomed with adult males

Table 14.2 Grooming activity of six adult males in camp during 1978.

Adult male[a]	No. of adult individuals male grooms (a) or is groomed by (b)				Percentage of total grooming time male grooms (a) or is groomed (b)				Percentage of total time in camp during which male grooms (a) or is groomed (b)	
	Males		Females		Males		Females			
	(a)	(b)	(a)	(b)	(a)	(b)	(a)	(b)	(a)	(b)
Humphrey	4	6	4	6	86.5	76.0	13.5	24.0	7.5	13.0
Evered	4	5	12	11	25.0	24.0	75.0	76.0	34.5	36.0
Figan	5	5	9	12	52.0	54.0	48.0	46.0	15.0	22.0
Satan	6	6	13	14	36.0	30.5	64.0	69.5	36.0	41.0
Jomeo	6	6	7	7	85.0	78.0	15.0	22.0	16.0	24.0
Sherry	6	5	13	11	55.0	43.5	45.0	56.5	15.0	19.0

Note: By the Wilcoxon matched-pairs signed-ranks test (two tailed), none of the differences in percentage of time spent grooming and being groomed by males or females is significant; however, the difference in percentage of total time in camp during which males groom and are groomed is significant ($p < 0.05$). That is, males are groomed significantly more than they groom.
a. Males are listed in decreasing order of age.

in preference to females. Satan, equally clearly, preferred females—not only those who cycled during the year, but those who were anestrous. Figan's grooming time was distributed fairly evenly between males and females, but his involvement with females was mostly when they were in estrus.

Of the three Satan was the most enthusiastic groomer; his involvement in grooming was high even back in 1970 (Bygott, 1974). Jomeo and Figan, whose overall scores are remarkably similar, were not prepared to invest as much time in grooming as Satan (although, as we shall see, Jomeo groomed more in other years).

All three males tended to groom higher-ranking or older male partners more than they were groomed by them. Analysis of grooming data for the other adult males (not shown) revealed that Goblin, the youngest, groomed all the older males more than they groomed him, and he was not groomed at all by the two oldest, Humphrey and Evered. The elder of two late-adolescent males, Mustard, was groomed only by Satan and Goblin; the other, Atlas, was not seen to be groomed by any of them.

Table 14.2 summarizes data from the profiles of the three males and compares their grooming behavior with that of three others: Evered, Sherry, and Humphrey. Evered was a prolific groomer, and both he and Sherry showed the same "females most" pattern as Satan. These three males all groomed more different females, and groomed females longer, than the others. Humphrey, the oldest and by 1978 the lowest ranking of the males, had the least involvement with females.

Differences in willingness to reciprocate in the grooming context vary from individual to individual even within the different age-sex classes. Satan and Evered, both avid groomers, not only respond readily to invitations to groom, but also maintain their enthusiasm throughout long sessions and make efforts to stimulate flagging partners. During 1978 either Evered or Satan was involved in 20 of the 28 sessions that lasted 40 minutes or longer; in 10 of the 12 that were

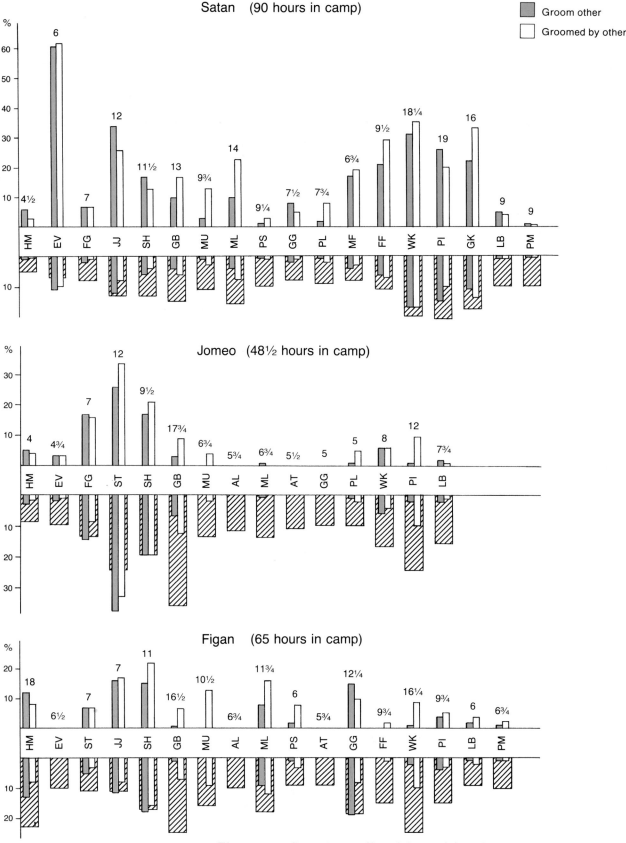

Figure 14.1 Grooming profiles of three adult males during one year, 1978, based on data from camp records. The columns *above* the initials show for each of the three males the percentage of total *time with each other adult* that he groomed, or was groomed by, that individual. The number of hours each dyad was together in camp is given at the top of the column. (Scores for dyads who spent less than four hours together have been omitted.) *Below* the initials the wide hatched columns show the percentage of total *time in camp* that each of the three males was present with each of the other adults. The narrow green and white columns show the percentage of total grooming time directed toward and received from each other individual.

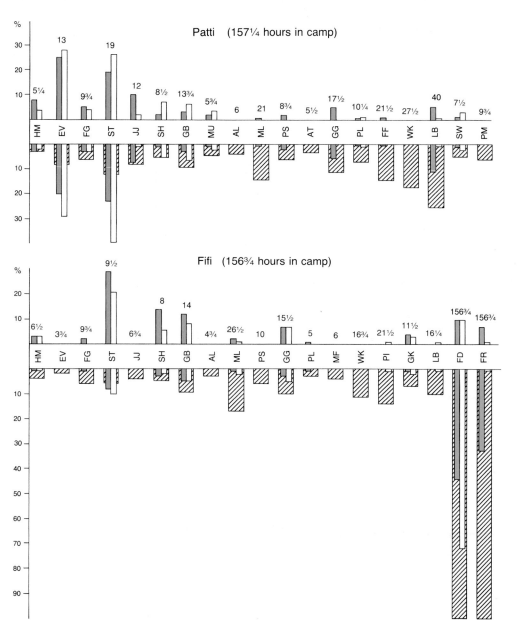

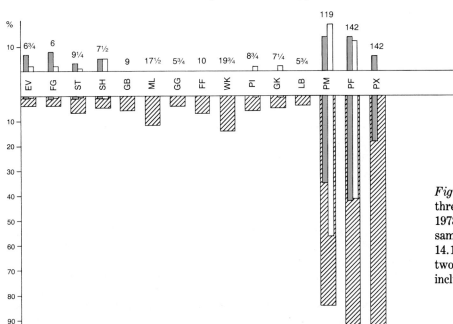

Figure 14.2 Grooming profiles of three adult females during one year, 1978. These histograms provide the same kind of information as Figure 14.1, except that offspring of the two mothers, Fifi and Passion, are included also.

Late-adolescent male Evered embraces, then grooms, Melissa during a reunion.

longer than 50 minutes; and in 6 of the 7 that were longer than 60 minutes. Evered *and* Satan took part in one of the two longest sessions (80 minutes), and Evered in the other (78 minutes). This is undoubtedly why, relative to association time, they rank high in the yearly scores of many individuals, male and female alike. Nor is it surprising that, over a seven-year period (1976 through 1982), each was the other's most preferred partner (except during one year when Satan, though still *groomed* most by Evered, himself groomed Humphrey slightly more). Males like Jomeo, Figan, and Humphrey tend to leave most of the work to their companions, who after a while give up and either stop grooming altogether or groom a more responsive partner— or sometimes themselves.

Figure 14.2 shows the grooming patterns of three females. The youngest, Patti, gave birth early in the year but lost the infant after one week. She resumed cycling, then became pregnant again in August. Thus she represents a young, childless female. The other two are mothers. Fifi throughout the year was accompanied by two offspring, Frodo and Freud; at the start of 1978 they were one and a half and seven years old respectively. Passion, the oldest female, was also in constant association with two sons, Pax (one year) and Prof (seven years). Passion was with her adult daughter, Pom, for 119 of the 142 hours that she was in camp during the year.

Both mothers devoted most of their grooming time to their families; Passion had three offspring to attend to and groomed Pom and Prof about equally and Pax rather less, while Fifi spent more time grooming Freud than infant Frodo. Both mothers received the greatest proportion of *total* grooming from these same elder offspring. The two infants did not groom their mothers at all except for brief attempts, each lasting a few seconds.

If we exclude family grooming, Fifi's graph is quite similar to that of Patti: both groomed adult males for up to 30 percent of the time they were together. Unlike Fifi, Patti received more grooming than she gave from five of her seven male partners, probably because she cycled during the year. Of the five fully mature male partners, it was

Pom grooms her mother.

two of those with the "females most" pattern, Satan and Evered, who spent most time grooming with her.

The graphs of all three females show quite convincingly how little time was devoted to grooming with nonrelated females. Fifi's most preferred female partner was Gigi; and the two of them groomed with each other for only 7 percent of their total association time. (Gigi, probably on account of her childless status, has over time been involved with more females in the grooming context, and groomed with them longer, than any other female.) Fifi's other grooming partner of note was Gilka; Fifi groomed her for a mere 4 percent of the time they spent together, and was groomed even less. Patti's two favorite female partners were Gigi and Little Bee; she groomed them both for 5 percent of the time they spent together and they did not groom her at all. Passion groomed no nonrelated females and was groomed by only two; in each case for rather less than 2 percent of association time.

As mentioned earlier, one of the reasons for the very low levels of grooming among the Gombe females probably lies in their reluctance to reciprocate during their rare bouts. The following incident took place in 1979. Patti, who had just arrived in camp, approached Fifi, turned, and sat, thus presenting her back for grooming. Fifi ignored her. Patti glanced around, then scratched vigorously down one shoulder. At this Fifi did respond and began to groom her. After one and a half minutes she stopped, got up, moved in front of Patti, and presented *her* back. Patti groomed Fifi for just over two minutes. She then stopped and, in her turn, moved in front of Fifi and scratched. Fifi, grooming her own leg, paid no attention; after thirty seconds she repositioned herself in front of Patti. The maneuvers were repeated—once more by Fifi, twice by Patti—so that eventually, when Patti ended the somewhat unsatisfactory performance by moving off, the pair had progressed a couple of meters from their starting point.

Table 14.3 summarizes the data for the three females of Figure 14.2 and provides comparative data for three others: Winkle, Little Bee, and Melissa. Winkle, pregnant for the last eight months of the year and with a six-year-old son, had a grooming pattern much the same as Patti's. Little Bee, about the same age as Patti, but with a year-old infant, spent very little time grooming with anyone.

Melissa, about the same age as Passion, had a very different grooming pattern in 1978: she devoted a much greater proportion of her time to grooming with adult males. This may have resulted partly from her fear of the infant killer, Passion, which caused her to stay close to the males whenever possible, and partly from the males' fascination with the unique phenomenon of Melissa's twins. (As a rule, grooming by adult males of females with new infants tends to increase over prebirth grooming rates.) I have therefore included Melissa's 1983 grooming scores in the table as well. These show a profile that is quite similar to Passion's in 1978.

Figure 14.3 summarizes the sex differences in grooming behavior that emerge from the male and female grooming profiles of Figures 14.1 and 14.2. The contrast between Satan and the childless Patti in the proportion of total camp time spent in grooming is striking. There is rather less difference in the scores for Patti and Figan, particularly

Table 14.3 Grooming activity of six adult females in camp during 1978.

Adult female[a]	No. of adult individuals female grooms (a) or is groomed by (b)						Percentage of total grooming time female grooms (a) or is groomed (b)						Percentage of total time in camp during which female grooms (a) or is groomed (b)	
	Males		Females		Offspring		Males		Females		Offspring			
	(a)	(b)	(a)	(b)	(a)	(b)	(a)	(b)	(a)	(b)	(a)	(b)	(a)	(b)
Melissa, 1978	5	6	6	9	4	2	20.5	12.0	7.0	4.5	72.5	83.5	23.5	18.5
Melissa, 1983[b]	5	4	3	1	3	3	7.0	10.5	1.5	0.5	91.5	89.0	21.0	14.0
Passion	5	4	1	3	3	2	3.9	2.5	0.1	0.1	96.0	97.0	37.0	27.0
Fifi	5	4	5	6	2	2	18.0	17.5	6.0	10.0	76.0	72.5	22.0	14.0
Winkle	7	7	4	5	1	1	49.0	62.5	3.5	11.0	47.0	27.0	16.4	11.2
Patti	7	7	9	4	—	—	60.0	86.5	40.0	13.5	—	—	10.0	8.0
Little Bee	4	6	8	5	1	0	7.0	15.0	16.0	85.0	77.0	0	11.2	6.8

Note: By the Wilcoxon matched-pairs signed-ranks test (two tailed), the difference in percentage of total time in camp during which females groom and are groomed is significant ($p < 0.05$). That is, females groom significantly more than they are groomed.
a. Females are listed in decreasing order of age.
b. Melissa's grooming pattern was atypical in 1978, so her 1983 scores are given here also.

with regard to time spent grooming, although Figan received three times as much grooming as she did. Perhaps most surprising is the fact that Satan actually spent a slightly higher proportion of his camp time in grooming and being groomed than did Passion, despite the fact that he was alone for almost 20 percent of the time whereas Passion was always in the company of at least one potential grooming partner.

Alleviation of Stress

While much social grooming in chimpanzee society takes place during peaceful and relaxed rest periods, often when participants are replete after a hearty meal, it occurs too in a variety of situations when the chimpanzees are tense, anxious, or fearful. A frightened subordinate may approach a superior after an agonistic incident, then reach out and groom him (or her). Often this is merely a few movements, a "token" that has become incorporated into the repertoire of submissive gestures already described (and from which, most probably, some forms of reassurance touching have derived). Grooming of this sort enables a subordinate to establish physical contact in time of stress. Similar perfunctory token grooming may be the response of an individual to a highly frustrating situation, as when a female briefly grooms the male from whom she has been unsuccessfully begging for ten minutes or so.

Grooming is often the response of a dominant individual to the fearful approach of a subordinate seeking reassurance. When two individuals (particularly adult males) are *both* in a highly aroused state and *both* seeking reassurance, they often start to groom mutually (sometimes after first embracing). This grooming may become a prolonged session, during which both participants gradually relax.

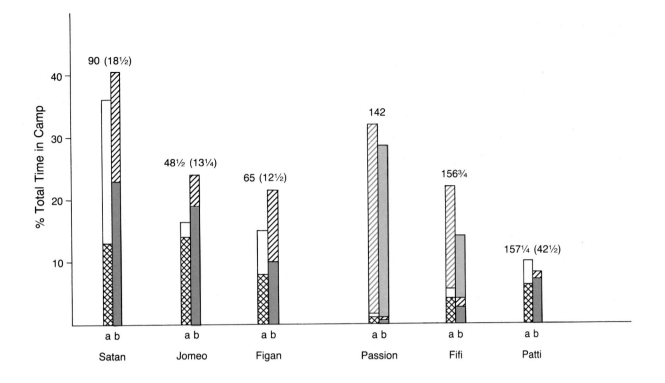

90 (18½)

48½ (13¼)

65 (12½)

142

156¾

157¼ (42½)

40

30

20

10

% Total Time in Camp

a b — Satan
a b — Jomeo
a b — Figan
a b — Passion
a b — Fifi
a b — Patti

⊠ Groom adult ♂ ♂
■ Groomed by adult ♂ ♂
☐ Groom adult ♀ ♀
▨ Groomed by adult ♀ ♀
▨ Groom offspring
▨ Groomed by offspring

a: % time spent grooming others
b: % time spent being groomed by others

Figure 14.3   The percentage of total time in camp during 1978 that three adult males and three adult females spent in grooming and being groomed. The total number of hours in camp appears at the top of each column; of those hours the number spent completely alone is given in parentheses. The bars show the percentage of time devoted to grooming with adult males, with adult females, and (for the two mothers) with offspring.

The chimpanzee is exposed to the reassuring influence of grooming from earliest infancy. Along with embracing, it is the immediate response of the mother to almost all forms of childhood anxiety, fear, or pain. Even a seriously hurt youngster will become calm under the deft, relaxing manipulations of his mother. When Mandy's three-month-old infant appeared with a torn and bleeding arm from which the bone protruded, her pain was obvious; she held her head rigid with eyes open, glazed and staring. But as Mandy groomed her she temporarily relaxed, and her eyes closed. Once the infant Flint was bitten by an enormous ant, which remained clinging to his brow; he whimpered and rubbed his face against Flo's breast. She was apparently unable to detect the cause of his distress, but groomed him from time to time. Each time she did so he quieted, only to whimper and rub his face again when she stopped.

The early-adolescent male, when he begins to spend time away from his family in association with adult males, is increasingly liable to provoke threats or attacks from his superiors. After he has been through a particularly difficult time, he leaves the males, apparently to search for his mother. The long grooming session that invariably follows his reunion with her undoubtedly helps to relax him after a time of considerable tension and stress. Of course, as the youngster matures and spends more time with individuals outside the family circle, they too groom him in moments of stress—just as he will eventually reassure his subordinates in the same way.

Maternal grooming, however, continues to calm and provide reassurance even when the recipient is fully grown. When twenty-three-year-old Figan screamed loudly after hurting his wrist in a dominance conflict, his mother, Flo, was about half a kilometer away. Old as she was (it was not long before her death), she ran to him. Figan, still screaming, approached her and she began at once to groom him. Gradually his screaming died away until he was quiet (Goodall, 1984).

Grooming that is directed *up* the hierarchy, which may function to reduce the anxiety of a subordinate, may at the same time calm the more dominant recipient. And grooming that is directed *down* the hierarchy, which may function to reassure a fearful subordinate, may in addition relax the groomer himself. It is not surprising to find that social grooming plays a vital role in patching up temporary rifts in normally friendly relationships, in improving those that are poor to start with, and in forging new friendly links (as when youngsters move into the adult world or when immigrant females seek acceptance). In this way, grooming is essential to the maintenance of harmony within the community as a whole.

The importance of grooming as a means of overcoming the tension of a reunion after separation is suggested by the fact that, of the 122 greetings recorded in 1979 that included some sort of friendly interaction, grooming occurred in 70.5 percent. As Figure 14.4 shows, it was particularly prevalent in greetings between pairs of males, occurring in 87 percent of the 46 observed; in half of these it was the more dominant of the two who groomed his subordinate. In half of the male-female greetings too, it was the more dominant individual—the male—who groomed. During the year analyzed, only eight friendly greetings that involved more than a soft grunt were observed between females. Both of the greetings between mothers and their adult daughters involved eager mutual grooming that resulted in prolonged sessions.

Bauer (1979) found that a male who displayed during a reunion sequence was more likely to become involved in a grooming session than one who did not, and that an individual was more likely to groom a chimpanzee he had just met than one with whom he had already been associating.

Grooming is an important component of sexual relationships, as we shall see in Chapter 16. Males such as Satan and Evered who groom females a good deal are likely to be most successful in the consort situation (Tutin, 1979). Brief token grooming is the typical response of the consort male to the fearful approach of a female he is trying to force to follow him. She thereby becomes less anxious, so that it is easier for him to lead her; furthermore, it is less stressful for her to follow, in that his grooming gives proof, as it were, of his fundamentally friendly intent. Throughout a consortship the male almost always grooms the female more than she grooms him.

While absolute levels of grooming between friendly males at Gombe may be high, the percentage of grooming relative to association time may be lower than between two males whose relationship is less

As Charlie starts to pant-hoot (his high level of arousal
can be inferred from his erect penis), adolescent Sniff
glances quickly at him, scratches (revealing tension),
then grooms the adult male.

Figure 14.4 The proportion of
friendly greetings in 1978 that included grooming between adult
males, between males and females,
and between females. The total
number of friendly greetings observed in each category is given at
the top of the column.

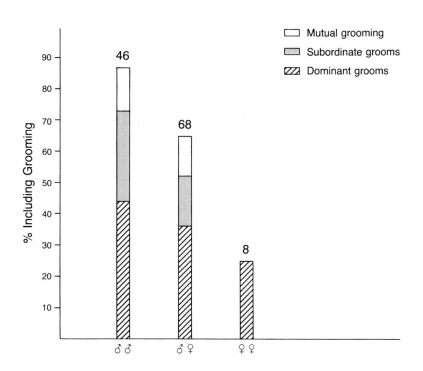

When Figan pauses in the grooming of his brother, Faben raises one arm, a request that the session continue. (L. Goldman)

friendly (provided they are not actually hostile). In 1974, while the brothers Figan and Faben had the highest absolute grooming scores, Figan groomed more with other males, relative to association time (Riss and Busse, 1977). The same was true of Figan and Humphrey in 1978. Absolute levels are high between males with a close relationship because they spend so much time together; each is available to the other as a grooming partner more often than any other male. Even if one member of such a pair is "busy" when the other approaches for grooming, he may be willing, at least for a while, to humor his friend. Wrangham (1975) describes a delightful incident involving Hugh and the younger and more dominant Charlie, thought to be brothers. Both males had carried food-laden branches to the ground to eat in comfort. Hugh finished first; he approached Charlie, gently took his branch away, laid it down out of his brother's reach, and began to groom him. Charlie reciprocated good-naturedly, but during the next twelve minutes slowly edged around until he could reclaim his property. He resumed eating and the session ended as amicably as it had begun.

Between companions such as these, levels of aggression are low and, for much of the time, neither has much need of reassurance. When two males have a less friendly relationship involving a good deal of antagonism, tension runs high. After an aggressive incident the subordinate may approach the other with submissive gestures; grooming is a frequent response and serves to calm both.

During times of instability in the upper levels of the male dominance hierarchy, the tensions generated as high-ranking males challenge one another may affect many members of the community. Even pairs of friendly males groom each other more often, seeking mutual comfort and reassurance as aggression erupts around them, particularly if one or both face aggressive challenges from a third. The result is apt to be a general escalation of grooming levels among the adult males. In the Arnhem colony, de Waal (1982) noted significant increases in time spent grooming by the three adult males, both during periods of instability in their dominance relationships and also when females came into estrus (when there was sometimes nine times as much grooming as during less stressful periods). At Gombe, young Goblin wreaked havoc with the male hierarchy from mid-1979 through 1980, as we shall see in the next chapter, and this created much tension among the senior males. At the same time, the Kalande community was pushing aggressively to the north, invading Kasakela territory. Probably because of the combination of these events, grooming levels among the adult males tended to be higher in 1980 (see Figure 14.5). Moreover, the overall grooming pattern of two of the males was somewhat affected: during 1979 Satan and Evered no longer showed the clear-cut "females most" pattern that had characterized their grooming behavior for the previous three years; instead, their grooming was distributed more or less equally between the sexes. Then, in 1980, they both swung over to the "males most" pattern that previously had been typical only of Humphrey and Jomeo. Figan showed the same trend. (Of course, we must remember that this change was due in part to the fact that there were so few cycling females during both 1979 and 1980.)

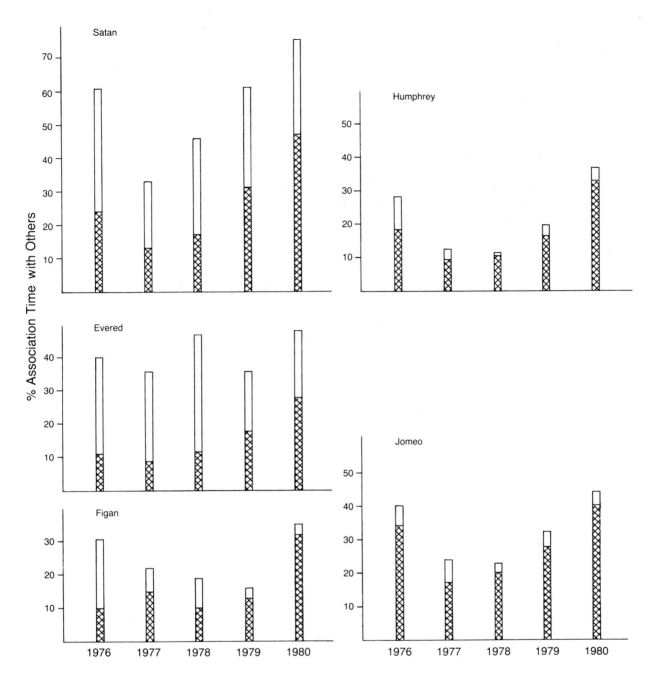

% Association Time with Others

Figure 14.5 Changes in the proportion of time spent grooming adult males as opposed to adult females, for five adult males over a five-year period, 1976 through 1980. The height of the column represents the time spent grooming expressed as a percentage of the total time spent in camp with others.

Groom adult ♂ ♂

Groom adult ♀ ♀

A Manipulative Strategy

Grooming is used in several rather different kinds of situations as a peaceful means of furthering the immediate goal of the groomer. The following are a few examples.

(a) Freud, after he had several times been prevented by his mother from touching his new sibling, began to groom her. Fifi relaxed and began to groom herself. Freud, by gradually working his way toward the baby, was able after three minutes to fondle Frodo's hand. In a similar ploy an infant during weaning may groom around the mother's breast, working his way ever closer to the nipple, then seizing a quick suckle. This strategy of diverting the attention of a more dominant individual is fairly common among the primates; it is often seen, for example, among the Gombe baboons.

(b) Before approaching Athena, who was in estrus, Sherry walked over to alpha male Figan and groomed him for twenty seconds. He then copulated with Athena (about 7 meters distant from Figan). Grooming in this context of placating the senior male is quite often observed at Gombe (Tutin, 1975) and also in the Arnhem colony (de Waal, 1982).

(c) Pax had been begging vigorously, with whimpers, for a share of his mother's fruits. Passion first attempted to hold her food out of his reach, then tried moving away—to no avail. Finally she held Pax firmly in her lap and groomed him with one hand. He became quiet and began to suckle, allowing Passion to finish her meal in peace. In this example the infant—that is, the subordinate individual—was pacified.

Chimpanzees, without doubt, gradually learn the calming effect that their own grooming behavior is likely to have on others. Grooming can then be used with *intent* as a manipulative tool. Over and above this, chimpanzees at times show some understanding of the significance of grooming interactions between others. Both at Gombe and in the Arnhem colony, adult males who were striving for alpha rank tried to break up grooming sessions involving their rivals.*

In the Arnhem colony Luit, challenging the current alpha, Yeroen, actively interfered almost every time his rival began a grooming session with another chimpanzee. By breaking these friendly contacts, Luit gradually ostracized Yeroen from potential allies and was able to take over the alpha position. When Nikkie subsequently challenged Luit, he displayed vigorously whenever Luit and Yeroen began to groom, while he himself groomed often with Yeroen and established a strong alliance with the older male that enabled him to maintain his position at the top (de Waal, 1982). At Gombe, Goblin practiced the same technique, displaying repeatedly toward and around pairs of senior males whenever they groomed. The situation was harder for Goblin, in that there were more senior males and he was unable to isolate them from one another. In fact, Goblin's behavior functioned to *increase* grooming and bonding among his rivals: the senior males often began to groom each other as soon as Goblin arrived. Presumably owing to the calming effect of their occupation, they were able

---

* Similar behavior was seen during the gradual formation of groups of unfamiliar gelada baboons. The dominant females consistently interfered with all friendly contacts between their male and other females (Kummer, 1974).

to ignore Goblin's tempestuous performances for longer than might otherwise have been possible. Studiously they avoided looking at him, instead peering with exaggerated interest at each other's coats. Only when he rushed by very close, or actually hit one of them, did they break off grooming to threaten him jointly.

Unequivocal evidence that chimpanzees can predict the effect of mutual grooming by others is provided by the females of the Arnhem colony when they act as mediators between rival males. This phenomenon is described in Chapter 19.

The importance of grooming in maintaining and improving relationships and in reducing levels of arousal has been stressed in many studies of nonhuman primates (for example, Carpenter, 1942; Yerkes, 1943; Sade, 1965; Oki and Maeda, 1973; Anderson and Chamove, 1979; Seyfarth, 1980). In a series of experiments designed to search for rules underlying behavior, Kummer (1974) found that social grooming was the end point in a predictable sequence of four behaviors between pairs of unfamiliar gelada baboons. (The same was true for hamadryas baboons; see Stammbach and Kummer, 1982.) Two males first worked out their relevant social rank by fighting. Some never got beyond this stage, but those that did progressed through presenting and mounting to grooming. Of particular interest is the fact that it was the winner of the fight who typically initiated the friendly behaviors, even when this meant that he risked losing his dominant status. Once the grooming stage was reached, the baboons apparently were able to relax. Males paired with females reached the grooming stage quickly and without fighting.

Pertinent to the above is a series of observations on pairs of captive hamadryas females. Those with a previously established friendly relationship groomed less, when separated from their groups, than those whose relationships were poor. The opposite was true when these females were in their social groups: there was more social grooming between pairs of friendly individuals, perhaps in order to protect the relationship in a competitive situation (Stammbach and Kummer, 1982). It might also be argued that when two animals (including humans) with a weak relationship find themselves together, separated from others, feelings of tension arise. These may be overcome, or at least reduced, by establishing friendly contact—grooming in nonhuman primates, speech in humans. On the other hand, if two animals who happen to be together have a satisfactory, relaxed relationship, there will be no tension and the need for grooming (or talking) will be diminished or absent. Perhaps if the friendly hamadryas females were exposed to some frightening stimulus during their period of togetherness, they would groom each other more.

Reynolds (1981) has pointed out that the social grooming networks of nonhuman primates share many properties with networks of material-object exchange in human societies: the *token* that is exchanged is different, but there is a similarity of *process*. The act of grooming, clearly pleasurable for the recipient, can be offered or requested, accepted or withheld. It can be reciprocated—immediately or on a subsequent occasion. Moreover, the process is volitional, given or withheld in accordance with the prediction of the actor about the likely consequences of his act. Grooming may be directed toward another with the intention of producing a desired effect; if that effect is not realized, grooming may be withheld in favor of other types of interaction or it may be intensified. De Waal describes social grooming as the price that may have to be paid, even by an alpha male, for favors received. Thus Nikkie, prior to copulating with a female, often first groomed both of the males subordinate to him. If, despite this, they displayed a little or pant-hooted as he approached the female, he returned and groomed some more, "thus raising the price" (de Waal, 1982, p. 179).

15. Dominance

(H. van Lawick)

July 1964 Mike, the new alpha, rests in the shade of a tree. A sudden crashing in the undergrowth heralds the arrival of Goliath, recently deposed from the top position. Mike does not move as Goliath charges flat out toward him, dragging a huge branch. At the last moment Goliath turns aside, swings into a nearby tree, and sits motionless. Only now does Mike begin to display, swaying the vegetation, hurling a few rocks, then climbing into Goliath's tree and swaying branches there. When he stops, Goliath displays again, leaping ever closer to his adversary until Mike responds. For a few moments both are wildly swaying foliage within 2 meters of each other; but there is no fight. They swing to the ground and charge off through the undergrowth, running parallel, then sit staring at each other. Goliath stands upright and rocks a sapling; Mike hurtles past, throwing a large rock. For the next twenty-three minutes the performance continues, and during the whole episode the only physical contact between them is when one is hit by the end of a bough swayed by the other. Finally, after a three-minute pause, Goliath moves rapidly toward Mike, crouches beside him with loud, submissive pant-grunts, and begins to groom him vigorously. For half a minute Mike ignores him, then turns and grooms his vanquished rival with equal intensity. For more than an hour they groom until both are relaxed and peaceful.

Ever since Schjelderup-Ebbe (1922) described the peck order in domestic chickens, the concept of dominance as an explanatory principle in the ordering of animal societies has had its ups and downs. As Hinde (1978) points out, it is necessary to distinguish between dominant and subordinate status as applied to, first, the relationship between a pair of individuals A and B (in which A bosses or is "superior to" B in a variety of contexts) and second, the patterning of relationships in a group of more than two individuals. In a linear hierarchy A will boss B, C, and D; B will boss C and D; and so on down the line. In many species, particularly in the higher primates, the dominance relationship between B and C is often complicated by the presence or absence of A, where A is an ally of C. If B and C are on their own, B can boss C; but if A (who is dominant to B) is nearby, then B will *not* be able to boss C, who may in fact be able to boss *him* (or her). This is the principle of *basic,* as opposed to *dependent,* rank (Kawai, 1958).

The relative status of A and B may initially be determined by fighting. But dominance must not be confused with aggression; once A has become dominant to B, and their relationship is clear-cut, A will not necessarily be more aggressive than B, and in fact there will be little need for overt aggression between them. Nevertheless, the fact that B may direct submissive behaviors toward A, even in the absence of overt aggressive signals, suggests that B is aware of A's *potential* for the expression of his superior status through aggression. If B oversteps the mark—if, for example, he approaches too closely as A feeds—a slight threat is usually enough to remind B of his relative rank and he will avoid further irritating A. As mentioned, the principle of dominance is beneficial not only for A, who exerts himself to the minimum, but also for B, who instead of wasting energy in a fight he would probably lose, simply moves off to find food elsewhere (Lack, 1966).

In chimpanzee society the dominance relationship between A and B, determined initially by an aggressive encounter (or through observation of aggression and the traditions of the community), may, as A gets older and weaker, persist beyond the time when B would, if it came to a fight, be physically able to defeat A and so reverse their relative status. In other words, B continues to defer to A through habit.

In the natural habitat the hierarchy, the patterning of all the different relationships within the chimpanzee society, is never static. Changes occur as youngsters compete during play sessions, learning the strengths and weaknesses of their companions; as games become aggressive; as allies step in to help on both sides, and win or lose; as young males challenge, first, the females of the community and then, one by one, the senior males; as young immigrant females compete with resident females for food; as males fight over females in estrus; as each individual watches the interactions among all the others. Nevertheless, for weeks or months at a time the ordering of relationships may be relatively stable, at least among the older individuals. When it is at its most stable, overall levels of aggression are likely to be low; each individual, in a manner of speaking, "knows his place" relative to each other. Threats in many contexts take the place of attacks. Thus a hierarchy, while it cannot be said to develop in order to control aggression within a society, often functions in exactly this way. We need only observe the escalation of levels of fighting during times of instability. In 1976, after Figan lost his clear-cut alpha status, there was one attack by an adult male (on any victim) per 103.5 hours during follows of adult males (a total of 889 hours); in 1978, when Figan had regained his alpha position, the figure dropped significantly to one per 261.5 hours (during 1,187 hours).

An observer encountering a group of chimpanzees for the first time will, if familiar with their behavior, be able to rank many of the individuals by watching their interactions upon reunion. As we have seen, greeting behavior typically serves to reaffirm the relative social status of the individuals involved. The subordinate shows submissive patterns, to which the other may respond with reassurance gestures.

High-ranking Charlie approaches alpha male Mike (*foreground*), who has just arrived and performed a charging display. He still shows some hair erection. Charlie first pant-hoots, then, as he gets closer, starts to scream. He is obviously very tense as he seeks reassurance contact. Mike turns his head away, a sign that he is unlikely to direct aggression toward his subordinate. (B. Gray)

The actual postures and gestures displayed, and their intensity, depend on the age, sex, personality, relative status, and state of arousal of the individual concerned. De Waal (1978, 1982) and Bygott (1979) found, when they analyzed patterns of dominance in chimpanzees, that the pant-grunt (de Waal's *rapid ohoh*) was the best indicator of relative status. A chimpanzee, during intense social excitement, particularly when in the presence of an ally, may display toward or hit a superior (to whom he would normally pant-grunt); but the latter will not pant-grunt to him. Presenting and crouching are also indicators of relative status, particularly among adult males. The soft bark of mild threat, uttered by confident individuals, is a call that is directed only *down* the hierarchy.

In chimpanzee society in the wild, males are the dominant sex. A normally healthy adult male is able to dominate all the females of his community, even when a number of them are together. In some captive situations, coalitions of females can put fully grown males to flight (de Waal, 1982) but this almost never happens in the wild, either at Gombe or at Mahale (Nishida, 1979).

The Male Hierarchy

When Bygott (1974) studied male dominance at Gombe between 1970 and 1972, there were fifteen socially mature males, and there was no precise linear hierarchy. For example, although Humphrey at that time certainly dominated more individuals than did Charlie, Humphrey was very wary of Charlie and avoided him whenever possible. Some of the males interacted so infrequently that meaningful conclusions could not be drawn about their relevant social rank. Bygott (p. 133) thus proposed a "hierarchy of levels." There was a clear-cut *alpha*—Mike; three *high-ranking* males—Humphrey, Hugh, and Charlie; six *middle-ranking* males—Figan, Evered, Hugo, Leakey, Faben, and Dé; and five *low-ranking* males—Godi, Goliath, Willy Wally, Jomeo, and Satan. After the community division in 1972, when there were fewer males in the Kasakela community, they could most of the time be ranked in a linear hierarchy (Riss and Goodall, 1977).

Figure 15.1 illustrates changes in dominance rank among the males of the Kasakela community as measured by the directionality of pant-grunting (1970 to 1974) and of pant-grunting and displaying toward in 1976 and from 1978 to 1980. The 1970 to 1974 data are from camp records only (the 1970 figures are from Bygott, 1974); the remaining four years combine in-camp and out-of-camp information for all males, from all records. These are raw data and do not take into account the varying times that different dyads were observed together. Nevertheless, they provide an overall picture which, along with other measures of dominance (such as winning of fights, fleeing, screaming, and so on), are helpful in monitoring changes in relative dominance rank from year to year. The combined pant-grunt and display data show how subordinate males, even when still pant-grunting to superiors, start to challenge them by means of displays. As their display frequency toward a particular individual increases, they tend to pant-grunt less to him. One or more years after these challenges are initiated, it is the defeated male who may begin to pant-grunt to his

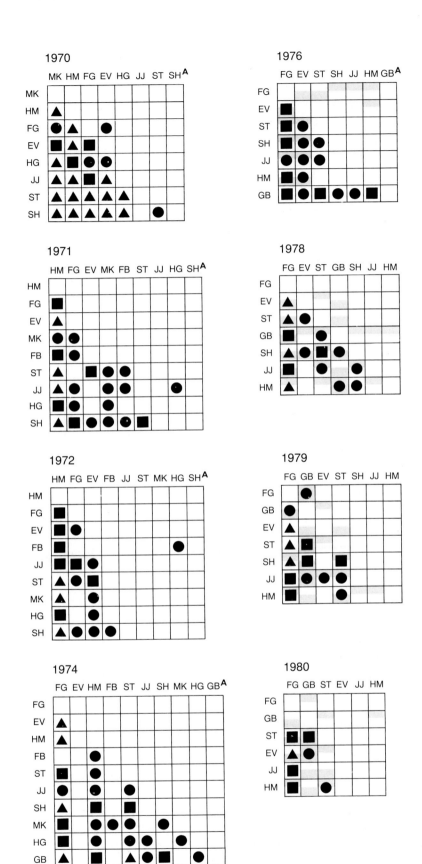

Figure 15.1 The directionality of pant-grunting (row to column) and displaying toward (column to row) among the adult and late-adolescent Kasakela males over eight years. From 1970 to 1974, data are from in-camp records only; from 1976 to 1980, data are from in-camp and out-of-camp records.

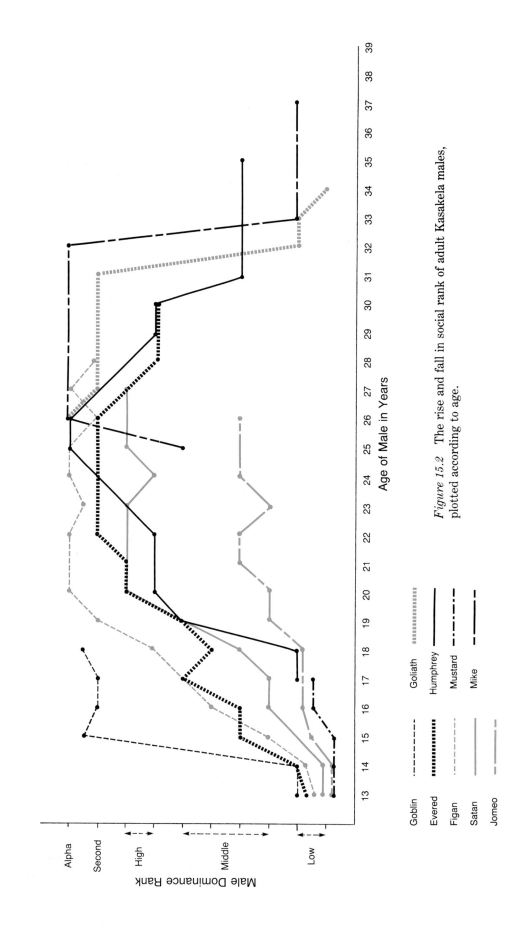

Figure 15.2 The rise and fall in social rank of adult Kasakela males, plotted according to age.

onetime subordinate (although the senior males are often seemingly reluctant to pant-grunt to young "upstarts" such as Goblin). At the same time the frequency of displays by the younger male toward his now-defeated rival will suddenly decrease.

Figure 15.2 shows the rise and fall in the dominance hierarchy of nine adult males, plotted by age. The males were ranked, on the basis of directionality of pant-grunts, within Bygott's hierarchy of levels: first and second positions are distinguished, and approximate positions within high-, middle-, and low-ranking classes. The figure suggests that although there is much individual variation, males tend to reach their zenith in the hierarchy between age twenty and twenty-six years; the median for six males was twenty-two years. Goblin, who reached top status on a one-to-one basis at the age of fourteen, was exceptional, and his case will be discussed in some detail. Before the community split it was much harder for young males to work their way up the hierarchy because there were twice as many senior males to defeat before they could reach a high position. Once a male reaches about thirty years of age his status drops, gradually or suddenly. Old males such as Goliath, Mike, and Hugo were very low ranking during the last few years of life.

Factors other than age which determine the position of a male in the dominance hierarchy include physical fitness, aggressiveness, skill at fighting, ability to form coalitions, intelligence, and a number of personality factors such as boldness and determination. Yerkes (1943, p. 47) observed that some dominance struggles were like "a contest of wills in which self-confidence, initiative, resourcefulness and persistence" seemed to be important. At Gombe some males strive with much energy to better their social status over a period of years; others work hard for a short while, but give up if they encounter a serious setback; a few seem remarkably unconcerned about their social rank.

Although the hierarchy of the adult males may be relatively stable for months, even years, at a time, there will always be some young, low-ranking males on the lookout for opportunities to better their status, ready to take advantage of a senior should he show signs of ill health or aging, or if he loses an ally. For the most part a reversal of dominance between two males takes place over an extended period of time during which the lower-ranked individual repeatedly challenges the other and eventually wins more and more frequently. In such cases there may be a long period during which it is difficult to ascertain which of the two holds the higher rank. Other changes occur more quickly as the result of one or a few severe, decisive fights.

The Charging Display and Fighting
Early in my study it became apparent that the charging display was of major significance to the male as he worked his way up, or maintained his position in, the hierarchy (Goodall, 1971). Bygott's analysis of the 1971 data showed that high-ranking males displayed with greater frequency than low-ranking ones and that, in general, a male did not display more often than an individual who was clearly dominant to him when in the presence of that individual. This was particularly true of the nonvocal charging display, which is more closely associated

with the tendency to attack than is the vocal charging display. When a male is obviously subordinate to and fearful of another, he seldom performs *any* nonvocal displays in the other's presence, and is likely to stop displaying if the other male suddenly arrives on the scene. Conversely, a male sometimes displays unusually often in the presence of a specific other individual. For example, after Figan had finally defeated Evered and become alpha male, he nevertheless continued to emphasize his position. During the fifty-day follow the next year Figan directed a higher percentage of his displays at Evered than at any other male; Evered, for his part, displayed in Figan's presence only once, when *all* the males were displaying during meat-eating excitement (Riss and Busse, 1977).

A senior male when challenged by the display of a younger rival often stands firm and may threaten or display himself. Then, at least in the early days of a dominance conflict, the subordinate may flee. There are times, however, when the higher-ranking individual avoids a particularly vigorous display and this response, without doubt, provides psychological reinforcement to the challenging male. Often the senior male will ignore (or appear to ignore) the displays of a young male who may then repeat his performance again and again, passing ever closer to the "victim" until eventually the senior is forced to take notice. Then he responds with threat, chasing, or—sometimes—avoiding. As we saw in the last chapter, the displays of a rival can more easily be ignored if the victim is able to groom with a companion. De Waal (1982) describes similar behavior in the Arnhem colony when the older male, Yeroen, studiedly ignored the prolonged and noisy displays of the young Nikkie.

Sometimes two rival males, close in the hierarchy, will display toward or near each other time and again, first one and then the other. I have already described how Evered and Figan behaved in this way during the tense months of their relationship. Goliath and Mike did the same both before and after Mike took over as alpha. And I have related how at the end of one long display-conflict of this sort Humphrey threw no less than twenty-two rocks, causing Satan, his rival, to leap for safety into the branches above.

Bygott found that there were differences among the males in the ratio of displays to physical attacks. Mike, who was alpha at the time, was observed to display eight times for every one fight, whereas the number two male, Humphrey, had a ratio of only three displays for one fight. Humphrey was a large, heavy male and fighting was probably less risky for him than for smaller males. The three who had the highest display-to-attack ratios were all medium sized: Mike was top ranking, Figan dominated the middle-ranking males, and Godi the low-ranking ones. This emphasizes the importance of the charging display as a strategy for the male chimpanzee in his struggle to achieve or maintain high rank; it enables him to succeed even without the advantages of large build or body weight, attributes that have been considered important for attaining high social status in many primate species (Trivers, 1972).

Our long-term study has emphasized the fact that while threats (among which the charging display is particularly effective) serve, for the most part, to *maintain* or challenge the existing social order, *reversals* in rank among the males have, in almost all cases, been the result of fights. As Scott (1958) pointed out, fear is quickly learned and hard to extinguish, so that a single decisive loss can have a long-lasting effect. Moreover, fighting ability itself may be diminished by failure, just as it may be enhanced by success (Kuo, 1967).

Dominance conflicts involve two kinds of fighting. First, the "hit-and-run" variety, as when a young male displays past a rival, slapping or stamping on him briefly as he does so, or when he seizes the opportunity offered by an ongoing fight to pound quickly on one (or even both) of the contestants, then hastens away. A combination of repeated attacks of this sort, during which the aggressor is seldom hurt, together with avoidance responses to his vigorous charging displays, gradually increases the confidence of a young male to the point where he will either dare to initiate a more severe attack or will stand firm and retaliate in the face of aggression directed toward him by his rival. This is the second and far more serious kind of dominance fighting—the kind that may lead to a permanent reversal of rank between two males. Sometimes it takes the form of a single decisive attack: it seemed, for example, that Mike lost his alpha status as a result of Humphrey's one observed attack. More often, rank reversals occur after a series of significant fights, such as the many attacks by Figan and Faben on Evered, and those between Figan and Humphrey, described in Chapter 8.

One decisive fight between Sherry and Satan, observed in 1975, seems to have influenced Sherry's subsequent career. Sherry was seen for the first time in 1974 to attack one of the mature males. The victim was Mike, so old and frail by then that his defeat represented only a psychological victory for the young male (it was soon after this that Sherry was seen, for the first time, joining a grooming session with the alpha, Figan). Subsequently Sherry began to challenge some of the senior males with frequent and vigorous displays and occasional attacks. The fight with Satan occurred when Sherry displayed toward the older male, who turned and attacked him severely. When Sherry, screaming loudly, managed to escape, he was bleeding from bad wounds on one shoulder, both hands, his back, head, and one leg. It seemed to be a turning point in Sherry's life, for it marked the end of his serious attempts to dominate the other males. Afterward, although he did sometimes initiate attacks on senior males (and occasionally won), it was in contexts such as meat eating or sexual rivalry—when there was some immediate reason for fighting.

To date there are no examples, at either Gombe or Mahale, of dominance fights (or any kind of fights *within* a community) that have resulted in the death of the loser. This fact does not mean that no such deaths occurred; a number of adult males have vanished (and were presumed dead, of unknown causes). The fight initiated by Goblin, which led to a gang attack on him by five males, resulted in serious wounds. The series of fights that took place over several months between the alpha and the second-ranking male of the K group at Mahale left both rivals with bad injuries (Nishida, 1983).

In two captive groups rivalry between males has led to the death of the loser. At the Arnhem colony three adult males (Yeroen, Luit, and Nikkie) had vied for dominance over several months. Eventually

Dé mounts Godi, and the two males confront Humphrey, the new alpha, during feeding excitement. This is an opportunistic coalition; all three are highly aroused. (D. Bygott)

J.B. and Mike, coalition partners, perform a joint charging display. (H. van Lawick)

Nikkie became alpha, through coalition with the former alpha, Yeroen; but the relationship remained tense. One night an unobserved fight took place in the sleeping quarters. In the morning Luit was found dead. Several toes and his scrotum had been torn off, and he had sustained many other wounds (de Waal, 1985). The incidents that led to the death of an adult male at Lion Country Safaris are described in Word Picture 15.1.

Coalitions

The ability of a male chimpanzee to enlist support during conflicts is perhaps the most crucial factor in attaining and maintaining high rank. At Gombe there are *stable coalitions*, based on long-term mutually supportive relationships or friendships (often involving siblings) and *opportunistic coalitions*, when two or more males (whatever their relationship) temporarily join forces against a common rival. Senior males often form opportunistic coalitions against a young male who is striving for high rank.

The alliance between Figan and Faben, which enabled Figan to attain alpha status, is a perfect example of a stable coalition. First apparent in 1967 (after Figan became dominant to his elder brother), it persisted until Faben's disappearance and presumed death in 1975. Faben almost always supported Figan's aggressive initiatives during this period. Even if he did not, on only one occasion was he observed to join forces against his brother, and never after 1973. Other stable coalitions involved Jomeo and Sherry (siblings); Hugh and Charlie, Leakey and Worzle, and Mike and J.B. (suspected siblings); McGregor and Humphrey (suspected uncle-nephew); Goliath and David Greybeard (thought not to be closely related); Figan and Humphrey, and Figan and Evered (known not to be closely related). Jomeo and Sherry had a very close supportive relationship from 1968 (when Sherry, as

A Dominance Struggle Ending in Death

For three months in 1971–72 a detailed study was made of the social behavior of a large chimpanzee group living on an island at Lion Country Safaris, Florida. At the start of the observations Old Man was clearly alpha male. On the night of 10–11 December 1971, severe fighting broke out on the island (the observer slept on the site, in a trailer). The commotion lasted for about forty minutes, from 0350 hours to 0430. In the morning Old Man and a younger male, Harold, both had bad wounds, and Old Man was submissive to Harold. Although Old Man gradually regained something of his former position, the dominance hierarchy was certainly disturbed. Conflicts broke out occasionally over the next six weeks—including one series of fights which, since they occurred at night, could not be seen but sounded severe (and three of the four adult males had slight wounds in the morning). Two weeks after this, a dominance conflict was for the first time actually observed: it involved fierce fighting between Old Man, Harold, and Black Knight. Old Man sustained a very severe wound in one foot, almost certainly inflicted by Black Knight. The dominance hierarchy afterward was very unsettled, and three days later Harold attacked and decisively defeated Black Knight, who had a severe wound on one foot and lacerations on his arms and head. After the fight he was heard, for the first time, to pant-grunt to Harold. As Harold swaggered around, displaying and hair bristling, Black Knight sat close to Old Man and the pair groomed as Harold continued to display. Obviously he had become alpha; even Old Man pant-grunted to him (Gale and Cool, 1971). Unfortunately, the Florida study came to an end soon after this episode.

About a year later Black Knight was found curled up in the center of the island, covered with wounds. Part of one ear was gone, the other torn, and his hands and feet were badly mauled with several bones exposed and some fingers and toes missing. A gash stretched from one shoulder to the opposite hip. Ten days after surgery, he escaped from his cage and was found sitting at the edge of the moat, gazing over at the other chimpanzees. The following day he died. An autopsy showed extensive adhesions in the pleural and peritoneal cavity and at least five broken ribs. There were still signs of massive bruising on his chest, back, and abdomen (T. Wolfe, personal communication).

a young adolescent, first began to leave his mother and travel with his elder brother) until 1977. That year Sherry was observed to challenge and dominate Jomeo, after which they traveled together less often. Neither was ever observed to support any other male against his brother.

Stable coalitions are characterized by high levels of association between the partners. This means that when one of the pair is in trouble, his ally often happens to be present and thus *able* to help. Moreover, the very fact that they are together often inhibits aggression from other males. Bygott (1974) noted that the only time he ever saw Humphrey direct aggression toward Faben was when Figan was not present.

Typically, when one member of a stable coalition displays (whether against a rival or not), his partner joins him and the two display in unison. When Figan and Faben were together during the fifty-day follow of Figan in 1974 (Riss and Busse, 1977), 60 percent of Figan's displays were accompanied by a Faben display. Hugh and Charlie had a relationship that was as close and supportive as the one between Figan and Faben, and their parallel displays, as described, were impressive and intimidating. In 1970, before the community split, Humphrey (then number two male) had the highest overall display frequency and displayed about twice as often as any other male who was in his party *except* when he was with Hugh and Charlie—who then displayed twice as often as Humphrey. When Humphrey met Hugh and Charlie together (and they usually were), they displayed in parallel at Humphrey, who avoided them; he seemed more subdued in their presence than at any other time (Bygott, 1974).

Even when one member of a coalition does not actually join in a conflict, it seems that his proximity provides "moral support" for the other. Goliath, for instance, took courage from the presence of his ally, David Greybeard. When Figan initiated the fight that gave him final dominance over Humphrey, his ally Faben, although he did not help, nevertheless was close by. After Faben's death Figan derived reassurance and comfort from the presence of old Humphrey, even though Humphrey seldom actively supported him.

Figure 15.3 illustrates the coalition network among the adult males of the Kasakela community over the seven-year period 1976 through 1982. There are two gaps in the data presented—1977 and the last four months of 1981 (the relevant information has not yet been extracted from the Kiswahili reports). Nevertheless, the figure gives a graphic representation of changes and stability of alliancing behavior over the years. I have only included interactions when two or more males actively joined forces to challenge one or more other males, or to retaliate in the face of a challenge directed against them. I have not included instances when one male simply protected another, as Figan sometimes protected Goblin from 1976 to the end of 1978. Only if Goblin subsequently joined Figan in threatening his defeated adversary is the interaction included.

Immediately apparent is the strong alliance between the brothers Jomeo and Sherry, which persisted until Sherry's presumed death in 1979. A network for 1973–74 would show a similar alliance between Figan and Faben. A comparable relationship developed between Figan and Humphrey, which lasted from 1978 until Humphrey's death in May 1981 (Figure 15.3b–d). These males were *never* seen to join coalitions against each other after 1976.

It was during 1976 that the other males repeatedly ganged up on Figan in apparent attempts to overthrow their alpha (Figure 15.3a). Evered, Satan, Jomeo, and Sherry on many occasions collectively challenged Figan; only Humphrey, for the most part, kept out of these interactions (the two he joined were coalitions of four against one, as we shall see in Figure 15.5). Humphrey also had ambivalent relationships with most of the other males. Of undoubted help to Figan as he strove to regain his alpha position was the fact that the relationships among the other males were also unstable. There was a good deal of antagonism, on a one-to-one basis, between his two principal opponents, Evered and Satan. The only reason they were occasionally

The presumed brothers Hugh and Charlie were close companions and made a strong and powerful team. Here they arrive in camp together, uttering pant-hoots in unison. (B. Gray)

allies (against Figan) was probably purely opportunistic, in that the desire of each to challenge Figan was stronger than their mutual rivalry. Satan, in addition, was still in the final stages of dominating Humphrey, and Humphrey was also being challenged by Sherry (sometimes supported by Jomeo) and Goblin. Even Evered still seemed to find it necessary to reemphasize his dominance over the previous alpha. Thus Humphrey turned increasingly to Figan; and Figan, for his part, found great psychological comfort in the older male's proximity.

By 1978 (Figure 15.3b) the males had calmed down. Figan had regained his unequivocal alpha position (although he never again had the complete power he had exercised when Faben was alive). He had not yet begun to form alliances with the other males because there was little need: his close relationship with Humphrey was established and he was almost never challenged.

In 1979 the older males, particularly Jomeo and Sherry, were beginning to team up against Goblin. And after the "great fight" in November, when Figan and four other males attacked Goblin, the senior males allied even more frequently than before. All these coalitions were in response to aggression initiated by Goblin, as we shall see. Goblin himself had no allies. Only Satan once joined him in order to challenge Figan, and that, as mentioned, was probably for his own opportunistic ends.

Figure 15.3e shows the coalition network after Humphrey died in early 1981. Figan and Evered continued to join forces against Goblin, as did all the other males. But Figan was no longer clear-cut alpha, and once again there was a good deal of ambivalence among the older males. After Figan's disappearance in May 1982 Goblin concentrated

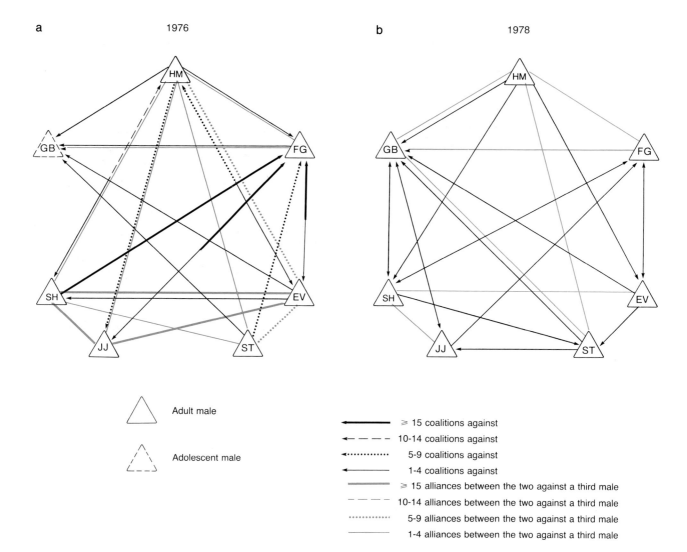

a 1976   b 1978

Adult male

Adolescent male

⟶ ≥ 15 coalitions against
← − − − 10-14 coalitions against
←·········· 5-9 coalitions against
⟵ 1-4 coalitions against
⟶ ≥ 15 alliances between the two against a third male
− − − − 10-14 alliances between the two against a third male
·········· 5-9 alliances between the two against a third male
⟶ 1-4 alliances between the two against a third male

Figure 15.3   Coalition networks illustrating the overall pattern of alliances among the adult Kasakela males from 1976 through 1982. Some networks cover an entire year, others are for a portion of a year; allowance should be made for the difference in the number of months.

When two individuals are linked by a green line only (for instance, Sherry and Jomeo in 1976), neither was observed during that year to join alliances against the other. Thicknesses of the lines indicate frequency of alliance; Sherry, for example, was observed to ally with his brother Jomeo more often in 1976 than with Satan. The green lines do not have arrows, for the most part, because the interactions (two or more individuals joining forces to threaten or attack one or more others) are mutually supportive. In the case of Figan and Goblin in 1976 and 1978, the black arrow on the green line indicates that whereas Figan was seen to help Goblin chase off senior males, Goblin was *never* seen to support Figan in any situation.

Black lines illustrate antagonistic relationships; arrows indicate the direction of the aggression. In some instances the lines are unidirectional, but in 1978, for example, Sherry was seen in a coalition against Goblin and Goblin against Sherry. Where the number of coalitions is greater in one direction than in the other (as for Humphrey and Evered in 1976), the thickness of the line changes at the middle.

Individuals linked by green *and* black lines displayed varying degrees of ambivalence. For instance, in 1976 Humphrey formed an alliance *with* Figan and also joined coalitions *against* Figan.

422 Dominance

c   January-November 1979 (11 months)

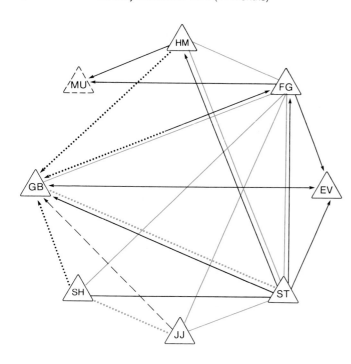

d   December 1979-May 1981 (6 months)

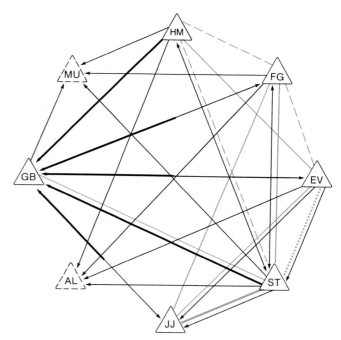

e   June-August 1981 (8 months)
     January-May 1982

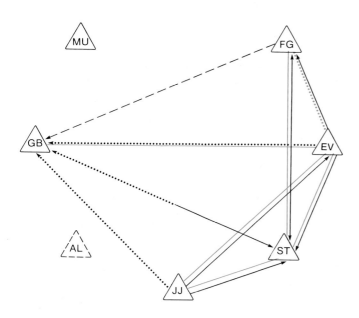

f   June-December 1982 (7 months)

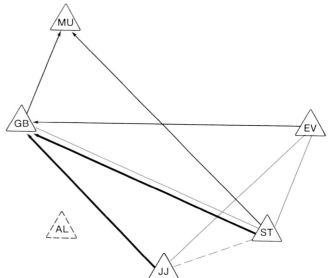

423 Dominance

his aggression on Satan and Jomeo, seeming not to bother with Evered, who by then looked quite old. Satan and Jomeo responded, as Figure 15.3*f* shows, by repeatedly forming alliances against Goblin.

In the Arnhem colony the adult females participated fully in the dominance conflicts of the three adult males. When Luit began to challenge the aging alpha, all the females supported Yeroen, in an effort to maintain the old order. Systematically Luit tried to break up contacts between Yeroen and the females, displaying toward them if they groomed or even sat near him, sometimes attacking them. Gradually the older male spent less and less time interacting with his would-be allies, and eventually Luit became alpha. Changing his strategy, Luit formed an alliance with his defeated rival and zealously prevented interactions between Yeroen and the third adult male, Nikkie. Here, however, Luit failed: Yeroen established a strong coalition partnership with Nikkie, and Luit was defeated. Nikkie then became alpha, but relied heavily on his supportive relationship with the oldest male. Yeroen, in turn, worked hard to prevent any friendly contacts between the other two males. Three years later the Yeroen-Nikkie alliance was still strong and, as described earlier, they (probably jointly) killed Luit during the night.

In the K group of the Mahale Mountains only three adult males remained in 1976. One of these, Kasonta, had been alpha since 1967, but he was defeated when a younger rival, Sobongo, formed a coalition with the third male, Kamenanfu. Kamenanfu occasionally changed sides during the prolonged and fierce periods of fighting between the other two. Nishida (1983) speculates that this "manipulation" gave Kamenanfu considerable advantage, since neither of the others could afford to antagonize him; both needed his support to win.

Motivation and Alpha Status
The alpha is the highest-ranking male of the community. This does not mean that he is able to control *all* situations, but instances when he is intimidated by other males (including coalitions) are rare. Some males devote much more time and energy than do others to the improvement or maintenance of their social status. Figure 15.4 shows which males over a twenty-one-year period attained, or made determined efforts to attain, the alpha position. In 1961 Goliath was alpha: we do not know how he achieved the position, but he maintained it by high levels of aggression and spectacular, fast charging displays. In 1964 Mike challenged and defeated Goliath, through the intelligent incorporation of noisy cans into his charging displays. He remained alpha, with a high display rate, for the next six years; at various times he was able to dominate between nine and fourteen other adult males. In early 1971, when Mike was old, Humphrey overthrew him—probably with a single fight; no special strategy was involved. Humphrey reigned for only twenty months, after which there was a six-month period when Humphrey, Evered, and Figan were vying for top rank and there was no unequivocal alpha. Figan took over, with Faben's help, in May 1973. Two years later, when Faben disappeared, Figan became increasingly unsure of himself. There was a nine-month period when, again, there was no unambiguous alpha. Figan, however, was able to reestablish himself, and his position was unchallenged until

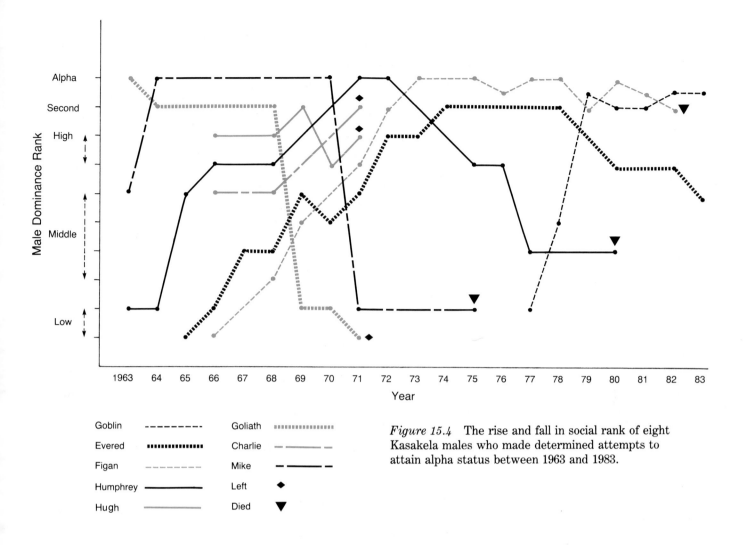

Male Dominance Rank

Alpha
Second
High

Middle

Low

1963 64 65 66 67 68 69 70 71 72 73 74 75 76 77 78 79 80 81 82 83

Year

Goblin — — — — —   Goliath ▪▪▪▪▪▪▪▪▪▪▪▪▪

Evered ▪▪▪▪▪▪▪▪▪▪▪   Charlie — — — —

Figan — — — — —   Mike — ▪ — ▪ —

Humphrey ——————   Left ◆

Hugh ——————   Died ▼

Figure 15.4   The rise and fall in social rank of eight Kasakela males who made determined attempts to attain alpha status between 1963 and 1983.

mid-1979, when young Goblin made a determined bid to overthrow him. Another period ensued when, for five months, there was no obvious alpha. Once again Figan worked his way back to the top, but nine months before his disappearance, Figan finally lost control. Once again there was no clear-cut alpha, this time for just over two years. On a one-to-one basis Goblin was clearly top, but he was frequently unable to intimidate pairs of senior males. In 1984 Goblin finally qualified as alpha male.

With the exception of Humphrey, these alpha males shared one characteristic, an intensely strong motivation to dominate their fellows. Humphrey got to the top because he was large and aggressive. Prior to his decisive attack on Mike he had become respected, if not feared, by most other individuals of the community. As Bygott (1974) indicated, because of Humphrey's size and strength he could attack with relative impunity. Mike, much smaller, with worn teeth and broken canines, would almost certainly have lost out to Humphrey several years prior to the final showdown had the younger male dared to challenge. Instead, Humphrey continued to show extreme deference to the old alpha throughout the year before the fight, an excellent example of how dominance status can be maintained through habit. And even when the time came, Humphrey might not have attacked Mike except for the presence of Faben, at that time his frequent companion and ally.

Humphrey, after becoming alpha, maintained a high frequency of both display and attack, yet was at the top for only twenty months. Throughout this period, on those occasions when the southern males (those who were in the process of forming the Kahama community) moved into the northern (Kasakela) part of the range, Humphrey was nervous and avoided them, particularly the Hugh-Charlie coalition. If the community had not divided, it is almost certain that Humphrey would never have become alpha: he showed none of the persistence and determination that characterized the dominance conflicts of Goliath, Mike, Figan, and Goblin.

Mike and Figan were both relatively small males, and both had low attack-to-display ratios. Mike, in fact, was never observed to attack another male at all during his four-month struggle to attain alpha status. For both Mike and Figan intelligence and ingenuity in display patterns were instrumental in getting them to the top. Mike's use of empty kerosene cans in his charging displays has been described (Goodall, 1968b, 1971). It must again be emphasized that while *all* of Mike's comtemporaries had the opportunity to use these cans, and most of them occasionally did so, only Mike profited from the experience and used it to further his own ends. (Although on two separate occasions Figan, then an early adolescent, was observed by himself in the bushes, performing practice displays with one of Mike's discarded cans.)

Mike pant-hoots as he starts a charging display with two kerosene cans in 1964. (H. van Lawick)

Mike's deliberate planning was a striking aspect of his rise to alpha status. Once, for example, as a group of six adult males groomed about 10 meters away, Mike, after watching them for six minutes, got up and moved toward my tent. His hair was sleek and he showed no signs of any visible tension. He picked up two empty cans and, carrying them by their handles, one in each hand, walked (upright) back to his previous place, sat, and stared at the other males, who at that time were all higher ranking than himself. They were still grooming quietly and had paid no attention to him. After a moment Mike began to rock almost imperceptibly from side to side, his hair very slightly erect. The other males continued to ignore him. Gradually Mike rocked more vigorously, his hair became fully erect, and uttering pant-hoots he suddenly charged directly toward his superiors, hitting the cans ahead of him. The other males fled. Sometimes Mike repeated this performance as many as four times in succession, waiting until his rivals had started to groom once more before again charging toward them. When he eventually stopped (often in the precise spot where the other males had been sitting), they sometimes returned and with submissive gestures began to groom Mike.

Whereas most directed displays are performed in silence, Mike's were almost always accompanied by pant-hoots; it is as though he was emphasizing his identity in connection with the noise of the racketing cans. Since he learned to keep as many as three banging and bounding ahead of him as he charged, flat out, toward his rivals, it is scarcely surprising that his bluff was so effective. After we removed the cans,

Figan, alpha male for ten years, 1973 to 1983. (C. Tutin)

Mike made determined efforts to secure other human artifacts to enhance his displays—chairs, tables, boxes, tripods, anything that was available. Eventually we managed to secure all such items; he then made extensive use of natural objects. Once, for example, Mike came charging down a slope toward other males dragging *two* palm fronds in one hand. Suddenly he paused, seized a third, and with great difficulty continued his display with all three. Mike maintained the highest display rate of all the males throughout his tenure as alpha (Bygott, 1974). By 1970 he was old in appearance and some of the younger males (such as Figan and Evered) were already challenging him. It was probably only his keen social awareness that enabled him to maintain his position as long as he did. Bygott comments that Mike tended to approach camp silently and, if he saw other males there, would burst upon them in a vigorous display, startling and scattering them.

Figan too showed intelligence in the timing and positioning of his displays. Sometimes, as he approached an unsuspecting group of males, he moved quietly around until he reached a point on the slope above them, then charged down suddenly. This gave his display an element of surprise that was extremely effective (Bygott, 1974). Another tactic was his early-morning or late-evening arboreal display, performed while his companions were in their sleeping nests. Again, the element of surprise caused much confusion. It was during a series of late-evening displays, after the rest of the group had nested for the night, that Figan finally gathered his courage and attacked Humphrey, twice; and it was these attacks that led to the status reversal between the two.

Without doubt it was the support of his brother, Faben, that enabled Figan to attain alpha status. As described in Chapter 8, Faben's help was crucial in the final defeat of Evered. Yet I have the distinct impression that Figan would have managed to reach the top even without Faben's help (just as I suspect Mike would have done so even without the cans). These are idle speculations, but the determination to better their rivals, and the ingenuity shown in doing it, by both Figan and Mike, are facts. Figan, from early adolescence on, had seized every available opportunity to get the better of his superiors—when they became sick, were injured, or showed any sign of weakening—even if it only gave him a temporary advantage (Goodall, 1971; Bygott, 1974). He even subjected his elder brother to a series of aggressive displays when Faben lost the use of one arm during the polio epidemic. After that Figan was able to dominate Faben for the rest of his life. Figan began to display at Mike at least a year before Humphrey finally dared to challenge the old male; and Mike was intimidated by Figan's threats before he heeded those of any other male.

Even though Figan lost overall control after Faben's death, he managed to regain alpha status despite the fact that *all* the other males united against him from time to time. Figure 15.5 shows that during 1976 twenty observed coalitions were directed against Figan by various combinations of the other males. Fourteen times three of the males joined forces against him, and twice four of his rivals challenged him at the same time.

Figure 15.5 The various combinations of adult male alliances against Figan that were observed during 1976. For example, Evered, Satan, and Sherry challenged Figan three times; Evered, Jomeo, and Sherry challenged him eleven times (the triple bars indicate high frequency), and so on. A total of twenty coalitions are shown in this way.

No. times coalition formed	Humphrey	Evered	Satan	Jomeo	Sherry
3		△————	△		—————△
11		△══════		══△══	══△
1		△————		————△	
1		△————			————△
1				△———	———△
1		△————		————△	
2	△———	△————		————△	———△
20 coalitions	2	19	3	16	18

There were four separate factors, I believe, that enabled Figan not only to cling to his shaky position during the period following Faben's death, but subsequently to strengthen it, so that during 1977 he regained much of his previous control.

(a) Even though he sometimes fled, screaming, from the other males, they never pursued their advantage to its logical conclusion; they chased Figan, but never attacked him. Instead, when they caught up with him, all those involved began screaming loudly, approached, embraced, and began to groom one another with much excitement.

(b) When Figan next encountered one of his adversaries separated from any possible ally, he typically displayed toward him with vigor and was seen to attack each of them at least once that year. He himself was not seen to be even mildly attacked during the year.

(c) The temporary loss of a clear-cut alpha seemed to have an unsettling effect on relations among the contenders themselves. Only Jomeo and Sherry were really staunch allies—and Sherry was barely out of his adolescence, while Jomeo (as we shall see) had at that time little or no motivation to better his status.

(d) Figan strengthened his relationship with Humphrey.

Mike showed a similar ability to withstand formidable coalitions. The most striking example was an incident that took place in 1964, soon after he had become alpha. Led by David Greybeard, who was quickly joined by Goliath, an alliance of five senior males advanced threateningly on Mike. He turned and fled, seeking refuge up a tree. The five followed, displaying and uttering fierce waa-barks, and it seemed that Mike would lose his newly acquired status. Instead, as they pursued him into the branches, he suddenly turned on them—and all five retreated to the ground! Mike's victory was particularly impressive in that even his suspected sibling, J.B., joined in against him. The fact that as many as five angry adult males can be intimidated by one determined individual, quite on his own, is another example of the importance of psychological factors in chimpanzee dominance interactions. It implies also that the lone male who dares to face such opposition is either stupid (cannot imagine the possible consequences) or has rather a large share of boldness—a quality that perhaps comes close to courage. Mike, Goliath, Figan, and (as we shall see) Goblin all showed this characteristic in a variety of situations. None of them could be described as lacking in intelligence.

These males were, in addition, characterized by persistence. Bygott observed that Figan displayed repeatedly at older males, charging toward them again and again. The first time he was observed to use intimidation tactics on Faben, he displayed toward and around him nine times in fifteen minutes before finally shaking him out of his tree. In 1971 he displayed nine times in quick succession toward the high-ranking Hugh and then, as Hugh continued to ignore him, he finally dared to attack him briefly (Bygott, 1974). Mike showed similar perseverance in his displays with the cans, challenging the senior males, particularly the alpha Goliath, day after day until he achieved success. Goliath, for his part, made valiant attempts not only to hold onto his top rank in the face of Mike's challenge, but to regain it after his defeat. Goblin perhaps had this quality of persistence in greater measure than any of the others. When challenging the adult females, for example, he displayed not once or twice in succession, but ten times or more (his record was twelve times in fifteen minutes). Even after being chased and beaten up by pairs of infuriated females, he was quite ready to try again the next time he encountered them. And, as he got older, Goblin showed the same tenacity when challenging the senior males, displaying time and time again until he had provoked some kind of response. And if the result was a fight that Goblin lost, it made little difference to his determination to force the issue when next he met the older male. Even when he was badly wounded during dominance conflicts, Goblin never gave up; indeed, he seemed to renew his challenges with increased vigor once his injuries had healed.

Goblin's rise to alpha position is unusual in many ways. Commentary 15.2 provides a summary of the events surrounding his accomplishment.

Loss of Rank

As shown in Figure 15.2, some males begin to drop in the hierarchy at a much younger age than others; some show a gradual loss of rank, and others lose status rapidly. The response of a male to his loss of rank—whether because of ill health, defeat by another male, loss of an ally, or any other factor—will be determined in large part by his age at the time. Thus when an old male (over thirty years of age) suffers a serious setback, he is likely to drop precipitously through the hierarchy—as did Mike when, after being attacked by Humphrey, he lost his alpha status. This is understandable: prior to the incident Mike's social position had been maintained for some time through habit alone. Humphrey's submissive behavior before the decisive attack was obviously because he remembered Mike as a powerful and aggressive male; and Mike perpetuated this unrealistic picture for an amazingly long time by skillful tactics—such as keeping out of the way when it seemed that he might become involved in fighting, and continuing his impressive and well-timed displays. But his threats were pure bluff. Long before Humphrey's attack Mike would have lost a test of physical strength to almost any of the other adult, or even late-adolescent, males. And some of the younger males such as Figan and Evered seemed to sense this fact, for they began to ignore Mike's displays before the showdown.

In marked contrast to Mike's sudden drop in rank, Goliath and Humphrey (both of whom were toppled from the alpha position at about age twenty-six) remained high ranking for several more years (five and three respectively). When Goliath was about thirty-one he became very sick. After a three-week absence he reappeared in an emaciated condition, weighing only 31.5 kilograms (as compared with 35 kilograms three months later). He was immediately submissive to the other males and, like Mike, remained very low ranking for the rest of his life.

A younger male is more resilient and can move up the hierarchy again (provided he is motivated to do so) even after quite a serious setback. When Faben and Pepe, as young males rising fast in social rank, were crippled by polio, their loss of status was immediate and unambiguous: when threatened by others they were submissive and fearful. After adjusting to their physical disabilities, however, both began to rise once more in the hierarchy. Another male of similar age, Willy Wally, was also afflicted with polio. He was very low ranking to start with and it made little difference to his status (Bygott, 1974); he never made any subsequent attempts to move up the hierarchy.

When a male has been severely and decisively beaten during a fight with a rival, he may, especially if he is young (like Goblin in 1979) or highly motivated (like Figan in the same year), tackle his opponent again. Even if he does not regain his previous status, he may nevertheless maintain a relatively high position in the hierarchy. But if he is defeated again, once or several times, he may to all intents and purposes drop out of the struggle. Sherry appeared to lose his initiative in matters of rank after his severe beating by Satan and made no further serious attempts to better his status.

Evered, in response to the repeated persecutions of the powerful Figan-Faben alliance, began to spend less and less time in the core area of the community range. Admittedly he had occasionally absented himself for periods of a month or so even *before* these attacks, but his peripheral wandering peaked in the 1973–74 period. Kasonta, the old K-group male at Mahale who lost his position to a younger rival, wandered, mostly alone, for over a year in the northern part of his range. He is thought to have associated with the other two males for only one five-day period during the entire time. Interestingly, he then returned to the core area and regained his alpha status (Nishida, 1983). In the M group the former alpha Kajugi, after being defeated during conflicts with the other males, also moved out to the periphery of the community range and was still wandering there alone after two years (Hiraiwa-Hasegawa, Hasegawa, and Nishida, 1983).

Commentary 15.2

Goblin: A Case History

Goblin's meteoric rise through the male dominance hierarchy resulted at least in part from his unusual relationship with Figan. From the time when he first began to travel independently of his mother, Goblin was attracted by Figan. He spent more time with him than with any other adult male (Pusey, 1977), and this close association continued throughout his adolescence. Goblin was very respectful of his "hero," followed him around, watched what he did, and often groomed him. Figan occasionally was mildly aggressive, but on the whole he was remarkably tolerant.

When Goblin began to challenge the adult females of the community, Figan was never seen to help him even when, as quite often happened, the higher ranking and more aggressive of them turned on Goblin and beat him up. But Goblin, by the age of twelve, had managed to dominate them all. At the end of 1976, therefore, he turned his attention to the adult males (Figure 15.6). And almost every time that his displays provoked an aggressive response from one of them, Figan charged over to help Goblin. After a while Figan's mere presence seemed to inhibit the other males, so that Goblin, like the child of a high-ranking female, repeatedly "got away with it" when he challenged his seniors.

It was Humphrey, oldest and by 1977 lowest ranking of the senior males, who was Goblin's first target. Figan never supported Goblin in these conflicts; obviously his close relationship with Humphrey was too important. But he did not help Humphrey; at most he displayed between the two, thus ending the incident. Goblin, however, succeeded in intimidating Humphrey without assistance, and by the beginning of 1978 Humphrey was heard, for the first time, to pant-grunt submissively to Goblin. Meanwhile, Figan continued to back Goblin in his conflicts with the other males, and during 1978 Goblin seldom challenged any of them unless Figan was present. In the last four months of that year, for example, Goblin was observed to display toward other males twenty-nine times; Figan was present on all but nine of these occasions. Twice, when the males concerned turned on Goblin, Figan charged over to his young friend's defense and the aggression stopped. When the older males themselves initiated aggression against Goblin (seen a mere thirteen times during the whole year), he retaliated *only* when Figan was present (six times): twice Figan supported him.

By early 1979 Goblin, on a one-to-one basis, could intimidate all of the senior males except Figan. And in May he was seen, for the first time, to challenge Jomeo and Sherry when they were together. Four times he displayed past the two as they groomed. He went a little closer each time, finally hitting Jomeo. The brothers were enraged and chased him off. But Goblin did not give up. Four months later came a dramatic conflict, which Goblin won despite being attacked by both brothers—each heavier than himself. From then on he was very obviously able to dominate them, even when they were together.

In June 1979 came the first sign of change in Goblin's relationship with Figan. One day, instead of hurrying over to greet the alpha on arrival as he had always done before, Goblin ignored him. After this he very rarely pant-grunted to Figan, who became increasingly tense; when Goblin appeared unexpectedly, Figan often grinned and squeaked, then hastened to one of the older males for reassurance.

In August, when Figan was lame from sores on his fingers, Goblin began to challenge him in earnest, displaying toward him again and again, sometimes hitting him as he ran past. Figan always rushed for support to one of

Figure 15.6 The various combinations of adult male alliances against Goblin that were observed during six periods. Double bars between coalition partners indicate a higher frequency of coalition than single bars, and triple bars indicate a still higher frequency.

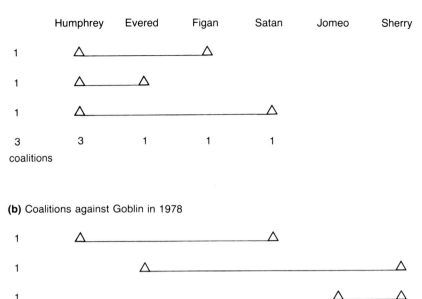

(a) Coalitions against Goblin in 1976

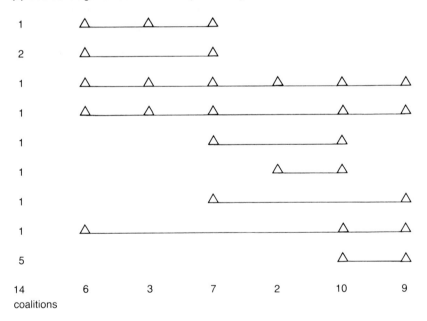

(b) Coalitions against Goblin in 1978

(c) Coalitions against Goblin in 1979 (Jan.- Nov.)

the senior males. As Figure 15.6c shows, in 1979 all of these males formed coalitions with Figan against young Goblin. Goblin, having turned against his long-time supporter, now had no ally. Instead, he relied on the devastating effect of vigorous displays repeated over and over again. It was obvious that he had not only profited from Figan's support through the long years of their close association, but he had also learned some of Figan's techniques, such as early-morning displays and surprise tactics (hiding in the undergrowth as others were heard, then charging out with startling suddenness).

(d) Coalitions against Goblin Dec.1979 to May 1981 (after the great fight, until Humphrey's death)

	Humphrey	Evered	Figan	Satan	Jomeo	Sherry
6	△———————————△					dead
6	△—————————————————△					
1	△	△		△		
2	△	△	△	△	△	
7		△———————△				
1		△———————————————————△				
1			△————△			
1			△——————————————△			
12				△=======△		
4		△———————————————△				
41	15	15	17	26	16	
coalitions						

(e) Coalitions against Goblin Jun.- Aug. 1981, Jan.- May 1982 (8 months) until Figan's death

	Humphrey	Evered	Figan	Satan	Jomeo
1	dead	△———————△———————△			
2			△———————△		
2			△———————————————△		
2		△———————△			
7		3	7	3	2
coalitions					

(f) Coalitions against Goblin Jun.- Dec. 1982 after Figan's death

	Evered	Figan	Satan	Jomeo
1	△————dead————△			
1	△———————————————△		△	
14			△=======△	
16	2		16	15
coalitions				

Early-adolescent Goblin grooms alpha male Figan in 1973. He misinterprets a sudden movement by Figan (perhaps to swat a fly) as a threat and leaps back, screaming, then reaches out for reassurance. (C. Packer)

At the end of September 1979 the first serious fight between Figan and Goblin was seen: it was a decisive win for Goblin. Figan, kicked from the tree into which he had fled, fell some 10 meters to the ground and ran off screaming. Later that month, after Goblin had displayed at him several times in succession, Figan again took refuge in a tree; Goblin sat below, calmly grooming with Mustard, for a whole hour. Figan became increasingly agitated and began whimpering. But he did not dare to climb down until Goblin had moved away.

Thus by October 1979 Goblin, only fifteen years of age, had the highest basic rank of all the males. He was unable to control situations when two of the others were together, but this did not prevent him from repeatedly displaying around them, causing much confusion.

In mid-November Goblin was defeated in the "great fight" I have mentioned. Figan, joined by Humphrey, Evered, Satan, and Jomeo, attacked Goblin after an aggressive meat-eating session. Goblin (who had himself initiated the incident) was badly wounded before he managed to escape. After this he was very uneasy in Figan's presence, and Figan regained confidence. In December Figan once sat for twenty minutes below a tree in which Goblin was trapped and apprehensive; the tables were turned.

By February 1980 Figan had reestablished himself as alpha, though a rather weak alpha. Goblin, once he had recovered from his wounds, continued to challenge the other males (except Figan). He devoted most of his energy to resolving his position with Satan, finally and unequivocally. After the great fight, however, the senior males repeatedly and enthusiastically supported one another against Goblin, as shown in Figure 15.6d.

Gradually, during 1981, Goblin regained his confidence and by the end of the year had begun to challenge Figan again. As Figure 15.6e shows, Figan, who became more and more afraid of Goblin, formed coalitions against him with each of the three remaining senior males. Despite this, and even before Figan's disappearance in mid-1982, Goblin once more became top-ranked male. Still he could not dominate the senior males when any two of them were together, but with Figan gone he displayed at the others with renewed vigor, disrupting peacefully resting or feeding groups again and again with his endless displays. He went out of his way to terrorize Jomeo (Sherry had vanished in 1979). Once, after Goblin had displayed at and around Jomeo three times, finally kicking him hard, Jomeo frantically appealed first to Mustard and then to Atlas for support. When both young males refused to help, Jomeo threw a violent tantrum and left. Indeed, unless he was with one of the other senior males, he began to avoid Goblin's company.

Throughout 1983 Goblin persisted in his attempts to dominate the seniors. They, for their part, groomed one another with persistent concentration when Goblin began his displays. In this way they were sometimes able to ignore him for as long as twenty minutes—but in the end he invariably goaded them into action. Here is a typical incident from mid-1983.

It began when Goblin joined a group that included Satan and Jomeo. As soon as he appeared, the two seniors approached each other and began to groom. Goblin, after displaying around them seven times, attacked Satan, leaping into the tree above and stamping down onto the older male's head. Satan and Jomeo, screaming loudly, charged Goblin together. And despite the fact that his adversaries weighed 49 and 47 kilograms respectively, Goblin, who is only 37 kilograms, stood his ground and tackled them both. For just over a minute they grappled and hit at each other—then Satan and Jomeo fled. Goblin chased them, throwing rocks. Five minutes later he attacked Satan again and then left the group altogether.

Finally, sometime in 1984, Goblin became undisputed alpha male, able to control almost any social situation. And today his closest companion is Jomeo. The two travel together, groom each other, and share meat after a successful kill. Things have calmed down among the adult males, and Goblin's display rate has probably dropped by half.

It should be noted that Evered, during his absences to the north, is thought to have spent at least some of the time in consort with females; in that way he was able to turn his temporary "exiles" to his reproductive advantage. Sherry, after giving up in 1976, also spent long periods away from the center of the community range; perhaps he too was consorting with stranger or peripheral females. The defeated K-group male at Mahale sometimes associated with the females of his community, at least during the early part of his fifteen-month exile (Nishida, 1983).

Lack of Motivation

One male, Jomeo, has always been remarkable for his lack of preoccupation with matters of rank. He was first identified in 1964 when he was about eight years old. His mother, Vodka, was still alive but was shy and seldom seen. In 1965 Jomeo, like other early-adolescent males, began to challenge the community females. Although he was often attacked and chased off, as are all youngsters at that age, he once swaggered and bristled so effectively that he was able to withstand the efforts of the high-ranking Passion when she tried to chase him from a box of bananas. In the end, they shared the fruits.

In 1966 Jomeo appeared one day with a badly wounded foot. He was unable to put it to the ground for almost a month, and to this day all the toes are curled under. We do not know what happened, but it marked an abrupt end of his attempts to intimidate adult females. In 1968, after Jomeo had mildly attacked her infant, Passion turned on him; he fled, screaming, with the female in hot pursuit. Even in 1971, when he was about fifteen years old and weighed some 44 kilograms, Passion was still able to put him to flight. At the end of that year, again in unknown circumstances, Jomeo once more received severe injuries. This time he sustained deep wounds on his back, hands, feet, and head.

In 1972 Jomeo's young brother, Sherry, began to challenge the community females and Jomeo invariably went to his rescue when necessary. This often involved him in fights which, with his heavy build and large size (as well as the backing of Sherry), Jomeo had no trouble in winning. Rapidly he gained more confidence and the females, suddenly learning his strength, began to treat him with respect.

At the same time Jomeo's display frequency rose, but his technique was poor. Once, as he charged downhill, he tripped over a tree root and fell. Another time he tried to enhance his display by rolling a large rock; when it failed to move, he stopped and began to pull and heave, quite destroying the effect of his display. It was the same when a sapling he seized remained firmly rooted; he stopped, heaved, and tugged at it until finally he tore it from the ground, then tried to continue his display. But the sapling was so large that he sometimes had to move backward, using both hands.

Very rarely has Jomeo been seen to *initiate* aggressive interactions with community males. He may retaliate if provoked, and he sometimes seizes the opportunity for a hit-and-run attack during ongoing aggression. In 1977, when Humphrey was already a low-ranking male, Jomeo did once display toward him and hit him a few times. But while he got the better of him on that occasion, four years later an even

Mustard, one of the few males who exerts very little effort in attempts to better his dominance status.

older Humphrey was able to chase Jomeo away during group excitement.

One other male, Mustard, appears to be developing in the same way as Jomeo. In 1981, when he was sixteen years old, Fifi was able to put him to screaming flight on five separate occasions. He was seen to challenge females only a very few times until he was seventeen. We know something of Mustard's family background and early life; are there clues there to his curious lack of "dominance drive"?

Until he was five years old, Mustard had many opportunities to interact with other youngsters because his mother, Nope, regularly joined the large banana-feeding gatherings in camp. After the change in feeding methods, however, she became increasingly solitary. Because of Nope's somewhat atypical reproductive cycle, both Mustard and his younger sister were suckled for seven years. Mustard was extremely dependent as an early adolescent: he seldom left his mother, except for very brief periods, until he was eleven years old. His lack of social experience during the impressionable years of early adolescence was mentioned in Chapter 4. Any of these factors, as well as the inherited genetic makeup of the individual, might underlie the unusual lack of dominance drive. As far as Jomeo is concerned, we do know that his disinterest in social rank does not stem from lack of aggressive tendency per se; he took part in all five of the observed attacks on Kahama individuals and was one of the more brutal assailants. Not until we are able to document the genealogy and early history of other males who, as adults, show behavior similar to that of Jomeo and Mustard, will it be possible to probe more deeply into this fascinating issue.

Dominance and the Female

De Waal (1982, p. 186) found that in the Arnhem colony the female hierarchy was stable and was based on "respect from below rather than intimidation and a show of strength from above." He ascertained that most pant-grunts (which he referred to as "greetings") were "spontaneous" and felt that, for the female, the *acceptance* of dominance was probably more important than the *proving* of dominance. He therefore suggested that Rowell's (1974) "subordinance hierarchy" might be applied meaningfully to female chimpanzees.

However, de Waal was not present when the Arnhem colony was first formed. It is possible that the "respect from below" was in fact based on the outcome of aggressive interactions at the time of first acquaintance. Yerkes (1943) found that when adult females were placed together, they went through the same process of settling their relative status as did the adult males.

At Gombe, although there is a good deal of aggressive interaction among the females, they cannot be ranked in a clear-cut dominance hierarchy. This is partly because some of them associate only rarely; partly because, even when they do spend time together, agonistic interactions between some dyads are seldom observed; and partly because the relative status of any two females is affected (perhaps even more than for males) by the presence or absence of particular others—principally family members. There are, however, females who are clearly very high ranking (one of whom often emerges as alpha) and others who rank very low; the remainder fall into a middle-ranking

High-ranking female Circe threatens Melissa during banana feeding. (H. van Lawick)

As higher-ranking Flo passes, Olly touches her in greeting. (H. van Lawick)

class. Thus Bygott's "hierarchy of levels" can be meaningfully applied to the females.

In the days of heavy banana feeding (1964 to 1968), when the chimpanzees gathered together frequently in large parties, the females associated with one another far more than they do today. It seemed (although the data are not yet analyzed) that there was a more clearly defined rank order among them. Certainly until the end of 1965 Flo was at the top. She was able to win fights with all the other females unless they were backed by late-adolescent sons. By 1966, however, Flo was looking very old and gradually began to lose her high status. The robust female Circe, who had a strong supportive relationship with Nope, increasingly threatened Flo during banana feeding and twice that year attacked her. Usually during these encounters Flo showed submissive behavior, but occasionally she retaliated and then sometimes Circe gave way. Indeed, early in 1967 Flo actually attacked Circe and won the fight. But by mid-1967 Circe was dominant over Flo in agonistic contexts except when the latter was actively supported by either of her sons, Figan or Faben.

Circe died early in 1968 of unknown causes. Two females then moved to the top, Gigi and Passion. Pending fuller analysis it is not evident which dominated the other at that time; possibly they were equal. It is not surprising that Gigi should have high status among the females because her behavior in so many ways is like that of a male. As we have seen, she is an atypical female: large, aggressive, and sterile, she has spent more time with the adult males than any other female. She often displays along with the males during patrols or other group excitement, she performs streambed displays, and she has the top hunting scores of any female. By 1975 it was clear that Gigi ranked higher than Passion. After one conflict between the two (the start of which was not observed) Gigi chased Passion, who was screaming loudly, for at least 200 meters. (Observer: R. Barnes.)

De Waal (1982) found that the only reliable measure for ranking the females of the Arnhem colony was the directionality of pant-grunts. To some extent this is true also for the Gombe females, but there are many cases where *both* members of a dyad pant-grunt to each other, usually simultaneously, during a reunion. Moreover, a female may pant-grunt to another to whom she is clearly dominant in other agonistic measures (such as attack and avoid) during times of excitement, or when she is particularly nervous during the early weeks after parturition.

The approximate rank order of the central Kasakela females during 1978 and 1979 has been worked out, based on directionality of pant-grunting. Gigi (who pant-grunted to no other female) was highest. Passion, who pant-grunted only to Gigi, ranked second. She was pant-grunted *to* by far more females than was Gigi, suggesting that they were more afraid of Passion. In view of her infant-killing tendencies, which led to violent aggression with many of the females, this fact is scarcely surprising.

Below these two highest-ranking females came Fifi, Melissa, and Miff. Each was seen to pant-grunt to only three other females. Next came Winkle, Athena, and Patti. And finally, the lowest ranking were Pallas and Little Bee.

Gilka was in very poor health during these two years and became increasingly solitary. Her dominance position was unclear: she was not seen to pant-grunt to many individuals, but this was partly because so few interactions were seen. Subjectively, she seemed to be the lowest ranking of all the females. When she had the support of her brother, Evered, however, she was able to assert herself. Once, as Fifi and Gilka sat together licking a mineral block set out in camp, Fifi hit at Gilka (who got in her way), Gilka hit back (this is typical, even for low-ranking females, during squabbles with other females), and Fifi at once attacked her. Gilka, screaming, rushed off a short distance, then returned and held out her hand to Fifi, who touched it. Both females resumed licking. Suddenly Gilka gave a loud waa-bark and again began to scream. To my surprise she hurled herself at Fifi, hitting and grappling. Then I noticed that Evered had arrived. He stood surveying the scene, hair slightly bristling. Fifi, who presumably saw Evered also, pulled away from Gilka and retreated; now *she* was screaming. Gilka remained at the salt, directed a few waa-barks at Fifi, then licked along with Evered. After a suitable interval (three and a half minutes) Fifi quietly approached the siblings, groomed Evered for a few moments, then joined in the licking, keeping Evered between herself and Gilka.

One factor that is of overwhelming significance for the rank of a female at Gombe, outweighing all other variables except extreme sickness or extreme old age, is the nature of her family—and which family members are with her when she encounters another female.

For the young female, the rank of her mother is of crucial importance, for a high-ranking, aggressive female will almost always support her daughter during agonistic interactions with other females. In 1964, when Flo was top-ranking female, her juvenile daughter, Fifi, was able to threaten or even attack females much older than herself; if they dared to retaliate (which they seldom did), Flo would at once charge to her daughter's defense. In addition, Fifi had the advantage of two elder brothers, Figan and Faben, both of whom would sometimes support her. There were only three juveniles at that time: Fifi

As Fifi and Gilka lick the mineral block at camp, Fifi threatens Gilka (in the center), who screams, then seeks reassurance. Suddenly Evered, probably in response to his sister's screams, appears with hair bristling. Gilka immediately flies at Fifi and the females fight, both screaming. Gilka later shares the block with Evered and utters defiant waa-barks at Fifi, who continues to scream in rage.

was responsible for *all* of the seventeen attacks made by her age class that year (Goodall, 1968b). Her victims often showed conflict behavior when Fifi threatened them. Once, for example, she approached Melissa (about eight years older than Fifi herself). Melissa, who was chewing a banana skin, mildly threatened Fifi who, immediately bristling, glanced over at Flo. Melissa grinned and, extending her arm, placed the back of her wrist appeasingly to Fifi's mouth. In response, Fifi merely gave a waa-bark and *wrist-shook* at Melissa (this was Fifi's idiosyncratic version of the arm-raise threat). Melissa, still grinning, hit at Fifi but at the same time presented to Figan, who was nearby. This ended the incident and Fifi departed. In more recent years Pom, Moeza, and Gremlin, taking advantage of the high status of their respective mothers, have also been able to threaten females considerably older than themselves.

Fifi, as we have seen, is one of the higher-ranking females today. Like Flo before her, Fifi has a high display rate: during an eighteen-month period from mid-1976 through 1977 sixty-six charging displays by females were observed, thirty-nine of which were directed at other females during agonistic encounters. Fifi was responsible for 41 percent of the directed displays. Moreover, her displays (like those of Flo) were extremely vigorous and included stamping, drumming, and throwing. In 1981, as mentioned, she was observed displaying toward the sixteen-year-old Mustard on five separate occasions; each time he fled. Once, after chasing him away, she displayed, drumming and throwing and giving deep pant-hoots, for over two minutes. Another time, when Melissa threatened Fifi's juvenile son, Freud (who had made her own son scream during play), Fifi displayed vigorously toward her and Melissa rushed out of the way. And once, when she was traveling with a small female party, Fifi displayed seven times in fifty minutes and attacked Melissa, her daughter Gremlin, and the low-ranking Little Bee. There was no obvious reason for any of these attacks. The following morning, in the same wild mood, she displayed at and attacked Little Bee again, this time fiercely, leaving her bruised, bleeding, and muddy from being dragged through a stream.

It should not be surprising that Fifi has become a high-ranking female. Not only was she able to score many wins during early dom-

inance conflicts, owing to the support of family members (thus improving her fighting ability through success; Kuo, 1967) but her two eldest offspring are both males. As we have seen, offspring of either sex will usually help their mothers in aggressive interactions; this is why the outcome of a conflict between two females of similar rank may be largely determined by the age and sex of the accompanying offspring. Fifi, when she conclusively defeated both Melissa and her adolescent daughter, was with her ten-year-old son. While he did not actually attack on behalf of his mother (he did not need to), he several times displayed when Fifi did and thereby enhanced the effectiveness of Fifi's challenge.

One incident, observed in late 1964, further illustrates these principles. It started when the old mother Marina threatened Fifi (then a juvenile). Flo at once ran to her daughter's aid and the two elderly mothers fought, clinching and tumbling over and over. Marina's late-adolescent son, Pepe, who had been feeding in a palm tree nearby, quickly swung down and charged toward the battling females. Flo saw him coming and fled; Marina and Pepe together chased Flo and Fifi away. Two years previously Flo had not only been able to dominate Marina on a one-to-one basis, but had dominated Pepe as well. By 1964, however, when Marina was even older and more frail, Pepe was larger and stronger and the tables were turned—but only when he was present. Often, of course, he was away from his mother, traveling with the adult males; then Flo could win. If Flo and Marina had fought when *all* of their offspring were present, Flo would probably have won, for she had two adolescent sons whereas Marina had only the one.

While coalitions between mothers and members of their immediate families are the most usual, adult females who are not closely related do join forces from time to time on an opportunistic basis. Flo and Olly several times jointly attacked Circe when she first appeared (presumably as an immigrant) in the community range, and an example of an opportunistic coalition of three Kasakela females against the immigrant Patti is described in Chapter 17. Quite often two females will form a temporary alliance against an adolescent male when he persists in challenging one or both.

High-ranking Flo (juvenile Fifi in background) starts a charging display in 1969. Notice her bristling hair and compressed-lips face. (P. McGinnis)

What Profit High Rank?

The immediate advantage of high rank in the female hierarchy lies mainly in increased ability to appropriate desirable food. As we saw in Chapter 12, many aggressive interactions that take place between females are in the context of feeding. Theoretically, increased food intake produces healthier females, who give birth more often and raise offspring who are also fitter since they too have more food. In some primate species, such as the gelada baboon, high-ranking females are indeed more successful mothers (Dunbar, 1980). The same may be true of chimpanzees, but there are not yet enough data to be certain. In some parts of Africa, though, where the environment is more harsh than it is at Gombe, the ability to compete successfully for food may be critical. We do know that the offspring of high-ranking females tend to become high ranking themselves. If, therefore, it can be shown that a high-ranking female has a reproductive advantage, it will clearly be significant to her overall genetic fitness. It means that her daughters will, in their turn, be more productive and successful as mothers, and her sons will probably have a reproductive advantage also.

The benefits of high rank—reproductive advantage and priority of access to desired resources—seem, for the most part, less significant for chimpanzee males than for the males of many other primate species. Feeding competition was discussed in Chapter 10: when food is short, the chimpanzees tend to fragment and the alpha male will have no greater advantage than the top-ranked individual in each of the small groups. In competition for meat, the *older* males are often more successful than those who have highest rank. The extent to which high rank affects the reproductive success of the male will be discussed fully in the next chapter.

A high-ranking male has less reason to fear aggression from other males, and he is the frequent recipient of deferential behavior from many subordinates. These benefits will increase as he works his way up the hierarchy—but, at the same time, he will have to work harder to maintain his position. Perhaps chimpanzees are among the few creatures who are prepared to expend energy and run risks to attain high rank not only for the material advantages it may bring, but for the psychological benefits as well.

16. Sexual Behavior

(C. Gale)

July 1963   Flo arrives in camp early, accompanied by David Greybeard and Goliath. It is the second day of her full sexual swelling, and both males copulate with her before taking a single banana. Each time, juvenile Fifi rushes up and pushes at her mother's suitor. Things calm down and the chimpanzees sit quietly, eating. Suddenly we see a movement in the bushes on the far side of the camp clearing. Through binoculars I see Mr. McGregor, then Mike and J.B. There too are Huxley, Leakey, Hugo, Humphrey, and a number of adolescent males. Most have never visited camp before: they have followed Flo. Soon she moves into the bushes; in the next fifteen minutes, she is mated by each male, and Fifi races up to interfere every time. Once, as McGregor copulates, Fifi jumps right onto Flo's back and pushes so forcefully that he loses his balance and tumbles back down the slope.

For the next week Flo is followed everywhere by her male retinue. If she sits or lies down, several pairs of eyes swivel in her direction; when she gets up, the males are on their feet in no time. Whenever there is any kind of excitement, the adult males one after the other copulate with Flo. There is no fighting; each takes his turn. Once, as David Greybeard mates Flo, quick-tempered J.B. swaggers around the pair and sways a branch so that its end crashes down on David's head. But he does not attack. David merely presses closer to Flo, shuts his eyes, and carries on.

The young female begins to travel away from her mother with adult males during her first estrous periods. For the duration of adolescent sterility—one to three years—she mates not only with the males of her own community, but with at least some from a neighboring social group. When she becomes pregnant, between ages twelve and fourteen, she typically makes a commitment to one community. Often at Gombe she remains in her natal group, with her mother, to give birth and raise her offspring, the first of whom may well have been sired by a male from a neighboring community. The young male, by contrast, always remains in his natal group and competes with other resident males for mating rights. Together the males protect their female resources from neighboring males and try, themselves, to recruit young females from adjacent communities.

The menstrual cycle of the female chimpanzee, both in captivity (Graham, 1981) and at Gombe (McGinnis, 1973; Tutin, 1979; Tutin and McGinnis, 1981), has been described in detail. Its mean length for adult nonpregnant females is thirty-six days. The cycle can be divided into four major phases easily recognizable in the field: *inflation*, or *tumescence*, when the swelling gradually enlarges (about six days); *maximal tumescence*, when the swelling is described as full (about ten days); *detumescence*, when the skin of the anogenital area rapidly becomes flabby (five to six days); and a period when the sex skin is quiescent or *flat* (about fourteen days). *Menstruation* (about three days) occurs some nine days after the start of detumescence.

The stages in the sexual cycle are largely regulated by the gonadal hormones: the development of the swelling is associated with increasing secretion and excretion of estrogens in the follicular phases; detumescence, with decreased levels of estrogens and increased levels of progesterone in plasma and urine. Ovulation is most likely to occur on the last day of maximal tumescence or early on the first day of detumescence, but can occur earlier (Graham et al., 1972; Graham, 1981). In this chapter the last four days of tumescence are referred to, collectively, as the periovulatory period, or POP. Because viable sperm may be present in the fallopian tubes for up to forty-eight hours, a male who copulates on the first of the four days could theoretically be the father of an infant conceived on the third day (Speroff, Glass, and Kase, 1983; Thompson-Handler, Malenky, and Badrian, 1984).

Both interindividual and intraindividual variation in cycle length occur. Age is an important variable; adolescents are apt to have prolonged periods of maximum swelling (Tutin and McGinnis, 1981). In captivity, too, adolescents tend to have longer cycles (Graham, 1981). Irregularities, as we have seen, may be caused by injury or illness. Conversely, a variety of social factors may stimulate tumescence, and there is some evidence for cycle synchrony among the chimpanzees both at Gombe and at Mahale (McGinnis, 1973; Nishida, 1979).

The changes in the sex skin are significant to the observer because they reflect internal hormone changes and the probable time of ovulation. The full swelling, large and pink, is highly conspicuous. Estimates of its size range from 938 cubic centimeters (Erikson, 1966) to 1,400 cubic centimeters (Yerkes and Elder, 1936). It corresponds with behavioral estrus, when the female is most receptive to the sexual advances of adult males. All studies of sexual behavior in chimpanzees, both in the laboratory and in the field, have reported that most copulations between adult males and females take place during periods of maximal genital swelling. The turgescence of the sex skin may facilitate copulation: Allen (1981) found that a male achieved ejaculation with fewer thrusts when a female was at least half swollen than when she was flabby or flat. (The level of sexual arousal of the male at the time could, of course, also be a factor.)

Copulation is not, however, entirely restricted to the fully swollen phase. This is particularly true in captivity, where animals are less concerned with the pressures of day-to-day living than in the natural state and have more time for nonadaptive experimentation. Yerkes

(1943) found that a dominant and aggressive male could coerce a young and fearful female into accepting his sexual advances at any stage of her cycle. And sometimes nonswollen matings take place with the willing consent of a female, particularly certain individuals (Lemmon and Allen, 1978; Allen, 1981). Even at Gombe, copulations sometimes, though rarely, occur in semiswollen or flat phases of the female's cycle. During a five-year period adult males were observed to copulate 1,475 times, of which 96.2 percent were with females who were fully or almost fully swollen. Of the remaining 3.8 percent, 23 copulations were with half-swollen females, 13 with quarter-swollen females, and 20 with completely flat females. Two males between them were responsible for 73 percent of these sexual interactions with partially swollen or flat females. One of these was Goblin, whose sexual behavior, as we shall see, was unusual in other ways. The other was Evered, and all of his 27 such copulations took place during consortships—16 of them were with the completely flat Athena and Dove.

Infants, juveniles, and adolescents typically copulate with females during inflation and detumescence as well as during the fully swollen phase. This practice is to the advantage of youngsters, as they can learn much about sexual behavior at times when the older males are not particularly interested in the females and thus are highly tolerant.

The sexual swelling is a visual aid to the male chimpanzee, who can at a glance tell a good deal about the reproductive condition of a female. Although the swelling looks virtually the same throughout its maximum stage and although, as described in Chapter 5, precise olfactory cues regarding the period of ovulation are not detectable in vaginal smears, male chimpanzees nonetheless seem able to determine the POP with some precision. A review of the Gombe data showed that when females in estrus were in parties with several males present, either (a) there was an *increase* in copulation rate (with mature males) during the four days prior to the final day of full swelling or (b) they were mated *less often* because the alpha male, Figan, was showing possessive behavior and inhibiting the sexual advances of other males. Thus it seems, as mentioned earlier, that adult males are more attracted to females, and court and mate with them most often (unless inhibited by their superiors) during the POP, when fertilization is most likely to take place (Tutin, 1979).

Females showing presumed anovulatory swellings (during pregnancy or early postpartum nonbleeding cycles) are usually not popular among the older males. They are most likely to be mated when they first arrive in a party where there is at least one adult male; copulation may then be a part of the reunion. A similar response toward less sexually desirable females was described by Allen (1981). He found that the one adult male of his colony almost always copulated with a female who was reintroduced to his cage after an absence, even if he subsequently ignored her. This "strange-female effect," as he called it, held even when the female concerned was anestrous. At Gombe it is perhaps the novelty of immigrant females that makes them so attractive to community males (as a consequence of which the young female seems to prefer "new" males to those of her own natal group; Pusey, 1980).

It is quite obvious, however, that female attractiveness—sex appeal—is not *only* dependent upon hormonal and physiological factors or levels of arousal in the male; personality factors are also highly significant. Adult males of a variety of macaque species (rhesus, pigtail, stumptail, and Japanese monkeys) and yellow, chacma, and olive baboons, show clear-cut preferences for particular females, which may be independent of cycle stage. The literature has been reviewed briefly by Allen (1981). Similar partner preferences, independent of hormonal influences, are clearly of major significance for chimpanzees. Yerkes (1939) concluded that, in the laboratory, individual and social factors contributed more than hormonal ones in determining the behavioral interactions between male and female chimpanzees.

From the early days of the research at Gombe, it was evident that some females stimulated more intense sexual excitement among the adult males than did others. Tutin (1979) made a careful study of partner preference. She found that on thirty-eight occasions when two maximally swollen females were approximately the same distance from a mature male prior to a copulation, the older of the two was chosen thirty times. Certainly it is true that some old females are sexually very popular. This was made abundantly clear when Flo, aged over thirty, came into estrus in 1963. She was followed everywhere by as many as fourteen adult males, many of whom braved the novelty of our camp in order to keep close to the object of their desire. It was the same when she resumed cycling again in 1967 (when she was involved in fifty copulations during one day). To some extent this may be attributed to the novelty factor; both times Flo had been anestrous for periods of four years or so. But there is clearly more to sex appeal than this. Some young females have stimulated far more sexual interest among the adult males than others. Fifi and Pom were both equally attracted to the adult males, hurrying toward them the moment they showed any signs of sexual interest, or even before. Both spent periods of time sitting near adult males, staring at their penes. Yet the males mated Fifi with much more enthusiasm, and with higher frequency, than they did Pom. Among the various personality traits that might have been responsible for this difference, the most influential, perhaps, was the nature of the relationship between the young female and the adult males. Fifi, in general, had an easygoing relationship with them; as an adolescent, she had frequently joined them in vigorous play sessions. When males courted her, she usually approached quite calmly and crouch-presented for copulation without fuss. Pom, on the other hand, was invariably nervous with adult males, bobbing and pant-barking loudly, reaching out to dab at their faces, leaping away if they reached out to reassure her. Like Fifi, she quickly approached in response to their courtship, but tended to be much more tense as she crouch-presented. It is undoubtedly significant that while both these young females had high-ranking and aggressive mothers, Flo was highly popular and had a relaxed relationship with the adult males; Passion, like her daughter, tended to be tense and nervous. Finally, Fifi had two elder brothers and was therefore much more used to male company than Pom, who was a firstborn (Goodall, 1984).

Courtship and Copulation

The male chimpanzee, like the female, is endowed with impressive external genitalia. His erect penis is large compared with that of a gorilla or an orangutan, though small compared with that of a human. One captive chimpanzee had an erect penis measured at 8 centimeters long, compared to the 3-centimeter penis of a fully adult gorilla, the 4-centimeter penis of an orangutan, and the 13-centimeter mean length in the human male (Short, 1979). The chimpanzee's penis is thin and tapers to a point, presumably to assist in penetration of the female's swelling during copulation. When the penis is flaccid, it is usually invisible, concealed by the prepuce, the orifice of which is flush with the body wall. When erect, it shows up clearly against the pale skin of the thighs and lower abdomen. The *male-invite* posture (Tutin and McGinnis, 1981) exposes the penis, bright pink against the pale skin of the thighs and lower abdomen. Sometimes the signal value of an erect penis is enhanced by a sudden flicking movement, which may be repeated several times.

If the penis of the male chimpanzee is considered relatively large, the pendulous scrotum must be considered enormous. Schultz (1938) found the testes to account for 0.269 percent of body weight for three captive males—which is high for primates. The corresponding percentages for gorillas and humans are 0.017 and 0.079 respectively (Short, 1979). The size of the chimpanzee testes reflects a large spermatogenic capacity; presumably, as Short points out, it is an adaptation to meet the demand for high copulatory frequencies in a promiscuous society, where females typically outnumber males, where young females cycle for rather long periods, and where mating takes place throughout the year.

Copulation (or mating) is almost always preceded by a male courtship display that signals the sexual arousal of the male and attracts the attention of the female. The display, *always* accompanied by a penile erection, may include one or more of the following as well: direct gaze, hand on branch, hair erect, branch shaking, one or both arms stretched toward the female, bipedal swagger, sitting hunch. The male may rock from side to side, sometimes swaying vegetation, stamp with one foot, or hit the ground with the knuckles of one hand. Almost all these patterns may also occur in purely aggressive contexts, so that the male-invite posture directing attention to the erect penis is probably important in helping the female to assess the sexual intent of the male. It was the second most frequently recorded element in 200 male courtship sequences, scoring 126 and topped only by penile erection (Tutin and McGinnis, 1981). A penile erection by itself can be a confusing signal because it occurs in a variety of nonsexual contexts also, such as reunion or other kinds of social excitement. Tutin (1979) observed fifteen occasions when females approached and presented to males who had penile erections but had shown no other courtship behaviors; copulation occurred only three times. McGinnis (1973) reported one successful mating when a young female backed onto the erect penis of a feeding male and herself achieved intromission. Older and more experienced females typically wait for additional signals to clarify the male's sexual motivation.

Charlie displays the *male-invite* courtship accompanied by scratching, and thumping the ground with his knuckles; Fifi, in response, crouch-presents. Fifi's infant sibling, Flint, interferes, placing his hands on Charlie's face. This preoccupation with the *face* of the mating male is typical. (P. McGinnis)

When the female becomes aware of a courtship display, she usually glances at the male and then, either calmly or fearfully, approaches and crouch-presents, with her rump toward him. The male usually copulates in a squatting position. For the most part he cannot be said to *mount* the female. He may lean forward along her back and perhaps grasp her with one hand, but often he copulates without touching her except at the point of sexual contact. There is considerable variation in posture: sometimes a female barely flexes her legs and the male mates in an almost upright posture; occasionally, when a female ignores a male's courtship, he approaches and tries to achieve intromission while she reclines on the ground. A variety of postures also occur when mating takes place in the trees, structured by the position of available branches. Sometimes the female crouch-presents in a position that makes copulation awkward for the male; he may then move to a more suitable location and, if she does not immediately follow him, recommence his courtship.

At Gombe the adult male typically takes the initiative in sexual encounters with adult females. But adolescent females are sometimes insatiable (as is the case also in the Arnhem colony; de Waal, 1982) and repeatedly solicit males of all ages, even going so far as to tweak flabby penes. Usually a male ignores such approaches; sometimes he moves away. Occasionally the "wrong" female (usually an adolescent) responds to the courtship of an adult male. Typically he ignores her and continues to summon the other. Or two females may arrive at the same time, in which case he usually mates the one he was courting and pays no attention to the second.

Ejaculation occurs after an average of 8.8 pelvic thrusts (Tutin and McGinnis, 1981). Once, when a female leaped away after only three thrusts, ejaculate (or at least preejaculatory fluid) was already squirting into the air. During copulation the male often utters copulation pants. These sounds crescendo, probably as ejaculation takes place. The female typically gives characteristic high-pitched copulation screams (or squeals). The question of whether or not female nonhuman primates experience orgasm has not yet been decisively answered (see for instance Burton, 1971), but it seems likely that they can. Certainly the home-fostered Lucy, who masturbated almost daily after attaining sexual maturity, appeared to achieve orgasm; at the climax of a bout of self-stimulation she laughed, looked happy, and stopped suddenly (Temerlin, 1975, p. 108). Adolescent females at Gombe sometimes finger and tickle their genitals, laughing softly as they do so. In view of the fact that some females (mostly adolescents) show a keen desire for copulation, even in situations when males are showing no sexual interest in them, sexual contact seems, at the very least, to be far from unpleasant.

Copulation may be terminated by the female's darting forward (sometimes prior to ejaculation) or the male's stepping backward. Sometimes, perhaps when he has not achieved ejaculation, the male initiates a second, even a third, copulation within five minutes or so. After a successful copulation traces of fresh semen can be detected on the penis. The male, after peering down at it, often picks a handful of leaves and wipes himself. The female only occasionally wipes her bottom with leaves after mating, but she may subsequently pick out and eat the congealed semen, or vaginal plug.

Male chimpanzees are capable of ejaculating at a remarkably high rate. One captive male did so three times in succession after intervals of only five minutes (Allen, 1981). In the wild, Tutin (1979) found that there was a significant correlation between age and hourly rate of adult male copulation. Sherry, youngest male in the sample, had the highest rate (0.72) and the two oldest, Mike and Hugo, had the lowest rates (0.26 and 0.33 respectively). The other males fell between these extremes. Before the community split, however, when mixed-party groups tended to be larger, Bygott (1974) found that young males had lower copulation rates than their higher-ranking companions. The greater degree of tension in large parties is likely to inhibit the sexual behavior of subordinate males.

Both in the wild (Goodall, 1968b; Tutin, 1979) and in captivity (Allen, 1981) copulation rate tends to peak early in the morning. Thus at Gombe there is likely to be a flurry of mating activity when a female in estrus first leaves her nest. Indeed, an ardent old male once leaped into the nest of a young female and courted her vigorously before she had even sat up. Tutin, analyzing the rate at which females copulated during the day, found that there was an average of between five and six copulations per female per hour in the early morning, after which the rate dropped gradually to about two per hour in midmorning, rose very slightly during the afternoon, and tapered off to one per hour in the evening. The exact number will depend, of course, on how many males are in the party at the time. The copulation rate of one adult male, Evered, the only individual present throughout the estrous period of one female, bears out the reality of the morning peak. Taken over a nine-day period his scores were 0700–0900 hours, 8; 0900–1100, 6; 1100–1300, 1; 1300–1500, 6. No copulations were observed after 1500.

Tutin suggests that relatively high levels of circulating plasma testosterone in the morning, coupled with nighttime abstinence, may be a factor in sexual activity early in the day. (In fact, copulations have been seen twice during bright moonlight, but there have not been enough vigils at night to know how often this happens.)

To some extent, frequency of copulation can be correlated with social excitement and generally high levels of arousal in the adult males. When, for example, a party arrives at a food source or hears other chimpanzees calling nearby, or when two parties join, then there may be a burst of mating along with other indications of heightened arousal such as hair erection, pant-hooting, and charging displays. It is at such times that a female in estrus may be courted and mated by a succession of extremely excited adult males. I watched one party as it arrived at a new food source: the attractive female climbed into the tree along with eight bristling males, each of whom copulated with her in quick succession in a period of five minutes. In the days when large numbers of chimpanzees congregated in camp for long periods, this phenomenon was particularly apparent. If one or more males joined a party that included a swollen female, not only the newcomer males, but also many of the others, were likely to mate with her.

Late adolescent Willy Wally copulating with late adolescent Bumble. (C. Gale)

When we provided more bananas to individuals who had been peacefully resting and grooming, there was often a sudden burst of copulatory activity along with the feeding activity. Social facilitation evidently plays an important role in promiscuous mating of this sort. Often as a male watches a nearby copulation his penis becomes erect, and he may then approach and mate with the same female immediately afterward.

All this means that when a female in estrus is traveling in a large party with many males, she is likely to be involved in many copulations. When there were between twelve and fourteen fully mature males in the Kasakela-Kahama community before the split, as many as fifty copulations were counted during one day when the old female Flo was at the peak of estrus (McGinnis, 1973), and twenty to thirty copulations per day were common for a popular female at that time. On the other hand, when a female is traveling with one male only, she is unlikely to be mated more than four or five times in a day.

Mating Patterns

A reproductively mature female experiences six rather different sexual situations, described in detail by Tutin (1979).

(a) During the early stages of tumescence she will usually be ignored by adult males, but mated by infants, juveniles, and early adolescents, a situation that may prevail throughout an infertile cycle. On the other hand, some sexually popular females, particularly during their first swelling after a long anestrous period, may be mated by adult males even before they reach maximum tumescence.

(b) Once a female's swelling reaches full size, particularly if she is undergoing a fertile cycle, she typically becomes the nucleus of a party comprising many or all of the community males. In this situation mating is initially promiscuous (or opportunistic) and her sexual partners will probably include every male in the party with the exception (almost always) of any adult son or (sometimes) adult maternal sibling.

(c) At some stage during her period of swelling, usually in the second half of maximal tumescence, one male may become *possessive*, following her closely, grooming her frequently, and while copulating with her himself, trying to inhibit copulations by other males (possible only when they rank lower than he does).

(d) When the top-ranking male in the party behaves in this way he can, to a certain extent, *monopolize* the female—by inhibiting or aggressively preventing other males from copulating with her.

(e) At some stage of her cycle (including anestrous periods) one of the adult males may attempt to lead her away from the others, to the periphery of the home range. If he is successful, the pair forms a *consortship* (McGinnis, 1973) and may remain together for up to three months.

(f) Finally, the female, particularly as a late adolescent, may wander toward or be recruited by males of a neighboring community, and be mated by them. She may remain in this community as an immigrant or return to her natal group—sometimes pregnant.

During a single cycle a female may experience one, a few, or, successively, all of the first five of these situations. It will depend, among other things, on whether her cycle is fertile or anovulatory and on her popularity as a sexual partner—which may, in turn, depend on whether other females are in estrus at the same time and, if so, who they are and whether or not *they* are sexually popular. To some extent also, her sexual experience will be affected by the current state of the male dominance hierarchy; a female who is highly attractive to a powerful alpha male is likely to be monopolized by him during successive estrous periods and is less apt to be taken off on consortships.

Of major significance to an understanding of reproductive behavior is the very marked variation, from one year to the next, in the number of females who are sexually available to the males. In order to give some idea of this variability, the data from an eight-year period, 1976 to 1983, have been summarized and are presented in Appendix Table E1. These data are not absolutely precise. In some cases I have *assumed* that a given female developed a swelling, even when this was not observed—if, at a time when she was *expected* to come into estrus, she vanished for two weeks or more and returned flat. In other cases there was simply not enough information on the female in question to make any assumptions about her reproductive state during a long absence. Some periods of swelling, therefore, have undoubtedly been omitted. Nevertheless, I believe the appendix provides a reasonably accurate picture of the extent to which the various females contributed to the sexual activity of the community at different times. During this eight-year period a total of twenty-two habituated females (eight of them peripheral) showed among them *at least* 276 periods of estrus. An unknown number of visitors also contributed to the sexual satisfaction of the Kasakela males, but we have no idea how many times they were sexually receptive.

The rather dramatic decrease in numbers of sexually active females from 1979 to 1982 was almost entirely due to the infant-killing behavior of Passion. She is thought to have cannibalized all but one (Frodo) of

the infants born during the three-and-one-half-year period 1974 to mid-1977. When during 1977 she stopped attacking newborns, practically all the community females were either pregnant or had small babies; none of these became sexually active for at least the next three years, and many not for five.

The Group Situation

When a mature female undergoing a fertile cycle is associating with a number of mature males, she is likely to be involved in a flurry of sexual activity when she attains full swelling, particularly if it is her first cycle after a long period of lactational anestrus. Even so, there is usually a striking lack of overt aggression on the part of the males present, particularly during the first half of maximal tumescence. At this stage the female seems to be regarded as a common resource to be shared by any males present. If she is highly popular, then some adolescent and low-ranking males may be sexually inhibited, but even they can usually sneak a copulation or two out of sight of their superiors. It is during this early stage of full swelling that one most often sees a succession of males copulate with the female, one after the other, without overt signs of hostility.

At some point during this period one of the males may start to show possessive behavior toward the female. If he ranks low in the dominance hierarchy he cannot, as mentioned, prevent higher-ranked rivals from copulating with the female. When they do, he usually does nothing at all and, as Tutin (1979) points out, it is sometimes difficult to demonstrate objectively that he has even noticed. An alpha male, or the highest ranked among those present, however, is able to monopolize the female to some extent. His close proximity to her will often serve to inhibit his subordinates; if not, he will prevent their sexual advances by aggressive interference. This usually takes the form of threat. If he actually attacks, it is the female who is likely to be the victim; because if he fights with his rival, a second male may take the opportunity to copulate with the female. By attacking his female instead, the possessive male not only stops her mating with his rival, but also delivers a warning that she should avoid the sexual advances of other males.

In the late 1960s Mike, as alpha male, was sometimes observed to monopolize a female, and Bygott (1974) found that Humphrey was easily able to do so when he was top ranking in 1971. Figan took over the alpha position in 1973 and typically claimed access to females around the time of ovulation. At Mahale the alpha males of K group and the large M group both showed possessive behavior toward females in estrus. Indeed, in K group, when there were not more than three adult males, the alpha was able to monopolize several estrous females at the same time (Nishida, 1983).

Similar monopolization of females in estrus was observed by de Waal (1982) at the Arnhem chimpanzee colony, where the alpha male had clear control of mating rights and was "incredibly intolerant" when lower-ranking males tried to mate any female in the colony. If he noticed a clandestine copulation, it was, as at Gombe, the female who was apt to bear the brunt of his displeasure. He was very likely to attack her, so for the most part the females refused to comply with courtship demands of lower-ranking suitors—although they did so

This sequence starts with Leakey's courtship of Gigi, in camp. As Goliath approaches the couple, Gigi gives nervous pant-grunts. Goliath, squeaking, takes over after Leakey has completed copulation. When both males have finished, Gigi escapes up a tree. Sequential copulations of this sort are common during food excitement and other forms of social excitement such as reunions. (P. McGinnis)

eagerly, even through bars, when the alpha was temporarily shut outside or in a different cage.

The upshot is that when the male dominance hierarchy is relatively stable, and most particularly when there is a clear-cut alpha, there is not likely to be much fighting over the female, even when she is in the most fertile periovulatory period and surrounded by many adult males. Levels of aggression per se will not necessarily be low, however: the presence of the female, and the frustration imposed on most of the males by the inhibition of the sexual urge, mean that levels of arousal are high and the threshold for aggression correspondingly low. Fights will occur as a result of redirected aggression, and they will be triggered by a variety of petty causes—to the discomfiture of lower-ranking members of the party, male and female alike. Yet this is nothing compared to the bedlam that ensues in a similar situation when there is no unambiguous alpha male, and when as a result the entire hierarchy is unstable. This was the case in November 1976 when Figan, after the disappearance of his brother, had not yet regained his position at the top of the male hierarchy. The events that took place when the female Pallas was fully swollen are summarized in Word Picture 16.1. An almost identical replica of this chaotic sexual gathering took place in March 1982 when Nope was in estrus and almost certainly conceived; again there was no clear-cut alpha male.

The Consortship

A far more reliable mating strategy for a male, but one that may require a good deal of skillful social manipulation, excellent timing, and sometimes a fair amount of brutality, is the formation of a consortship. If successful, the consortship enables him to maintain absolutely exclusive mating rights over a female during the period of ovulation when, all things being equal, he may be able to impregnate her. A good many consortships, however, do not fall into this category. Some occur at times when a female is physiologically unable to conceive: (a) when she is anestrous (either between fertile swellings or when pregnant) or (b) when she is undergoing an anovulatory swelling (during adolescent sterility or during pregnant or early postpartum cycles). In addition, many attempts to establish consortships fail owing to lack of cooperation by the female and/or early interference by other males.

Of the total number of consortships formed during a given year, only a very small number are actually observed, even in part. Often, however, a relationship can be inferred. McGinnis (1973), by examining camp attendance data, was able to infer a number of consortships based on the coincidental absence of a given male and female. When such absences coincided with an estrous period of the female concerned, and when all other mature males were observed at some point during the absence of the pair in question, it was reasonable to suppose that a consortship was in progress. As camp attendance fell after 1970, it was no longer possible to use this method in its original form. However, by this time the chimpanzees were being followed more frequently out of camp, so that by using a combination of in-camp attendance data and out-of-camp travel and association chart data, it was possible to make similar inferences about probable consortships (Tutin, 1979).

The Period of Maximal Swelling of Adult Female Pallas,
22 to 29 November 1976

During this period Pallas almost certainly conceived and was therefore highly fertile. She was attended throughout by all but one of the Kasakela adult males, Satan, and even he eventually turned up after returning from a consortship in the north with another female. The daily size and composition of this sexual gathering is illustrated graphically in Figure 16.1, and the travel route followed over six of the days is shown in Figure 16.2.

These events occurred at a time when the dominance hierarchy of the adult males was unstable: Figan, after the death of Faben, was still in the process of reestablishing his position as alpha. During those portions of the eight days that the gathering was followed, 169 aggressive incidents, including 20 actual attacks, were recorded. Thirteen of these attacks were between mature males. This gives a frequency of 1.7 aggressive acts and 0.2 attack per hour. During the same period 84 copulations were observed, by five adult males and two late adolescents. For most of the time Pallas stayed up in the trees, feeding and resting, while the adult males, when they were not also feeding, sat below on the ground and gazed from time to time at Pallas. When one of the males started to climb toward her, hair and penis erect, his progress frequently triggered an outburst of extraordinary social excitement and aggression, with the adult males displaying toward and around one another, attacking one another, and then engaging in a frenzy of tactile reassurances—embracing, kissing, and patting. Throughout these interactions there were loud screams and waa-barks not only from the rival males, but also from others present. One typical episode, chosen from many similar incidents, illustrates the behavior. These events took place on 27 November, four days before the start of Pallas' detumescence.

At 1023 Evered climbs toward Pallas and courts her. Figan, hair bristling, moves toward the tree. Pallas crouch-presents but Evered, presumably inhibited by Figan, does not copulate. Instead he turns and attacks Goblin (who, as usual, is staying very close to Pallas). Evered pulls Goblin to the ground, then displays away. Figan, along with Humphrey, Jomeo, and Sherry, all display and rush toward Pallas. As Figan moves toward her, Jomeo jumps on him, stamps once, and displays away. Meanwhile Humphrey and Evered fight. Figan turns on Sherry and attacks him. Satan does not join in this confusion, but sits by himself. Jomeo, screaming loudly (as are Figan, Humphrey, and Sherry) leaps down from the tree, charges toward Goblin, and attacks him. Sherry then runs to Evered and the two embrace. On a different branch Humphrey and Figan embrace and kiss. Things calm down. (Observer: E. Mpongo.)

During the whole eight-day period, Figan achieved only 8 percent of the observed 84 copulations. The oldest male present, Humphrey, did best at 31 percent. Sherry and Jomeo were observed to copulate at about the same frequency as Figan; Evered, slightly older, scored a little higher. Late-adolescent Goblin showed interesting behavior, remaining close to Pallas throughout. Typically he sat near her up in the trees while the older males remained on the ground. He was observed to copulate 12 times (14.3 percent of the total observed), and may well have done so more often during the commotion caused by the displays and fights of the adult males. (Goblin behaved in exactly the same way in relation to Pallas in estrus in November 1973, when Tutin was observing the group. Unfortunately the events of the two periods are not comparable; Pallas was three months pregnant in 1973,

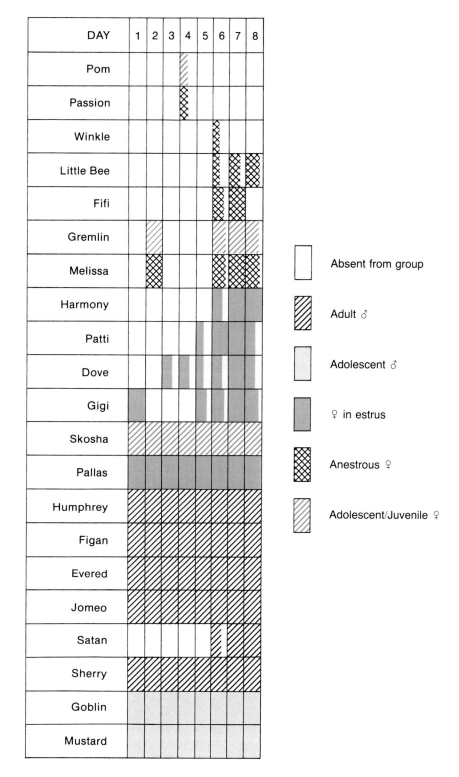

Figure 16.1 Daily fluctuations in the number of individuals associating in a large sexual party, or gathering, that was followed for eight days in November 1976. Nucleus of the gathering was Pallas, the sexually popular female, in estrus.

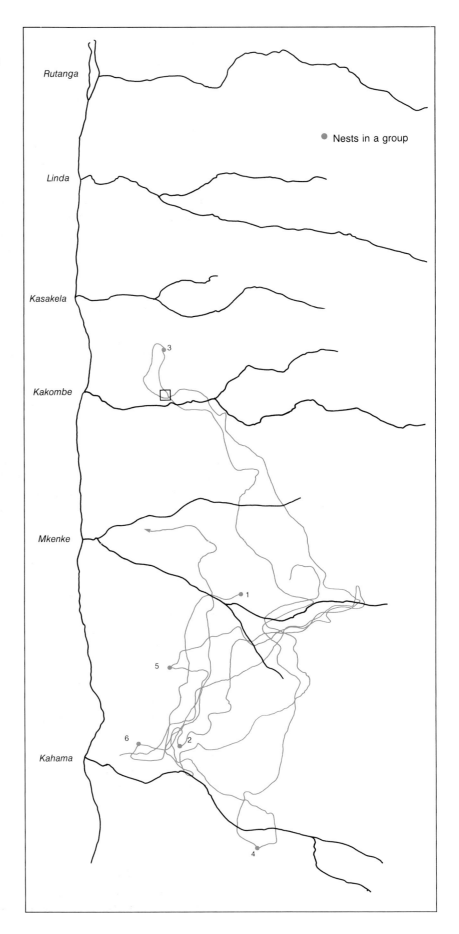

Figure 16.2 Travel route of the gathering of November 1976, over six of the eight days that the group was followed. Nests are numbered in consecutive order; the open square shows the location of camp.

Rutanga

Linda

Kasakela

Kakombe

Mkenke

Kahama

Nests in a group

and although copulation rates were high, the males for the most part were less interested sexually than in 1976.)

Between 1966 and 1983, 258 consortships of three days or longer were fully or partially observed (33), or else inferred. Table 16.1, a summary of this information, shows that the number of consortships per year varied—mainly, of course, in accordance with the number of mature males and the number of cycling females who were available. Only about half (43.7 percent) of these consortships could have resulted in conceptions—those with females who were undergoing fertile cycles and those that lasted throughout the crucial POP. At other times the females were infertile or else the relationship terminated before the fertile period of their cycles. The percentage of the total consortships that may in fact have resulted in conceptions rose from about 8 percent before the division to 23 percent afterward.

Some males tended to favor brief relationships initiated just prior to the POP of the female; others tended to lead their females off at different stages of the cycle, for much longer intervals. Very high-ranking males, or males undergoing tense dominance relationships, usually favored short consortships or gave up consorting altogether, as we shall see.

INITIATION In order to establish a consortship, a male must first persuade the female to accompany him away from other mature males of the community—that is, to areas near the periphery of the home range. He does this with signals of courtship: gaze toward, hair erection, branch shaking, rocking, and arm stretching. When the female responds to one or more of these patterns by approaching, he gets up and moves away, looking back over his shoulder to make sure she follows. If she does not, he stops and repeats his summons. If she continues to show reluctance, his actions become increasingly violent and eventually he may display toward and all around her—he may also attack her. Behavior of this sort I term *herding*, after Kummer (1968, p. 36), "punishing a companion for being too far away."

If a male is able to establish exclusive mating rights with a female undergoing a fertile cycle, and be with her as the only male during her time of ovulation, he has a good chance of fathering a child. If he can lead her away during maximal swelling, he need only be away from the center of the community range for a few days. This is to his advantage in three ways.

(a) His task of maintaining proximity with the female and ensuring she does not elude him will be less exacting, partly because the time span is shorter, partly because a swollen female often seems more willing to follow a male than one who is anestrous.

(b) There is less chance that the pair will encounter males from a neighboring community who might attack the consort male and make off with his female.

(c) The danger of his facing aggression from other community males on his return (sometimes the response to an individual when he

Table 16.1 Number of consortships (observed, partially observed, or inferred) from 1966 to 1971, before the community division, and from 1972 to 1983 afterward. The number of reproductively mature (RM) males in the community (end-of-year count) is shown. The percentage of consortships that included the estrous peak (EP) and the number that may have resulted in conceptions are also given.

Year	No. RM males	No. periods of swelling	No. consortships	% consortships that included EP	Consort conceptions	
					Probable	Possible
1966	13	a	15	40.0	1	0
1967	14	43	29	41.4	0	0
1968	12	42	36	38.9	0	1
1969	13	33	24	41.7	1	0
1970	12	32	19	52.6	1	1
1971	13	43	17	47.1	0	0
Mean 1966–1971	*12.8*	*38.6*	*23.3*	*43.6*		
1972	8	50	15	13.3	0	1
1973	8	37	15	40.0	2	1
1974	8	31	6	33.3	0	1
1975	5	50	13	69.2	1	0
1976	5	47	10	50.0	0	1
1977	6	55	18	50.0	2	0
1978	6	34	13	30.8	3	0
1979	5	17	2	100.0	1	0
1980	6	22	5	20.0	0	0
1981	6	25	4	20.0	0	0
1982	5	21	7	28.6	0	0
1983	6	59	10	70.0	1	0
Mean 1972–1983	*6.2*	*37.3*	*9.8*	*43.8*		

Note

Numbers of consortships per year vary in relation to both the number of males and the number of periods of swelling shown (collectively) by late-adolescent and RM females. No distinction is made here between cycles that were presumed fertile and those that were presumed infertile.

a. Only two months' data available for analysis.

Leakey shows *arm-stretch* and *male-invite* postures as he initiates a consortship with Olly. Note his flaccid penis. Juvenile Gilka, Olly's daughter, approaches alongside her mother. (H. van Lawick)

appears after a prolonged absence) is minimized.

However, it is not always possible for a male to lead away a female at this optimal time. The closer she is to her POP, the more intense the competition from other males is likely to be, and the would-be consort male will have problems of a different sort.

When a female remains as part of a sexual party throughout maximal swelling and ovulation, this may be (a) because no male *wants* to form a consortship (as seems to have been the case with some young females, such as Gremlin) or (b) because no male is *able* to do so—because the alpha male is showing possessive behavior, because the female refuses to cooperate, or because (as was almost certainly the case with Pallas and Nope, as well as Flo in 1963 and 1967) virtually all the males are so competitive and alert, and levels of arousal so high, that consorting opportunities do not arise. It is probably for this reason that so many consortships are initiated during the period of inflation, or when the female is quite flat, or even when she is flabby after her full swelling has subsided. Of the total 258 consortships (1966–1983) 39.5 percent were initiated with maximally swollen females, 20.5 percent with females who were completely flat, 16 percent during various stages of tumescence, 12 percent with those who were flabby, and 12 percent were with females who were already pregnant.

A male's technique for initiating a consortship will depend on the cycle stage of the female, who else is present at the time, his own particular personality, and the extent to which the female is willing to be led. When the female is swollen, and especially if she is going through a fertile cycle, she is almost certain to be part of a sexual party. In these circumstances the would-be consort male must exercise ingenuity if he is to succeed—and he is only likely to do so if the female herself is prepared to cooperate. It is in this context that possessive behavior shown by a low-ranking male acquires significance; by keeping close to the female and grooming her frequently, he is not only able to copulate with her at a higher frequency than might otherwise be possible (as was the case with late-adolescent Goblin and Pallas) but, if he carefully monitors the situation in the group, he can take advantage of an appropriate opportunity to try to lead her away.

Tutin (1975) describes one occasion when Satan, who had been showing possessive behavior toward Miff, in estrus, was able to lead her away from other males. The opportunity came after a period of rest. Miff did not move until all of the adult males, with the exception of Satan, were out of sight. She then got up and followed the others. Satan immediately moved ahead of Miff and traveled in front of her with frequent backward glances. After thirty-eight minutes there was a burst of pant-hoots from the other males—500 meters away and no longer ahead of them. Miff looked toward the calls, then at Satan, who sat shaking branches. After a moment she approached him and followed his lead over the ridge into the next valley. Tutin speculates that Miff may not have been aware of Satan's intentions until she heard the direction of the others' calling. This was where she demonstrated her willingness to cooperate with Satan; had she called out or refused to follow, the consortship would probably have ended. Her calls would almost certainly have attracted the other males, as would her screams had Satan punished her disobedience by attacking.

On two occasions Satan took advantage of the fact that females in estrus tend to make their nests later than other members of their party. After the other males had retired for the night, Satan shook branches at the female concerned, as she fed. When she eventually followed, he led her away and they nested close together. The following morning he rose early, shook branches at his female until she left her nest, then led her in the opposite direction from the other chimpanzees. Again, it was only because the female cooperated that Satan achieved his goal (Tutin, 1975). Satan showed the same evening and early-morning technique when in 1984 he persuaded Fifi (anestrous between swellings) to follow him on a consortship.

Once a male has succeeded in leading a female *in estrus* away from other chimpanzees, he usually proceeds as rapidly as she will follow until the pair is out of earshot of the others. The response of the female to calling from the party they have left is critical. Miff only looked toward the sounds, then followed Satan in the opposite direction. But on four occasions females who had been following consort males traveled rapidly toward others heard calling nearby; their males allowed them to go without trying to salvage the consortship (Tutin, 1975). At other times, consort females responded by themselves giving pant-hoots. Twice when males appeared in response to such calls, they then attacked the consort male (McGinnis, 1973). This may explain the three occasions when swollen females called out in response to nearby pant-hoots, and their consort males left them and moved rapidly away.

It is during the early stages of a consortship, when the male is trying to lead his female away from the core area of the community range, that she is most likely to try to escape and the male may be at his most aggressive. When Evered initiated a consortship with Winkle that ultimately would lead to conception, she was half swollen. During the five hours it took him to lead her across Linda Valley (from Kasakela to Rutanga, as shown in Figure 16.3) he displayed at her repeatedly and attacked her six times, twice quite severely. Tutin (1975, 1979) describes unusually high levels of aggression when Faben forced Pallas to accompany him beyond Rutanga Valley, and Word Picture 16.2 describes Humphrey's attempts to coerce Nope into accompanying him. Both Pallas and Nope were in the flat phase of the estrus cycle. At this time the male does not have to be concerned about competitive reactions from other males, but he often has to work hard to force the female into following him at all.

Tutin (1975, 1979) found that there were clear-cut individual differences in the leading behavior of the various males. Hugo and Faben tended to be impatient when their females delayed during travel and were quick to shake branches and display. Figan and Satan were more patient. Once, for example, Figan waited calmly for forty minutes while Patti, his consort female, continued to fish for termites. Five years later, in 1978, Figan was still the most tolerant of the males in the consortship situation. Other males fell between these extremes.

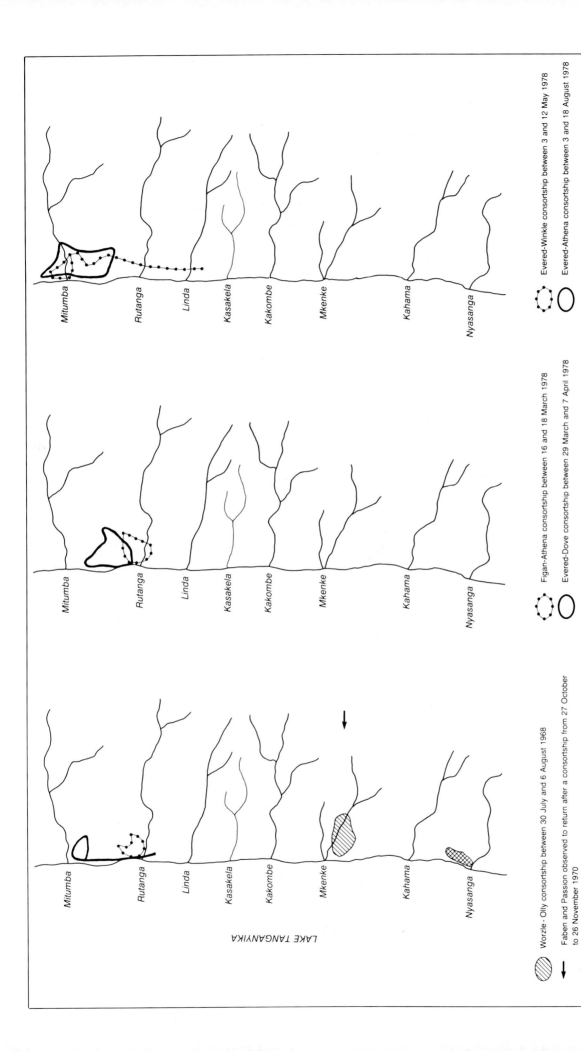

LAKE TANGANYIKA

Mitumba
Rutanga
Linda
Kasakela
Kakombe
Mkenke
Kahama
Nyasanga

Worzle - Olly consortship between 30 July and 6 August 1968

Faben and Passion observed to return after a consortship from 27 October to 26 November 1970

Figan-Gigi consortship between 22 January and 2 February 1972

Faben-Pallas consortship between 27 June and 8 July 1973

Sniff-Little Bee safari 21 May 1974

Figan-Athena consortship between 16 and 18 March 1978

Evered-Dove consortship between 29 March and 7 April 1978

Evered-Winkle consortship between 3 and 12 May 1978

Evered-Athena consortship between 3 and 18 August 1978

Figure 16.3 The location of several of the consortships that were observed between 1968 and 1978.

Humphrey Attempts to Initiate a Consortship with Nope, November 1979

On 1 November 1979 Sherry and Nope reappeared in the central part of the Kasakela community range after an inferred consortship of ten days, during which Nope almost certainly conceived; she was fully swollen at the outset and returned flat. That same day Humphrey forced her to follow him northward, to Rutanga Valley (whence she had, almost certainly, just come). The relatively short journey from camp to the southern slopes of the valley (approximately 2 kilometers) took eight hours. During this period there were fifteen separate branch-shaking sessions, some of which lasted for ten minutes, before she reluctantly approached. Three times Humphrey displayed vigorously around her, flailing her wildly with the surrounding branches, until eventually she rushed toward him, crouched with loud pant-grunts and squeaks, and kissed his belly or thigh. He had the greatest difficulty in forcing her to cross the various ridges between valleys and gulleys. Each time they arrived at one of these more open, elevated places, Nope stopped and gazed back toward the core area of the community range, ignoring Humphrey's frantic branch shakings and swaggerings as long as she dared. Some of the delays in travel that triggered the wildest episodes of branch waving from Humphrey were when Nope was waiting for her seven-year-old daughter, Lolita. In one instance the child was fishing for termites. Nope had left her own fishing in response to Humphrey's demands, but after traveling—reluctantly and with many pauses—for 100 meters, she stopped and looked back toward Lolita. As Humphrey's swaggerings became violent, Nope began to whimper; and when it seemed he could no longer be prudently resisted, she ran up to him, screaming, then followed, whimpering and repeatedly looking back over her shoulder. Presently, Lolita's loud crying was heard in the distance as she searched for her mother. Nope started back toward her, stopped, looked at Humphrey, than ran to him screaming loudly and embracing him. When Lolita finally arrived, she leaped into her mother's arms and, despite her age, began a lengthy suckling session. Throughout this interval Nope completely ignored Humphrey's summons, intently grooming the child and not looking at him until the session was over.

At 1745 hours, as Humphrey fed on blossoms, Nope added some leafy branches to an old nest and lay down. Presently Lolita joined her mother, stretched out beside her, and began to suckle. At 1800 Humphrey climbed to the ground, where he sat looking up at Nope. Once he shook a little clump of grasses, then lay down and watched her. At 1815 he made his own nest, 5 meters away from his consort female, in which he settled for the night. (The pair could not be followed the next day, but Nope must have escaped soon afterward because three days later Humphrey was back with the other Kasakela males.)

Mr. Worzle alternated between long spells of patient waiting and sudden bouts of impatient aggression. On one occasion, as he persuaded Olly to follow him to their consort range in the south, he waited quietly on the ground while she dozed in a day nest above him. From time to time he glanced up and shook a handful of grass, rocking slightly from side to side, but Olly merely opened her eyes, grunted softly, then dozed off again. Suddenly, after twenty-three minutes, Worzle's hair began to bristle. Leaping up, he swaggered from foot to foot, swaying branches vigorously; soon he was sweating profusely in the hot sun. Olly pant-grunted but remained in her nest. Three minutes later Worzle swung up and attacked her, causing a frenzy of screams and pant-barks—and instant, obedient following for the next hour (P. McGinnis, unpublished field data).

One male, Leakey, twice tried to take *two* females with him. Once as he sat feeding, Fifi, his most frequent partner (McGinnis, 1973), appeared and he at once began to shake branches at her. Before he had succeeded in leading her away, Olly, his second most frequent partner, appeared. He at once began shaking branches at *her!* He then attempted to lead both females, shaking branches and gazing first at one and then the other. Both were reluctant to follow him. He became increasingly tense and suddenly attacked Fifi, even though she had just run toward him. Olly hastily escaped. When Leakey noticed, he instantly raced to the place where he had last seen her and began to search, climbing trees and peering in all directions. Fifi took the opportunity of slipping quietly away. Leakey spent ten minutes searching for the two females, then gave up. On the second occasion events proceeded in a similar way, but then only Olly managed to escape, and Leakey went off with Fifi.

THE EXCLUSIVE RELATIONSHIP The period of consortship, whether the female led away is swollen or flat, fertile or infertile, represents an unusually close relationship between two adult non-related chimpanzees of opposite sex, one that may last for a considerable period of time. The first detailed account of consort behavior describes the relationship between Mr. Worzle and Olly in 1968. Patrick McGinnis followed the pair for six days of a consortship that continued for another three (unpublished field data). Since then, seven other consort pairs have been followed for six days or more, and twenty-five others for shorter periods. Only that of Faben and Pallas in 1973 was observed in its entirety, although the consortship of Evered and Winkle was observed throughout her period of maximal swelling before the pair was lost to the observer's view (to reappear, two days later, with Winkle flat).

The relationship between each male-female pair was typically relaxed and tolerant once the consort range had been established. The female became more tractable—partly, no doubt, because she was in an unfamiliar area and felt the need to stay close to the male. During sudden alarms (such as the calls of strangers or the sound of passing fishermen) the female often ran to the male and the two embraced and kissed. The male, once he had succeeded in coercing the female out of earshot of community males, usually became much calmer and was prepared to adjust his own movements to those of the female, at least for much of the time.

Mr. Worzle, after leading Olly to the south, waited patiently as she fed in the late evening, lying on his back watching her, shaking a branch above his head from time to time. Only when it began to get dark, nearly an hour after he had first summoned her, did he become really insistent, rushing up the tree with hair bristling. Olly followed then, but nested in a very low palm, the only suitable nesting tree within 15 meters. Worzle, in order to maintain proximity, was forced to sleep on the ground below her, using some of the fronds from her tree. This is the only report at Gombe of a chimpanzee's spending the night on the ground (except when too sick to climb).

During the consortships there were many sessions of relaxed social grooming when the male almost always groomed for longer than his female. If the female was accompanied by members of her family, they often joined the grooming sessions too—and occasionally there was a play session involving the male. Despite the fact that the mother's preoccupation with a nursing child sometimes made her more disobedient, no instances of male aggression toward the youngsters were recorded.

In only one of the eight well-observed consortships was the female, Winkle, fully swollen. Her copulations with her male, Evered, were infrequent, four in one day being the highest number recorded. Of particular interest were copulations between Evered and two of his other consort females, Dove and Athena, both of whom were anestrous and pregnant. Evered, as mentioned earlier, copulated with other flat females during consortships. For the most part anestrous females, when faced with (rare) sexual demands from adult males—usually during the excitement of reunion—screamed and showed fearful submissive behavior. The consort females, particularly Athena and Dove, were calm. Their behavior reflected the relaxed relationship that typifies the successful consortship.

It seems that the Faben-Pallas consortship described in detail by Tutin (1975) was unusual, as their relationship was tense almost all the time and Faben was very aggressive toward his female. Pallas was anestrous throughout and extremely reluctant to follow Faben: in order to keep her close he resorted to relatively high levels of aggressive herding throughout the thirteen-day period. He did groom Pallas quite frequently but unlike other consort males he became increasingly insistent that she groom him in return. During the last three days of this consortship Faben became more and more demanding, insisting on very close proximity, and by the final day Pallas was unwilling to move in *any* direction. Faben sometimes had to initiate travel by pushing her along with his one good arm around her.

THE CONSORT RANGE One of the dangers faced by a consorting couple is the risk of encountering hostile neighbors. Figure 16.3 shows the location of eight known consortships. Only Worzle and Olly remained well inside the community range (at that time); the other pairs were in areas where the Kasakela community range overlapped that of the Mitumba community in the north (Faben and Passion were observed returning from an inferred consortship in the southeast, as indicated in the figure). Evered and Figan both established consort ranges close to Mitumba fishing huts near the lakeshore; because unhabituated chimpanzees do not venture so close to human dwellings, theirs was a very safe location.

Consorting pairs typically maintain vocal silence, and while this functions to conceal their presence from members of their own community, it may be more important that they do not reveal themselves to neighbors. In fact, calls presumed to have been made by chimpanzees of the Mitumba community were heard quite frequently during some consortships. When these calls were from locations estimated as more than 700 or 800 meters distant, the chimpanzees usually gazed or glanced toward the sounds, then paid no further attention; sometimes they did not even look up. When the calls were closer, however, the pair traveled silently and rapidly in the opposite direction—toward their own core area.

Only during the Evered-Athena consortship were encounters with strangers actually observed. Unhabituated mother-infant pairs were seen on four separate occasions, and twice their presence led to attacks by Evered. Once he attacked very fiercely, and the female was lame and bleeding heavily when she escaped. On another occasion one of the two females encountered, chased and attacked eleven-year-old Atlas before she saw the observers and ran off. The high frequency of encounters observed during this consortship resulted from the fruiting of a large fig tree near the beach; the chimpanzees met when they went there to feed.

That consort behavior at the periphery of the home range can, indeed, be a risk is discussed in the next chapter. Dé was near the periphery of the Kasakela-Kahama border when he was fatally attacked by Kasakela males. He may have been forming a consortship with Little Bee, with whom he was traveling at the time. The immigrant female Harmony, during an absence from the Kasakela core area, was encountered near the Mitumba-Kasakela periphery in the north, apparently in consort with a stranger male. Kasakela males displayed toward the pair; the male ran off but Harmony, in estrus at the time, was attacked and firmly herded back to the south.

TERMINATION   The thirteen-day consortship of Faben and Pallas ended when, in response to her screams (as Faben attacked her), pant-hoots were heard nearby from three Kasakela males. When they presently called again, Faben attacked her once more, but made no attempt to lead her away from the approaching males, one of whom was his brother Figan. The five chimpanzees remained together for the rest of the day. Faben stayed close to Pallas much of the time and waited for her, but did not attempt to lead or herd her. The presumed consortship of Harmony and the Mitumba male also ended with the arrival of other adult males, in this case from a different community.

Once, in 1963, Goliath appeared leading an unhabituated female, in estrus, toward camp. She was understandably terrified and hastily retreated when she saw me and my tent. Goliath rushed after her and attacked her, trying in vain to lead her toward the bananas he knew were waiting; but she refused to follow. Goliath was torn between conflicting desires: for five minutes he alternately ran down the slope toward camp, then turned and rushed back up to make sure his female was still there. When he left her long enough to gather an armful of fruit, he pushed his luck too far—she took the opportunity to slip away quickly. McGinnis and Tutin both observed females who stole away when their males were busy feeding, thus ending consort attempts.

Visiting

At Gombe visitors are shy and we know little about their mating pattern. Typically they seem to be sexually popular (the "strange-female effect" of Allen, 1981) and are often the nucleus of sexual groups. Moeza, however, when she visited the Mitumba community as a late adolescent, was several times encountered fully swollen and quite on her own. Females, as we have seen, sometimes form consortships with neighboring males: Evered was observed apparently in consortship with a Mitumba female; and we have seen that Harmony, after she had immigrated into the Kasakela community, was encountered traveling with a Mitumba male (presumably the pair was consorting). It is not yet clear in cases such as these whether the female is visiting the male's community, or he hers.

Incest

In Japanese macaques (Imanishi, 1965; Enomoto, 1974), rhesus monkeys (Sade, 1968; Missakian, 1973), and olive baboons (Packer, 1979), mating between close kin who remain together as adults is rare. The same is true for chimpanzees (Goodall, 1968b; Tutin, 1979; Pusey, 1980).

Table 16.2 shows that incestuous matings between sexually mature males and their mothers are extremely uncommon. Figan and Faben, who were with their mother for at least some part of her five periods of estrus, were never observed even *trying* to mate her—during her 1963 and 1967 periods of estrus they were the only males of any age who did not do so. In fact, they never showed the slightest sign of sexual arousal in her presence. The same was true of Evered and his mother, Olly.

Tutin observed Satan copulating with his mother, Sprout. She tried to escape from him but he followed her to the top of a tall tree. Although she submitted, she screamed loudly throughout and leaped away prior to ejaculation. Goblin quite frequently showed sexual interest in his mother, Melissa, when he was nineteen years old. His first observed copulation was during her first postpartum cycle after the birth of the twins, two or three days before deflation. When he summoned her she refused to approach; eventually, after repeated branch shaking and two short bouts of chasing, Goblin stamped on her back three times and soon afterward gave up. The following day he was again observed to chase his mother. This time she stopped, screaming, and crouched for copulation; but after he had delivered a few thrusts she leaped away before he had ejaculated. Another mild attack resulted, but she escaped, hitting out at him as she did so, then took refuge up a tree. He gazed up and shook branches at her, but she climbed very high and after a further minute he gave up. During Melissa's next period of swelling Goblin was once more observed as he summoned her, but he did not persist when she ran from him.

The following year, after a miscarriage, Melissa resumed cycling. She was followed daily during both her periods of swelling. Goblin was observed to summon her only once during the first estrus and he quickly gave up when she avoided him. A month later, however, he copulated with her, apparently successfully, after displaying and chasing her up a tree.

Over the years we have collected data on the sexual relationships of five late-adolescent or mature females with their elder known or

Table 16.2 Frequency of copulation between mothers and their adult sons.

Mother	Son	Age of son (years)	No. of cycles seen together	No. of copulations	
				Seen	Attempted
Flo	Faben	16	2	0	0
		20	1	0	0
		22	2	0	0
	Figan	14	1	0	0
		16	2	0	0
Olly	Evered	15	4	0	0
Sprout	Satan	18	2	1[a]	0
Melissa	Goblin	19	2	1[a]	3
		20	2	1	1
Nope	Mustard	16	2	0	0

a. Mother protested violently, screamed, pulled away prior to ejaculation.

assumed brothers: Miff with Pepe, Gilka with Evered, Gigi with Willy Wally, and Fifi with both Figan and Faben. Copulations between these brother-sister pairs were observed very infrequently. On the one occasion that Miff was seen copulating with her elder brother she was quite calm; mostly, he showed no sexual interest in her. Gigi once attacked her presumed elder brother, Willy Wally, when he persisted in trying to mate with her, but another time she accepted him without fuss. Gilka was never seen to be mated by her brother, Evered, despite the fact that, when swollen, she was followed extensively. Three times he showed low-key sexual interest in her, but did not follow when she moved away. Pusey (1980) has shown that these three females all associated significantly less often with their brothers after first estrus.

Fifi was mated by all late-adolescent and mature males during her first full swelling *except* her two brothers, Figan and Faben. Figan (aged thirteen) was not observed to show interest in her at this stage, but Faben (about nineteen) was twice seen to approach her with hair and penis erect. Each time Fifi hurried away, screaming, although she was eager to respond to the sexual advances of most suitors (Goodall, 1971). During her eighth period of estrus, however, both Figan and Faben were seen copulating with her (Figan once, Faben twice). She did not try to escape from Faben, but when Figan approached and courted, she screamed and tried to jump from the tree. Figan pursued her and caught up, but she did not present; he copulated, as best he could, as she hung, screaming, from a branch. During the two years of Fifi's adolescent sterility, Figan was seen to mate with her only four times, Faben seven.

After giving birth and after five years of lactational anestrus Fifi once more began to cycle. She was observed during parts of four periods of estrus: Figan (still alpha) was seen to copulate or try to copulate with her on seventeen different occasions. Five times she resisted these courtships so persistently that he gave up. Twice, even though she refused to cooperate, he persisted until he did in fact achieve intromission. Once this was after courting her vigorously for over a minute, during which time he swayed the vegetation so wildly that at the end she was virtually imprisoned beneath a layer of

Faben, eyes half closed, attempts to ignore Flint and Goblin, who both interfere as he copulates with his sister, Fifi. She is uttering a copulation scream. (H. van Lawick)

branches; these he held down over her back during mating! Fifi was not observed to reject any other male in this way.

The relationship between the sixth sibling pair, Goblin and Gremlin, was unusual. He was the second mature male seen to copulate with her during her first adult swelling; she responded to his courtship with the typical crouch-present and showed no signs of protest. Over the next seven months, during which Gremlin cycled regularly, Goblin was seen to court his sister an additional twenty-eight times. On twenty-four of these occasions she accepted him calmly, as she had the first day, but four times she became very upset, screaming and trying to avoid his sexual advances. Once he managed to copulate regardless, but only after chasing and attacking her quite severely (one of the few instances of "rape," and similar to his behavior when he tried to mate with his mother). The other three times Gremlin managed to avoid him. Gremlin also resisted some of the sexual advances of other males, but not so frequently: she resisted (though not always successfully) 15.4 percent of Goblin's copulation attempts as compared with 12.0 percent of Satan's; and she resisted the advances of Evered and Jomeo once each.

Table 16.3 provides some raw data on the frequency with which four of these brother-sister pairs were observed to copulate, and compares these scores with the average number of copulations observed per reproductively mature male during each of the time periods reviewed. We see that Figan copulated (or tried to) with Fifi more

Table 16.3 Relative frequency of observed copulations between maternal siblings versus mean number with non-related reproductively mature males during selected periods.

Male		Female		Observed female copulations	Mean no. per reproductively mature male	Copulations with sibling		
Name	Age (years)	Name	Age class			Number observed	Number resisted	Number resisted successfully
Figan	14	Fifi	Late adolescent	250[a]	14.0	1	0	0
	22		Between 1st and 2nd offspring	38	6.2	1	0	0
	27		Between 2nd and 3rd offspring	154	22.0	12	7	5
Faben	20	Fifi	Late adolescent	250[a]	14.0	2	0	0
	28		Between 1st and 2nd offspring	38	6.2	2	0	0
Pepe	16	Miff	Late adolescent	100[b]	6.0	1	0	0
Goblin	16	Gremlin	Late adolescent	128	18.3	26	4	3

a. Four months during 1967 and 1968 (random).
b. Three months during 1967 and 1968 (random).

often when he was twenty-seven than he had in earlier years. The table also shows that the observed number of Goblin-Gremlin copulations was actually *greater* than the male average. (In fact, as is shown in Table 16.4, Goblin's observed number of copulations was virtually the same as the score for Satan, slightly higher than that of Mustard, Atlas, and Humphrey; and considerably higher than that of the other three senior males.)

Copulations between fathers and daughters and between paternal siblings are unlikely to be inhibited, for the individuals concerned do not "know" their relationship. There is no close bonding between them and they do not achieve the high level of familiarity that presumably underlies incest avoidance between mothers and sons and maternal siblings. Tutin (1979) and Pusey (1980) found that young females were sometimes reluctant to respond to the courtship of much older males, and they suggest that this could be another mechanism for minimizing incestuous matings. Recently, however, it has been possible to observe the sexual behavior of two young females, Gremlin and Moeza, who were certainly sired by Kasakela males (Gremlin surely and Moeza possibly during consortships). Gremlin's presumed father, Evered, was seen to copulate with her eight times during eight periods of maximal swelling. She was observed in the group situation for most of the estrus period during which she is thought to have conceived; Evered was seen to mate her successfully four times. Moeza was observed during parts of six periods of estrus; Figan, who might have sired her, was seen to copulate with her four times. Neither of these young females showed aversion to being mated by the older males Evered, Figan, and Jomeo (Gremlin mildly rejected Evered and Jomeo once each, then quickly gave in), but the latter did not seem to be very interested in them. Table 16.4 summarizes the copulation data for these two females.

Table 16.4 Number of times two late-adolescent females were observed to copulate (during maximal tumescence) with each of the adult males of the community, including those who were possibly their fathers, and the number of times they were observed to resist the courtship displays of these males.

	Gremlin			Moeza		
Male	No. copulations observed	No. copulations resisted	No. resisted successfully	No. copulations observed	No. copulations resisted	No. resisted successfully
Goblin[a]	26	4	3	14	0	0
Satan	25	3	2	16	0	0
Mustard	20	0	0	2	0	0
Atlas	18	0	0	7	1	1
Humphrey	17	0	0	[d]	—	—
Evered[b]	8	1	0	9	0	0
Jomeo	6	1	0	9	0	0
Figan[c]	7	0	0	4	0	0
Total	127	9	5	61	1	1

a. Gremlin's brother.
b. Probably Gremlin's father.
c. Possibly Moeza's father.
d. Died before Moeza came into estrus.

No consortships, either observed or inferred, have involved mothers and sons or maternal siblings. No male has ever been observed to *try* to take his mother or sister on a consortship. Because the father-daughter relationship is not known, there are likely to be occasions when such pairs go off together. But again, the fact that older males tend to be less sexually interested in young females of their own community can reduce the likelihood of incestuous consortships of this sort. In fact, of the 258 consortships that were observed or inferred from 1966 to 1983, in only 18 percent was the male old enough to have been the father of his partner. No male in the consort situation is thought to have impregnated a female young enough to have been his daughter (Appendix Table E2), but the numbers as yet are too small to know whether there is a real reluctance on the part of older males to consort very young females. Certainly Leakey had no such inhibition: when he was an estimated thirty-one years old, late-adolescent Fifi was his favorite consort partner.

One mechanism for avoiding incestuous matings is, of course, female transfer (Goodall, 1968b; Nishida, 1979; Tutin, 1979; Pusey, 1980). Even though at Gombe most of the young females known or believed to have been born in the Kasakela community did not transfer out permanently, almost all of them are thought to have at least visited neighboring communities during late adolescence, and some Kasakela infants are thought to have been fathered by noncommunity males.

Although the likelihood of father-daughter and paternal sibling inbreeding is reduced by the mechanism of female transfer, it is not altogether eliminated even for those females who transfer permanently. Thus a young female who was fathered during a visit by her mother to a neighboring community is highly likely to be mated by her father (provided he is still alive) if she in turn visits or transfers into the community where she was conceived: inevitably the older males will *all* be sexually stimulated by young visitor or immigrant females, whereas, as we have seen, they are less aroused by females who grew up in the same community.

Paternity

If the birthdate of an infant is known fairly precisely and if the gestation period is taken as an average of 229.4 days, it is usually possible to infer during which period of swelling conception occurred. Often, too, females show obvious irregularities in the cycle that immediately follows fertilization, and some show *no* further swellings once they are pregnant. A female may conceive (a) when she is surrounded by most or all of the community males, in the sexual party or group situation; (b) when she is on a consortship with a single male; or (c) during a visit to a neighboring community (either in a group situation or on a consortship).

Appendix Table E2 shows the known or assumed situation of females during POPs that almost certainly resulted in conception from 1967 to 1984. Sixteen times females were probably impregnated in the group situation, thirteen times during consortships, and six times during visits to neighboring communities. (In the other instances there was insufficient information on the females to infer their situations; they are listed as uncertain.) It would seem that a female has roughly equal chances of becoming pregnant in the group and the consort situation; but a more careful analysis of the data shows that this is not, in fact, the case.

Let us turn again to Appendix Table E1. It lists for each Kasakela female (from 1976 to 1983) not only the number of cycles she showed per year, but also her situation, when known, during the peak of estrus of each full swelling. We find that over the eight years 47.5 percent of estrous peaks were in the group situation, 20.3 percent in the consort situation, and 10 percent during visits. In sixty-one instances (22.2 percent) the situation of females during their estrous peaks could not be determined from the available data. These figures are derived from a consideration of *all* cycling females during those years—including those known or assumed to have shown anovulatory swellings when, therefore, they were physiologically incapable of conception. If we are interested in the reproductive efficiency of the different mating patterns, we should try to eliminate all such infertile swellings.

Table 16.5, covering the same time period, shows the situation of females assumed to be undergoing fertile cycles at the time of their estrous peaks. These data reflect the same general trend that emerged from the pooled data for all females; during a high proportion of POPs (51.7 percent) the females were surrounded by males in the group situation. Of the POPs, 22.5 percent occurred when the females were on consortships and 15.0 percent when they were visiting neighbors. In the remaining cases the females' situation during the POP was uncertain. All twelve of the females considered in this sample conceived: nine did so once, two twice, and two three times each. Approximately half of these nineteen conceptions (52.6 percent) resulted from promiscuous matings in the group situation, compared with about

Table 16.5 The situation of twelve cycling females over the eight-year period 1976–1983 during the periovulatory period (POP) of maximal swelling: in the group situation, on consortship, or visiting neighboring males. These figures are taken from the larger sample presented in Appendix Table E1, after exclusion of nonfertile cycles.

Year	No. of cycling females	Total no. of POPs	Percent of POPs				No. of conceptions		
			Group	Consort	Visit	Uncertain	Group	Consort	Visit
1976	7	25	64.0	16.0	12.0	8.0	2	1[a]	0
1977	6	27	40.7	33.3	11.2	14.8	1	2	1
1978	4	9	66.7	33.3	0	0	2	2	0
1979	1	2	0	100.0	0	0	0	1	0
1980	1	3	100.0	0	0	0	1	0	0
1981	1	6	83.0	0	0	17.0	1	0	0
1982	3	14	28.5	14.3	42.9	14.3	2	0	0
1983	8	34	50.0	20.6	17.6	11.8	1	1	1
Total	31	120	51.7	22.5	15.0	10.8	10	7	2

Note: The differences among the categories of conceptions are significant by the Friedman two-way analysis of variance test (one tailed), $p < 0.001$.

a. Conception in group *or* consortship.

a third (36.8 percent) from exclusive mating in the consortship situation. However, a significantly greater proportion of the twenty-seven consortships led to conceptions (25.9 percent) than was the case for the sixty-two group situations (16 percent).

Success in the Group Situation

As we have seen, there may be a good deal of competition among males in the group situation. And although the highest-ranking of the males gathered around a sexually popular female can to some extent prevent his subordinates from copulating with her, he is not always in control. His task will be easier if his top status is undisputed. When Figan was clear-cut alpha during 1973 and 1974, he repeatedly monopolized the females of his choice during their POPs and at least sometimes was the *only* adult male to copulate at such times (Tutin, 1975, 1979). After the loss in 1975 of his brother and ally, he was never again able to monopolize females as effectively as he had during the height of his power.

The ability of the alpha to defend exclusive mating rights depends also on the number of other adult males in the social group at the time. There were many more adult males in the community prior to the division, when Mike and Humphrey were alphas, than there were after 1972. Perhaps this is why they rarely monopolized females in estrus. At the time when the alpha male of K group at Mahale so easily defended his mating rights, he had only two other adult males to contend with: the same was true for the various alpha males of the Arnhem colony.

Even if an alpha male *is* powerful enough to maintain exclusive mating rights in the group situation, there are drawbacks to this strategy, as Tutin (1979) has pointed out. For one thing, the alpha must maintain constant vigilance. If he becomes distracted—as during a hunt, a meeting with strangers, an outburst of social excitement—or if the chimpanzees travel through very thick bush and he is not sufficiently close to his female, his competitors may seize the opportunity for rapid, clandestine matings. On one occasion when Figan's attention was elsewhere (watching monkeys), three other males hastened up and successfully mated with his female, Pallas; a fourth quietly led her away (Riss and Busse, 1977). Tutin (1979) and de Waal (1982) describe similar incidents. The best that even a powerful alpha male can, realistically, hope to do is to ensure that *most* of the copulations around the time of ovulation are his.

Analysis of data collected in the group situation at Gombe, from 1976 to 1983, shows that at some point during the four-day POPs of most females, they copulated with most or all of the reproductively mature males of their community. Five times observations were made during some portion of those POPs when the females concerned almost certainly became impregnated. These data, set out in Table 16.6, show that even when females were thus at the very peak of fertility, mating was quite promiscuous. Unless there happens to be an all-powerful alpha at the time, it seems that a male of any age and rank has a chance of being the one to sire a "group baby." Even twelve-year-old Goblin could have fathered Pallas' 1976 baby.

Success in the Consortship Situation

The male who successfully initiates and maintains a consortship with a fertile female probably has a better chance of fathering her child than he would in the group situation, even if he were alpha. Not only because he will hold exclusive mating rights, but also because the consort situation itself may be more conducive to fertilization—when the pair is away from the excitement and raised levels of aggression that typically prevail in the sexual party and that inevitably subject the female to some stress.

Eight different males have almost certainly fathered thirteen "consort infants" between 1967 and 1984. (This total does not include *possible* fathers.) These data are given in Appendix Table E2. At the time the infants were conceived, their assumed fathers ranged in age from fourteen to thirty-one. Four times the male was between five and ten years older than the mother, four times he was four to nine years younger, and in the remaining five instances the pair were almost the same age (within about three years). Six times the males were high, four times middle, and three times low in rank.

One characteristic shared by most of the successful consortships was their duration; couples who spend longer together, it seems, are more likely to be successful. Thus only three of the thirteen conception consortships were less than ten days, while seven were fifteen days or more; of these, two lasted for over a month.

Although the numbers are still small, and paternity at best is based on assumption rather than fact, it does seem that male chimpanzees are not equally successful at impregnating females in the consortship situation. Table 16.7 provides some information on overall consortship patterns shown by the seven males for whom we have the most longitudinal data. Evered and Sherry scored the highest rates of conception per observation year. Jomeo is not thought to have sired an infant at all, and Figan may not have either. For Sherry, Evered, and the third most successful male, Faben, the percentage of total consortships that resulted in conception (probable and possible combined), and the proportion that included the crucial POP, was high. Evered and Faben tended to lead their females away for relatively

Table 16.6 Copulations observed between seven adult males and five females during the periovulatory periods (POPs) when conception occurred.

| Female | Number of observed copulations | | | | | | | Portion of POP observed |
	Humphrey	Evered	Figan	Satan	Jomeo	Sherry	Goblin	
Pallas (1976)	5	2	3	1	5	5	4	Last two days
Patti (1977)	0	2	1	3	0	2	1	Parts of last two days
Fifi (1980)	3	2	3	0	2	Dead	4	Last two days
Gremlin (1981)	Dead	1	4	6	2	Dead	5	Last three days
Nope (1982)	Dead	5	6	3	1	Dead	1	First, third, fourth days

Table 16.7 Consort behavior of the seven adult males for whom the most longitudinal data are available (over sixteen years).

Measure	Humphrey	Faben	Evered	Figan	Satan	Jomeo	Sherry
No. years male observed[a]	11	9	16	15	12	15	7
Total no. of consortships	34	23	33	24	30	15	16
% including POP	38.2	47.8	45.5	54	33.3	21.4	43.8
% initiated when female							
fully swollen	51.5	25	35.3	40.9	55.2	30.8	43.8
Mean duration (days)	5.6	14.6	18.4	14.9	5	9.3	6.5
% > 10 days	35.3	50	45.5	54.1	15	20	33
% > 26 days	6	13	18	0	0	0	0
Longest consortship (days)	28	45	99	26	18	26	21
No. of consort conceptions							
Probable	2	1	4	0	1	0	2
Possible	0	1	1	2	1	0	1

a. From year first consortship was seen or inferred.

long periods of time (two to three weeks). Sherry, in marked contrast, had a low score for this measure; he, Humphrey, and Satan typically consorted for less than a week. These three males tended to initiate consortships with females who were maximally swollen, but while Sherry usually was able to maintain the relationship throughout the POP, the other two (particularly Satan) often failed.

Prior to her first pregnancy and between successive ones, a female, as we have seen, is likely to show a number of cycles. Does she, during these periods, consort more often and for longer with the male who fathers her infant than with other males? To try to answer this question I tallied the number and length of the consortships that each of the mothers was known or thought to have had prior to the birth of her baby, and the proportion of these that were with the assumed father. The results are shown in Table 16.8. Humphrey, Evered, and Satan all scored high on both measures, whereas Sherry and Faben, two of the most successful consort fathers, had remarkably low scores for both.

We should not lose sight of the fact that the preference of females for individual males may be crucial to the success or failure of a consortship. High levels of association and grooming with maximally swollen females in the group situation, as well as willingness to share food with them, were found to be significantly positively correlated with frequency of male participation in consortships (Tutin, 1979).

Table 16.8 Frequency of association and amount of time spent in association with females whose infants may have been sired by six adult males, compared with the overall consortship involvement of the mothers.

Measure	Combined scores of females impregnated by—					
	Evered	Satan	Humphrey	Figan	Faben	Sherry
No. females impregnated[a]	5	2	2	2	2	3
Combined female consortships	15	4	5	7	18	9
No. with father[b]	10	3	3	3	4	3
Combined female consortships including POP	10	2	3	3	6	8
No. with father[b]	8	2	3	3	2	3
Total consort days	193	23	85	97	214	213
% with father[b]	75.5	69.6	69.4	44:3	24.3	19.2

a. Females *possibly* impregnated by the particular male are included.
b. Probable and possible fathers. These differences in percentages are significant by the χ^2 test (one tailed), $df = 5$, $p < 0.001$.

Evered, Sherry, and Satan had high scores in these affiliative measures. The frequency of aggressive punishment administered to "disobedient" females may also be important; the least punitive male was Jomeo.

For a male like Evered who specializes in long consortships, each one represents a considerable investment in time and energy and may expose him to the double risk of jeopardizing his position in the male dominance hierarchy and encountering hostile neighbors. At the same time even an outwardly successful consortship (which included the POP) is no guarantee that the female has conceived. If she has not, and if some other male impregnates her during her next estrus, he will have lost heavily on his investment. His best strategy, therefore, would be to form another consortship with her himself. If she is not already pregnant, he will have another chance to rectify matters.* Serial consortships, therefore, are probably an efficient strategy even if the female is already pregnant during one or more of them.†

Evered consorted with Passion for two weeks apparently without impregnating her, then after ten days took her off again on an inferred consortship for thirty-nine days, during which time she conceived (she did not come into estrus again before giving birth). He took Dove off and not only impregnated her but maintained the association for an interval sufficient to include one if not two cycles after she had become pregnant.

Evered's technique did not always work. During the summer of 1979 he is thought to have been on a marathon consortship with Nope:

* Even if she is, he may possibly help to safeguard his unborn infant if, by consorting with her, he keeps her from multiple copulations in a sexual party during her next (infertile) cycle. Newly pregnant Kikuyu and Masai women of Kenya sometimes engaged in frequent intercourse at the time when the next ovulation was due, as a way of aborting an unwanted fetus (Leakey, 1977). A similar mechanism seems to be recognized by the !Kung bushmen of the Kalahari (Shostak, 1981).

† The all-time record for serial consortships must go to Leakey, who in 1968–69 took Fifi away six times in succession, and when she reappeared from an extended absence, perhaps in the south, he took her yet again. He did not, however, impregnate her. He did succeed in getting Olly pregnant during the first of a series of three long consortships; and after an interlude in which Goliath stole her away, Leakey took her for two more successive consortships. It is unfortunate that after all this effort the infant was stillborn.

Leakey shaking branches at Olly during the initiation of consortship. (H. van Lawick)

neither was seen in the central part of the Kasakela range for ninety-nine days, and three times they were encountered near the lakeshore in the Mitumba Valley area (by field staff going to and from their village). Yet it seems that upon her return Nope was not pregnant. She cycled again and was taken off by Sherry, who almost certainly fertilized her—during a mere ten days!

Thus there is no single recipe for success: different males specialize in different techniques, and some are better, or luckier, than others. Evered typically set off with a partially swollen or flat female and was remarkably successful in keeping her with him until, during, and sometimes after her period of maximum fertility. Sherry, on the other hand, was most likely to lead off a fully swollen female, maintain the relationship during her POP, and then allow her to wander off.

What of the less successful males? Figan scored high on many of the measures that led to success for Evered and Sherry, yet we cannot with certainty ascribe even one consort baby to him. Satan's problem has been lack of success in maintaining consortships through the POP. His first observed consortship was in 1970, when he was about fifteen years old; it was initiated when the female, Flo, was fully swollen. Four days later the pair reappeared among the other males. Flo was still fully swollen and had presumably escaped. This became an oft-repeated pattern. Eight of his next eleven consortships over the next four years were initiated with fully swollen females; he kept only one of them throughout her POP. The other seven reappeared, like Flo, and were mated by other males. From 1975 on his technique improved somewhat. He continued to initiate consortships with fully swollen (eight times) or inflating (eight times) females, and he managed to keep half of them to himself.

Jomeo's scores are the lowest in all measures of consorting behavior. He is thought to have taken females on fifteen consortships only, during as many years. Only five of these included the POP of the female concerned. Seven were with females already pregnant with the offspring of other males. It should also be mentioned that Jomeo's performance in the promiscuous group situation was equally poor; his frequency of copulation was usually low. Thus it would appear that Jomeo has been unable to compete successfully with other males in any reproductive sphere.

One final point should be made: there are some occasions when males consort with females who are either peripheral community members unknown to human observers or who belong to neighboring communities. After Evered was defeated in the dominance conflicts of 1973, he began to spend longer and longer (up to seventy-five days) in the north of the community range. It was assumed that the violent aggression that his reappearance usually triggered from the powerful Figan-Faben alliance was responsible for the increasing length of his absences. While this was undoubtedly true to some extent, he was observed during one such absence in Mitumba Valley, apparently in consort with a stranger female in estrus. From this incident we assume that during some, maybe all, of his visits to the north he was consorting with unhabituated females. After Faben's disappearance Evered spent less time wandering in the north and once again became a central community male. But he continued to make occasional mysterious visits outside the central part of the Kasakela range.

Evered is not the only male who has, from time to time, vanished for more than ten days. Faben was sometimes away for periods of up to forty-one days in the early 1970s, and Sherry too was away five times between 1975 and 1977. Figan had seven unexplained absences scattered throughout the fifteen years of his reproductive life (none during his two years of supreme power). Thus infants may sometimes be conceived during consortships that go unrecorded, with females who are unhabituated and unknown to the research staff at Gombe.

Male Lifetime Reproductive Strategies

Natural selection has provided the male chimpanzee with an exceptionally efficient mechanism for production, storage, and serial ejaculations of viable sperm. If he is to compete effectively against his rivals, he must be prepared to make the most of any opportunity to impregnate a female, whether this be by snatching quick copulations, or monopolizing a female in the group situation, or persuading or forcing a female to establish a consortship with him. It is of special interest to find that some males are able to adopt different strategies at different periods of their lives. Figan and Humphrey were observed before, during, and after they became alphas and it has been possible to correlate some clear-cut changes in their reproductive strategies with the changes in social rank.

Figan early demonstrated his ability to initiate and maintain potentially successful consortships (those including POPs). In 1967, for example, when he was about fourteen years old, he was responsible for five of the eleven POP consortships recorded for the community males. He went on four the following year and perhaps sired an infant (although as Miff returned from their consortship still fully swollen and was mated by other males during the last two days, it is highly uncertain). Figan's mating strategy changed after he had risen to top rank. He gave up consorting altogether and instead did his best, often very successfully, to maintain exclusive mating rights over a series of females at the height of estrus. Two years later, when he lost his extremely powerful position, he changed his strategy for a third time: he monopolized females to the extent that this was possible for him in the group situation, showed high frequencies of promiscuous copulations, and occasionally risked taking females on consortships. (There is a possibility that he sired Little Bee's infant in 1976, but she, like Miff, returned from the consortship fully swollen and was mated by at least one other male prior to detumescence.) Thus Figan showed a good deal of flexibility in his efforts to maximize his reproductive success. In view of this it seems strange that we cannot attribute even *one* infant to him with any degree of certainty; there is a distinct possibility that he may have been sterile, but unfortunately this is something we shall never know.

Humphrey showed similar versatility. Using more aggressive techniques than Figan, he led females away on a number of potentially successful consortships prior to his rise to the alpha position in 1971. During his tenure as top male Humphrey, unlike Figan, did not entirely give up consorting, but he never stayed away long enough to jeopardize his alpha status. The longest of his six consortships at this time was eight days: the average length (5.1 days) was considerably less than that of his earlier consortships (11.2 days). After his fall from the top Humphrey consorted more frequently again, and the average length increased to 13.0 days (from 1974 until his death seven years later). Humphrey as alpha did not consistently monopolize females in this context as Figan had, but his copulation frequency in the group situation was high and remained so until his death.

None of the other males have shown major changes in reproductive strategy.

Male Dominance and Reproductive Success

The ultimate selective force that drives the male chimpanzee to vigorous and aggressive attempts to better his status should be the maximizing of his own reproductive potential at the expense of his competitors. Yet, as we have seen, only a really secure alpha male can assert his mating rights in a group situation. At the same time, low-ranking and even crippled males who have highly developed strategic skills in consort formation are able to sire infants without the need to compete with higher-ranked rivals in the group situation. They are thus able to pass their genes on to the next generation even if—or perhaps *because*—they opt out of the adult male power struggle. Evered is a perfect example of a male who, having clearly lost in his bid for top rank, devoted himself with considerable success to increasing his reproductive output. Sherry, also defeated in matters of rank, and Faben, unable to compete because of his paralyzed arm, showed similar tendencies.

Thus it would almost seem that there is no overwhelming need, in connection with reproductive success, for the sometimes huge expenditure of energy and the high degree of risk in a male's attempts to rise to the top of the hierarchy. Why, then, did this intense dominance drive evolve? Perhaps to cope with different problems at a time when chimpanzees lived in a less fluid social group—one more similar to the multimale society of the savanna baboon. In this sort of situation, as we have seen, an alpha male chimpanzee can sometimes monopolize a female at the height of estrus and sometimes maintain exclusive mating rights over more than one female. And even if he is not all-powerful, he can still gain an edge over rivals by achieving the highest number of matings in the promiscuous situation.

Once, for whatever reasons, group structure began to change—once community members gained in independence and started to wander off alone or in small parties—then the whole question of reproductive strategy must have changed. The alpha male was not always around when a female came into her peak of estrus (though he often was, because of the prolonged swelling period). Chimpanzees not only show pronounced flexibility in their behavior, they have an undoubted ability to capitalize on chance learning experience. Once a low-ranking male (especially one with the intelligence of a Mike or a Figan) discovered the delights of uninterrupted copulation with an attractive female far from the watchful eye of his alpha, it is more than likely that he would actively try to reproduce this halcyon situation on future occasions. After all, the basic courtship behavior of branch shaking is a signal that commands a female to approach; it is only one additional step from summoning a female to a secret copulation behind a tree-trunk to demanding that she leave the group entirely and follow her suitor wherever he wishes. This behavior, with its attendant reproductive advantages, would be selected for.

It is not impossible, in fact, that the chimpanzees are presently undergoing a gradual evolutionary change in reproductive strategy. Although at first it might seem highly improbable that a species should be in the process of a major behavioral change, on careful reflection it need not be so surprising. After all, the human species is still in the midst of changes in social behavior that began at the dawn of humanity and show few signs of abating. Our own dizzy progress has depended on cultural evolution; we should not forget that chimpanzees also have flexible behavior and that, in their society as in ours, innovative behaviors can be passed on to the next generation through observational learning, imitation, and practice.

Female Choice

Yerkes and Elder (1936) concluded from their study of sexual behavior in chimpanzees that if the male and female concerned were mature, sexually experienced, and familiar to each other, the female tended to control mating. At such times copulations were restricted to a relatively brief period of maximal genital swelling seldom more than one third of the total length of the cycle. On the other hand, if the male was highly dominant and sexually assertive, whereas the female was young, timid, inexperienced, and/or unfamiliar to the male, he was able to coerce her into accepting his sexual advances in the absence of heightened female responsiveness. Mating was then likely to take place for a relatively longer period and irrespective of the phase of the female's cycle.

Females of the Arnhem colony were, for the most part, "free to choose whether or not to have sex"; if they were not prepared to accept a male, "that was the end of it." If a male persisted he was likely, eventually, to be chased by the female, who was often then joined by other females. In that colony the individual preference of the female was important and did not always tally with the rank of the male, so that the females themselves to a great extent "engineered the evasion of the rules which existed among the males" (de Waal, 1982, p. 175).

The extent to which a Gombe female can avoid an undesired copulation depends in part on the situation (the composition of the party) and in part on the tolerance of the male concerned. During one five-year period, 1977 to 1981, adult males were observed courting adult, fully swollen females 1,475 times. Only 61 times (4.1 percent) did the females fail to respond within one minute.* Forty-six percent of these

---

* Tutin (1979) observed 209 uncompleted copulatory sequences, of which 86 percent occurred when females did not respond to male courtship. However, she included males of all ages in her sample and only differentiated between those under and over three years of age. There are many occasions when females ignore the courtship of older juveniles and adolescents, but far fewer cases of adult females refusing to submit to the sexual demands of adult males.

As Humphrey approaches late-adolescent Fifi and begins to court her, she looks around, then away. Humphrey rocks slightly and stamps his foot on the ground. At this Fifi looks toward him again, crouch-presents, and backs onto his erect penis—at which she appears to look, through her legs. During the copulation Flint approaches and interferes. Humphrey, less tolerant than many males, threatens Flint. After the copulation Fifi demonstrates her relaxed relationship with adult males, reclining and self-grooming close to Humphrey. (P. McGinnis)

refusals, or attempted refusals, involved the sterile Gigi and 21 percent occurred when Figan and Goblin were courting their sisters (see the earlier section on incest).

On sixteen of these occasions the rebuffed males persisted until the females gave in. Twelve times they emphasized their courtship by aggressive swaggering displays or brief chases before the females submitted, and once (Goblin and his sister) copulation was preceded by an attack. Mostly, however, the males were remarkably tolerant and patient. In December 1980 Kidevu, who had copulated with Figan on her arrival in his party ten minutes before, reclined and looked in the other direction when he began to court her a second time. After shaking a branch intermittently for three minutes, Figan moved closer, hair bristling; standing bipedally, he swayed branches vigorously. She did not respond. Still bristling, he pushed her—without result. Finally, using both hands, he raised her rump and proceeded to copulate with her. Afterward he released her and stepped back. Kidevu darted away, but then lay down again to rest and did not seem unduly disturbed. (A similar copulation technique was once used by Goblin in order to achieve intromission with a flat female, Fifi.)

Another example of male tolerance was provided by Humphrey in 1969. After branch shaking at Fifi, who was reclining under a palm tree, he approached her with bristling hair, swaggering. Fifi got up rather slowly and moved away from him around the tree. He followed, still with all his hair erect. As she continued to avoid him, she gradually went faster and faster around the tree, and the two of them circled an incredible six times before he finally broke away and displayed off down the slope. Given his highly aggressive temperament and his frequent attacks on females in other contexts, his restraint was remarkable.

As Tutin (1979) points out, a male who is consistently avoided by a particular female may give up and seek his sexual gratification elsewhere. But sometimes he persists. In the mid-sixties Gigi often rebuffed Humphrey—the sound of her barks and screams as she avoided his ardent and repeated courtship became a familiar sound around camp when she was fully swollen. Humphrey did not give up and at the time of his death had consorted with her more often than any other male (Appendix Table E3). Almost always, unless he is crippled or very old, an adult male can coerce an unwilling female into copulating with him and he can, at least on some occasions, force her to embark on a lengthy consortship.

When a low-ranking male shows possessive behavior toward a swollen female in the group context, she can usually avoid him (Tutin, 1979). However, if it is the highest-ranking male present who shows sexual interest in her, there is little she can do other than steal an occasional clandestine copulation when he is not looking.

If a low-ranking male tries to force a female to follow him on a consortship and if she is in estrus at the time, she may, as we have seen, be able to bring the undesired relationship to an end (by calling out so that nearby males are alerted to her whereabouts and may "rescue" her). But if she is flat or only partially swollen, her calling will probably not bring other males to the scene and she may be unable to escape. There are times when females are obviously extremely unhappy in the consortship situation. Once, for example, Leakey insisted that Fifi, who was flat and menstruating, follow him away from Flo. Fifi was particularly reluctant because Flo had just produced an infant, a source of great fascination for Fifi. She paused and looked back several times as she trailed after the big male, but followed—whimpering every few minutes—for the next hour. (The episode ended happily for Fifi; Leakey met other males, lost interest in consorting, and Fifi was allowed to leave.)

This raises an interesting question: Can we assume, when a female unresistingly follows a male on a consortship, that she does so from choice? Certainly it would appear so if, when she hears other males nearby, she remains silent and makes no attempt to escape. But we must remember that, as McGinnis (1973) points out, her obedience may have been shaped by attacks administered as a direct consequence of her calling on a previous occasion. It is in this context, perhaps, that we should consider some of the punishing assaults perpetrated on flat or partially swollen females when they refused to follow male consorts. I saw the first of these in 1961 when an old mother, with a child on her back, refused to follow a male from one valley to the next. In the space of three minutes he attacked her fiercely twice, after which she followed. Even if other males do come along in response to the screams of a female in this situation, her male loses little; competition for access to an anestrous female has never been seen. And the female herself may go through a traumatic learning experience—in the future she will know she should either avoid that particular male or, if she cannot, she should obey his signals.

In 1976 Evered tried to lead Passion, one-quarter swollen, to the north (the direction opposite to her preferred area). She persistently refused to approach when Evered branch shook for her, and he attacked her five times in just over two hours. These attacks were all severe, and during the fourth her hand was badly hurt so that she was unable to use it as she walked. This, of course, slowed her down still more. Evered continued to shake branches and display and, as she limped after him, she whimpered continually. Despite her obviously lame condition, Evered attacked her again. This time her screams did attract a group of males. Evered went off with them, abandoning Passion without a backward glance. The following day an inferred consortship began between Evered and Passion, which lasted for fourteen days. And, just ten days after that, another inferred consortship began during which her infant was certainly conceived. Evered and Passion (with her family) were away for thirty-eight days.

Perhaps we should consider here the seemingly unprovoked attacks, often very severe, that may be perpetrated by adult males on adult females for no reason obvious to human observers. The attacks typically occur when the actors have for some time been in peaceful association, and when there is no change in the social or environmental situation. During one three-year period, 1975–1977, fifty-one severe attacks of this type were recorded: 83 percent of them were directed against cycling females who were *not* fully swollen at the time. The incidents are puzzling. Does the male give some subtle signal to which the female fails to respond? Is he suddenly overcome by a desire to copulate (owing to the fact that her sex skin, though not full, provides some kind of stimulation), so that he perceives her lack of receptivity as an act of implicit disobedience? Whatever the cause, these attacks certainly function to increase the fearful respect of the females for the males concerned; they learn that they must either totally avoid a particular male, or quickly respond in a positive way to his requests.

Female Sexual Strategies

The sexual swelling is a conspicuous feature of chimpanzee reproductive behavior. At maximum size it appears to cause some discomfort to the female; she reclines rather than sits on it or, when in a tree, positions the protuberance carefully to one side of a branch. The swelling, particularly vulnerable if she becomes involved in aggressive interactions, is often ripped open. As a result it may subside rather rapidly. Because ovulation occurs at the end of the swollen period, sudden detumescence could well delay conception. In the face of these obvious disadvantages, there must be a major evolutionary reproductive advantage for the female or the swelling could not have been selected for.

Pom in estrus.

Sexual swellings are also found in the bonobo (*Pan paniscus*) and some of the monkeys, being most common in *Papio*, *Macaca*, and *Cercocebus*, all of which are organized into multimale societies. The swelling, which coincides with behavioral estrus, appears several days *before* the fertile periovulatory period. so that most or all of the group's males will be aware of the female's reproductive state by the time she ovulates. If we argue that the higher-ranking and more assertive males make the best sires, then in many cases the swelling will be to the female's advantage; for the most part, these are the males who will, if warned early, have the best chance of impregnating her. In chimpanzee society this function may be particularly crucial, because females tend to be relatively solitary and community males may be scattered.

Even more crucial for the chimpanzee female is the signal value of her swelling during *intercommunity* interactions. Neighboring males will know of her desirable reproductive state before they get too close, and sexual rather than hostile behavior will be stimulated. We have observed no instances of brutal attacks on stranger *swollen* females at Gombe. Intergroup signaling may also be an adaptive function of the sexual swelling in other primate species. At Gombe male baboons frequently transfer from one troop to another. Prior to the change, the male typically surveys the new troop from a tree overlooking part of its range; from such a vantage point female sexual swellings are readily visible. Some males typically choose to transfer into neighboring troops with more cycling females than their own (Packer, 1979).

Thus, in this context alone, by facilitating her chances of mating with neighboring males and providing new genetic material for her offspring, the sexual swelling may be considered highly adaptive for chimpanzee (and baboon) females.

In addition, the swollen state is very definitely associated with a variety of privileges for the female concerned. This was first demonstrated by the experiments of Yerkes. Pairs of male and female chimpanzees were tested in a situation where choice food was delivered in such a way that only one member of the pair could feed at a time. He found that "female sexual status proves to be of outstanding importance and under certain circumstances the sexual relation of the mates may supplant natural dominance to an extent which enables the subordinate mate (the female) to act for a time as though privileged to assume dominance . . . This competitive food-getting test reveals that control passes from male to female at the beginning of sexual receptivity, and from female to male when receptivity ends" (Yerkes, 1943, p. 75).

Gombe females also become more assertive during maximal swelling. Flo, when she developed her first swelling after four or five years of lactational anestrus, pushed in among the big males for bananas—which she had not done before. It was even more striking when Olly, upon resuming her estrous cycles, showed the same confident behavior, for she was normally excessively nervous when with the adult males (Goodall, 1968b). Another female, Nope, usually extremely timid of humans (other than myself), reclined at her ease within a few feet of the field assistants during two of her rare periods of genital swelling. And there are clear-cut benefits too. For one thing, swollen females are groomed more often by males. For another they are more successful when begging. Reviewing the results of his tests, Yerkes wrote: "It becomes the privilege of the female . . . to take control of the competitive situation and reap whatever benefits accrue in exchange for sexual accommodation or potential accommodation of the mate" (1943, p. 76). When a female in estrus is begging meat from a male at Gombe, it is not at all unusual to see the male, carcass clutched in one hand, pause in his feeding to mate her—after which she is usually allowed to share his prey. I have even seen females, during copulation, reach back and take food from the mouth of the male.

In the female chimpanzee the swollen state has been prolonged well beyond the biological need for female receptivity and attractiveness to the males around the time of ovulation. For one thing, the ten to sixteen days of swelling is far longer than the one to five that would be necessary for strictly procreative purposes. For another, the young female not only cycles during an extended period of adolescent sterility, but maximal swelling during this time is unusually long. She also may continue to show irregular swellings during the first months of pregnancy, some of which may be extended for up to three weeks. Although most females do not resume estrous cycles for three to four years after parturition (the average for eight mothers was exactly three and a half years), some show (nonfertile) swellings as early as one year after giving birth.

These infertile cycles benefit the female socially and biologically— she becomes increasingly integrated into adult society, travels with males to peripheral areas, learns new food sources, and perhaps becomes better aware of the dangers of intercommunity conflict; she also gets more grooming and more frequent shares of meat. The child of an older female will also benefit as a result of his mother's increased sociability. Her swelling, in a way, serves as a sexual bargaining point; the price she has to pay is the occasional undesired consortship, forcible recruitment by neighboring males, the risk of injury during promiscuous group mating which, in turn, may delay fertilization— and a good deal of minor discomfort.

The female bonobo has progressed even further along the evolutionary road toward more extensive periods of receptivity.* She is fully or partially swollen almost all the time; only during extreme old age does she show complete detumescence comparable to the common chimpanzee in anestrus. Moreover, genital swelling not only continues through much of pregnancy, but is *typically* resumed about a year after parturition in females of all ages (Thompson-Handler, Malenky, and Badrian, 1984). There are many ways in which the social behavior of the common bonobo differs from that of the chimpanzee (Kuroda, 1980), and some of these may well be a consequence of the almost continual receptivity and attractiveness of the female. The initiation of copulation between a mature male and female is the prerogative of

---

* In the hamadryas baboon, a young male typically starts his one-male unit by kidnapping a one-year-old female from a neighboring group. He treats her in a protective "maternal" way. Perhaps as a result of being his sole female companion, she may start to show irregular sexual swellings when she is only two years old—some three years earlier than is usual for the savanna baboon (at least at Gombe). The adult female hamadryas also benefits from her periods of swelling, because at such times she is often able to lead her male (Kummer, 1968).

either sex, and the sexual act itself shows greater variety of positions than does that of the common chimpanzee, including the ventroventral embrace favored especially by the female. Affiliative bonds between male and female bonobos are strong. The two sexes commonly share vegetable food, and there are descriptions of females clambering onto the shoulders of males to reach overhead delicacies. Sometimes a male will even sleep with a female at night, curled up in the same nest. Finally, male aggression and dominance rivalry is minimal. It sounds like a utopian society—and viewed against this, it would seem that the Gombe chimpanzees have a long way to go.

Discussion

Sex plays a powerful role in chimpanzee social life. Because of its unique fluid nature, chimpanzee society, perhaps more than that of any other primate, is affected by the presence or absence of receptive females. For one thing, party size tends to be greater when a sexually attractive female is present, and party size in itself affects the frequency of aggression and a variety of other social interactions. The tendency for females to show their first postpartum swellings toward the end of the dry season results in huge gatherings of socially stimulated individuals—sexual jamborees in which, some years, more than half of the community's females, each flaunting her pink flag, contribute to the sexual gratification of the males. These gatherings influence many aspects of social life. Individuals who do not see one another for weeks at a time for much of the year have an opportunity to move about and feed together and thus reaffirm the bonds that, however loose, bind them into a single community. Infants and juveniles, as a result of the increased sociability of their mothers, have rich opportunities for more frequent interactions with peers and elders—particularly meaningful for those whose mothers are at the asocial end of the scale.

Of special interest is the flexibility of the chimpanzees' mating patterns—the different options that are available to the male for the maximizing of his reproductive success, and the fact that at least some males are able to vary their mating strategies with respect to their position in the male dominance hierarchy at different times of their lives. Some males, as we have seen, can even turn failure to succeed in the struggle for high rank to their own reproductive advantage, leaving the core of the community range and associating with either peripheral females of their own community or those of a neighboring social group.

The calm and relaxed nature of the relationship between the male and the female during a successful consortship, which may be maintained for well over a month, suggests that chimpanzees have the potential for developing more permanent ties with nonrelated adults of the opposite sex. And, in this context of stronger and more meaningful pair bonding, the relatively high incidence of copulations during the flat or partially swollen stages of the female cycle is particularly notable.

It is becoming increasingly apparent that sex not only plays a major role in shaping social patterns within the community, but affects,

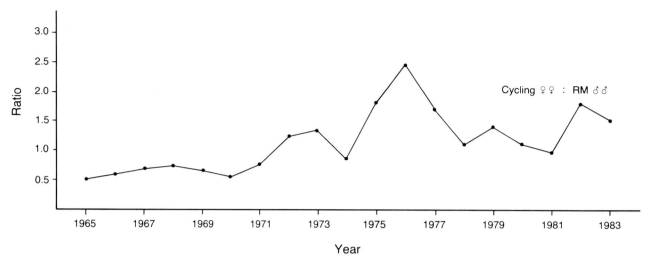

Figure 16.4 Fluctuations in the ratio of cycling females to reproductively mature males in the Kasakela community over the nineteen-year period 1965 to 1983.

equally strongly, relations between neighboring communities. Adult males, perhaps visiting peripheral areas in search of adolescent females for recruitment into their own social group, may come upon and brutally attack older females with young, or encounter neighboring males with resultant hostility on one or both sides. Consort pairs in particular, because they tend to establish their range in overlap zones, run considerable risk of being surprised by a stealthy patrol of neighboring males.

The fluctuating ratio of cycling females (including late adolescents and visitors) to adult males, as shown in Figure 16.4, may have led to aggressive incidents far more severe than competition between males for mating rights within the community. The number of sexual partners available to the males in 1970, prior to the community division, seems very low (although there may have been other unhabituated females in peripheral areas of their community range). The need to recruit young cycling females from the neighboring Kalande community in the south may well have influenced the ranging patterns of the breakaway males at that time.

The long-term nature of our research has yielded unique data regarding the influence that individual chimpanzees have had in determining the course of events—in making chimpanzee history. Three females in particular should be cited.

Gigi, infertile and cycling year after year, has been the core of countless sexual parties. Once community males have been drawn together by the magnetic swelling of a female, they are far more likely to move out to peripheral areas of their range to patrol the boundaries. Gigi's influence was especially pronounced during years when there were only a few other cycling females: in 1974, for example, Gigi, when in estrus, was the nucleus of eleven of the seventeen parties (about 65 percent) that were followed as they patrolled in the south of the Kasakela range. And it is during patrols, as we shall see in the next chapter, that strangers may be savagely attacked. It is at such times, too, that young females may be actively recruited from neighboring social groups.

Gigi (L. Goldman)

Mandy was cycling in 1971 when the original KK community was beginning to divide. Each time she came into estrus she attracted all the Kahama males, and it was when they were gathered together that they most often moved, in a compact group, into the northern part of their range—the area that was soon to become Kasakela territory. The Kahama males, thus united, were able to intimidate Humphrey and the other northern males. Had this situation continued, the course of the Kahama-Kasakela intercommunity conflict might have been very different.

But in 1974, when the first attacks on Kahama males were observed, it was another Kahama female who was cycling—late-adolescent Little Bee. She was not very attractive to the males of her own community but was popular with the Kasakela males, who in fact made a number of excursions to the south in apparent attempts to recruit her. Her appeal for them may have been heightened as a result of the decrease in the number of cycling females available in their own community that year.* Thus Little Bee not only failed to act as a cohesive force in the Kahama community, but actually attracted the hostile Kasakela males into the area of conflict.

A good deal of further research and fortuitous observation is needed before we can hope to clarify all the factors that determine the frequency and nature of interactions between members of adjacent communities. There can be little doubt, however, that the availability of sexually attractive females is one of the more crucial elements—as we shall see in the next chapter.

* The next sharp drop in numbers of cycling females in the Kasakela community was between 1979 and 1981. During this period not only were a number of unhabituated cycling females seen from time to time in association with Kasakela males, but two new females, Kidevu and Candy, joined the habituated community. It is quite possible that they were recruited in the same way as Little Bee.

17. Territoriality

(H. van Lawick)

November 1974   The three Kahama males Charlie, Sniff, and Willy Wally have captured a piglet. As they feed, another adult male appears on the far side of a narrow ravine. The observer has never seen him before; he is obviously a member of the powerful Kalande community that ranges over the hills to the south. Charlie, upon seeing the stranger, drops his meat and runs northward, closely followed by Sniff and Willy Wally. The observer hastens after them and presently finds Charlie in a tree, screaming loudly. Other chimpanzees are calling up ahead, so the observer leaves Charlie and runs on. A few minutes later he catches up with Sniff. Hair bristling, the young male is performing a spectacular display along a trail partway down a steep-sided ravine. Charlie is still screaming back in the forest, but Sniff is uttering deep, fierce-sounding roar pant-hoots. Below, hidden in the vegetation, other chimpanzees are heard calling loudly and charging through the undergrowth. Sniff, as he displays, repeatedly picks up large rocks and hurls them into the ravine so that they land near the strangers beneath. He hurls at least thirteen before he moves out of sight. The chimpanzees below are throwing too; every so often a rock or stick flies up from the undergrowth but, falling far short of Sniff, the missiles roll back harmlessly into the ravine. The observer determines that at least three of the strangers are adult males. They see him and retreat southward. (Observer: H. Matama.)

Sade (1972) has suggested that we should consider as the basic unit of social organization not a single group, but rather a *system of social groups* interacting in a given area. It has become clear, from the pattern of intercommunity interactions that has emerged at Gombe over the years, that this approach would be very meaningful. Unfortunately, until very recently we have only been able to habituate one social group of chimpanzees (although it divided, so that for a while there were two groups). As yet we know almost nothing about the community to the north—the Mitumba community—or the one to the south—the Kalande community. The unhabituated chimpanzees from these groups usually flee whenever they see humans. When females have transferred into our study group, we have known nothing of their background; when our known females left, we were unable to follow their subsequent careers. Nevertheless, it has been possible to piece together an overall picture of the relations of Gombe chimpanzees with members of neighboring communities, a picture that accords well with the description of intercommunity (or interunit-group) interactions at Mahale (Nishida, 1979; Nishida et al., 1985).

Three facets of chimpanzee social organization, all of which have been described in previous chapters, are crucial to the understanding of interactions between neighboring communities. First, chimpanzees neither travel in stable groupings nor follow predictable travel routes from day to day. A lone male may suddenly encounter several males of a neighboring community, a party of males may surprise a single female, and so on. This random pattern contrasts with intertroop encounters in many primate species. The baboons at Gombe, for example, can anticipate where they are likely to encounter neighbors. And when they do meet them, they know that even if the whole troop is not visible, the other members are unlikely to be far away. Second, male chimpanzees remain in their natal community, whereas females may transfer out. Third, female chimpanzees are so distributed throughout the community range that some will travel relatively often in the overlap zones between neighboring communities. Keeping these facts in mind, we can now examine the role played by adult male chimpanzees in maintaining, defending, or enlarging the range they share.

Excursions and Patrols

It must be emphasized that although streams or ravines sometimes appear to serve as boundaries of a community range (a party of males may advance with relative confidence to a stream, but behave with extreme caution after crossing it), for the most part there are no well-defined boundary lines as such (although there are at Mahale; Nishida, 1979). The peripheral areas of a home range are usually overlap zones between neighboring communities.

Community members make periodic visits to peripheral parts of their range for a variety of reasons:

(a) After forming a *consortship* with a female, a male tries to lead her to a place where they are less likely to encounter other community males. Such a pair (accompanied by any dependents the female may have) may remain in a tiny part of an overlap zone for periods of up to a month.

(b) *Excursions* may be made for the purpose of feeding on an abundant seasonal food located in a peripheral area. Often large parties, including females and young, travel to such food sources. Before moving in to feed, they usually pause on a ridge overlooking the valley; sometimes the males will give pant-hoots and perform charging displays. Then they remain silent, obviously listening for a response. If all is quiet they continue their journey. At other times a party may travel to the same area with apparent unconcern, as if they "know" it is safe that day.

(c) Males visit peripheral areas to *patrol* or monitor them. They may be accompanied by females, usually females in estrus who happen to be traveling with them at the time.

Figan and Mike gaze toward neighboring territory. (Gombe Stream Research Center)

Sometimes males start to exhibit patrolling behavior after a large party has moved to a peripheral area to feed—possibly when they detect signs suggesting that neighbors are, or have recently been, in the area. When this happens, anestrous females and young who happen to be with the males (for feeding) often tag along. At other times two or more adult males together leave the core area of their range and travel purposefully toward the periphery. Anestrous females, especially mothers, almost always drop out on such occasions; those in estrus usually remain with the males.

A patrol is typified by cautious, silent travel during which the members of the party tend to move in a compact group. There are many pauses as the chimpanzees gaze around and listen. Sometimes they climb tall trees and sit quietly for an hour or more, gazing out over the "unsafe" area of a neighboring community. They are very tense and at a sudden sound (a twig cracking in the undergrowth or the rustling of leaves) may grin and reach out to touch or embrace one another.

During a patrol the males, and occasionally a female, may sniff the ground, treetrunks, or other vegetation. They may pick up and smell leaves, and pay particular attention to discarded food wadges, feces, or abandoned tools on termite heaps. If a fairly fresh sleeping nest is seen, one or more of the adult males may climb up to inspect it and then display around it so that the branches are pulled apart and it is partially or totally destroyed.

Perhaps the most striking aspect of patrolling behavior is the silence of those taking part. They avoid treading on dry leaves and rustling the vegetation. On one occasion vocal silence was maintained for more than three hours. A male may perform a charging display during which he drums on a treetrunk, but he does not utter pant-hoots. Copulation calls are suppressed by females, and if a youngster inad-

Table 17.1 Details of visits to northern and southern boundary areas at Gombe over a six-year period, 1977 to 1982.

Boundary visits	1977	1978	1979	1980	1981	1982
Total days males were followed for six or more hours	109	224	155	93	100	77
Percentage of those days that parties visited boundaries	17.5	14	23	23	11	20
Total number of visits to boundaries[a]	19	32	35	21	11	16
Percentage of those visits that —						
Party visited southern boundary	47	78	88	62	63	62
Party visited northern boundary	53	22	12	38	37	38
Party patrolled	26	31	31	33	27	13
Party fed	37	44	57	57	54	38
Stranger females were encountered	16	19	34	29	18	13
Stranger males were seen or heard nearby	[b]	28	20	33	45.5	56

Note: By the Mann-Whitney U test (one tailed) there were significantly more visits to the southern boundary than to the northern boundary ($p < 0.003$).

a. A number of visits occurred when female targets were followed when traveling with parties of males. These are included in the table.

b. Most encounters with stranger males during 1977 were with Kahama male Sniff.

vertently makes a sound, he or she may be reprimanded. By contrast, when patrolling chimpanzees return once more to familiar areas, there is often an outburst of loud calling, drumming displays, hurling of rocks, and even some chasing and mild aggression between individuals. Particularly spectacular are displays along streambeds or around waterfalls, which may last as long as ten minutes. Possibly this noisy and vigorous behavior serves as an outlet for the suppressed tension and social excitement engendered by journeying silently into unsafe areas.

During excursions or patrols to peripheral areas the chimpanzees often hear, smell, or actually see individuals of the neighboring community. Responses vary from quiet avoidance or flight to aggressive chasing or fighting, and depend to a large extent on the relative size and age-sex composition of the two parties. Table 17.1 shows how often chimpanzees of the Kasakela community were observed to visit their northern and southern boundaries during a six-year period, 1977 to 1982, and the proportion of those visits when patrolling behavior was seen, when stranger females were encountered, and when stranger males were thought to be nearby.

Contacts between Males of Neighboring Communities

When males hear pant-hoots and drumming from an obviously *larger* number of unhabituated adult males, they sometimes stare toward the sounds in silence, then hastily retreat. When the number of males in the two parties appears to be similar, members of both sides usually engage in vigorous displays with much drumming and throwing, interspersed with pant-hoots, roar pant-hooting, and waa-barks. After a wild outburst the participants stand or sit in silence, apparently waiting to see if the other party will reciprocate. If it does, another outburst ensues. Vocal challenges of this sort are common and usually end with one or both parties withdrawing, noisily, to the core areas of their respective home ranges.

There have been very few occasions when encounters between males of neighboring communities have been relatively well observed—when, that is, at least some of the strangers were actually *seen* by field observers. And even these are incomplete, because the unhabituated individuals invariably flee once they see the humans. The most complete observations were those described below.

Kasakela and Northern Males

Three encounters have been observed between Kasakela males and males from the Mitumba community to the north.

(a) In 1974 six Kasakela males suddenly noticed six Mitumba males sitting in a tree some distance away. The Kasakela males stared toward the strangers, hair erect, then began to move slowly in their direction. But when the Mitumba group charged, displaying, toward them, the Kasakela males turned and fled southward. (Observer: H. Mkono.)

(b) In 1979 Kasakela males encountered one Mitumba male, apparently in consort with the Kasakela female Harmony. They charged him and, when he fled, attacked Harmony and moved back with her to their core area. (Observer: H. Mpongo.)

(c) In 1980 a party of five mature Kasakela males (with Fifi, in estrus) saw a Mitumba male with a female and infant, north of Rutanga Stream. Instantly the Kasakela males turned and fled south. Pant-hoots and waa's broke out from the Mitumba party, indicating that more males were present than the one who had been visible to the field assistant. (Observer: H. Mkono.)

Kahama and Kalande Males

Three such encounters took place at the end of 1974, between the three surviving Kahama males (Charlie, Sniff, and Willy Wally) and males of the Kalande community to the south (also see Goodall et al., 1979).

(a) One of these resulted in the only attack that has been seen on an unhabituated male. The three Kahama males suddenly noticed two strangers, a male and a female, and charged the pair. Both fled, but the male, partially paralyzed in one leg, was caught and briefly attacked by Charlie and Sniff, while Willy Wally displayed nearby. When the stranger escaped he ran off screaming; the victors chased him a short way, then stopped. (Observers: E. Mpongo and A. Bandora.)

(b) A second incident, involving Sniff, Charlie, and Willy Wally is described at the beginning of this chapter.

(c) The third observation took place when a party of seven to nine strangers was suddenly observed south of Kahama Stream. Charlie and Willy Wally fled northward. Sniff remained behind and did not follow his companions until two of the Kalande males charged to within 25 meters of him. (Observers: A. Pierce and C. Chiwaga.)

Kasakela and "Eastern" Males

An encounter with stranger males in the extreme east of the Kasakela range was observed early in 1975. It seems likely that these were, in fact, Kalande males who had moved northward along the high eastern areas of the Kahama community range (or even through country outside the national park to the east of their range). The episode began when four Kasakela males heard a sudden burst of pant-hoots and drumming high in Kakombe Valley. They became very excited, called loudly, and ran toward the calls. The two youngest, Sherry and late-adolescent Goblin, were in the lead and eventually came upon a party of three adult males. Both parties displayed toward each other, waving branches and hurling rocks. When all four Kasakela males charged in unison, the strangers retreated. The Kasakela males continued to display and call, but did not give chase. (Observer: P. Leo.)

Other encounters were between the Kasakela and Kalande males after the annihilation of the Kahama community. They were also highly aggressive in nature and will be described in a later section of this chapter. Thus it is clear that interactions between males of neighboring communities are typically hostile, a finding that has emerged also from the study at Mahale (Nishida, 1979).

Interactions between Males and Unhabituated Females

The response of adult males to a "stranger" female—that is, a female unknown to the human observers and presumed to be from a neighboring community—depends to a large extent on her age and reproductive state. Thus we should consider separately (a) young nulliparous females, estrous and anestrous; and (b) anestrous mothers with dependent young. When they are in estrus, mothers with infants can probably be considered as part of the (a) category.

Young Nulliparous Females

Usually well tolerated by neighboring males, females of this category may move freely between natal and neighboring communities, although they may encounter aggression from resident *females* of the new community. When adult males encounter one of these young visitors, the greeting may be relatively calm, or the female may flee—in which case she is likely to be chased and mildly attacked. These attacks may be compared with attacks in the context of consort formation and serve a function similar to the neck bite in the hamadryas baboon—that is, causing the female to remain with the aggressor (Kummer, 1968). Here again, behavior is probably influenced by the presence of the observer, in that an unhabituated female will usually flee from humans, thereby causing the males to chase and attack her. There is evidence that adult males will, at least on some occasions, actively recruit young females into their community. The shyness of unhabituated females would explain why more examples of such behavior have not been observed.

Anestrous Mothers and Infants

The response of Kasakela males to strangers of this category is very different. During the eight-year period 1975 to 1982 Kasakela males were observed when they encountered such females on twenty-five occasions. Seventy-six percent of these encounters were aggressive in nature, involving chases or attacks. Moreover, in contrast to the brief attacks on young nulliparous females, all but one of the fifteen actual attacks on the older females were extremely severe. Three times they resulted in the death of the females' infants—twice the infants were eaten, or partially eaten, by the aggressors (Goodall, 1977). After ten of these attacks it was possible to see the victim: each time she was bleeding heavily from wounds on the limbs and/or back, and in at least eight cases, on the face or head.

Prior to 1975 three fierce attacks on older females had been recorded: one of these had also led to the killing and partial eating of an infant (Bygott, 1972; Goodall et al., 1979). A further observation reported by Wrangham (1975) is relevant here. In 1974 a party of Kahama males made a detour, apparently to inspect the body of an old Kalande female in the south of their range. It seems that she had died from a number of wounds, the most severe being deep punctures on her back. The only animals that could have inflicted these wounds, other than chimpanzees, are leopards or baboons. The body was not eaten, nor were there any lacerations of the sort that would be expected if the attack had been made by a leopard. And baboons have never been known to inflict serious injury on chimpanzees (even during hunting episodes when chimpanzees attack infant baboons). It seems likely, therefore, that this female also was a victim of chimpanzee aggression. The Kahama party paused briefly near the body and one male smelled and groomed it; then they moved on. (Observer: Y. Selemani.)

The worst attacks on stranger females are summarized in Table 17.2, and Figure 17.1 shows where each of these episodes took place. Case 2 provides the best example of the persistence with which chimpanzees may *hunt* a stranger female. A large party of Kasakela chimpanzees, including one adolescent and nine adult males, suddenly heard a single chimpanzee call. Instant pandemonium broke out and, with pant-hoots, waa-barks, and screams, they raced toward the sound. Presently they fell silent, and traveled in a fairly compact group toward the place where the call had originated. There they sat for fifteen minutes. They traveled on a short distance, stopped again, and sat in silence for half an hour. Afterward they moved on, still silently, penetrating the home range of the northern Mitumba community. They began smelling the ground and treetrunks, but then seemed to give up and began to feed. Suddenly, three hours and twenty minutes after the original call had been made, they raced to the east (retracing their steps); a few minutes later there was a tremendous outburst of calling. When the observers caught up, they saw Humphrey, Evered, and Satan jointly attacking an old anestrous female, who was crouched to the ground and screaming. She managed to escape and ran off with the Kasakela party in hot pursuit; the observers were quickly left behind. The quantity of blood at the site of the attack suggested that the victim had been badly hurt (see also Goodall et al., 1979).

A second attack (case 3) is described fully in Word Picture 17.1. Again a large party of Kasakela chimpanzees was involved, and on that occasion the adult males cooperated to surround their victim rather as they may during a baboon hunt. In attacks 6 and 7 the males showed similar behavior. Case 16 provides another example of an infant snatched from the mother, flailed against the ground as its captor displayed, then released. The infant was not seriously hurt and was able to return to its mother.

Case no.	Date	Specifics of attack	
1	September 1971	*Victims:*	Mother and 2½-year-old infant
		Present:	*Mike, Humphrey, Figan, Satan, Jomeo*
		Type:	Level 3
		Duration:	2 minutes
		Wounds:	Mother loses half an ear. Much blood seen. Infant killed and eaten.
		Comments:	Males race forward and jointly attack mother-infant pair. Probably Humphrey seizes infant: when confusion dies down, he is holding the body. Infant alive and calling for 4 minutes while being eaten. Eleven days later a female, thought to be the same, is encountered by Humphrey, Faben, Jomeo, and Figan. Humphrey attacks her, then all four males chase her as she runs off. They catch up with her, and Humphrey, Faben, and Jomeo attack her jointly but do not follow when she runs off a second time.
		Observer:	D. Bygott (both incidents)
2	May 1973	*Victim:*	Old female; offspring not seen
		Present:	Mike, Hugo, *Humphrey, Evered,* Faben, Figan, *Satan,* Jomeo, Sherry, Goblin, Gigi, 2 cycling females
		Type:	Level 3
		Duration:	6 minutes
		Wounds:	Much blood on ground and vegetation afterward
		Comments:	Description in text
		Observers:	C. Tutin, C. Kakuru, R. Bambanganya
3	November 1975	*Victims:*	Mother and 1½- to 2-year-old female infant; also juvenile (sex undetermined)
		Present:	*Humphrey, Figan, Satan, Jomeo, Sherry,* Goblin, Atlas, 2 females and their offspring
		Type:	Level 4 (two attacks)
		Duration:	First attack, 10 minutes; second attack (30 minutes later), 10 minutes
		Wounds:	Much blood seen afterward. Infant has deep gash in groin and other wounds on face, hands, feet, back.
		Comments:	Full description in Word Picture 17.1
		Observers:	E. Mpongo and K. Selemani
4	July 1976	*Victims:*	Mother and infant; also second stranger, not seen clearly
		Present:	*Evered,* Figan, *Satan,* Goblin, Gigi
		Type:	Level 3
		Duration:	4 minutes
		Wounds:	Blood seen afterward

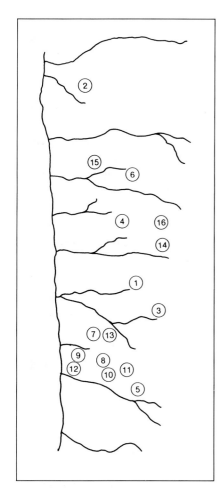

Figure 17.1 Sites of attacks by adult males on unhabituated females. The numbers correspond to the case numbers of Table 17.2.

Table 17.2   (continued)

Case no.	Date	Specifics of attack	
		Comments:	Evered, Satan, and Figan race toward victims. Female gives three waa-barks and runs. Evered and Satan attack, Figan runs toward second stranger. Goblin and Gigi display. Victim escapes.
		Observers:	K. Selemani and H. Mpongo
5	November 1976	*Victims:*	Anestrous female; also 1 or 2 others, not seen
		Present:	*Evered*, Figan, *Jomeo, Sherry, Goblin*, Mustard, 2 females
		Type:	Level 3
		Duration:	2 minutes
		Wounds:	Not observed
		Comments:	Scream heard. Figan, Jomeo, and Sherry grin, embrace, race toward. Jomeo and Sherry attack together; Evered joins in. Figan and Goblin display. Males leave victim and race toward unseen others. Female crouches behind vegetation; Goblin attacks briefly. She flees.
		Observers:	E. Mpongo and H. Mpongo
6	March 1977	*Victims:*	(a) Mother and 5-year-old (b) Female with 5- to 6-year-old male juvenile, who escapes up tree
		Present:	*Figan, Satan, Sherry, Goblin, Melissa*, 2 other females
		Type:	(a) and (b) Level 3
		Duration:	(a) 4 minutes; (b) 5 minutes
		Wounds:	(a) Many, worst on face, back, and above anus; bleeding all over. Moves very slowly after attack. (b) Not seen clearly, but blood observed on female and on vegetation.
		Comments:	(a) Sherry first to race toward strangers. He attacks and Gigi displays close by, dragging branch and swaying vegetation. All males join in vicious gang attack, biting, pulling at victim's hair, dragging. After 4 minutes second stranger waa's about 100 meters away; males leave first victim and race toward second. (b) All males attack at once. Melissa joins in, biting stranger's ear. Stranger bites Melissa's bottom. Males continue to bite, beat, stamp. Victim escapes five times; each time attackers seize her and continue fighting. After 4 minutes Figan and Satan stop; Evered and Sherry attack again.
		Observers:	J. Athumani and R. Bambanganya

Table 17.2   (continued)

Case no.	Date	Specifics of attack	
7	August 1978	*Victims:*	Mother and 5-year-old
		Present:	*Humphrey*, Figan, *Satan, Sherry, Goblin, Mustard*, 1 female
		Type:	Level 3
		Duration:	1 to 2 minutes
		Wounds:	Not observed
		Comments:	Males see stranger, embrace, race toward. Humphrey, Satan, Sherry attack vigorously. Infant escapes and runs off. Humphrey and Satan attack a second time. Female escapes and rushes up tree. Goblin follows and attacks; Mustard joins in. When victim tries to escape, males prevent by surrounding her. She leaps to another tree and drops to ground. Males all chase; Satan grabs and attacks, Humphrey joins in. She escapes.
		Observers:	J. Athumani and H. Matama
8	October 1978	*Victims:*	(a) Mother and infant (b) Middle-aged female (c) Mother and infant
		Present:	*Humphrey, Figan, Satan, Jomeo, Sherry, Goblin*, Gigi (cycling)
		Type:	(a) Level 2 (b) and (c) Level 3
		Duration:	(a) Brief; (b) 4 minutes; (c) 3 minutes
		Wounds:	(a) None (b) Very bad on rump, rectum penetrated; also on face and hands (c) Not seen, but much blood
		Comments:	Three strangers have colobus monkey kill; Kasakela party rushes forward. Observation poor because of thick undergrowth. Second two attacks very fierce; all males participate.
		Observers:	H. Mpongo and Y. Alamasi
9	November 1978	*Victim:*	Middle-aged anestrous female; others present, not seen
		Present:	Figan, *Satan, Jomeo, Sherry, Goblin*, Mustard
		Type:	Level 4
		Duration:	5 minutes
		Wounds:	One foot badly wounded; victim cannot use. Much blood. Not clearly seen.
		Comments:	Sound heard; Figan, Jomeo, and Sherry embrace and grin. Move on cautiously. Thirty minutes later Satan, leading, races forward. Others embrace, grin, follow. Satan is heard attacking stranger, pounding and slapping as she screams. Stops after 1 minute. Victim tries to run. Satan

Table 17.2   (continued)

Case no.	Date	Specifics of attack	
			and Goblin chase and grab her by wounded foot, pull her down. All males display up to her and slap her and display away. One minute later Jomeo attacks again, hits and stamps continually, as others waa-bark, roar, display. Goblin chases as victim tries to run; she crouches, screaming. Goblin seizes her and starts to attack. She escapes; he chases, pounds on her. Other males race off, seemingly after another stranger. Return after 1 minute, pant-hooting and roar pant-hooting. Chase her for 3 minutes, then leave.
		Observers:	Y. Mvruganyi and S. Rukemata
10	January 1979	*Victims:*	Mother and infant
		Present:	*Humphrey*, Figan?, *Jomeo*, Sherry?, *Goblin*, Gigi
		Type:	Level 4
		Duration:	5 minutes
		Wounds:	Not seen, but a great deal of blood
		Comments:	Attack not well seen, because in thick undergrowth. Three separate attacks by Humphrey, Jomeo, and Goblin. All males and Gigi display, pant-hoot, roar. Infant runs from victim and is also attacked; screams a lot. Victims escape; males continue to display and roar pant-hoot.
		Observers:	Y. Mvruganyi and K. Selemani
11	February 1979	*Victims:*	Mother and infant
		Present:	*Figan, Satan, Jomeo, Goblin,* 1 female
		Type:	Level 3
		Duration:	2 minutes
		Wounds:	Not seen
		Comments:	All males rush and attack vigorously, then display away. Poor observation in thick undergrowth.
		Observers:	H. Mpongo and Y. Alamasi
12	February 1979	*Victims:*	Mother and infant
		Present:	Humphrey, Figan, Satan, *Jomeo, Sherry*, Goblin
		Type:	Level 3
		Duration:	3 minutes
		Wounds:	Not seen, but much blood
		Comments:	Jomeo and Sherry attack jointly and viciously, pounding and hitting; others display back and forth, pant-hoot and roar pant-hoot.
		Observers:	H. Mpongo and Y. Alamasi
13	March 1979	*Victims:*	Mother, tiny baby, and juvenile (sex undetermined)

Table 17.2 (continued)

Case no.	Date	Specifics of attack	
		Present:	Humphrey, Evered, Figan, Satan, Jomeo, Sherry, Goblin, Mustard, Atlas, Gigi (cycling), 1 female
		Type:	Level 3
		Duration:	2 to 3 minutes
		Wounds:	Raining too hard to see mother; infant seized, killed, eaten.
		Comments:	Pouring rain. Mustard first to notice strangers; rushes forward and attacks juvenile. All adult males leap onto mother and attack, pounding and stamping on her. Sherry seizes infant; all males share flesh. Female and juvenile vanish.
		Observers:	H. Mkono and H. Matama
14	May 1979	*Victims:*	Middle-aged mother and 3-year-old infant; also female adolescent (about 9 years old)
		Present:	Humphrey, Figan
		Type:	Level 4
		Duration:	Total encounter 15 minutes, attack at least 6 minutes
		Wounds:	Many; worst on back, hands, face. Infant wounded on one leg during first attack (possibly other wounds). Much blood all along trail.
		Comments:	Mother pant-grunts as Figan and Humphrey approach. They rush at her, do not attack at once. While Figan displays, pant-hoots, and roar pant-hoots, Humphrey seizes mother and half drags, half rolls her 5 meters down rocky gully. She escapes, bleeding. Males follow. Observers follow blood along trail; hear another attack, 6 minutes after first. Humphrey attacks, Figan displaying around. Humphrey again pushes and rolls her down gully, toward stream; Figan joins in. Both youngsters follow, screaming. Attack lasts 1 minute. Victim tries to run; Figan grabs her and attacks again. Both males display, drumming and calling, and leave.
		Observers:	H. Matama and Y. Alamasi
15	December 1980	*Victims:*	Mother and 4- to 5-year old infant; also young cycling female
		Present:	Figan, Satan?, Jomeo, Goblin, Mustard?, Atlas, adolescent male, 3 females
		Type:	Level 3
		Duration:	2 minutes
		Wounds:	Not seen; much blood
		Comments:	Figan, Jomeo, and Goblin race toward mother and start to attack. She

Table 17.2   *(continued)*

Case no.	Date	Specifics of attack	
			manages to escape. All males chase her; sounds of severe attack heard (from sounds, assault judged to be one of worst).
		Observers:	H. Mkono and H. Matama
16	January 1982	*Victims:*	Mother and 2-year-old infant; also juvenile (sex undetermined)
		Present:	*Evered, Figan, Satan, Jomeo, Goblin, Mustard,* Atlas, 3 adolescent males, *Gigi,* 1 cycling female
		Type:	Level 4
		Duration:	Encounter 15 minutes, attack during most of time
		Wounds:	Great deal of blood
		Comments:	Mother seen in tree; leaps down, returns for infant. Goblin first to rush forward and attack fiercely. Juvenile runs off. Satan, Goblin, and Mustard attack together. Other males and Gigi display vigorously, join in attack. After 10 minutes Evered seizes infant and races off, displaying and flailing infant. Gigi and Atlas follow. Infant alive, screaming and grabbing at vegetation. Evered hurls infant away, displays back, and attacks mother. After 4 more minutes she escapes and climbs tree. Goblin chases and attacks her in tree; Gigi rushes up, hits and pounds. All display away as mother vanishes.
		Observers:	E. Mpongo and R. Fadhili

Possible Attacks on Kasakela Females

The presence of human observers makes it unlikely that attacks by unhabituated males on Kasakela females will be seen. So far we have only indirect evidence that such aggression does, in fact, occur. This will be described in a later section of this chapter, along with the other Kasakela-Kalande interactions that took place after the annihilation of the Kahama community in 1977.

Response of Community Females to Strangers

For the most part, patrolling the boundaries and repelling strangers from the community range are male activities. Females seldom accompany males on patrols unless they are in estrus, and even then they usually trail along in the rear, not often showing the alert behavior typical of males. During attacks on stranger females, Gigi, the female most often present during intercommunity conflict, typically displayed around the scene of action and once became so aroused that she charged after the two observers, slapping and hitting them (case 4 in Table 17.2). Melissa, in estrus, joined the males when they attacked victim 6 and bit her ear. In response the female managed to bite Melissa's swelling, making a wound. Once when a Kasakela consortship pair encountered strangers in the north, one of them, an anestrous mother, charged at and attacked Atlas, the adolescent son of the consort female. Unfortunately (for our records, though not for Atlas) she saw the observers and moved away.

Females of the Kasakela community have sometimes taken part in chases that preceded male attacks on unhabituated females. Once, after a large party had raced noisily in the direction of a single call, the unhabituated female (Harmony) who was found was attacked by only one member of the Kasakela party: adult female Fifi kicked the stranger some 30 meters to the ground after chasing her through the branches. In fact, young immigrant females (and Harmony was already transferring at that time) face a good deal of hostility from resident females. This has been recorded on all occasions when it was possible to collect information (some immigrants have been too shy for observation until they became accepted members of their new community).

The most dramatic account was provided by Caroline Tutin when she followed the immigrant Patti in 1973 (described by Pusey, 1977). On 1 December Patti was traveling with the young male Satan. They joined a group with three resident females and other males. One of the females at once climbed toward Patti, who screamed and avoided. Two others, Winkle and Athena, approached, embraced, then attacked Patti severely. The fight was broken up by one of the adult males. Patti, keeping close to her protector, threatened the two aggressive females with waa-barks and raised arm. The following day, as Patti was feeding in a tree, two females, Winkle and Miff, approached; standing below the newcomer's tree, they gazed up at her. Patti at once began to move off through the branches. The two resident females followed below. As Patti leaped for the next tree, Miff and Winkle raced to cut off her escape; meanwhile, a third female, Athena, ran to join the others and Patti was surrounded. Slowly these three moved toward Patti, who went as high as she could in her tree. There she clung, screaming, unable to escape. Two adult males, presumably in response to her cries, suddenly charged toward the scene. Patti hurried to them, following with waa-barks, as one displayed at her aggressors. After this she remained very close to one of the big males throughout the rest of the observation. Gradually, with repeated contacts, these aggressive responses of the chimpanzees faded and a month later Patti was seen grooming with her erstwhile persecutors.

It is possible that resident females of other communities were responsible for unobserved attacks on three nulliparous Kasakela females (Nova, Pom, and Moeza). Each of them, after an absence from the center of the Kasakela community range when she was thought to have been visiting a neighboring community (in 1966, 1977, and 1983 respectively), reappeared with deep wounds on her back, head, face, hands, and feet; Nova, in addition, had lost many handfuls of hair from her back. The injuries inflicted by Passion on Melissa (Chapter 12) show that females are capable of causing considerable damage when sufficiently motivated.

Attack by Kasakela Males on a Stranger Female and Her Infant (location shown as no. 3 in Figure 17.1)

The incident began when a large Kasakela party—five adult males, two adult females, and four immatures—encountered an unknown number of strangers in the south of the community range. Most of the Kasakela males raced ahead, and two attacks, heard but not seen, took place in thick bush. Meanwhile a stranger female, with a female infant clinging in the ventral position and followed by a juvenile, climbed a nearby tree. The Kasakela chimpanzees not involved in the attacks gave loud waa-barks as they saw this female, and Fifi climbed the tree after her, stamping on the trunk and continuing to utter waa-barks. Soon the Kasakela males returned and some of them joined Fifi in the tree. The stranger juvenile climbed down and was not seen again.

Two of the males, Jomeo and Sherry, displayed toward the female and attacked her, seizing and stamping on her, one after the other. The Kasakela chimpanzees then quieted. Most of them sat in the tree with the stranger, some staring at her, some actually feeding. Presently the stranger approached one of the males, presented, and twice extended her hand submissively. She received no response. After this it seemed as though she tried to leave, moving through the branches toward another tree. Immediately some of the males climbed down and moved along the ground, looking up, while others followed through the branches. When she stopped, most of them again climbed into her tree.

Thirty minutes after she was first attacked, she again approached one of the males, Satan, and presented, twice reaching to touch him. Satan actively rejected these contacts—the second time he picked a large handful of leaves and scrubbed his leg where her hand had rested. Immediately after this she was attacked again, first by Figan, then by Sherry, and finally by Satan. During this violence her infant fell and was at once seized by Jomeo, who was below. The mother instantly dropped to the ground, and there followed eight minutes of confusion as Jomeo and the mother battled for possession of the infant while the other males leaped down and together continued their attack on the female. Finally, as the mother escaped (bleeding heavily), Jomeo raced up a tree with her infant. Figan followed, seized the youngster from Jomeo, and, holding her by one leg, leaped through the tree, smashing the infant against the branches and trunk as he did so. He jumped to the ground and continued to flail his victim against the rocks as he ran. Finally, after charging for some 40 to 50 meters he flung her from him.

She was not dead. Satan approached, picked her up, groomed her, then gently put her down. She was next "rescued" by a four-year-old male (Freud, son of Fifi), who carried her for over an hour, constantly supporting her since she was too badly wounded, and probably too shocked, to cling unaided. He was still carrying her when Satan approached, threatened Freud (by shaking a branch at him, which knocked him over), and took the infant himself. For the next hour and a half Satan carried her, holding her while he fed, supporting her as he traveled, grooming her from time to time while he rested. Finally he left her on the ground, where she was picked up by eleven-year-old Goblin. He carried her for only ten minutes, then left her in some thick bushes. (She died later of her wounds.)

The Birth and Annihilation of a Community

In 1972 we recognized the existence of a new community at Gombe—the Kahama community, which had previously been part of the large KK study community. By the end of 1977, after an existence of five years, the Kahama community was no more. The events that led to the death (or dispersal) of the Kahama individuals are, I believe, unique in the history of primate field research. Unfortunately the factors leading up to the community division are, to some extent, confused by the banana-feeding regimen and it will be many years before we have the opportunity to observe a similar split. To give as complete an account as possible, let me start at the beginning of my study.

In the summer of 1960 the crop of figs in Kakombe Valley was exceptionally good, and two other fruits were available in large quantities in Linda Valley. I made many observations from the peak overlooking Kakombe Valley. Often I saw parties of chimpanzees traveling from Linda, in the north, down into Kakombe to feast on the figs. Frequently chimpanzees also arrived from the south. Several times I observed excited and noisy reunions as a large party from the south charged down the slope into the valley to join those who had arrived from the north, or vice versa. After a period of feeding, during which members of both parties mingled peacefully, the large gathering sometimes divided again, one group moving off to the north, the other to the south. As I did not recognize individuals at that time, I was unable to ascertain whether the chimpanzees who went to the south were those who had arrived from that direction or whether there had been an exchange of members. Sometimes virtually all the chimpanzees moved in a large party to the north to feed on the fruits in Linda Valley. At the time I was of the opinion that I was observing two separate social groups, which frequently encountered each other without undue hostility. This was what led me initially to suppose that "only a geographic barrier would constitute a limiting factor on the size of a community, although individuals living at opposite ends of the range might never come into contact" (Goodall, 1968b, p. 215).

Between 1962 and the end of 1965 a total of nineteen mature and adolescent males, all of whom seemed to know one another, gradually discovered the banana-feeding area at camp. By 1966 it was clear that the males could be divided into two groups: those who spent more time south of Kakombe Valley (where camp is situated) and those who spent more time to the north. We referred to these as the northern and southern "subgroups." Three adult females seemed to associate more closely with the southern males than with the northern males.

In 1971 the southern males visited the northern part of the KK community range less and less often, while the northern males traveled less frequently in the south. Bygott (1979) has described the gradual decrease in frequency of meetings and associations between males of the two subgroups during that year. He has also documented the gradual decrease in frequency of visits to camp by the southern subgroup. The timing coincided with the reduced banana feeding, but this is not thought to have been a significant factor in the change of ranging pattern, for when these males did come to camp, they were always fed (Bygott, 1974).

At this time there were eight fully mature males in the northern subgroup: six prime (Humphrey, Evered, Figan, Jomeo, Satan, and crippled Faben), and two old (Mike and Hugo). In the southern subgroup were six fully mature males: four prime (Charlie, Godi, Dé, and crippled Willy Wally), one past prime (Hugh), and one old (Goliath). An adolescent male (Sniff) was also part of this group. The Kahama group was smaller, but, as mentioned previously, the two top-ranking males, Hugh and Charlie, were able to intimidate Humphrey, alpha male of the Kasakela group. One female who previously had visited camp regularly (Mandy) and another who had been a fairly frequent visitor (Madam Bee) allied themselves with the southerners, as did the rarely observed immigrant, Wanda.

When the southern males moved north toward camp, they tended to do so in a compact group—and the northern males usually traveled in parties containing at least five of their number when they moved to the south. A pattern developed whereby parties of males from one subgroup left their core area and traveled a kilometer or so until they met males of the other subgroup. When meetings occurred, the southern males tended to perform parallel charging displays, which invariably caused the northern males to scatter. After this initial excitement, interaction was usually peaceful until the "invading" party returned to its own preferred area.

Thus the situation in 1971 was strongly reminiscent of that which I had observed in the early sixties—except that the "meeting place" had moved one valley to the south. There is thus the possibility that the KK community had begun to divide in the early 1960s, that the process was interrupted by our daily banana feeding, then resumed (perhaps because of the large number of males) ten years later.

During 1972 the northern and southern males met only occasionally. Some peaceful interactions were seen, but in general the males increasingly avoided one another, and the southern males stopped visiting camp as a group toward the end of that year. Only one old male, Goliath, and Madam Bee, with her daughters, visited camp (a couple of times) after 1972.

By the beginning of 1973 two separate communities were recognized, the northern or *Kasakela* community, based on Kakombe and Kasakela valleys, and the southern or *Kahama* community, based on Kahama Valley (Figure 17.2*a*). Wrangham (1975), when he analyzed ranging data for 1973, confirmed that the areas used by males of the two communities were geographically separate. Sleeping Valley, shown on the map, was the only significant overlap zone. Although a few nonaggressive interactions were observed that year between males of the two new communities, they involved only the three oldest members: Mike and Hugo of Kasakela, Goliath of Kahama. All had associated closely for many years. Indeed, Goliath and Mike had become very friendly from 1967 until the community division, and Goliath's decision to move south was puzzling.

The characteristic response of the other males during north-south—Kasakela-Kahama—encounters is illustrated by the following observations.

(a) In January 1973 a party of Kasakela males traveled to a ridge overlooking the overlap area and sat in silence. Only when they heard

Figure 17.2 In *a*, the community range and core area of the Kasakela and Kahama communities are shown in 1973, just after the division. Sleeping Valley is the overlap zone between the two communities. In *b*, the extended range of the Kasakela community at the end of 1977 is indicated. Note particularly the expansion of the core area.

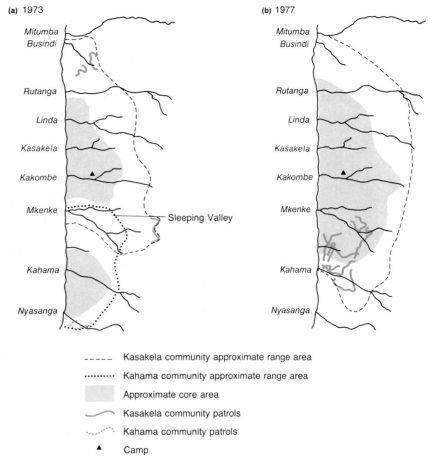

Mike grooms Goliath, who was seen in peaceful association with Kasakela males only five months before their savage assault.

calls from Kahama males in the distance did they move down to feed in Sleeping Valley.

(b) In May of the same year four Kasakela males patrolled to the south. They sat overlooking Kahama Valley for some time, apparently saw and heard nothing, and returned to their core area.

(c) The following month seven Kasakela males heard pant-hoots to the south as they moved down into Mkenke Valley. They at once called and displayed, then ran silently toward the stream. The Kahama males continued to call on the opposite side. For the last few hundred meters, only four Kasakela males continued to run forward. They reached the stream and sat silently while the Kahama males continued calling and displaying. Soon the Kasakela party retreated, maintaining silence for some time before breaking into pant-hoots and drumming.

Early 1974 saw the first of the series of violently aggressive episodes that led to the complete annihilation of the Kahama community. The Kasakela males began a southward movement that culminated in 1977 with annexation of the Kahama community range. By the end of 1974, the estimated range of the Kasakela community was 15 square kilometers; the range of the Kahama community, estimated at about 10 square kilometers in early 1973, had shrunk to about 3.8 square kilometers. During the 1974–75 wet season, the four Kahama males who remained seldom left an area that was only 1.8 square kilometers, and even this was sometimes entered by parties of Kasakela individuals.

Adult males, accompanied by a female in estrus and her infant, sit silently and scan the territory of the Kahama community. (C. Busse)

Moreover, the large, unhabituated Kalande community to the south was also encroaching on the small remnant of the Kahama community range (Pierce, 1978). By the end of 1977, the last Kahama male had vanished. The Kasakela males, along with their females and young, began feeding and nesting in the southern part of their community range. This led to a marked increase of the Kasakela core area, as shown in Figure 17.2b. The total range at this time was estimated at approximately 17 square kilometers.

The attacks on the members of the Kahama community that took place between 1974 and 1977 were all consistently brutal and protracted, and each one is described in some detail in the following pages. Figure 17.3 shows where each of the attacks took place.

The Attack on Godi

Godi (P. McGinnis)

In January 1974 a large mixed party of Kasakela individuals traveled southward. (Observer: H. Matama.) At 1415 hours six adult males (Hugo, Humphrey, Faben, Figan, Jomeo, and Sherry), an adolescent male (Goblin), and a female in estrus (Gigi) began to travel more purposefully southward. The others stayed behind. From time to time calls were heard from the south, and the chimpanzees began to travel quickly and silently in that direction. Suddenly they came upon Godi, who was feeding in a tree. He leaped down and fled. Humphrey, Jomeo, and Figan were close on his heels, running three abreast; the others followed. Humphrey grabbed Godi's leg, pulled him to the ground, then sat on his head and held his legs with both hands, pinning him to the ground. Humphrey remained in this position while the other males attacked, so that Godi had no chance to escape or defend himself.

Figan, Jomeo, Sherry, and Evered beat on Godi's shoulder blades and back with their hands and fists; Hugo bit him several times. Gigi raced around and around, screaming loudly. Goblin kept out of the way.

Finally Humphrey released his victim and the others stopped their attack, which had lasted ten minutes. Hugo, screaming loudly, stood upright and hurled a large rock at Godi; it fell short. The attacking party left and moved rapidly to the south, uttering pant-hoots and displaying. Throughout the attack all had been screaming loudly. Later, calls were heard farther south. The Kasakela party hurried toward them, then stopped, eventually returning to their core area.

After the attack Godi remained motionless for a few moments, then as his attackers moved off, he slowly got up and looked after them, screaming. He was very badly wounded: a great gash extended from his lower lip down the left side of his chin, and his upper lip was swollen. He was bleeding from his nose and from cuts in the side of his mouth. There were puncture marks on his right leg and between his ribs on the right side, and he had a few small wounds on his left forearm. Godi was never seen again, despite the fact that research staff continued to work in the Kahama area until 1978.

The Attack on Dé

This incident took place seven weeks later, at the end of February 1974. (Observers: E. Mpongo and A. Bandora.) The Kasakela party consisted of Evered, Jomeo, Sherry, and Gigi (anestrous). At 0845 they began to travel purposefully to the south. Upon arrival in Sleeping Valley they moved slowly, pausing often to stare around and listen. Suddenly they tensed, hair erect, and stared into a tree. Two baboons appeared and the chimpanzees grunted softly and relaxed. They continued their slow travel until suddenly, at 0915, they all raced forward. A moment later there was a confusion of screaming, waa-barks, and roar pant-hooting as though there had been a predation. The observers, catching up, found the three Kasakela males attacking Dé. Displaying and calling nearby were Charlie and Sniff. Little Bee, in estrus, was near Dé.

After about two minutes Evered and Jomeo charged southward toward Charlie and Sniff. Sherry continued to attack Dé, stamping on him and biting him. Dé tried to pull away but was unable to. Perhaps he had already been badly hurt by the gang attack, or perhaps he was still weak from an illness he had suffered the previous year. He soon stopped struggling and sat hunched over, uttering squeaks. Sherry continued the assault for another two minutes, and then, as Dé seemed to bite him, ran off. Dé climbed a tree. Sherry returned and renewed the attack, and Dé began to scream loudly again.

Soon after this Jomeo and Evered charged back. Jomeo at once joined his brother in the attack. Dé jumped to another tree, but the brothers leaped after and continued hitting and biting him. Dé, again screaming, once more jumped away, but landed on a branch that cracked, so that he was left hanging close to the ground. Jomeo leaped from the tree, seized Dé's leg, and pulled him down. Evered joined in and Dé lay flat on the ground, no longer even trying to escape, as the three prime males assaulted him, all screaming. Little Bee was watching from 5 meters away. Then Gigi began a display, stamping and slapping the ground, and finally joining in the attack. This became even more vicious as all four Kasakela chimpanzees hit and stamped

Dé, two months after the attack. (J. Moore)

again and again on the prostrate victim and dragged him, faintly squeaking, along the ground. One or more of the aggressors tore skin from Dé's legs with their teeth.

Again Evered and Jomeo charged off after the retreating Charlie and Sniff, waa-barking and waving branches. Sherry continued to attack Dé, who again screamed loudly. Little Bee moved to the south, in the direction taken by the Kahama males (who could no longer be heard). At this, Evered and Jomeo followed her and only then, twenty minutes after the start of the attack, did Sherry finally leave Dé and follow the others. (Gigi was left near the victim and was not seen again that day.) Little Bee tried to continue southward, but the males displayed at her and she stopped. After a short rest the party turned back to the north. As they traveled, Little Bee frequently hesitated; each time, one of the males displayed at her, shaking branches, and she screamed, presented, then moved on with them.

Two months later Dé was encountered by himself in Kahama. Emaciated, with his spine and pelvic girdle protruding, he had a bad unhealed gash on his left inner thigh that prevented normal locomotion (perhaps where strips of skin had been removed). Nails had been torn off his fingers, and one toe was partly severed. He had lost part of one ear. His scrotum had shrunk to about one-fifth normal size. He was followed for five consecutive days by D. Riss, during which it was noted that he had much difficulty climbing. He was never seen again despite intensive searching.

The Attack on Goliath
This episode occurred in February 1975. (Observers: E. Bergmann-Riss and A. Bandora.) During the afternoon five adult males (Humphrey, Faben, Figan, Satan, and Jomeo) and one adolescent male (Goblin) left a large gathering and traveled south. Faben led. They moved slowly and cautiously, pausing to listen intently. Humphrey turned back after fifteen minutes and was not seen again that day. Faben led on across Bare Tree Ravine. The party climbed into a tree and sat for forty-eight minutes staring toward Kahama. The only sounds were made by Goblin as he fed, rustling the vegetation.

Suddenly the males climbed to the ground, sat for a moment, then Faben gave pant-hoots, displayed, and raced forward about 25 meters to where Goliath, possibly hiding, sat in a tangle of low bushes. Faben started to attack, leaping at the old male and pushing him to the ground, his functional hand on Goliath's shoulder. Goliath was screaming, the other males giving pant-hoots and waa-barks and displaying. Faben continued to pin Goliath to the ground until Satan arrived. Both aggressors then hit, stamped on, and pulled at the victim who sat, hunched forward. Jomeo, screaming, joined in. Once Goliath was lifted bodily from the ground and dropped back again. During the attack, which lasted for about twenty minutes, Goblin repeatedly ran in, hit Goliath, and raced off again. As the three males continued to pound and kick their victim, Figan displayed past at least eight times, hitting the old male each time and once biting his thigh.

The other males continued to beat up their victim without pause, using fists and feet. Goliath initially tried to protect his head with his arms but, as Dé had done, he soon gave up and lay quite still. Faben

Goliath being groomed by Jomeo, one of the males who later helped to kill him. (H. Bauer)

took one of his arms and dragged him about 8 meters over the ground. Satan dragged him back again. Faben leaped onto Goliath and repeatedly stamped on him as he lay stretched out, face down. Faben then began hitting him with his hand; Satan and Jomeo did the same. With very rapid movements Jomeo began to drum with his hands on Goliath's shoulder blades while Faben sat on the old male's back, took one of his legs and, with his one good arm, tried to twist it around and around.

Eighteen minutes after the start of the attack Jomeo left Goliath, followed by Satan and Faben. The Kasakela party, incredibly excited, displayed with drumming, pant-hooting, and roar pant-hooting. Faben displayed back toward the victim and charged right over him. Only the faintest squeak indicated that the inert male was still alive. After this the aggressors set off at a fast run back north. There was much vocalization, including loud screaming, and the males drummed on trees as they traveled to rejoin the gathering they had left earlier.

At the time of this attack Goliath already looked extremely old. His head and back were partially bald, his ribs and spine protruded, his teeth were worn to the gums. In the attack he was, inevitably, very badly hurt. He had one severe wound on his back, low on the spine; another behind his left ear, which was bleeding profusely; and another on his head. When the aggressors had gone, he several times tried to sit up but fell back, shivering all over. He was holding one wrist in his hand as though it was broken. Like Godi, Goliath, despite intensive searching by all research personnel and field staff, was never seen again.

The Death of Charlie

In May 1977 fishermen living near the Nyasanga Research Camp heard the sounds of fierce fighting. Shortly afterward five large male chimpanzees passed their huts—completely habituated and fearless, they were obviously Kasakela males. Two days later the body of Charlie was found, close to where the sounds of the fighting had been

Charlie (C. Busse)

heard. He was lying, face down, at the edge of Kahama Stream. He had died as a result of multiple wounds—on his head and neck, rump, scrotum, legs, arms, hands, and feet. The injuries were similar to those sustained by victims of intercommunity fights, and there can be little doubt that Charlie suffered a fate similar to that of Godi, Dé, and Goliath.

The Attack on Sniff

By mid-1977 it seemed fairly certain that Sniff was the only remaining Kahama male. Field assistants working in the south often recognized his pant-hoots, yet they never heard Willy Wally respond. At 1615 on 11 November all six Kasakela adult males (Humphrey, Evered, Figan, Satan, Jomeo, and Sherry), as well as late-adolescent Goblin, began to patrol near Kahama Stream, leaving three anestrous females behind. (Observers: K. Selemani and Y. Alamasi.) At 1640 Sherry drummed, without calling, and the other males listened, tense and bristling. There was no response and after five minutes they displayed; this time all gave pant-hoots. After listening again, they crossed the stream and moved rapidly and silently toward Nyasanga Valley: for fifteen minutes they were lost to sight as the observers were left behind.

At 1715 pant-hoots and drumming were heard. The six males appeared, heading northward from Nyasanga and in a state of considerable excitement. Displaying, dragging branches, drumming, and roar pant-hooting, Figan, Satan, Humphrey, and Goblin moved to the north. (Humphrey probably went on northward; he was not seen again that day.) The other three continued to display near Kahama Stream. At 1722 sounds of a fierce attack were heard from the direction of Figan's party. Jomeo (followed by field assistant Alamasi) ran toward the frenzied screaming, and they encountered Sniff, bleeding from a recent attack and still screaming, running southeastward. Jomeo ran toward him and Sniff, trying to escape, ran straight into Evered and Sherry. All three males converged on Sniff, who crouched to the ground as Jomeo and Evered hit and pulled at his limbs while Sherry bit at his face and head. The attack was still in progress when at 1728 Figan, Satan, and Goblin arrived and joined in. All the males were screaming loudly and uttering waa-barks. Sniff by now was bleeding from wounds on the mouth, nose, forehead, and back, and his left leg seemed to be broken. As he crouched, Satan grabbed him by the neck and sucked blood from his nose. Satan and Sherry then each grabbed a leg and, screaming, dragged him down the slope.

At this point the males calmed down a little. Satan, Figan, and Jomeo sat about 10 meters away. (Evered had already left.) When, giving shaky waa-barks, Sniff sat up, Goblin at once approached and hit his nose several times. Sherry watched intently, then at 1736 he too began hitting at Sniff. Goblin moved away. Figan, Jomeo, and Satan left, heading back north, calling loudly and displaying. Goblin went with them a short distance, then returned and sat about 12 meters from Sniff, watching as Sherry continued to hit him. At 1740 Sherry stopped his attack, and he too sat and watched, from 4 meters away. Goblin left, following Figan's party. When Sniff got up and moved southward very slowly, his leg trailing after him, Sherry fol-

Willy Wally (N. Pickford)

Sniff (H. van Lawick)

lowed closely. He did not attack, and each time Sniff paused, Sherry sat, then followed as Sniff moved on. At 1800, some thirty-five minutes after he first began his attack, Sherry finally left the crippled male, pant-hooting and displaying eastward, along the stream. Sniff, who had given a few small waa-barks while Sherry followed him, now moved on in silence. When he moved into thick undergrowth, the observers left, as he seemed disturbed by their presence.

Sniff was seen the following day, barely able to move. We decided that it would be merciful to end his suffering—but he could not be found again. Five days later there was a strong smell of putrefaction, the source of which could not be discovered.

Madam Bee (H. Bauer)

The Attacks on Madam Bee

The crippled Kahama female Madam Bee and her daughters Little Bee (born in 1960) and Honey Bee (born in 1965) were subjected to a *series* of attacks by the Kasakela males, but only the mother was hurt. Figure 17.3 shows the location of these aggressive incidents.

The community split had not had a marked effect on the ranging patterns of this family, because Madam Bee, like the Kahama males, had always tended to roam in the southern part of the KK community range. Her visits to camp, never very frequent, became gradually less so. Since her preferred range was in the overlap zone between the two new communities, she continued to have some association with Kasakela individuals during 1973. Even as late as August 1974 the family visited camp—their first visit in two years. They encountered the old male Hugo and several females, but there was little social interaction and no aggression. (A week later Madam Bee tried to visit camp again, but each time she started down the opposite slope, Willy Wally, who was in consort with Little Bee, displayed at her and eventually she gave up.)

FIRST ATTACK   In September 1974 the entire Bee family was attacked by a party of five Kasakela males (Evered, Faben, Figan, Satan, and Jomeo) and one female (Gigi). (Observers: A. Bandora and H. Matama.) Only Madam Bee, who was attacked very severely first by Jomeo and then by Figan, was injured; she sustained a deep gash in one leg. Little Bee was attacked successively by Evered, Faben, and Gigi; Honey Bee was attacked by Satan. At the time Madam Bee was showing the first signs of an estrus swelling. Little Bee was fully swollen and accompanied the aggressors when they returned to the north. The following month, after Little Bee had returned to her family, circumstantial evidence suggests that when she was again in estrus she was once more led back north from her home range by Kasakela males (Goodall et al., 1979).

SECOND ATTACK   In February 1975 another attack on Madam Bee was seen. (Observer: E. Bergmann-Riss.) Only Figan, Faben, and Gigi were present, and only Faben attacked her. It was not a serious attack and no wounds were visible afterward.

THIRD ATTACK   Three months later, in May, Madam Bee was set upon by three Kasakela males (Figan, Satan, and Jomeo) when, along

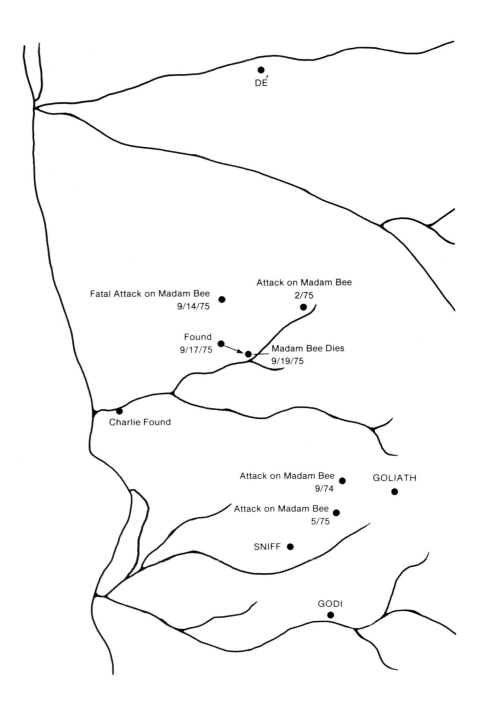

Figure 17.3 Sites of the attacks by Kasakela males on Kahama individuals, 1974 to 1977. The map covers roughly the area from Sleeping Valley on the north to Nyasanga Valley on the south.

with a female in estrus (Patti), they were patrolling in the south. (Observer: R. Bambanganya.) The patrol began at 1215. At 1244 the males suddenly ran forward as they saw Madam Bee and Honey Bee some 40 meters ahead. Jomeo led the attack, but it was Satan who took over and delivered the worst punishment, wounding Madam Bee badly in one thigh. During these attacks Figan chased Honey Bee. He followed her up a tree, grappled with her, and she fell to the ground. She was then attacked, mildly, by Satan and Jomeo. The entire incident lasted five minutes, after which the Kasakela males, who had uttered loud pant-hoots during the attacks, displayed back northward. Possibly they were after Patti, who had vanished during the aggression.

UNOBSERVED ATTACK(S) Four months later, in September, Madam Bee was encountered during a follow, in very poor health. She was walking slowly and with difficulty and had several unhealed wounds; the two worst were on her head and on one shoulder. Clearly she had been subjected to one or more attacks.

THE FATAL ATTACK Six days later, when observers H. Matama and E. Mpongo were searching for Madam Bee to monitor her progress, they heard sounds of fierce conflict. Hurrying to the source, they found Figan, Satan, Jomeo, and Sherry displaying around and attacking the old female. Goblin and four Kasakela females (including Little Bee, who by this time had transferred into the Kasakela community) were watching the action. As the observers arrived, Jomeo, dragging Madam Bee by one arm, displayed down the slope, then turned and stamped on her and slapped her. As he paused, Figan displayed up, stamped on the victim, dragged her for 3 meters, and charged away. All four males then displayed, screaming; the Kasakela females rushed out of the way into the trees. To add to the confusion a troop of baboons was milling around. Madam Bee tried to stand; she was shivering all over. At once Satan displayed up, threw her to the ground, stamped on her, and dragged her a few meters. Figan then attacked her with ferocity, hitting and stamping on her again and again. She was too much hurt (perhaps winded) to scream. When she stopped moving, he displayed away. Jomeo returned, pulled her inert body toward him, half picked her up and slammed her down, stamped on her, and rolled her over and over along the slope. Finally he stopped and sat a few meters away, watching her.

Madam Bee tried to stand, but fell and lay still. Again she tried to move; this time she started back up the slope, screaming and heading for a thicket. Satan displayed up, smashed her to the ground, pulled her toward him, pushed her away, then pounded on her with hands and feet continuously until she collapsed and lay completely motionless for two minutes. The observers thought she was dead. Satan, his hair fully erect, stood by and swayed branches at her. Goblin also sat close, watching intently. Satan continued to wave branches until finally she moved again. Goblin left at this point, joining the other three males who were displaying nearby. Satan remained for another minute, following and watching as she very slowly moved into the dense undergrowth. She disappeared fifteen minutes after the start of the assault, and Satan left.

When the Kasakela party moved noisily away, one of the observers crawled into the bush after Madam Bee—but was unable to find her. Nor could I, when some thirty minutes later I was brought to the scene. Despite intensive searching from morning to evening, she was not found until the third day. Her presence then was revealed by Honey Bee, who was moving about in a tree above. Madam Bee was scarcely able to move. During the attack she had received deep wounds in the following places: left ankle, right knee, left wrist, right hand, back (several wounds), and big toe of left foot (hanging by a strip of skin). She died on the fifth day after the attack.

Thus, after an existence of five years, the breakaway community had been virtually exterminated. The two Kahama females, Mandy and Wanda, may have joined the Kalande community—although it seems more likely, in view of the hostile attitude of male chimpanzees at Gombe to stranger mothers, that they and their offspring met the fate of Madam Bee.

Little Bee, as mentioned, transferred permanently into the Kasakela community in 1975. Four months after her mother's death Honey Bee appeared in the Kasakela home range and for the next three years she was seen, every so often, when in estrus. Unlike Little Bee, the younger sister did not transfer permanently into the community of the aggressor males. In 1979 another young female, Kidevu, estimated as being twelve to thirteen years old, immigrated into the Kasakela community from the south. In facial appearance she strongly resembled the Kahama female Mandy—particularly when compared with photos of Mandy taken when she was about the same age as Kidevu. Mandy's daughter Midge, born before the community split, would have been thirteen years old in 1979. It seems almost certain that Kidevu and Midge are one and the same.

Invasion from the South

The assault on Sniff in November 1977 marked the final stage in the annihilation of the Kahama community as such. In 1978, as we have already seen, the victorious Kasakela males, along with their females and young, began to sleep as well as feed in the area which, for the previous five years, had been the heart of the Kahama core area (Figure 17.4a). But this situation did not last for long. The Kahama community had apparently acted as a buffer between the Kasakela chimpanzees and the powerful Kalande community (comprising at least nine fully mature males) in the south. This community, which had already begun to push northward in 1975, now began to shift its core area still farther to the north.

During the 1978–79 wet season, when great stands of the two seasonal fruits *Landolphia* and *Saba florida* were ripening in Kahama and Mkenke valleys, Kasakela individuals repeatedly encountered parties of Kalande chimpanzees, mostly males, north of Kahama Stream. Areas where the Kasakela males showed patrolling behavior (Figure 17.4a and b) clearly indicate the northward penetration of their Kalande neighbors.

In 1979 Kasakela males were followed twenty-six times when they crossed Bare Tree Ravine, moving south toward Kahama. On 54 percent of these occasions they heard or actually saw strangers, assumed to be members of the Kalande community. And so the Kasakela males moved southward with ever-increasing caution and apprehension, as gradually their newly annexed land began to shrink (Figure 17.4b). Early in 1980 one of the Kasakela females, Passion, whose

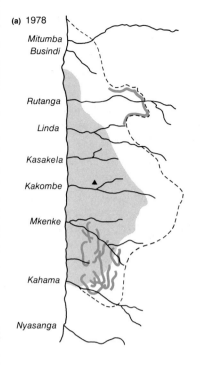

(a) 1978

Mitumba
Busindi
Rutanga
Linda
Kasakela
Kakombe
Mkenke
Kahama
Nyasanga

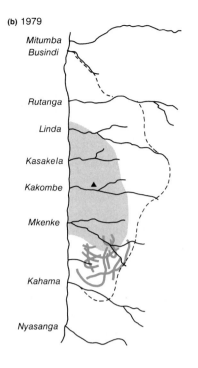

(b) 1979

Mitumba
Busindi
Rutanga
Linda
Kasakela
Kakombe
Mkenke
Kahama
Nyasanga

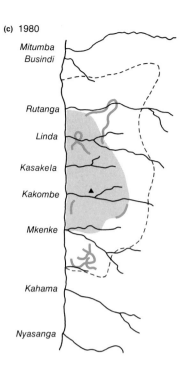

(c) 1980

Mitumba
Busindi
Rutanga
Linda
Kasakela
Kakombe
Mkenke
Kahama
Nyasanga

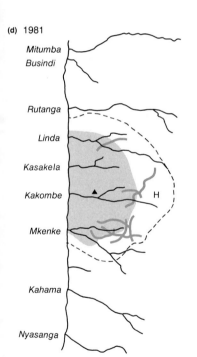

(d) 1981

Mitumba
Busindi
Rutanga
Linda
Kasakela
Kakombe
Mkenke
Kahama
Nyasanga

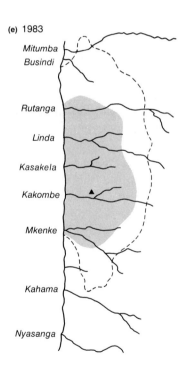

(e) 1983

Mitumba
Busindi
Rutanga
Linda
Kasakela
Kakombe
Mkenke
Kahama
Nyasanga

- - - - -   Kasakela community approximate range area

    Approximate core area

    Kasakela community patrols

▲   Camp

H   Humphrey's skull found

Figure 17.4 Changes in community range between 1978 and 1983 that resulted from the invasion of the powerful Kalande community in the south. Patrolling behavior by the Kasakela males indicates the presence of Kalande individuals in community territory. (Patrolling in 1983 is not shown, as the data have not yet been analyzed.)

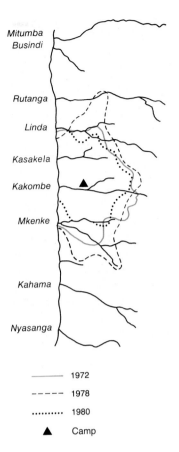

Mitumba
Busindi

Rutanga

Linda

Kasakela

Kakombe ▲

Mkenke

Kahama

Nyasanga

——— 1972

- - - - - 1978

.... 1980

▲ Camp

Figure 17.5 Passion's year range in 1972, 1978, and 1980. Her range in the south, which enlarged in 1978 (after the annihilation of the Kahama community) shrank after Passion almost certainly was attacked by Kalande males. Her 1980 range in the north is almost identical to that plotted by Wrangham (1975) in 1972. (In 1978 she ranged more widely than usual in the north when she accompanied her daughter, in estrus, as explained in Chapter 9.)

preferred area had always been south of camp, appeared with extremely bad wounds. Her three-year-old infant son was also severely wounded. They arrived in camp from a southerly direction and lay in the undergrowth, where Passion dabbed at her lacerated bottom. The other members of her family, who had become separated, eventually rejoined her. From that day on Passion and her family were no longer observed to use the southern part of their range. Figure 17.5, which compares her observed year range in 1972 and 1978 with her observed range in 1980, shows this change clearly. It is highly probable that Passion had been the victim of intercommunity male aggression.

By 1981 the Kasakela community range was down to about 9.6 square kilometers—half its 1977 size—and, in fact, the chimpanzees seldom moved outside their core area of some 5.7 square kilometers. The males were only observed crossing Mkenke Stream on two occasions, and even in the northern part of Mkenke Valley they often moved with caution and sometimes showed patrolling behavior (Figure 17.4*d*)—a sure sign that strangers were, or had recently been, in the vicinity. Early that year Humphrey disappeared; his skull was subsequently found near the eastern periphery of the community range. Then during September and October two Kasakela females lost infants who were between one and three years old. One of these mothers, Nope, was seen with her infant on 15 October and appeared without her six days later; Nope had bad wounds on her face, one hand, and one leg. The second mother, Dove, is a peripheral female and her infant vanished during a two-month period when she was not seen at all. Any or all of these casualties could have been the result of conflict with Kalande males.

With only five fully mature males (from May on, after Humphrey's disappearance) the Kasakela community not only lost ground in the south, but seemed barely able to hold its own in the north. The total range shrank to around 6 square kilometers, scarcely enough to support the eighteen adult females and their families.

At the start of the 1981–82 wet season the Kalande community again moved northward in strength to feast on the rich crops of *Saba florida* and *Landolphia* fruits in Mkenke Valley. At times they even moved to the southern slopes of Kakombe Valley and from camp could be heard calling. On a never-to-be-forgotten day in 1982 four Kalande males actually appeared in camp. For twenty minutes two Kasakela females, Fifi and Melissa, had been staring over the valley to the south and uttering waa-barks. Then as the strangers appeared, Fifi and her family ran northward, but Melissa rushed up a palm tree and was chased by a Kalande male and mildly attacked. Her four-year-old son, Gimble, encountered a second male—but was only sniffed. The four males left after this, moving back toward the others of their group, who were still calling. Melissa, screaming, climbed down, gathered up her infant, and followed Fifi. Perhaps it was the strangeness of the camp setting that saved Melissa from a worse attack.

Shortly after the above incident observer E. Mpongo, fishing on the lake on his day off, heard Kalande males calling from the Mkenke-Kahama ridge and, perhaps in reply, Mitumba males calling from the Linda-Kasakela ridge. Kasakela individuals remained completely silent, and for the next three days were not heard to make any loud

Three Kasakela males startle during a patrol: Figan,
Satan, and late-adolescent Mustard.

calls at all. I arrived two days after Mpongo had heard the calls from
the lake. When Evered appeared in camp, alone, he seemed tense,
startling at any sudden sound and continually scanning the southern
slopes of Kakombe Valley. I followed him as he traveled with extreme
caution, moving very silently, to the top of the Kakombe-Mkenke
ridge. During the fifteen-minute climb he repeatedly stopped to listen.
When he reached the top he sat in complete silence, facing first south,
then west, then east, for a total of twenty minutes. He next retraced
his steps, moving very fast indeed, and began to feed on *Saba florida*
fruits on the valley floor. The field assistants later told me that prior
to this instance no Kasakela chimpanzee had been seen feeding from
these richly laden vines. The reason, the assistants suggested, was
that the rushing noise of Kakombe Stream would make it impossible
for chimpanzees feeding there to hear the approach of their "enemies,"
and they feared a surprise confrontation. Evered's behavior bore out
this interpretation.

During the 1982 dry season the Kasakela males began once more
to extend their range to the north. As a result, their overall range at
the end of the year had increased to an estimated 11 to 12 square
kilometers (Figure 17.4*e*). They managed to maintain this area
throughout 1983. Indeed, in the 1983–84 wet season, with Atlas swelling the ranks of the mature males and five late adolescents joining in
the patrols, the Kasakela males became more assertive and the Kalande males seemed to become more cautious. No Kalande individuals
were seen north of Mkenke Stream, and Kasakela patrols once more
began to move, quite regularly, as far south as Kahama Stream.

Individual Participation

Without doubt most adult male chimpanzees, particularly young prime individuals, are strongly motivated to travel to peripheral areas of their home range. In many instances their response to the sight or sound of presumed strangers (especially the frenzied rush toward females), as well as the long periods spent in stealthy stalking of victims, suggests that these encounters are highly attractive to the participants. To some extent all adult males show a keenness to participate in these exciting incidents, although old males take part less often as it becomes harder for them to travel long distances. Even among the prime males, however, there are distinct individual differences.

Sherry, between 1974 and his disappearance in 1979, was almost always in the forefront of the patrols and several times was the leader when parties traveled to peripheral areas to feed. Once when Sherry was patrolling with five other males south of Mkenke Stream, calls were heard (from Kahama individuals) farther to the south. The Kasakela party called and displayed, then most of them, led by Figan and Humphrey, retreated northward. Sherry, however, ran toward the calls—followed, after a moment, by Faben. It was ten minutes before they gave up and rejoined the others. On three other occasions Sherry was left near strangers when the rest of his party headed for home. Twice Sherry remained with Kahama victims—Dé and Sniff—and continued to attack after the other Kasakela males had displayed away.

The youngest Kahama male, Sniff, showed similar enthusiasm for encounters with his neighbors. He was in the vanguard of the attack on the male with the paralyzed leg, and after other Kahama males had run from an encounter with Kalande chimpanzees, Sniff on his own engaged in the spectacular rock-hurling episode against at least three stranger males. Another time, after he, Charlie, and Willy Wally had retreated from a party of Kalande individuals, Sniff went back by himself—as though he simply had to have another look. There was one occasion in September 1975 when Sherry led Figan, Satan, and Jomeo on a patrol to the south. Suddenly the four males raced forward, a bedlam of sound broke out ahead, and then, a moment or so later, all four Kasakela warriors came racing back—pursued by Sniff, with Charlie bringing up the rear! All six were lost to view, so that the outcome is not known.

Jomeo, Satan, and Evered all participated about equally. Jomeo was violent and brutal in *all* attacks on Kahama individuals, and because he has always been the largest and heaviest of the Kasakela males, he was probably responsible for many of the worst wounds, including Sniff's broken leg. Faben, despite a completely paralyzed right arm, often led patrols, was in the forefront of chases, and took part in two of the attacks on the Kahama males. He was, if anything, the most brutal of Goliath's assailants. Figan, while he took part in many patrols, tended to travel in the rear and it was often he who turned back first if neighboring males were heard or seen. And Humphrey, although he eagerly took part in many patrols and encounters in the north, very rarely patrolled to the south. Indeed, he often left parties that began to move purposefully toward Kahama. He took part in only one attack on a Kahama individual, that on Godi. Humphrey's reluctance—fear, even—to travel in the south dates from 1971 when, even though he was alpha male of the Kasakela community, he was, as we have seen, easily intimidated by the Charlie-Hugh coalition on those occasions when the Kahama males moved northward.

Other Populations

The evidence that has been gradually accumulating at Mahale, the only other long-term study site, points to a similar aggressive relationship between males of different communities—or different unit-groups (Nishida, 1979; Nishida et al., 1985). I have already noted in Chapter 9 the annual movement to the north of the powerful M group, which in turn led to the displacement to the north of the smaller K group—often on the very day that the main body of M group arrived. Nishida describes how chimpanzees of K group typically retreated rapidly and avoided making any calls when they heard calls from M group during the times both groups were in the overlap zone. When K-group males were in their core area, however, they sometimes responded with calling and displaying if M-group calls were heard from the overlap zone. And if M group was represented by only a very few individuals, K-group males occasionally showed aggressive behavior even if *they* were in the overlap zone and the M-group members were in their own exclusive range.

Nishida (1979) states that the unit-groups at Mahale maintain their ranges with well-delineated and "traditional" boundaries. These boundaries are not ecological (such as Mkenke Stream between the Kasakela and Kahama communities at Gombe), but rather are "invisible barriers." For seven years no adult male of K group was seen to cross the community boundaries, and adult females seldom did so. Similar boundaries were recognized for M group, although an M-group party was once seen to penetrate the exclusive area of K group.

Mahale males also show patrolling—or "scouting"—behavior and occasionally direct charging displays toward "enemy groups." Nishida comments that it is mainly the size and composition of the party, especially the number of adult males, that decides the issue during an intercommunity encounter. Meetings when adult males were present in both parties always included aggressive elements. There is only one report of fighting when (in 1974) a lone M-group male, Mimikire, was chased for 200 meters by three K-group males, then seized and attacked by one of them, Kasonta. The two wrestled bipedally "until Kasonta forced Mimikire to the ground and bit him on the right thigh. Mimikire retreated little by little on his buttocks," with Kasonta stamping on his back. The victim escaped after five minutes. Kasonta chased him a short way, then returned, with much displaying, to the rest of his party (Nishida, 1979, p. 89).

While this attack showed none of the brutality of the gang attacks described at Gombe, more recently all the adult K-group males over six years old have disappeared, one after the other. Nishida and his colleagues (1985) speculate that intercommunity conflict may have been responsible for at least some of these disappearances. The larger M group is now in the process of taking over the K-group females.

At Mahale, in 1974, K-group males were observed killing and cannibalizing the infant of an M-group female. In another observation, in 1976, M-group males fiercely attacked a K-group female (Wantendele) and seized, killed, and ate her infant (Nishida, Uehara, and Nyundo, 1979). Four years after that attack, when Wantendele's next youngster was just over four years old, the mother was again subjected to a fierce attack by M-group males (Nishida and Hiraiwa-Hasegawa, 1985). This was in January 1981, when the K-group males were greatly reduced in number. It is thought that she was encountered by an M-group patrol as she and her son traveled in the overlap zone between the two communities. When researchers arrived, alerted by screaming and other calls, Wantendele was being attacked; already she was badly wounded on her head, back, and fingers. Her son had also been hurt, but less severely. After the attack Wantendele remained with the aggressors another two hours, during which she was attacked twenty times (seven times her infant was clinging ventrally). Most of these attacks were mild, as were the three directed against her son when he was on his own. When another very severe assault broke out, in which three of the four M-group males took part, the observers, fearing for the chimpanzee's life, intervened and chased away the aggressors. Wantendele and her son then fled back toward their own core area.

Ten months later the same mother-infant pair revisited M-group range. This time Wantendele was in estrus, though not fully swollen. She was observed interacting with M-group chimpanzees for four successive days, and it was obvious that she was attempting to transfer. Some friendly interactions, as well as many mildly aggressive ones, were seen. After this the pair disappeared, presumably returning to the remnants of their community. A month later, again in estrus, again not fully swollen, Wantendele once more appeared in M-group range. This time she and her infant were subjected to a brutal gang attack of the type seen so many times at Gombe. Eleven M-group chimpanzees joined the assault: six adult and two immature males and three adult females. Wantendele was badly wounded on her head, ears, arm, fingers, toes, and sexual swelling. Her son's wounds, while not severe, were also extensive. After seven minutes she managed to pull away. The four observers, again afraid Wantendele's life was in jeopardy, formed a protective cordon around the mother-infant pair; amazingly, things calmed down almost at once. Wantendele remained with M group for the rest of the day (Nishida and Hiraiwa-Hasegawa, 1985). She visited again a few times during 1982 and transferred permanently the following year, after the last of the K-group males had disappeared (Nishida et al., 1985).

The disappearance of all the K-group males has led to an influx of K-group females into M group. The result is an unusually high ratio of adult females to males; in 1982 it was 1:3.55 (Hiraiwa-Hasegawa, Hasegawa, and Nishida, 1983). No infant of these transfer females had, by the end of 1982, survived longer than three months. Three of the females are known to have become pregnant twice. Of the four resulting infants who were actually seen by observers, one was killed and cannibalized by M-group males at three months of age (Kawanaka, 1982a), two others are thought to have met a similar fate, and one died of unknown causes at one month of age. In the remaining two cases the pregnant females appeared without infants after about a month's absence (Hiraiwa-Hasegawa, Hasegawa, and Nishida, 1983). More recently a fourth immigrant female reappeared with a tiny infant after a six-month absence. She was subjected to a brutal attack by three M-group males, during which the alpha male seized, killed, and ate the infant—which, like the other six, had almost certainly been sired in his own unit-group (Takahata, 1985).

Relevant to this discussion is a series of nighttime raids that were made by wild chimpanzees of Niokolo-Koba National Park in the Senegal on a group of chimpanzees being rehabilitated there. The rehabilitant group has been described by Brewer (1978), and I am grateful to her for permission to discuss these observations here.

At the time of the events referred to, which were witnessed by Brewer's assistants, the eldest chimpanzee in the group, Tina, had a two-year-old infant sired by one of the rehabilitant males. Aggression from the wild chimpanzees toward the ex-captives had been observed, on and off, since the early days of the study but intensified during 1977, when on several occasions parties of wild chimpanzees chased rehabilitants from the forest to the outskirts of camp. One night Tina was chased from her nest right into camp by a wild adult male, who left only when a torch was shone in his face. The following month, during the full moon, a group of five to eight wild adult chimpanzees displayed intermittently near camp, pant-hooting and drumming all night. Seven months later a wild male again chased Tina into camp during the night, while a second individual displayed and called on the outskirts. Two nights later came the incident that led to termination of the rehabilitation program in the Senegal.

At 0200 hours there was an outburst of screaming and waa-barks. The assistants rushed out and saw four adult wild chimpanzees right in camp, clearly visible in the bright moonlight. Four of the rehabilitants were huddled, screaming, at the door of the house. One of them, a young female, had been attacked badly; her brow and forehead were gashed deeply and there was a severe wound in one thigh. The raiders retreated when humans appeared, but stayed nearby displaying for the next hour. At 0600 the displaying began again at the edge of the camp clearing. At 0630, when Tina approached a wild male showing submissive behavior, he viciously attacked her and was quickly joined by two others. By the time the assistant chased them off, Tina had been badly hurt and almost certainly would not have survived without medication. She lost half her nose, a good part of her eyebrow, half an ear, and half an index finger. She had a two-inch slit on her upper lip and muzzle. Her hands were so torn that she could not use them for days, and her feet were almost as bad. Her arms and legs were punctured, and she had deep gashes on the inside of one thigh and one calf. For several days she could not climb and simply lay in the grass in the camp clearing. Her infant escaped unscathed.

Why these violent incidents after several relatively peaceful years? The rehabilitant group was located in part of the home range of the wild chimpanzee community. The year of the raids there was a drought, and at the height of the dry season the only known source of running water was that close to Brewer's camp. This, moreover, was one of the few locations where *tabbo* trees produced fruit that year, a particularly important resource because many fruit crops had failed or yielded poorly. Finally, the wild chimpanzees were the focus of a study initiated by W. McGrew and C. Tutin in mid-1976 (McGrew, Baldwin, and Tutin, 1981) and by 1977 had begun to lose some of the fear of humans that previously would probably have kept them away from Brewer's camp.

As a result of the aggression, the rehabilitant group was relocated elsewhere. In some ways the situation is comparable to that at Gombe where, as a result of their repeated attacks on Kahama individuals, Kasakela community members regained access to an area which, before the community division, had been an integral part of their home range.

<table>
<tr><td>

Discussion

</td><td>

Quite clearly adult male chimpanzees, and to some extent adult females, may behave with extreme brutality toward members of other communities. The information presented in this chapter raises some interesting questions. What is the function of the brutal assaults on multiparous stranger females? Are chimpanzees territorial? Do they attack adult conspecifics with intent to kill? And does their behavior throw any light on the origins and evolution of that uniquely human practice—warfare?

The Assaults on Stranger Females—An Unsolved Puzzle

The attacks on unhabituated females at Gombe led on four occasions to the death of their infants; three times those infants were eaten or partially eaten. And twice at Mahale adult males are thought to have seized and then eaten the infants of females of a neighboring unit-group.

There is increasing evidence from research on a variety of mammalian species to suggest that infanticide may be a common male reproductive strategy. It was first described for langurs by Sugiyama (1965) and has subsequently been observed in a variety of species (review by Hrdy, 1977). Typically a male takes a female from the group of a rival male to start his own group or fights and takes over all females: in either case he kills any infants there may be. The advantage of infanticide to the male is high in any species in which the birth interval is typically long and when the loss of a nursing infant causes the female to resume estrus quickly: (a) the female becomes available to the new male for mating, (b) offspring bearing his genes may be produced earlier than if he had not killed the previous infants, (c) the reproductive success of his defeated rival will be diminished, and (d) he will not waste energy protecting the progeny of that rival.

Although a female chimpanzee does indeed come into estrus rapidly after losing an infant (as much as five years sooner than she might if it had survived, depending on its age at the time of death), it does not appear that the attacks on mother-infant pairs by male chimpanzees can be fitted into this framework. In the cases at Gombe the aggression was clearly directed against the *mothers*, not the infants.

</td></tr>
</table>

In the twelve observed gang attacks on mother-infant pairs, twice infants were seized, flailed, and eaten; twice they were seized, flailed, and hurled away; twice they left their mother and climbed a tree, where they were ignored; once an infant left the mother and was attacked, then left; and on the other five occasions they were left clinging to their mother during the fights. I say "they were left" advisedly—two females, Passion and Pom, were able to wrest Melissa's infant from her despite her valiant resistance, and there is no question that the males could have taken the infants had this been their goal. During the attacks on Wantendele at Mahale, described above, it was clearly the mother who was the prime target of the males' aggression: her son was seldom attacked when he was on his own (six times in all) and this aggression was mild, whereas she was subjected to thirty-two separate attacks during the same period of time. In the savage assault on the rehabilitant female in the Senegal, the mother was almost killed while her infant was completely unharmed; this would not have been the case had the three attacking males been after the infant.

At Mahale, the two mothers whose infants were killed by neighboring males remained in their own social groups after the attacks; they were *not* mated by the aggressor males when they resumed estrus (Kawanaka, 1981). Admittedly Wantendele eventually transferred, but the reason almost certainly was that her own community males all had gone. At Gombe, there is no evidence to suggest that any of the four mothers who were attacked, and whose infants were killed (or died), subsequently became available to the Kasakela males for mating; they were, if anything, more likely to have died themselves of their injuries. Moreover, Passion and her family actively avoided the presumed location of their attack by Kalande males and stopped using part of what had been their preferred area (Figure 17.5).

If, then, the assaults are not directed at infanticide and subsequent recruitment of the female victims, why *are* they perpetrated? Probably we do not need to search for any causal explanation beyond the fact that the victims were all members of neighboring communities: chimpanzees, it seems, have an inherent aversion to strangers. I have already described in Chapter 12 the fierce attack on the small, harmless female newcomer who was introduced to Köhler's Tenerife group. Not only were the female victims at Gombe and Mahale strangers,* they were for the most part encountered in overlap zones where the chimpanzees, fearful of neighboring males, are often nervous. We have seen how the tension that builds up during travel in these unsafe areas may lead to wild displays and occasional attacks on scapegoats within the party, once the males have returned to their own safe core area. Thus when patrolling males encounter an anestrous female (who poses no threat and who is not at the time sexually desirable), she is likely to be attacked. And because of the heightened level of arousal the aggression is apt to be fierce.

* These females may, in fact, have been encountered previously (perhaps several times) during patrols and excursions to peripheral areas. They may even have visited the aggressor males as late adolescents during periods of estrus. But they were still highly unfamiliar, and their social bonds were with members of another community—which, of course, included "dangerous" males. Köhler's female had actually been in full view of her aggressors for several weeks before her introduction to the colony.

Of particular interest here are the descriptions of interactions between Mahale's Wantendele and the M-group males during periods when she visited in an almost fully swollen state. The adult males apparently found themselves in an unusually difficult conflict situation, because their natural aversion to her "stranger" status was counteracted by their attraction to her "sexual partner" status. Their behavior was ambiguous: friendly interactions were interspersed with aggressive threats and attacks. Two males broke off grooming sessions to jump on and suddenly bite the mother or her son, yet twice lower-ranking males interceded on behalf of the strangers (Nishida and Hiraiwa-Hasegawa, 1985). With tension running so high, the sparking of the fierce gang attack (in November 1981) is not really surprising.

One other fact may be pertinent to our discussion. At Gombe, as I have repeatedly stressed, the bond between mothers and grown daughters is remarkably close and enduring. At Mahale, no such bonds existed between *any* of the K-group senior females and their daughters, although recent data suggest that in the larger M group two relationships between mothers and their adult daughters are similar to those that prevail at Gombe. I showed in Chapter 7 that Gombe mothers tend to remain with their daughters when the latter travel with adult males during late adolescence; mothers may even accompany their daughters on consortships (providing, as far as the male is concerned, most unwelcome chaperones).

All but one of the violent attacks on Madam Bee took place during the time when the Kasakela males were actively recruiting her daughter Little Bee (as, in fact, did one of the attacks on Kahama male Dé). Shortly after the last, fatal attack on Madam Bee, her ten-year-old daughter, Honey Bee, appeared in the Kasakela community range. For the next three years Honey Bee appeared from time to time, in estrus, in association with Kasakela males. True, she vanished again three years later and did not transfer permanently as her elder sister had done, but in the interim she provided an additional cycling female for the males. Two of the young females who immigrated, Harmony and Jenny, were carrying young infants (assumed to be orphan siblings) when they were first seen with Kasakela males. This suggests that their mothers had recently died. Another immigrant, Winkle, first appeared as a very young adolescent, at an age when females, even at Mahale, are usually still associating closely with their mothers. Thus she too may have been an orphan. The immigrant female Kidevu, as mentioned earlier, is almost certainly the daughter of Kahama female Mandy; and I have already suggested that Mandy may have met the same fate as Madam Bee.

Perhaps the recruitment of new females into the community at Gombe is facilitated by repeated brutal attacks on their mothers, which serve (as apparently with Little Bee) to weaken the mother-daughter bond or (as with Honey Bee and perhaps Jenny, Harmony, Winkle, and Kidevu) to break it altogether. If we examine Table 17.2, we find that two of the female victims *were* in association with adolescent females, one estimated at nine, the other at eleven years of age. Two mothers were accompanied by seven- or eight-year-old youngsters whose sex was not determined (and one, not included in the table because she was attacked by a single male only, had a seven-year-old daughter). Five other victimized females had youngsters four to six years old; theoretically these females might have had older offspring, as might the three middle-aged and old victims. The fact that older offspring were not observed does not mean they did not exist. Honey Bee, for instance, was not seen at all during the fatal attack on her mother. Moreover, during three of the fights (cases 4, 5, and 9) the aggressors suddenly stopped fighting and rushed toward sounds of another stranger, rather as they left Dé to chase after and recruit Little Bee. For the present, this whole area must remain speculative. More facts are badly needed.

Are Chimpanzees Territorial?

If a fixed area is *occupied* by one or more individuals, we may talk about a home range (Burt, 1943). Any part (or all) of this overall range that is *defended* against other conspecifics, or even against individuals of other species, is a territory (Noble, 1939). Noble was an ornithologist, and in fact a vast amount of the literature on territorial behavior relates to birds (for example Hinde, 1956). In many bird species territory owners are single males, or a male-female pair. A territory owner, having established his boundaries, often by fighting with neighboring males, stays in his domain, usually proclaiming his presence by means of vocal and/or visual displays. After the initial aggressive demarcation of boundaries, neighboring territory owners typically respect one another's property and further overt aggression is rare. If an intruder moves in, he will be challenged and chased off or attacked. A territory owner tends to be confident and easily roused to aggression inside his domain, but becomes progressively less assertive the farther away from it he ventures. Thus if he does move into neighboring territory, he is likely to retreat hastily if challenged. Any aggressive incidents that do occur between neighboring territory owners are, therefore, likely to occur along the boundaries. Often territories are set up during the breeding season and abandoned when the young have been successfully raised.

Patrolling males. (C. Busse)

Hinde (1956) has stressed the heterogeneous quality of the behaviors that constitute territoriality and the multiple functions they are likely to serve. Those most relevant to the present discussion are the spacing out of individuals (or groups of individuals) so that territory owners can forage and hunt, mate or raise their young, without interference, and the maintenance of an area that provides an adequate food supply for the young or for the subsistence of the territory owners. The inhabitants of a territory also have the advantage of familiarity with the topography but, as Wilson (1975) points out, this benefit results from staying in one place, whether that place is defended (territory) or not (home range).

Territorial behavior is widespread in the animal kingdom. Among the nonhuman primates two species that adhere closely to the classical picture of territoriality are the gibbons (Carpenter, 1940) and the South American *Callicebus* (Mason, 1968). Both form groups comprising an adult male-female pair accompanied by any immature offspring. Their territories are relatively small, so that the boundaries can be visited daily, and there is little overlap. For the most part, ritualized threatening displays serve to maintain the integrity of the territories. Male gibbons, for example, leap vigorously about in the trees near boundaries and utter loud calls, which serve to indicate their whereabouts to neighbors. Some aggressive chasing and attacks may occur, even between established neighbors—usually in the small overlap zones and frequently when rival groups are trying to feed in the same tree (Ellefson, 1968).

When a group occupies a *large* home range, it is impossible to monitor and defend all the boundaries. Most species of Old World primates with extensive home ranges, such as baboons, have considerable overlap with neighboring groups and are not considered territorial. They do not monitor or mark the boundaries of their home range or of their core area, they have no brightly colored markings to proclaim their presence, they do not exchange vocal displays with neighboring groups (as, for example, the howler monkey does; Carpenter, 1934), and intertroop conflicts are rare. Often there is such a clear-cut dominance relationship among troops that one typically gives way at the approach of another. By contrast, troops of vervet monkeys (*Cercopithecus aethiops*), at least in some areas,* meet one another frequently near their common boundaries, where the males engage in ritualized aggressive displays. They often sit in tall trees near their boundaries looking out over neighboring territory; their blue scrota and red penes are very conspicuous (McGuire, 1974). Kummer (1971) comments that while aggression may determine the initial distribution of space among territorial primates, the members of adjacent groups soon learn where the boundaries lie, and at least most of the time respect them. Conflicts are thus reduced to a minimum and defense of territory is apt to be limited to harmless advertisement or demonstration. This again complies with the classical territoriality first described for so many bird species.

Against this background we can now review chimpanzee intercommunity interactions. In a number of respects chimpanzees also comply with classical territoriality:

(a) Conspecifics from neighboring social groups who intrude into the home range are aggressively expelled (except for those females who are recruited);

(b) Boundaries are visited frequently and monitored;

(c) Parties traveling to peripheral areas show tense, nervous behavior and are much less confident than when in their core area;

(d) Auditory displays, loud pant-hoots, and drumming may be exchanged between parties of adult males of neighboring communities, followed by ritualized aggressive display, and members of both sides may retreat without conflict;

(e) Boundaries may be respected over a number of years. For two years after the community split Kasakela males recognized Mkenke Stream as a boundary between themselves and the Kahama community; at Mahale, members of the powerful M group, over a seven-year period, seldom moved into the exclusive area of the smaller K group.

In three important ways, however, chimpanzee behavior does *not* comply with classical territoriality.

(a) Both at Gombe and Mahale it is the relative size and composition

* Vervets, like some other primate species, are territorial in certain parts of their range and not in others (Gartlan and Brain, 1968).

of the two neighboring parties that determine the outcome of an encounter, rather than the geographic location. A small patrol will turn and flee if it meets a larger party, or one with more males, even *within* its own range; whereas if a large party, traveling *out* of its range, meets a smaller party of neighbors, it is likely to chase or attack. When parties are approximately the same size, with similar numbers of adult males, visual and auditory display exchanges without conflict typically result.

Almost without exception, species that show territorial behavior move about their domains in stable groups (or alone). Thus when a male gibbon and his female travel to one of their boundaries and meet their neighbors, they will always meet the same neighboring pair. When a group of vervets moves to feed near one of its territorial borders, the adult males are not likely to be surprised by an encounter with a neighboring troop that has suddenly doubled (or halved) in size.

An interesting exception, among mammals considered to be territorial, is the spotted hyena (*Crocuta crocuta*). These predators live in clans and, at least in resource-rich areas such as Tanzania's Ngorongoro crater, each clan occupies a territory the boundaries of which are regularly patrolled and scent-marked (Kruuk, 1972). Within the clan, members travel about in small subgroups, the composition of which is continually changing. Provided three or more of the high-ranking females (the dominant sex) are present, a party may set off on a patrol, monitoring the clan boundary. If two parties from neighboring clans, of similar size and composition, encounter each other near their common boundary, there will be aggressive displays, whooping, growling, and bristling manes, followed by noisy withdrawal on both sides. But a patrol almost always gives chase if it encounters a lone hyena, or a pair, even if these neighbors are in their own territory. A captured individual may be mauled and left to die. The significance of size and composition of party is most clearly demonstrated when a prey animal is chased by one clan across a territorial boundary and killed in the domain of a neighboring clan. Hyenas are noisy, and as the hunters feed, their whooping, giggling calls alert other individuals within earshot, friend and foe alike. After a while the territory owners begin to gather in a growling cluster facing the trespassers. When there are enough of them, they charge. The hunters, whose appetite has been partly assuaged (at least that of the high-ranking females, who push in first), retreat and the carcass changes ownership. Then it is the hunters who stand at a distance, bristling and calling. Meanwhile other hyenas continue to arrive. The trespassers become bolder *as their ranks swell;* they charge again and may once more take over their kill. I watched one incident of this sort, which took place across the territorial boundary, in which the kill changed sides no less than five times before it was finally consumed (Goodall and van Lawick, 1970).

The social structure of hyenas is similar in many ways to that of chimpanzees. The fact that in both species individuals of the community (or clan) travel around in parties of *variable* size and composition means that there will be opportunities for intimidating neighbors seldom vouchsafed to other species. Because hyenas and chimpanzees are intelligent, as well as being hostile to neighbors, they can and do take advantage of such opportunities when they arise.

(b) Chimpanzees have a large home range with *considerable overlap* between neighboring communities. In most species described as territorial the overlap zone is not extensive and the territory is small enough for the occupants to monitor the boundaries frequently. Wilson (1975, p. 265), however, proposed a rider to the definition of territoriality: A resident "can guard its entire territory all of the time, or it can defend only those portions of the territory within which it happens to encounter an intruder at close range." Lions are considered territorial. There may be considerable overlap in the ranges of neighboring prides, but whenever and wherever intruders are encountered, they are expelled—or killed (Pusey and Packer, 1983). Chimpanzees actually expend considerable energy in *creating* opportunities to encounter intruders at close range. The adult males move out to peripheral areas, on average, once every four days.

(c) It is perhaps in the *violence* of their hostility toward neighbors that chimpanzees, like hyenas and lions, differ most from the traditional territory owners of the animal kingdom. Their victims are not simply chased out of the owners' territory if they are found trespassing; they are assaulted and left, perhaps to die. Moreover, chimpanzees not only attack trespassers, but may (as Figure 17.4d shows) make aggressive *raids* into the very heart of the core area of neighboring groups. Admittedly, boundaries may be respected for some of the time, but during the long-term studies at Gombe and Mahale there have been three major "invasions": Kasakela males took over Kahama range, Kalande males pushed deep into Kasakela range, and M group moved into K-group range. During all these invasions adult males (and some females) were killed or disappeared. Even if it is argued that the Kasakela males were merely trying to *reclaim* an area to which they had previously had free access, the assertion does not explain the northward thrust of the Kalande community or the takeover by the M group at Mahale.

On the basis of the facts presented in this chapter, particularly if we accept Wilson's rider to the definition of territoriality, I believe that chimpanzees should be considered territorial. But theirs is a form of territoriality that has shifted away from the relatively peaceful, ritualized maintenance of territory typical for many nonhuman animals, toward a more aggressive type of behavior. In the chimpanzee, territoriality functions not only to repel intruders from the home range, but sometimes to injure or eliminate them; not only to defend the existing home range and its resources, but to enlarge it opportunistically at the expense of weaker neighbors; not only to protect the female resources of a community, but to actively and aggressively recruit new sexual partners from neighboring social groups.

Do Chimpanzees Show Intent to Kill?

Madam Bee died on the fifth day after the last assault on her by the Kasakela males, and Charlie's battered body was found two days after the sounds of a fierce battle. Goliath and Sniff could in no way have survived after the observed attacks—Sniff had a broken leg as well as other injuries, and Goliath, in addition to being terribly mauled, was old. Dé and Godi would almost certainly have been seen during the intensive research in the south from 1975 to 1977 (Dé's pitiable condition, two months after the observed assault, has been described). That the bodies of the four males were not found is not surprising; Madam Bee, despite two days of searching in the immediate vicinity of where she had in fact been lying, was only discovered because of her daughter's presence in the trees above. Gombe's undergrowth is extremely thick and tangled in parts, and chimpanzees can keep very quiet.

Even given that all the Kahama victims died (and probably some of the unhabituated females, too), we can tell nothing about the "intentions" of the aggressors. The actual patterns of assault were very similar in the five observed incidents: (a) the attacks were all long—the shortest lasted at least ten minutes, and three continued more than twice as long; (b) all were gang attacks, during which the aggressors sometimes assaulted the victim one after the other, or two to five assailed the victim simultaneously; (c) all the victims were, at some point, held to the ground by one or more of the aggressors while the others hit and pounded; (d) all the victims, in addition to being hit, stamped on, and bitten, were dragged first in one direction, then another; (e) in each case the victims (even if in the initial phases of the conflicts they had made some attempt to defend themselves or escape) gave up and crouched or lay passively until the onslaught was over; (f) after the attacks each victim was not only wounded but more or less immobilized; and (g) during each incident the observers, all thoroughly experienced in chimpanzee behavior, *believed* that the aggressors were trying to kill their victims.

I asked the field assistants why they thought so. They said it was because the attackers showed some of the patterns which, while commonly seen during the killing of large prey, have not been seen during *intra*community fighting—as when one of Goliath's legs was twisted, when a strip of flesh was torn from Dé's thigh, or when Satan drank the blood pouring from Sniff's nose. Moreover, in all cases the attacks continued until the victims were incapacitated.

Similar battering to insensibility was, incidentally, seen on two occasions when chimpanzees attacked civets (*Civettictis civetta schwarzi*). The two individuals, late-adolescent male Mustard and adult female Patti, seized their victims, displayed with and flailed them, then beat and stamped on them until the creatures were immobile. Although not completely dead at the time, they died later of their injuries. A wounded serval (*Felis serval hindei*) was also hit and stamped on, but left by the assailants when it dragged itself into thick vegetation (Teleki, 1973b). Thus chimpanzees may attack and cause death in situations other than hunting for food.

When chimpanzees are hunting, they seldom deliberately kill small prey animals (such as infant monkeys). Usually the victim dies as a result of being eaten—and since the brain is apt to be eaten first, quick death ensues. A larger prey, however (such as an adult monkey), presents a problem—it can (and does) struggle to escape, bite, and generally make it difficult for the captor to feed in peace. In such a case the hunter bites at or tears off limbs, breaks legs, drags, displays with, and pounds the victim until it is incapable of further annoying the captor. This state of incapacity is almost certainly a more significant criterion for the chimpanzee than physiological death. Quite often a large prey dies slowly, while it is being consumed. If it makes sounds (which it usually does), these are ignored; but if it hits out or struggles violently, another outburst of hitting or flailing is triggered.

After being attacked, Goliath and Dé were left lying almost motionless on the ground, uttering faint squeaks. Sniff was still moving; Sherry remained, watching him closely as he dragged himself laboriously toward cover. When Madam Bee "fainted," however, and lay *totally* motionless for two minutes, Satan displayed around her and waved branches until she moved again. Then he followed her as she crept toward a thicket. It was almost as though he could not be sure, until he saw her move, that the attack had been "successful"—that she had been truly incapacitated. There is evidence, too, that the aggressors return at a later time to the "scene of the crime." One and a half hours after the attack on Dé the Kasakela males, on their way back north (with Little Bee), revisited the place where they had left their victim and searched around for thirty minutes. Five weeks after the assault on Sniff, the aggressor males were followed as they journeyed to the place of the attack; they peered about and smelled the ground for ten minutes. Nor should we forget the Kahama party that made a detour apparently to inspect the dead body of an old female, probably their victim. They exhibited no surprise or alarm when they approached her body; clearly they already knew she was there. It is as though the aggressors check up on the results of their attacks.

In summary, then, it seems almost certain that the Kasakela males were making determined attempts, through wounding and battering, to incapacitate the Kahama chimpanzees. If they had had firearms and had been taught to use them, I suspect they would have used them to kill.

The Precursors of Warfare

"At no period in human history was there a golden age of peace," writes Quincy Wright (1965, p. 22). Wright defines war as "*armed* conflict between groups" (italics added). Perhaps, even more important, we might describe warfare as *organized* conflict. In either case, it is uniquely human behavior and appears to be an all but universal characteristic of human groups (Eibl-Eibesfeldt, 1979). Wars have, of course, been fought over a wide range of issues, including culturally and intellectually determined ideological ones. In essence, wars have functioned, at least ecologically, to secure living space and adequate resources for the victors. To some extent, too, they have served to reduce population levels and conserve natural resources (Russell and Russell, 1973). Since warfare involves conflict between groups of people, rather than between individuals, it has, through genocide, played a major role in group selection. This fact, first pointed out by Darwin (1871), has been elaborated on by others, notably Keith (1947), Bigelow (1969), Alexander (1971), and Eibl-Eibesfeldt (1979).

Warfare is, of course, a product of cultural evolution (Tinbergen, 1968) and, as Pitt (1978) points out, most authors have discussed its influence on human evolution during historical, or near-historical, times. Alexander and Tinkle (1968) and Bigelow (1969), however, postulate the existence of a primitive form of warfare in the early hominids (the "dawn warriors," as Bigelow picturesquely labels them), and stress the important role it may have played in spreading such valued human qualities as altruism and courage (see also Campbell, 1972). Moreover, the early practice of warfare would have put considerable selective pressure on the development of intelligence and of increasingly sophisticated cooperation among group members. This process would escalate, for the greater the intelligence, cooperation, and courage of one group, the greater the demands placed on its enemies (Wilson, 1975). The powerful pressure that warfare almost certainly exerted on the development of the human brain was suggested by Darwin and Keith. Alexander and Tinkle, Bigelow, and Pitt go further, believing warfare to have been the *principal evolutionary pressure* that created the huge gap between the human brain and that of our closest living relatives, the anthropoid apes. Stone Age groups with inferior brains, who could not win wars, were exterminated long ago (see also Sagan, 1977).

Granted that destructive warfare in its typical human form (*organized, armed* conflict between groups) is a cultural development, it nevertheless required preadaptations to permit its emergence in the first place. The most crucial of these, as pointed out by Alexander and Tinkle and by Bigelow, were probably cooperative group living, group territoriality, cooperative hunting skills, weapon use, and the intellectual ability to make cooperative *plans*. Another basic preadaptation would have been an inherent fear of, or aversion to, strangers, expressed by aggressive attack (Eibl-Eibesfeldt, 1979). Early hominid groups possessing these behavioral characteristics would theoretically have been capable of the kind of organized intergroup conflict that could have led to destructive warfare.

Chimpanzees not only possess, to a greater or lesser extent, the above preadaptations, but they show other inherent characteristics that would have been helpful to the dawn warriors in their primitive battles.

(a) The killing of adult conspecifics is not common among mammals, for such conflicts can be dangerous for aggressors and victims alike. It is often emphasized that it has been necessary to train or shape human warriors by cultural methods: glorifying their role, condemning cowardice, offering high rewards for bravery and skill on the battlefield, and emphasizing the worthiness of practicing "manly" sports during childhood (see for example Tinbergen, 1968; Pitt, 1978; Eibl-Eibesfeldt, 1979). But if the early hominid males were *inherently* disposed to find aggression attractive, particularly aggression directed against neighbors, this trait would have provided a biological basis for the cultural training of warriors. I describe in this chapter how a young male chimpanzee is often strongly attracted to intergroup encounters, even to the extent of approaching a number of potentially dangerous neighbors on his own. This characteristic is shown by young males of other nonhuman primate species. A male gibbon, for example, may leave his female and young in order to watch or actually participate in a dispute between two other groups (Ellefson, 1968). Young male langurs may leave their troops and travel as much as 500 meters to engage in whooping and branch-shaking encounters with neighbors

(Ripley, 1967). And on Cayo Santiago island parties of male rhesus sometimes actively seek contact with another troop and initiate hostile encounters; these may escalate and lead to the involvement of the older and higher-ranking males of both groups (Morrison and Menzel, 1972). Intertroop encounter is also attractive to some silverback male gorillas (Fossey, 1979).

(b) In our own species cultural evolution permits *pseudospeciation* (Erikson, 1966)—the transmission of individually acquired behavior from generation to generation within a particular group, leading to the customs and traditions of that group. This process is analogous to the formation of species through genetic inheritance. Pseudospeciation in humans means, among other things, that the members of one group may not only see themselves as different from members of another, but also behave in different ways to group and nongroup individuals. In its extreme form pseudospeciation leads to the "dehumanizing" of other groups, so that they may be regarded almost as members of a different species (LeVine and Campbell, 1971). This process, along with the ability to use weapons for hurting or killing *at a distance*, frees group members from the inhibitions and social sanctions that operate within the group and enables acts that would not be tolerated within the group to be directed toward "those others." As Eibl-Eibesfeldt (1979) stresses, this lack of inhibitions is a prime factor underlying the development of destructive warfare.

Thus it is of considerable interest to find that the chimpanzees show behaviors that may be precursors to pseudospeciation in humans. First, their sense of group identity is strong; they clearly differentiate between individuals who "belong" and those who do not. Infants and females who are part of the group are protected even if the infants were sired by males from other communities. Infants of females who do *not* belong may be killed. This sense of group identity is far more sophisticated than mere xenophobia. The members of the Kahama community had, before the split, enjoyed close and friendly relations with their aggressors. By separating themselves, it is as though they forfeited their "right" to be treated as group members—instead, they were treated as strangers.

Second, nongroup members may not only be violently attacked, but the patterns of attack may actually differ from those utilized in typical intracommunity aggression. The victims are treated more as though they were prey animals; they are "dechimpized."

There are two further aspects of chimpanzee behavior that are of interest in relation to the evolution of behavior associated with human intergroup conflict.

(a) Cannibalism, in humans, has been reported from almost all parts of the world, and evidence from paleoanthropology suggests that it dates back at least to the mid-Pleistocene. Many fossil skulls from that time on are characterized by a "careful and symmetric incising of the periphery of the foramen magnum," which on the basis of comparative evidence is assumed to have been made for the purpose of extracting the brain for eating (Blanc, 1961, p. 131). The motives behind the eating of enemies have varied historically: a preference for the taste of human flesh; revenge—by eating an enemy he is completely destroyed; a magical belief that the victor can acquire some of his enemy's qualities of courage or strength from the flesh (Eibl-Eibesfeldt, 1979). Until comparatively recently, cannibalism was thought to be another behavior "by which the human animal may be sharply distinguished from other primates" (Freeman, 1964, p. 122). In the chimpanzee, we have seen that cannibalism may follow intergroup conflict with neighboring females. The bizarre behavior directed at the corpse by some adult males could well, with a little more intellectual sophistication, evolve into ritual.

(b) Warfare, in humans, almost always leads to acts of great *cruelty* (by no means, of course, confined to warfare). Freeman (1964, p. 121), writes that "the extreme nature of human destructiveness and cruelty is one of the principal characteristics which marks off man, behaviorally, from other animals." Implicit in the Oxford dictionary definition of cruelty as "delight in or indifference to another's pain" is a certain level of cognitive sophistication. In order to be cruel, one must have the capability (1) to understand that, for example, the detaching of an arm from a living creature will cause pain and (2) to empathize with the victim. It is because we humans unquestionably have these abilities that we are able to be cruel. Both human and chimpanzee children may mutilate insects or small animals; human children, at least in most Western cultures, are taught that this is cruel. If a group of humans behaved in the same way as the gangs of Kasakela males when they attacked their Kahama victims, the behavior would be described as cruel; so would the slow killing of large prey animals. Of course, chimpanzees are intellectually incapable of creating the horrifying tortures that human ingenuity has devised for the deliberate infliction of suffering. Nevertheless, they are capable to some extent of imputing desires and feelings to others (Woodruff and Premack, 1979) and as we have seen in Chapter 13 they are almost certainly capable of feelings akin to sympathy. The Premacks' Sarah consistently chose photographs of her "enemy" strewn with cement blocks, suggesting that she may possess some precursor of sadism. On the other hand, her motivation may have been nothing more than mischief making.

On the Threshold of War?

"War among primitives," writes Eibl-Eibesfeldt (1979, p. 171), "is often limited to raids in which they stalk or creep up to the enemy, using tactics reminiscent of hunting." However "primitive," raids of this sort are the result of careful planning—and if the enemy somehow finds out about these plans, he will counter with a plan of his own. This, of course, is the major difference between primitive warfare and the stealthy, cooperative raids of the Kasakela males into Kahama territory, which sometimes culminated in brutal assault. While fully admitting the major role that warfare has probably played in shaping our uniquely human brain, I submit that until our remote ancestors acquired *language*, they would not have been able to engage in the kind of planned intergroup conflicts that could develop into warfare—into organized, armed conflict.

The chimpanzee, as a result of a unique combination of strong affiliative bonds between adult males on the one hand and an unusually hostile and violently aggressive attitude toward nongroup individuals on the other, has clearly reached a stage where he stands at the very threshold of human achievement in destruction, cruelty, and planned intergroup conflict. If ever he develops the power of language—and, as we have seen, he stands close to that threshold, too—might he not push open the door and wage war with the best of us?

18. Object Manipulation

David Greybeard fishing for termites. This termite mound provided the first-ever observation, back in 1960, of chimpanzee tool *making* in the natural habitat. Our early hominid ancestors undoubtedly used twigs and sticks long before they made the first stone implement. (J. D. Waters, née Goodall)

Two extracts from my diary

4 November 1960 Just to the south of the Kasakela cliff is a termite hill . . . My path passes about 90 meters south of this. 0815: Saw a black object in front of termite hill. Peered through vegetation. It was a chimp. Quickly dropped down and crawled through the sparse dry grass until I reached a tree with greenery sprouting at the base. Through these leaves could see the chimp about 45 meters away. Visibility was poor owing to (a) grass and (b) a tree between chimp and me, close to the termite hill. Made out that he was picking things from the mound and eating. He had his back to me. After a few minutes he moved away completely out of sight behind the hill. Cautiously backed away and moved in from another angle. Had a slightly better view. He still had his back to me. He turned slightly and, very deliberately, pulled a thick grass stalk toward him and broke off a piece about 45 centimeters long. Then unfortunately, he turned his back on me again. After a few minutes he climbed over the hill and moved away on the far side. I identified him as David Greybeard.

6 November 1960 By the termite hill were two chimps, both male . . . I could see a little better the use of the piece of straw. It was held in the left hand, poked into the mound, and then removed coated with termites. The straw was raised to the mouth and the insects picked off with the lips along the length of the straw, starting in the middle.

The present chapter on object manipulation and tool-using behavior, and the one on social awareness that follows, have been left to the end of this volume because, taken together, they provide an illuminating picture of the undeniable intelligence of the chimpanzee in nature.

The ability to make and use tools was, for a very long time, thought to be unique to our own species (see for example Napier, 1971). The emergence of tool-using performances in our prehuman ancestors marked a crucial step in our evolution: when, for the first time, an apelike creature made a tool to "a regular and set pattern" he became, by definition, Man (Leakey, 1961, p. 2). For this reason tool-using performances in nonhuman animals have always compelled attention.

To be classified as a tool an object must be held in the hand (or foot or mouth) and used in such a way as to enable the operator to attain an immediate goal (Goodall, 1970). If we accept this definition, a variety of animal species, including some insects, qualify as users of tools. The chimpanzee emerges as the most superior of them; he uses more objects, for more different purposes, than any creature except ourselves. But the mere use of an object as a tool is not, of itself, particularly remarkable. It is the cognitive aspects of tool-using performances that are of interest. The chimpanzee, with his advanced understanding of the relations between things, can modify objects to make them suitable for a particular purpose. And he can to some extent modify them to "a regular and set pattern." He can pick up, even prepare, an object that he will subsequently use as a tool at a location that may be quite out of sight. Most important of all, he can use an object as a tool to solve a completely novel problem.

Table 18.1 lists the various objects that chimpanzees, at Gombe and in other parts of Africa, have been seen to utilize as tools, and the contexts—feeding, body care, investigation, and intimidation—in which the tool-using performances have been observed.

The Feeding Context

In this context the chimpanzees at Gombe use more tools, more frequently, than in any other. The same appears to be true for chimpanzees in other parts of their range.

Termite Fishing

Termites (*Macrotermes bellicosus*) are obtained almost exclusively through the use of tools. At certain times of year, when the chimpanzees spend long periods of time fishing for and eating these insects, tool using is a normal part of the daily routine. During the main season from October to December, when heaps are visited frequently, a chimpanzee often pauses during travel, selects a grass stem or other material, and holding this in his mouth, proceeds to a termite mound. The mound may be out of sight when the tool is picked, even as far off as 100 meters, although it is usually much closer.

Passages into a termite mound are narrow and not completely straight, so the materials used must be smooth and fairly pliable if they are to be effective. Tools are fashioned from grasses, vines, bark, twigs, or palm frondlets. Sometimes a chimpanzee will pick up almost any suitable material that is nearby, including the discarded tools of others who have worked the mound previously. At other times clumps of grass, tangles of vines, and so on are carefully inspected before a tool is selected; a length may be picked, then discarded immediately before it has been used, and another choice made. To some extent the procedure reflects individual differences, but dry-season terminating calls for more skill and more care in the choice of material than wet-season fishing, when (a) the insects are near the surface and (b) the soldiers are on the defensive and quick to bite at any foreign material inserted into the nest.

One hundred forty-five wet-season tools used at Gombe had a median length of 28 centimeters (range 7 to 100; McGrew, Tutin, and

Table 18.1 Tool-using behaviors of wild chimpanzees, including (where known) the average length of the tools used at various sites. All references are given in the text.

Object and purpose	Gombe	Mahale[a]	Bossou, Guinea	Tai, Ivory Coast	Other locations
			Typical applications at—		
Leaves					
Sponging water	Used crumpled		Used unmodified		
Mopping food	Inside skull and strychnos	Carpenter ants (K)			
As brush	For bees, driver ants				
As fishing probe (midribs of the leaves)		20 cm (K)			
As napkin	For fruit juice, feces, etc.	As at Gombe			
As container	To catch feces				
Grass or small stems					
Termite fishing	28 cm	51.5 cm (K,M,B)			Kasakati Basin Senegal—30 cm Gabon—38 cm
Ant fishing		21.4 cm (K)			
Honey fishing					Kasakati Basin Central Africa
Investigation	Termite mound, holes in dead wood, etc.	Ant nests			
Leafy twigs					
As fly whisk					Uganda
Small sticks					
Ant fishing		*Camponotus* (K)			
Expelling, stirring up	Ants, bees	Ants (K)			
Perforating termite mound					Mbini (Río Muni)—52 cm
As toy	Tickling self				West Cameroon
As prod for termites			5–15 cm		
As prod for resin			10–20 cm		
Investigation	Holes in trees, etc.	As at Gombe (K)			
Large sticks					
Investigation	Holes in trees, feared objects, etc.	As at Gombe			
Dipping driver ants	15–113 cm				
Enlarging nest entrances	Of birds, bees				
Hooking branches			With hooked end		
As missile	Against chimpanzees, baboons, humans, etc.	As at Gombe	Up to 120 cm—against humans		Zaire, Benin; against stuffed leopard
As club	Against chimpanzees, baboons, humans, etc.	As at Gombe	Against stuffed leopard		Zaire, Benin; against stuffed leopard
Short, thick sticks					
As hammer				For *Coula* nuts	
As weapon	Against chimpanzees, baboons, humans, etc.	As at Gombe	Against humans		
Stones or rocks					
As hammer			For palm nuts and hard fruits	For *Coula* and *Panda* nuts	Liberia—for palm nuts Central Africa— for hard fruits
As missile	Against chimpanzees, baboons, humans, etc.	As at Gombe			
As toy	Tickling self				

a. The unit-group in which the behavior is observed is indicated in parentheses.

Juvenile Prof watches closely as his adult sister, Pom, inserts a long grass tool into a termite tunnel. After Pom leaves, Prof immediately takes over the tunnel. His choice of material (a thin strip of bark) is different, but its length is similar to that of Pom's tool. Both chimpanzees fished to a depth of approximately 45 centimeters.

William selects a vine tool for termite fishing.

Baldwin, 1979). Dry-season tools have not yet been systematically collected and measured, but I suspect the median length will prove to be greater, because the termites are in lower levels of the nest. Certainly, at this time of year the chimpanzees tend to use very long grasses or lengths of vine.

During a session tools become gradually shorter, as the chimpanzees remove pieces at the end that become frayed or excessively bent. When they become too short, replacements are selected. Most new tools are taken from suitable clumps of grass or other vegetation within 5 meters or so of the heap. One female several times climbed 5 meters into a tree near the mound to pick lengths of vine from the lower branches. Gremlin as a late adolescent once moved out of sight of the mound, a distance of some 7 meters, to pick lengths of supple, durable vine. She did this three times during a two-hour session and on each occasion returned with three or four lengths. It often happens that several tools are picked at once; those not put to immediate use are tucked into the groin or laid on the ground nearby.

Some material, such as thin grass (green or dry) or a smooth stem or vine, is suitable for use as is. Other material must be modified before it can be used efficiently. Leaves must be stripped from small twigs, leaflets from a main leaf rib, and slender fibrous lengths from bark, thick stems, or frondlets of palm. Sometimes grass must be thinned down, and the chimpanzee removes blades from each side of the midrib.

Chimpanzees show individual variation in termiting ability: some select more suitable tools, or show more persistence; some either obtain termites for longer periods than others working on the same heap, or appear to capture more. Two young females, Pom and Gremlin, are acknowledged by all of us at Gombe as the current champion termite fishers. Both tend to select very long tools. Gremlin had one dry-season tool that measured just over 1.5 meters, of which she inserted more than two thirds into the passage she was working. This required considerable skill, including a dexterous rotation of the wrist.

When I tried to use her tool after she had left, I only managed to get half of it into the passage.

It was not uncommon, during the difficult out-of-season termite sessions, to see Pom or Gremlin patiently watching as their more dominant mothers worked the only productive tunnel on a mound. When the older female gave up, the daughter moved in, often succeeding where her mother had failed. Once during a March session Passion noticed that Pom had found a productive site. At once she moved across and displaced her daughter, who then sat and watched as Passion tried, with little success, to work the passage. After twenty minutes Passion gave up and left the heap. Pom immediately returned and once more fished very successfully. After eight minutes Passion, who had been feeding nearby but out of sight, reappeared with a grass stem in her mouth. Once again she displaced Pom. Once again she failed where Pom had succeeded. After fifteen minutes she moved off, leaving the site to Pom. This time the young female worked continuously for twenty minutes, catching some termites (although fewer than at first). Passion reappeared, once more with a tool in her mouth, and made a final attempt to replicate her daughter's success. Failing again, she left for good. Pom persisted for another thirteen minutes, after which she went off after the rest of the family.

Ant Dipping

The "wands" used when chimpanzees dip for driver ants are more uniform in size and appearance than the termite-fishing tools. McGrew (1974) collected and measured thirteen wands, which ranged from 15 to 113 centimeters, with a median of 66 centimeters. Tools that are too short cannot catch many ants per dip. I watched a young adult male, Goblin, return on three successive days to the same driver ant nest. Each time he used an inappropriate tool, only about 13 centimeters long. By the time the ants had swarmed almost up to his hand, forcing him to withdraw his stick, he had captured only about 20— compared with the 292 captured and counted by McGrew in a mass that looked similar to that on the wand of a competent chimpanzee. (On other occasions, both before and after, Goblin has shown competent tool use when collecting these ants.) A tool must not be too long, either, or it will be awkward to maneuver; it must not be rough with protruding side branches, or it cannot be swept through the hand; and it must not be too thin, or it may bend or even break.

After use the tool is usually left near the site of the nest, and it will frequently be picked up and used by the next chimpanzee to arrive— who may, in fact, be the original manufacturer making a return visit.

Probing

During termite fishing a chimpanzee often uses a blade of grass or a stem as an investigatory probe, inserting it into a passage, withdrawing it, then carefully sniffing the end. Subsequently he may continue to enlarge the opening, or he may move off immediately—as though informed by olfactory cues that this particular passage is not productive.

Probes are also used to investigate holes in dead wood; the chimpanzee inserts the tool, then smells the end. This seems to convey information about the occupancy of the hole, for the chimpanzee either

Little Bee (in estrus) inserts her thin, peeled stick into a nest of driver ants. She begins to draw the tool rapidly through her right hand, her mouth open in readiness to receive the ant mass. She takes the ants from her hand to her mouth and chews them with rapid gnashing movements of her jaws.

drops the probe and moves off or he continues to investigate the hole. Sometimes the wood is broken open, usually to disclose grubs of some sort, which are eaten. Occasionally, of course, the nest has been abandoned and the chimpanzee, despite his careful investigation, tears open the wood for nothing. Infants and juveniles are more likely to use sticks to investigate holes than are adults.

Breaking into Nests

Other tool-using performances involving sticks were seen only occasionally in the feeding context. Stout sticks were sometimes used to enlarge the openings of underground bees' nests: three times the chimpanzees stood upright and vigorously moved the sticks back and forth a few times, then discarded them and took honey with their hands. And once two females and several youngsters prodded a nest without success. One female, Miff, used a stick to enlarge the opening of a hole in a tree, then reached in and took a fledgling, probably a hornbill. The stick perhaps broke down the clay these birds use to seal the entrance to their nests.

Seven times I saw chimpanzees (six times a juvenile and once an adult female) use sticks to try to force openings in the very hard, football-sized arboreal nests of *Crematogaster* ants. None of these individuals succeeded (one was watched intently by other youngsters, three of whom then tried, unsuccessfully, themselves). It is not clear if they were trying to prey on the ants; in 1961 I saw a group of chimpanzees poking long sticks into one of these nests and presumed that they were eating the ants. However, the only time I saw a

Pax trying to break open a *Crematogaster* (ant) nest.

Olly tries to pry open a banana box with a strong stick, which she fetched from some 15 meters away. Her daughter, Gilka *(right)*, and Fifi watch intently. Subsequently both infants work at the box with tiny twigs. (H. van Lawick)

chimpanzee actually *eating* these ants, she broke off a large piece of dead wood that contained a nest and picked off the insects with her lips. Richard Wrangham (cited by Nishida and Hiraiwa, 1982) saw chimpanzees picking *Crematogaster* ants from dead wood on two occasions. Possibly the 1961 group was poking at something else that had moved into an abandoned nest: once a chimpanzee opened a nest of this type (by hand) and extracted two woodpecker fledglings (R. Wrangham, personal communication).

Soon after the introduction of banana boxes at Gombe many chimpanzees used sticks to try to open them. Sometimes they showed skillful trimming techniques to fashion the end of a stick as a "chisel" to push into narrow openings.

Rousing the Occupants of Holes

Another use for sticks is to disturb and expel the inhabitants of holes. Nishida (1973) was the first to describe this usage. At Gombe the technique was mostly seen when a chimpanzee was investigating a hole in a tree: he broke off a branch, pushed it into the opening, and moved it rapidly backward and forward. Then the tool was removed and the chimpanzee peered into the hole. Twice ants swarmed out and were eaten; once the insects that appeared seemed to be termites, and they also were eaten. Once, however, the chimpanzee (an adult female) merely gazed at some ants that emerged and made no attempt to consume them. Three times young adolescents disturbed bees in this way and then ran off. Often nothing at all came out of the hole. In many instances the tool users were infants or juveniles. It is possible that they enjoy watching the frenzied activity of insects disturbed in this way: Kohts's Ioni often took a piece of straw and poked at cockroaches in the slits of his cage. "He apparently found much amusement in witnessing the flight of the insects and would invariably start his little game afresh on seeing them recover from their first panic" (Kohts, 1935, p. 533). At Gombe infants sometimes investigate trails of ants during solitary play, occasionally prodding them with twigs.

Drinking
Sometimes rainwater collects in a hollow "basin" in a tree. When a chimpanzee cannot reach this water with his lips, he picks a handful of leaves, chews them briefly (thus crumpling them and making them more absorbent), then inserts the sponge into the water, withdraws it, and sucks out the liquid. This process may be repeated until the water is gone or until the individual is satisfied. During one three-year period, 1978 to 1980, mature individuals were seen to use sponges on fourteen occasions (eleven times females, three times males) in the typical way, and Jomeo used a sponge when drinking from a stream. Thirty additional instances of sponging (fifteen in streams) were contributed by infants and juveniles during the three years. Wrangham (1975) observed another adult male, Hugo, using a sponge in a stream and I saw an adult female, Patti, do so in 1977.

Miscellaneous
An adult male, Evered, once removed the last traces of fruit from the inside of a strychnos shell with a dead leaf, which he then sucked (Wrangham, 1977); Hugo used leaves to clean the inside of a baboon skull in the same way (Teleki, 1973c). Chimpanzees sometimes use leaves to clean sticky fruit juice from their hands, faces, or other parts of their bodies.

The adult female Miff used a large handful of leaves to brush bees from the surface of their hive, prior to helping herself to comb and honey. The same female two years later used leaves for a similar purpose, this time to brush away the driver ants that were swarming up the sapling from which she was "dipping." In both cases the use of the leaves probably saved her hand from some painful stings or bites.

Other Chimpanzee Populations
Chimpanzees in the Mahale Mountains feed on termites, using a fishing technique almost identical to that used at Gombe. B group feeds in this way on *Macrotermes*, and K group on *Pseudacanthotermes* (Uehara, 1982). In the Kasakati Basin area, the chimpanzees apparently have the same skill; two chimpanzees once ran from a termite

mound, leaving a couple of typical fishing tools behind them (Suzuki, 1966). The method seems to be used too by a group of chimpanzees in the Mount Asserik region of the Senegal (McGrew, Tutin, and Baldwin, 1979). In addition, the captive group rehabilitated by Brewer developed, without tutelage or example, a very similar technique. These individuals had spent their early years in the wild (in Guinea), one of them until about the age of three; she probably was replicating the behavior of her original social group (Brewer, 1978).

Chimpanzees at three of four sites in Mbini, once known as Río Muni (Jones and Sabater Pi, 1969), and at a site in West Cameroons (Struhsaker and Hunkeler, 1971) also use tools when feeding on *Macrotermes*, but both the type of tool and the technique employed are different. The behavior has not been directly observed, but the tools left scattered around the mounds (sometimes protruding from them) are sticks, rather than fishing twigs. Apparently the chimpanzees use these to "perforate" or break into the nests. After exposing the termites, the chimpanzees probably feed on them without further use of tools. (At the fourth site in Mbini it seems that if termites are eaten at all, they are eaten without the use of tools, for none could be found despite intensive searching; Jones and Sabater Pi, 1969.)

At Bossou in Guinea, Sugiyama and Koman (1979) observed another method of tool use in feeding on an unidentified species of termite. Twice male chimpanzees broke off and peeled small sticks, then pushed them into the nests made by termites in small hollows in treetrunks where dead branches had fallen off. The males pounded the sticks up and down for a few moments, then withdrew them with a few (usually mangled) termites adhering to the end. Both spent about thirty minutes working in this way, but their termite yield was low; the method seems highly inefficient.

The chimpanzees at Bossou used what appears to be the same technique for gathering resin from hollows in *Carapa procora* trees. For this purpose the technique is effective: the resin is glutinous and adheres to the end of the prodding tool.

Chimpanzees of K group and M group (B group not yet known) in the Mahale Mountains use tools when feeding on several species of arboreal ants, particularly *Camponotus* (Nishida, 1973; Nishida and Hiraiwa, 1982). The tools vary in size according to the width of the entrance to the nest. Sometimes a small side branch is picked and poked directly into a nest. At other times, when the entrance is small, tools similar to those used in termite fishing are prepared. If no ants cling to the tool, the chimpanzee often shakes it violently to and fro in the "expelling" technique, which then usually causes the ants to emerge. (If they do not, the chimpanzee may leave his perch and stamp on the trunk of the tree, thus shaking the nest.) Ants crawling on the branches are picked off with the lips and tongue or mopped with the back of the hand, as in termite fishing. In one incident, when a great many ants were swarming around the entrance, a female picked a big handful of leaves, wiped up a mass of ants, and ate them. Sticks were also used as investigatory or olfactory probes when chimpanzees tested a newly discovered site (Nishida and Hiraiwa, 1982).

K-group chimpanzees occasionally show a similar fishing technique when feeding on the honey of two kinds of bees (Nishida and Hiraiwa,

Anting at Mahale. (T. Nishida)

1982) and chimpanzees in the Cameroons were once observed poking sticks into an underground bees' nest and eating the honey (Merfield and Miller, 1956).

At Bossou a fascinating use of branches was observed. During a period of just over two weeks the chimpanzees daily visited a ripe fig tree, which because of its wide, smooth trunk could not be climbed. They therefore went as high as possible in a different tree, the topmost branches of which almost reached the lowest ones of the fig. From this elevation they tried, but initially failed, to climb into the fig tree. Different males, one after the other, broke off branches, stripped them of leaves and twigs, and holding one end, tried to pull the nearest fig branch within reach. In some cases one or two side branches were left on the tool and these served as hooks. The chimpanzees, standing upright, sometimes beat at the branch above, sometimes pressed it downward with a long hooked tool while reaching for it with their free hand. Between bouts of tool using, they bounced and rocked the branch they were standing on, trying to get enough swing to reach the fig tree. One male persisted in these attempts for as long as fifty-one minutes before attaining his goal—at which point other individuals hooted loudly while he displayed (as though in excitement) through the fig tree. As the days went by, it became increasingly difficult to get the figs; gradually all the suitable branches of the access tree were removed, and so too were many of the lowest fig branches. No chimpanzee was seen to carry a stick up into the access tree to use as a tool (Sugiyama and Koman, 1979).

The leaf sponge for drinking water was used by the Gambia group. Here too the behavior appeared spontaneously, without tutelage (Brewer, 1978). At Bossou, one chimpanzee was seen to dip an unchewed, uncrumpled leaf into water and then lick off the drops. In the Budongo Forest of Uganda, a chimpanzee dipped his *hand* into a water bowl and licked the water off his fingers. (This behavior, seen also at Gombe, does not necessarily mean that Budongo Forest chimpanzees never use sponges.)

In West Africa, at sites in Liberia (Beatty, 1951), Ivory Coast (Savage and Wyman, 1843–44; Struhsaker and Hunkeler, 1971; Rahm, 1971; Boesch and Boesch, 1981), and Guinea (Sugiyama and Koman, 1979), chimpanzees have been observed using hammer and anvil techniques to open oil-nut palm seeds or other small, hard-shelled items. The systematic study of Boesch and Boesch at Tai Forest, Ivory Coast (now in its sixth year), shows that during the main nut season chimpanzees engage in nut cracking at least as often as the eastern chimpanzees fish for termites or ants. Five types of nuts are cracked, mostly those of *Coula edulis* and *Panda oleosa*. For all nut cracking the chimpanzee needs a hard surface (such as a rock or tree root) as an anvil and a stone or stout stick as a hammer. *Coula* nuts are cracked in trees as well as on the ground, which means that the chimpanzee must take a hammer with him when he climbs. The *Panda* is a much harder nut and must be cracked with a stone hammer; careful positioning and a precise hammering technique are essential if these nuts are to be opened without smashing the contents. Stones are scarce on the floor of the rain forest and sometimes must be transported for *hundreds* of meters. These observations (which are continuing) present some of the most sophisticated examples of chimpanzee technology yet discovered. Similar hammer and anvil techniques were utilized by the chimpanzees at Bossou, who opened the nuts of the oil-nut palm to extract the kernel. In this area too, stones had to be transported to cracking sites (Sugiyama and Koman, 1979).

Body Care

Chimpanzees are quite fastidious, and if their bodies become soiled with dirt (feces, urine, mud, and so forth) they often use leaves to wipe themselves. They also use leaf napkins to dab at bleeding wounds and (occasionally) to rub themselves during or just after heavy rain. Table 18.2 shows the frequency with which napkins were seen to be used for different purposes over a six-year period, 1977 to 1982.

By far the most common context was penis wiping by males after mating. To give some idea of the frequency of this behavior, the number of times that individual males wiped was calculated as a percentage of the number of occasions when they were seen to mate. The results show that some males are more fastidious than others:

	Matings	% wipes
Adults:		
Humphrey	226	5.8
Jomeo	108	3.7
Satan	193	3.6
Atlas	113	3.5
Goblin	216	2.8
Sherry	128	1.6
Figan	257	1.2
Evered	211	0.5
Early adolescents:		
Wilkie	91	5.5
Freud	210	2.9

Juvenile Fifi uses a handful of leaves to wipe blood from her clitoris, bitten during a fight. (H. van Lawick)

Other immature males also wiped their penes, but because far fewer matings were seen, these observations have not been included in the analysis.

Bottom wiping by females after copulation, by contrast, was observed only rarely. In part this is because the male's penis is visible to him after the sexual act and he can see the remains of ejaculate clinging to the tip. In addition, during sexual contact the male may become contaminated with feces from the female's bottom. We once saw Gigi approach and present in response to Hugo's vigorous courtship; staring at her diarrhea-smeared bottom, he seemed to change his mind and moved away without mating her. Two males on other occasions carefully wiped Gigi's dirty bottom with leaves when she presented; one subsequently mated her, the other did not.

The Gombe chimpanzees, in fact, seem to have an almost instinctive horror of being soiled with excrement and only very rarely have been seen to touch feces (their own or another's) with their bare hands. If a chimpanzee accidentally becomes smeared with the feces of another, the offending substance is wiped off carefully with leaves. Once Satan, in the midst of attacking a female, got spattered with her fear dung.

Table 18.2 Chimpanzee use of leaf napkins over a six-year period, 1977 to 1982.

Usage	Number of observations	Substance removed				
		Feces	Urine	Semen	Blood	Other
To wipe self[a]						
Own penis	77	?		•		
Own bottom						
After defecation	37	•				
After urination	5		•			
After mating	6			•		
When menstruating	1				•	
To remove—						
Feces of other	31	•				
Urine of other	11		•			
Fruit juices	15					•
Mud	1					•
Stranger's touch[b]	2					•
Rain	3					•
To dab wound	22				•	
To wipe other						
Wound	10				•	
Bottom	7	•				
Nose	1					•
Urine	1		•			
Total	230	75	17	83	33	22
Percent of total		33.6	7.5	36	14	9

a. Ten additional instances of wiping were observed when the substance removed was not identified.
b. Once the "stranger" was human.

Immediately he stopped chasing her and scrubbed himself seven times, each time with a new napkin. Mothers usually clean themselves at once if they are accidentally dirtied by the excrement of their infants. Individuals with diarrhea may wipe themselves with leaves. One female, Melissa, was particularly fastidious in this respect; she accounted for thirteen of the thirty-seven observed incidents. Her daughter, Gremlin, also wiped her bottom often (but five of her eleven wipes were after she had urinated).

Köhler (1925, p. 75) observed a similar fastidiousness and made a further interesting observation. Like virtually all captive chimpanzees, the individuals in his colony practiced coprophagy and in this context they would unhesitatingly pick up feces with their bare hands. But when they became *accidentally* soiled—if for example "one of them steps in excrement, the foot cannot, as a rule, tread properly after . . . The creature limps off till it finds an opportunity of cleansing itself" (with rags, straw, pieces of paper, and so on). Coprophagy is rare in the wild; when it does occur the chimpanzee almost always picks food material from the feces with his lips, and only infants have been observed to handle excrement. One old male, Mike, who had eaten a good deal of meat the evening before, picked a couple of large leaves, placed them on his hand, and defecated onto them. He picked out pieces of undigested meat with his lips, then dropped the leaves and the remains of the feces to the ground below his tree. (Observer: H. Bauer.)

If a chimpanzee is accidentally sprinkled with urine (by a companion above him, for example), it too may be wiped off with leaves, but the behavior is not so frantic and sometimes the victim merely glances up and calmly moves out from under the remainder of the shower.

As we have seen, chimpanzees sometimes use leaves to wipe sticky fruit juice from themselves. The eating of unripe strychnos fruits causes copious salivation, and it was in this context that eleven of the fifteen instances of food wiping were seen during the six years under review. Interestingly, all of these incidents involved members of one family: Passion, Pax, Prof, and Pan. During earlier years, however, other individuals were seen wiping themselves in the feeding context. Fifi once scrubbed her chest and belly repeatedly after carrying an armful of overripe bananas.

Chimpanzees often dab at bleeding wounds with leaves, which they then lick; they may repeat the process many times. Young Gremlin once got a bad cut on her bottom. For two days she picked large handfuls of leaves rather frenziedly *while* she was urinating and pressed them to her bottom as soon as she had finished; probably the urine hurt or stung the wound. An adolescent female wiped her bottom with leaves when she was first observed to menstruate.

In two instances wiping was in response to contact with strangers. The first occurred when a stranger female was surrounded by Kasakela individuals (she became the victim of one of the very severe attacks described in the last chapter). Pant-grunting nervously, she approached Satan, reached out, and touched his arm submissively. Satan immediately moved away, picked some leaves, and wiped the place of contact. The second incident took place in 1968 after an infant female, Pom, swung playfully over a visitor (Robert Hinde) and investigated him by stamping on his head with her foot. She then sniffed her foot, picked some leaves, and wiped it vigorously.

Occasionally a chimpanzee uses leaves to wipe a companion. During the six years this was seen nineteen times and involved only family members. One infant, Frodo, wiped his mother's bottom when she was in estrus, removing smears of feces. Another infant, Michaelmas, twice wiped his mother's bottom (she too was in estrus) *after* he had copulated with her. Melissa four times used leaves to wipe her twins after one of them had defecated. Once she wiped *both*, though only one was dirty. Ten times youngsters dabbed at the wounds of others: six times Prof gently wiped the severe wound of his infant sibling, and four times Gimble dabbed at a bleeding wound sustained by his mother. During a play session with his two-year-old brother, Freud paused to urinate. His penis was erect (often the case during vigorous play) and the urine, glinting in the sun, formed an arc the end of which made contact with young Frodo. Both brothers watched, seemingly fascinated, until the stream ended. Freud then picked six big leaves and vigorously scrubbed Frodo where he had been wet. The last case was seen when infant Pax sneezed. His brother, Prof, gazed intently at the thick mucus that dribbled from his nostrils, picked some leaves, and carefully wiped it away.

A few more examples of wiping others come from earlier years. Infant Flint was seen wiping the dirty bottom of his mother and his sister Fifi (once each), removing mud from his mother's back, and

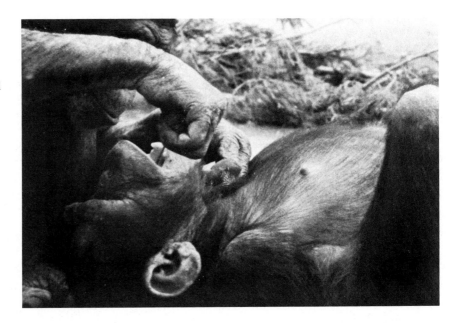

Chimpanzee dentistry: adolescent female Belle (one of the Menzel chimpanzees) works at the loose deciduous tooth of a companion. On one occasion she actually performed a successful extraction. (W. C. McGrew)

cleaning mud from the head of an adult male. Miff was seen to wipe the bottom of her infant, Michaelmas, in 1973, five years before he was seen to wipe hers. And when Miff's elder daughter, Moeza, was an infant, she too was seen to wipe her mother's bottom. I have already mentioned the two males who wiped diarrhea from the backside of a female.

Chimpanzees in captivity sometimes use objects as tools in this cooperative manner. McGrew and Tutin (1972) describe the use of twigs in dental grooming by a member of the Menzel group. Not only did the "dentist," Belle, clean the teeth of a young male with twigs, she actually performed an extraction, removing a loose deciduous premolar in one and a half minutes. (These chimpanzees also poked at and removed their own teeth with tools; and a chimpanzee at Gombe was observed picking with a twig at her teeth where, it seemed, something was stuck.) D. Fouts (1983) describes how the youngster Loulis used a willow twig to probe a sore in the foot of a young female; but his foster mother, Washoe, repeatedly took away the tool and groomed the patient in more traditional fashion.

Captive chimpanzees sometimes use sticks to scratch themselves (see for instance Köhler, 1925, and a lovely photograph in Kummer, 1971, p. 148). This behavior was seen only once at Gombe, but one young female, Pom, during very heavy rain repeatedly poked a short, firm twig through the hair on her head, almost certainly in response to the tickling of the rainwater as it seeped through her hair. (She was also making the most extraordinary facial grimaces, presumably for the same reason.)

Investigation

The use of grasses and stems as olfactory probes during feeding, especially to test passages during termite fishing, has already been described. In addition, chimpanzees use twigs and sticks to investigate objects that they cannot otherwise reach or that they are afraid to

Flint investigates his newborn sibling. (H. van Lawick)

touch with their hands. When a dead python was placed at the feeding area, eight-year-old Fifi, who had been staring at it for some time, smelled the end of a long palm frond on which the snake had lain. She then pushed the frond, hand over hand, until its tip touched the python's bloodied head, withdrew it, and sniffed the end. A juvenile, Flint, prevented by his mother from touching his newborn sibling, contacted the baby gently with a stick and then sniffed the end. In the early days of banana feeding I sometimes hid a fruit on my person in order to smuggle it to a youngster when the big males were not looking. Fifi once approached and tried to feel in my pocket; when I prevented her from doing so (because adults were close by), she picked a long piece of grass, poked it into my pocket from a distance, then smelled the end of her tool. Her suspicions were apparently confirmed, for she followed me, whimpering, until I was able to hand over my offering. In 1965 two infants used small twigs to inspect the genital area of females (normally done by touching, then sniffing the finger). In 1971 this behavior was seen in another infant. Sometimes youngsters use sticks to investigate running water, poking and swirling while watching the effect intently. Frodo, as an infant, once poked around in his dung; this was at a time when we were examining feces on a regular basis and he may have been imitating us.

Intimidation

The male chimpanzee typically enhances his charging display by dragging, waving, or hurling branches, and by rolling or throwing rocks. Even when these performances are not directed at specific other individuals, they nevertheless serve to make the performer seem larger and more dangerous than he actually may be. One male, Humphrey, on two separate occasions threw more than twenty rocks in

Mike learned to keep two, or even three, empty kerosene cans clattering ahead of him during his charging displays. This strategy so intimidated his rivals that over a four-month period he rose from a low position in the hierarchy of fourteen males to become alpha; no physical attack by Mike was ever observed. (H. van Lawick)

the general direction of another male (each time it was Satan) during protracted dominance conflicts. Satan eventually was forced to seek refuge up a tree while Humphrey continued to charge about and hurl rocks below. I have described how Sniff of the Kahama community threw at least thirteen rocks into a ravine where males of the unhabituated Kalande community were displaying. Sometimes very large rocks are uprooted and rolled during charging displays (one was estimated to weigh about 6 kilograms); they can be extremely dangerous for chimpanzees (or humans or any other creatures) who happen to be downslope.

A convincing demonstration of the purposeful incorporation of objects into charging displays was Mike's use of empty 15-liter kerosene cans, as described in Chapter 15. Most impressive was the calm manner in which he selected his cans prior to commencing his display. When (after having been hit a few times) we hid all the cans, Mike went to extreme lengths to take other human artifacts, and when we managed to secure everything against his depredations, he dragged and threw natural objects with a very high frequency over a period of several months.

Whipping, Flailing, and Clubbing

Whipping occurs when a chimpanzee takes hold of a growing branch or sapling and swings it vigorously up and down, hitting the victim. In one protracted dominance conflict between Mike and Goliath whipping was the only aggressive contact between the rivals. Males have been seen whipping females who refused to crouch for mating, and occasionally rival males have whipped a copulating couple. I was sometimes whipped in the early days when the chimpanzees responded aggressively to my presence.

The term *flailing* is used when a chimpanzee picks up a stick or palm frond, or actually breaks one off a tree, and, usually in an upright position, brandishes his weapon at the opponent. This is highly effective intimidation: a fully adult male baboon, who may ignore the threatening gestures of an adult male chimpanzee, is likely to avoid the approach of a female or even a juvenile chimpanzee who is flailing a large stick.

When a detached stick or frond is used to hit or beat an opponent, this is referred to as *clubbing*. Over a six-year period (1977 to 1982) 22 percent of 188 observations of flailing ended in clubbing.

Köhler (1925) noted an additional use of sticks as weapons—jabbing—which has not been observed at Gombe. Chimpanzees of his colony used sticks or pieces of wire to jab unsuspecting humans, dogs, or domestic chickens. Sometimes a chimpanzee would lure a chicken within range by throwing out pieces of bread (another example of tool use!) and then suddenly jab the unfortunate fowl. Occasionally it was a cooperative effort; one chimpanzee threw the bread and another jabbed.

Aimed Throwing

In addition to the generalized hurling of objects during displays, chimpanzees throw stones, rocks, or sticks at definite objectives such as

Juvenile Flint threatens a male baboon. (H. van Lawick)

conspecifics, baboons, humans, or a variety of other species. Aimed throwing may be overhand or underhand: larger missiles are more often thrown underhand, and sometimes launched with both hands. Chimpanzees have good aim, but the missiles often fall short of their targets.

The establishment of the feeding area at Gombe clearly affected the observed frequency and efficiency (in terms of choice of missile) of aimed throwing, at least by the adult males, as the following tabulation shows:

	1963	1964	1968	1977	1980
Total number of males	16	15	14	6	6
Number of males seen to throw	5	8	9	3	4
Number of missiles	9	32	113	9	11
Number of hits	1	4	5	2	0

The 1963, 1964, and 1968 scores are from camp records only; the 1977 and 1980 scores, however, include observations of throwing in camp and out of camp and during *all* follows. These numbers cannot be treated statistically but they give, nevertheless, some idea of the changing picture.

The increase in the number of missiles thrown during the height of competition for bananas (1966 to 1969) reflects both the increased tensions generated by the close proximity of many adult males at the feeding area and competition for bananas among the chimpanzees themselves and between chimpanzees and baboons. When this competition and crowding declined, there was less need for the males to throw (although throwing continued with high frequency among some immatures, as we shall see).

The throwing of stones and rocks, like the brandishing of sticks, is an extremely effective method of intimidation. True, weapons are seldom used in serious fighting (the chimpanzees usually inflict punishment by biting, hitting, and stamping), but there are undoubtedly occasions when the use of sticks or rocks, by intimidating the victim, prevents aggressive incidents from becoming more violent. During the 1977–1982 period at Gombe, 412 instances of aimed throwing were observed, sometimes involving more than one missile. When young chimpanzees were engaged in aggressive play with baboons, the exact number of stones or branches thrown was not always recorded, so I have listed these as single bouts. Thus the actual number of objects thrown, as far as youngsters are concerned, was considerably greater than indicated.

Table 18.3 shows the observed frequency of flailing, clubbing, and aimed throwing in chimpanzees of various age-sex classes over the six-year period. Mature males threw more than they flailed and clubbed, whereas mature females had a tendency to flail more than they threw. Immature females were seen to throw and club at similar frequencies. Immature males, clearly, were the major weapon users. In fact, as we shall see, it was one or two individuals per year who were responsible for the high rates of both throwing and flailing. Immature, like mature, males showed a tendency to throw more than they flailed and clubbed.

Probably the most significant fact to emerge from the data is that there is a clear-cut sex difference in patterns of aimed throwing by adults. A greater percentage of adult males throw in comparison with adult females. During the six years under consideration, for example, five of the seven adult males were observed to throw at least twice, whereas of the twelve frequently observed females (Appendix C) only three were observed to throw at all. (The five males were seen to throw a total of 39 missiles, the three females all told threw 12.) In 1968 *all* of the nine males who regularly visited camp threw at least once, but only two of the twelve females did so. The males were seen to throw 113 objects, the females 2. No adult female has ever been seen to throw more than five times in a given year, while the record for an adult male is over fifty. Moreover, although some young females do throw quite frequently, they do not do so to nearly the same extent as young males.

Contexts for Weapon Use
Table 18.4 shows, for mature (thirteen years and over) and immature chimpanzees of each sex, the targets at which they directed both aimed throwing and flailing over the six-year period. For throwing I have added information from 1968. With the exception of that year, when there was fierce chimpanzee-baboon competition over the bananas in camp, adult male chimpanzees threw and flailed most often at conspecifics. Mature females threw more often at baboons, but flailed more often at conspecifics. For immature males there was little difference; the high rate of throwing at humans in 1968 was caused by one youngster, Flint. In the years under consideration there were very few observations of weapon use by immature females. As an eight-year-old, however, Fifi had thrown frequently: six of her fourteen throws recorded in a four-month period in 1967 were at baboons

Table 18.3 Observed incidents of weapon use in adult and immature males and females over the six-year period 1977 to 1982.

Usage	1977	1978	1979	1980	1981	1982	1977	1978	1979	1980	1981	1982
	Mature males						Mature females					
Throwing	9	18	6	5	1	0	0	9	2	0	1	0
Flailing	2	0	1	0	0	0	5	3	5	2	1	3
Clubbing	1	0	0	0	0	0	2	0	1	0	1	1
	Immature males						Immature females					
Throwing	33	53	72	23	78	85	5	0	5	1	4	2
Flailing	38	34	42	13	13	9	6	2	1	2	3	2
Clubbing	5	4	7	4	5	6	1	1	0	1	1	0

Table 18.4 The percentage of throwing incidents and flailing incidents that were directed by mature and immature males and females at different targets over the six-year period 1977 to 1982. For throwing only, comparable data are given for 1968.

	Percentage of total targets that were—				Total no. incidents	No. individuals who threw or flailed
	Chimpanzees	Baboons	Humans	Other		
Throws, 1968						9
Mature males	33	56.5	9	2	113	
Mature females	50	50	0	0	2	2
Immature males	21.5	24	54.5	0	37	5
Immature females	33.3	33.3	33.3	0	3	2
Throws, 1977–1982						
Mature males	64	32	2	2	39	7
Mature females	17	83	0	0	12	5
Immature males	28	40	29	3	344	9
Immature females	35	35	30	0	17	6
Flailings, 1977–1982						
Mature males	75	25	0	0	4	5
Mature females	63	37	0	0	19	8
Immature males	46	37	14	3	149	8
Immature females	37	37	13	13	16	7

and four at other chimpanzees. And Pom, when she was between nine and ten years old, also threw quite often—and flailed more—but the detailed data have not yet been extracted from the reports.

Young chimpanzees may start to flail or throw during play sessions with young baboons. At this point the baboons tend to run off and the chimpanzees display after them, brandishing sticks or throwing stones. The behavior is described as aggressive play, the degree of aggression depending on the age and sex of the chimpanzee concerned and judged by the response of the baboon playmate.

Adult males threw rocks at rivals during dominance conflicts, and at females and youngsters during other aggressive incidents. Adolescent males threw at females to enhance their intimidation displays. Thirty-five percent of the 129 objects seen to be thrown by two young males, Atlas and Freud, between the ages of eight and ten, were directed at older females in this context. The technique is highly effective: on thirty-two of the forty-five occasions the females avoided or showed submissive behavior. Sometimes the young males only needed to pick up and brandish rocks, without throwing them, to elicit submissive behavior in young females. Older or higher-ranking females, if they saw young males reaching for or holding rocks, occasionally approached them and took the weapons away. Three times (throughout the entire study period) a male chimpanzee, after avoiding a rock hurled at him by another male, was seen to pick up a missile himself and hurl it back.

Some chimpanzees were more inclined than others to throw rocks at humans. The most frequent exponents have been Humphrey, Flint, Atlas, Freud, and, most recently, Frodo. Flint in 1968 was observed throwing thirty times: 60 percent of these missiles were hurled at humans. And Frodo in 1981 threw 74 percent of his eighty-nine missiles at humans. I found ten reports of Humphrey throwing in 1974 and six times the objective was a human. Although he threw fewer missiles, he selected larger rocks than the youngsters and threw much harder—and with better aim.

The Bossou chimpanzees also threw at human observers. They picked branches and threw them, mostly underhand, from the trees above. The adult males showed good control and threw large missiles up to 120 centimeters long and 3.2 kilograms in weight. Sometimes the humans were hit, and the incidents "actually represented severe attacks for the authors" (Sugiyama and Koman, 1979, p. 516).

Youngsters sometimes throw or flail when they encounter monitor lizards or other small creatures. Both Fifi and Gremlin chased slow-moving snakes for about 10 meters while brandishing sticks and periodically whipping the reptiles with one end of the branch. Köhler's young chimpanzees used sticks to investigate small animals such as lizards and mice: if one of these creatures made a rapid movement toward the chimpanzee, the stick became a weapon and the victim was hit hard.

Four times, from 1970 on, chimpanzees were seen to throw large rocks at adult bushpigs.

(a) I described in Chapter 12 how the adolescent male Atlas routed a group of pigs who had displaced him and his family from a patch of fallen fruits.

(b) During a hunt adult male Mike threw a melon-sized rock at an adult pig that was protecting its young (Plooij, 1978). The missile was probably intended to make the prey run. Chimpanzees have also been observed to flail at prey animals in this context.

(c) Humphrey stood upright and hurled a rock, estimated as weighing at least 5 kilograms, at an adult pig he encountered while traveling.

(d) A party of mothers and young encountered a boar who grunted and charged. All the females rushed up trees, but eight-year-old Freud remained below, displaying and hurling many rocks and branches at the pig until (perhaps because of the human observers) it ran off.

During one baboon hunt one of the six participating males threw many rocks at the male baboons who were attacking the hunters

Flint hurls a rock at the camera.
(H. van Lawick)

(Chapter 11). The late-adolescent female Gremlin broke off and threw a large dead branch at a male baboon who was threatening her mother (who had just seized the baboon's kill). Both females flailed vigorously on this occasion.

Two unusual throwing incidents involved the same adult male, Hugo. In the first he hurled a number of large rocks (up to 15 centimeters in diameter) toward the dead body of Rix, the male who broke his neck when he fell from a tree; none of them hit (Teleki, 1973a). In the second episode he stood upright to throw an extremely large rock, estimated to weigh more than 5 kilograms, at the motionless body of Godi, one of the victims of intercommunity aggression. Had it hit, the rock could have inflicted substantial damage.

Kortlandt (1962, 1963, 1967) and his colleagues (Albrecht and Dunnett, 1971) carried out a number of field experiments that involved placing a stuffed leopard (its head was electrically wired to move from side to side) in an area where chimpanzees came to feed on papaya (at Benin) or grapefruit (at Bossou). Many of the adult chimpanzees displayed around this dummy, often dragging, flailing, or throwing sticks. Frequently the leopard was clubbed. On the basis of these observations, Kortlandt suggests that chimpanzees may use sticks as clubs during encounters with leopards.

Very few encounters between chimpanzees and large predators have been observed in the wild. There are four reports of meetings between chimpanzees and leopards, all in Tanzania. Two are from Gombe (Goodall, 1968b; Pierce, 1975), one from Kasakati Basin area (Izawa and Itani, 1966; Itani, 1970), and one from Mahale (Nishida, 1968). Only in two cases were sticks used in intimidation displays; they were not used as clubs. At Gombe an adolescent female, Honey Bee, broke off and threw a branch at a lame leopard as it moved beneath her. A young male at Mahale threw small branches and bits of dead vine at a leopard in the undergrowth below, but did not seem to be very disturbed.

A small group of chimpanzees, captured at juvenile and subadult ages in a savanna area of West Africa where leopards abound, were kept in a large enclosure at the Institut Pasteur. Adriaan Kortlandt confronted them with a stuffed leopard having a baby chimpanzee doll on its paws. The apes fiercely attacked the dummy and scored four hits with the clubs that had been laid ready for them. In the final attack the club swung by the mother carrying the infant grazed the belly of the leopard and hit the ground with a speed of 50 mph, measured on 16-mm film. The bending of the stick shows the force of the blow.
(Copyright © A. Kortlandt)

There are two reports of chimpanzee encounters with lions. One of these, which took place near Kigoma in the 1950s, also involved the use of sticks as missiles. It was described to me graphically by a local Tanzanian, Mzee Mbrisho. Five adult chimpanzees screamed and threw down branches at a young male lion, which ran off as Mbrisho and his companions appeared. The second report comes from the Ugalla region of southwest Tanzania. Kano (1972) came upon a group of about fifteen chimpanzees calling in a very agitated manner. Suddenly a lion ran out of the brush below them (disturbed by the approach of Kano—as was he by the approach of the lion!). It was 1100 hours and the chimpanzees were still in the trees where they had slept; obviously they had not dared to leave.

Individual Differences

There are obvious differences among individuals with regard to the frequency of aimed throwing (and, to a lesser extent, of flailing and clubbing). Some individuals throw often, others very rarely; a good many females have never been observed to throw at all. Since 1964 there have only been four adult males who were observed to throw at relatively high frequency year after year—Hugo, Mike, Worzle, and Humphrey. Among immature males throwing became a frequent activity for four juvenile and adolescent males—Flint, Atlas, Freud, and Frodo. Two of these youngsters, Atlas and Freud, threw most frequently between the ages of seven and ten. Flint died when he was eight; Frodo at the time of writing is at a very high peak in his eighth year. The only two females who excelled at throwing, Fifi and Pom, did so when they were between nine and ten years, after which the behavior was seen only seldom. The more recent data are summarized in Table 18.5, where I have also included the information from 1968. Of the ten individuals who threw enthusiastically, four are members of the same family (Fifi, Flint, Freud, and Frodo) and two (Humphrey and Atlas) are suspected of being father and son. Flint was four years old when his sister Fifi was at her peak of throwing; he perhaps learned by watching her behavior. Freud was only fourteen months old when Flint died, and it seems unlikely he could have imitated throwing at that age. But he was between three and six when Atlas was at his peak and might have learned from him. Frodo watched his elder brother intently from the age of two (when Freud was almost seven). Thus for each of the three the behavior could have been acquired as a result of watching an older "throwing champion."

In 1974, when Atlas was seven years old, he began to throw frequently at human observers. He selected quite large rocks (up to 3 kilograms) and since he threw from very close quarters, he often hit his target. The behavior was effective—there were very few follows on his family that year! During 1975 and 1976 he continued to throw often, particularly at baboons in camp, where he frequently flailed and clubbed also. He was easily able to drive adult male baboons away from the mineral lick we sometimes set out, even when mature male chimpanzees had been intimidated (by the baboons). He also began to throw with increasing frequency at conspecifics, incorporating the behavior into the intimidation displays that he directed at adult females. During the last three months of 1975 he was seen to throw at

Table 18.5 Individuals who were responsible for the majority of throwing incidents during seven years, 1968 and 1977 to 1982, classified by age and sex. Figures in parentheses are numbers of objects actually seen to be thrown.

| | Number of objects thrown | | | | | | | | Number of individuals who threw as fraction of number in age-sex class | |
| | 20 or more | | 10–19 | | 5–9 | | 1–4ᵃ | | | |
Year and age category	♂	♀	♂	♀	♂	♀	♂	♀	♂	♀
1968										
Mature	2: HM(25), WZ(48)		2: MK, HG		1		4	2	9/9	2/11
Adolescent							2	1: FF	2/16	1/6
Juvenile or infant			1: FT				2	1: PM	3/4	1/4
1977										
Mature					1: HM		2		3/6	0/10
Adolescent			2: AL, FD					3: PM	2/3	3/2
Juvenile or infant							4		4/6	0/5
1978										
Mature			1: HM				3	3: FF, PM	4/7	3/10
Adolescent	1: FD(26)		1: AL						2/2	0/3
Juvenile or infant							1		1/3	0/6
1979										
Mature					1: HM			1: PM	1/7	1/11
Adolescent	2: AL(21), FD(49)								2/2	0/2
Juvenile or infant						1	1: FR		1/9	1/4
1980										
Mature					1: AL		3: HM		4/6	0/11
Adolescent					1: FD				1/4	0/3
Juvenile or infant						1	1: FR		1/8	1/5
1981										
Mature							2: HM, AL	2	2/7	2/11
Adolescent							2: FD		2/6	0/3
Juvenile or infant	1: FR(74)							2	1/7	2/5
1982										
Mature									0/5	0/10
Adolescent							1: FD		1/7	0/2
Juvenile or infant	1: FR(56)						3	1	4/6	1/4

a. Initials shown are those of individuals who threw more frequently in other years.

females seven times (they screamed and avoided), and once at an older adolescent male. He even hurled a missile at adult male Satan, who had just attacked him (it was perhaps fortunate for him that Satan did not notice). In 1981, however, when he was fourteen years old, Atlas was only observed to throw once, and the following year not at all. Freud threw most often in his seventh and ninth years.

Freud's younger brother, Frodo, is the most zealous thrower yet. Like Atlas, his performances can be extraordinarily effective; he has succeeded in intimidating at least this human observer! Having carefully selected a very large rock (occasionally he has to give up because he cannot carry the missile of his choice), he approaches closely, sometimes to within a meter, and hurls it with great force. From this distance he can hardly fail to hit, and I have been bruised on a number of occasions. In 1983 he began to dislodge extremely big rocks, weighing at least 8 kilograms, and push them toward victims who were downslope. Once he rolled no less than five in quick succession: it was not clear whether the target of this bombardment was a female chimpanzee or myself—we both fled!

Fifi's throwing as an adolescent was sometimes amusing: she would nonchalantly gather one or even two handfuls of small stones, wander toward her unsuspecting victim (usually a baboon or female chimpanzee), then from about 2 meters suddenly hurl the missiles at her adversary. At least that was presumably her intention; but her aim was poor and, especially since she usually turned and ran off as she threw, the stones often went up into the air and landed back near (once on) the thrower. Fifi is one of the few adult females who still occasionally throws. As adolescents both Pom and Fifi flailed and clubbed more often than they threw.

Köhler (1925, p. 80) also noted individual differences in his colony. The thrower par excellence was Chica, "who learnt to aim excellently and expressed her skill with equal delight against her fellow apes and us." At Bossou two individuals, an early-adolescent female and a juvenile male, between them accounted for 54 percent of the sixty-nine missiles thrown (Sugiyama and Koman, 1979).

Other Contexts

When an infant is unable to follow his mother from one tree to the next, he may proceed as far as he can along a branch and then stop and whimper. Sometimes a mother returns and finds she cannot reach her infant. After surveying the situation she climbs to where she can just reach the tip of a branch of the infant's tree, draws it closer, hand over hand, and holds it in place until the infant has navigated the crossing. This bridging action is not strictly tool use, of course, but like much of the object use in play, described below, it represents an awareness of the relations between things.

Youngsters utilize many objects during solitary play, demonstrating the extent to which infants make use of the objects in their environment—sometimes in a very inventive way. Fruit-laden twigs, strips of skin and hair from an old kill, or highly prized pieces of cloth may be draped over the shoulders or carried along in the neck or groin pocket (that is, tucked between the neck and shoulder or thigh and belly); stones or small fruits may be hit about on the ground, from one hand to the other, or thrown short distances into the air and retrieved.

Sometimes a large stone or a short stout stick is used in self-tickling—a performance that *can* be labeled tool use. Juvenile and adolescent females are particularly apt to show this behavior: the object is pushed and rubbed into those especially ticklish areas between neck and shoulder and in the groin. The activity may last for up to ten minutes and is often accompanied by loud laughing. Sometimes these tickling tools are carried up into a nest and the game proceeds there. Two young females (an infant and a juvenile) tickled their own genitals with sticks while laughing. And three infant males were observed to carry stones, fruits, or small rocks (once even a

The dead male flower cluster of an oil-nut palm is frequently used as a toy by young chimpanzees—in this case, infant Flint.

Flint with a strychnos toy. He is showing a play face. (H. van Lawick)

piece of dry dung) for short distances, set them down, and make thrusting movements on them with erect penes. Interestingly, they were all members of the same family (Goblin, Gimble, and Getty); they could not have learned from one another because in each case the behavior had long been extinguished in one before it appeared in the next.

A strychnos fruit is a very popular toy. It may be carried short distances, rolled on the ground, or rubbed over the body. A most impressive performance took place when Freud, age seven years, who had been playing in this way, not only threw the ball almost a meter into the air but actually caught it. He spent the next five minutes trying, without success, to repeat his accomplishment, retrieving the fruit three times when it rolled away from him.

Often, when initiating social play or during an ongoing session, one of the youngsters will break off a leafy twig or pick up some other object such as a piece of palm frond; with this in mouth or hand he approaches, then runs from, the chosen playmate. Usually a chasing response ensues—sometimes a tug-of-war.

Cultural Traditions

Table 18.1 lists the tool-using performances that have been observed at different sites. It must be emphasized that it is not a *comprehensive* compilation of tool-using patterns shown by chimpanzees throughout their range, but rather a list of the patterns that have been *observed*. Only at Gombe and Mahale (K group) have habituated chimpanzees been studied over a long enough period for us to be reasonably certain that all the commonly used patterns have been recorded. Incomplete as the picture is, it gives us some idea of the diversity of objects that chimpanzees use as tools and the variety of purposes to which they may be put. Patterns range from the inefficient prodding and crushing of termites in tree nests to the careful selection, preparation, and skillful insertion of suitable material into a mound; from poking a single unmodified leaf into water and sucking off the liquid to a crumpled, absorbent sponge used for the same purpose; from the smashing of a hard-shelled fruit against a fixed surface without the use of a tool to the sophisticated behavior shown by the nut crackers of Tai. Objects may be used as they are, or they may be carefully modified to make them more suitable for the purpose required; they may be picked up at the site where they are to be used, or they may be transported over considerable distances, sometimes from a point in space from which the chimpanzee cannot see the ultimate destination.

In all populations studied for any length of time there are important items in the diet, at least during some months of the year, that would be difficult or impossible to acquire without tools. Chimpanzees at Gombe spend up to 20 percent of their feeding time fishing for termites in November, and females fish throughout the year. It seems probable, too, that termites are extensively eaten in both the Senegal and Mbini (Río Muni). Chimpanzees at Mahale probe for arboreal ants almost daily, their bouts averaging about thirty minutes, during which they capture between two hundred and a thousand ants (Nishida and Hiraiwa, 1982). The Tai chimpanzees crack *Coula* nuts from November to March, with a peak in December, when they seem to do so almost all day. They crack *Panda* nuts from January to October, with a peak from February to April, and some *Parinari* nuts from June to October. Thus there are no months when they do not use the hammer and anvil technique, and there are four months when they do so very extensively (Boesch and Boesch, 1983).

The techniques used for termite fishing at Gombe and the Senegal, at opposite ends of the chimpanzees' range, seem to be similar, if one can judge by the tools used; this technique and that used by the Mbini chimpanzees, on the other hand, are very different. The fishing techniques of the Gombe and Mahale K-group chimpanzees seem much alike—although the species of insect captured is different. The chimpanzees of Tai and Bossou, separated by some 200 kilometers, both use hammer stones. There are a number of differences in technique, and whereas the Bossou group cracks oil palm nuts, the Tai group does not, although the nuts occur there (Boesch and Boesch, 1983). So far, however, there are no observations of eastern chimpanzees using hammer stones for anything, despite an abundance (at least at Gombe) of stones, rocks, and hard seeds.

These differences, as I have argued elsewhere (Goodall, 1970, 1973), can be regarded as culturally divergent traditions. Once a technique has become established in a given community, it probably persists virtually unchanged for countless generations.* Certainly there are no obvious differences between the tool-using behaviors shown by the young chimpanzees at Gombe today and those of the past generation. Young chimpanzees learn the tool-using patterns of the community during infancy, through a mixture of social facilitation, observation, imitation, and practice—with a good deal of trial and error thrown in

As Winkle enlarges the opening of a bees' nest with a stout stick, her daughter, Wunda *(left)*, and young Frodo watch. When Winkle gives up, Wunda picks up and uses her mother's tool.

* Granite anvils used by the Tai and Bossou nut crackers show depressions indicating their use over very long periods of time (Sugiyama and Koman, 1979; Boesch and Boesch, 1983).

Flint uses the termite-fishing technique in an artificially constructed water basin. Eventually he crumples the grass blade, and this forms a miniature sponge. (H. van Lawick)

(Goodall, 1973). The capacity to acquire a given technique through learning plays a role similar to that played by genetically controlled behavior in lower species in ensuring the continuation of some behaviors, tool use being one of them (Marais, 1969; Kummer, 1971).

The environment undoubtedly has a part in shaping the tool-using patterns of a given population (McGrew, Tutin, and Baldwin, 1979). The high annual rainfall of Mbini, for instance, means that termites continue work on their nests for most of the year, so that the walls are moist and porous and vulnerable to penetration. The termite mounds at Gombe and the Senegal become extremely hard through the dry months, and chimpanzees cannot break them open. As a result, the Mbini chimpanzees had more incentive to develop their destructive method of catching termites by breaking up the nests. Yet no special environmental factors seem responsible for some of the other differences, such as the capture of driver ants at Gombe but arboreal ants at Mahale, or the use of hammer stones in some areas and not in others. We should probably acknowledge that a crucial role has been played by certain individuals who take their place as the "wheel inventors" of the chimpanzee world.

Chimpanzees in captivity are known to be capable of insightful problem solving, or ideation. One clear-cut example was seen at Gombe. The adult male Mike was afraid to take a banana from my hand. He threatened me, shaking a clump of grasses. As he did so the end of one blade just touched the banana. He let go of the grasses, picked up a thin plant, dropped it immediately, and broke off a thicker stick. He then hit the banana to the ground, picked it up, and ate it. When a second banana was held out, he used the tool immediately. This particular piece of problem solving was not important for the Gombe chimpanzees. Nor (since chimpanzees rarely eat their dung) was Mike's creative use of leaves to catch his feces and thus avoid dirtying his hand. The point is that the chimpanzee is *capable* of performances of this sort and, having once achieved a goal by a particular means, will almost certainly be able to repeat the behavior.

Because chimpanzees are extremely curious and watch any unusual action with close attention, and because they are able to learn through observing the behavior of others, a novel performance of this sort may be passed on to others in the group.

In point of fact we should probably return to the infant or juvenile as the most likely candidate for stumbling upon a new tool-using technique. Infants, once they have acquired a new behavior such as termite fishing, use it repeatedly in novel contexts. Infant Flint "fished," as though for termites, in the hairs of his mother's leg—not exactly a useful new behavior! But the same infant, when he was four years old, used these fishing techniques when drinking water from a hollow in a tree. Initially he sucked the drops of water from the end of his piece of grass; as he repeated the action, the grass became increasingly crumpled until he ended up with a minute sponge. This is exactly the kind of performance that could have led to the "invention" of the water sponge in the first place. (Perhaps it was an infant trying to replicate the resin-gathering performance of his mother, but in the wrong place—that is, a termite nest—who acquired the first termites by the inefficient method seen in Guinea.)

Infants are more exploratory than adults and their behavior is more flexible. During one two-year period at Gombe sticks were used to investigate holes in trees on eleven occasions. Eight times the wielders were infants or juveniles; the other three were adolescent females. The most illuminating incident was when a juvenile male, Wilkie, poked his stick into an ant nest, causing a stream of fierce black ants to emerge. Wilkie avoided them. His mother, who had been watching, immediately approached and ate the ants. It seems most likely that these were a species of large carpenter ants, a favorite of the Mahale ant eaters. It is not difficult to see how investigative behaviors such as this could lead to the development of a new tool-using pattern in the community. Infants (particularly firstborn offspring, who have no siblings and must amuse themselves as best they can while their mothers feed) frequently play with ants, watching them, flicking them as they crawl up and down the trunk of a tree, and often poking them with tiny twigs.

Another observation is relevant here: the longest bout recorded when chimpanzees fed on termites *without* the use of tools was when a juvenile, Prof, broke open a termite heap from which the reproductives were emigrating. He and his family fed on the winged insects for over thirty minutes. This is the only example I could find of Gombe chimpanzees breaking open a termite mound. True, it did not involve the use of a stick; but given the propensity of juveniles to wield sticks and their fondness for flying termites, it is not difficult to see how a Mbini-type behavior might have evolved.

At Gombe the only instance of the use of a rock as a hammer was when infant Flint utilized one to repeatedly smash a small object (probably an insect) on the ground. Flint also hit at an insect with a wooden club. Thus the patterns necessary for the development of nut-cracking skills are already present in the Gombe chimpanzees, and the use of hammer stones at some point in the future is not an impossibility.

As we saw in Chapter 10, because females at Gombe spend much longer than males in searching for and feeding on termites (perhaps

Infant Flint clubs an insect, an innovative performance that he was not seen to repeat. (H. van Lawick)

driver ants as well), it follows that they also use tools in the feeding context more frequently than males. As yet, however, there is no evidence that they are more *skillful* in their use of tools. Of interest are the recent findings at Tai, where females not only crack nuts more frequently than males, but show more dexterity in their manipulation of hammer stones (Boesch and Boesch, 1981). This sex difference is particularly significant with relation to the perpetuation of the tool-using cultures of a given community. Females spend long periods using tools; thus their infants—the tool users of the future—have ample time to learn the necessary skills. Moreover, in chimpanzee society it is the female who typically transfers from one community to the next. Not only does she thereby widen the gene pool of the neighboring group, she may also enlarge its cultural repertoire. For if she brings a new tool-using technique with her, she will, at the very least, pass it on to her offspring and thus begin the dissemination of the behavior through the new group. If the chimpanzee studies presently under way, and those about to begin, can be continued into the future, we shall learn more about tool-using traditions throughout the range and eventually perhaps be able to record the appearance and spread of new techniques both within and among communities.

19. Social Awareness

Freud and his infant sister, Fanni.

February 1982 Fifi and her family—Freud, Frodo, and one-year-old Fanni—are resting. Presently Freud sits up, glances at Fifi, then gathers his infant sister into the ventral position and sets off to the north. Fifi stops grooming Frodo and follows, but after taking a few steps sits again; Fanni is already coming back. Freud, after eating, leaves for five minutes in a desultory manner, returns, and sits near his mother. Presently Fifi and Frodo set off to the south; Fanni totters behind. Freud hurries after them, gathers up Fanni, and heads in the opposite direction. Fifi stops. She gazes after her son, then turns and follows. Thirty meters away Fanni struggles down. At once Freud pushes her ahead and with little shoves keeps her moving in his chosen direction. They proceed for a couple of meters, then, as Fanni tries to escape, Freud grabs her ankle, draws her close, and begins to groom her. Fanni becomes quiet and Fifi, who has caught up, stands watching. After a couple of minutes Freud stands to go, facing north—but before he can take a step, Fifi seizes Fanni and pulls gently. Freud resists for a moment, then gives in. Fifi travels rapidly to the south. Freud looks after her, then follows. Later that evening, as the family feeds, they hear excited calling from chimpanzees to the east. Freud at once starts toward the sounds, but Fifi is still feeding. He gathers Fanni and sets off in the direction of the others. Fifi follows. After 70 meters Fanni returns to her mother, but this time Fifi goes with Freud and the family nests with a large party.

Almost every chapter of this book has emphasized the complexity of the fusion-fission society in which the chimpanzee lives and has provided examples of the way in which he must cope with this ever-changing social scene. In the natural habitat heavy demands are placed upon his cognitive abilities; if he cannot meet the challenge, he will fare less well than his more intelligent companions. He must be able to sort out and correctly respond to information from a wide variety of stimuli. His social environment may change at any moment from a peaceful party of two or three individuals to a large and excited gathering, and he must be able to adjust his behavior accordingly.

Let us examine, as an example, the complexity that can surround a seemingly simple communicative act. An adult male sits with an

Dé copulates with Fifi
(P. McGinnis)

erect penis, glances toward a fully swollen female, and shakes a branch. The female responds by approaching him, crouching with her bottom oriented toward him, and copulation takes place. This sequence has been observed thousands of times over the past twenty-five years—certainly often enough for us to label the shaking of the branch, when it occurs together with the glance and the penile erection, as a signal that has the goal of initiating copulation. When we see the male signaling thus, we can—all things being equal—predict the response of the female. Frequently, however, all things are not equal. For example:

(a) The female may ignore, or at least not respond to, the male's signal. She may do so because he is not a preferred sexual partner. Or because of the presence of a higher-ranking male, who might attack her if copulation took place with a subordinate male. Or she may have failed to notice the signal. (If the female continues to ignore his courtship, the male may repeat his signal more vigorously. He may move closer and show increasingly aggressive behaviors, which may culminate in attack. He may finally succeed in attaining his goal. Or he may give up.)

(b) The female, in response to the signal, may squeak, scream, or actively avoid the male rather than approach. She may do this because, as before, she does not want sexual contact with him but is afraid of his aggression should she fail to respond. Or because she is afraid of a higher-ranked male and *his* response. Or (rarely) because she has misinterpreted the signal as an aggressive one. The male will have the same options as in (a) above.

(c) The female may respond appropriately, approaching the male and crouching. But the male may not copulate, perhaps because he is inhibited by the presence of a higher-ranking male. (Or, occasionally, because the *wrong* female has approached and presented, if two are in estrus at the same time.)

(d) A male may court a female on one occasion but not on another. The reason may be a difference (not detectable to the observer) in her sexual attractiveness. Or because a higher-ranking male, present both times, shows possessive behavior on one occasion and not the other. Or because an ally of the male was present on the previous occasion, and/or an ally of the rival is present on this one.

The above examples are by no means exhaustive, but they give some idea of the variables that must be taken into account when interpreting social interactions. After years of experience it becomes easier to make sense of some of the complex, fast-moving behavior sequences (see de Waal, 1982, p. 31); yet there are times when we are humiliatingly inadequate at grasping meanings that to the chimpanzee actors are obvious. On the other hand, we sometimes have the satisfaction of making a correct interpretation in a situation where our chimpanzee subject has quite certainly made a wrong one; when, for example, an adult male brushes a fly from his nose and an adolescent male nervously grins and jumps away, mistaking the movement for a threat.

If an observer is to make valid interpretations, it is vital that he or she know as much as possible about the relative position of the various individuals in the dominance hierarchy, and about their relationships—

friendly or unfriendly, ambivalent or neutral; for complex interactions involving three or more individuals are an integral part of chimpanzee social life. To illustrate: The juvenile female Pooch approaches high-ranking Circe and reaches for one of her bananas. Circe at once hits out at the youngster, whereupon Pooch, screaming very loudly indeed, runs from camp in an easterly direction. Her response to the rather mild threat seems unnecessarily violent. After two minutes the screams give way to waa-barks, which get progressively louder as Pooch retraces her steps. After a few moments she reappears; stopping about 5 meters from Circe, she gives an arm-raise threat along with another waa-bark. Following behind Pooch, his hair slightly bristling, is the old male Huxley (who had left camp shortly before in an easterly direction). Circe, with a mild threat gesture directed toward Pooch and a glance at Huxley, gets up and moves away. Pooch has used Huxley as a "social tool." This little sequence can be understood only because we know of the odd relationship between the juvenile and the old male who served on many occasions as her protector and was seldom far away. In order to behave in an appropriate fashion, it was, of course, necessary that *Circe* also know the facts of the relationship.

To what extent is a monkey or ape able to make complex cognitive assessments of the rank order and relationships among the other members of its social group? Can it predict the effect of its own behavior on other individuals? Can it choose one course of action over another on the basis of such predictions? Until quite recently, as Kummer (1982) points out, ethologists have tended to take a parsimonious view of complex behavior. However, as descriptions of intricate social interactions have accumulated from field and laboratory studies on a variety of primate species, it has become necessary to revise many of the early oversimplistic explanations.

In 1971, discussing complex interactions involving three individuals, Kummer wrote, "Success requires that a monkey know and integrate the status of group members present and their alliances and antagonisms towards him and among each other" (p. 148). Subsequently he and his colleagues (Bachmann and Kummer, 1980) carried out a series of elegantly conceived experiments with captive hamadryas baboons, showing that these primates are not only aware of the relationships among others but are able to use this knowledge to guide their own behavior. Thus a male hamadryas baboon was more likely to respect the bond between a rival male and the rival's female if he had previously watched the two together; when, that is, he had information about their relationship. If he *did* make a bid to take over the female, he was more likely to do so if the preference of the female for her "owner" (as determined by prior choice tests) ranked low. That is, he made a "decision" on whether or not to fight based on the *quality* of their relationship. Moreover, he was more likely to fight if the rival male was familiar to him; in other words, when he was in a position to make a prediction about the outcome of a fight and the effect it might have on their relationship (Kummer, 1982).

Mason (1982, p. 138), describing the social behavior of a group of captive rhesus monkeys, writes that they "not only know their status *vis à vis* other members of the group, but also know the status of other group members in relation to each other, and can use this information to their advantage." He illustrates his statement with the following example. The monkeys are competing for access to water: a subordinate, A, threatens a more dominant individual, B, who is drinking. At the same time, A glances toward a third, C, who is dominant over both A and B. C begins to threaten B, who abandons the water; A quickly approaches and drinks. A has "used his knowledge of the power relations between B and C to displace B."

Hinde (1984, pp. 68–69) is thus able to write, "Laboratory studies have often grossly underestimated the intellectual capacities of non-human primates . . . evidence is accumulating that at least some . . . may engage in more complex manipulations of their social environment than they have hitherto been given credit for."

There is no doubt whatsoever that a chimpanzee is capable of assessing the complexities of the ever-changing social environment and planning his own behavior accordingly. Before substantiating this claim, let me speculate on the manner in which an infant chimpanzee, growing up in the Gombe community, acquires the kind of knowledge he needs for the development of complex social skills.

Acquisition of Social Knowledge

The infant is not born with built-in responses that will dictate his behavior in complex social situations. To be sure, many of the sounds, gestures, and postures with which the chimpanzee expresses himself are genetically coded, but he must learn how and when to use them appropriately (Menzel, 1964). He learns by trial and error, social facilitation, observation and imitation, and practice. This knowledge is not acquired overnight, and he often makes mistakes that in many instances result in reprimand.

The inherent complexities of social interactions surround the chimpanzee infant from the time he is born. Initially, however, he can scarcely be viewed as a separate individual: he is part of a package that includes his mother also. She shapes and cushions his first interactions with other individuals. If he totters up to an adult male who is resting peacefully, his mother watches but permits the contact. If, however, the male shows hair erection and other signs of anxiety or annoyance, she hurries after her infant and removes him. If another infant approaches and plays gently with him, the contact is often tolerated by the mother; but if the other is at all rough, the mother will remove her child, or threaten the playmate, or both. During this period the infant gradually becomes familiar with the signs that proclaim sex, age, individuality, and mood in those around him. His own immediate family—his mother and siblings—he will of course know best of all.

As he gets older, his mother increasingly permits contact with other individuals and he has more opportunity to experiment with behaviors that previously he has only watched. He may sometimes be mildly rebuffed, but his vigilant mother usually removes him from situations where punishment might be too severe. Moreover, other adults are for the most part highly tolerant of small infants. His first serious rejections are often administered by his mother herself, and she is seldom overpunitive. Thus his confidence—and knowledge—grow.

As a result of gradually accumulating experience, along with increasing motor skills and independence, he finds out that a given individual, B, is liable to behave differently toward him when he, A, is close to his mother than when he is farther away from her. He discovers that there are things he cannot do—such as taking food from B—unless his mother is nearby, in which case she will protect him from retaliation by B. And so he learns to keep close to her in potentially unsafe social situations.

He next finds out that if B's ally, C, is present, he may have to be more cautious even if his own mother is there, *unless* she is higher ranking than B and C combined. Having mastered this fact, he may discover that when his mother is with *her* close ally, D (perhaps an adult son or daughter), this coalition may be able to intimidate B + C. He sees that other individuals (such as siblings) may stand in for his mother. He learns *their* relationship with others, and how this too varies relative to the distance from his (and their) mother; her rank, of course, is still crucial. And so, stage by stage, radiating out from the focal relationship with his mother, he learns more and more about his society.

As a result of these *direct* learning experiences, our youngster should eventually be in a position to respond appropriately in a variety of social situations and to predict the probable effect of his own behavior (and that of his allies) on various individuals. But he must also learn from his role as an *observer*, from watching interactions between others which, even though neither he nor his all-important mother is involved, may at any moment affect *him*. If, for example, C attacks D, D may turn and attack A, our chimpanzee pupil, in redirected aggression. If A is able to predict an outcome of this sort, he may be able to avoid such an attack. Moreover, if A can learn, from watching an interaction between C and D, that C is dominant over D, he has obtained some helpful information: he will then know that C will be more useful to him as an ally against D than will D against C (although it may be that A and D, acting together, can intimidate C). Some social events, such as a dominance reversal, can have significant ramifications, and the more A can learn from careful observation of the behavior of others, the greater will be his ability to manipulate his fellows to his own advantage.

In order to acquire and make reasoned use of the kind of information outlined above, the young chimpanzee must have an attention span sufficient to follow long-drawn-out social maneuvers, and he must be able to remember all that he has learned from one occasion to the next. He also needs to know how to integrate information received through different sensory modalities, and how to form concepts. For example, when he hears a pant-hoot in the distance, he needs to relate the sound in some way to the individual who made it and, moreover, have some kind of symbolic representation of the message conveyed. If he has an understanding of the wants, purposes, and emotional attitudes of his companions—a "theory of mind" (Premack and Woodruff, 1978)—he will be able to anticipate their future behavior and better plan his own. Finally, he must be able to combine his isolated pieces of knowledge, drawing on past experience and surveying the present scene, if he is to respond appropriately to each new encounter.

Predicting the Consequences of Behavior

A chimpanzee (like any higher mammal) varies his behavior in accordance with the age-sex class—and the identity—of the individual with whom he is interacting. Quite clearly he is well aware of the nature of his relationship with the other, and whether it is appropriate for him to be assertive or cautious. If a young male were not concerned about the possible consequences of copulating with a female under the eyes of his seniors, he would not bother to lead her behind a tangle of vegetation before doing so; he does not lead a female into the bushes when there is no superior male around. Chimpanzees in some cases are even able to single out individual elements of a behavior sequence that must be inhibited or concealed if they are to avoid trouble from their superiors.

It is equally obvious that a chimpanzee has a sophisticated understanding of the relationships between *other* individuals and the probable outcome of interactions between them. An illustration is the triadic interaction of Circe, Pooch, and Huxley, described earlier. Pooch threatened Circe only after the arrival of Huxley, which suggests that she knew not only that the old male would hasten to her defense if necessary, but also that Circe, ranking lower than Huxley, would be unlikely to harm her when her ally was present. And Circe certainly recognized the supportive nature of the relationship between Pooch and her ally.

When a chimpanzee bystander, watching an interaction between two or more others in which he has no part, shows by his behavior that he can predict the outcome of the event, this is a cognitive achievement of a higher magnitude. The following is an example. In 1975 Freud was often rough during play with other youngsters. When, as frequently happened, he made the younger Prof scream, there was sometimes a series of reprisals that involved all members of both families (Prof's sister threatened Freud, Freud screamed, Fifi threatened Pom, Passion ran to support Pom). Twice the upshot was a fight between Fifi and the higher-ranking Passion (which Fifi lost). After giving birth to a new infant later in the year, Fifi usually tried to avoid such conflicts. On two different occasions I saw her get up and move rapidly away when play between Freud and Prof seemed liable to erupt into aggression. Presumably she knew that trouble was imminent and that she was likely to become involved if she stayed. (Once her prediction of discord was quite correct; the second time Freud followed after his mother.) Another mother with a small infant behaved in exactly the same manner when her adolescent son, Atlas, began threatening Pom, swaggering around her with rocks and sticks. At the Arnhem colony one female, Tepel, solved a similar problem in a different way. When her child began to squabble with a playmate, Tepel glanced nervously at the other mother (who sat nearby). She then went over to the most dominant of the females, who was asleep. Tepel poked her a few times, then gestured at the squabbling children. The dominant female took a few steps toward them and gave a mild threat, which restored peace (de Waal, 1982).

Even more sophisticated was the behavior sometimes seen in the Arnhem colony during times of instability in the male dominance hierarchy. After aggressive conflicts, levels of arousal in the colony tended to remain high until the rival males had become reconciled. At such times females were seen to act as mediators. Two unreconciled males, for instance, sat apart, each avoiding the other's gaze. After a while an adult female approached one of them, groomed him briefly, and then presented. After he had inspected her genital area, she began to move slowly toward the second male; the first followed, continuing to sniff at her backside from time to time. The female sat between the two males, both of whom began to groom her. When after a short interval she discreetly withdrew, the two rivals simply continued to groom—each other! *All* the adult females of the colony were observed to act as mediators, although some were more skillful than others. The behavior was certainly purposeful, for the mediator, as she moved toward the second male, kept looking around to be sure the first was following. If he was not, she would sometimes pause and tug at his arm. The Arnhem females were also seen to remove rocks quietly from the hands of males before they could be employed as missiles (de Waal, 1982). Females at Gombe sometimes take rocks or branches from swaggering early-adolescent males before they can hurl them, but the targets would have been the females themselves; at Arnhem the female pacifists were more concerned that *others* should not be hit. Obviously they knew that the result would have been discord among the adult males and a disruption of the peace.

Manipulation and Deception

Some communication sequences are simply statements, such as a greeting when a subordinate, A, concedes the dominant status of a superior, B, who in turn acknowledges A's presence and indicates that all is well between them. Although the "goal" is simply to reaffirm the relevant status and maintain the relationship, the greeting functions also to preserve the social order. At other times the goal of a communicative signal is to change the behavior of the recipient in accordance with the signaler's wants or needs. A dominant individual can emphasize his meaning with commands or threats, and if these fail he may attack. A subordinate, however, must rely on requests or appeals; if these fail he must either give up or use more devious tactics.

In Dominant Individuals

There are many occasions when aggression toward a subordinate is inappropriate—when the dominant chimpanzee, in order to get his way, must resort to persuasion. Good examples can be found in the context of social grooming. Although, as mentioned in Chapter 14, one male occasionally mildly threatened a companion who was slow to reciprocate, such behavior was unusual. By and large, grooming interactions are peaceful in nature. The following is a detailed account of one sixty-minute session involving the adult female Pom and her two immature brothers. Pom used five different techniques obviously intended to deflect Prof's grooming activity from their small brother, Pax, to herself. Four of these were successful. Twice she reached out and touched his face; once he responded and turned to groom mutually, but the second time he ignored the signal. Pom then put her hand under his chin—he was bent diligently over Pax—and forced his face up until their eyes met; he then groomed her. Once she seized a handful of leaves, moved 3 meters away, and began to groom them frantically. Prof and Pax both followed and stared. Pom then dropped the leaves and groomed Prof, who immediately reciprocated. Twice Pom, after looking at Prof and scratching vigorously to no avail, suddenly began to groom him with frenzied movements and loud clacking of her teeth. Both times Prof turned, startled, to watch what she did, but he did not then groom her. (On other occasions I have seen this ploy work.) During the same session, Pax once tried to interrupt, by presenting his back, requesting grooming when Prof actually *was* working on his sister. Pom instantly moved, interposing herself between the two brothers, and Prof continued to groom her. On two occasions when there was a lull in grooming activity, Pom presented her back to Prof; both times he at once moved in front of her and they groomed mutually.

The act of grooming itself, as described in Chapter 14, may be used by higher-ranking individuals in peaceful attempts to persuade subordinates to comply with their wishes—as when a male is trying to lead a female off on a consortship, or a mother, pestered by her begging infant, is trying to feed in peace.

During the sometimes traumatic period of weaning, mothers quite often try to divert their offspring's attention from suckling by initiating bouts of either grooming or play. Three mothers (Pallas, Little Bee, and Patti) initiated extremely vigorous play at such times, laughing loudly even while their infants were whimpering. Pallas (but not the others) *always* succeeded in eliciting a play response, thus putting a temporary stop to her daughter's demand to nurse. It was Pallas, too, who so often began a play session on occasions when her daughter refused to follow—when, for example, she was enjoying a game with another youngster. After a few moments of energetic play Pallas, with a final playful push, moved off quickly; her infant usually followed. Other mothers may make a few tickling movements in similar situations, then follow up by dragging their infants along the ground behind them. The youngsters apparently consider this great fun, for they may laugh as they bump along and usually scamper off, ahead of their mothers, when they are released. A mother may also start to groom or play with a youngster (usually her own) who is trying to touch her small infant.

In Subordinates

Many times the presence of a higher-ranking individual, B, interferes with a goal-directed activity of A. In such circumstances A has two options: he (or she) may refrain from performing the desired activity, at least while B is there; or he may try to attain his goal despite B's presence. Let us look at the tactics A can employ to "get his own way" in spite of B. There are two rather different kinds of situation involved: in the first, A's goal is some kind of interaction with B (A wants something from B, or wants him to do something or stop doing something); in the second, A's goal has nothing whatever to do with B, but B's presence interferes with the execution of the goal (which might be copulating with a female who is near B). The means by which A can attain his goal are often similar in these two contexts—which I shall refer to, for convenience, as *A with B* and *A despite B*.

When Freud repeatedly touches his newborn sibling, his mother, Fifi, eventually starts to play with him, bite-tickling his hand. Freud laughs and is distracted from the baby.

(1) *A can enlist the help of a third individual.* A can sometimes achieve a goal such as making B go away (an *A with B* situation) by enlisting the help of an ally, C. C's rank need not be higher than that of B, provided the combined forces of A and C can intimidate B. Incidents of this sort are seen most often when A's goal appears to be an attack on B, which by himself A does not dare to carry out. Miff, for instance, upon encountering Passion (who shortly before had tried to snatch Miff's infant), ran off screaming—only to return with two adult males who intimidated Passion on Miff's behalf.

An ally may also be enlisted in *A despite B* situations. Once, for example, Sherry was trying to eat a large piece of meat but was unable to do so because the higher-ranking Satan repeatedly displayed around him, tried to seize the prey, and once attacked him. Finally Sherry ran over to alpha male Figan, kissed and embraced him, then sat close beside him and ate in peace. (Figan himself had meat as well.) Another example involves seven-year-old Frodo, who wanted to play with Sparrow's infant. Every time he attempted to do so, Sparrow threatened him. Finally Frodo persuaded the infant to follow him to *his* mother, Fifi, who ranked higher than Sparrow. Frodo was then able to play with the infant without interference. Sometimes a youngster will try to *drag* an infant close to his own mother. If he succeeds, and provided she ranks higher than the mother of his playmate, he can play for as long as the infant stays with him.

In the above examples the ally (active or passive), by enabling A to achieve a goal otherwise not possible, may be regarded as a *social tool*, as first proposed by Chance (1961) and subsequently elaborated upon by Kummer (1982). Of course, A's strategy may fail (because B is too strong for A and his ally, C; or because C refuses to cooperate). So too a twig picked for termite fishing may break. But it still can be called a tool even though it never serves the purpose for which it was intended.

Another example of social tool use (in *A with B* situations) occurs when an older offspring, usually a male, tries to persuade his mother

Flo watches as Fifi makes off with infant sibling Flint. (H. van Lawick)

Figan sets off with his infant sibling, Flint, in an attempt to persuade his mother, Flo, to follow. But Flint slides down and returns to Flo.

Faben persuades Flint to cling ventrally as he too tries to lead Flo from camp.

to follow him while she is still feeding or resting. He may, when normal leading behavior has failed, carry off his infant sibling, as described in the anecdote at the start of this chapter. If the infant is very young, the mother will almost always get up and follow. Of course, the infant may escape and run back to her, or she may retrieve him and return to what she was doing. Often, though, this maneuver is highly successful. It was seen most often in the Flo family: both Figan and (occasionally) Faben used infant Flint in this way; later Freud used infant brother Frodo, and both Freud and Frodo made use of infant sister Fanni.

A different type of social tool use occurs when one chimpanzee utilizes another to help him attain a *nonsocial* goal, which he could not readily accomplish by himself. The earliest accounts of this behavior were those of Köhler (1925) and Crawford (1937). Köhler describes how one chimpanzee would try to pull another one beneath fruit suspended from the ceiling and endeavor to obtain the lure by climbing onto his companion. Obviously not a very effective strategy, it often ended with "a struggling group of chimpanzees, who all gripped each other, and lifted their feet to climb, but none of whom wanted to be footstool" (p. 50). In an attempt to investigate cooperation, Crawford designed an experiment in which two young chimpanzees, if they pulled together on a rope, could draw a tray of food into their cage. Because the dominant member of the pair almost always took the food, it was not easy for him to persuade his partner to help! Yerkes (1943, p. 190) describes the behavior: "The observer may see both solicitational and directive gestures . . . One subject may take up its rope and, instead of beginning to pull, look toward the other enquiringly, reach out and touch it as if to attract attention to the work in hand and, if these reminders do not suffice, push or turn the animal towards its rope and await the moment for a joint pull."

The most commonly observed examples of this type of behavior at Gombe are the manipulations of mothers by their infants. The goal of the infant may be any task that he cannot easily accomplish by himself—reaching the top of a thick trunk, crossing a stream or moving from one tree to another, obtaining food from a hard shell, and so on. It is when the infant reaches two or three years of age and can, if he tries hard, accomplish these things by himself that he begins to make determined efforts to manipulate his mother—by reaching toward her, whimpering, screaming, begging, and so on. If he is successful, he makes life easier for himself.

Winkle's two offspring were both very skillful in using tactics of this sort. Wilkie's goal was usually to persuade his mother to accompany him when he wanted to join other chimpanzees. The following took place when he was being weaned at age five years. As Winkle set off in her chosen direction, Wilkie declined to follow. Instead he sat, whimpering softly. After proceeding some 10 meters Winkle stopped, looked back, then returned to her son, gathered him into the ventral position, and moved off again (heading, almost certainly, to a large fruiting fig). After a few moments, however, Wilkie detached himself and set off in a different direction. Winkle stopped, gazed after the infant, scratched, and gave a few soft grunts. Wilkie, when he saw that she was not following, also stopped and again began to whimper—louder this time. After half a minute Winkle again approached him, once more took him ventral, and now proceeded in *his* chosen direction. After three minutes she turned and again headed for the fig tree. Almost at once Wilkie detached himself and set off the other way, whimpering as he traveled, and repeatedly looking around at his mother. This time she gave in and followed him. They joined a small party where Wilkie played boisterously with other youngsters. After one and a half hours, during which Winkle fed intermittently on leaves or sat grooming herself, she moved away from the others. Wilkie followed for some 15 meters but then turned back to the group, whimpering. Winkle gazed after him, then (probably with a sigh) trailed behind him, back to the chimpanzees they had left. That night they all nested together.

Wunda, as an infant, was seen to manipulate her mother most often in the termite-fishing context. She begged (usually successfully) for almost every tool that Winkle began to use; she took over (whimpering) each tunnel that Winkle began to work. At this time she was six years old. The following year she frequently refused to follow when Winkle gave up fishing and wanted to move on. Wunda continued to work, whimpering and repeatedly gazing toward where her mother waited, for periods of up to an hour.

It is not uncommon for a female to take over, from her adolescent or adult daughter, a productive tunnel that has been opened up only after patient searching, or a tool that has been carefully selected and newly prepared for use. Gremlin in particular suffered from frequent depredations of this sort. Another incident in which a mother used her daughter as a tool occurred when I was visiting the Zurich zoo. The chimpanzee group there, which includes an old female Lulu and her two offspring, had been given wooden logs, with holes drilled through and stuffed with raisins. The innovation was a recent one and Lulu, who almost immediately appropriated her twelve-year-old daughter's log, seemed to have difficulty balancing on her perch: she had a log in one hand, another in one foot, and a tool (a long twig) in the other hand. As she worked on her own log, Sita, her daughter, sat very close, watching and begging. To my astonishment, after five minutes Lulu suddenly handed Sita her tool. Soon the reason was clear: with one hand freed, Lulu exchanged the position of the logs, putting the almost-finished one in her foot and the untouched one (her daughter's) ready to work in her hand. Then, with complete assurance, she reached out for her twig—which Sita handed over without a murmur.

(2) *A can divert B's attention.* This occurs in *A with B* situations and is most often observed when A wants something from B. For example, if A wants to examine B's infant or a piece of B's food, he or she will approach cautiously, sit very close, and begin to groom B. This tactic, as we have seen, will calm B and will also serve to divert attention from A's real goal. A number of nonhuman primate species use this technique, often successfully. A chimpanzee infant who is being weaned may, when he wants to suckle, first groom the mother around her nipples for a few minutes, then try to nurse (Clark, 1977). And a juvenile may groom his way closer and closer to a new infant sibling and finally touch the baby once the mother is relaxed. I have seen a youngster touch her infant sibling with one hand while grooming her mother with the other. Once, as Goliath (then alpha male) sat eating a piece of meat, David Greybeard began to groom him intently; after a few minutes, as he continued to groom with one hand, the other cautiously moved closer and closer to a fallen scrap of meat. Once he had secured his prize, David stopped grooming and moved off to eat.

(3) *A can throw a tantrum.* The temper tantrum seems to be an uncontrolled, uninhibited, and highly emotional response to frustration. It appears as a last resort, when A has failed to get his way. It is observed most frequently in youngsters who are going through the peak of weaning, after they have begged, whimpered, and cajoled their mothers for an opportunity to suckle, but to no avail. In some ways it seems absurd to think that such a spontaneous outbreak could be a deliberate strategy for achieving a goal. Yet Yerkes (1943, p. 30) wrote, "I have seen a youngster, in the midst of a tantrum, glance furtively at its mother . . . as if to discover whether its action was attracting attention." And de Waal (1982, p. 108) says, "It is surprising (and suspicious) how abruptly children snap out of their tantrums if their mothers give in."

At Gombe a mother almost always does give in. The tantrum seems to make her tense and even nervous. She hastens to embrace the screaming child—who, of course, begins instantly to suckle. As he does so, the mother often gives a soft bark of threat. Her behavior, roughly translated, might read "Anything for peace!"

Tantrums, primarily observed in infants, occur in older individuals too, particularly in adolescent males when their plea for reassurance contact after an attack is ignored. We have seen that Mr. Worzle, after begging persistently and unsuccessfully for a share of meat, threw a tantrum so violent that he almost fell out of his tree. Goliath, the higher-ranking possessor of the carcass, immediately tore the prey apart and gave half to his screaming companion. De Waal (1982) describes the extremely violent tantrums occasionally thrown by two of the Arnhem adult males when they were defeated in dominance conflicts. He writes of one of the males, then thirty years old, "We found it difficult to take Yeroen's despair completely seriously: it seemed so exaggerated and *posed*" (p. 108; italics added). After "regaining some of his self-composure," Yeroen would run to the females, his potential allies, and beseech their support. If they refused, he would once again throw a violent tantrum. "It was as if Yeroen was trying to arouse pity and mobilize his sympathisers [against his rival]" (p. 107).

(4) *A can attain his goal without B's knowledge.* This can be achieved in several ways: by diverting B's attention, as described above; by operating carefully and quietly while B is sleeping or looking in another direction; or by hiding. We have already seen how an adolescent or low-ranking male, whose goal is to copulate with a female, will move behind a rock or tangle of vegetation, walking quietly and glancing back toward the female. He also glances at least once toward B, the high-ranking male from whom he is attempting to hide his purpose. If the female is willing to take part in this clandestine operation, the strategy is usually successful.

A can also withhold relevant information. The experiments of Premack and Woodruff (1978), designed to demonstrate systematically whether chimpanzees are capable of deception, were described in Chapter 2. Four young chimpanzees learned *not* to reveal the whereabouts of food (by gestures, glances, and so on) when they knew the experimenter would eat the reward, but eagerly indicated its position to a more generous human. The deception most often observed at Gombe is concealment of interest. There was an occasion when Figan, as an early adolescent, had somehow acquired a dead adult colobus monkey. He was feeding on the flesh, occasionally allowing his infant sister to take small pieces. Below him in the tree was his mother, Flo. I was surprised to see that she appeared to have no interest in obtaining a share since she, of all the females, was usually avid for meat. On this occasion, however, she was not even watching. I had been observing for about five minutes when Flo, very casually, moved a little closer to Figan. Still she paid no attention to the meat. She stopped, just beyond arm's reach, and again sat and idly groomed herself. Seven minutes later she once more began to climb and all at once made a lightning grab at the monkey's dangling tail. Figan, however, had clearly been anticipating just such a move, and he leaped even more quickly than she did. Flo repeated her maneuver on and off for the next hour but without success. As an adult, Figan himself became a past master at pretending disinterest and then successfully seizing meat from others.

When a juvenile wants to play with a new infant, in addition to the distraction technique described he (or she) may pretend disinterest—reaching out to the baby while looking elsewhere. Melissa was unusually protective when she gave birth to twins, and it was common to see her juvenile daughter, studiously looking away from the babies, reach back behind herself to touch them. Sometimes she would lie on her back, gazing into the sky, and touch the prohibited infants with her toes.

Gremlin, lying on her back, gently touches one of the "forbidden" twins with her toes.

Once as I was watching four females and their youngsters feed in a tree, I inadvertently disturbed a francolin (a grouselike bird), which flew up with loud characteristic calls. A moment later the most subordinate of the females, Little Bee, climbed slowly down. She stood close to me and I could see, from the direction of her gaze and her eye movements, that she was *visually* searching the ground where the bird had been. After thirty-five seconds she suddenly moved rapidly forward and, without hesitation, reached out and picked up two eggs. Almost before she had put them into her mouth, the other three females had rushed down and gathered around her, peering at her mouth and feeling about in the area of the nest. If Little Bee had searched manually instead of standing and using her eyes, it is quite possible that one of the higher-ranking females would have found the eggs before she did.

In 1965 we designed remote control boxes for banana feeding. To open a box, one had to unscrew a nut and bolt and release a handle. This caused wire, threaded through a buried pipe, to slacken and the metal lid of the box then fell open. Two young chimpanzees learned to unscrew the nut and bolt. One of them, Evered, released handle after handle, each time running to the box with loud food grunts. The trouble was that the adult males all sat around waiting for Evered to feed them, and he very seldom reaped a reward for his skillful manipulations. Figan, on the other hand, after performing a similar service for his seniors on a few occasions, quickly learned to modify his behavior when adult males were in camp. Very nonchalantly, and with an apparent lack of purpose, he wandered to a handle. There he sat and performed the entire unscrewing operation with one hand, never so much as glancing at what he was doing. Thereafter he simply sat, gazing anywhere but at the box, one hand or one foot resting on the handle so that the lid remained shut. There he outwaited the big males, sometimes for as long as thirty minutes. Only when the last one had gone did he release the handle and (silently) run to claim his well-earned reward.

One year we occasionally hid bananas up in the trees, so that while the big males fed from boxes, the females and youngsters had a chance to get some too. One day Figan (aged then about ten years) spied a leftover banana some time after the feeding orgy had ended. It was directly above one of the high-ranking males, Goliath, who sat peacefully grooming. After glancing at Goliath, Figan moved away and for the next half hour remained where he could not see the banana. The moment Goliath left, Figan quietly returned and took his prize.

A very similar observation was made by de Waal (1982). The Arnhem chimpanzees were shut inside their building while a box full of grapefruit was half buried in the sand of the enclosure, with small yellow patches left visible. The chimpanzees had seen the box full; they saw it empty as de Waal returned, and when they were released they rushed off and searched diligently for the fruit. Some of them, including a young male, Dandy, went right by the spot but none stopped. That afternoon, when the group had settled to rest, Dandy got up quietly, made a beeline for the grapefruit, dug them up, and ate them.

Sometimes a chimpanzee learns to conceal a particular element of a behavior sequence that might give him away. When Figan was about nine years old, there was no formalized method of handing out bananas at Gombe; often the big males got most of the fruits. One day Figan, who had had no bananas at all, waited behind when the rest of the group moved off. When they were out of sight, we handed him some fruits. Greatly excited, he gave loud food barks. The whole group immediately raced back and Figan lost all but one of his bananas. The following day he again waited and got bananas. This time, although we could hear faint, choking sounds in his throat, he remained virtually silent and ate his allotment undisturbed. Never again did Figan call out in this kind of situation.

During clandestine copulations females typically contribute to the deception by inhibiting the copulation scream or squeal. De Waal (1982) describes one adolescent female who used to scream particularly loudly. When she did so during surreptitious matings, the alpha male always charged up and interrupted the forbidden activity. By the time she was almost adult she had learned to stifle the call during clandestine copulations, although she continued to vocalize when mated by the alpha himself.

As described earlier, the Gombe chimpanzees, when traveling near the periphery of the home range, move very quietly and suppress vocal sounds. In fact, the males may even try to suppress calls made by others. On two separate occasions the adolescent Goblin vocalized during patrols. Once, as already mentioned, he was hit; the second time, embraced. During another period of noiseless travel an infant got loud hiccoughs; her mother became extremely agitated, repeatedly embracing the child until eventually the sounds ceased. A human observer who is too noisy on such occasions may be threatened.

These observations are very interesting, for they suggest that the chimpanzees have some concept of the need for silence at such times. Similar evidence comes from the Arnhem colony: once, when the screams of her infant seemed likely to bring male retribution, Tepel rushed to him and clamped a hand over his mouth (de Waal, 1982).

Two additional types of signal concealment were observed among the chimpanzees of the Arnhem colony. As a young male surreptitiously courted a female, a senior male suddenly came around the corner and the subordinate quickly covered his penis with his hands. This gesture was seen several times. It certainly seems that these young males were aware that an erect penis would give them away (de Waal, 1982). Even more remarkable was the signal concealment first observed in the old male Yeroen when he was being challenged by Luit: he seemed to try to hide signs of uncertainty from his rival. After a conflict he invariably moved away from Luit with "an expressionless face" and only began to grin and yelp when he was some distance away and had his back turned. Luit, when he in his turn was being challenged by Nikkie, showed similar behavior. After one of their conflicts Luit sat on the ground below Nikkie, facing away from him. When Nikkie began to pant-hoot, Luit, clearly tense and nervous, grinned widely but "immediately put his hand to his mouth and pressed his lips together" (p. 133). Twice more Luit hid his fear in this way. Then, after Luit had moved away and was out of earshot, Nikkie began to grin and very softly squeak. In these incidents the withholding of information enabled each of the males to appear less fearful than he really was. Perhaps the same explanation applies to Goliath: when he was striving to retain his top rank against Mike's challenge, he occasionally turned his back on his rival and sat facing away as Mike, on arrival, started a tumultuous charge toward him. And maybe it was for the same reason that Goblin, during his many fights with males much heavier than himself, usually battled in silence, at times when his opponents were screaming loudly. Even when he lost he did not often scream, but displayed away without calling.

(5) *A can provide B with false information.* The eldest of the four chimpanzees in Premack and Woodruff's 1978 experiments, for instance, learned to give wrong information to the "selfish" investigator (the one who ate all the food once the baited container was pointed out to him).

Menzel (1974) describes how a subordinate female, Belle, who had been shown the whereabouts of hidden food, tried in various and ever more sophisticated ways to withhold this information from the dominant male, Rock (because if she led him to the place, he invariably took all the food himself). Rock quickly learned to see through her various subterfuges. If she sat on the food, he learned to search beneath her. When she began sitting halfway toward the food, he learned to follow the direction of her travel until he found the right place. He even learned to go in the opposite direction when she tried to lead him *away* from the food. And since she would sometimes wait until he was not looking, Rock learned to feign disinterest, but was ready to race after her once she began to head for the prize. Sometimes a single piece of food was hidden in a different place from the large pile. Belle would lead Rock to this piece and, while he ate it, run to the pile. When Rock learned to ignore this decoy and continued to keep an eagle eye on Belle, she threw temper tantrums.

The events described took place over a period of several months. Initially, Belle was merely concealing information; only when Rock began consistently to outmaneuver her did she begin to *lie*. Rock learned to respond in kind, even wandering off in another direction until, just as she was about to take the food, he turned and raced over to prevent her.

Another example of lying comes from Roger Fouts (cited in Davis, 1978). The young chimpanzee Bruno began to play with a hose; the bigger Booee soon took it from him. Suddenly Bruno went to the door of the hut and gave a loud waa-bark, at which Booee dropped the hose and rushed outside. Bruno quickly resumed playing with his stolen toy. The entire sequence was repeated three times. Fouts tells me that previously he had distracted Bruno in this way; the chimpanzee had, it seems, penetrated the deception and subsequently practiced the technique for his own ends.

Fouts once crossed swords with another chimpanzee liar. On this occasion it was the Temerlins' Lucy, who had regular lessons in ASL. The incident occurred after Lucy, when no one was looking, defecated in the middle of the living room. Fouts arrived shortly thereafter. Temerlin (1975, p. 122) gives this verbatim report of their signed conversation:

Roger What's that?
Lucy Lucy not know.
Roger You do know. What's that?
Lucy Dirty, dirty.
Roger Whose dirty, dirty?
Lucy Sue's.
Roger It's not Sue's. Whose is it?
Lucy Roger's.
Roger No! It's not Roger's. Whose is it?
Lucy Lucy dirty, dirty. Sorry Lucy.

Another example of deception is provided by de Waal (1982). During a fight with the alpha male, the elderly male Yeroen hurt his hand. Throughout the following week he limped very badly—but *only when in the alpha's sight*. At other times he walked quite normally. De Waal speculates that perhaps in the past the alpha had been easier on him when he was limping, and he had learned from this experience.*

The Gombe chimpanzees also can lie. I have already described how Figan, even as an adolescent, was able to initiate group travel by setting off with a brisk and purposeful walk. He did this when other

* I have to mention here that one of my dogs, after hurting his paw badly, received a good deal of attention and sympathy. The paw was still sore when I had to leave. I returned two weeks later and was shocked to see that he was still keeping his paw off the ground. It was only after I had knelt to examine the proffered foot, with proper sounds of concern, that I saw the amazed faces of my family; his wound had healed over a week before, and he had been walking quite normally until I came into the house.

chimpanzees had been loitering at camp for long periods during which he himself had had no bananas. Five to ten minutes later he returned—alone. Of course he got some fruit. The first time, I thought the sequence was a coincidence: the second time, I began to wonder; the third time, it was evident that his maneuver was intentional. On one occasion Figan found on his return that a high-ranking male had arrived during his absence. Then, like Belle, he threw a tantrum and, still screaming, ran after the group that he had led away so jauntily a few minutes before.

Figan's purposeful departure indicated to the others that he was headed for a good food source and they followed him toward it. In fact, he almost certainly *was* headed for one. The chimpanzees have a far too sophisticated knowledge of the whereabouts of ripe food to follow a youngster in an unprofitable direction. Figan's "lie" was to travel in the direction opposite to *his* food goal. On the first occasion he had, in fact, probably intended to feed elsewhere; then, when the others followed, he had seized the opportunity to return—and, subsequently, profited from that experience.

It is possible that infants may learn to make use of their mothers' protective responses. My first observation of this behavior occurred as I followed Fifi and her four-year-old son, Frodo, who was being weaned. After he had twice tried to climb onto his mother's back and twice been rejected, he followed slowly with soft hoo-whimpers. Suddenly he stopped, stared at the side of the trail, and uttered loud and urgent-sounding screams, as though suddenly terrified. Fifi, galvanized into instant action, rushed back and with a wide grin of fear gathered up her child and set off—carrying him. I was unable to see what had caused his fear response. Three days later, as I followed the same mother-infant pair, the entire sequence was repeated. And, a year later, I saw the same behavior in a different infant, Kristal, who was also being weaned.

Were these infants lying? Or was their fear real; were they suddenly *frightened* of maternal rejection? Obviously more observations are necessary, but I am of the opinion that they were intentionally manipulating their mothers. A similar behavior was shown by Fifi herself when, on her arrival in camp one day, she encountered the aggressive adult male Humphrey. As she appeared, he started toward her, hair bristling. Instead of rushing forward, pant-grunting and crouching, or running off (either of which is a common response), Fifi appeared to ignore Humphrey altogether. She moved past him, climbed into a low tree, stared intently into the undergrowth, and uttered two loud waa-barks. Humphrey (as well as Fifi's two offspring) hastened to join her. As her companions all gazed at the place she had indicated, Fifi herself turned and calmly began to groom Humphrey. I stared through binoculars—and after a while went to investigate. I found nothing. It is, of course, possible that she had seen a snake . . .

The foregoing has done only partial justice to the skill of the Gombe chimpanzee in interpreting and manipulating the behavior of others. Not until the more subtle aspects of communication are recorded reliably and analyzed fully, not until we have a better appreciation of the quality of information the chimpanzee is able to convey with his calls, will we be in a position to map the full extent of his abilities in this sphere. Observations of captive groups such as the Arnhem colony have assuredly made meaningful contributions to our understanding of the complexities of chimpanzee social awareness. No one has revealed these complexities in more depth and with greater understanding than Frans de Waal. The careful and patient observations he and his co-workers have made reveal striking examples of social intelligence. Some of these performances, such as female mediation and the hasty hiding of the penis by a junior male surprised during illicit courtship, have never been seen at Gombe. We must ask, therefore, whether the Arnhem chimpanzees do, in fact, interact in a more complex way than those at Gombe. Or is it, as de Waal himself suggests, simply that he and his colleagues have been able to observe the chimpanzees in much finer detail than is possible in the wild, and as a result have uncovered subtleties of behavior overlooked at Gombe?

Certainly the conditions for observation are superior when the subjects are in an enclosure. They can be watched continuously, day after day, with little to obstruct the view; events that are crucial to the understanding of changing relationships are likely to be recorded; and subtle changes in these relationships can be documented on a daily basis. And the more complete the history of complex events, the more complex the behavior of the various actors is likely to appear. Nevertheless, I suspect that the difference between the Arnhem and Gombe chimpanzees is real, and that it is largely the result of a captive versus a wild group. The large field enclosure is, in a sense, a halfway house between the laboratory and the field. In the controlled conditions of the laboratory, it is the *investigator* who, by skillful manipulation of testing procedures, can encourage the chimpanzees to perform at high levels and thereby obtain evidence of cognitive sophistication. In the field enclosure, it is the *conditions of captivity* that, by their pressure on the existing repertoire of social skills, encourage the chimpanzees to develop increasingly elaborate performances.

There are three ways in which the captivity of the Arnhem chimpanzees may have affected their social interactions. First, confined primates have considerably more time available for social pursuits. In the African forest, chimpanzees cannot afford to devote the whole of their considerable intellectual abilities to competing with rivals or improving relations with friends. They must expend a good deal of energy, particularly during the dry season, in finding and processing food. Life in the wild carries with it at all times an element of uncertainty, often of excitement. At any moment the chimpanzees may encounter a party of hostile, potentially dangerous neighboring males. There may be a strenuous hunt, a stimulating encounter with baboons, and so on. In other words, much of the chimpanzees' mental skill is occupied with day-to-day living. In marked contrast, captive chimpanzees are provided with food and shelter and their ills are attended to. They do not have to seek food and spend long hours in its preparation, nor sit for hours hunched and shivering in the rain. They have no complex decisions to make about the direction in which to travel

Males of the group at Lion Country Safaris in Florida. As the alpha (*right*) starts an aggressive charging display, the second-ranking male shows a bipedal swagger and play face. By trying to initiate play in this situation, he is often able to distract his superior and thus prevent disruption of the peace. (C. Gale)

or the company to keep. By the same token, they miss out on the excitement and tension of patrols and hunting and the like. There is no element of danger to spice their lives. And so they can devote themselves almost entirely to their position in the hierarchy and their relations with others. This freedom from survival pressures may lead to novel social behaviors in other primates (Kummer and Goodall, 1985). Kummer and Kurt (1965) recorded nine communicative signals in a troop of zoo hamadryas baboons—signals that had not been observed in any wild group. In addition, the protected threat sequence was more sophisticated and effective in the zoo group than the version observed in the wild.

Furthermore, the conditions of captivity result in *stability of the social environment*. At Gombe individuals are relatively free to associate with whomever they choose, and the number and identity of their companions change continually. Captive chimpanzees have no such freedom of choice, and they do not have to cope with the complexities inherent in the comings and goings of their wild counterparts.

Since all members of their group are always present, they are far more familiar with one another than are most adults at Gombe and thus are better able to predict one another's behavior. Allies are continually at hand, and social strategies that rely on cooperative support can be developed into polished performances that are not readily attainable in the natural habitat.

Finally, the conditions of captivity may actually create a *need* for more sophistication in the social sphere. At Gombe, when levels of aggression become too high, a chimpanzee is able to leave the group and move off with a companion of his choice or by himself. We can imagine the male who has lost a dominance conflict inwardly fuming as he stalks through the forest, and we know he may vent his frustration on a lower-ranking individual unfortunate enough to cross his path. But with the passing of time his tension will lessen. Admittedly, he will not forget his grudge, and when he next meets his rival there may be a renewed outbreak of aggression; but this time he may have an ally with him, or the rival may *not* have his ally. The female at Gombe is even more likely to slip off by herself, or with her family, and thus avoid the tempestuous conflicts of rival males.

The Arnhem chimpanzees have no such freedom. They are captives not only in the literal sense, but figuratively, in the web of their society. They can move to the other side of their enclosure, but they cannot leave the group entirely—alone or with chosen companions. A male cannot lead a female away from his rivals and in a distant place mate with her in peace; he cannot avoid the persistent persecution of a powerful rival. Nor does he have the benefit of the occasional cooperative foray into hostile territory, where fear of a common enemy will create temporary solidarity among community males. Thus for the captive male there is a new need for improved social maneuvering, particularly when this involves the concealment of intention, the maintenance of close bonds with allies, and, above all, reconciliation after conflict. De Waal comments that after disputes the opponents were "drawn to each other like magnets" (p. 40), and tense relations prevailed until some sort of reconciliation was effected. Indeed, the males seldom retired to their sleeping cage for the night until they *had* become reconciled, or at least "declared a truce." There is, after all, evidence that when his tactics fail a male may pay with his life. Nor is it only at Arnhem that captive males have risen to this challenge: the third-ranking male of the Lion Country Safaris group used play as a diversionary tactic against the aggressive displays of the alpha, and I do not doubt that a whole variety of sophisticated innovations of this sort will be revealed as more data accumulate from various groups in captivity.

It seems that the behavior of female chimpanzees may be affected by confinement even more than that of the males. Captive females, unlike those at Gombe, form close and enduring friendships (Köhler, 1925; de Waal, 1982). And, perhaps because they cannot leave when their situation becomes too demanding, they are apt to play a more active role in the ordering of their society. Their alliances with the males often swing the outcome of dominance conflicts in favor of the males of their choice. Their mediation between unreconciled rival males restores peace to the group, a harmony that Gombe females can find by moving off and leaving the males to their own devices.

It is fascinating to contemplate the changes that would occur if the Gombe chimpanzees were deprived of their freedom—or if the Arnhem chimpanzees were given theirs.

Conclusion

Fifi fishing for termites.
(H. van Lawick)

In Chapter 2 I reviewed the perceptual and intellectual world of the chimpanzee as revealed by careful testing in the laboratory and other conditions of captivity. Subsequent chapters have described, in some detail, how one community of chimpanzees copes with day-to-day life in the wild. Let us now return to the issue raised at the close of Chapter 2: to what extent do chimpanzees in the natural habitat *need* the higher cognitive abilities that we know they have?

Daily wanderings from one food source to another, excursions to peripheral areas, and establishment of consort ranges clearly demand of the chimpanzees some sophisticated mechanisms for displacement in space and time. Spatial memory is crucial. The chimpanzees' mental map is extensive and they can easily relocate food patches within the 8 to 24 square kilometers of the Gombe home range. Their spatial memory is rich in detail; they know not only the position of major foods—great stands of richly fruiting trees, for example—but also the whereabouts of solitary trees and individual termite mounds. They remember, at least over several weeks, the exact places where specific major events, such as intercommunity conflicts, took place. In the Tai Forest the chimpanzees not only remember the location of hammer stones and nut-bearing trees, but seem able to choose stones, when the tree is quite out of sight, in such a way as to keep transport distances to a minimum (Boesch and Boesch, 1984). In other parts of Africa, such as the Senegal, where the home range may run to several hundred square kilometers, even heavier demands must be made on their mental abilities in this sphere.

Each day brings with it the need for decision making. To join A, or B, or remain alone? To move east to the fig tree, or south to the palm-nut grove? Making decisions of this sort involves cognitive assessments of a whole variety of factors, such as the whereabouts of neighboring males (encountered perhaps one or two days previously) or of a higher-ranking rival (determined by recognition of his voice), the abundance of figs over palm nuts, and/or the degree of ripeness of one versus the other (calculated from information remembered from the day before).

That the chimpanzees do indeed make plans for the immediate future is obvious. Thus a chimpanzee may pick a blade of grass for use as a tool at a termite heap that is some distance away and quite out of sight. In one of my favorite anecdotes Goliath sets off with his last two bananas in one hand, then suddenly retraces his steps to a pile of the fruit and exchanges his slightly overripe bananas for two firmer ones, which we know he prefers, and takes those with him instead. Is there a need for the chimpanzee to plan still further ahead? Is it possible that the male who initiates a consortship with a female some time before her periovulatory period *knows* that she will sooner or later develop a full swelling, and plans accordingly? When Satan followed a female in estrus until she nested, then slept close beside her, was he planning the early-morning getaway next day? Or did he simply take advantage, each time, of the favorable circumstances in which he found himself in the morning? When Little Bee felt those figs in the darkening twilight, did she *plan* to rise at dawn and feast before competitors arrived? We do not know, but surely an individual who could make such plans would have an advantage over one who could not.

It is very clear that the chimpanzees need their premathematical skills: they must make accurate and sometimes rapid assessments of which branch has *more* rather than *fewer* ripe fruits, sometimes even as they race up the tree in order to claim the best site ahead of their competitors. The more precise the reckoning of an individual, the more bountiful will be his reward. The old female fishing for termites quickly notices when her daughter's success rate exceeds her own and often firmly insists that they change places.

The termite fisher also demonstrates the need for well-developed conceptual ability. When, for example, a passage yields but little to probing with the only material that is within reach, the chimpanzee may travel 5 meters or so in search of a more suitable implement. And it is evident that a certain type of tool is in mind: it may be vine, or grass, or bark, but whichever it is, a number of different pieces of that particular material (and no other) may be visually or manually inspected before a suitable one is selected. This process requires an ability to classify the various properties of the materials involved—straight versus bent, firm versus flexible, and so on. Exactly how do wild chimpanzees classify the objects and events around them? Presumably there are many simple categories such as *edible/inedible, leaflike/fruitlike, friend/foe, safe/dangerous, likely-to-attack/peaceful,* and so on. Some of their categories are probably, if anything, too rigid—objects classified as edible are *only* those that have been eaten from infancy—those that are part of the tradition of the particular community into which the chimpanzees are born. This is in marked contrast to their flexibility in social behaviors—but, of course, experimenting with new foods is more risky.

The capacity of chimpanzees for symbolic thought probably emerged as a result of the need for concept formation, for perceiving the relations between things. We can trace a pathway along which representations of, for example, a fig become progressively more distant from the fig itself. The value of a *fig* to a chimpanzee lies in eating it. It is important that he quickly learn to recognize, as *fig*, the fruit above his head in a tree (which he has already learned to know through taste). He also needs to learn that a certain characteristic odor is representative of *fig*, even though the fig itself is out of sight. Food calls made by other chimpanzees in the place where he remembers the fig tree to be located may also conjure up a concept of *fig*. Given the chimpanzees' proven learning ability, there does not seem to be any great cognitive leap from these achievements to understanding that some quite new and different stimulus (a symbol) can also be representative of *fig*. Although chimpanzee calls are, for the most part, dictated by emotions, cognitive abilities are sometimes required to interpret them. And the interpretations themselves may be precursors of symbolic thought.

All of the cognitive abilities mentioned above are called into play during social interactions—a sphere in which a chimpanzee, to be successful, needs a well-developed capacity for reasoned thinking. As we saw in the last chapter, a low-ranking individual can and frequently does gain a desired goal by skillful and devious means, despite the presence of a disapproving superior. If a male is to attain, and above all maintain, a position at the top of the hierarchy, he must be capable of planning his moves and manipulating his subordinates, many of whom may be physically stronger than he. And he will fare better if he has the ability to sustain a close, supportive relationship with another male. In chimpanzee society the social environment is continually changing, and each individual needs to be mentally alert, able to weigh the salient features of a given situation and make rapid adjustments to his behavior if necessary. Thus a young male, swaggering and displaying at a female one moment, suddenly stops and sleeks his hair if a higher-ranking male arrives on the scene. Because blustering aggression is no longer suitable, he must use more subtle tactics if he is to win the cooperation of the female. Sometimes lightning decisions must be made—to flee an aggressor or approach with appeasing gestures. And mistakes may be brutally punished.

If a creature is to be socially aware in a truly sophisticated sense, not only is it necessary that he impute purposes and wants to others, he should be able to learn simply by watching the experiences of others. "If Humphrey attacks Fifi when *she* takes his food, he will attack *me* if *I* do the same." We do not yet know to what extent the Gombe chimpanzees can reason in this way, but the ability to do so would clearly be advantageous. There is *need* in a society of this sort for some glimmering of a concept of self—an ability, as we noted in Chapter 2, "to assume the perspective of others towards itself," which enables an individual to perceive itself as an object. Like naive chimpanzees in captivity, the Gombe chimpanzees do not appear to recognize themselves in a mirror (Goodall, 1971); nevertheless, some individuals spend minutes at a time crouched over still water, apparently gazing at their reflection. Of interest in the context of self-awareness is the behavior of infant Goblin who, after watching his mother wipe her dirty bottom with leaves, picked more leaves and wiped his own clean bottom. Not only did he demonstrate a well-developed sense of the relation between things (which, of course, is a necessary prerequisite to any intelligent act of imitation), but recognition of the fact that he too had a bottom—a bottom that was separate from, but the same as, that of his mother: "*her* bottom, *my* bottom."

Faben threatening his mirror image.
(H. van Lawick)

And what of imagination? Do chimpanzees in the natural habitat have any *need* for a fantasy world? I believe that one of the ways in which imagination is useful for humans is that it enables us to practice in the *mind* behavior that we intend to perform in *reality*. I actually learned to jump a horse, between one riding lesson and the next, by practicing over imaginary hurdles every night for a week. And my uncle, a senior consulting surgeon, never went to sleep before an operating day without first reviewing in his mind every step he would take in each of the types of surgery on his agenda for the next day— including how he would react to the things that might go wrong. Practice of this sort can also be an immensely powerful tool for the improvement of strategy in tricky social negotiations that lie ahead.

Viki Hayes *imagined* the pull toy, but *enacted* the movements that gave substance to her make-believe. Most small human children act out their imaginary games in this way. At Gombe, on three separate occasions, adolescent males have been encountered as they performed charging displays in the forest, far from their companions. And Figan, it will be remembered, practiced Mike's display technique with an abandoned kerosene can, by himself, in the bushes. Who are we to say that these actions were not performed expressly to terrorize a host of imaginary chimpanzees? Once four-year-old Wunda watched intently from a safe distance as her mother, using a long stick, fished for fierce driver ants from a branch overhanging the nest. Presently Wunda picked a tiny twig, perched herself on a low branch of a sapling in the same attitude as her mother, and poked her little tool down—into an imaginary nest? And when she subsequently withdrew it, how do we know that it was not swarming with a record, though nonexistent, catch?

Live ants are fascinating too. Small infants, particularly firstborn offspring who have no elder sibling as a built-in playmate, may spend long periods watching, poking, flicking, and generally having fun with trails of small, harmless ants. Sometimes they use the termite-fishing technique at the nest entrance. It is precisely this sort of behavior—curiosity and the practicing of learned patterns in new contexts—that can, because of the chimpanzee's ability to profit from chance experience, lead to innovations. And because chimpanzees are captivated by unusual behavior in their companions, new patterns can be passed from one to another through the channels of observation, imitation, and practice. This is the dawn of *cultural* evolution; it means that the performance of a gifted individual can spread through a group and quite rapidly become part of its tradition. The kin-selection mechanism, which so jealously guards the genetic advantages of a given genealogy, is confounded and the whole group can profit, at the expense of a neighboring group, from the chance performance of one of its members. We have not yet recorded the incorporation of a new behavior, either technical or social, into the Gombe community (although some innovations were imitated and practiced by a few other individuals for short periods). But obviously each of the cultural traditions that we observe today stems from the innovative behavior of some chimpanzee genius of the past.

We know that there are exceptionally gifted chimpanzees in the wild. As Köhler's Sultan shone in the realm of technological problem solving, so Mike and Figan stand out among the Gombe males as masters of reasoned thinking and skillful tactical and social manipulation. Then there are those, like Pom and Gremlin, who excel at termite fishing. Each chimpanzee has his or her own special characteristics. Indeed, the marked differences in personality from one to another are rivaled only by individual variation in our own species. In chimpanzee society, genetic diversity is encouraged by the flexible mating patterns that allow *all* males (including cripples and those of low rank) to sire infants, as long as they have the social skills to persuade females to follow them on consortships. The more adept males have an even better chance of passing on their genes, so that intelligence, to the extent that it is hereditary, can win out over high rank and aggressiveness.

It is clear, I think, that the higher cognitive abilities of the chimpanzee *are* called upon in his natural habitat, even though some are brought into play much more frequently than others. Thus we are now in a position to superimpose, one on the other, two portraits of the chimpanzee: the one that is emerging from studies in captivity, and the one that is gradually being painted from observation in the natural habitat. The result? A whole that is even more fascinating, even more complex, than the sum of its parts. Of course, the picture is not yet finished; and it may well be that in some places the reality we are trying to capture has been slightly distorted, that some of the brush strokes are a little crooked or even in the wrong place. But with the quest for understanding joined by scientists from many disciplines, we have a picture that is much closer to reality than was the case twenty-five years ago when the Gombe study first began.

One striking finding from all studies of the chimpanzee—in laboratory, home, and forest—is the sometimes uncanny similarity between certain aspects of chimpanzee and human behavior: the long period of childhood dependency, the postures and gestures of the nonverbal communication system, the expressions of emotion, the importance of learning, the beginning of dependency on cultural tradition, and the startling resemblance of basic cognitive mechanisms. Our own success as a species has been due entirely to the explosive development of the human brain. Our intellectual powers are so superior to those of even the most gifted chimpanzee that attempts made by scientists to spell out the similarity of mental processes in man and chimpanzee have largely been met with ridicule or outrage.* When Köhler and Yerkes first published their findings on insightful problem solving in chimpanzees, these were instantly and abusively denounced as a "pernicious, I should say disgusting, tendency to depart from the truth" (Pavlov, 1957, p. 557).† But results of this sort were replicated, and the evidence for sophisticated mental performances became ever more convincing. One by one, the attributes once believed unique to man have been found to exist in "lowlier" forms of life.

Nevertheless, we must not forget for an instant that even if we do differ from the apes not in kind, but only in degree, that degree is still overwhelmingly large. Knowledge of the ways in which our behavior is *similar to* that of the chimpanzee, combined with knowledge of how it is *different*, helps us, I believe, to pinpoint what it is that makes man unique. This is not the place to discuss such issues in detail, but I should like at least to mention three points.

First, we have developed a complex symbolic language. We can not only *show* our children how to do things, we can also *tell* them how and *explain* why. As Lorenz (1977, p. 161) says, this "enables tradition to become free of objects." We can discuss events that happened in the remote past and make complex contingency plans for a future that may be close or distant. We can, in addition to planning an aggressive foray against our neighbors, determine how best to defend ourselves. Words give substance to abstract thoughts. The interaction of mind with mind broadens ideas and sharpens concepts—even as competitive interactions among chimpanzees can bring about more sophisticated social strategies.

---

* Most recently the controversy has raged around the language acquisition studies. Speaking for myself, it seems a reasonable assumption that creatures so like us in genetic makeup should show the kind of prelinguistic ability that must have existed in our prehuman ancestors.

† In 1933 I. P. Pavlov acquired two chimpanzees, Raphael and Rosa, and began a series of extensive problem-solving tests at the Koltushi Biological Station. One of these tests was designed to replicate Köhler's box-piling problem. Pavlov confirmed Köhler's findings but maintained that insight is achieved *progressively* as a result of the chimpanzees' problem-solving behavior, in contradiction to Köhler's interpretation of insight learning as a *sudden new perception* of the relationship between the different elements of the problem (Windholz, 1984).

Second, I believe that we are able, as no other creature is, to overcome by conscious choice the "selfish gene" (Dawkins, 1976) of our biological heritage. Our acts of altruism are *not* always selfish; and by the same token our acts of violence are *not* inevitable.

Third, while our basic aggressive patterns are not so different from those of a chimpanzee, our comprehension of the suffering we may inflict on our victims is of an entirely different order of magnitude.

Are the chimpanzees at the end of their evolutionary trail? Or are there pressures in their forest habitat that might, given time, push them farther along the path taken by our own prehistoric ancestors, producing apes that would become ever more human? It seems unlikely, for evolution does not often repeat itself. Probably chimpanzees would become slowly more *different*—perhaps developing the right side of the brain at the expense of the left, for example.

Such questions are purely academic. They could not be answered for countless thousands of years, and even *now* it is clear that the days of the great African forests are numbered. If the chimpanzees themselves survive for a while in freedom, it will be in a few isolated patches of forest begrudgingly conceded, where opportunities for genetic exchange between different social groups will be limited or impossible; this is already the case at Gombe. And it seems not unlikely that the day will come when the only chimpanzees are in laboratories and zoos. (At that point if any evolution goes on, it will be *un*natural selection at work!)

Meanwhile there is much to learn about chimpanzee behavior in the wild—before it is too late. It is fortunate indeed that a number of long-term studies are under way or planned in different geographic locations. In the Tai Forest of the Ivory Coast, Christophe and Hedwige Boesch have already habituated many of the chimpanzees and hope to carry on for many more years; in Lope National Park of Gabon, Caroline Tutin has begun the arduous task of habituation and plans a long-term study; Yukimaru Sugiyama and his co-workers return as often as possible to Bossou, Guinea; in the Kibale Forest Research Station of Uganda, Richard Wrangham will join Isabiriye Basuta to establish another long-term study on animals already habituated. And, of course, the work in Tanzania's Mahale and Gombe national parks will continue. With so many groups under observation, and more researchers prepared to gather their data under terms dictated by the chimpanzees themselves, our understanding will progress by leaps and bounds. There will be an increased chance of seeing rare behaviors, such as intercommunity conflicts, with more people watching. And as information is pooled and compared, our knowledge of cultural variation across the range will increase.

Let us hope that this new comprehension of the chimpanzees' place in nature will bring some relief to the hundreds who presently live out their lives as prisoners in our laboratories and zoos. Let us hope that, even as our greed and shameless destruction of the natural world gradually take from yet more chimpanzees their forests, their freedom, and often their lives, our knowledge of their capacity for affection and enjoyment and fun, for fear and suffering and sadness, will lead us to treat them with at least the compassion we would accord fellow

Ham, the first chimpanzee to journey into space, made history in January 1961 with his ride in a Mercury Redstone rocket. His "reward"? Over sixteen years alone in a zoo cage. Happily, he found a measure of freedom during the last two and a half years of his life in the beautiful outdoor exhibit at the North Carolina Zoological Park. Here his favorite female, Maggie, watches him. (Courtesy of the Asheboro *Courier-Tribune* and the North Carolina Zoological Park)

humans. Let us hope that if we continue to use chimpanzees for painful or psychologically distressing experiments, we shall have the honesty to label our actions what they are—the infliction of torture on innocent victims.

If only those responsible for the care of captive chimpanzees (and other animals too) could experience with me some of the more intimate moments at Gombe. For so often they generate overwhelming shame for the behavior of our own species—our arrogant assumption that *our* needs, *our* pleasure, *our* wishes must inevitably come first. Let me try to share, now, one from the countless hundreds of such moments.

The sun is setting over Lake Tanganyika. Melissa and her daughter, Gremlin, have made their nests some 10 meters apart. Melissa's son Gimble still feeds on *msongati* pods, his lips white with the sticky juices. Gremlin's infant, Getty, dangles above his mother, twirling, kicking his legs, and grabbing at his toes. From time to time Gremlin reaches up, idly tickling his groin. After a few minutes he climbs away through the branches, a tiny figure outlined against the orange-red of the evening sky. When he reaches a small branch above Melissa's nest he suddenly drops down, plop, onto her belly. With a soft laugh his grandmother holds him close and play-nibbles his neck and his face until it is his turn to chuckle. He escapes, climbs up, and plops down again for more. Then again, and again.

After a time he goes back to his mother and lies beside her, suckling, one arm on her chest. Melissa reaches out for a handful of leafy twigs, arranges them carefully under her head and shoulder, and relaxes. As the quick, tropical dusk settles, Gimble stops feeding and stands at the edge of his mother's nest, seeking permission to join her for the night. She reaches to him, and he climbs in and lies beside her.

Suddenly from the far side of the valley come the melodious pant-hoots of a single male: Evered, probably in his nest too. It is Gimble who starts the answering chorus, sitting up beside Melissa, his hand on her arm, gazing toward the adult male—one of his "heroes." Melissa joins in next, still lying on her back; then Gremlin too. And Getty last of all, interrupting his meal to add his infant calls to the evening song. "Seven o'clock on a fine, dry evening and all's well."

At least during their lives, Gombe—in conservation-minded Tanzania—will still be safe.

(R. Wrangham)

A. Data Collection at Gombe

Inevitably, over a period of twenty-five years *recording* methods in the field have changed; basic *observation* techniques, however, have not. The chimpanzees gradually became habituated to first one, then several, then finally *any* human observer. Since that time the data have been collected on foot by one or two people who are careful not to frighten or irritate their subjects: they move quietly, keep a distance of at least 5 meters, and avoid abrupt movements that might be interpreted as threat.

Data Sources

The long-term records of the Gombe Stream Research Center are compiled from a number of sources.*

* A microfilm of all records between 1963 and May 1975 has been made, with copies housed in California and at Cambridge University in England. Originals (or paper copies) of all data are stored at my home in Dar es Salaam.

In 1973–1975, when the Gombe Stream Research Center was affiliated with Stanford University, a computer program had been tailored to the needs of the long-term project at Gombe and to those of the newly established Stanford Outdoor Primate Facility; computerizing of the data and statistical analysis were under the direction of Helena Kraemer. The punching of the Gombe data was found to be extremely costly, in terms of both money and time. Ways to streamline the process were being investigated at the time Stanford withdrew its support from the Gombe project in 1975 (chimpanzee research at the Outdoor Primate Facility came to an end soon afterward).

During the years immediately following the 1975 kidnapping, it became very difficult for me to find funds to keep the relatively large-scale project at Gombe operational (at the level of ongoing data collection, and training and supervision of field staff). Needless to say, I had neither the funds nor the time to investigate methods of computerizing incoming data.

Many of the long-term records (all those prior to 1968 and almost all after 1975) are in the form of written notes. Analysis of this kind of information is, of course, extremely time-consuming. I have tried to extract and organize the data from these reports in such a way as to standardize them over the ever-increasing period of time involved, and to make retrieval more efficient. Even so, in order to review certain aspects of behavior I have occasionally had to undertake marathon searches through file after file, coding the data as I went along. Thus, while I have sometimes presented in this book data drawn from many years (particularly when a long-term view is important) in other cases I have extracted information from only a few years, or even a single year. In these instances, however, I have substantiated the data by referring to observations on the same behavior made in other years (even if they were collected in a slightly different way). In other words, within the limitations of changes in recording meth-

In-Camp Data

The following data are recorded daily (and have been for some behaviors since 1963, when Flo came into estrus and attracted into camp most of the adult males of the study community):

(a) Time of arrival and departure of each individual, composition of party, and direction from which it came and toward which it went.

(b) Reproductive state of each adult female—that is, anestrous, inflating, fully swollen, or deflating.

(c) Illness or injury, with comments on recovery.

(d) Arrival and departure of baboon or baboon troop (with identification of troop).

(e) Grooming and play (who grooms or plays with whom, including play with baboons); recorded on the minute every two minutes, entered on check sheets since 1969.

(f) Behavioral interactions, also individual charging displays and vocalizations—as many as can be seen, in as much detail as possible. From 1970 until 1975 these data were recorded by students on check sheets; after the students left in May 1975, the same data were handwritten in Kiswahili by the Tanzanian field assistants (and later transcribed onto check sheets—except for the attendance data, which were still entered directly onto printed forms).

(g) Descriptive notes on any unusual events.

(h) Weight of the chimpanzees; at least twice a month we try to lure them up a rope attached to a spring balance by placing a banana in a tin near the top of the rope. We then read the weight with binoculars. This method was initiated by Richard Wrangham in 1970. It is not possible to accurately weigh mothers with small infants, because the latter are constantly clinging to them.

One observer is responsible for maintaining this record. He is equipped with check sheets, lined paper, binoculars, and (since 1982) a two-minute-interval timer. At midday there is a change of observer. Obviously, if there are many chimpanzees in camp at the same time, one person cannot see all the animals at once. If certain individuals are not being observed, this is recorded on the attendance sheet, so that the data can subsequently be used as accurately as possible.

Follows of Target (or Focal) Individuals

A selected individual is followed from his or her nest in the morning—or from camp, or from some location to which the observer has been attracted by calls or where there is a known temporary food source—until the target nests that night, or until the required number of hours has been obtained, or until the chimpanzee is lost to view.

Between 1968 and 1975 much of the information at Gombe was collected by undergraduate and graduate students, along with some postgraduates (mostly from Europe and the United States). Each student was accompanied by one of the Tanzanian field assistants.

The following data are recorded during a follow:

---

odology and data coding over the years, I have tried to use the long-term records in such a way as to present as accurate and longitudinal a picture as possible of the behavior of the Gombe chimpanzees.

(a) Behavior of the target chimpanzee

The behaviors noted and the precise method of recording have varied somewhat in accordance with the specific questions being asked by different researchers. David Bygott, for instance, was especially interested in agonistic interactions between adult males; Caroline Tutin followed females in estrus and recorded information pertaining to mating patterns; Richard Wrangham collected data on the time budgets, feeding behavior, and ranging patterns of adult males; and so on. The data were recorded on tape and subsequently transcribed—in the form of written notes or onto check sheets—or they were entered directly onto check sheets in the field. All researchers have been required to contribute information (including qualitative descriptive reports) for the long-term records, particularly for the "character files" of each chimpanzee.

From 1975 on reports have been written in Kiswahili by Tanzanian field assistants. The accounts describe the major activities of the target individual. Time is noted whenever the chimpanzee changes activity—for instance, when he stops feeding and starts to rest, or stops resting and starts to travel. All interactions involving the target are described.

Since 1970 the food eaten by the target, and the length of time spent feeding, have been recorded (a procedure initiated by Richard Wrangham).

After 1978 we began recording, on the minute every five minutes throughout a follow, all grooming and play in which the target is involved (and that of the offspring if the target is a mother).

(b) Composition of the target's party

On the minute every fifteen minutes the names of other individuals in the target's party are checked on the standard association chart. (A party, as defined in 1972, is made up of all individuals within 100 meters of the target. In very thick vegetation, or when the chimpanzees are in a large gathering, there may be errors in the recording of party composition.) Since 1978 it has been the practice to mark with a dot, instead of a check, any chimpanzees who are known to be within this distance but who cannot be seen by the observer from a position close to the target (and, in most cases, probably not by the target either).

From 1975 on the estrus condition of all females seen has been noted on the chart.

(c) Other information on the association chart

Encounters with other animal species such as baboons, monkeys, bushbucks, and bushpigs are recorded.

It is on this chart that the type of food eaten by the target is noted.

(d) Travel route

The route is plotted on a standard map sheet. The position of the target is indicated by a number written every fifteen minutes; this number corresponds to one entered every fifteen minutes on the association chart. If strangers are seen or heard, their position is plotted on the map. Wrangham (1975) compared nesting areas recorded by different observers between one evening and the next morning. He found, in a sample of thirty, that the median distance between two records of the same nesting area was 70 meters. Such data were considered acceptably reliable (analysis of ranging data uses squares with sides of 100 meters).

(e) Other events

Other significant events that occur in the target's party are noted, even if the target is not directly involved: fights, hunts, anything that seems unusual.

Since 1971 the Tanzanian field assistants have been almost totally responsible for the standardized association and travel chart data; from mid-1975 on they have collected the majority of the information at Gombe (because I am there for only about a third of every year, and the chimpanzees are observed daily). One observer writes the reports while a second is responsible for the association chart and travel map—and, for the time sampled, grooming and play data. Appendix B is a translation, more or less word for word, of the reports made by two of the field assistants, accompanied by examples of the travel map and the association chart (from which the Feeding column has been deleted and a Comments column added, for illustrative purposes).

Mother-Infant Data

In 1969 Robert Hinde helped to design a check sheet suitable for recording mother-infant interactions, based on his research on captive rhesus monkeys. Mother-infant data have been recorded continuously since then, and although the check sheet has been modified over time, the categories remain much the same. Each record is accompanied by a descriptive account.

Incidental Observations

It has always been the practice of Gombe personnel to write reports of any interesting events they happen to observe when not specifically on the job.

Rainfall and Temperature

Measured daily.

Banana Feeding

Wrangham (1974) has described in detail the various phases of banana feeding up to 1973, after which time the method of distribution has remained essentially the same. Each chimpanzee is given five to eight bananas once every ten days or so—when alone or with a compatible individual (or individuals), so that aggression is unlikely. Records are maintained of when each chimpanzee receives bananas.

Selection of Targets

My basic rationale is to monitor the behavior of as many individuals as possible. Appendix C shows that some individuals have been followed far more than others. With respect to the females this is because some, even after many years, are still relatively unhabituated (Dove, Nope, Sparrow, Sprout). Some other females who appear to have been followed less were part of the mother-infant study, so that more is known about them than the appendix indicates.

To some extent the chimpanzees themselves dictate the emphasis of the research from month to month, year to year. When an individual is going through a turbulent period of change (such as Goblin's meteoric rise in the hierarchy), or when unexpected and unique behavior occurs (Passion's infant killing, Melissa's delivery of twins), we try to collect a great deal of information so as to get as complete a picture as possible.

It would be pointless to deny that sometimes a target is selected simply because the observer happens to prefer that particular chimpanzee over another. If I set out to look for a target for a mother-infant follow and come upon two groups comprising Melissa and her family and Fifi and her family, and if I need data on both, I select Melissa. If I follow Fifi, I know that I shall be hit, banged, and generally given a bad time by Frodo. (On the other hand, if I have not collected my quota of data on Fifi, I select her, Frodo notwithstanding!) And just as I prefer some individuals to others, so too do the field assistants.

The more hours spent in the field, the better the likelihood that critical events will be seen. An incredible number of such events *have* been observed (for example, Pom's infant being blown from a tree in a gale, many intercommunity attacks, episodes in the dominance conflict between Figan and Humphrey); hundreds more, over the years, have been missed.

Training of Field Assistants
The first field assistant, Hilali Matama, was employed in 1968—initially in order to accompany students, because national park regulations required that observers work in pairs. As additional Tanzanians were employed, it was decided in 1970 to train them to collect simple standardized data on party composition and behavior of the target individual. The travel and association charts have already been discussed. Between 1972 and 1975 many students were at Gombe working on their doctoral dissertations. Because each one relied on his or her field assistant to provide reliable information about party composition, association, and travel routes, considerable effort was put into the training of these men. Those who were unreliable, or who proved incapable of improving their performance, were dismissed. Richard Wrangham, in particular, devoted much time to this informal but intensive training program.

In 1974 Larry Goldman accepted the position of senior scientist at Gombe. He and graduate student Donna Anderson put in many hours to further train the field assistants by enlarging the categories of behavior that they recorded during follows, by initiating the use of check sheets, and by performing reliability tests. In addition, I began a series of weekly seminars on primate behavior (in Kiswahili) for all field staff.

This emphasis on training the field assistants was fortunate in view of the events of the following year. Immediately after the kidnapping, when all non-Tanzanians had left Gombe, data collection (in camp) was continued after a break of only two days, despite the fact that the field staff had been, as it were, abandoned. Even though I myself was only able to come to Gombe for a few days at a time during 1975 and part of 1976, observations continued. When I did visit Gombe, I spent all my time working with the field assistants—going through their reports, word by word, explaining where not enough information had been recorded, answering their questions, and giving encouragement. I went into the field with them and checked the accuracy of their reports. By 1977 these men had reached a very high standard indeed—as I believe the reports translated in Appendix B show.

When a new field assistant is employed, he goes through a period of training. Initially he accompanies one of the experienced men and is taught how to fill in the association chart and travel map. During this period (usually three to six months) he gradually learns to identify the various chimpanzees, their behaviors, their food plants, and the topography. When the senior staff are confident that he knows these things, he writes his first report. The senior person who is with him also writes a report, and the two are compared. Only when the new assistant has been through a training period of six months is it decided whether or not he should be employed permanently. During my own stays at Gombe I not only go through his reports, comparing them with those of the experienced staff, but I do some record taking with him myself.

Recording Equipment

The observer responsible for the report on the target carries paper and pen (or pencil), and wears a reliable watch. He has a pair of Minolta binoculars in his pocket. The second observer carries the charts on which he records association and travel patterns, paper for notes, and a timer.* Both men carry torches for finding their way home after dark. During follows, in addition to binoculars and notebook, I carry an Olympus camera† in my pocket; at other times I use a Nikon with various lenses. The observer in camp has two clipboards for his various charts and check sheets (on attendance, health, bananas, grooming, and play) and some blank sheets for notes. He has a pair of binoculars and a timer.

Thus our equipment is simple. Most important is health, patience, determination—and endless fascination for, empathy with, and respect for our subjects.

Organization of Data

Our large volume of data, some of which (in the form of daily journal notes) goes back to 1960, is organized as follows:

(a) Each chimpanzee has a character file in which information thought to be of interest is included as part of an ongoing sorting of raw data. These files represent a unique and ever-growing collection of life histories.

(b) Various files on particular aspects of behavior such as tool using, mating, begging, fighting, and intercommunity interactions are also maintained throughout the year.

* This timer, designed especially for us by Ian Curington, times intervals of one, two, and five minutes. Tested in temperature and humidity conditions that simulated those at Gombe, it has proved completely reliable over four years.

† A similar camera has recently been added to the equipment of the field assistants.

(c) All association data are analyzed, by hand, on a month-by-month basis. This allows us to compile for each individual an index of associations both in camp and out of camp. I have made extensive use of this information in Chapters 7 and 8, where the value and shortcomings of the index are discussed. Appendix D gives association matrixes of target males and females during two sample years, 1978 and 1979.

The Observers

In this book I make extensive use of the long-term records just described, records that have been compiled by field assistants, undergraduate and graduate students, and researchers—from the United States, Europe, and Tanzania. Thus it is important to examine thoughtfully the performance and outlook of all these workers—particularly of the Tanzanians, who have collected the bulk of the data from 1975 to the present. None of the current team of field assistants have received academic training beyond the high school level, and some of them have had even less formal education. Am I justified, then, from the point of view of quality of data, in making scientific use of their work?

Before I address this question, let me indicate the qualities that seem to me to be fundamental in *any* observer who is to collect reliable ethological field data. First of all, he (or she) must appreciate and respect the animal being studied, in order that the all-important observer-subject relationship be built up and maintained. He must be able to make rapid, accurate identification of all the individuals of the study group. He must be completely familiar with the various behaviors (the ethogram) of the animal species he is studying. He must be able to understand and interpret these behaviors well enough to make an acceptably detailed report of a complex and fast-moving interaction, without wasting time in noting irrelevant details. He must be aware of the value of recording negative information—for example, that a chimpanzee joins a party and does *not* greet any other individual, or that conditions were such that the observer was *unable* to see whether or not a greeting took place. He must transcribe his field notes as soon as possible, so that information perhaps only partially scribbled in the field can be immediately elaborated upon. He must be aware of the need for scrupulous integrity and record only facts of which he is certain. He must (insofar as possible) be objective and not allow personal bias to distort the data. Intuitive interpretations, while they may be of crucial importance to the understanding of a given sequence of events, must not be confused with facts. Finally, the observer must be highly motivated to collect information; field conditions are often far from ideal, and he will fall short of his task if his enthusiasm does not overcome, at least most of the time, any personal discomfort.

Observer Performance
Against this background of necessary qualities we can now evaluate the performance of the research assistants and other scientists who have contributed to the long-term records at Gombe.

Appreciation and Respect

In order to maintain the animal-human relationship of mutual respect, tolerance, and trust that was built up during the early years at Gombe, I laid down certain rules of conduct for the field workers. Observers, whenever possible, were to maintain a distance of about 5 meters between themselves and their subjects; they were not to move between two chimpanzees, particularly a mother and members of her family, if it could possibly be avoided; they were not to talk loudly or make sudden movements; if approached and "pestered" by a young chimpanzee, they were to either move quietly away or try to ignore the behavior and thereby discourage it; if a chimpanzee seemed excessively nervous during a follow, they were to drop back, then if the animal was still upset, discontinue observations; and if sick they were not to work with the chimpanzees for fear of passing on infections.

Almost all the American and European research personnel who have worked at Gombe came because of an interest in (and often love of) animals. In some cases, however, when individuals were highly motivated academically to collect as much information as possible (for doctoral dissertations, for example), they disobeyed some of the rules. For instance, they would move between a mother and her older child in order to record, in as much detail as possible, her interactions with her infant.

Tanzanians in general have a keen appreciation of animals of various species. Often I have watched those who observe the chimpanzees. Their responses to interesting or amusing behaviors are, for the most part, very similar to my own. When I return to Gombe after an absence, the men spontaneously tell me the most recent noteworthy news about the chimpanzees (and about the baboons); by and large, they pick out the incidents I would have selected myself.

There are, of course, major differences in the way people relate to the chimpanzees. The best observers are quiet people; the worst, those who are insensitive and/or excitable. These characteristics are not peculiar to any cultural group. My long-term Tanzanian field assistants all have an interest in chimpanzee behavior and are unobtrusive and perceptive. Those who were not have been screened out at an early stage—not only because I, or senior members of the staff, noticed their shortcomings, but because they themselves found the work distasteful and left.

Identification of Chimpanzees

Human observers joining the research center have taken between one and three months to become familiar with the chimpanzees of the study group. For short-term assistants (from the United States or Europe) this time requirement was sometimes a serious drawback; but from 1968 on, when it became customary for research personnel to work in pairs, newcomers were able to follow chimpanzees in the company of an "old hand" and the difficulty was to some extent overcome.

There is another big advantage to the practice of working in pairs: two observers, particularly when they watch from slightly different vantage points, can often piece together the sequence and participants of a fast-moving, complex event far more accurately than could a single person.

No Tanzanian can become a permanent member of the research team today if, after his training period, he is not able to identify accurately every individual in the study group. (He is also required to know at least the major food plants and the topography of the community range.)

The value of working in the field, day in, day out, year after year, is brought home to me each time I return to Gombe after an absence of three months or so. For the first week I often have difficulty making split-second identifications, particularly of infants seen without their mothers, who have grown or changed in the interim. And identification of males displaying past through the undergrowth is even more difficult. The Tanzanians are superb in this respect, as I have witnessed time and again.

Familiarity with the Ethogram
Students and research assistants from abroad have been required to read all the available literature prior to their arrival. A *Gombe Glossary* of terms was prepared and this, too, became required reading. Between 1973 and 1975 students and research assistants from Stanford University received detailed training on the chimpanzee ethogram (and data-recording techniques) at the Stanford Outdoor Primate Facility, where they collected data with methods very similar to those in use at Gombe. Even so, only experience in the field during their first few months enabled newcomers to become truly familiar with chimpanzee behavior in the natural habitat and with the idiosyncrasies of posture, gesture, and voice of the various individuals. However, provided they were accompanied by one of the regular field staff, the students were usually able to collect reliable data after a month.

The Tanzanians, when they first join the research team, have read no literature (there is none in Kiswahili). Nevertheless, extensive discussions with new recruits after the first three months of basic training showed that those who were subsequently employed had quickly learned to understand the various elements of the behavioral repertoire. Some of the Tanzanians have only begun to write as adults and have not yet acquired speed in doing so. Sometimes the man responsible for target information is still recording one interaction when the target chimpanzee has embarked upon another. This is clearly a drawback, but is largely compensated for by the fact that two men work together; the second is able to watch the new interaction and subsequently relate it for inclusion in the report. This slowness in writing also means that certain events are not recorded in enough detail (even when transcribing field notes the following day, when the events are still fresh, the men may be reluctant to devote the necessary amount of time to filling in the details). I want to stress, however, that most of these men, perhaps because they have until very recently been forced to rely on the spoken word for the storing and passing of information, have phenomenal memories. When asked for details of an incident written too sketchily in their report, they are usually able to supply an incredible wealth of additional information, weeks or even months afterward. That these remembered facts are for the most part accurate, I have verified on many occasions by asking about incidents that took place when I too had observed the episode. Their almost-total recall includes individuals present, their position at the time, the order in which they participated, the precise components of behavior, and often the time of the various happenings to within one or two minutes. If an unusual event was observed (such as a cannibalism or an intercommunity attack), it has been possible to interview the observers separately and glean a good deal of additional information. (We have very recently begun to use small tape recorders at Gombe, so that significant events can be recounted on tape at the time of occurrence.)

Every few months I check the reliability of the reports of the different field assistants (often without telling the men concerned) simply by making simultaneous recordings and then comparing the results. When only a few animals are under observation, our comments are usually identical except that I sometimes have noted more fine details of behavior. When several chimpanzees are together or complex events take place, again I find that I tend to have noted more separate elements of behavior—provided observation conditions are reasonably good. But if such events take place in tangled vegetation, especially if they are high up in the treetops, the Tanzanians often observe more than I do. The point is that what the men write down is accurate, and it is extremely rare that a major occurrence is omitted from their records.

After I have compared reports in this way, I have long discussions with the men involved, going through each and every discrepancy in detail. This has resulted in marked improvement in the quality of the reports over the last few years. No field assistant who has consistently *omitted* important data, or (worse) written *incorrect* sequences of behavior, has been kept on as a member of the research team.

Personal Bias and Objectivity

Objectivity in recording is the goal of all behavioral scientists. This is not to say that a certain amount of empathy with one's subjects is not admissible—indeed, I personally believe it to be desirable—but the report itself must separate objective fact from subjective interpretation. The Tanzanian field staff have been taught to write down what they actually see, then, in parentheses, their interpretation of unusual or interesting happenings. Often these interpretations, based on intuitive understanding when in the presence of the animals themselves, provide me with valuable insights that otherwise would be lost.

Scientists and students from the West who came to Gombe may be divided into three categories: those for whom chimpanzees were merely preprogrammed machines, those who were more sentimental and tended to be too anthropomorphic, and those who steered a middle course. Almost all, in fact, were in the third category. There were very few in the first, and even if they arrived at Gombe with this mechanistic outlook, they were usually somewhat changed by the time they left. The Tanzanians almost without exception have fallen into the third category. Several, after working with us for a while, moved into the second category.

All observations are likely to be influenced, to a greater or lesser extent, by the particular biases of the observer. These may be cultural in origin or they may relate to the particular perspective, interests, or even sex of the person concerned. Many people are fascinated by the dramatic and obviously aggressive displays of adult males; only a few (such as myself) will be anxious to observe the behavior of mothers and infants during such agonistic events. Members of the all-male Tanzanian field staff for the most part prefer to follow the males. Only one or two take pleasure in spending long hours with the mothers and infants.

Sometimes an observer is biased by his scientific theory. It may (often subconsciously) influence him to see most clearly those behaviors that seem relevant to his hypothesis and to reject those that appear unimportant or even destructive to it. (Particularly dangerous, although not a problem at Gombe, is the research design planned *prior to* experience with the animals in the field, because it may be based on too narrow a theoretical view.) When I set out for the field in 1960, I was quite without hypothesis and I certainly had no ax to grind. Only when I had observed behavior sequences again and again did I begin to try to explain what I saw, and not until I came under the influence of Robert Hinde at Cambridge University did I begin to formulate any theories. Louis Leakey had looked upon my naive outlook as a point in my favor, and in retrospect I believe he was right.

Integrity in Data Collection
Some inaccuracies in data collection are the result of overconfidence or sloppiness, both of which may lead an observer to guess the start or end of an incompletely observed behavior sequence, or the identity of a half-glimpsed chimpanzee, and record it as fact. Also, people who have been in the field for two months or so may *think* they know the identity of a given chimpanzee, or *think* they understand the meaning of a given behavior, when in fact they are mistaken. With the incorporation of long-term field assistants into the research team, this source of inaccuracy has been largely eliminated.

It is probably extremely rare that an observer deliberately falsifies data, although if a scientist did so the falsification would probably be very hard to detect. I have been told of an African field assistant employed in a field research project in Kenya who, having quickly learned the kind of data he was supposed to collect, proceeded to provide it simply by using his imagination. There was one such case at Gombe. Thinking to impress, the individual in question provided quite a detailed report concerning the behavior of a chimpanzee he had encountered while off duty. Unfortunately for him the particular chimpanzee was at the time being observed by one of the senior field staff! When I next returned to Gombe I was immediately told, with much indignation, of his shocking behavior. He had in fact, because of social pressure, already resigned.

Perhaps the best testament to the intrinsic honesty of the field assistants is the regularity with which they write down their own shortcomings: "We arrived too late at the nest and Figan had already left. We searched for one and a half hours and eventually found him feeding." It would be exceedingly simple to *invent* Figan's behavior during the first hour and a half of daylight. Of course, the fact that the men normally work in pairs serves as a control on deliberate dishonesty, since both would have to connive to falsify the reports.

Enthusiasm and Motivation

When an individual is collecting data for his own study, he is likely to be highly motivated to fit in as many hours of observation as possible, even when this means undergoing considerable personal hardship and discomfort (follows in the rain, for example). It took me about eighteen months before I could get within 50 meters of most of the chimpanzees, and during this period there was scarcely a day that I did not rise at 0530 and remain in the field until dark (often sleeping out at night in order to be close by in the morning, or to see if anything happened in the moonlight). There were two reasons for my dedication: one was an overwhelming desire to learn about the chimpanzees; the other was my reluctance to let Louis Leakey down.

Working with the habituated chimpanzees now is a very different kind of experience, with its special rewards and its special difficulties. Because the chimpanzees would not tolerate my following them in the early days, I was unable to watch at close quarters the fascinating interactions within a family group—but neither did I have to spend a half hour on my stomach wriggling fast (or as fast as possible) under thorny vines. Most of the students and research assistants who have been at Gombe have been enthusiastic and dedicated. Many of them came, often in the face of severe competition, because they desperately wanted field experience. Even so, there were some who seized on the slightest excuse to abandon a follow because of rain or some other threat to their comfort, who jumped at opportunities to go into town on an errand and thereby enjoy lavish meals and cold drinks provided by kind friends.

The Tanzanian field assistants are in a different category. For them the collection of data is a job. Only those who are exceptionally dedicated (and I must single out Hilali Matama, the senior man) work beyond the call of duty—although all of them go out of their way to check the identity of a chimpanzee they may see on the way to their village or the market, or on their days off (some interesting information has been gained in this way). Nevertheless, many of the men lack motivation, which leads to pleas of ill health during the uncomfortable rainy months. Output of work tends to fall when I am away for too long at this time, and even bonuses offered for extra hours in the field are sought by few. Still, the work that they produce during their hours of duty is accurate and conscientious; and productivity, along with morale, increases magically during the dry months when following can be a real pleasure.

In summary, each category of observer at Gombe has over the years contributed to the long-term records in a different way. I have tried to use the records selectively, to omit or treat with caution statements made by observers whose standards were perhaps not quite high enough. By and large, however, our long-term records are an outstanding example of enthusiasm, scrupulous accuracy, dedication, and patience, contributed by many people from different disciplines and different cultures, all helping in our efforts to understand better the behavior of our closest living relatives.

B. Two Follows of the Adult Male Jomeo

(both exact translations from the Kiswahili)

Report from the Field

Target: Adult male Jomeo, 9 May 1981
Field Assistants: Yahaya Alamasi and Godfrey Kahela

0645 Jomeo in nest.
0650 Jomeo leaves nest, climbs, feeds on *viazi pori** fruit.
0716 Wanders along, feeding on *budyankende* fruits.
0808 Stops feeding, climbs, and feeds on *viazi pori* fruits again.
0835 Travels.
0840 Feeds on *budyankende* fruits as he walks.
0903 Feeds on termiting earth, then travels on slowly.
0910 Climbs and feeds on *nkonzi* leaves.
0920 Lies in nest.
0957 Travels and feeds on *budyankende* fruits.
1006 Rain, not very heavy. Jomeo continues feeding.
1120 End of rain. Jomeo continues feeding.
1128 Travels.
1135 Climbs and feeds on *kihololo* fruits.
1157 Travels.
1200 Climbs and feeds on *msiloti* flowers.
1255 Lies in nest.
1353 Feeds on *msiloti* flowers.
1426 Travels.
1433 Climbs and feeds on *mmanda* fruit.
1435 Gigi (½)† and Michaelmas appear, climb a different tree. Then Gigi and Michaelmas move into Jomeo's tree but on a different branch. Gigi pant-grunts to Jomeo. Jomeo, hair out a little. Gigi feeds, at a distance of about 10 meters. Michaelmas is by himself; that is, on his own branch.

* The plant names in this report and the next are those used to identify the foods at Gombe; some are in Kiswahili, some in Kiha.

† (½) refers to size of sexual swelling, in this case half swollen; (1) is fully swollen; (0) is anestrous.

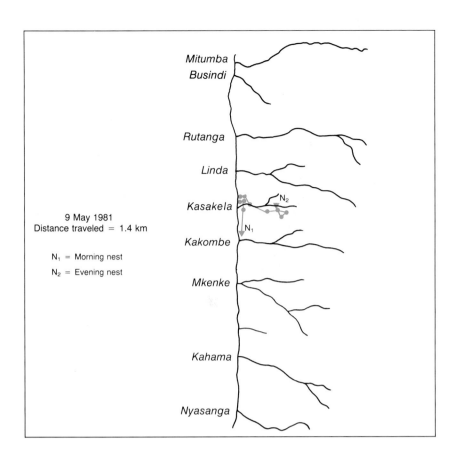

9 May 1981
Distance traveled = 1.4 km

N₁ = Morning nest
N₂ = Evening nest

1555 Jomeo rests.
1600 Jomeo travels, leaves Gigi and Michaelmas.
1610 Reaches Kasakela Stream. Climbs and feeds on *nkonzi* leaves. Gigi reappears.
1625 Gigi, who has been following Jomeo, grunts and continues travel.
1634 Jomeo travels.
1645 Jomeo crosses KK11* and continues on.
1650 Jomeo climbs and feeds on *msiloti* blossoms.
1723 Jomeo grunts, climbs, makes a nest.
1730 Sleeps, alone.

The association chart and travel map presented here show Jomeo's activities and route in detail.

* KK11 means the eleventh Kasakela Karongo (or ravine), counted west to east as they run down to join Kasakela Stream.

Time	Humphrey	Figan	Evered	Satan	Jomeo	Goblin	Mustard	Atlas	Beethoven	Passion/Pax	Pom	Prof	Melissa/Gimble	Gremlin	Nope	Athena/Apollo	Aphro	Pallas/Kristal	Skosha	Gigi	Miff/Mo	Moeza	Michaelmas	Fifi/Fanni	Freud	Frodo	Winkle/Wunda	Wilkie	Patti/Tapit	Little Bee/Tubi	Dove	Sparrow	Joanne	Jageli	Sprout/Spindle	Kidevu	Caramel/twins	9 May 1981 Target (Jomeo) Activity in Each 1/4 Hour	# Target Calls	Map Location
0700					N₁ ✓																																	Feed	-	N₁
																																						Feed	-	▼
																																						Feed, Travel	-	
																																						Feed, Travel	-	•
0800																																						Feed, Travel	-	
																																						Feed	-	
																																						Feed	-	
																																						Feed, Travel	-	•
0900																																						Feed, Travel	-	
																																						Feed, Travel	-	
																																						Feed, Nest	-	
																																						Nest	-	•
1000																																						Nest, Travel, Feed	-	
																																						Feed	-	
																																						Feed	-	
																																						Feed	-	•
1100																																						Feed	-	
																																						Feed	-	
																																						Feed, Travel	-	
																																						Feed	-	•
1200																																						Feed, Travel	-	
																																						Feed	-	
																																						Feed	-	
																																						Feed	-	•
1300																																						Feed, Nest	-	
																																						Nest	-	
																																						Nest	-	
																																						Nest	-	•
1400																																						Nest, Feed	-	
																																						Feed	-	
																																						Feed, Travel	-	
																			Skosha				Michaelmas															Travel, Feed	-	•
1500																																						Feed	-	
																																						Feed	-	
																																						Feed	-	•
																																						Feed	-	
1600																																						Feed, Rest	-	
																																						Travel, Feed	-	
																																						Feed	-	
																																						Travel	-	•
1700																																						Travel, Feed	-	
																																						Feed	-	▼
					N₂																																	Feed, Nest	1	N₂
																																						Nest		

N = Nest

▢ = Easily visible to observers

▨ = Partially swollen ♀

Report from the Field

Target: Adult male Jomeo, 14 October 1982
Field Assistants: Eslom Mpongo and Gabo Paulo

0617   Jomeo, Satan, Beethoven, and Freud in nests at KK10.

0634   Climb down. Beethoven presents, pant-grunts to Satan; Satan, hair out, stamps on Beethoven, who screams and falls 3 meters to the ground. Jomeo and Freud pant-grunt. Jomeo climbs a tree and sits.

0640   Satan feeds on *mitati* leaves. Beethoven watches, then feeds on the same.

0655   Jomeo feeds on *mitati* leaves too. They continue feeding and slowly head southeast.

0703   They stop feeding and travel. Freud does not feed.

0708   Freud pant-hoots, all pant-hoot, and Jomeo stamps. These pant-hoots are in response to calls from KK9 and KK8.

0710   The group arrives at KK9 and climbs, feeds on *kirukia* fruit. Goblin, who was feeding at KK9 before the arrival of the group, displays. Beethoven pant-grunts. Jomeo and Satan, no response. Goblin displays around, swaying, hitting saplings.

0715   Goblin climbs, feeds with the others, and food-grunts. Freud vanishes.

0725   They stop feeding, climb down. Goblin pant-hoots; the others are silent.

0726   All climb, feed on fruits at KK9 with pant-hoots.

0738   They stop feeding, climb down. Goblin displays, silently, not toward anyone. They travel southeast.

0743   Goblin displays toward Jomeo and Satan, who waa-bark. Beethoven also waa-barks. Then they climb *mtobogolo* tree and feed on figs with loud pant-hoots and food calls.

0745   Jageli joins, pant-hoots, and climbs to feed with the others.

0748   Jomeo, Goblin, Satan, Beethoven, and Jageli, all pant-hoot.

0841   Satan, Goblin, Jageli stop feeding, climb down and rest. Jomeo and Beethoven continue feeding.

0843   Freud arrives and rests near the others.

0850   Jomeo and Beethoven stop feeding and climb down. Jomeo leads northeast and the others follow.

0855   Jomeo and Satan rest, groom each other. Goblin displays past, and then near them for about 50 meters, hurling dead branches and dragging them. Jomeo, Satan watch. Jageli, Beethoven pant-grunt. Beethoven moves closer to Satan, and sits near him to rest.

0857   Goblin pant-hoots. Jomeo, Satan, Freud, Beethoven, Jageli pant-hoot.

0902   Beethoven joins the grooming session. Satan stops grooming and lies down. Jomeo and Beethoven groom mutually.

0906   They stop grooming and travel east. They are at Dung Hill. Satan leads the travel.

0908   Joanne and her infant Jimi join. Goblin displays, throwing branches, but not toward Joanne.

0910   They arrive at Kakombe Falls. There is a troop of colobus; they look up and start to run along the ground below the troop.

0917   Jageli, Beethoven climb to chase in the trees. The colobus males chase them and they leap down. Jomeo continues moving below.

0927   Satan, Goblin, Beethoven, Freud, and Jageli chase a female colobus with infant. Three male colobus rush toward the chimps, barking; the chimps waa-bark and return to the ground.

0930   They stop hunting and continue to travel east.

0945   Jomeo, Satan, Goblin, Beethoven, Jageli, Freud, and Joanne arrive at KK6 and meet Evered, Mustard, Atlas, Prof, Pom, Pax, Sprout (¼), Spindle, Miff (1), Michaelmas, Mo, Athena (¼), Aphro, Apollo, Wilkie, and Nope. Evered and the others are all feeding on *mashindwi* fruit, and all pant-grunt, food call. Goblin starts to display toward Evered, who gives loud pant-barks, climbs higher in his tree. Goblin continues to display around Evered's tree. Satan displays as well, and many now pant-hoot with much noise.

0950   Freud climbs toward Miff (1), branch shakes. Miff presents, Freud copulates. Jomeo is resting on the ground.

0957   Melissa, Gremlin, Gimble, Getty, Fifi, Frodo, Fanni, Little Bee, and Tubi join the group. The youngsters are playing.

1000   Jomeo climbs and starts to feed on *msiloti* leaves.

1006   Jomeo stops feeding, climbs down. Observation on the group is poor. Satan mates Miff (1), who squeals.

1010   Goblin mates Miff (1). Mo interferes, Goblin pushes her. Jomeo is resting on the ground. Aphro lies in front of Jomeo, presenting her ribs, and Jomeo grooms her. Frodo follows Aphro to Jomeo, and pushes her head with his hand three times. Aphro does not move away. Frodo picks up a rock, holds it in his hand, and hits Aphro's head. Jomeo reaches out his hand and prevents this. Aphro moves away and continues to play with Frodo.

1016   Jomeo grooms Beethoven for four minutes, then stops.

1021   Satan displays, dragging and swaying saplings, hair out. Jomeo, hair out, displays after Satan and they display toward Goblin; both Jomeo and Satan are grinning. They do not attack Goblin, but stop and embrace. Jomeo then goes toward Melissa. The chimpanzees are making much noise at this point: they are spread out and now most move off west. Jomeo, Satan, and Goblin do not follow.

1023   Jomeo and Satan groom mutually. Goblin displays near them and runs off 60 meters. Jomeo and Satan ignore.

1029   Goblin displays toward and past them again. They continue to groom. Goblin sits about 30 meters away from them.

1031   Goblin displays toward them and now they waa-bark and chase him. Goblin goes 40 meters and stops. Jomeo and Satan stop chasing after 10 meters and groom mutually.

1032   Goblin approaches and stands just uphill of them. The rest of the group is not near now.

1034   Goblin displays toward Jomeo and Satan and they hold each other, arm raise, and waa-bark. Satan charges and stamps once on Goblin's rump, Satan and Jomeo chase Goblin and waa-bark. Goblin again sits about 30 meters away, but after thirty seconds or so he returns.

1036   Goblin displays very close. Satan and Jomeo watch.

1042   Goblin pant-hoots. Jomeo and Satan also pant-hoot and the group pant-hoots from a distance. Jomeo and Satan groom again.

1053   Goblin approaches Jomeo and Satan with hair out. Jomeo and Satan look at Goblin. Goblin holds out a hand to Jomeo, who reciprocates, and they touch hands slowly. Then Goblin joins the grooming.

1110   Jomeo, Satan, Goblin are still grooming. Suddenly they startle and waa-bark once each, loudly. Jomeo stops grooming. Satan and Goblin continue. Cannot see what startled them.

1113   Beethoven arrives; Jomeo and Beethoven groom. Now Nope and Mustard appear, with pant-hoots.

1116   Jomeo, Goblin, Mustard, Nope, and Beethoven pant-hoot and waa-bark loudly, prolonged. The group, just west, replies from KK7.

1119   Goblin and Nope groom.

1123   Jomeo and Beethoven stop grooming, Satan and Mustard groom. Jomeo and Beethoven lie resting.

1142   Mustard stops grooming with Satan. Mustard and Goblin groom mutually and Satan and Nope groom mutually.

1143 Jomeo starts off northwest, and the others all follow.

1152 Jomeo and companions feed on *mashindwi* fruit, all pant-hoot. The rest of the group pant-hoots to the west.

1200 Jomeo and companions stop feeding, climb down and head toward the northwest.

1205 Now Skosha and Kristal, Miff (1), Mo and Michaelmas, Melissa, Gremlin, Gimble and Getty, Prof and Pax, Sprout, Spindle, Jageli, Wilkie, and Freud all appear with pant-grunts.

1206 Goblin mates Miff (1). Mo and Kristal approach to interfere, but Goblin arm-raises (right arm) and thus stops them. Miff gives copulation squeals. Then all travel north, upper KK7.

1210 Satan displays, chases after Kristal, and attacks her, swatting once with his hand. Then he seizes her and drags her 14 meters. The spot is rocky with no earth visible and Kristal screams a lot, and the others including Skosha waa-bark.

1212 Jomeo climbs to feed on *mashindwi* fruit, a little apart from the others. Observation on the group is poor. Jageli climbs to feed near Jomeo.

1234 Jomeo and Jageli stop feeding, travel north.

1238 Jomeo and Jageli climb to feed on *ngongo* leaves. The group is near.

1243 Jomeo and Jageli pant-hoot and the group replies with pant-hoots.

1255 Melissa, Gremlin, and their infants (Gimble, Getty), Goblin, Skosha, Kristal, Wilkie, Beethoven, Miff (1) and Mo, and Michaelmas join, with many food calls and pant-grunts.

1303 Gigi (¼), Fifi, with Freud and Fanni, Little Bee and Tubi, Atlas, Sprout and Spindle, Patti and Tapit, Kidevu and Konrad, Satan, and Evered now join, and all feed together. Some feed on *ngongo* leaves, others feed on *muhandehande* fruits. The youngsters play. Satan is ahead of the group. Observation on the group is poor.

1307 Jomeo stops feeding, climbs down. Gigi (¼) pant-grunts to him, Jomeo hair out and displays toward her. Gigi climbs a tree, Jomeo follows her up the tree. Gigi leaps to the ground, presents, Jomeo inspects, and then wipes his finger on the ground. Jomeo travels on and all follow northeast along the hillside.

1310 The group feeds on *muhandehande* and *magusuhande* fruits.

1318 Jomeo summons (courts) Miff (1) and mates her. Miff squeals. Mo interferes. The group is nearby.

1325 Jomeo feeds, moving ahead of the group on his own.

1405 Jomeo feeds and travels west toward the others and rejoins them. Some are feeding, some resting on the ground, and others resting in the trees. Jomeo climbs a tree to join Evered, Satan, Fifi, Freud, Frodo, and Fanni. Spindle is there too. Satan and Evered groom mutually and Jomeo sits nearby.

1412 Evered and Jomeo groom mutually. The males keep gazing north toward Upper Linda. Fanni and Tapit play.

1430 Jomeo and the group move off north.

1438 They arrive at Linda Stream. There Freud and Atlas display. Atlas displays toward Fifi and Gigi (¼). Freud displays toward Little Bee. The females waa-bark and climb trees. Evered, Satan, Mustard, Goblin display. Freud follows and attacks Little Bee. The males drum. Atlas displays past field assistant Mpongo, slapping him. There is a cacophony of sound, of waa's, pant-hoots, and screams. All head north.

1443 Jomeo and the group climb to feed on *msiloti* leaves in Linda Valley. Gimble, Tubi, and Tapit feed and play.

1452 Goblin mates Miff (1), who is silent. Mo watches.

1455 They stop feeding and move to the north. The group climbs the hill between Linda and Rutanga streams. Evered and Satan are ahead.

1506   The group stops and feeds on *mashindwi* leaves. Some of them feed on *magusuhande* fruits.

1528   They climb down and head northeast, feeding on fallen *magusuhande* and *muhandehande* fruits.

1535   They continue to feed, now on *muhandehande*, climbing trees and gazing toward Upper Rutanga Stream.

1546   They stop feeding and head northwest, all traveling together.

1616   They arrive at Rutanga Stream and drink. They are very silent and the youngsters do not play.

1622   Evered, Satan, Jomeo, Goblin, Mustard, Atlas, Jageli, Beethoven, Wilkie, Freud, Gigi (¼), Sprout (¼) travel north of Rutanga. The others drop off at this point, heading back south. Jomeo and his group start to patrol.

1633   They sniff the ground along the trail, but do not show signs of fear. They move calmly on to the north.

1638   They come to the edge of the valley, arriving at Busindi, and head northwest, traveling fast.

1645   They inspect a termite heap, sniff around, gaze at tools left there by others, pick them up, and sniff them. Then they move on.

1650   They drink again, and sniff around another termite heap, without fishing. They travel on, now moving very cautiously, peering into the vegetation, pausing to listen from time to time. They are all close together.

1655   Satan, Jomeo, and Mustard climb to pick fruits. The others pick up fallen ones. They feed.

1659   They travel on, chewing wadges, northwest in Busindi.

1702   They rest, listening and gazing north.

1706   They move on, still heading northwest.

1708   They see nine bushpigs. There are two large young, two small young, and three very tiny piglets. Two adults, both female. The pigs snort. The chimps move around the area. The adult pigs are very fierce, snorting loudly, while all the young keep close together. The chimps run from the pigs. Goblin, Atlas, Sprout, and Gigi sway the undergrowth, climb and waa-bark at the pigs, and shake branches at them. Now the other chimps also threaten and show signs of hunting. One of the pigs charges Satan and Atlas.

1715   Goblin, Satan, Atlas, and Gigi are very close together facing an adult pig. They are giving very loud waa-barks. Sprout, Jageli, and Beethoven are above, also giving waa-barks. This is the end of the patrol.

1730   Jomeo and the others climb trees and rest.

1745   They come down and now head east following the path used by fishermen.

1750   Jomeo and Beethoven travel together, ahead of the others, who go south and vanish. They pant-hoot but Jomeo and Beethoven remain silent. Satan, Jageli, and Wilkie follow behind Jomeo. Jomeo and Beethoven pant-hoot.

1803   Jomeo and Beethoven arrive at Rutanga Falls. Jomeo displays, throws rocks, drags branches, and drums. Beethoven follows behind and also displays and drums once. Jomeo, hair out, gives roar pant-hoots as he displays.

1808   Jomeo stops displaying and they head south.

1810   Jomeo pant-hoots. Beethoven pant-hoots too. Satan, Jageli, Wilkie reply from behind.

1820   Jomeo and Beethoven climb and feed on *msalasi* fruits. Satan, Jageli, Wilkie join in feeding, climbing a different tree. They are in Rutanga Valley.

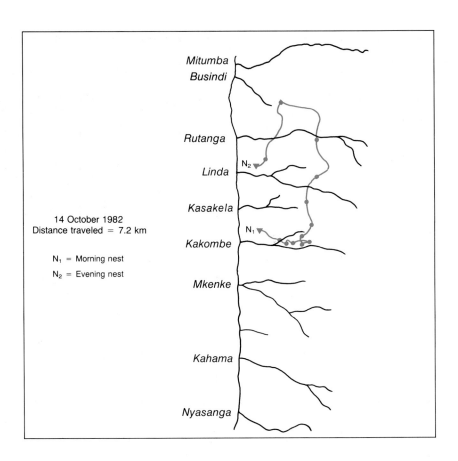

14 October 1982
Distance traveled = 7.2 km

N₁ = Morning nest

N₂ = Evening nest

1826   Wilkie and Jageli leave Satan and approach Jomeo. They climb and feed with Jomeo and Beethoven.

1830   Jomeo, Beethoven, Wilkie, and Jageli feed on *msiloti* leaves. Satan moves off alone.

1834   Jomeo, Beethoven, Wilkie, and Jageli climb down and head south to LK8 (Linda Karongo 8).

1841   They arrive at LK8 and head south.

1843   They climb to feed on fig fruits in LK8.

1845   They stop feeding, climb down, and head for LK7. Wilkie leaves Jomeo and heads toward LK7 whence come loud pant-hoots and other calls of the big group. Satan's and Goblin's calls are distinguished. Jomeo, Beethoven, and Jageli head west.

1851   They climb to feed on *msiloti* leaves in LK7.

1853   They pant-hoot in reply to the big group.

1915   They stop feeding, cross into other trees, and make their nests.

Mwisho (The End)

Once again, the association chart and travel map show Jomeo's activities and route in detail.

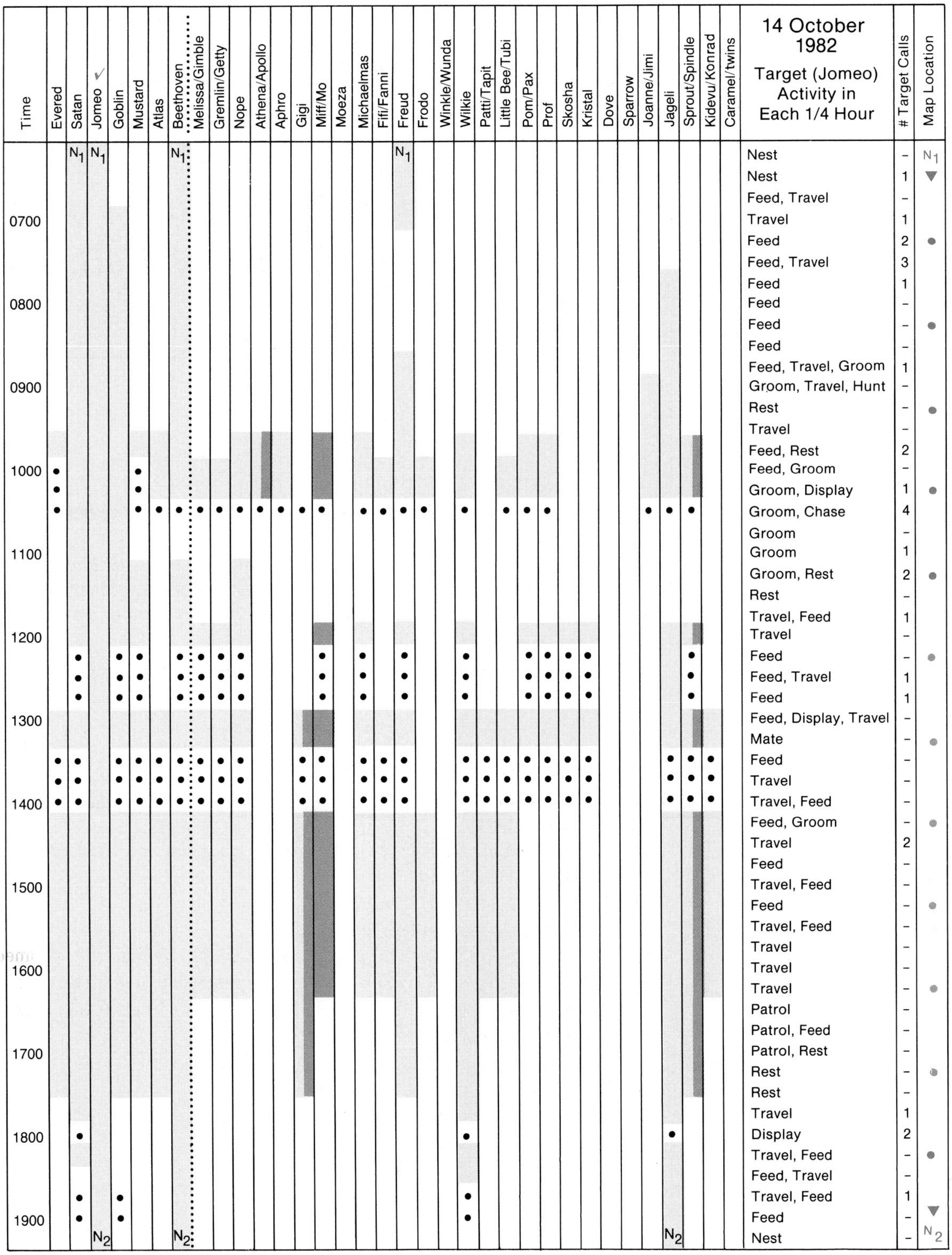

14 October 1982

Target (Jomeo) Activity in Each 1/4 Hour

Time	Activity	# Target Calls	Map Location
	Nest	–	N₁
	Nest	1	▼
0700	Feed, Travel	–	
	Travel	1	
	Feed	2	•
	Feed, Travel	3	
0800	Feed	1	
	Feed	–	•
	Feed	–	•
	Feed	–	
0900	Feed, Travel, Groom	1	
	Groom, Travel, Hunt	–	
	Rest	–	•
	Travel	–	
1000	Feed, Rest	2	
	Feed, Groom	–	
	Groom, Display	1	•
	Groom, Chase	4	
1100	Groom	–	
	Groom	1	
	Groom, Rest	2	•
	Rest	–	
1200	Travel, Feed	1	
	Travel	–	
	Feed	–	•
	Feed, Travel	1	
1300	Feed	1	
	Feed, Display, Travel	–	
	Mate	–	•
	Feed	–	
1400	Travel	–	
	Travel, Feed	–	
	Feed, Groom	–	•
	Travel	2	
1500	Feed	–	
	Travel, Feed	–	•
	Feed	–	
	Travel, Feed	–	
1600	Travel	–	
	Travel	–	•
	Travel	–	
	Patrol	–	
1700	Patrol, Feed	–	
	Patrol, Rest	–	
	Rest	–	•
	Rest	–	
1800	Travel	1	
	Display	2	•
	Travel, Feed	–	
	Feed, Travel	–	
1900	Travel, Feed	1	▼
	Feed	–	N₂
	Nest	–	

Column headers (individuals): Evered, Satan, Jomeo ✓, Goblin, Mustard, Atlas, Beethoven, Melissa/Gimble, Gremlin/Getty, Nope, Athena/Apollo, Aphro, Gigi, Miff/Mo, Moeza, Michaelmas, Fifi/Fanni, Freud, Frodo, Winkle/Wunda, Wilkie, Patti/Tapit, Little Bee/Tubi, Pom/Pax, Prof, Skosha, Kristal, Dove, Sparrow, Joanne/Jimi, Jageli, Sprout/Spindle, Kidevu/Konrad, Caramel/twins

Nest markers: N₁ (Satan, Jomeo, Beethoven, Freud) at top; N₂ (Satan, Beethoven, Jageli) at bottom.

Legend

N = Nest

☐ = Easily visible to observers

⊡ = Within 100 meters of target

■ (dark) = ♀ in estrus

▨ (light) = Partially swollen ♀

C. Observation Hours over Five Years of Follows

For each of the adult males and for the most frequently followed adult females between 1978 and 1982 are given (a) the number of target hours, (b) the number of hours observed when in groups with other targets, and (c) the number of hours seen in camp.

Year and hours of observation	Males							Females											
	HM[a]	FG[b]	EV	ST	JJ	SH[c]	GB	PS[d]	PM	ML	GM[e]	AT	MF	PL[f]	FF	WK	GG	LB	PI
1978																			
No. target hours	207	360	444[g]	216	220	198	242	377	90	368		138	230	122	236	218	83	186	
No. hours seen with other targets	495	566	419	436	361	327	457	196	542	200		588	158	154	222	433	491	324	
No. hours seen in camp	51	65	81	90	49	101	170	142	133	215		39	76	86	157	179	78	112	
Total observation hours	753	991	944	742	630	626	869	715	765	783		765	464	362	615	830	652	622	
1979																			
No. target hours	112	374	108	137	142	157	272	154	67	193		138	100	106	215	461	53	112	89
No. hours seen with other targets	372	368	257	322	328	240	379	108	202	180		168	88	65	162	181	263	203	260
No. hours seen in camp	81	159	84	78	102	136	221	109	119	185		63	92	105	161	209	128	155	175
Total observation hours	565	901	449	537	572	533	872	371	388	558		369	280	276	538	851	444	470	524
1980																			
No. target hours	56	234	96	115	133		219	208	60	170		70	116	82	216	126	31	105	94
No. hours seen with other targets	246	403	389	368	331		390	81	216	238		85	129	40	279	170	289	142	203
No. hours seen in camp	36	67	74	52	40		102	51	49	76		24	21	45	83	96	44	15	47
Total observation hours	338	704	559	535	504		711	340	325	484		179	266	167	578	392	364	262	344
1981																			
No. target hours	45	122	135	95	139		185	60	40	200	205	46	118	101	448	164	20	40	92
No. hours seen with other targets	74	212	241	239	164		217	67	91	184	336	73	109	36	152	126	255	84	122
No. hours seen in camp	21	92	133	148	91		241	61	60	147	144	30	56	38	212	178	71	32	78
Total observation hours	140	426	509	482	394		643	188	191	531	685	149	283	175	812	468	346	156	292
1982																			
No. target hours		95	116	133	92		201		272	186	157	108	76	102	269	135	46	92	67
No. hours seen with other targets		128	266	332	264		291		158	203	286	113	109	20	261	162	309	219	145
No. hours seen in camp		25	100	109	70		180		84	194	193	14	36	64	253	156	42	63	95
Total observation hours		248	482	574	426		672		514	583	636	235	221	186	783	453	397	374	307

a. Died in May 1981.
b. Disappeared in May 1982.
c. Died in October 1979.
d. Died in February 1982.

e. Adolescent until 1981.
f. Died in September 1982.
g. Includes follows during EV-WK, EV-AT, and EV-DO consortships.

D. Association Matrixes

The tables that follow present the raw data on which a number of the figures in Chapters 7 and 8 are based.

First are shown the 1978 and 1979 association matrixes for those adult and adolescent males and adult females who were targets for at least 75 hours during the year (Appendix C). For each individual the percentage of total target hours that were spent in the same party as each other individual is shown. The total number of target hours and the percentage of time spent alone are also shown.

	HM	FG	EV	ST	JJ	SH	GB	MU	AL	PS	PM	ML	AT	MF	PL	FF	WK	GG	LB	NP	PI	SW	HR	DO	Number of target hours	Percentage of target hours spent alone
1978																										
HM		61	33	42	34	36	37	46	51	2	16	7	37	16	12	8	23	31	23	7	9	8	17	11	207	21
FG	49		23	33	30	18	33	26	37	8	11	8	33	8	7	10	18	28	12	6	17	8		8	360	15
EV	27	36		32	13	20	20	17	28	8	8	10	18	7	9	10	22	21	15	4	12	7	15	13	127	12
ST	44	46	26		24	26	35	28	26	6	10	11	23	13	6	13	20	22	19	4	18	11	13	6	216	25
JJ	33	44	13	35		30	36	31	27	2	7	8	25	8	5	5	18	25	12	2	10	5	4	7	220	33
SH	36	41	30	36	26		37	28	28	13	17	9	29	7	6	7	26	38	7	2	6	7		4	198	30
GB	20	29	10	21	22	15		20	17	3	4	15	18	2	2	3	11	14	6		9	4		2	242	42
PS	5	7	5	6	4	5	4	4	6		97	9	6	2	3	7	9	7	5	*	3	5			377	66
PM	7	7	6	3	7	6	6	5	7	95		6	7	1	9	5	4		5	2	7	8			90	69
ML	6	10	9	7	7	6	16	7	9	6	6		8	4	7	11	12	7	22	3	12	5			368	53
AT	19	30	39	8	18	14	18	18	89	5	5	3		12	1	4	1	11	18	5	9	6		2	138	26
MF	12	22	7	11	11	9	9	9	19	2	2	4	12		2	9	8	20	7	6	8	6		2	230	46
PL	6	6	1	5	4	4	4	2	4	3	6	11	1	6		6	3	5	12	2	11	4			122	65
FF	9	15	6	12	6	8	13	7	14	4	3	10	11	9	*		8	16	17	*	17	7	6	1	236	48
WK	10	11	14	6	8	5	5	15	6	6	4	2	6	3	5	7		9	7		6				218	62
LB	9	12	4	10	4	3	10	5	12	3	4	20	10	14	6	20	6	17			20	9			186	27
1979																										
HM		45	23	14	11	1	10	19	21	3	4	4					4	11		7	8		6	1	111	29
FG	53		20	26	41	27	38	28	27	1	2	8	4	4	2	4	6	24	3	3	12	4	13		379	13
EV	24	45		27	24	23	15	29	31	5	6	5	4	4	8	9	6	22	1	13	7	4	15		107	17
ST	9	27	11		14	12	30	15	18	2	5	5	12	3			9	17	7	4	8	4	12	3	177	33
JJ	17	34	18	17		16	21	20	19	1	1	8	2	3	3	6	10	21	4	5	7	7	14	3	146	35
SH	8	16		10	21		18	6	7	2	1	4	1	2	3	2	4	9	1		5	1	1	4	159[a]	55
GB	15	16	11	25	12	16		15	14	1	2	9	6	6	2	8	7	16	6	2	9	2	5	2	277	41
PS	2	1	*	1	1		2	1	3		35	1	1	*	1	1	1	*				*			154	88
PM	3	4	*	1	3		4	*	5	38		6	3			1	5	3	2		3	8			78	70
ML	2	12	6	7	10	7	16	7	11	5	9		9	2	3	10	8	10	20	*	18	4		5	183	41
AT	2	8	6	4	3		7	3	4	1	1	4			1	3	9	12	14	16	11	1		5	141	39
MF				4	1		1	*	*								3	5		14	3				102	72
PL	*	*	2	3		*	2	1	2			3	3	4		4		3	2	5		7	1		115	80
FF	3	5	4	4	2	4	6	2	6	2	3	13	4	6	5		7	10	17	1	13	3	2	3	262	67
WK	7	7	5	6	3	3	5	4	3	1	2	3	3	2	1	5		5	3	1	3	5	2	2	465	75
GG	22	35	22	29	16		22	25	29	2	4	5	25	17	3	14	22		11	3	28	10			116	35
LB	1	1	4	1	1		36	1	11	3	3	13	12	2	*	13	8	3			5	2			112	58
PI	5	12	3	4	3	31	37	14	4	1	5	18	5	6	5	18	3	11	23	1					92	21

Note

* = <0.5% target hours association time.

a. To October 1979.

Below are the out-of-camp association indexes for adult and adolescent males and females during 1981. The matrix was compiled by totaling for each individual the number of quarter-hours (or portions thereof) that he or she was seen with each other community member. For each dyad (A and B), association time out of camp is expressed as a percentage of total time each was seen without the other plus time they were seen together. That is,

$$\frac{A + B \text{ together}}{(A \text{ without } B) + (B \text{ without } A) + (A \text{ and } B \text{ together})} \times 100.$$

	FG	EV	ST	JJ	GB	MU	AL	PS	PM	ML	GM	NP	AT	MF	MZ	PL	SS	FF	FD	WK	WL	GG	SW	PI	LB	BE	DO	KD	JG	JO	SP	SY
FG																																
EV	40																															
ST	21	18																														
JJ	32	23	16																													
GB	20	22	35	20																												
MU	27	21	19	20	16																											
AL	26	24	15	17	16	21																										
PS	4	3	2	4	2	3	5																									
PM	4	3	2	5	2	3	6	82																								
ML	6	5	8	4	7	5	6	3	2																							
GM	7	5	13	4	9	7	6	3	2	54																						
NP	4	4	4	3	3	6	5	—	—	1	1																					
AT	13	9	10	8	7	14	21	5	4	6	7	5																				
MF	12	9	8	8	7	10	13	5	5	5	8	5	8																			
MZ	9	7	6	5	5	7	9	4	4	4	7	3	5	62																		
PL	3	2	1	1	1	2	2	2	2	3	3	*	1	1	1																	
SS	3	2	1	1	1	2	2	2	2	3	3	*	2	1	1	70																
FF	8	7	10	4	9	5	4	2	1	9	11	1	3	4	3	1	1															
FD	14	14	12	9	14	9	9	2	2	8	12	2	4	4	3	1	1	33														
WK	6	5	7	4	5	6	7	3	3	7	8	2	6	9	6	2	2	7	7													
WL	9	7	11	6	7	8	10	4	4	8	9	2	7	10	7	2	2	7	9	34												
GG	27	19	13	17	15	16	22	5	5	6	7	2	11	13	12	3	3	5	9	9	8											
SW	5	3	6	3	4	5	7	1	1	3	3	3	12	4	2	2	2	2	3	2	4	2										
PI	9	7	7	6	9	7	11	3	3	8	8	1	7	8	8	2	2	7	8	7	6	31	2									
LB	5	3	4	3	4	4	3	5	5	11	9	2	4	5	2	2	2	6	6	7	7	8	*	13								
BE	48	35	23	30	23	27	27	4	4	6	8	4	18	12	7	3	3	7	13	6	7	24	7	10	5							
DO	3	2	3	3	2	3	5	4	4	3	2	3	5	2	2	2	2	1	1	1	1	1	14	1	1	3						
KD	10	9	16	6	10	13	11	4	4	5	5	1	13	3	2	2	3	2	3	2	4	4	11	3	1	11	5					
JG	37	29	26	25	23	27	21	4	4	6	8	2	14	6	4	3	3	8	13	6	8	18	6	10	5	23	3	15				
JO	1	1	*	1	1	2	2	1	1	*	*	—	*	*	*	—	—	—	—	*	*	1	—	1	—	2	—	—	15			
SP	8	6	5	8	12	11	13	1	1	1	1	4	6	7	4	—	—	2	5	2	2	10	—	7	3	9	—	*	9	—		
SY	3	3	4	4	4	7	6	1	1	*	1	6	2	7	4	—	—	1	2	2	2	5	—	3	3	4	—	—	3	—	20	

* = <0.5% hours spent together.

E. Supporting Data on Sexual Behavior

The three tables that follow provide additional information relative to Chapter 16, Sexual Behavior.

Table E1 (overleaf) The sexual situation of Kasakela females during the last four days of maximum swelling, 1976 to 1983. The number of swellings recorded for each female and the situations in which conceptions are thought to have occurred are indicated.

When a female is known to have been in association with two or more Kasakela males, she is shown in the Group situation (48.5 percent); when known or inferred to have been on a consortship with a Kasakela male, in the Consort situation (13.0 percent); when absent during a period that did not coincide with the absence of *any* Kasakela male, in the Away situation, for she was almost certainly visiting with males of a neighboring community (15.5 percent). A female is indicated as Uncertain when, during a given month, there was insufficient information in the records to permit assigning her to any of the first three situations (23.0 percent).

The data for the fourteen central females are the most precise; for the eight peripheral females they are less so, particularly for the last three, who were rarely observed and during some months were not seen at all.

Year		PS	ML	PL	MF	FF	NP	WK	LB	PI	PM	GM	MZ	GG	AT	SW	DO	HR	KD	CD	SP	CA	JO	Total no. of swellings
1976	*Total*	3	1	4	1			4	3	9	5			8	1		3	5						47
	Group	1	1ᵃ	2ᵃ	1			3	1	7	4			5			3	1						29
	Consort	1								2ᵇ	1			1				1						5
	Away	1		1				1						1				2						6
	Uncertain			1						1	1			1	1			2						7
1977	*Total*	4	1		3			5		7	10			8	3	7	4	*3*						55
	Group	2			1			2		2ᵃ	4			4	2	4	2	3						26
	Consort	2ᵃ	1ᵃ					2		3					1	1								10
	Away				1ᵃ						3			3		2	1							10
	Uncertain				1			1		2	3			1			1							9
1978	*Total*						1	5		1	2			9	6	1	4	2	3					34
	Group							3			2ᵃ			4	5	1ᵃ	3ᵃ	2						20
	Consort							2ᵃ		1ᵃ							1							4
	Away													5										5
	Uncertain						1								1ᶜ				3					5
1979	*Total*						2							8				*3*	4					17
	Group													8										8
	Consort						2ᵃ																	2
	Away																							—
	Uncertain																	3	4					7
1980	*Total*				5							1		6					5			*2*	*3*	22
	Group				3ᵃ							1		4					1					9
	Consort				2									2										4
	Away																							—
	Uncertain																		4			2	3	9
1981	*Total*											6	2	5					6		*3*	*3*		25
	Group											5ᵃ	2	4										11
	Consort													1					1					2
	Away																							—
	Uncertain											1							5		3	3		12
1982	*Total*				3		3				3			6	1	2	3				*2*			21
	Group						2ᵃ							4	1	1ᵃ	1							8
	Consort				1		1							1										3
	Away				2						2					1	2							6
	Uncertain										1			1							2			4
1983	*Total*		4		2			4	4	5	7		7	8			6			6	4			59
	Group		2ᵃ					4	1	2			3	6			2			4				25
	Consort		1							1ᵃ	4						1							7
	Away				2ᵃ			2	1	2		4					2			1				15
	Uncertain		1					1	1	1				2			1			1	4			12

Note: Numbers in italics = *at least* this number of swellings.

a = conception thought to have taken place.

b = conception in group *or* consortship.

c = conception when situation of female uncertain.

Table E2 Situation between 1967 and 1984 of each adult or late-adolescent Kasakela female during the periovulatory period when conception is thought to have occurred (based on a gestation period of 229.4 days). In some cases females returned from a consortship just before deflation. If they were then mated by other Kasakela males, the consort male may or may not have been the father of the infant (for example, infants Plato, Moeza, Wilkie, Sandi, Tubi). In two cases the estimated conception date fell midway between two periods of swellings: Otta could have been sired in either the group or the consort situation. The abbreviation m/sb indicates miscarriage or stillbirth.

| Female | | Infant | | Situation | Consort babies | | | |
| | | | | | Father | | | No. days of |
Name	Age	Name	Date of birth	of female	Name	Age	Rank	consortship
Olly	31	m/sb	September 1967	Consort	Leakey	31	Middle	27
Flo	39	Flame	August 1968	Group				
Melissa	19	m/sb	November 1969	Consort	Faben	22	Middle	15
	20	Gremlin	November 1970	Consort	Evered	18	Middle	18
	26	♂	January 1976	Uncertain				
	26	Genie	November 1976	Group				
	27	Twins	October 1977	Consort	Satan	22	High	4
	34	m/sb	April 1984	Group				
Passion	20	Prof	October 1971	Group				
	26	Pax	December 1977	Consort	Evered	25	High	39
Nope	20	Lolita	June 1973	Away				
	27	Hepziba	June 1980	Consort	Sherry	18	Middle	10
	30	Noota	November 1982	Group				
Athena	14	Atlas	September 1967	Group				
	19	Aphro	June 1973	Group				
	25	Apollo	February 1979	Uncertain				
Nova	15	Skosha	March 1970	Uncertain				
Pallas	15	Plato	September 1970	Consort/Group	(Faben)	23	Middle	7
	18	Villa	April 1974	Consort	Humphrey	27	High	13
	20	Banda	February 1976	Uncertain				
	21	Kristal	August 1977	Group				
Miff	12	Moeza	January 1969	Consort/Group	(Figan)	15	Middle	7
	17	Michaelmas	October 1973	Consort	Humphrey	27	High	18
	21	Mo	April 1978	Away				
	27	Mel	January 1984	Away				
Winkle	13	Wilkie	October 1972	Consort/Group	(Evered)	20	High	5
	19	Wunda	December 1978	Consort	Evered	26	High	16
Dove	13	Dominie	November 1972	Away				
	19	Dapples	November 1978	Consort	Evered	26	High	35
Fifi	12	Freud	May 1971	Away				
	17	Frodo	June 1976	Consort	Sherry	14	Low	17
	22	Fanni	March 1981	Group				
Sparrow	13	Sandi	October 1973	Consort/Group	(Satan)	18	Middle	18
	18	Barbet	September 1978	Group				
	23	Sheldon	May 1983	Group				
Patti	17	♂	April 1978	Group				
	18	Tapit	February 1979	Consort	Goblin	14	Low	3
	23	Tita	January 1984	Consort	Mustard	19	Low	7
Little Bee	15	m/sb	April 1976	Group				
	16	Tubi	August 1977	Consort/Group	(Figan)	23	High	23
	23	Darbie	March 1984	Group				
Gilka	13	Gandalf	June 1974	Away				
	14	Otta	July 1975	Consort/Group	(Sherry)	13	Low	14
	16	Orion	August 1976	Uncertain				
Pom	13	Pan	October 1978	Group				
Gremlin	11	Getty	May 1982	Group				

Table E3 Consort partner preferences from 1966 to 1983. Number in green in each cell shows the number of consortships between the specific male and female; number in black in each cell shows the total number of consortships in which the *female* was involved during the reproductive life of the male.

In each cell below, values are given as **green / black** (specific consortships / total consortships); a single value is the green (specific) number only.

	HM	FB	FG	EV	ST	JJ	SH	GB	MU	AL	HX	LK	HG	DV	*GOL*	MK	RX	*HH*	WZ	*WW*	*CH*	*GI*	PP	*DE*	Total ♀ consortships
PS	5/1	2	5	5/3[a]	4	5	3	2				2	2/1		2	2		2	2	2	2	2		2	5
ML	12	11/6[a]	12/3	14/4[a]	6/3[a]	14	1	2	2	2	1	11	11	5	11	11	5	11	5	11	11	11	6	11	16
NP	5	3	6	8/1	8	8/4	5/1[a]	5/2				3	3		3	3		3		3	3	3		3	8
AT	17/5	7/3	17/2	18/4	17/2	18/1	10	11	1	1	1	1	7	1	7	7	1	7/1	1	7	7/1	7	1	7	19
GG	39/11	36/4	43/7	43/8	29/4	43/1	6	4		1	1	22/1	36/2	4	28/3	36/1	6	28	14	28	28	28	1	28/1	43
PL	20/3[a]	19/6[b]	20	20	9/1	20/1	8/2				1	12/2	16/1	6	12/1	16/1	4	12	4	12/2	12	12	1	12	20
MF	11/2[a]	9	11/2[b]	12	6/3	12	4	3	1	1	2	6	9	3	6/1	9	3	6/1	6	6	6	6/1	5	6/2	12
WK	12	6/1	12	12/4[a,b]	12/1	12/4	12	3				3	6/1		6	6/1		6		6	6	6/1		6	13
DO	3		3	4/2[a]	4	4	3/2	1	1	1															4
FF	18/5	12	20	20/2	9/1	20/2	6/1[a]	2				12/7	12		12	12/1	4	12	6	12	12	12/1	3	12	20
SW	5	4/2	6	7/1	7/2[b]	7	5/1	3		1/1		4			4										7
PI	13/2	4/1	13/1	14/1	14/3	14	13/3	6/1	1/1[a]	1		3/1			3										14
LB	7	4/1	7/2[b]	7	7/2	7/2	7	1				3			3										7
GK	9/3	9	9	9	9/3	9	9/2[b]					9			2	9/1				2	2	2		2	9
PM				5	5/3	5	5	5/1	5/1																5
MZ			1	1	1/1	1	1	1	1																1
KD	2		3	3/1	3/2	3	3																		3
FLO	5	5	5	5	2/1	5						5/2	5	1	5/1	5	1	5/1	3	5	5	5	1	5	5
OL	17	17	17	17		17					1	17/6[a]	17/5	8/2	17/1	17	1	17/3	17	17	17	17	1	17	17
S	7	7	7	7		7					5/2	7	7	7	7/1	7	5/1	7	7	7	7	7	5/1	7	5
MN	3	3	3	3/1		3/1					3	3	3	3	3	3/1	3	3	3	3	3	3	3	3	3
BM	3	3	3	3		3					1	3	3	3	3	3	3	3/1	3/1	3/1	3	3	3	3	3
NV	13/3	13	13/1	13/2		13/2					2	13	13	3	13	13/1	3	13	10	13/2	13/1	13/1	2	13	13
P	6	6/5	6	6		6/1					1	6	6	6	6	6	6	6	6	6	6	6	1	6	6
Total ♂ consortships	34	24	24	34	31	15	16	4	2	2	2	16	13	2	10	5	2	4	3	5	2	4	1	3	258

Note: Individuals in box at upper left are those for whom the most longitudinal data are available. Those shown in italics separated to form the Kahama community.

a. Probable conception during consortship.

b. Possible conception during consortship.

References

Albrecht, H., and S. C. Dunnett. 1971. *Chimpanzees in Western Africa.* Munich: Piper-Verlag.

Alexander, R. D. 1971. The search for an evolutionary philosophy of man. *Proc. Roy. Soc. Victoria*, 84:99–120.

Alexander, R. D., and D. W. Tinkle. 1968. A comparative book review of *On Aggression* by Konrad Lorenz and *The Territorial Imperative* by Robert Ardrey. *Bioscience*, 18:245–248.

Allen, M. 1981. Individual copulatory preference and the "strange female effect" in a captive group-living male chimpanzee (*Pan troglodytes*). *Primates*, 22:221–236.

Altmann, S. A., ed. 1967. *Social Communication among Primates.* Chicago: University of Chicago Press.

Anderson, J. R., and A. S. Chamove. 1979. Contact and separation in adult monkeys. *S. Afr. J. Psychol.*, 9:49–53.

Andrew, R. J. 1963. The origins and evolution of the calls and facial expressions of the primates. *Behaviour*, 20:1–109.

——— 1972. The information potentially available in mammal displays. In R. A. Hinde, ed., *Non-Verbal Communication*, pp. 179–203. Cambridge: Cambridge University Press.

Argyle, M. 1972. Non-verbal communication in human social interaction. In R. A. Hinde, ed., *Non-Verbal Communication*, pp. 243–267. Cambridge: Cambridge University Press.

Bachmann, C., and H. Kummer. 1980. Male assessment of female choice in hamadryas baboons. *Behav. Ecol. Sociobiol.*, 6:315–321.

Baldwin, P. J., W. C. McGrew, and C. E. G. Tutin. 1982. Wide-ranging chimpanzees at Mt. Asserik, Senegal. *Int. J. Primatol.*, 3:367–385.

Bandura, A. 1970. The impact of visual media on personality. In E. A. Rubenstein and G. V. Coelho, eds. *Behavioral Sciences and Mental Health: An Anthology of Program Reports*, pp. 398–419. Washington, D.C.: Government Printing Office.

Barnett, S. A. 1958. The nature and significance of exploratory behaviour. *Proc. Roy. Phys. Soc. Edinburgh*, 27:41–45.

Bartlett, A. D. 1885. On a female chimpanzee now living in the Society's gardens. *Proc. Zool. Soc.* (London), 673–676.

Bauer, H. R. 1976. Ethological aspects of Gombe chimpanzee aggregations with implications for hominization. Ph.D. diss., Stanford University.

———— 1979. Agonistic and grooming behavior in the reunion context of Gombe Stream chimpanzees. In D. A. Hamburg and E. R. McCown, eds., *The Great Apes*, pp. 395–404. Menlo Park, Calif.: Benjamin/Cummings.

Bauer, H. R., and M. Philip. 1983. Facial and vocal individual recognition in the common chimpanzee. *Psych. Rec.*, 33:161–170.

Beach, K., R. S. Fouts, and D. H. Fouts. 1984. Representational art in chimpanzees. *Friends of Washoe*, 3:2–4; 4:1–4.

Beatty, H. 1951. A note on the behavior of the chimpanzee. *J. Mammal.*, 32:118.

Bielert, C., and L. A. Walt. 1982. Male chacma baboon sexual arousal: mediation by visual cues from female conspecifics. *Psychoneuroendocrinology*, 7:31–48.

Bigelow, R. S. 1969. *The Dawn Warriors: Man's Evolution towards Peace*. Boston: Little, Brown.

Blanc, A. C. 1961. Some evidence for the ideologies of early man. In S. L. Washburn, ed., *Social Life of Early Man*, pp. 119–136. Chicago: Viking Fund Publications in Anthropology no. 31.

Blodgett, F. M. 1963. Growth retardation related to maternal deprivation. In A. J. Solnitz and S. A. Provence, eds., *Modern Perspectives in Child Development*, pp. 83 ff. New York: International University Press.

Boehm, C. 1981. Parasitic selection and group selection: a study of conflict interference in rhesus and Japanese macaque monkeys. In A. B. Chiarelli and R. S. Corruccini, eds., *Primate Behavior and Sociobiology*, pp. 161–182. Berlin: Springer-Verlag.

Boesch, C. 1978. Nouvelles observations sur les chimpanzés de la forêt de Tai (Côte d'Ivoire). *Terre et Vie*, 32:195–201.

Boesch, C., and H. Boesch. 1981. Sex differences in the use of natural hammers by wild chimpanzees: a preliminary report. *J. Hum. Evol.*, 10:585–593.

———— 1983. Optimization of nut-cracking with natural hammers by wild chimpanzees. *Behaviour*, 83:265–286.

———— 1984. Mental map in wild chimpanzees: an analysis of hammer transports for nut cracking. *Primates*, 25:160–170.

Boreman, T. 1739. *A Description of Some Curious and Uncommon Creatures*. London.

Bowlby, J. 1973. *Attachment and Loss*. Vol. 2, *Separation*. London: Hogarth Press.

Brewer, S. 1978. *The Forest Dwellers*. London: Collins.

Brink, A. S. 1957. The spontaneous fire-controlling reaction of two chimpanzee smoking addicts. *S. Afr. J. Sci.*, 241–247.

Brodkin, A. M., D. Shrier, R. Angel, E. Alger, W. A. Layman, and M. Buxton. 1984. Retrospective reports of mothers' work patterns and psychological distress in first-year medical students. *J. Am. Acad. Child Psych.*, 4:479–485.

Brown, G. W., M. Bhrolchain, and T. Harris. 1975. Social class and psychiatric disturbance among women in an urban population. *Sociology*, 9:225–254.

Brown, J. L. 1975. *The Evolution of Behavior*. New York: W. W. Norton.

Buchanan, J. P., T. V. Gill, and J. T. Braggio. 1981. Serial position and clustering effects in a chimpanzee's "free recall." *Mem. Cog.*, 2:651–660.

Buffon, G. L. L. 1766. *Histoire naturelle, générale et particulière*, vol. 14, chaps. 1–3. Paris.

Burt, W. H. 1943. Territoriality and home range concepts as applied to mammals. *J. Mammal.*, 24:346–352.

Burton, F. 1971. Sexual climax in female *Macaca mulatta*. *Proc. Third Int. Cong. Primatol.*, 3:180–191.

Busse, C. D. 1976. Chimpanzee predation on red colobus monkeys. Manuscript.

———— 1977. Chimpanzee predation as a possible factor in the evolution of red colobus monkey social organization. *Evolution,* 31:907–911.

———— 1978. Do chimpanzees hunt cooperatively? *Am. Nat.,* 112:767–770.

Butler, R. A. 1965. Investigative behavior. In A. M. Schrier, H. F. Harlow, and F. Stollnitz, eds., *Behavior of Non-Human Primates,* vol. 2, pp. 463–490. New York: Academic Press.

Bygott, J. D. 1972. Cannibalism among wild chimpanzees. *Nature,* 238:410–411.

———— 1974. Agonistic behaviour and dominance in wild chimpanzees. Ph.D. diss., Cambridge University.

———— 1979. Agonistic behavior, dominance, and social structure in wild chimpanzees of the Gombe National Park. In D. A. Hamburg and E. R. McCown, eds., *The Great Apes,* pp. 405–428. Menlo Park, Calif.: Benjamin/Cummings.

Campbell, D. T. 1972. On the genetics of altruism and the counter-hedonic components of human culture. *J. Soc. Issues,* 28:21–37.

Carpenter, C. R. 1934. A field study of the behavior and social relations of the howler monkeys (*Alouatta palliata*). *Comp. Psychol. Monogr.,* 10:1–168.

———— 1940. A field study in Siam of the behavior and social relations of the gibbon (*Hylobates lar*). *Comp. Psychol. Monogr.,* 16:1–212.

———— 1942. Sexual behavior of free-ranging rhesus monkeys *M. mulatta.* *J. Comp. Psychol.,* 33:113–162.

Chamove, A. S., H. Harlow, and G. Mitchell. 1967. Sex differences in the infant-directed behavior of pre-adolescent rhesus monkeys. *Child Dev.,* 38:329–335.

Chance, M. R. A. 1961. The nature and special features of the instinctive social bond of primates. *Viking Fund Publ. Anthropol.,* 31:17–33.

Cheney, D. L., and R. M. Seyfarth. 1980. Vocal recognition in free-ranging vervet monkeys. *Anim. Behav.,* 28:362–367.

Clark, C. B. 1977. A preliminary report on weaning among chimpanzees of the Gombe National Park, Tanzania. In S. Chevalier-Skolnikoff and F. E. Poirier, eds., *Primate Bio-social Development: Biological, Social, and Ecological Determinants,* pp. 235–260. New York: Garland.

Clutton-Brock, T. H. 1972. Feeding and ranging behaviour of the red colobus monkey. Ph.D. diss., Cambridge University.

Collins, D. A., C. D. Busse, and J. Goodall. 1984. Infanticide in two populations of savanna baboons. In G. Hausfater and S. B. Hrdy, eds., *Infanticide: Comparative and Evolutionary Perspectives,* pp. 193–216. New York: Aldine.

Crawford, M. P. 1937. The cooperative solving of problems by young chimpanzees. *Comp. Psychol. Monogr.,* 14:1–88.

Crook, J. H. 1966. Gelada baboon herd structure and movement: a comparative report. *Symp. Zool. Soc. Lond.,* 18:237–258.

Cullen, J. M. 1972. Some principles of animal communication. In R. A. Hinde, ed., *Non-Verbal Communication,* pp. 101–121. Cambridge: Cambridge University Press.

Darwin, C. 1871. *The Descent of Man.* London: John Murray.

———— 1873. *The Expression of Emotions in Man and Animals.* London: John Murray.

Davenport, R. K. 1979. Some behavioral disturbances of great apes in captivity. In D. A. Hamburg and E. R. McCown, eds., *The Great Apes,* pp. 341–356. Menlo Park, Calif.: Benjamin/Cummings.

Davenport, R. K., and C. M. Rogers. 1970. Inter-modal equivalence of stimuli in apes. *Science*, 168:279–280.

———— 1971. Perception of photographs by apes. *Behaviour*, 39:318–320.

Davenport, R. K., C. M. Rogers, and S. Russell. 1975. Cross-modal perception in apes: altered visual cues and delay. *Neuropsychologia*, 13:229–235.

Davis, F. 1978. *Eloquent Animals*. New York: Coward, McCann and Geoghegan.

Dawkins, R. 1976. *The Selfish Gene*. New York: Oxford University Press.

Doering, C. H., P. R. McGinnis, H. C. Kraemer, and D. A. Hamburg. 1980. Hormonal and behavioral response of male chimpanzees to long-acting analogue of gonadotropin-releasing hormone. *Arch. Sex. Behav.*, 9:441–450.

Döhl, J. 1968. Über die Fähigkeit einer Schimpansin, Umwege mit selbständigen Zwischenzielen zu überblicken. *Z. Tierpsychol.*, 25:89–103.

Dollard, J., L. W. Doob, N. E. Miller, O. H. Mowrer, and R. R. Sears. 1939. *Frustration and Aggression*. New Haven: Yale University Press.

Dooley, G. B., and T. V. Gill. 1977. Acquisition and use of mathematical skills by a linguistic chimpanzee. In D. M. Rumbaugh, ed., *Language Learning by a Chimpanzee: The Lana Project*, pp. 247–260. New York: Academic Press.

Dunbar, R. I. M. 1980. Determinants and evolutionary consequences of dominance among female gelada baboons. *Behav. Ecol. Sociobiol.*, 7:253–265.

Dunbar, R. I. M., and E. P. Dunbar. 1975. *Social Dynamics of Gelada Baboons*. Contributions to Primatology, no. 6. Basel: Karger.

Eaton, G. G., and J. A. Resko. 1974. Plasma testosterone and male dominance in a Japanese macaque (*Macaca fuscata*) troop compared with repeated measures of testosterone in laboratory males. *Horm. Behav.*, 5:251–259.

Eibl-Eibesfeldt, I. 1971. *Love and Hate*, trans. Geoffrey Strachan. London: Methuen.

———— 1979. *The Biology of Peace and War*. New York: Viking.

Eisenburg, J. F., and R. E. Kuehn. 1966. The behavior of *Ateles geoffroyi* and related species. *Smithsonian Misc. Collect.*, 151:1–63.

Ellefson, J. O. 1968. Territorial behavior in the common white-handed gibbon, *Hylobates lar Linn*. In P. C. Jay, ed., *Primates: Studies in Adaptation and Variability*, pp. 180–199. New York: Holt, Rinehart and Winston.

Enomoto, T. 1974. The sexual behaviour of Japanese monkeys. *J. Hum. Evol.*, 3:351–372.

Erikson, E. H. 1966. Ontogeny of ritualization in man. *Phil. Trans. Roy. Soc. Lond.*, Ser. B, 251:337–349.

Falk, J. L. 1958. The grooming behavior of the chimpanzee as a reinforcer. *J. Exp. Anal. Behav.*, 1:83–85.

File, S. K., W. C. McGrew, and C. E. G. Tutin. 1976. The intestinal parasites of a community of feral chimpanzees. *J. Parasitol.*, 62:259–261.

Fletcher, R. 1966. *Instinct in Man*. New York: Schocken Books.

Flint, M. 1976. Does the chimpanzee have a menopause? Paper presented at the forty-fifth annual meeting of the Association of Physical Anthropology, St. Louis.

Flynn, J., H. Vanegas, W. Foote, and S. Edwards. 1970. Neural mechanisms involved in a cat's attack on a rat. In R. E. Whalen, R. E. Thompson, M. Verzeano, and N. Weinberger, eds., *The Neural Control of Behavior*, pp. 135–173. New York: Academic Press.

Fossey, D. 1979. Development of the mountain gorilla (*Gorilla gorilla beringei*): the first thirty-six months. In D. A. Hamburg and E. R. McCown, eds., *The Great Apes*, pp. 139–186. Menlo Park, Calif.: Benjamin/Cummings.

——— 1983. *Gorillas in the Mist.* Boston: Houghton Mifflin.

——— 1984. Infanticide in mountain gorillas (*Gorilla gorilla beringei*) with comparative notes on chimpanzees. In G. Hausfater and S. B. Hrdy, eds., *Infanticide: Comparative and Evolutionary Perspectives*, pp. 217–236. New York: Aldine.

Fouts, D. H. 1983. Loulis tries his hand at surgery. *Friends of Washoe*, 3:4.

Fouts, R. S. 1973. Talking with chimpanzees. In *Science Year, the World Book Science Annual*, pp. 34–49. Chicago: Field Enterprises Educational Corp.

——— 1974. Language: origins, definitions and chimpanzees. *J. Hum. Evol.*, 3:475–482.

——— 1975. Communication with chimpanzees. In G. Kurth and I. Eibl-Eibesfeldt, eds., *Hominisation and Behaviour*, pp. 137–158. Stuttgart: Gustav Fischer Verlag.

Fouts, R. S., and R. L. Budd. 1979. Artificial and human language acquisition in the chimpanzee. In D. A. Hamburg and E. R. McCown, eds., *The Great Apes*, pp. 374–392. Menlo Park, Calif.: Benjamin/Cummings.

Fouts, R. S., D. H. Fouts, and D. J. Schoenfeld. 1984. Sign language conversational interactions between chimpanzees. *Sign Lang. Stud.*, 34:1–12.

Fouts, R. S., A. D. Hirsch, and D. H. Fouts. 1982. Cultural transmission of a human language in a chimpanzee mother-infant relationship. In H. E. Fitzgerald, J. A. Mullins, and P. Gage, eds., *Psychobiological Perspectives*, pp. 159–193. Child Nurturance Series, vol. 3. New York: Plenum.

Fox, G. J. 1982. Potentials for pheromones in chimpanzee vaginal fatty acids. *Folia Primatol.*, 37:255–266.

Freeman, D. 1964. Human aggression in anthropological perspective. In J. D. Carthy and F. J. Ebling, eds., *The Natural History of Aggression*, pp. 119–130. Institute of Biology, symposium no. 13. New York: Academic Press.

Frost, S. W. 1959. *Insect Life and Insect Natural History*, pp. 356–369. New York: Dover.

Galdikas, B. M. F. 1979. Orangutan adaptation at Tanjung Puting Reserve: mating and ecology. In D. A. Hamburg and E. R. McCown, eds., *The Great Apes*, pp. 195–234. Menlo Park, Calif.: Benjamin/Cummings.

Gale, C., and W. Cool. 1971. A two month continuous study of a large chimpanzee group at Lion Country Safaris, Florida. Manuscript.

Gallup, G. 1970. Chimpanzee: self-recognition. *Science*, 167:86–87.

——— 1977. Self-recognition in primates. *Am. Psychol.*, 32:329–338.

Gardner, B. T., and R. A. Gardner. 1980. Two comparative psychologists look at language acquisition. In K. E. Nelson, ed., *Children's Language*, vol. 2, pp. 331–369. New York: Halsted Press.

Gardner, R. A., and B. T. Gardner. 1969. Teaching sight language to a chimpanzee. *Science*, 165:664–672.

——— 1978. Comparative psychology and language acquisition. In K. Salzinger and F. E. Denmark, eds., *Psychology: The State of the Art*, pp. 37–76. New York: Annals of New York Academy of Sciences.

——— 1983. Early signs of reference in children and chimpanzees. Manuscript.

Gartlan, J. S., and C. K. Brain. 1968. Ecology and social variability in *Cercopithecus aethiops* and *Cercopithecus mitis*. In P. C. Jay, ed., *Primates: Studies in Adaptation and Variability*, pp. 253–292. New York: Holt, Rinehart and Winston.

Ghiglieri, M. P. 1984. *The Chimpanzees of Kibale Forest: A Field Study of Ecology and Social Structure.* New York: Columbia University Press.

Gill, T. V. 1977. Conversations with Lana. In D. M. Rumbaugh, ed., *Language Learning by a Chimpanzee: The Lana Project*, pp. 225–246. New York: Academic Press.

Gillan, D. J. 1982. Ascent of apes. In D. R. Griffin, ed., *Animal Mind—Human Mind*, pp. 177–200. Berlin: Springer-Verlag.

Gilula, M. F., and D. N. Daniels. 1969. Violence and man's struggle to adapt. *Science*, 164:396–405.

Goldfoot, D. A., and K. Wallen. 1978. Development of gender role behaviors in heterosexual and isosexual groups of infant rhesus monkeys. In D. C. Chivers and J. Herbert, eds., *Recent Advances in Primatology*, vol. 1, pp. 155–160. New York: Academic Press.

Goodall, J. 1963. Feeding behaviour of wild chimpanzees: a preliminary report. *Symp. Zool. Soc. Lond.*, 10:39–48.

———— 1965. Chimpanzees of the Gombe Stream Reserve. In I. DeVore, ed., *Primate Behavior*, pp. 425–447. New York: Holt, Rinehart and Winston.

———— 1967. Mother-offspring relationships in chimpanzees. In D. Morris, ed., *Primate Ethology*, pp. 287–346. London: Weidenfeld and Nicolson.

———— 1968a. Expressive movements and communication in free-ranging chimpanzees: a preliminary report. In P. Jay, ed., *Primates: Studies in Adaptation and Variability*, pp. 313–374. New York: Holt, Rinehart and Winston.

———— 1968b. Behaviour of free-living chimpanzees of the Gombe Stream area. *Anim. Behav. Monogr.*, 1:163–311.

———— 1970. Tool-using in primates and other vertebrates. In D. S. Lehrman, R. A. Hinde, and E. Shaw, eds., *Advances in the Study of Behavior*, vol. 3, pp. 195–249. New York: Academic Press.

———— 1971. *In the Shadow of Man.* Boston: Houghton Mifflin; London: Collins.

———— 1973. Cultural elements in a chimpanzee community. In E. W. Menzel, ed., *Precultural Primate Behaviour*, vol. 1. Karger: Fourth IPC Symposia Proceedings.

———— 1975a. Chimpanzees of Gombe National Park: thirteen years of research. In I. Eibl-Eibesfeldt, ed., *Hominisation und Verhalten*, pp. 74–136. Stuttgart: Gustav Fischer Verlag.

———— 1975b. Patterns of behaviour: the chimpanzee. In V. Goodall, ed., *The Quest for Man*, pp. 130–169. London: Phaidon Press.

———— 1977. Infant-killing and cannibalism in free-living chimpanzees. *Folia Primatol.*, 28:259–282.

———— 1979. Life and death at Gombe. *Natl. Geo.*, 155:592–621.

———— 1983. Population dynamics during a fifteen-year period in one community of free-living chimpanzees in the Gombe National Park, Tanzania. *Z. Tierpsychol.*, 61:1–60.

———— 1984. The nature of the mother-child bond and the influence of the family on the social development of free-living chimpanzees. In N. Kobayashi and T. Berry Brazelton, eds., *The Growing Child in Family and Society*, pp. 47–66. Tokyo: University of Tokyo Press.

———— Forthcoming. Social rejection, exclusion and shunning among the Gombe chimpanzees. In R. Masters and M. Gruter, eds., *Ostracism: A Social and Biological Phenomenon*.

Goodall, J., and J. Athumani. 1980. An observed birth in a free-living chimpanzee (*Pan troglodytes schweinfurthii*) in Gombe National Park, Tanzania. *Primates*, 21:545–549.

Goodall, J. van Lawick, and H. van Lawick, 1970. *Innocent Killers*, pp. 149–207. Boston: Houghton Mifflin; London: Collins.

Goodall, J., A. Bandora, E. Bergmann, C. Busse, H. Matama, E. Mpongo, A. Pierce, and D. Riss. 1979. Inter-community interactions in the chimpanzee population of the Gombe National Park. In D. A. Hamburg and E. R. McCown, eds., *The Great Apes*, pp. 13–53. Menlo Park, Calif.: Benjamin/Cummings.

Gordon, T. P., R. M. Rose, and I. S. Bernstein. 1976. Seasonal rhythm in plasma testosterone levels in the rhesus monkey (*Macaca mulatta*): a three year study. *Horm. Behav.*, 7:229–243.

Gordon, T. P., R. M. Rose, C. L. Grady, and I. S. Bernstein. 1979. Effects of increased testosterone secretion on the behaviour of adult male rhesus living in a social group. *Folia Primatol.*, 32:149–160.

Goy, R. W. 1966. Role of androgens in the establishment and regulation of behavioral sex differences in mammals. *J. Anim. Sci.*, 25:21–31.

Graham, C. E. 1970. Reproductive physiology of the chimpanzees. In G. Bourne, ed., *The Chimpanzee*, vol. 3, pp. 183–220. Basel: Karger.

────── 1979. Reproductive function in aged female chimpanzees. *Am. J. Phys. Anthropol.*, 50:291–300.

────── 1981. Menstrual cycle physiology of the great apes. In C. E. Graham, ed., *Reproductive Biology of the Great Apes*, pp. 286–303. New York: Academic Press.

Graham, C. E., D. C. Collins, M. Robinson, and J. R. K. Preedy. 1972. Urinary levels of estrogens and pregnanediol and plasma levels of progesterone during the menstrual cycle of the chimpanzee: relationship to the sexual swelling. *Endocrinology*, 91:13–24.

Gunderson, V. M. 1982. The development of intra-modal and cross-modal recognition in infant pigtail macaques (*M. nemestrina*). *Abstract International*, B, 43:907–908.

Haldane, J. B. S. 1955. Population genetics. *New Biol.*, 18:34–51.

Halperin, S. D. 1979. Temporary association patterns in free ranging chimpanzees: an assessment of individual grouping preferences. In D. A. Hamburg and E. R. McCown, eds., *The Great Apes*, pp. 491–500. Menlo Park, Calif.: Benjamin/Cummings.

Hamburg, D. A., G. R. Elliott, and D. L. Parron, eds., 1982. *Health and Behavior: Frontiers of Research in the Biobehavioral Sciences*, pp. 293–302. Washington, D.C.: National Academy Press.

Hamilton, W. D. 1964. The genetical evolution of social behaviour, pts. 1, 2. *J. Theor. Biol.*, 7:1–52.

Handler, P. 1970. *Biology and the Future of Man*. New York: Oxford University Press.

Harlow, H. F. 1965. Sexual behavior in the rhesus monkey. In F. A. Beach, ed., *Sex and Behavior*, pp. 234–265. New York: Wiley.

Hartman, C. G. 1931. Relative sterility of the adolescent organism. *Science*, 74:226–227.

Hasegawa, T., and M. Hiraiwa. 1980. Social interactions of orphans observed in a free-ranging troop of Japanese monkeys. *Folia Primatol.*, 33:129–158.

Hasegawa, T., M. Hiraiwa, T. Nishida, and H. Takasaki. 1983. New evidence on scavenging behavior in wild chimpanzees. *Curr. Anthropol.*, 24:231–232.

Hayes, C. 1951. *The Ape in Our House*. New York: Harper and Brothers.

Hayes, K., and C. Hayes. 1951. The intellectual development of a home-raised chimpanzee. *Proc. Am. Phil. Soc.*, 95:105–109.

——— 1952. Imitation in a home-raised chimpanzee. *J. Comp. Physiol. Psychol.*, 45:450–459.

——— 1953. Picture perception in a home-raised chimpanzee. *J. Comp. Physiol. Psychol.*, 46:470–474.

Hayes, K., and C. H. Nissen. 1971. Higher mental functions of a home-raised chimpanzee. In A. M. Schrier and F. Stollnitz, eds., *Behavior of Non-Human Primates*, vol. 4, pp. 59–115. New York: Academic Press.

Hebb, D. O. 1945. The forms and conditions of chimpanzee anger. *Bull. Can. Psychol. Assoc.*, 5:32–35.

Herrnstein, R. J., and D. H. Loveland. 1964. Complex visual concept in the pigeon. *Science*, 146:549–551.

Hinde, R. A. 1956. The biological significance of the territories of birds. *Ibis*, 98:340–369.

——— 1966. *Animal Behavior: A Synthesis of Ethology and Comparative Psychology*. New York: McGraw-Hill.

——— 1976. Interactions, relationships and social structure. *Man*, 11:1–17.

——— 1978. Dominance and role—two concepts with dual meanings. *J. Soc. Biol. Struct.*, 1:27–38.

——— 1979. *Towards Understanding Relationships*. New York: Academic Press.

———, ed. 1984. *Primate Social Relationships: An Integrated Approach*. Sunderland, Mass.: Sinauer.

Hinde, R. A., and Y. Spencer-Booth. 1971. Effects of brief separation from mothers on rhesus monkeys. *Science*, 173:111–118.

Hiraiwa-Hasegawa, M., T. Hasegawa, and T. Nishida. 1983. Demographic study of a large-sized unit-group of chimpanzees in the Mahale Mountains, Tanzania. Mahale Mountains Chimpanzee Research Project, Ecological Report no. 30.

Hladik, C. M. 1973. Alimentation et activité d'un group de chimpanzés réintroduits en forêt gabonaise. *Terre et Vie*, 27:343–413.

——— 1977. Chimpanzees of Gombe and the chimpanzees of Gabon: some comparative data on the diet. In T. H. Clutton-Brock, ed., *Primate Ecology*, pp. 481–501. New York: Academic Press.

van Hooff, J. A. R. A. M. 1967. The facial displays of the catarrhine monkeys and apes. In D. Morris, ed., *Primate Ethology*, pp. 7–68. London: Weidenfeld and Nicolson.

——— 1973. The Arnhem Zoo Chimpanzee Consortium: an attempt to create an ecologically and socially acceptable habitat. *Int. Zoo Yearbk.*, 13:195–205.

Hoppius, C. E. 1789. *Anthropomorpha. Amoenitates academicae (Linné), Erlangae*. (1st ed., 1760.)

Hrdy, S. B. 1977. Infanticide as a primate reproductive strategy. *Am. Sci.*, 65:40–49.

Hunsperger, R. W. 1956. Affektreaktionen auf elektrische Reizung im Hirnstamm der Katze. *Helv. Physiol. Acta*, 14:70–92.

Imanishi, K. 1965. The origin of the human family: a primatological approach. In S. A. Altmann and Yerkes Regional Primate Center, eds., *Japanese Monkeys: A Collection of Translations*, pp. 113–140. Published by the editors; printed at the University of Alberta.

Itani, J. 1965. Social organization of Japanese monkeys. *Animals*, 5:410–417.

——— 1970. *Chasing Wild Chimpanzees*. Tokyo: Chikuma-Shobo.

Izawa, K. 1970. Unit groups of chimpanzees and their nomadism in the savanna woodland. *Primates*, 11:1–46.

Izawa, K., and J. Itani. 1966. Chimpanzees in Kasakati Basin, Tanganyika (I). Ecological study in the rainy season 1963–1964. *Kyoto Univ. Afr. Stud.*, 1:73–156.

Jolly, A. 1966. Lemur social behavior and primate intelligence. *Science*, 153:501–506.

Jones, C., and J. Sabater Pi. 1969. Sticks used by chimpanzees in Río Muni, West Africa. *Nature*, 223:100–101.

Kano, T. 1971a. The chimpanzees of Filabanga, Western Tanzania. *Primates*, 12:229–246.

—— 1971b. Distribution of the primates on the eastern shore of Lake Tanganyika. *Primates*, 12:281–304.

—— 1972. Distribution and adaptation of the chimpanzee on the eastern shore of Lake Tanganyika. *Kyoto Univ. Afr. Stud.*, 7:37–129.

Katchadourian, H. A. 1976. Medical perspectives on adulthood. *Daedalus*, 105:29–56.

Kaufman, I. C., and L. A. Rosenblum. 1969. Effects of separation from the mother on the emotional behavior of infant monkeys. *Ann. N.Y. Acad. Sci.*, 159:681–695.

Kaufmann, J. H. 1962. Ecological and social behavior of the *coati Nasua narica*, on Barro Colorado Island, Panama. *Univ. Calif. Publ. Zool.*, 60:95–222.

Kawabe, M. 1966. One observed case of hunting behavior among wild chimpanzees living in the savanna woodland of western Tanzania. *Primates*, 7:393–396.

Kawai, M. 1958. On the rank system in a natural group of Japanese monkeys, pts. I, II. *Primates*, 1:111–112, 131–132.

Kawamura, S. 1959. The process of sub-culture propagation among Japanese macaques. *Primates*, 2:43–60.

Kawanaka, K. 1981. Infanticide and cannibalism in chimpanzees with special reference to the newly observed case in the Mahale Mountains. *Afr. Stud. Monogr.*, 1:69–99.

—— 1982a. Further studies on predation by chimpanzees of the Mahale Mountains. *Primates*, 23:364–384.

—— 1982b. Association, ranging, and social unit in chimpanzees of the Mahale Mountains, Tanzania. Mahale Mountains Chimpanzee Research Project, Ecological Report no. 22.

Kawanaka, K., and M. Seifu. 1979. The third case of cannibalism among wild chimpanzees and the case of non-agonistic encounter between chimpanzees of different unit-groups observed in the Mahale Mountains. Mahale Mountains Chimpanzee Research Project, Ecological Report no. 6.

Keeling, M. E., and J. R. Roberts. 1972. Breeding and reproduction of chimpanzees. In G. Bourne, ed., *The Chimpanzee*, vol. 5, pp. 127–152. Basel: Karger; Baltimore: University Press.

Keith, A. 1947. *A New Theory of Human Evolution*. Gloucester, Mass.: Peter Smith.

Kellog, W., and L. A. Kellog. 1933. *The Ape and the Child*. New York: McGraw-Hill. Rev. ed., New York: Hafner, 1967.

Keverne, E. B., R. R. Meller, and A. M. Martinez-Arias. 1978. Dominance, aggression and sexual behavior in social groups of talapoin monkeys. In D. J. Chivers and J. Herbert, eds., *Recent Advances in Primatology*, vol. 1, pp. 533–548. New York: Academic Press.

King, N. E., V. J. Stevens, and J. D. Mellen. 1980. Social behavior in a captive chimpanzee (*Pan troglodytes*) group. *Primates*, 21:198–210.

Klein, L. L., and D. J. Klein. 1973. Social and ecological contrasts between four types of neotropical primates (*Ateles belzebuth, Alouatta seniculus, Saimiri sciureus, Cebus apella*). Paper presented at the ninth annual meeting of the International Congress of Anthropology and Ethnological Science, Chicago.

Köhler, W. 1921. Aus der Anthropoidenstation auf Teneriffa. *Sitzber. preuss. Akad. Wiss., Berlin*, 39:686–692.

────── 1925. *The Mentality of Apes*. London: Routledge and Kegan Paul. Reprint ed., New York: Liveright, 1976.

Kohts, N. 1923. *Untersuchungen über die Erkenntnisfähigkeiten des Schimpansen*. Museum of Darwinianum, Moscow, p. 453.

────── 1935. *Infant Ape and Human Child (Instincts, Emotions, Play, Habits)*. Scientific Memoirs of the Museum of Darwinianum, Moscow, no. 3. (In Russian, with English summary, pp. 524–591.)

Konner, M. 1982. *The Tangled Wing*. New York: Holt, Rinehart and Winston.

Kortlandt, A. 1962. Chimpanzees in the wild. *Sci. Am.*, 206:128–138.

────── 1963. Bipedal armed fighting in chimpanzees. *Proc. 16th Int. Cong. Zool.*, 3:64.

────── 1967. Handgebrauch bei freilebenden Schimpanzen. In B. Rensch, ed., *Handgebrauch und Verständigung bei Affen und Frühmenschen*, pp. 59–102. Bern: Huber.

Kraemer, H. C., C. H. Doering, P. R. McGinnis, and D. A. Hamburg. 1980. Hormonal and behavioural response of male chimpanzees to a long-acting analogue of gonadotropin-releasing hormone. *Arch. Sex. Behav.*, 9:5.

Kruuk, H. 1972. *The Spotted Hyena: A Study of Predation and Social Behavior*. Chicago: University of Chicago Press.

Kuhn, H. J. 1968. Parasites and the phylogeny of the catarrhine primates. In B. Chiarelli, ed., *Taxonomy and Phylogeny of Old World Primates with Reference to the Origin of Man*, pp. 187–195. Turin: Rosenberg and Sellier.

Kummer, H. 1957. Soziales Verhalten einer Mantelpavian-Gruppe. *Beiheft Schweiz. Z. Psychol.*, 33:1–91.

────── 1968. *Social Organization of Hamadryas Baboons: A Field Study*. Bibliotheca Primatology, no. 6. Chicago: University of Chicago Press.

────── 1971. *Primate Societies: Group Techniques of Ecological Adaptation*. Chicago: Aldine.

────── 1974. Rules of dyad and group formation among captive gelada baboons (*Theropithecus gelada*). *Symp. 5th Cong. Int. Primatol. Soc.*, pp. 129–159.

────── 1979. Intra- and intergroup relationships in primates. In M. von Cranach, K. Foppa, W. Lepenies, and D. Ploog, eds., *Human Ethology: Claims and Limits of a New Discipline*, pp. 381–434. Cambridge: Cambridge University Press.

────── 1982. Social knowledge in free-ranging primates. In D. R. Griffin, ed., *Animal Mind—Human Mind*, pp. 113–130. Berlin: Springer.

Kummer, H., A. A. Banaja, A. N. Abo-Khatwa, and A. M. Ghandour. 1981. Mammals of Saudi Arabia—primates: a survey of hamadryas baboons in Saudi Arabia. *Fauna of Saudi Arabia*, 3:441–471.

Kummer, H., and J. Goodall. 1985. Conditions of innovative behaviour in primates. *Phil. Trans. Roy. Soc. Lond.* B, 308:203–214.

Kummer H., and F. Kurt. 1965. A comparison of social behavior in captive and wild hamadryas baboons. In H. Vagtborg, ed., *The Baboon in Medical Research*, pp. 1–16. Austin: University of Texas Press.

Kuo, Z. Y. 1967. *The Dynamics of Behavior Development*. New York: Random House.

Kurland, J. A. 1977. *Kin Selection in the Japanese Monkey*. New York: Karger.

Kuroda, S. 1980. Social behavior of the pygmy chimpanzees. *Primates*, 21:181–197.

Lack, D. 1966. *Population Studies of Birds*. Oxford: Clarendon Press.

Langer, S. 1957. *Philosophy in a New Key*. Cambridge, Mass.: Harvard University Press.

van Lawick-Goodall, J., *see* Goodall, J.

Leakey, L. S. B. 1961. *The Progress and Evolution of Man in Africa*. London: Oxford University Press.

——— 1977. *The Southern Kikuyu before 1903*, vol. 2. New York: Academic Press.

Lemmon, W. B., and M. L. Allen. 1978. Continual sexual receptivity in the female chimpanzee (*Pan troglodytes*). *Folia Primatol.*, 30:80–88.

LeVine, K. A., and D. T. Campbell. 1971. *Ethnocentrism: Theories of Conflict, Ethnic Attitudes, and Group Behavior*. New York: Wiley.

Lieberman, P., E. Crelin, and D. Klatt. 1972. Phonetic ability and related anatomy of the newborn and adult human, neanderthal man, and the chimpanzee. *Am. Anthropol.*, 74:287–307.

Lindburg, D. G. 1971. The rhesus monkey in North India. In L. A. Rosenblum, ed., *Primate Behavior*, vol. 2, pp. 1–106. New York: Academic Press.

Lorenz, K. 1963. *On Aggression*, trans. M. K. Wilson. New York: Harcourt, Brace and World.

——— 1977. *Behind the Mirror: A Search for a Natural History of Human Knowledge*. New York: Harcourt Brace Jovanovich.

MacKay, D. M. 1972. Formal analysis of communicative processes. In R. A. Hinde, ed., *Non-Verbal Communication*, pp. 3–26. Cambridge: Cambridge University Press.

Mackinnon, J. R., and K. S. Mackinnon. 1978. Comparative feeding ecology of six sympatric primates in west Malaysia. In D. J. Chivers and J. Herbert, eds., *Recent Advances in Primatology*, vol. 1, pp. 305–321. New York: Academic Press.

Marais, E. 1969. *The Soul of the Ape*. New York: Atheneum.

Marler, P. 1969. Vocalizations of wild chimpanzees, an introduction. *Proc. 2nd Int. Cong. Primatol. (Atlanta, 1968)*, 1:94–100.

——— 1976. Social organization, communication, and graded signals: the chimpanzee and the gorilla. In P. P. G. Bateson and R. A. Hinde, eds., *Growing Points in Ethology*, pp. 239–280. New York: Cambridge University Press.

Marler, P., and W. J. Hamilton III. 1966. *Mechanisms of Animal Behavior*. New York: Wiley.

Marler, P., and L. Hobbett. 1975. Individuality in a long-range vocalization of wild chimpanzees. *Z. Tierpsychol.*, 38:97–109.

Marler, P., and R. Tenaza. 1976. Signaling behavior of wild apes with special reference to vocalization. In T. Sebeok, ed., *How Animals Communicate*, pp. 965–1033. Bloomington: Indiana University Press.

Martin, D. E. 1981. Breeding great apes in captivity. In C. E. Graham, ed., *Reproductive Biology of the Great Apes*, pp. 343–375. New York; Academic Press.

Martin, D. E., C. E. Graham, and K. G. Gould. 1978. Successful artificial insemination in the chimpanzee. *Symp. Zool. Soc. Lond.*, 43:249–260.

Mason, W. A. 1965. The social development of monkeys and apes. In I. DeVore, ed., *Primate Behavior*, pp. 514–543. New York: Holt, Rinehart and Winston.

———— 1968. Use of space by *Callicebus* groups. In P. C. Jay, ed., *Primates: Studies in Adaptation and Variability*, pp. 200–216. New York: Holt, Rinehart and Winston.

———— 1979. Environmental models and mental modes: representational process in the great apes. In D. A. Hamburg and E. R. McCown, eds., *The Great Apes*, pp. 277–293. Menlo Park, Calif.: Benjamin/Cummings.

———— 1982. Primate social intelligence: contributions from the laboratory. In D. R. Griffin, ed., *Animal Mind—Human Mind*, pp. 131–144. Berlin: Springer-Verlag.

McCormack, S. A. 1971. Plasma testosterone concentration and binding in the chimpanzee: effect of age. *Endocrinology*, 89:1171–77.

McGinnis, P. R. 1973. Patterns of sexual behaviour in a community of free-living chimpanzees. Ph.D. diss., Cambridge University.

———— 1979. Sexual behavior in free-living chimpanzees: consort relationships. In D. A. Hamburg and E. R. McCown, eds., *The Great Apes*, pp. 429–440. Menlo Park, Calif.: Benjamin/Cummings.

McGrew, W. C. 1974. Tool use by wild chimpanzees in feeding upon driver ants. *J. Hum. Evol.*, 3:501–508.

———— 1975. Patterns of plant food sharing by wild chimpanzees. *Proc. 5th Cong. Int. Primatol. Soc. (Nagoya, Japan)*, pp. 304–309. Basel: Karger.

———— 1979. Evolutionary implications of sex differences in chimpanzee predation and tool use. In D. A. Hamburg and E. R. McCown, eds., *The Great Apes*, pp. 440–463. Menlo Park, Calif.: Benjamin/Cummings.

———— 1981. The female chimpanzee as a human evolutionary prototype. In F. Dahlberg, ed., *Woman the Gatherer*, pp. 35–73. New Haven: Yale University Press.

———— 1983. Animal foods in the diets of wild chimpanzees (*Pan troglodytes*): why cross-cultural variation? *J. Ethol.*, 1:46–61.

McGrew, W. C., and C. E. G. Tutin. 1972. Chimpanzee dentistry. *JADA*, 85:1198–1204.

———— 1978. Evidence for a social custom in wild chimpanzees? *Man*, 13:234–251.

McGrew, W. C., P. J. Baldwin, and C. E. G. Tutin. 1981. Chimpanzees in a savanna habitat: Mt. Asserik, Senegal, West Africa. *J. Hum. Evol.*, 10:227–244.

McGrew, W. C., C. E. G. Tutin, and P. J. Baldwin. 1979. Chimpanzees, tools, and termites: cross-cultural comparisons of Senegal, Tanzania, and Río Muni. *Man*, 14:185–214.

McGrew, W. C., C. E. G. Tutin, P. J. Baldwin, M. J. Sharman, and A. Whiten. 1978. Primates preying upon vertebrates: new records from West Africa. *Carnivore*, 1:41–45.

McGuire, M. T. 1974. The St. Kitts vervet (*Cercopithecus aethiops*). *J. Med. Primatol.*, 3:285–297.

McReynolds, P. 1962. Exploratory behavior: a theoretical interpretation. *Psychol. Rep.*, 11:311–318.

Mead, G. H. 1934. *Mind, Self and Society*. Chicago: University of Chicago Press.

Meddin, J. 1979. Chimpanzees, symbols, and the reflective self. *Soc. Psychol. Quart.*, 42:99–109.

Mellen, S. L. W. 1981. *The Evolution of Love*. Oxford: W. H. Freeman.

Menzel, E. W., Jr. 1964. Patterns of responsiveness in chimpanzees reared through infancy under conditions of environmental restriction. *Psychol. Forsch.*, 27:337–365.

———— 1971. Communication about the environment in a group of young chimpanzees. *Folia Primatol.*, 15:220–232.

———— 1973. Leadership and communication in a chimpanzee community. In E. W. Menzel, Jr., ed., *Precultural Primate Behavior*, pp. 192–225. Basel: Karger.

———— 1974. A group of young chimpanzees in a one-acre field. In A. M. Schrier and F. Stollnitz, eds., *Behavior of Non-Human Primates*, vol. 5, pp. 83–153. New York: Academic Press.

———— 1975. Communication and aggression in a group of young chimpanzees. In P. Pliner, L. Krames, and T. Alloway, eds., *Nonverbal Communication of Aggression*, pp. 103–133. New York: Plenum.

———— 1978. Cognitive mapping in chimpanzees. In S. H. Hulse, H. Fowler, and W. K. Honig, eds., *Cognitive Processes in Animal Behavior*, pp. 375–422. Hillsdale, N.J.: Lawrence Erlbaum Associates.

Menzel, E. W., Jr., D. Premack, and G. Woodruff. 1978. Map reading by chimpanzees. *Folia Primatol.*, 29:241–249.

Menzel, E. W., Jr., E. S. Savage-Rumbaugh, and J. Lawson. 1985. Chimpanzee spatial problem-solving with the use of mirrors and televised equivalents of mirrors. *J. Comp. Psychol.*, in press.

Merfield, F. G., and H. Miller. 1956. *Gorillas Were My Neighbours*. London: Longmans.

Midgley, M. 1978. *Beast and Man: The Roots of Human Nature*. Ithaca: Cornell University Press.

Miles, W. R. 1963. Chimpanzee behavior: removal of foreign body from companion's eye. Paper presented at the 100th annual meeting of the National Academy of Sciences, Washington, D.C.

Missakian, E. A. 1973. Genealogical mating activity in free-ranging groups of rhesus monkeys (*Macaca mulatta*) on Cayo Santiago. *Behaviour*, 45:225–241.

Morris, D. 1962. *The Biology of Art*. London: Methuen.

Morris, D., and R. Morris. 1966. *Men and Apes*. New York: McGraw-Hill.

Morris, K., and J. Goodall. 1977. Competition for meat between chimpanzees and baboons of the Gombe National Park. *Folia Primatol.*, 28:109–121.

Morrison, J. A., and E. W. Menzel, Jr. 1972. Adaptation of a free-ranging rhesus monkey group to division and transportation. *Wildl. Monogr.*, 31:1–78.

Moynihan, M. 1955. Some aspects of reproductive behavior in the black-headed gull (*Larus ridibundus ridibundus L.*) and related species. *Behaviour*, suppl. no. 4:1–201.

Nagel, U., and H. Kummer. 1974. Variation in cercopithecoid aggressive behavior. In R. Holloway, ed., *Primate Aggression, Territoriality, and Xenophobia*, pp. 159–185. New York: Academic Press.

Napier, J. 1971. *The Roots of Mankind*. London: George Allen and Unwin.

Nicolson, N. A. 1977. A comparison of early behavioral development in wild and captive chimpanzees. In F. E. Poirier, ed., *Primate Biosocial Development: Biological, Social, and Ecological Determinants*, pp. 529–560. New York: Garland.

Nishida, T. 1968. The social group of wild chimpanzees in the Mahale Mountains. *Primates*, 9:167–224.

———— 1973. The ant-gathering behaviour by the use of tools among wild chimpanzees of the Mahale Mountains. *J. Hum. Evol.*, 2:357–370.

———— 1976. The bark-eating habits in primates, with special reference to their status in the diet of wild chimpanzees. *Folia Primatol.*, 25:277–287.

———— 1979. The social structure of chimpanzees of the Mahale Mountains. In D. A. Hamburg and E. R. McCown, eds., *The Great Apes*, pp. 73–122. Menlo Park, Calif.: Benjamin/Cummings.

———— 1980. The leaf-clipping display: a newly discovered expressive gesture in wild chimpanzees. *J. Hum. Evol.*, 9:117–128.

———— 1983. Alpha status and agonistic alliance in wild chimpanzees (*Pan troglodytes schweinfurthii*). *Primates*, 24:318–336.

Nishida, T., and M. Hiraiwa. 1982. Natural history of a tool-using behaviour by wild chimpanzees in feeding upon wood-boring ants. *J. Hum. Evol.*, 11:73–99.

Nishida, T., and M. Hiraiwa-Hasegawa. 1985. Responses to a stranger mother-son pair in the wild chimpanzee: a case report. *Primates*, 26:1–13.

Nishida, T., and S. Uehara. 1983. Natural diet of chimpanzees (*Pan troglodytes schweinfurthii*): long-term record from the Mahale Mountains, Tanzania. *Afr. Study Monogr.*, 3:109–130.

Nishida, T., S. Uehara, and R. Nyundo. 1979. Predatory behavior among wild chimpanzees of the Mahale Mountains. *Primates*, 20:1–20.

Nishida, T., R. W. Wrangham, J. Goodall, and S. Uehara. 1983. Local differences in plant-feeding habits of chimpanzees between the Mahale Mountains and Gombe National Park, Tanzania. *J. Hum. Evol.*, 12:467–480.

Nishida, T., M. Hiraiwa-Hasegawa, T. Hasegawa, and Y. Takahata. 1985. Group extinction and female transfer in wild chimpanzees in the Mahale National Park, Tanzania. *Z. Tier psychol.*, 67:284–301.

Nissen, H. W. 1931. A field study of the chimpanzee. *Comp. Psychol. Monogr.*, 8:1–121.

Nissen, H. W., and M. P. Crawford, 1936. A preliminary study of food-sharing behavior in young chimpanzees. *J. Comp. Psychol.*, 12:383–419.

Noble, G. K. 1939. The role of dominance in the social life of birds. *Auk*, 56:263–273.

Norikoshi, K. 1982. One observed case of cannibalism among wild chimpanzees of the Mahale Mountains. *Primates*, 23:66–74.

———— 1983. Prevalent phenomenon of predation observed among wild chimpanzees of the Mahale Mountains. *J. Anthropol. Soc. Nippon*, 91:475–479.

Oki, J., and Y. Maeda. 1973. Grooming as a regulator of behavior in Japanese macaques. In C. R. Carpenter, ed., *Behavioral Regulators of Behavior in Primates*, pp. 149–163. E. Brunswick, N.J.: Bucknell University Press.

Packer, C. R. 1977. Reciprocal altruism in *Papio anubis*. *Nature*, 265:441–443.

———— 1979. Inter-troop transfer and inbreeding avoidance in *Papio anubis*. *Anim. Behav.*, 27:1–36.

Parkel, D. A., R. A. White, and H. Warner. 1977. Implications of the Yerkes technology for mentally retarded human subjects. In D. M. Rumbaugh, ed., *Language Learning by a Chimpanzee: The Lana Project*, pp. 273–286. New York: Academic Press.

Patterson, F. 1979. Talking gorillas as informants: questions posed by Jane Goodall regarding wild chimpanzees. *Gorilla*, 2:1–2.

Patton, R. G., and L. I. Gardner, 1963. *Growth Failure in Maternal Deprivation*. Springfield, Ill.: Charles C Thomas.

Pavlov, I. P. 1955. *Selected Works*. London: Central Books.

———— 1957. *Experimental Psychology and Other Essays*. New York: Philosophical Library.

Perachio, A. A. 1978. Hypothalamic regulation of behavioral and hormonal aspects of aggression and sexual performance. In D. C. Chivers and

J. Herbert, eds., *Recent Advances in Primatology*, vol. 1, pp. 549–566. New York: Academic Press.

Pierce, A. H. 1975. An encounter between a leopard and a group of chimpanzees at Gombe National Park. Manuscript.

—— 1978. Ranging patterns and associations of a small community of chimpanzees in Gombe National Park, Tanzania. In D. C. Chivers and J. Herbert, eds., *Recent Advances in Primatology*, vol. 1, pp. 59–62. New York: Academic Press.

Pitt, R. 1978. Warfare and hominid brain evolution. *J. Theor. Biol.*, 72: 551–575.

Plooij, F. X. 1978. Tool use during chimpanzees' bushpig hunt. *Carnivore*, 1:103–106.

Plotnik, R. 1974. Brain stimulation and aggression: monkeys, apes, and humans. In R. L. Holloway, ed., *Primate Aggression, Territoriality, and Xenophobia*, pp. 389–416. New York: Academic Press.

Poirier, F. E. 1974. Colobine aggression: a review. In R. L. Holloway, ed., *Primate Aggression, Territoriality, and Xenophobia*, pp. 123–158. New York: Academic Press.

Polis, E. 1975. A comparison of two aged wild chimpanzees at Gombe. Manuscript.

Popovkin, V. 1981. The monkey buys a banana. *Moscow News Weekly*, 6:10.

Premack, A. J., and D. Premack. 1972. Teaching language to an ape. *Sci. Am.*, 227:92–99.

Premack, D. 1976. On the study of intelligence in chimpanzees. *Curr. Anthropol.*, 17:516–521.

Premack, D., and G. Woodruff. 1978. Does the chimpanzee have a theory of mind? *Behav. Brain Sci.*, 1:515–526.

Prestrude, A. M. 1970. Sensory capacities of the chimpanzee: a review. *Psychol. Bull.*, 74:47–67.

Pryor, K. 1984. *Positive Reinforcement*. New York: Simon and Schuster.

Pusey, A. E. 1977. The physical and social development of wild adolescent chimpanzees. Ph.D. diss., Stanford University.

—— 1979. Intercommunity transfer of chimpanzees in Gombe National Park. In D. A. Hamburg and E. R. McCown, eds., *The Great Apes*, pp. 465–480. Menlo Park, Calif.: Benjamin/Cummings.

—— 1980. Inbreeding avoidance in chimpanzees. *Anim. Behav.*, 28: 543–552.

—— 1983. Mother-offspring relationships in chimpanzees after weaning. *Anim. Behav.*, 31:363–377.

Pusey, A. E., and Packer, C. 1983. Once and future kings. *Nat. Hist.*, 8: 54–62.

Rahm, U. 1971. L'emploi d'outils par les chimpanzés de l'ouest de la Côte-d'Ivoire. *Terre et Vie*, 25:506–509.

Ransom, T. W. 1972. Ecology and social behavior of baboons in the Gombe National Park. Ph.D. diss., University of California, Berkeley.

Rapaport, L., M. Yeutter-Curington, and D. Thomas. 1984. The influence of estrus on social behavior in a group of captive chimpanzees (*Pan troglodytes*): a preliminary report. Paper presented at the annual meeting of the Animal Behavior Society, Washington, D.C.

Reinhart, J. B., and A. L. Drash. 1969. Psychosocial dwarfism: environmentally induced recovery. *Psychosom. Med.*, 31:165–172.

Reite, M. 1979. Towards a pathophysiology of grief. Paper presented at the annual meeting of the American Psychosomatic Society, Dallas.

Rensch, B., and J. Döhl. 1967. Spontanes öfnen verschiedener Kistenverschlüsse durch einen Schimpansen. *Z. Tierpsychol.*, 24:476–489.

————— 1968. Wahlen zwischen zwei überschaubaren Labyrinthwegen durch einen Schimpansen. *Z. Tierpsychol.*, 25:216–231.

Reynolds, P. C. 1981. *On the Evolution of Human Behavior.* Berkeley: University of California Press.

Ripley, S. 1967. Intertroop encounters among Ceylon grey langurs (*Presbytis entellus*). In S. A. Altmann, ed., *Social Communication among Primates*, pp. 237–254. Chicago: University of Chicago Press.

Riss, D. C., and C. Busse. 1977. Fifty day observation of a free-ranging adult male chimpanzee. *Folia Primatol.*, 28:283–297.

Riss, D. C., and J. Goodall. 1977. The recent rise to the alpha rank in a population of free-living chimpanzees. *Folia Primatol.*, 27:134–151.

Rohles, F., and J. V. Devine. 1966. Chimpanzee performance on a problem involving the concept of middleness. *Anim. Behav.*, 14:159–162.

————— 1967. Further studies of the middleness concept with the chimpanzee. *Anim. Behav.*, 15:107–112.

Rose, R. M., T. P. Gordon, and I. S. Bernstein. 1972. Plasma testosterone levels in the male rhesus: influences of sexual and social stimuli. *Science*, 178:643–645.

Rose, R. M., J. W. Holaday, and I. S. Bernstein. 1971. Plasma testosterone, dominance rank and aggressive behavior in male rhesus monkeys. *Nature*, 231:366–368.

Rose, R. M., I. S. Bernstein, T. P. Gordon, and S. F. Catlin. 1974. Androgens and aggression: a review and recent findings in primates. In R. L. Holloway, ed., *Primate Aggression, Territoriality, and Xenophobia*, pp. 275–304. New York: Academic Press.

Rowell, T. 1974. The concept of social dominance. *Behav. Biol.*, 11:131–154.

Roy, A. D., and H. M. Cameron. 1972. Rhinophycomycosis entomophthorae occurring in a chimpanzee in the wild in East Africa. *Am. J. Trop. Med. Hyg.*, 21:234–237.

Rumbaugh, D. M. 1974. Comparative primate learning and its contributions to understanding development, play, intelligence, and language. In B. Chiarelli, ed., *Perspectives in Primate Biology*, pp. 253–281. New York: Plenum.

Rumbaugh, D. M., and T. V. Gill. 1977. Lana's acquisition of language skills. In D. M. Rumbaugh, ed., *Language Learning by a Chimpanzee: The Lana Project*, pp. 165–192. New York: Academic Press.

Rumbaugh, D. M., and J. L. Pate. 1984. The evolution of cognition in primates: a comparative perspective. In H. L. Roitblat, T. G. Bever, and H. S. Terrace, eds., *Animal Cognition*, pp. 569–587. Hillsdale, N.J.: Lawrence Erlbaum Associates.

Rumbaugh, D. M., T. V. Gill, and E. C. von Glasersfeld. 1973. Reading and sentence completion by a chimpanzee (*Pan*). *Science*, 182:731–733.

Russell, C., and W. M. S. Russell. 1973. The natural history of violence. In C. M. Otten, ed., *Aggression and Evolution*, pp. 240–273. Lexington, Mass.: Xerox College.

Saayman, G. S. 1971. Behaviour of adult males in a troop of free-ranging chacma baboons (*Papio ursinus*). *Folia Primatol.*, 15:36–57.

Sade, D. S. 1965. Some aspects of parent-offspring and sibling relations in a group of rhesus monkeys, with a discussion of grooming. *Am. J. Phys. Anthropol.*, 23:1–18.

————— 1968. Inhibition of son-mother mating among free-ranging rhesus monkeys. *Sci. Psychoanal.*, 12:18–38.

————— 1972. Life cycle and social organization among free-ranging rhesus monkeys. Paper presented at meetings of American Association of Anthropologists, Washington, D.C.

Sagan, C. 1977. *The Dragons of Eden.* New York: Random House.

Sapolsky, R. M. 1982. The endocrine stress-response and social status in the wild baboon. *Horm. Behav.*, 16:279–288.

―――― 1983. Endocrine aspects of social instability in the olive baboon. *Am. J. Primatol.*, 5:365–376.

Savage, T. S., and J. Wyman. 1843–44. Observations on the external characters and habits of the *Troglodytes niger*, Geoff., and on its organization. *Boston J. Nat. Hist.*, 4:362–386.

Savage-Rumbaugh, E. S., D. M. Rumbaugh, and S. Boysen. 1978. Symbolic communication between two chimpanzees (*Pan troglodytes*). *Science*, 201:641–644.

Schaller, G. B. 1963. *The Mountain Gorilla: Ecology and Behavior.* Chicago: University of Chicago Press.

―――― 1972. *The Serengeti Lion.* Chicago: University of Chicago Press.

Schiller, P. H. 1952. Innate constituents of complex responses in primates. *Psychol. Rev.*, 59:177–191.

Schjelderup-Ebbe, T. 1922. Beiträge zur Sozialpsychologie des Haushuhns. *Z. Psychol.*, 88:225–252.

Schultz, A. H. 1938. The relative weight of the testes in primates. *Anat. Rec.*, 72:387–394.

Scott, J. P. 1958. *Aggression.* Chicago: University of Chicago Press.

Scott, J. P., and E. Fredericson. 1951. The causes of fighting in mice and rats. *Physiol. Zool.*, 24:273–309.

Scott, J. P., and J. L. Fuller. 1965. *Genetics and the Social Behavior of the Dog.* Chicago: University of Chicago Press.

Seyfarth, R. M. 1980. The distribution of grooming and related behaviours among adult female vervet monkeys. *Anim. Behav.*, 28:789–813.

Seyfarth, R. M., D. L. Cheney, and R. A. Hinde. 1978. Some principles relating social interactions and social structure among primates. In D. J. Chivers and J. Herbert, eds., *Recent Advances in Primatology*, vol. 1, pp. 39–52. New York: Academic Press.

Short, R. V. 1979. Sexual selection and its component parts, somatic and genital selection, as illustrated by man and the great apes. *Adv. Stud. Behav.*, 9:131–158.

Shostak, M. 1981. *Nisa: The Life and Words of a !Kung Woman.* Cambridge, Mass.: Harvard University Press.

Silk, J. B. 1978. Patterns of food sharing among mother and infant chimpanzees at Gombe National Park, Tanzania. *Folia Primatol.*, 29:129–141.

Simpson, M. J. A. 1973. The social grooming of male chimpanzees: a study of eleven free-living males in the Gombe Stream National Park, Tanzania. In R. Michael and J. H. Crook, eds., *The Ecology and Behavior of Primates*, pp. 411–502. New York: Academic Press.

Smythies, J. R. 1970. *Brain Mechanisms and Behavior.* New York: Academic Press.

Sollereld, H. A., and M. J. van Zwieten. 1978. Membranous dysmenorrhea in the chimpanzees (*Pan troglodytes*): a report of four cases. *J. Med. Primatol.*, 7:19–25.

Speroff, L., F. Glass, and N. Kase. 1983. *Clinical Gynecological Endocrinology and Fertility*, 3rd ed. Baltimore: Williams & Wilkins.

Spitz, R. A. 1946. Anaclitic depression: an inquiry into the genesis of psychiatric conditions in early childhood. In *The Psychoanalytical Study of the Child*, vol. 2, pp. 53–74. New York: International University Press.

Sroufe, L. A. 1979. The coherence of individual development: early care, attachment and subsequent developmental issues. *Am. Psychol.*, 34:834–841.

Stammbach, E., and H. Kummer. 1982. Individual contributions to a dyadic interaction: an analysis of baboon grooming. *Anim. Behav.*, 30:964–971.

Stenhouse, D. 1973. *The Evolution of Intelligence: A General Theory and Some of Its Implications.* London: George Allen and Unwin.

Struhsaker, T. T. 1977. Infanticide and social organization in the redtail monkey (*Cercopithecus aethiops*). *Anim. Behav.*, 19:233–250.

Struhsaker, T. T., and P. Hunkeler. 1971. Evidence of tool-using by chimpanzees in the Ivory Coast. *Folia Primatol.*, 15:212–219.

Sugiyama, Y. 1965. Behavioral development and social structure in the two troops of Hanuman langurs (*Presbytis entellus*). *Primates*, 6:213–247.

Sugiyama, Y., and J. Koman. 1979. Tool using and making behavior in wild chimpanzees at Bossou, Guinea. *Primates*, 20:513–524.

Suzuki, A. 1966. On the insect eating habits among wild chimpanzees living in the savanna woodland of western Tanzania. *Primates*, 7:481–487.

——— 1969. An ecological study of chimpanzees in a savanna woodland. *Primates*, 10:103–148.

——— 1971. Carnivority and cannibalism observed among forest-living chimpanzees. *J. Anthropol. Soc. Nippon*, 79:30–48.

Takahata, Y. 1985. Adult male chimpanzees kill and eat a male newborn infant: newly observed intragroup infanticide and cannibalism in Mahale National Park, Tanzania. *Folia Primatol.*, 44:121–228.

Takahata, Y., T. Hasegawa, and T. Nishida. 1984. Chimpanzee predation in the Mahale Mountains from August 1979 to May 1982. *Int. J. Primatol.*, 5:213–233.

Takasaki, H. 1983. Mahale chimpanzees taste mangoes—towards acquisition of a new food item? *Primates*, 24:273–275.

Teleki, G. 1973a. Group response to the accidental death of a chimpanzee in Gombe National Park, Tanzania. *Folia Primatol.*, 20:81–94.

——— 1973b. Notes on chimpanzee interactions with small carnivores in Gombe National Park, Tanzania. *Primates*, 14:407–412.

——— 1973c. *The Predatory Behavior of Wild Chimpanzees.* E. Brunswick, N.J.: Bucknell University Press.

——— 1981. The omnivorous diet and eclectic feeding habits of chimpanzees in Gombe National Park, Tanzania. In R. S. O. Harding and G. Teleki, eds., *Omnivorous Primates: Gathering and Hunting in Human Evolution*, pp. 303–343. New York: Columbia University Press.

Temerlin, M. K. 1975. *Lucy: Growing up Human.* Palo Alto, Calif.: Science and Behavior Books.

Thomas, D. K. 1961. The Gombe Stream Game Reserve. *Tanganyika Notes Rec.*, 56:34–39.

Thompson-Handler, N., R. K. Malenky, and N. Badrian. 1984. Sexual behavior of *Pan paniscus* under natural conditions in the Lomako Forest, Equateur, Zaire. In R. L. Sussman, ed., *The Pygmy Chimpanzee: Evolutionary Biology and Behavior*, pp. 347–368. New York: Plenum.

Thorpe, W. H. 1956. *Learning and Instinct in Animals.* London: Methuen.

Tinbergen, N. 1968. On war and peace in animals and man. *Science*, 160:1411–18.

Tinklepaugh, O. L. 1932. Multiple delayed reaction with chimpanzees and monkeys. *J. Comp. Psychol.*, 13:207–243.

——— 1933. Corrections to "A diet for chimpanzees and monkeys in captivity." *J. Mammal.*, 14:68–69.

Traill, T. S. 1821. Observations on the anatomy of the orangutan. *Mem. Wernerian Nat. Hist. Soc.*, 3:1–49.

Trivers, R. L. 1971. The evolution of reciprocal altruism. *Quart. Rev. Biol.*, 46:35–57.

—— 1972. Parental investment and sexual selection. In B. Campbell, ed., *Sexual Selection and the Descent of Man*, pp. 136–179. Chicago: Aldine.

Tulp, N. 1641. *Observationum Medicarum*. Amsterdam.

Turnbull-Kemp, P. 1967. *The Leopard*. London: Bailey Brothers & Swinfen.

Tutin, C. E. G. 1975. Sexual behaviour and mating patterns in a community of wild chimpanzees (*Pan troglodytes schweinfurthii*). Ph.D. diss., University of Edinburgh.

—— 1979. Mating patterns and reproductive strategies in a community of wild chimpanzees (*Pan troglodytes schweinfurthii*). *Behav. Ecol. Sociobiol.*, 6:39–48.

Tutin, C. E. G., and P. R. McGinnis. 1981. Chimpanzee reproduction in the wild. In C. E. Graham, ed., *Reproductive Biology of the Great Apes*, pp. 239–264. New York: Academic Press.

Tutin, C. E. G., and W. C. McGrew. 1973. Chimpanzee copulatory behavior. *Folia Primatol.*, 19:237–256.

Tyson, E. 1699. *The Anatomy of a Pygmie*. London.

Uehara, S. 1982. Seasonal changes in the techniques employed by wild chimpanzees in the Mahale Mountains, Tanzania, to feed on termites. *Folia Primatol.*, 37:44–76.

de Waal, F. B. M. 1978. Exploitative and familiarity-dependent support strategies in a colony of semi-free-living chimpanzees. *Behaviour*, 66:268–312.

—— 1982. *Chimpanzee Politics: Power and Sex among Apes*. New York: Harper & Row.

—— 1985. Scapegoating in primates: a double-edged sword. (In press.)

de Waal, F. B. M., and J. A. Hoekstra. 1980. Contexts and predictability of aggression in chimpanzees. *Anim. Behav.*, 28:929–937.

de Waal, F. B. M., and A. Roosmalen. 1979. Reconciliation and consolation among chimpanzees. *Behav. Ecol. Sociobiol.*, 5:55–66.

Weisbard, C., and R. Goy. 1976. Effect of parturition and group composition on competitive drinking order in stumptail macaques. *Folia Primatol.*, 25:95–121.

Wickler, W. 1969. *The Sexual Code*. London: Weidenfeld and Nicolson.

Wilson, A. P., and S. H. Vessey. 1968. Behaviour of free-ranging castrated rhesus monkeys. *Folia Primatol.*, 9:1–14.

Wilson, E. O. 1975. *Sociobiology: The New Synthesis*. Cambridge, Mass.: Belknap Press of Harvard University Press.

Wilson, W. L., and A. C. Wilson. 1968. Aggressive interactions of captive chimpanzees living in a semi-free ranging environment. *Rep. 6571st Aeromed. Res. Lab.*, Holloman Air Force Base, N. M.

Windholz, G. 1984. Pavlov vs. Köhler: Pavlov's little-known primate research. *Pav. J. Biol. Sci.*, 19:1, 23–31.

Woodruff, G., and D. Premack. 1979. Intentional communication in the chimpanzee: the development of deception. *Cognition*, 7:333–362.

Wrangham, R. W. 1974. Artificial feeding of chimpanzees and baboons in their natural habitat. *Anim. Behav.*, 22:83–93.

—— 1975. The behavioral ecology of chimpanzees in Gombe National Park, Tanzania. Ph.D. diss., Cambridge University.

—— 1977. Feeding behavior of chimpanzees in Gombe National Park, Tanzania. In T. H. Clutton-Brock, ed., *Primate Ecology*, pp. 503–538. New York: Academic Press.

———— 1979. Sex differences in chimpanzee dispersion. In D. A. Hamburg and E. R. McCown, eds., *The Great Apes*, pp. 481–490. Menlo Park, Calif.: Benjamin/Cummings.

Wrangham, R. W., and E. van Z. Bergmann-Riss. In press. Frequencies of predation on mammals by Gombe chimpanzees, 1972–1975.

Wrangham, R. W., and T. Nishida. 1983. *Aspilia* spp. leaves: a puzzle in the feeding behavior of wild chimpanzees. *Primates*, 24:276–282.

Wrangham, R. W., and B. Smuts. 1980. Sex differences in behavioural ecology of chimpanzees in Gombe National Park, Tanzania. *J. Reprod. Fert. (Suppl.)*, 28:13–31.

Wright, Q. 1965. *A Study of War*. Chicago: University of Chicago Press.

Yerkes, R. M. 1925. *Almost Human*. New York: Century.

———— 1939. Sexual behavior in the chimpanzee. *Hum. Biol.*, 11:78–111.

———— 1943. *Chimpanzees: A Laboratory Colony*. New Haven: Yale University Press.

Yerkes, R. M., and J. H. Elder. 1936. Oestrus, receptivity, and mating in the chimpanzee. *Comp. Psychol. Monogr.*, 13:1–39.

Yerkes, R. M., and A. Petrunkevitch. 1925. Studies of chimpanzee vision by Ladygina-Kohts. *J. Comp. Psychol*, 5:99–108.

Yerkes, R. M., and A. W. Yerkes. 1929. *The Great Apes: A Study of Anthropoid Life*. New Haven: Yale University Press.

Acknowledgments

(C. Boehm)

My thanks must begin with the Tanzanian government, which gave me permission first to start and then to pursue my research in the Gombe Stream area. I express my profound gratitude to our former president, Mwalimu Julius Nyerere, lover of trees, conserver of forest habitats, and a botanist in his own right. I thank too the many government officials, in particular the various regional commissioners and district development directors of the Kigoma region, who have been unfailingly helpful and supportive over the whole twenty-five-year period. When I began the project, Gombe was a game reserve and I received much help from the Tanzania Game Department. Subsequently the area was designated a national park, and I am grateful to members of Tanzania National Parks for the help and support they have given me over the years, especially the chairman of the board of trustees, Adam Sapi Mkwawa.

Many foundations, institutions, and individuals have contributed funds over the years: my sincere thanks to them all. It was Leighton Wilkie of the Wilkie Foundation who contributed the first crucial seed money, and for that I am forever indebted. The National Geographic Society funded the entire research program for many years and to the present day continues to make grants and support the work in various other ways. I shall never forget the enthusiasm and support of the late Leonard Carmichael and Melville Bell Grosvenor; they were true friends of Gombe. Over the years I have been helped and encouraged by so many members of the organization. In particular, Melvin Payne has been tremendously supportive, and a close personal friend as well. The same is true of Neva Folk, Robert Gilka, Joanne Hess, Mary Smith, Edwin Snider, and Frederick Vosburgh. My warmest thanks to them.

Substantial aid was given over a five-year period by the W. T. Grant Foundation of New York; Philip Sapir and the late Douglas Bond were unfailingly helpful. In addition, generous grants have been made by the L. S. B. Leakey Foundation. I want to make special mention of the late Allen O'Brien, as well as George Jagels and Coleman Morton; and an especially warm thank-you goes to Tita Caldwell and Joan Travis, not only for their friendship and hospitality, but for arranging my annual spring lecture tours, which for several years were virtually the sole support of the Gombe research. Other aid has come from the New York Zoological Society—I thank William Conway, Royal Little, Peggy and Charles Nichols, and Laurance Rockefeller—and the Wenner Gren Foundation—my warm thanks to Lita Osmundsen. Gombe has also received assistance from the Bothin Helpin Fund, the British Science Research Council, and the East African Wildlife Society.

Many individuals have graciously made donations to help me maintain the research at Gombe since foundation support ended soon after the 1975 kidnapping. They are too numerous to mention by name, but my heartfelt thanks go to every one, not only for major contributions, but also for the many smaller gifts that represent the same magnanimous spirit on the part of the givers. One of my most treasured donations came to me in Africa from a small boy who mailed a quarter, taped to a sheet of paper, with a note saying that there would be more when he could earn money himself.

I am very grateful to the companies that have donated goods over the years: Avery Implements, Fisons Limited, Imperial Chemical Industries Limited, and Nestlé SA. I extend warm thanks to James Caillouette, who has long assisted with medical supplies for the Tanzanian staff and has become a personal friend.

At this point I must insert a paragraph of special gratitude to those who made possible the Jane Goodall Institute for Research, Conservation, and Education, the tax-exempt organization into which all recent donations have been channeled. It was conceived by the late Ranieri di san Faustino and his wife, Genevieve. After his death Genie battled on with the help of other wonderful friends—Joan Cathcart, Bart Deamer, Margaret Gruter, Douglas Schwartz, Dick Slottow, and Bruce Wolfe—to make the institute a reality. So much effort, so much generosity in time, or money, or both—how can I ever thank them enough? Subsequently, other loyal supporters have joined the board: Larry Barker, Hugh Caldwell, the late Sheldon Campbell, Warren Iliff, Jerry Lowenstein, and Frank Talbot. To all these people, my warm appreciation of their help and friendship.

One of Gombe's most generous benefactors over many years has been Gordon Getty. In 1979 he agreed to become president of the Jane Goodall Institute, and in 1983 he and his wife, Ann, put up a fabulous challenge gift. It is difficult to find words to express my deep gratitude to Gordon and Ann—not only for the money, but for the faith in the research, and in myself, that it symbolized. Because of the beneficence of so many others, the gift was matched and has gone far to ease my financial worries for the future of the work. I am deeply appreciative of Melinda McGee's major effort in helping us to raise this large sum of money.

The Tanzanian town of Kigoma is the last stop in the "civilized" world en route to Gombe. Ever since 1960, when I arrived there for the first time, wonderful people have been incredibly welcoming, receiving wave-tossed researchers into their homes and sending them off refreshed for the return journey along the lake. They have been many, but special thanks to Blanche and Tony Brescia, Subhadra and Ramji Dharsi, Rahma and Christopher Liundi, Lois and Maurice Marrow, and Eester and Timothy Shindika for their assistance and warm hospitality. Ramji helped significantly with administrative problems in Kigoma before moving to Dar es Salaam, where he continues to assist the project. The Brescias from 1975 to 1984 volunteered to look after all on-the-spot administration for Gombe Research Center: their contribution was an important one, and they are sorely missed. Asgar Remtulla, with Kirit and Jayant Vaitha, have stepped into the breach and I am most appreciative.

Over the years many people have worked at Gombe in an administrative capacity, and my thanks go to them all: to Godfrey Kipuyo, Margaret and Nick Pickford, Gerald Rilling, Moshi Sadiki, Emmanuel Tsolo, and Emilie van Zinnicq Bergmann-Riss. And special gratitude to the national parks wardens who have helped me so much: Eli Kyambile, Etha Lohay, Rweyongeza Mwenera, Augustino Sayaloi, and Ami Seki. My warm appreciation, too, for the cooperation of Frank Silkiluwasha, appointed by the Tanzania Wildlife Institute as 1981 director of the research center.

Many others have worked hard over the years in supporting roles at the center. Dominic Charles went to Gombe as cook with my mother and me in 1960. In the days when up to twenty students worked at Gombe, there were many who kept the center operational—cooking, cleaning, and so on—and I am grateful to every one of them. Most especially to my old friend, Rashidi Kikwale: it was Rashidi who first accompanied me into the mountains and taught me some of the forest trails; it was Rashidi who stood firm at the time of the kidnapping, refusing to tell the invaders where the key to the gasoline store was kept, and getting beaten for his loyalty; and it is Rashidi who still makes sure that all goes well for me whenever I visit Gombe. It would not be the same without him.

None of the work described in this book would have happened had it not been for the foresight of the late Louis Leakey and his faith in my ability to succeed. Louis had the original idea; he found seed money so that it could become reality; he arranged for my enrollment at Cambridge University; and through those early months of frustration at Gombe his encouragement never faltered.

Now let me turn to the actual research. My deepest and most heartfelt thanks to Robert Hinde, my supervisor while I worked for my doctoral degree at Cambridge University. When I arrived at Cambridge, I was completely naive in matters of analysis and presentation of research results. Robert, with his keen and critical mind, guided my exuberant writing style into a mold more suitable for scientific expression and was, without doubt, the best teacher I could have found anywhere. Moreover, he was instrumental in obtaining a substantial grant from the British Science Research Council and has helped in significant ways with the organization of long-term data collection.

I am also deeply grateful to David Hamburg, who in 1972 negotiated an affiliation between Gombe and Stanford University and thus enabled a succession of talented and gifted students from the human biology program to work at Gombe as research assistants, each for six months to a year. All were first trained rigorously at Stanford, and they made a very substantial contribution to the long-term records. It was Dave, too, who was responsible for bringing the Gombe research project to the attention of the Grant Foundation, with the resulting major support.

I want to express my thanks to Abdul Msangi who, when he was dean of the Faculty of Science at the University of Dar es Salaam, arranged for a number of his zoology undergraduates to join our team at Gombe between 1973 and 1975. As current director of the Tanzania Scientific Research Council, he continues to provide encouragement.

I should like too to thank J. B. Gillet of the East African Herbarium and Bernard Verdcourt of the Royal Botanic Gardens at Kew, who identified food plant specimens; S. F. Barnett and E. Soulsby of the School of Veterinary Medicine at Cambridge, for parasitological examination of feces; and W. V. Harris and I. H. Yarrow of the British Museum of Natural History, for their identification of termite and ant specimens.

I first dreamed of setting up a research center at Gombe in 1963, and it was Hugo van Lawick who worked out the plans with me. He helped not only with the administrative details but also with the overall organization of the collection of data. Over and above this, of course, Hugo's principal contribution has been his brilliant and sensitive photography of the chimpanzees. He has built up a unique record; many of his photographs illustrate this book, and his ciné films on chimpanzee behavior have been viewed by millions of people the world over. He has in this way truly enriched our understanding of the Gombe chimpanzees.

Now comes the task of trying to thank adequately all those who have taken part in the collection of chimpanzee data at Gombe: the researchers, students, and field assistants who have contributed to the unique long-term records without which this book could not have been written. In the very early days Alice Sorem Ford, Sonia Ivey, and Edna Koning worked long, hard hours under incredibly difficult circumstances. At that time we allowed ourselves as free time *one evening* a week. For the rest, it was seven days of observation and recording, six evenings of writing late into the night. I have not forgotten their contributions, reaching back to those early times, and I am lastingly grateful for what they did.

Some assistants worked at Gombe a relatively short time, but they too added to our knowledge: Sally Avery, Janet Brooks, Pamela Carson, Sue Chaytor, June Cree, Sanno Keeler, Patti Moehlman, Sean Sheehan, and Neville Washington. From 1972 on came the succession of highly trained undergraduates from the human biology program at Stanford and the zoology undergraduates from Dar es Salaam: Donna Anderson, Molly Brecht, Curt Busse, Cay Craig, John Crocker, Paul Harmatz, Carrie Hunter, Clever Kakuru, Henry Klein, Dede and Mark Leighton, Susan Loeb, Addie Lyaruu, Peter Meic, Nancy Merrick, Kitt Morris, Alex Mwankupilis, Nancy Nicholson, David Riss, William Sheeja, Joan Silk, Sara Simpson, and Ann Vander Stoep Hunt.

Other research assistants stayed for a year or more helping to accumulate data for the records: Lori Baldwin, Cathleen Clark, Caroline Coleman, Ruth Davis, Carole Gale, Godfrey Kipuyo, Ann Pierce, Ami Seki, Ann Simpson, Dawn Starin, and Geza Teleki. There were those who, after working as research assistants, were accepted into graduate programs and collected data for their doctoral dissertations: Harold Bauer, David Bygott, Patrick McGinnis, Anne Pusey, Mitzi

The research team at Gombe in 1973.

Thorndal, and Richard Wrangham. A few went straight into their own doctoral work: Hetty and Frans Plooij, and Caroline Tutin. Some came as postdoctoral researchers: Helmut Albrecht, Larry Goldman, William McGrew, and Michael Simpson. Peter Marler in 1967 carried out a three-month study of vocalizations. The work of all these people has benefited our understanding of chimpanzee behavior and has been of inestimable value to me in the writing of this book. In particular, I would thank those who made special efforts to contribute to the organization of the long-term record: David Bygott, Alice Sorem Ford, Pat McGinnis, Anne Pusey, Geza Teleki, and Richard Wrangham; and to the training of the Tanzanian field staff: Donna Anderson, Harold Bauer, Larry Goldman, Mitzi Thorndal, and Richard Wrangham.

Several students had just commenced their field research at Gombe when the unfortunate kidnapping event of 1975 brought their work to an abrupt and untimely end: Richard Barnes, Emilie van Zinnicq Bergmann-Riss, Carrie Hunter, Emily Polis, Steve Smith, and Barbara Smuts. I extend, yet again, my deepest sympathies. It was an unthinkable event, and its ramifications were far flung.

Over the years a number of students have collected data on the baboons and red colobus monkeys: this is not the place to mention them by name, but their studies have often thrown light on chimpanzee behavior, and I want to acknowledge the contribution they have made. Anthony Collins, during his four-year sojourn at Gombe studying baboons, helped in so many ways that he must be singled out.

My debt of gratitude to Ruth Davis, who lost her life while collecting data, is immense. She made a significant contribution during her time at Gombe. She was the first person to collect systematic data on the behavior of the adult males, and it was while she was struggling to keep up with a fast-moving group, in the far south of their range, that she met with her fatal accident. My heartfelt sympathy is extended again to her parents.

Now I come to the field assistants, for whose hard work and dedication I have the highest respect. These men work at Gombe for many years, in some cases for life. After the 1975 kidnapping the research

The research team and Goblin at Gombe in 1985. (C. Boehm)

would have come to an end had it not been for this team of conscientious Tanzanians. I have described their responsibilities and capabilities in some detail in this book; here I simply express my most sincere thanks. Most especially to Hilali Matama, who began in 1968 and is now in day-to-day charge of the data collections, and to Hamisi Mkono and Eslom Mpongo, who have also been with me for over ten years. For their contributions I thank too Yahaya Alamasi, Jumanne Athumani, Rugema Bambanganya, Adriano Bandora, Cretus Chiwaga, Ramadhani Fadhili, Bruno Helmani, Lai Lukumay, Hamisi Matama, Jumanne Mkukwe, Haji Mpongo, Yusufu Mvruganyi, Yahaya Ntabiriho, Gabo Paulo, Sadiki Rukumata, and Kassim and Yassini Selemani and Selemani Yahaya.

It is difficult to express my indebtedness to my late husband, Derek Bryceson, for his help, support, and advice. Without him I doubt I could have kept the research going after the kidnapping of 1975. Derek, with his vast knowledge and understanding of Tanzania, was at that time director of the national parks. He helped me to train the field staff to its current high level of efficiency and to reorganize the data collection along its present lines. Derek was fascinated by the chimpanzees and, lame though he was (after being shot down in World War II), he climbed the steep path to the observation area every day and spent hours watching and photographing the chimpanzees there. Many were the discussions I had with him on puzzling aspects of chimpanzee behavior; his comments, proffered from the point of view of a farmer, were often penetrating and gave me new insights. His contribution was indeed great; even now, because he was so loved and honored in Tanzania, his name confers on me, his widow, a position I would never otherwise have attained.

Next I must try to thank the loyal friends in Dar es Salaam who sprang into the breach after Derek's death, helping and strengthening me in many ways: first and foremost, Derek's son, Ian, who was teaching marine biology at the university. And Clarissa and Gunar Bårnes, Jenny and Michael Gould, Frauke and Benno Haffner, Dimitri Mantheakis and his sons, Sigy and Ted McMahon, Mollie and David Miller, Zuberi Mnyekea, Zeno George Ng'anga, Nancy and Robert

Nooter, Trusha and Prashant Pandit, Judy and Adrian Taylor, Faye and Roger Taylor, and Marina and Dick Viets. It would have been difficult to keep myself and the research going without the help of these people and many others, and I am more grateful than I can say.

Then there are my faithful canine friends, Baggins and Ginger (deceased) and Cinderella. Certainly there were times, as I struggled with this volume, when it was their belief in my infallibility that kept me going!

This book has, indeed, been a good many years in the making, and many people have shared in its conception. To acknowledge all those whose ideas have helped to mold my thinking is not possible, for I would have to start with Hugh Lofting, who wrote the Doctor Dolittle books and bent my attention to the study of monkeys and Africa when I was eight, then work through all those with whom I have talked and all those whose works I have read in the forty-odd years since. Even casual questions asked at lectures have sometimes sparked new ideas, new trains of thought. I should mention some, though, with whom I have discussed certain specific aspects of this book and who have contributed freely of their advice and ideas. Among them are three ex-Gombe students—David Bygott, Anne Pusey, and Richard Wrangham. Caroline Tutin and Geza Teleki provided the information for the chimpanzee range map in Chapter 3. Lynn and Pat McGinnis ran statistical tests on some of the tables. I have had stimulating discussions with many of the Japanese scientists and students who have worked at Mahale, particularly Mariko and Toshisada Hasegawa, Junichero Itani, Kenji Kawanaka, and Toshisada Nishida. Nor must I overlook the valuable insights gained from those who have worked long years with captive chimpanzees, especially Stella Brewer, Marc Cusano, Debbi and Roger Fouts, Trixie and Allen Gardner, Nancy King, Jill Mellen, Emil Menzel, Dave Thomas, Marianne Yeutter-Curington, and, high on the list, Pal Midget—who has worked for so long with Emil's group, first at the Delta Regional Primate Center, then at Stanford, and now in Texas.

Quite apart from the conceptualization of the book, many people have helped with the tedious analysis, all done by hand in the humid heat of Dar es Salaam: Heta Bomanpatell, Rosie Fieth, Diana Francis, Carol Ganiaris, Neil Margerison, and Gudilla Tarimo. For three special people, three very special words of thanks: Jenny Gould, Trusha Pandit, and Judy Taylor. All of them spent hundreds of hours in analyzing data for the various chapters. Jenny struggled with the draft manuscripts I handed her and uncomplainingly deciphered the mistyped sheets, liberally covered with illegible midnight scrawls. Trusha compiled table after table and drafted many of the complex figures that appear in the book. She lives next door and often volunteered help at all sorts of odd hours, when I was frantic about some tricky piece of analysis. These three have all helped with Gombe administration, especially during times when I was lecturing abroad. Over and above this, they organized *me* and the chaos of my papers—without them I doubt this book would ever have seen the light of day.

I must now express my heartfelt appreciation for the patience of William Patrick, in 1984 editor for science and medicine at Harvard

University Press. His forbearance as I missed deadline after deadline, as pile after pile of revised manuscripts arrived, embarrasses me when I think of it.

Even deeper gratitude goes to Vivian Wheeler, senior editor, for her hours of painstaking work, checking tables and figures against text and making invaluable suggestions for the rewording of confusing passages. Vivian flew to Tanzania to speed up the final editorial process. She has become a personal friend. And I thank the many others at Harvard University Press who have devoted their energy and creativity to the production of this book: Nancy Clemente, David Foss, Jean Hammond, Elizabeth Hurwit, Rebecca Saikia-Wilson, Kate Schmit, and—most especially—Marianne Perlak, the Press's art director, for her ingenious design of these pages. David Minard has my unbounded admiration for his superb rendering of the many complex figures and maps that landed on his desk from Dar es Salaam.

The book is further illustrated with photographs taken by many individuals over the years and is immensely enriched by their addition. A special word of appreciation must go to Larry Goldman, who tracked down a number of elusive portraits of specific chimpanzees, and to Richard Wrangham, who generously allowed me to use the superb photograph that so vividly captures the mood at the conclusion of the book. Melvin Payne and Mary Smith of the National Geographic Society, who kindly agreed to print almost all of the photographs used here (many from color slides), have my profound thanks.

I feel particularly indebted to Hans Kummer. Initially I planned this book as a slightly expanded version of my 1968 monograph. Hans read some early draft chapters, and as a result of his penetrating comments and stern admonitions, I realized that he was right: the scientific community deserved a more thorough analysis of a quarter-century's worth of data. We had many long discussions, and my debt to him is large. More recently Christopher Boehm has read all of the chapters in their final form; his criticisms and comments have been of utmost value, particularly those concerning chimpanzee vocalizations. I am extremely appreciative of his hours of time and our many enlightening discussions.

I am grateful to my sister, Judy Waters, who joined me in the early days to take the very first photographs of chimpanzee tool using in the wild. I thank my son—still known to his family and close friends as Grub, but to the rest of the world as Hugo—for putting up with a mother who was always immersed in chimpanzees. And how can I find words to thank my mother, Vanne Goodall, for the staggering contribution she has made? Not only did she encourage my childhood dream of studying wild animals, she even accompanied me to Gombe in 1960. Her wisdom and advice over the sometimes stormy years between then and now have been invaluable. She has helped with fund raising; she has read and commented on manuscripts; she has been, always, a tower of strength. When my son went to school in England at the age of nine, I should have certainly left the research to be with him had Vanne not been there, providing him with the same warm and supportive home and the same guidance and discipline that I had as I grew up.

Finally, I must end with the chimpanzees themselves, all those unique, vivid personalities: Flo and her talented family, Melissa and Goblin, Passion and Pom, Mike and Goliath, and he who will always have first place in my affections, David Greybeard. I thank them all—The Chimpanzees of Gombe—for their tolerance and for their *being* and for their help in better understanding ourselves.

Index of Names

Abo-Khatwa, A. N., 155
Alamasi, Y., 74, 126n, 274, 281, 285, 293, 297, 334, 377, 497, 498, 499, 510, 609
Albrecht, H., 10, 242, 555
Alexander, R. D., 530, 531
Alger, E., 379
Allen, M., 444, 445, 446, 449, 465
Altmann, S. A., 119
Anderson, D., 308, 601
Anderson, J. R., 407
Andrew, R. J., 117
Angel, R., 379
Argyle, M., 119
Aristotle, 5
Athumani, J., 285, 287, 391, 496, 497

Bachmann, C., 567
Badrian, N., 444
Baldwin, P. J., 9, 10, 44, 230, 249n, 522, 538, 543, 562
Bambanganya, R., 285, 287, 298, 354, 355, 495, 496, 512
Banaja, A. A., 155
Bandora, A., 331, 336, 492, 494, 507, 508, 511
Barnes, R., 438
Barnett, S. A., 38
Basuta, I., 9, 10, 593
Bauer, H. R., 134, 154, 159, 244, 387, 402, 546
Beach, K., 26
Beatty, H., 263, 544
Bergmann-Riss, E. van Z., 269n, 270, 331, 492, 494, 508, 511
Bernstein, I. S., 339, 340
Bhrolchain, M., 203
Bielert, C., 139
Bigelow, R. S., 530, 531
Blanc, A. C., 532
Blodgett, F. M., 103
Boehm, C., 378, 380
Boesch, C., 9, 10, 263, 269n, 544, 561, 564, 587, 593

Boesch, H., 9, 10, 263, 544, 561, 564, 587, 593
Boreman, T., 6
Bowlby, J., 203
Boysen, S., 36
Braggio, J. T., 31
Brain, C. K., 526n
Brewer, S., 269n, 521, 543, 544
Brink, A. S., 23
Brodkin, A. M., 379
Brown, G. W., 203
Brown, J. L., 314, 379
Buchanan, J. P., 31
Budd, R. L., 11, 34
Buffon, G. L. L., 5
Burt, W. H., 525
Burton, F., 448
Busse, C. D., 54, 107n, 158, 167, 177n, 181, 211, 236, 240, 241, 244, 269n, 270n, 272, 273, 285, 287, 299, 301, 304, 308, 331, 404, 416, 420, 473, 492, 494, 511
Butler, R. A., 26
Buxton, M., 379
Bygott, J. D., 120, 154, 178, 179, 209, 241, 283, 285, 287, 298, 316, 332, 336, 338, 341, 387, 393, 395, 412, 415, 416, 419, 420, 425, 427, 429, 430, 438, 449, 452, 494, 495, 503, 599

Cameron, H. M., 95
Campbell, D. T., 531, 532
Carmichael, L., 31
Carpenter, C. R., 407, 525
Catlin, S. F., 340
Chamove, A. S., 339, 407
Chance, M. R. A., 573
Cheney, D. L., 131, 133, 198
Chiwaga, P., 492
Clark, C. B., 54, 354, 370, 387, 392, 576
Clutton-Brock, T. H., 47, 49, 270n, 271
Collins, D. A., 285
Collins, D. C., 444
Cool, W., 144, 366, 419

Jolly, A., 39
Jones, C., 249n, 543

Kahela, G., 288, 294, 609
Kakuru, C., 495
Kano, T., 230, 269n, 557
Kase, N., 444
Katchadourian, H. A., 84
Kaufman, I. C., 101, 203
Kaufmann, J. H., 208
Kawabe, M., 269n
Kawai, M., 410
Kawamura, S., 266
Kawanaka, K., 230, 269n, 284, 299, 352, 520, 523
Keeling, M. E., 85
Keith, A., 530, 531
Kellog, L. A., 8, 17
Kellog, W., 8, 17
Keverne, E. B., 340
King, N. E., 375
Kipuyo, G., 285, 298, 354
Klatt, D., 11
Klein, D. J., 147, 155
Klein, L. L., 147, 155
Koehler, D., 92, 93
Köhler, W., 7, 8–9, 14, 17, 18, 19, 21–22, 23, 26, 27, 30, 33, 36, 39, 42, 118, 248, 322, 331, 374, 382, 385, 394, 523, 546, 548, 550, 559, 574, 585
Kohts, N., 8, 17, 118, 119, 248, 542
Koman, J., 10, 249n, 263, 543, 544, 545, 554, 559, 561n
Konner, M., 339
Koprowski, H., 93
Kortlandt, A., 9, 10, 144, 242, 334, 555, 556
Kraemer, H. C., 340
Kruuk, H., 285, 527
Kuehn, R. E., 155
Kuhn, H. J., 388n
Kummer, H., 39, 142, 147, 149, 155, 198, 199, 208n, 314, 318, 325, 338, 356, 406n, 407, 457, 484n, 493, 526, 548, 562, 567, 573, 584
Kuo, Z. Y., 417, 441
Kurland, J. A., 329
Kuroda, S., 484
Kurt, F., 584

Lack, D., 356, 410
Langer, S., 33
Latham, M., 256
van Lawick, H., 51, 282n, 527
van Lawick-Goodall, J. See Goodall, J.
Lawson, J., 35
Layman, W. A., 379
Leakey, L. S. B., 2–3, 9, 475n, 535, 607
Lemmon, W. B., 445
Leo, P., 493
LeVine, K. A., 532
Lieberman, P., 11
Lindburg, D. G., 338
Lorenz, K., 592

Loveland, D. H., 38
Lukumay, L., 285, 354

MacKay, D. M., 114n, 116
Mackinnon, J. R., 147
Mackinnon, K. S., 147
Maeda, Y., 407
Malenky, R. K., 444
Marais, E., 562
Marler, P., 119, 126, 128, 129, 130, 131, 134, 314
Martin, D. E., 85
Martinez-Arias, A. M., 340
Mason, W. A., 38, 122, 525, 567–568
Matama, H., 55, 126n, 256, 274, 277, 280, 285, 287, 291, 295, 307, 329, 331, 352, 354, 355, 488, 492, 494, 497, 499, 500, 506, 511, 513, 601, 608
Matama, S., 334
Mbrisho, M., 557
McCormack, S. A., 84
McGinnis, P. R., 84, 340, 387, 444, 447, 448, 450, 451, 453, 460, 463, 465, 481
McGrew, W. C., 9, 10, 30, 44, 54, 96, 130, 144, 145, 230, 249n, 251, 256, 261–262, 269n, 372, 374, 385, 390, 522, 536, 539, 543, 548, 562
McGuire, M. T., 526
McReynolds, P., 20
Mead, G. H., 33, 34
Meddin, J., 33, 34
Mellen, J. D., 375
Mellen, S. L. W., 382
Meller, R. R., 340
Menzel, E. W., Jr., 13, 20, 23, 30, 35, 141–142, 144, 176, 338, 532, 568, 580
Merfield, F. G., 544
Meunier, V., 6–7
Midgett, P., 23
Midgley, M., 39, 40, 380
Miles, W. R., 385
Miller, H., 544
Miller, N. E., 323
Missakian, E. A., 466
Mitchell, G., 339
Mkono, H., 287, 292, 293, 297, 492, 499, 500
Morris, D., 6–7, 24, 25, 26
Morris, K., 293, 309
Morris, R., 6–7, 24, 25
Morrison, J. A., 532
Mowrer, O. H., 323
Moynihan, M., 319
Mpongo, E., 126n, 131, 151, 285, 293, 307, 331, 354, 355, 357, 454, 492, 494, 495, 496, 500, 507, 511, 513, 516, 517, 612
Mpongo, H., 274, 277, 287, 288, 294, 492, 496, 497, 498
Mvruganyi, Y., 55, 126n, 281, 297, 498

Nadler, R., 81n
Nagel, U., 314, 325
Napier, J., 535

Nicolson, N. A., 391
Nishida, T., 9, 10, 86, 144, 159, 230, 237, 256, 262, 263, 264, 269n, 284, 296, 299, 352, 390, 412, 417, 424, 430, 436, 444, 452, 470, 489, 493, 519, 520, 521, 524, 541, 543, 555
Nissen, H. W., 8, 10, 42, 248, 375, 377
Noble, G. K., 525
Norikoshi, K., 269n, 284, 299
Nyundo, R., 269n, 284, 299, 352, 520

Oki, J., 407

Packer, C. R., 199, 282n, 380, 466, 483, 528
Parkel, D. A., 12
Parron, D. L., 203
Pate, J. L., 19
Patterson, F., 137
Patton, R. G., 103
Paulo, G., 151, 612
Pavlov, I. P., 39, 592
Perachio, A. A., 339, 340
Petrunkevitch, A., 26
Philip, M., 134
Pierce, A. H., 223, 331, 492, 494, 506, 511, 555
Pinneo, L., 13n
Pitt, R., 530, 531
Plooij, F. X., 136, 277, 285, 554
Plotnik, R., 335
Poirer, F. E., 338
Polis, E., 167
Popovkin, V., 37
Preedy, J. R. K., 444
Premack, A. J., 12, 29
Premack, D., 12–13, 20, 29, 33, 36, 37, 39, 386, 533, 569, 577, 580
Prestrude, A. M., 17
Pryor, K., 26, 388
Pusey, A. E., 54, 84, 87, 166, 168, 182, 337, 387, 392, 431, 445, 466, 467, 469, 470, 501, 528

Rahm, U., 544
Ransom, T. W., 279, 280, 282, 286
Rapaport, L., 394
Reinhart, J. B., 103
Reite, M., 101, 203
Rensch, B., 12, 31, 32
Resko, J. A., 340
Reynolds, F., 9, 10
Reynolds, P. C., 387, 407
Reynolds, V., 9, 10
Ripley, S., 532
Riss, D. C., 54, 107n, 158, 167, 177n, 178, 180, 181, 211, 236, 240, 241, 244, 330, 331, 393, 404, 412, 416, 420, 473, 492, 494, 508, 511
Roberts, J. R., 85
Robinson, M., 444
Rogers, C. M., 12, 28, 33
Rohles, F., 17
Roosmalen, A., 364

Rose, R. M., 339, 340
Rosenblum, L. A., 101, 203
Rowell, T., 437
Roy, A. D., 95
Rukemata, S., 286, 498
Rumbaugh, D. M., 12, 19, 27, 29, 36, 81n, 379
Runfeldt, S., 130
Russell, C., 530
Russell, S., 28, 33
Russell, W. M. S., 530

Saayman, G. S., 285, 338
Sabater Pi, J., 249n, 543
Sade, D. S., 407, 466, 488
Sagan, C., 531
Sapolsky, R. M., 336, 340
Savage, T. S., 376, 544
Savage-Rumbaugh, E. S., 35, 36
Schaller, G. B., 44, 137, 247, 285
Schiller, P. H., 27
Schjelderup-Ebbe, T., 409
Schmidt, M. J., 92
Schoenfeld, D. J., 24
Schultz, A. H., 447
Scott, J. P., 125, 314, 338, 354, 417
Sears, R. R., 323
Seifu, M., 284
Seki, A., 293
Selemani, Y., 494, 495, 496, 498, 510
Seyfarth, R. M., 131, 133, 198, 407
Sharman, M. J., 269n
Short, R. V., 447
Shostak, M., 475n
Shrier, D., 379
Silk, J. B., 372
Simpson, M. J. A., 387, 393
Sindimwo, A., 155, 285
Smuts, B., 154n, 159, 209, 231, 241, 245
Smythies, J. R., 339
Sollereld, H. A., 105
Spencer-Booth, Y., 101, 203, 379
Speroff, L., 444
Spitz, R. A., 203
Sroufe, L. A., 379
Stammbach, E., 407
Stenhouse, D., 19, 38, 39
Stevens, V. J., 375
Struhsaker, T. T., 249n, 285, 543, 544
Sugiyama, Y., 9, 10, 249n, 263, 522, 543, 544, 545, 554, 559, 561n, 593
Suzuki, A., 9, 10, 230, 269n, 284, 299

Takahata, Y., 263, 269n, 284, 299, 352, 489, 519, 520, 521
Takasaki, H., 296
Teleki, G., 45, 111, 269n, 279, 280, 286, 297, 300, 311, 330, 529, 542, 555
Temerlin, J., 11–12, 382–383
Temerlin, M. K., 11–12, 17, 23, 31, 33, 34, 382–383, 448, 581
Tenaza, R., 119
Thomas, D., 394
Thomas, D. K., 282

Index of Subjects

For ease of identification, main entries that are names of chimpanzees are given in green.

Abortion mechanism, 475n
Adolescence: defined, 81; association patterns, 166; follower relationships, 175, 202; separation from mother, 203–204; increased aggression, 353; ultrasubmissive, 368–369; grooming behavior, 392; sexual behavior, 443, 445, 450. *See also* Juveniles
Adoption, 101–103, 383–384
Aggressive behavior: intimidation, 1–2, 549–559; function, 3, 353–356, 480, 481–482; effect of feeding area, 52–54, 246; intercommunity, 111, 136, 176, 228, 317, 331–332, 491–517, 528, 530–534; threats and hostility, 120–123, 175–176, 314–315, 324–325; charging display, 123, 316, 415–418; calls, 128, 129, 136; lack, in leaders, 209; competition for plant food, 245–246; after kill, 299–301; attacks, 314, 317, 319, 329–330, 341, 493–502, 506–514; control and conflict resolution, 314, 317–318, 360–366, 407; herding, 314, 459; factors affecting arousal, 318, 332, 358–360, 407; coalitions, 318–319, 418–424, 427–428, 431–435, 441, 573–576; snowballing, 319; scapegoats, 320, 323–324, 333, 342, 367, 523; alliance breaking, 326; contagious, 329–330; interspecific, 331, 333–334; behavioral contexts, 332–333, 342, 344, 346–349; individual differences, 336–340; gender differences, 338–339, 341–353, 356; and intent to kill, 528–530; mediated by females, 570, 585–586; inappropriate, 571–572; effect of captivity, 583–586. *See also* Friendly behavior; Hierarchy, social; Hunting behavior; Relationships, individual
Aging signs, 81

Agonistic, defined, 314
Alarm calls, 132, 133, 135–136
Alcohol, liking for, 17
Alliances. *See* Coalitions
Ally, 33–34
Alpha status, 424–429, 472–473. *See also* Hierarchy, social
Altruism, 202, 376–381
Ameslan. *See* ASL
Analogy problems, 29
Anatomy, language and, 9–11
Anger, 118, 127. *See also* Tantrums
Annoyance, 118
Ant dipping, 251–254, 537, 539, 540, 543, 560–561
Anticipation, sense of, 31, 565–567
Ants. *See* Hymenoptera
Apprehension, 118, 127
Arnhem colony, 31; dominance fights, 121, 319, 324, 416, 417–418, 424, 577; alliance breaking, 326; friendly behavior, 364, 375; grooming, 404, 406; female hierarchy, 437, 438; sexual behavior, 452, 479, 580; feeding behavior, 579; signal concealment, 580; effects of captivity, 583–586
Art. *See* Drawing and painting
ASL (American Sign Language), 11–12, 18, 24, 25, 33–34
Aspilia, 237
Associations, group: patterns, 149–154, 170–171, 618–622; composition and size of parties, 154–158, 241–245; year-to-year changes, 158–159; changes over life cycle, 166–169; autonomy and social learning, 170–171; violence and intervention, 319, 325–328; cohesion, 357; effect of captivity, 583–586. *See also* Coalitions; Communities; Hierarchy, social; Relationships, individual

Genie, 74, 351
Genitals: abnormal development, 101, 103; presenting in submission, 360; male, 447. *See also* Sexual behavior; Sexual swelling
Gestures. *See* Communication; Reassurance; Submissive behavior
Gigi, 66–67, 68, 104–105, 165, 181, 184, 185; hunting and meat-eating behavior, 274, 277, 288–289, 305–311; sexual behavior, 486
Gilka, 58, 61, 64, 67–68, 94, 95, 103, 287, 289, 295; and Evered, 172; attacks on, 354; status, 439
Gimble, 74, 294–295
Goblin, 65, 68, 98, 99, 163, 167, 189–198, 210, 211, 215–216, 298, 406–407; and Figan, 177, 180, 184–185; hunting behavior, 286, 287, 302–304; lack of tolerance, 368; status, 425, 429, 431–435; unusual sexual behavior, 445, 468
Godi, 506–507, 528–530
Goliath (at Gombe), 69, 97, 98, 104; and David Greybeard, 61, 200, 201, 313; status, 424–425, 428, 429, 430; fatal attack, 508–509, 528–530
Goliath (at Washington Park Zoo), 375
Gombe National Park, 43–51
Gombe Stream Research Center, 3, 10, 43
Gorillas, 121, 137, 202, 285, 532
Greeting behavior, 366–368. *See also* Reunions
Gremlin, 150–151, 164, 210, 294–295; helping of family members, 377; copulation with brother, 468
Grooming, 26, 124–125, 133; with humans, 57–58, 388; calls and signals, 130, 131, 388–391; Gombe studies, 387; leaf grooming, 391, 572; mother and offspring, 391–392, 402; males and females contrasted, 392, 394–400; network, and hierarchies, 393–394, 400–405; as manipulative strategy, 406–408, 571–572, 576; dental, 548. *See also* Friendly behavior
Grosvenor, 67, 94; mother's care for, 383
Gua, 8, 17
Guinea, 8, 144, 263, 544. *See also* Bossou
Gyre, 74

Hair erection. *See* Piloerection
Ham, 593
Harmony, 88, 91, 308
Harold, 419
Hayes, Viki. *See* Viki
Health. *See* Disease(s); Injuries
Hearing, 16–17
Hector (baboon), 293
Hepziba, 76
Herding, 314, 457, 464
Hierarchy, social: reassurance and submission, 124–125, 360–366; preoccupation with, 184–185; benefits, 201–202, 205–206, 442, 478; dominance reversal,

324–325, 337–338, 415, 417, 429–430, 436; alliance breaking, 326; opportunistic attacks, 328, 418; and aggression, 336, 356, 410; factors affecting, 340, 356, 412, 414–415; grooming and, 388, 393–394, 400–405, 406; theory of, 409–412; changes at Kasakela, 412–415; charging display and fighting, 415–418; coalitions, 418–424, 427–428, 431–435, 441; alpha status, 424–429; lack of motivation, 425, 436–437; loss of rank, 429–430, 436; female, 437–442; reproductive success and, 452–453, 472–473, 478–479; interpreting observations, 566–567; manipulatory and deceptive behavior, 571–582; in captivity and in wild, 583–586. *See also* Aggressive behavior
Holloman Air Force Base, New Mexico, 24, 332
Honey, 131, 249, 255, 540, 543–544
Honey Bee, 72, 357, 513–514
Horace, 60
Hugh, 60–61, 99, 313, 329, 425–426, 429
Hugo (formerly Rodolf), 69–70, 104, 111, 209, 277, 280, 296, 297
Humans: relations with chimpanzees, 1–2, 40–41, 43, 54, 55–59, 61, 69, 70, 603–608; compared to chimpanzees, 3, 592–593; sensory apparatus, 15–16; as chimpanzee prey, 267, 268, 282–283; warfare, 530–534; rocks thrown at, 553, 554, 557–558. *See also* Evolution
Humidity, at Gombe, 47, 49
Humphrey, 57, 70, 96, 163, 167, 395, 398, 405, 462, 477–478; and Figan, 177, 178–179, 184; hunting behavior, 277, 278, 280–281, 287, 292, 302–304; cannibalism, 284, 298; aggressiveness, 329, 341, 376; status, 424–430; intercommunity aggression, 518–519
Hunting behavior: calls, 236–237; species hunted, 267–270; methods, 270–285; cooperation, 285–290; making the kill, 290–292, 529; piracy, 292–293, 294–295, 305–306, 309; scavenging, 293, 295–296; meat distribution, 299–301, 327–328; of males, 301–304; of females, 304–312; and aggression, 333–334, 529–530; of stranger females, 494; tool use, 554–555. *See also* Feeding behavior
Huxley, 111, 361
Hyena. *See* Crocuta crocuta
Hygiene, 231, 545–548
Hymenoptera: *Dorylus (Anomma) nigricans* (driver ant), 131, 249, 251–254, 260, 261–262, 264, 539, 540; *Apis mellifera* (honey bee), 131, 249, 252, 255; *Oecophylla longinoda* (weaver ant), 249, 251, 252, 253, 260, 261–262, 264; *Crematogaster clariventris* (ant), 249, 254, 264, 540–541; *Camponotus* (carpenter ant), 249, 254, 264, 543; *Monomorium afrum* (ant), 249, 264; *Blastophaga* (wasp), 249, 255; *Polistes* (paper wasp), 249, 254; *Trigona* (stingless

Hymenoptera *(continued)*
bee), 249; *Anthophoridae* (mining bee), 249
Hypothalmus, 338, 339

Imagination, 41–42, 590–591
Imitation, 23–24, 25, 145, 336–337
Immigrant females, 86–92, 331, 445, 451, 470–471
Incest, 68, 181, 197; mother-son, 466–467, 470; between siblings, 467–469, 470; father-daughter, 469–470; mechanisms for avoiding, 470–471
Individuality, 60n
Infancy: experience and learning, 26–28, 336–337, 369–370, 563, 568–569, 591; defined, 81; differential mortality, 113; tantrums, 118, 129–130, 135, 318, 374, 576–577; screams recognized by mother, 128–129, 131; and siblings, 176–177, 390, 565, 577; aggression, 336–337, 339, 353–356, 370; grooming, 390, 391–392; sexual behavior, 445, 450; infanticide, 522–523; tool use, 540, 559, 563, 564. *See also* Mother-offspring relations; Play
Infanticide, 112–113, 493–502, 522–523. *See also* Cannibalism
Injuries, 97–100; and pain, 17, 401; and medical intervention, 58–59; to eye, 71; from falling, 98, 100, 111–112; reproductive disorders and, 105; mortality, 110, 111; reaction of other chimpanzees, 121–122; during hunting, 279; fighting, 313, 334, 354–355, 551; compassion and, 382–386, 401; grooming and, 401; hygiene, 546, 547
Insects, at Gombe, 50; as food, 78, 248–262, 536–544; seasonality, 232, 235
Insight, 22–23
Instinct, 27, 38–39, 122–124, 391–392, 592
Institute for Primate Studies, Oklahoma, 11
Institut Pasteur, West Africa, 556
Intelligence: abstraction and generalization, 28–29; concept formation, 33–36, 588–599; definitions, 38; rationality and, 39–40; in apes and humans, 40–41, 592–593; imagination and, 41–42, 590–591; in display patterns, 426–427; in tool using, 535–536, 560–564; effect of captivity, 583–586, 591–592; spatial memory, 587; decision making, 587–588. *See also* Learning; Memory; Social problem solving
Intimidation displays, 1–2, 314–317, 549–559
Ioni, 8, 17, 542

Japanese macaque. *See Macaca fuscata*
Jealousy, 326–327
Jenny, 88, 91
Jessica, 87n
Joanne, 88, 91

Johannesburg Municipal Zoo, 23
Jomeo, 65, 96, 99, 163, 167, 284, 394–396, 398, 401, 405; and Sherry, 70–72, 377, 418–421, 428, 431; hunting behavior, 286, 287, 291, 302–304; lack of dominance drive, 436–437, 475, 476; intercommunity aggression, 518; detailed follows, 609–617
Julia, 12, 31, 32
Juveniles, 82; separation from mother, 203–204; teasing adults, 322; sexual behavior, 445, 450; tool use, 540, 563. *See also* Adolescence; Infancy

Kahama community, 46, 60–61, 69, 263, 492; formation, 84, 228, 229, 487, 503–504; annihilation, 223, 228, 229, 504–514
Kajugi, 430
Kalande community, 46, 228, 488, 492, 514–517
Kamenanfu, 424
Kasakati Basin, 537, 555
Kasakela community, 46, 263, 492–493; population changes, 80, 82; formation, 84; annexation of Kahama territory, 223, 228, 229, 299, 504–514; changes in male hierarchy, 412–415; invasion by Kalande community, 514–517
Kasonta, 424, 430, 519
Kibale Forest Research Station, Uganda, 9, 593
Kidevu, 87, 88, 90, 514
Kigoma, Tanzania, 9, 44–46, 49, 557
Kin selection, 202, 378–381, 591
Kissing, defined, 360
Kiswahili, 148n
Kristal, 77, 101, 103, 384

Lake Tanganyika, 45–46, 50, 51
Lana, 12, 29, 31, 34
Landolphia, 230, 242
Language: acquisition, 9–12; concept formation and, 28–29; symbol manipulation and, 33–34; power of, 534, 592. *See also* ASL; Yerkish
Leaf grooming, 391, 572
Leakey, 72, 97, 105, 297; and Fifi, 181–182; serial consortships, 463
Learning: problem solving and, 18–19, 21–24, 562, 589; definitions, 19; social interaction and, 19–21, 26–27, 568–569; imitation and, 23–24, 25; and teachability, 24–25, 478; motivation and, 25–26; early experience and, 26–28; and tool use, 562–564; effect of captivity, 583–586. *See also* Intelligence
Leopard. *See Panthera pardus*
Lepidoptera (caterpillars), 232, 249, 254, 262
Liberia, 263, 537, 544
Life cycle, stages defined, 79, 81
Lion. *See Panthera leo*
Lion Country Safaris, Florida, 144, 331, 365, 418, 419, 584, 585; drowning, 378

Mr. McGregor, 94, 274
Mr. Worzle, 95, 462–464; eye abnormality, 72, 105, 121, 122
Muggs, J. Fred, 25
Mustard, 164, 284, 291–292; lack of dominance drive, 437

National Geographic Society, 43, 51
Nest making, 2, 208
Nikkie, 406, 408, 416, 417–418, 580
Niokolo-Koba National Park. See Senegal
Noota, 76
Nope, 57, 76–77, 164, 308, 310, 462
Nova, 77, 95, 102

Object manipulation. See Tool using
Observer reliability, 51–52, 58–59, 601–602, 603–608
Oil-nut palm. See Elaeis guineensis
Old age: defined, 81; stresses, 86; disorders, 104; mortality, 110, 113; association patterns, 167, 169
Old Man, 419
Olfaction, 17, 137–139, 490, 539–540, 548–549
Olly, 61, 64, 67, 96, 103, 383, 462–464
Oor, 375
Orphans, 72, 101–104, 203, 379, 383–384

Pain: sense of, 17; effect on aggression, 335. See also Injuries
Painting. See Drawing and painting
Pallas, 165, 169, 289, 384; adoption of Skosha, 77, 102, 103, 384; hunting and meat-eating behavior, 304–307, 310, 311; attacked, 329–330; sexual popularity, 454–457, 464, 465
Pan, 111–112
Pan: troglodytes, 5; var. schweinfurthii, 49, 268; paniscus, 483, 484–485
Panda oleosa nuts, 537, 544, 561
Pant-grunt, 127, 129, 412, 438–439
Panthera: leo (lion), 50, 132, 334, 557; pardus (leopard), 50, 334, 555, 556
Papio anubis (olive baboon), 49, 132, 340, 380; effect of seasons on range, 107; communication, 132, 142; social structure, 147, 170; food competition with chimpanzees, 199–200, 236, 237, 334; range, 208; hunted, 268, 269, 270, 278–282, 283, 286–287; killing and eating of, 290, 292, 297; piracy from, 292–293, 294–295, 305–307, 309; play with chimpanzee young, 334, 552
Papio hamadryas (sacred baboon), 355–356, 407, 484n, 584
Papio ursinus (chacma baboon), cannibalism, 285
Parasites, 96–97, 255, 388n
Parasitic selection, 380n
Party, defined, 154
Passion, 68, 73, 77–78, 99, 100, 101, 102; association patterns, 164, 169, 337–338;

ranging patterns, 210, 216, 219–220, 223, 227; hunting and meat-eating behavior, 277, 287, 289, 290, 305–307, 310, 311; cannibalism, 283–284, 287, 289, 290, 299, 351–354; feeding behavior, 296–297, 299; grooming behavior, 397–400, 401
Patrols, 489–491, 518–519, 579–580
Patti, 68, 88, 90, 165, 397–400, 401; death of infant, 111; and Fifi, 183, 185; hunting and meat-eating behavior, 307, 308–309, 310, 311; victim of female aggression, 501
Pavlov Institute of Physiology, Russia, 13, 37
Pax: relations with siblings, 78, 383–384; injured, 100; death of mother, 101, 102; rescued by Prof, 384
Pedicularis schaefi (chimpanzee louse), 97, 388n
Pepe, 74, 94
Perception, chimpanzee, 16–19, 20, 137–139; and cross-modal transfer, 12, 28
Perception, human, 15–16
Periovulatory period (POP), defined, 444
Peter, 25
Pheromones, 137–139, 394
Photographs, response to, 28, 33, 37, 38
Physical contact. See Aggression; Friendly behavior
Piloerection, as mood communicator, 122, 139, 286, 315, 316, 334
Planning ability, 31, 32, 37, 533–534, 570–571, 587–588
Plato, 77
Play: imaginative, 41–42, 591; with humans, 57–58; injuries during, 100; calls, 130; solitary, 150, 559; social, 151, 369–372, 373; among siblings, 176–177; bonds, 182; with baboons, 334, 552; ending in aggression, 355, 370, 552; as distraction technique, 365–366, 572, 584; tool use, 559–560, 563
Poliomyelitis, 58, 67, 73–74, 82, 178; response to cripples, 121–122, 430
Pom, 73, 77–78, 99, 105, 164, 210, 216, 221, 296–297, 312, 446; relations with siblings, 102, 383–384; cannibalism, 283–284, 287, 289, 290, 299, 351–354
Pooch, 89, 105, 111, 361
POP. See Periovulatory period
Population dynamics, 79–113
Potamochoerus porcus (bushpig): hunted by chimpanzees, 49, 267, 268, 269, 270, 275–277, 287, 290, 293, 304–306, 554–555; attacks by, 100, 276–277; killing and eating of, 290, 296; food competition with chimpanzees, 334
Predators, 155; aggression toward, 334, 555–557
Predatory behavior. See Feeding behavior; Hunting behavior
Pregnancy, 85–86. See also Sexual behavior
Presenting, defined, 360
Problem solving, 18–19, 21–24, 25–26, 29. See also Intelligence; Learning; Social problem solving; Tool using

Sexual swelling *(continued)*
scribed, 444, 482; evolutionary advantage, 482–485. *See also* Estrus cycle; Females
Shadow, 145
Sharing, of food, 357, 369, 372–376, 386
Sherman, 35, 36
Sherry, 99, 163, 166, 167, 284, 395; and Jomeo, 70–72, 377, 418–421, 428, 431; hunting and meat-eating behavior, 272, 286, 287, 291, 301–304, 307; intercommunity aggression, 518
Sibling relations, 68, 71, 73, 74, 172; coalitions, 60–61, 64–65, 71, 418–419, 439; grooming, 64, 390, 404; use of younger sibling, 176–177, 565, 577; dominance reversal, 177, 178; social benefits, 204, 205–206; competition for plant food, 245; model for adult behavior, 337; and disease or injury, 383–384; incest, 467–470. *See also* Relationships, individual
Signaling, defined, 114n–115n. *See also* Communication
Sita, 576
Skosha, 90, 384; death of mother, 77, 101, 102, 103
Sleeping, size of parties, 154. *See also* Nest making
Smell, sense of. *See* Olfaction
Sniff, 272, 383, 488, 492, 518; adoption of infant sister, 102, 103, 383; fatal attack, 510–511, 528–530
Sobongo, 424
Social deprivation, 27–28. *See also* Orphans
Social excitement, 119, 127; aggression during, 332–333; friendly physical contact during, 358–360; and copulation, 449
Social facilitation, 20–21
Socialization, 166, 167–168, 568–569
Social problem solving, 36–38; intelligence and, 39; complexity of behavior and, 565–568; predicting consequences of behavior, 570–571; manipulatory behavior, 571–582; effect of captivity, 583–586
Social structure. *See* Communities, structure; Hierarchy, social
Social tools, 573–576
Soil eating, 246. *See also* Termite clay
Sophie, 102, 103
Sorema, 102, 103
Space travel, by chimpanzees, 24, 593
Sparrow, 88, 91, 164, 183, 307, 308
Spatial memory, 17–18, 237, 587
Speech, requirements for, 9–11
Sponge, for drinking. *See* Tool using
Sprout, 78, 87n; copulation with son, 466
Stanford Outdoor Primate Facility, California, 13n, 144, 152
Stone throwing, 69, 70, 315, 549–559
Strange-female effect, 445, 465–466
Stranger females, 86–89, 331, 451, 493–501, 522–525

Strychnos fruits, 238, 372, 537, 542, 547, 560
Submissive behavior, 122–125, 360–366. *See also* Aggressive behavior; Friendly behavior; Hierarchy, social
Sultan, 18, 23, 26, 27, 36, 39, 374–375
Sunday Times, London, 66, 79
Swahili. *See* Kiswahili
Symbolic representation, 33–34, 588–589. *See also* Language

Tai National Park, Ivory Coast, 9, 263, 537, 544, 560, 561, 564, 587, 593
Tantrums, 72, 118, 129–130, 135, 317, 318, 324, 374, 576–577
Tanzania, 44, 45, 230, 557. *See also* Gombe National Park; Gombe Stream Research Center; Kigoma
Taste, sense of, 17
Teasing, 322
Temperature: perception of, 17; at Gombe, 45, 47; effect on behavior, 335
Tenerife, 7
Termite clay, 249, 256
Termite fishing, 78, 237, 242, 248–251, 258–261, 536–539, 542–543, 563–564, 588
Termites. *See Macrotermes; Pseudacanthotermes*
Territoriality. *See* Communities; Ranging behavior
Testosterone, 339–340
Theropithecus gelada (gelada baboon), 325, 407, 442
Threats. *See* Aggressive behavior
Throwing. *See* Stone throwing
Tina, 521
Tolerance, 368–369
Tool using, 61; play and self-tickling, 150, 537, 559–560, 563; termite fishing, 237, 250, 251, 258–261, 535, 536–539, 542–543, 560, 561, 563–564, 588; ant dipping, 251–254, 537, 539, 540, 543, 560–561; tool defined, 536; investigation, 537, 539–540, 548–549; nest breaking, 537, 540–541, 563; hole probing, 537, 541–542; drinking, 537, 542, 544, 563; body care, 537, 542, 545–548; resin gathering, 537, 543; honey gathering, 537, 543–544; nut cracking and hammering, 537, 544–545, 560, 561, 563, 564; as weapon, 537, 549–559; tree climbing, 544; cultural traditions, 560–564; age and gender differences, 563–564
Toys, 559–560
Traditions. *See* Cultural traditions
Tragelaphus scriptus (bushbuck): as prey, 49, 268, 269; hunting and eating of, 270, 277–278, 292–293, 294–295, 296–297, 305–308
Travel. *See* Ranging behavior
Tschego, 26, 322, 374–375
Tubi, 73, 183–184, 355

Designer	Marianne Perlak
Compositor	DEKR Corporation
Printer	Halliday Lithograph
Text	10/13 Linotron Century Expanded
Display	Century Bold Italic
Paper	60-pound Warren Patina Matte
Binding	Joanna Arrestox B53500
Endsheets	Multicolor Oatmeal
Insert	Printed by John P. Pow Company on 80-pound Warren Lustro Gloss

Library of Congress Cataloging-in-Publication Data

Goodall, Jane, 1934–
    The chimpanzees of Gombe.

    Bibliography: p.
    Includes index.
    1. Chimpanzees—Behavior.   2. Mammals—Behavior.
3. Mammals—Tanzania—Gombe National Park—
Behavior.   4. Gombe Stream National Park (Tanzania).
I. Title.
QL737.P96G585   1986   599.88′440451   85-20030
ISBN 0-674-11649-6